Plate 1. Examples of mimetic fruits with colourful displays that may deceive avian frugivores (see Chapter 12). **(A)** Abarema brachystachya (Leguminosae), **(B)** Adenanthera pavonina (Leguminosae), **(C)** Pithecellobium sp. (Leguminosae), **(D)** Erythrina velutina (Leguminosae), **(E)** Rhyncosia phaseoloides (Leguminosae), **(F)** Ormosia arborea (Leguminosae), **(G)** Margaritaria nobilis (Euphorbiaceae), **(H)** Paeonia broteroi (Paeoniaceae). All photos by M. Galetti except Pithecellobium and Paeonia by C. Herrera. Printing costs supported by FUNDUNESP.

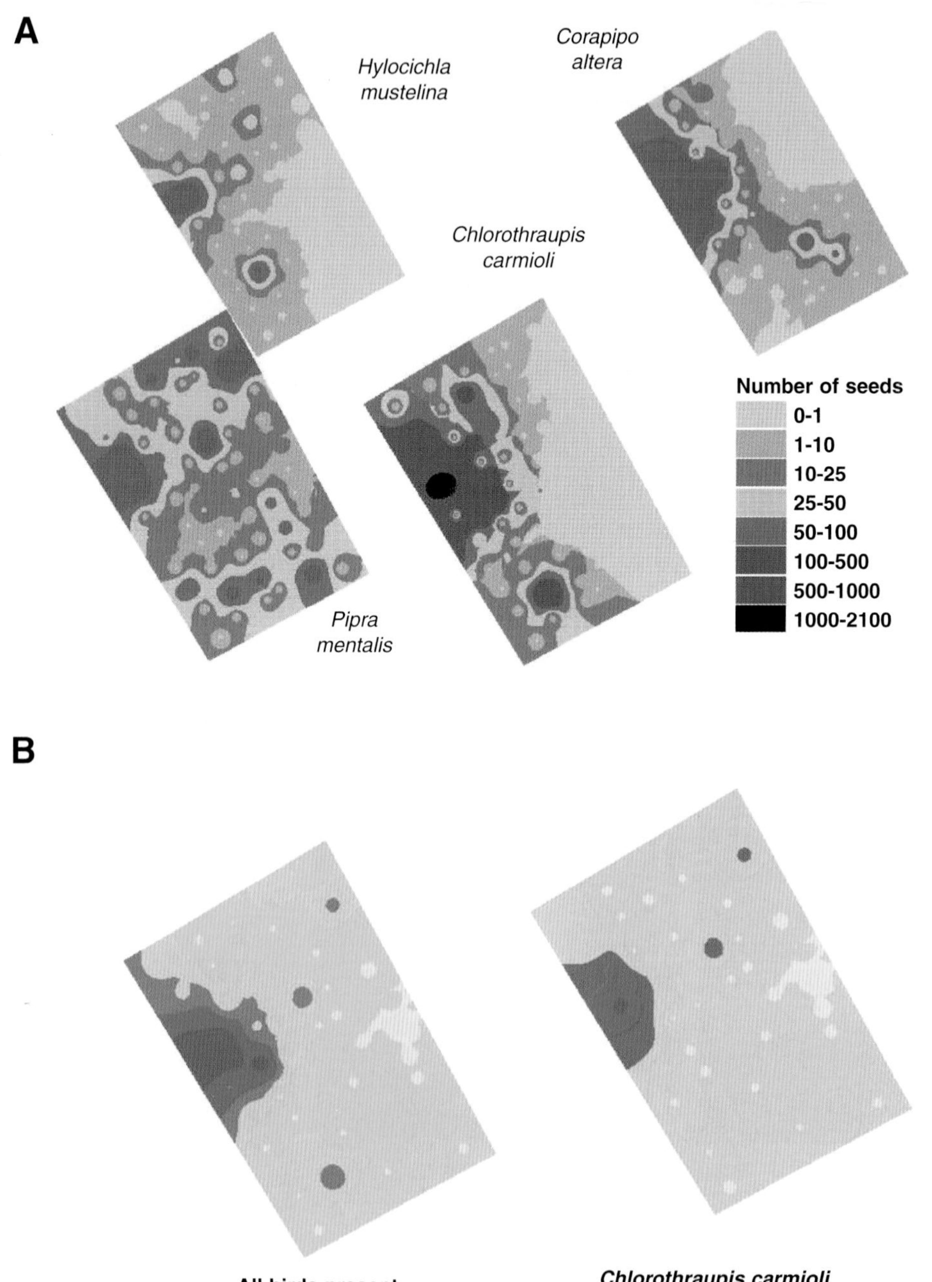

Plate 2. Seed shadows (see Chapter 26). **(A)** Hypothetical seed shadows of *Clidemia densiflora* are shown separately for each of four bird species. Seed shadows were generated using surface interpolation (inverse distance weighting) of seed numbers collected at 60 mist nets distributed uniformly across the approximate 10 ha study spot. **(B)** Alteration in seed shadows when the frugivore assemblage is complete and when *Chlorothraupis carmioli* has been removed. Hypothetical seed distributions at 60 points were derived using the HABITAT survival model. The seed surfaces displayed here were based on interpolation of modelled seeds at those 60 points using inverse distance weighting surface models. Plates 2A and 2B are included for illustrative purposes only. All statistical tests are based on the 60 point locations.

Seed Dispersal and Frugivory: Ecology, Evolution and Conservation

Dedication

We dedicate this book to Carlos Herrera and Henry (Hank) Howe, who led the development of this field and whose contributions have directly influenced all chapters of this book.

Seed Dispersal and Frugivory: Ecology, Evolution and Conservation

Edited by

Douglas J. Levey

Department of Zoology, University of Florida, Gainesville, Florida, USA

Wesley R. Silva

Departamento de Zoologia, Insitututo de Biologia, UNICAMP, Campinas, Brazil

and

Mauro Galetti

Departamento de Ecologia, Universidade Estadual Paulista, São Paulo, Brazil

CABI *Publishing*

CABI *Publishing* is a division of CAB *International*

CABI Publishing
CAB International
Wallingford
Oxon OX10 8DE
UK

Tel: +44 (0)1491 832111
Fax: +44 (0)1491 833508
Email: cabi@cabi.org
Web site: http://www.cabi-publishing.org

CABI Publishing
10 E 40th Street
Suite 3203
New York, NY 10016
USA

Tel: +1 212 481 7018
Fax: +1 212 686 7993
Email: cabi-nao@cabi.org

A catalogue record for this book is available from the British Library, London, UK.

Library of Congress Cataloging-in-Publication Data
Seed dispersal and frugivory : ecology, evolution, and conservation / edited by D.J. Levey, W.R. Silva, and M. Galetti.
p. cm.
Includes bibliographical references (p.).
ISBN 0-85199-525-X (alk. paper)
1. Seeds–Dispersal–Congresses. 2. Frugivores–Ecology–Congresses. 3. Frugivores–Evolution–Congresses. 4. Mutualism (Biology)–Congresses. 5. Coevolution–Congresses. 6. Animal-plant relationships–Congresses. I. Levey, Douglas John. II. Silva, W. R. (Wesley R.) III. Galetti, M. (Mauro) IV. International Symposium-Workshop on Frugivores and Seed Dispersal (3rd : 2000 : Sao Pedro, Sao Paulo, Brazil)

QK929.S444 2001
581.7′8–dc21 2001035222

ISBN 0 85199 525 X

Typeset in Melior by AMA DataSet Ltd
Printed and bound in the UK by Biddles Ltd, Guildford and King's Lynn

Contents

PLANT STRATEGIES

ANIMAL STRATEGIES

Acknowledgements

This book is the product of the Third International Symposium–Workshop on Frugivores and Seed Dispersal held in São Pedro (SP), Brazil, 6–11 August 2000. For financial support we thank the Conselho Nacional de Desenvolvimento Científico e Tecnológico (CNPq), Fundação de Amparo à Pesquisa do Estado de São Paulo (FAPESP), Fundação O Boticário de Proteção à Natureza, National Science Foundation (NSF), Universidade Estadual de Campinas (UNICAMP) and Universidade Estadual Paulista (UNESP, Rio Claro campus). Most authors provided comments on at least two manuscripts. Additional reviews were furnished by C. Bosque, R. Bustamente, M. Cain, J. Dalling, T. Engel, P. Feinsinger, T. Fleming, R. Green, C. Herrera, S. Hubbell, K. Kitajima, K. Silvíus, T. Theimer, D. Tilman and R. Ostfeld. Marco Aurélio Pizo and Ronda Green were valuable members of the Organizing Committee and helped with all aspects of planning. Verônica S.M. Gomes, Érica Hasui, William Zaca, Rafael Raimundo, Marina Fleury and Liliane Zumstein were members of the support staff; their efforts were not sufficiently appreciated by attendees because they worked so efficiently and effectively behind the scenes. Logistical support was provided by Marca Eventos, Scorpius Viagens e Turismo and the staff at Hotel Fazenda Fonte Colina Verde. Finally, we thank our children and wives, Lisa Wysocki, Celeste Silva and Carina Denny, for putting up with our absences and absent-mindedness while we prepared for the symposium.

Contributors

Juliann E. Aukema, *Department of Ecology and Evolutionary Biology, University of Arizona, Tucson, AZ 85721-0033, USA*

Michael L. Avery, *Wildlife Services, USDA-APHIS, National Wildlife Research Center, 2820 East University Avenue, Gainesville, FL 32641, USA (e-mail: michael.l.avery@aphis.usda.gov)*

Martijn Bartholomeus, *Tropical Nature Conservation and Vertebrate Ecology Group, Department of Environmental Sciences, Wageningen University, Bornsesteeg 69, NL – 6708 PD Wageningen, The Netherlands*

Bernardina Bello y Bello, *Instituto de Ecología, Universidad Nacional Autonoma de México, Mexico DF, Mexico*

John G. Blake, *Department of Biology and International Center for Tropical Ecology, University of Missouri – St Louis, 8001 Natural Bridge Road, St Louis, MO 63121-4499, USA (e-mail: blake@jinx.umsl.edu)*

Katrin Böhning-Gaese, *Institut für Biologie II, RWTH Aachen, Kopernikusstr. 16, 52074 Aachen, Germany (e-mail: boehning@bio2.rwth-aachen.de)*

Lynn A. Bohs, *Department of Biology, University of Utah, 257 South 1400 East, Salt Lake City, UT 84112, USA (e-mail: bohs@biology.utah.edu)*

Frans Bongers, *Silviculture and Forest Ecology Group, Department of Environmental Sciences, Wageningen University, PO Box 342, NL-6700 AH Wageningen, The Netherlands*

Osvaldo Calderón, *Smithsonian Tropical Research Institute, Apartado 2072, Balboa, Ancón, Republic of Panama*

Colin A. Chapman, *Department of Zoology, PO Box 118525, University of Florida, Gainesville, FL 32611, USA (email: cachapman@zoo.ufl.edu)*

Lauren J. Chapman, *Wildlife Conservation Society, 185th Street and Southern Boulevard, Bronx, New York, NY 10460, USA (e-mail: ljchapman@zoo.ufl.edu)*

Jerome Chave, *Department of Ecology and Evolutionary Biology, Princeton University, Princeton, NJ 08544-1003, USA*

Martin L. Cipollini, *Department of Biology, PO Box 430, Berry College, Mount Berry, GA 30149, USA (e-mail: mcipollini@berry.edu)*

Richard T. Corlett, *Department of Ecology and Biodiversity, University of Hong Kong, Pokfulam Road, Hong Kong, China (e-mail: corlett@hkucc.hku.hk)*

Paulo De Marco Jr, *Laboratório de Ecologia Quantitativa, Universidade Federal de Viçosa, 36570-000 Viçosa, Minas Gerais State (MG), Brazil*

Jan den Ouden, *Silviculture and Forest Ecology Group, Department of Environmental Sciences, Wageningen University, PO Box 342, NL – 6700 AH Wageningen, The Netherlands*

Rodolfo Dirzo, *Instituto de Ecología, Universidad Nacional Autonoma de México, Mexico DF, Mexico*

Donald R. Drake, *Botany Department, University of Hawaii, 3190 Maile Way, Honolulu, HI 96822, USA*

R. Scot Duncan, *Department of Zoology, PO Box 118525, University of Florida, Gainesville, FL 32611, USA (e-mail: duncan@zoo.ufl.edu)*

Jelmer A. Elzinga, *Centre for Terrestrial Ecology, Netherlands Institute of Ecology, PO Box 40, NL – 6666 ZG Heteren, The Netherlands*

Pierre-Michel Forget, *Muséum National d'Histoire Naturelle, Laboratoire d'Ecologie Générale, CNRS-MNHN UMR 8571, 4 avenue du petit Château, F-91800 Brunoy, France (e-mail: forget@mnhn.fr)*

Robin B. Foster, *The Field Museum, 1400 S. Lake Shore Drive, Chicago, IL 60605-2496, USA*

Mauro Galetti, *Plant Phenology and Seed Dispersal Research Group, Departamento de Ecologia, Universidade Estadual Paulista (UNESP), CP 199, 13506-900 Rio Claro, São Paulo, Brazil (e-mail: mgaletti@rc.unesp.br)*

J. Mauricio Garcia-C., *Apartado 1179-2100, Guadalupe, San Jose, Costa Rica*

José A. Godoy, *Estación Biológica de Doñana, CSIC, Apdo. 1056, E-41080 Sevilla, Spain*

Verônica S.M. Gomes, *Graduate Program in Ecology, IB UNICAMP, 13083-970 Campinas, São Paulo State, Brazil*

David S. Hammond, *Iwokrama International Centre for Rain Forest Conservation and Development, PO Box 10630, 67 Bel Air, Georgetown, Guyana*

Érica Hasui, *Graduate Program in Ecology, IB UNICAMP, 13083-970 Campinas, São Paulo State (SP), Brazil*

Henry S. Horn, *Department of Ecology and Evolutionary Biology, Princeton University, Princeton, NJ 08544-1003, USA*

Carol C. Horvitz, *Department of Biology, PO Box 249118, University of Miami, Coral Gables, FL 33124, USA (e-mail: carolhorvitz@miami.edu)*

Kazuhiko Hoshizaki, *Department of Biological Environment, Akita Prefectural University, Akita 010-0195, Japan (e-mail: khoshiz@akita-pu.ac.jp)*

Stephen P. Hubbell, *Botany Department, University of Georgia, Athens, GA 30602-7271, USA*

Philip E. Hulme, *NERC Centre for Ecology and Hydrology, Hill of Brathens, Banchory AB31 4BY, Scotland, UK (e-mail: pehu@ceh.ac.uk)*

Ido Izhaki, *Department of Biology, University of Haifa at Oranim, Tivon 36006, Israel (e-mail: izhaki@research.haifa.ac.il)*

Patrick A. Jansen, *Silviculture and Forest Ecology Group, Department of Environmental Sciences, Wageningen University, PO Box 342, NL – 6700 AH Wageningen, The Netherlands (email: Patrick.AJansen@btbo.bosb.wau.nl)*

Pedro Jordano, *Estación Biológica de Doñana, CSIC, Apdo. E-41080 Sevilla, Spain (e-mail: jordano@cica.es)*

Beth A. Kaplin, *Department of Environmental Biology, Antioch New England Graduate School, 40 Avon Street, Keene, NH 03431, USA (e-mail: bkaplin@antiochne.edu)*

Joanna E. Lambert, *Department of Anthropology, University of Oregon, Eugene, OR 97403-1218, USA (e-mail: jlambert@oregon.uoregon.edu)*

Josiane LeCorff, *Institut d'Horticulture, 2 rue le Nôtre, 49045 Angers, France*

Douglas J. Levey, *Department of Zoology, University of Florida, Gainesville, FL 32611, USA (e-mail: dlevey@zoo.ufl.edu)*

Simon A. Levin, *Department of Ecology and Evolutionary Biology, Princeton University, Princeton, NJ 08544-1003, USA*

Bette A. Loiselle, *Department of Biology and International Center for Tropical Ecology, University of Missouri – St Louis, 8001 Natural Bridge Road, St Louis, MO 63121-4499, USA (email: loiselle@umsl.edu)*

Janice M. Lord, *Department of Botany, University of Otago, PO Box 56, Dunedin, New Zealand (E-mail: jlord@planta.otago.ac.nz)*

Kim R. McConkey, *School of Biological Sciences, Victoria University of Wellington, PO Box 600, Wellington, New Zealand (e-mail: Kim.Mcconkey@vuw.ac.nz)*

Adrienne S. Markey, *Department of Botany, University of Otago, PO Box 56, Dunedin, New Zealand*

Jane Marshall, *Department of Botany, University of Otago, PO Box 56, Dunedin, New Zealand*

Carlos Martínez del Rio, *Department of Zoology and Physiology, University of Wyoming, Laramie, WY 82071-3166, USA (e-mail: cmdelrio@uwyo.edu)*

Karina Martins, *Laboratório de Ecologia Trófica, Departamento de Ecologia, Instituto de Biociências da Universidade de São Paulo, 05508-900 São Paulo, São Paulo State, Brazil*

Tarek Milleron, *Department of Rangeland Resources and the Ecology Center, Utah State University, Logan, UT 84322-5230, USA*

Kim Mink, *Department of Biology, PO Box 430, Berry College, Mount Berry, GA 30149, USA*

Susan M. Moegenburg, *Smithsonian Migratory Bird Center, National Zoological Park, Washington DC 20008, USA (e-mail: moegenburgs@nzp.si.edu)*

José Carlos Motta-Junior, *Laboratório de Ecologia Trófica, Departamento de Ecologia, Instituto de Biociências da Universidade de São Paulo, 05508-900 São Paulo, São Paulo State, Brazil (e-mail: mottajr@ib.usp.br)*

Helene C. Muller-Landau, *Department of Ecology and Evolutionary Biology, Princeton University, Princeton, NJ 08544-1003, USA (e-mail: helene@eno.princeton.edu)*

K. Greg Murray, *Department of Biology, Hope College, Holland, MI 49423, USA (e-mail: gmurray@hope.edu)*

Ran Nathan, *Department of Ecology and Evolutionary Biology, Princeton University, Princeton, NJ 08544-1003, USA (email: rann@bgumail.bgu.ac.il)*

Percy Núñez V., *Herbario Vargas, Universidad Nacional San Antonio de Abad de Cusco, Cusco, Peru*

Eric Paulk, *Department of Biology, PO Box 430, Berry College, Mount Berry, GA 30149, USA*

Carlos A. Peres, *School of Environmental Sciences, University of East Anglia, Norwich, Norfolk NR4 7TJ, UK (e-mail: C.Peres@uea.ac.uk)*

Nigel Pitman, *Center for Tropical Conservation, Duke University, PO Box 90381, Durham, NC 27708, USA*

Marco A. Pizo, *Departamento de Botânica – Universidade Estadual Paulista, Caixa Postal 199, 13506-900 Rio Claro, São Paulo State, Brazil (e-mail: pizo@rc.unesp.br)*

Carla Restrepo, *Department of Biology, University of New Mexico, Albuquerque, NM 87108, USA (e-mail: carlae@sevilleta.unm.edu)*

Sabrina E. Russo, *Department of Ecology, Ethology and Evolution, University of Illinois, Urbana, IL 61801-3707, USA*

Sarah Sargent, *Department of Environmental Sciences, Allegheny College, Meadville, PA 16335, USA (e-mail:ssargent@alleg.edu)*

Heather Schichter, *Center for Tropical Conservation, Duke University, PO Box 90381, Durham, NC 27708, USA*

Eugene W. Schupp, *Department of Rangeland Resources and the Ecology Center, Utah State University, Logan, UT 84322-5230, USA (e-mail: schupp@cc.usu.edu)*

Miles Silman, *Department of Biology, PO Box 7325, Wake Forest University, Winston-Salem, NC 27109, USA*

Wesley R. Silva, *Laboratório de Interações Vertebrados-Plantas, Depto. Zoologia, UNICAMP, 13083-970 Campinas, São Paulo State, Brazil (e-mail: wesley@unicamp.br)*

John Terborgh, *Center for Tropical Conservation, Duke University, PO Box 90381, Durham, NC 27708, USA (e-mail: manu@acpub.duke.edu)*

Raquel Thomas, *Iwokrama International Centre for Rain Forest Conservation and Development, PO Box 10630, 67 Bel Air, Georgetown, Guyana*

Anna Traveset, *Institut Mediterrani d'Estudis Avançats (CSIC-UIB), C/Miquel Marqués 21, 07190-Esporles, Mallorca, Balearic Islands, Spain (e-mail: ieaatv@clust.uib.es)*

Stephen B. Vander Wall, *Department of Biology and the Program in Ecology, Evolution and Conservation Biology, University of Nevada, Reno, NV 89557, USA (e-mail: sv@med.unr.edu)*

Marc van Roosmalen, *Departamento de Botânica, Instituto Nacional de Pesquisa da Amazônia, Caixa Postal 478, Manaus, Amazonas 69011-970, Brazil*

Sipke E. Van Wieren, *Tropical Nature Conservation and Vertebrate Ecology Group, Department of Environmental Sciences, Wageningen University, Bornsesteeg 69, NL – 6708 PD Wageningen, The Netherlands*

Miguel Verdú, *Centro de Investigaciones sobre Desertificación (CSIC-UV), Camí de La Marjal s/n, Apdo. Oficial, 46470-Albal, Valencia, Spain*

David M. Watson, *Environmental Studies Unit, Charles Sturt University, Bathurst, NSW 2795, Australia*

S. Joseph Wright, *Smithsonian Tropical Research Institute, Apartado 2072, Balboa, Ancón, Republic of Panama*

Preface

Plant–animal interactions have attracted the attention of evolutionary biologists and field ecologists since the time of Charles Darwin. In fact, Darwin himself laid the foundation for pollination biology (Darwin, 1877), a field that has flourished ever since. The evolutionary ecology of a second type of plant–animal interaction, herbivory, started a rapid rise to prominence in the 1960s (e.g. Ehrlich and Raven, 1964). Because frugivory does not have the long history of pollination biology and lacks the agricultural importance of herbivory, the field was slower to develop. A breakthrough occurred in the early 1970s when David Snow and Doyle McKey provided predictive evolutionary frameworks for fruit–frugivore interactions (Snow, 1971; McKey, 1975) and when Daniel Janzen and Joseph Connell proposed far-reaching consequences of seed dispersal for plant community structure (Janzen, 1970; Connell, 1971). Their theories attracted much attention; in approximately 10 years, the field had reached puberty, brash and confident. In 1985 Ted Fleming (University of Miami) and Alejandro Estrada (Los Tuxtlas Biological Station) organized an international symposium–workshop to assess the status of the field. Held at the Los Tuxtlas Biological Station in Mexico, it brought together a small but influential group of researchers (Estrada and Fleming, 1986). Six years later, a second symposium (also organized by Ted Fleming and Alejandro Estrada) attracted a much larger audience and resulted in a frequently cited volume (Fleming and Estrada, 1993a). Much changed in those 6 years (Fleming and Estrada, 1993b). Perhaps even more has changed in the 9 subsequent years. Now, the theories that led development of the field are largely viewed as ineffective or simplistic. More unsettling is a vague uncertainty about where the field is headed (Wheelwright, 1991; Leighton, 1995; Levey and Benkman, 1999). Whether this uncertainty is real or perceived remains to be seen. To tackle it, we decided to convene a third international symposium. The response was outstanding (> 250 attendees) and resulted in many new ideas (Fuentes, 2000).

We invited a wide range of participants, from many countries, of many ages, and working on diverse taxa from widely varying perspectives. Most of these participants contributed to this book. The result, we hope, is an easily accessible portrait of our field. We should not be content with such a snapshot, though; we must look forward. For the field to thrive, we must foster the next generation of biologists and help them see clearly which studies are most likely to define the future of the field. Thus, we have asked all authors to comment on avenues of especially promising research. In most cases these comments can be found at the end of chapters. Additional viewpoints on the field and its most prominent challenges were provided by a diverse panel of attendees, who led a round-table discussion on the last afternoon of the symposium. Their comments will be made available through Frugivory Updates, a listserve maintained by Ronda Green at Griffith University, Queensland, Australia (Ronda.Green@mailbox.gu.edu.au).

A distinct difference between this symposium and the previous two was the frequent emphasis on management and conservation. The desire to understand the evolutionary importance of plant–seed-disperser interactions has been largely replaced by the drive to document the ecological importance of the interaction. Clearly, there has been a shift away from studies of single species and towards interactions among them. Preserving a forest and its fruiting trees may ultimately prove misguided if seed-dispersers are not preserved, too (Redford, 1992). Given the rate at which species-rich forests continue to be lost worldwide, we urged all authors to speculate about the implications of their work for conservation and management.

Like its predecessors, this book is divided into sections on Historical and Theoretical Perspectives, Plant Strategies, Animal Strategies, and Consequences of Seed Dispersal. The distinctions between these sections are not as clear-cut as their titles imply; many chapters could have been placed in more than one section. Unlike its predecessors, this book includes a section on Conservation, Biodiversity and Management. In this section, authors present some of the first studies that directly address the ecological implications of seed-disperser extinction. The section also provides two chapters that focus on management of fruiting plants and management of fruit-eating birds.

Attendees of the symposium left reinvigorated. It is our hope that readers of this book can likewise catch the excitement and grasp the challenges presented at the symposium. The field is fertile, the seeds plentiful and the climate supportive. We hope the ideas expressed in this book will disperse, take root and grow!

Doug Levey
Wesley Silva
Mauro Galetti

References

Connell, J.H. (1971) On the role of natural enemies in preventing competitive exclusion in some marine animals and in rain forest trees. In: den Boer, P.J. and Gradwell, G.R. (eds) *Dynamics of Numbers in Populations.* Centre for Agricultural Publication and Documentation, Proceedings of the Advanced Study Institute, Osterbeek, Wageningen, The Netherlands, pp. 298–312.

Darwin, C. (1877) *The Various Contrivances by which Orchids are Fertilised by Insects.* J. Murray, London, 300 pp.

Ehrlich, P.R. and Raven, P.H. (1964) Butterflies and plants: a study in coevolution. *Evolution* 18, 586–608.

Estrada, A. and Fleming, T.H. (1986) *Frugivores and Seed Dispersal.* Junk, Dordrecht, The Netherlands, 392 pp.

Fleming, T.H. and Estrada, A. (1993a) *Frugivory and Seed Dispersal: Ecological and Evolutionary Aspects.* Kluwer, Dordrecht, The Netherlands, 392 pp.

Fleming, T.H. and Estrada, A. (1993b) General introduction. In: Fleming, T.H. and Estrada, A. (eds) *Frugivory and Seed Dispersal: Ecological and Evolutionary Aspects.* Kluwer, Dordrecht, The Netherlands, pp. xi–xii.

Fuentes, M. (2000) Frugivory, seed dispersal, and plant community ecology. *Trends in Ecology and Evolution* 15, 487–488.

Janzen, D.H. (1970) Herbivores and the number of tree species in tropical forests. *American Naturalist* 104, 501–528.

Leighton, M. (1995) Frugivory as a foraging strategy for ecologists. *Ecology* 76, 668–669.

Levey, D.J. and Benkman, C.W. (1999) Fruit–seed disperser interactions: timely insights from a long-term perspective. *Trends in Ecology and Evolution* 14, 41–44.

McKey, D.S. (1975) The ecology of coevolved seed dispersal systems. In: Gilbert, L.E. and Raven, P.H. (eds) *Co-evolution of Animals and Plants.* University of Texas Press, Austin, Texas, pp. 159–191.

Redford, K.H. (1992) The empty forest. *Bioscience* 42, 412–422.

Snow, D.W. (1971) Evolutionary aspects of fruit-eating by birds. *Ibis* 113, 194–202.

Wheelwright, N.T. (1991) Frugivory and seed dispersal: 'La coevolución ha muerto – !Viva la coevolución' *Trends in Ecology and Evolution* 6, 312–313.

1 Maintenance of Tree Diversity in Tropical Forests

John Terborgh[1], Nigel Pitman[1], Miles Silman[2], Heather Schichter[1] and Percy Núñez V.[3]

[1]*Center for Tropical Conservation, Duke University, PO Box 90381, Durham, NC 27708, USA;* [2]*Department of Biology, PO Box 7325, Wake Forest University, Winston-Salem, NC 27109, USA;* [3]*Herbario Vargas, Universidad Nacional San Antonio de Abad de Cusco, Cusco, Peru*

Introduction

Many tropical forests contain hundreds of tree species; some contain well over 1000. What are the mechanisms that allow such complex communities to persist over millennia? The question has no simple or agreed-upon answer; the only true answer is that we don't know.

Theoreticians have offered a great many suggestions, ten of which are listed in Table 1.1. The proposed mechanisms are extremely diverse. Some are based on abiotic processes, others on biotic processes. Some operate entirely via chance, whereas others depend on deterministic processes. For all these differences, the ten theories recall the Tower of Babel when considered collectively. Given the irreconcilable differences between some of them, it is highly unlikely that all of them are right. However, it is still possible that many of them contain a grain of truth, so it is best to keep an open mind until each can be tested with appropriate evidence.

To attempt a rigorous test of ten theories in a single chapter would be presumptuous, if not tedious. Here our purpose will be to present data from a series of empirical studies undertaken in western Amazonian forests at a wide range of spatial scales. We shall then comment briefly on each of the theories in light of the data presented. Some of the results provide direct tests of one or another of the theories, whereas other results may serve to inform new theories as yet unborn.

How do the processes that regulate tree species composition vary with spatial scale, from roughly 1 million square kilometres down to less than 1 m^2? We shall begin to answer the question by examining patterns of species composition at the very largest spatial scale and then work down to smaller scales.

Results and Methods

Pattern at the subcontinental scale

A commonly held view of Amazonian forests is that species composition is highly variable, even on spatial scales of a few hectares to a few square kilometres (Gentry, 1988). This idea has been given recent impetus by analyses of false-colour Landsat images, which reveal

Table 1.1. Ten theories of plant diversity.

Theory	Author
Broken stick	MacArthur, 1957
Niche pre-emption	Whittaker, 1965
Escape in space	Janzen, 1970; Connell, 1971
Intermediate disturbance	Connell, 1978
Community drift (non-equilibrium)	Hubbell, 1979
Lottery competition	Chesson and Warner, 1981
Resource limitation	Tilman, 1982, 1988
Spatial heterogeneity	Pacala and Tilman, 1994
Dispersal limitation (winner by default)	Tilman 1994; Hurtt and Pacala, 1995
Pathogens	Givnish, 1999

that *terra firme* (upland) forests with distinct reflectance properties form a complex mosaic over great expanses of the Amazon (Tuomisto *et al.*, 1995). Randomly oriented 30 km long transects overlying images of the Peruvian Amazon intersected a median of four distinguishable reflectance patches, suggesting hundreds of distinct *terra firme* forest types within the Peruvian Amazon alone.

To examine this proposition, we shall evaluate results from a series of 1 and 2 ha tree plots in south-eastern Peru and eastern Ecuador. The sites were selected to represent forest patches discernible as distinctive reflectance signals in Landsat images (Pitman *et al.*, 1999). The two regions are 1400 km apart, a vast distance in relation to the scale of patches in the reflectance mosaic revealed by Landsat images. The Peruvian network consisted of nine plots ranging in size from 0.875 to 2 ha and totalling 13.875 ha. The Ecuadorean network consisted of 15 1 ha plots (Pitman, 2000). In both sets of plots, all stems ≥ 10 cm diameter at breast height (dbh) were marked, mapped, measured and identified (or assigned to morphospecies).

In both regions, most tree species showed landscape-scale densities of fewer than one individual per hectare, but most individual trees in both networks belonged to a suite of common species. These common species combine high frequency with high local abundance, to form predictable oligarchic matrices over areas of at least several thousand square kilometres in each region (Pitman, 2000; Pitman *et al.*, 2001). So strong is the pattern that only 15% of the species in each region comprise > 60% of the individual trees in almost every plot.

Not only are the forests within each of the two regions surprisingly homogeneous; they are also remarkably similar to each other. More than two-thirds of the Peruvian species have been collected within the region sampled in Ecuador. Many of the most common species in Peruvian *terra firme* are also very common in Ecuadorean *terra firme*. The same handful of families (*Arecaceae, Moraceae, Myristicaceae, Violaceae*) have more common species than expected in both regions (see Terborgh and Andresen, 1998; ter Steege, 2000). Large-statured tree species are more likely to be common in both forests than small ones.

Notably, tree species recorded in the Peruvian and the Ecuadorean inventories show similar relative abundances in the two regions, even though the sampled areas are separated by 11° of latitude. For the 254 species shared by the two networks, abundance in Ecuador is positively and highly significantly correlated with abundance in Peru ($P < 0.0001$, $r^2 = 0.18$) (Fig. 1.1).

These results paint a surprisingly simple picture of how tree communities may be distributed over the Amazonian landscape. In contrast to the small-scale patches discerned in Landsat images, our inventories suggest that a relatively homogeneous but highly diverse tree community blankets a huge area that extends from Ecuador to south-eastern Peru, and perhaps beyond. A similar large-scale continuity of forest composition has recently been documented in Guyana (ter Steege, 2000).

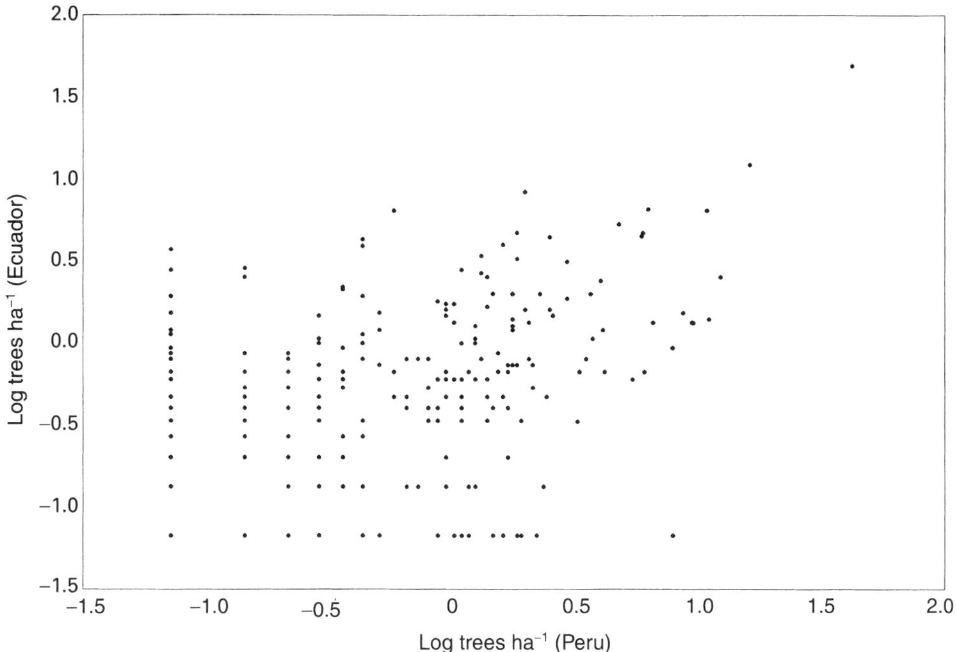

Fig. 1.1. Scattergram showing, for shared tree species, the abundance of each species in eastern Ecuador vs. south-eastern Peru.

Pattern at the regional scale

Just as notably, much of the uniformity observed at very large spatial scales is maintained when the scale is ratcheted down to the regional level. Many authors have emphasized habitat differentiation, especially as generated by edaphic gradients, as crucial to maintaining the diversity of plant communities (Tilman, 1982, 1988; Ashton and Hall, 1992; Tuomisto and Ruokolainen, 1994; Tuomisto *et al.*, 1995). Here we ask how south-west Amazonian trees respond to edaphic gradients. Al Gentry, who wrote extensively about Amazonian forests, was under the impression that beta diversity was characteristically high (Gentry, 1988). For example, two *terra firme* forest plots located only 2 km apart shared only about half their species, a fact he interpreted as supporting this view. But there remains the possibility that the extremely high diversity of these forests results in a sampling variance so great that compositional consistencies are effectively masked in small samples.

To examine this issue more closely, we consider data from 21 plots totalling 36 ha situated over an area of roughly 400 km^2 within the Manu Biosphere Reserve in south-eastern Peru. Each plot encompasses 1 or 2 ha, and the whole set of plots totals nearly 20,000 stems ≥ 10 cm dbh. The plots represent four edaphically distinct divisions of the landscape: upland (*terra firme*) forests, mature flood-plain forests, primary successional forests in river meanders and swamp forests (Pitman *et al.*, 1999).

Collectively, the plots contained 829 species and morphospecies. Nearly half of these taxa (45%) occurred in only one of the 21 plots. Little can be concluded about the edaphic requirements of such rare species, if only because many of them were represented by a single stem. If we take the 426 species that occurred in two or more plots, only 26% were confined to a single one of the four major forest types (Fig. 1.2). If we qualify the data even further and consider only the 365 species that occurred in three or more plots, then the proportion restricted to a single forest type drops to 15%. Clearly, if our sample of trees had been 40,000 or 100,000, instead of 20,000, the proportion restricted to a single forest type

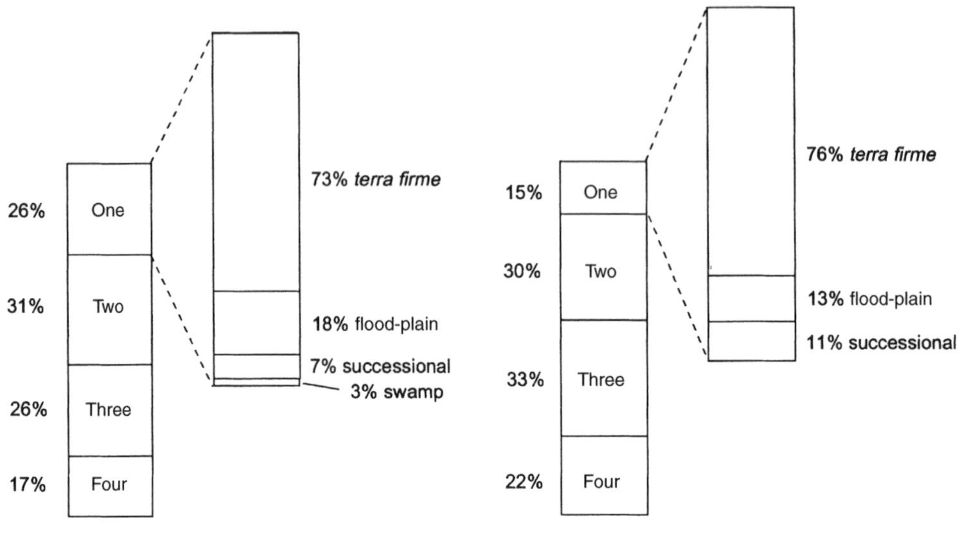

Fig. 1.2. Patterns of occurrence of tree species in 21 plots (36 ha) roughly evenly distributed over four major divisions of the landscape in south-eastern Peru: *terra firme* (uplands), mature flood-plain forest, primary successional forests and swamps. The left diagram includes 462 species and morphospecies that occur in two or more plots; the right diagram includes 365 species that occur in three or more plots. In both diagrams, the left-hand columns show the proportion of species occurring in one, two, three or four forest types; the right-hand columns show the proportions of habitat-restricted species in each of the four forest types. (Reprinted from Pitman *et al.*, 1999, with permission.)

would have dropped even further, probably to less than 10%. Of course, it is possible that the more common species are able to occupy a wider range of edaphic conditions than rare species (Brown, 1995), but, even so, increases in sample size would inevitably lower the fraction of edaphically restricted species. These results are not peculiar to south-eastern Peru. When sampling effort is standardized, data from the Ecuadorean plot network yield very similar results.

Pattern at the subhectare scale

At the spatial scale of a hectare or less, evolutionary–biogeographical influences and variation in physical factors are constrained to a minimum. Instead, we enter the realm of distances over which biotic interactions are presumed to be paramount. One such set of interactions is described by the Janzen–Connell mechanism, which proposes that the probability of survival of a seed or successful establishment of a seedling increases with distance from its parent tree. Seeds that fall relatively far from the parent enjoy enhanced survival, as they 'escape in distance' from predators, herbivores and/or pathogens (Janzen, 1970; Connell, 1971). There have now been dozens of attempts to test and evaluate the Janzen–Connell mechanism, using a variety of tree species and experimental designs (Clark and Clark, 1984).

Although the Janzen–Connell model was proposed 30 years ago, we know of no effort to investigate the consequences of the mechanism, as opposed to the processes driving it. We can do this, as it were, by looking back instead of forward. Rather than beginning with focal trees and studying the fates of seeds around them, we can reverse the process by starting with saplings and then asking how far each is to the nearest potential parent tree. To do this, we mapped all adult trees (defined as stems ≥ 10 cm dbh) in a 2.25 ha plot in mature flood-plain forest at the Cocha Cashu Biological Station in Peru's Manu National Park. Then we subsampled the nine central 30 m × 30 m subplots for small saplings, defined as those ≥ 1 m tall but < 1 cm dbh (Fig. 1.3). (Effectively, these are saplings ranging from 1 m to about 2.5 m in height.) By considering only those saplings growing in the central portion of the 150 m × 150 m adult tree plot, we ensured that all conspecific adults growing within 30 m of any sapling would have a known location. The central subplots were inventoried for saplings in 1996, 1997 and 1998. The locations of saplings were then related to the adult tree stand as it existed in 1990 to allow for the fact that most saplings were at least several years old at the time they were mapped.

For each sapling representing the 19 most common species in the adult tree stand ($n \geq 10$), we calculated the distance from the nearest conspecific adult that could have been its parent (Fig. 1.4). As predicted by the Janzen–Connell model, there were fewer saplings close to adults than somewhat farther away; the median sapling was 14 m from the nearest conspecific adult. It should be noted that the distances between conspecific adults of these common species were mostly in the range of 20–50 m, so that few saplings could be more than 10–25 m from

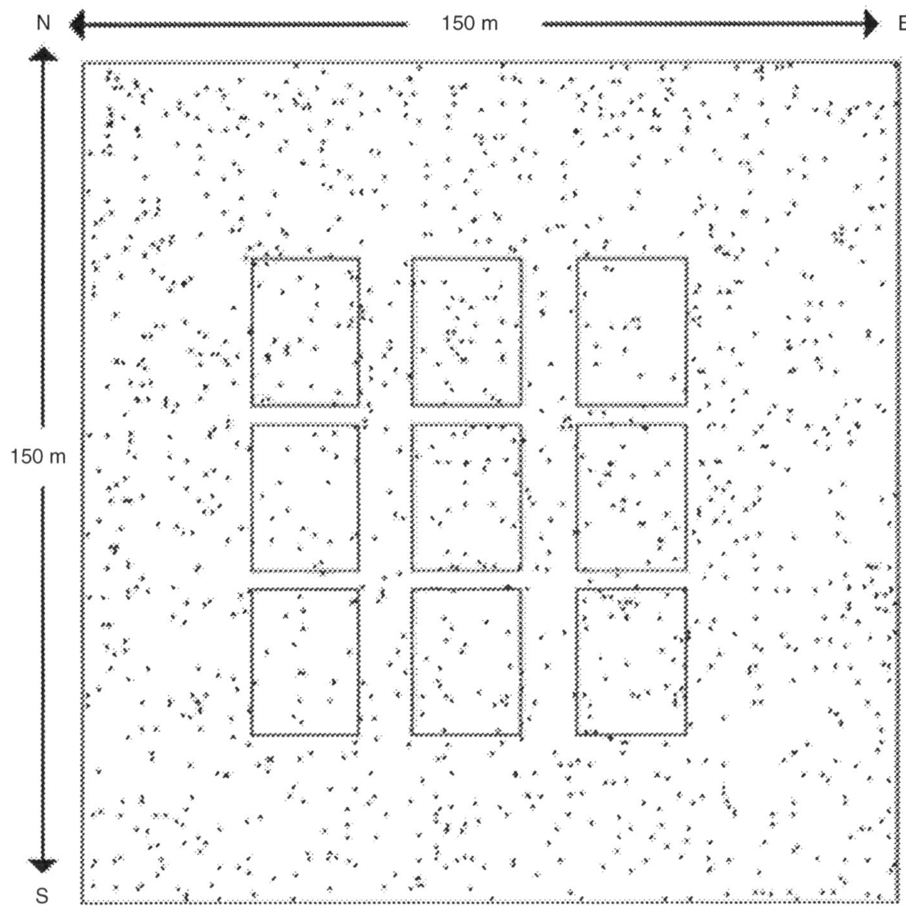

Fig. 1.3. Schematic diagram of the 2.25 ha adult tree plot at Cocha Cashu, Peru. The plot is subdivided into 25 30 m × 30 m subplots. Saplings ≥ 1 m tall and < 1 cm dbh were sampled in the nine central subplots to ensure that all saplings were ≥ 30 m from the nearest boundary of the adult tree plot.

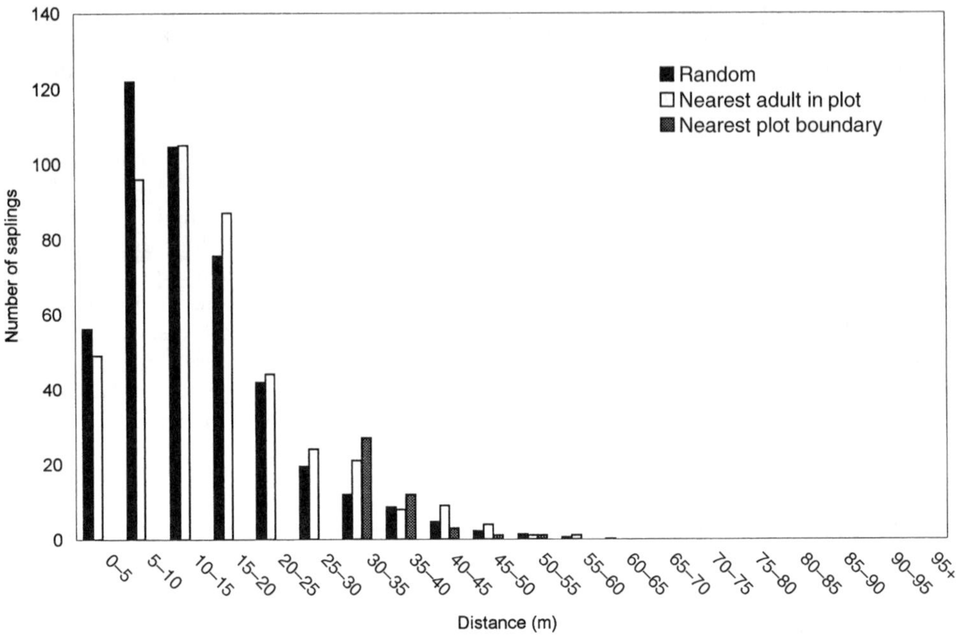

Fig. 1.4. Distribution of distances of saplings ≥ 1 m tall and ≤ 1 cm dbh to nearest conspecific adult for 19 common tree species in the 2.25 ha tree plot at Cocha Cashu, Peru. Black bars show the distribution for the same number of saplings of each species had they been randomly dispersed. White bars represent distances from saplings to the nearest conspecific adult in the plot. The differences between this distribution and randomly arrayed saplings are highly significant ($\chi^2 = 20.34$, $P < 0.01$). (Stippled bars represent distances from saplings to the nearest plot boundary in those cases where a plot boundary was closer than the nearest conspecific adult within the plot.) The species represented in Figs 1.4 and 1.5 were screened to eliminate inhomogeneities in the data set. Specifically, we did not include species that were (i) palms, (ii) small as adults (maturing at dbh < 10 cm) or (iii) considered to be gap pioneers.

the nearest adult, even if all stems were randomly arrayed in space. For perspective, we note that the mean nearest-neighbour distance between adult trees of any species is roughly 4 m.

If equal numbers of saplings of each species had been thrown down at random in the nine central subplots, the distribution of distances to nearest conspecific adults would be as shown by the black bars in Fig. 1.4. Comparison of the random vs. actual distributions reveals that somewhat fewer saplings than expected were close (≤ 10 m) to conspecific adults and somewhat more were further away (≥ 15 m), as would be predicted by the Janzen–Connell mechanism.

Now, we must introduce a complication. Some of the values shown for saplings that were > 30 m from the nearest conspecific adult are overstated, because the nearest conspecific adult to some saplings was likely to have been outside the adult tree plot and therefore invisible to the analysis. While we cannot eliminate this error, we can bound it by comparing the distance of each sapling from the nearest conspecific adult within the 2.25 ha plot with that of the nearest boundary of the plot, on the very conservative assumption that the nearest conspecific adult lay just outside the nearest boundary (stippled bars). The true values for nearest adults that were ≥ 30 m away therefore lie somewhere between the limits described by these two measures.

Next, we consider a group of 75 species that we shall label as 'less common' ($n \geq 1$, < 10). It would not be appropriate to call them 'rare', because more than half the tree species in the landscape occur at a density of fewer than one individual per hectare (see above). (The plot contains > 250 species of trees

≥ 10 cm dbh.) Among these less common species, there is a clear tendency to recruit closer to adults than to random points in the forest, although the median distance to nearest conspecific adults was substantially greater than those for common species (32 m) (Fig. 1.5).

If we could accurately measure the seed shadow of each species in the plot, it would be possible to make direct comparisons of the numbers of seeds falling vs. the numbers of saplings at different distances from adult trees. However, measuring dispersal has remained a daunting challenge at the empirical level because of the difficulties inherent in tracking seeds or their dispersers over large expanses of tropical forest (Wenny, 2000). The difficulty of measuring seed shadows is particularly acute in the case of large-seeded trees, because the size of their fruit crops tends to be small (Forget, 1992). Yet, in the forest at Cocha Cashu, large-seeded species predominate in terms of stand basal area (Silman, 1996).

We shall now look at two views of dispersal, one based on the method of using arrays of seed traps, and the other based on finding where seeds are eventually successful. We shall see that comparing the two approaches proves to be highly enlightening.

Within the same 2.25 ha permanent tree plot described above, Silman set up 40 0.5 m^2 seed traps at 20 stations arrayed in a rectangular grid, and monitored them for 2 years (Silman, 1996). At each sampling station, there were two traps located 5 m apart to control for small-scale sampling variation. The 20 m^2 of traps captured > 20,000 seeds of 195 species over the 2 years. However, half of the species were represented by only one or two seeds that fell into a single trap. Only 13 of the 195 species reached even 20% of the traps, and these were mostly wind-dispersed lianas (Fig. 1.6).

The surrounding flood-plain forest community is known to contain at least 905 species of trees, shrubs and lianas, of which the vast majority of species (88%) were invisible to the seed traps. Paradoxically, trees from early successional habitats near the river almost 1 km distant dispersed into the plot in surprising numbers: *Alchornea* (*Euphorbiaceae*) 20 seeds, *Sapium aereum* (*Euphorbiaceae*) 14

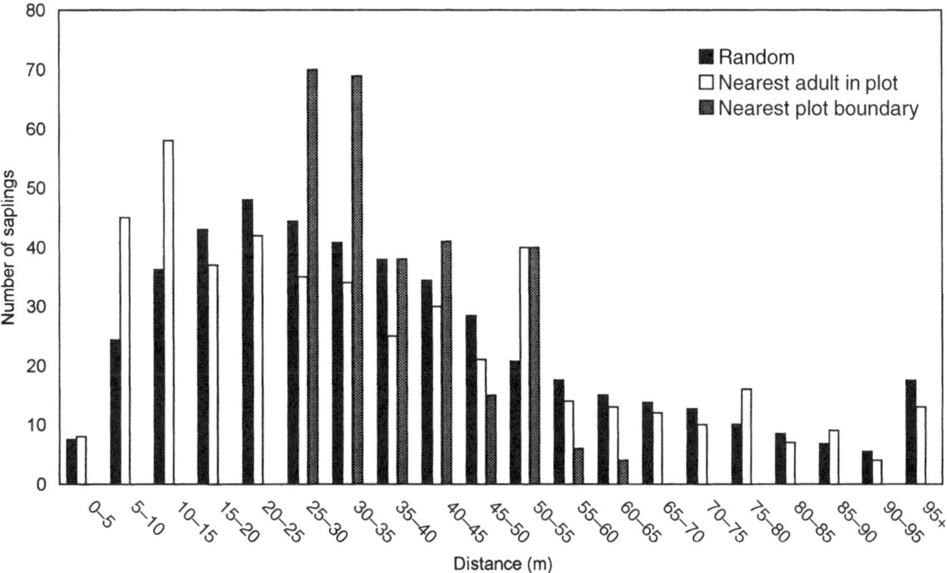

Fig. 1.5. Distribution of distances of saplings ≥ 1 m tall and ≤ 1 cm dbh to nearest conspecific adult for 75 less common tree species in the 2.25 ha tree plot at Cocha Cashu, Peru. Interpretation as in Fig. 1.4. The distribution of distances to nearest adult within the plot (white bars) and that of randomly arrayed saplings are highly significantly different (χ^2 = 68.13, P < 0.005).

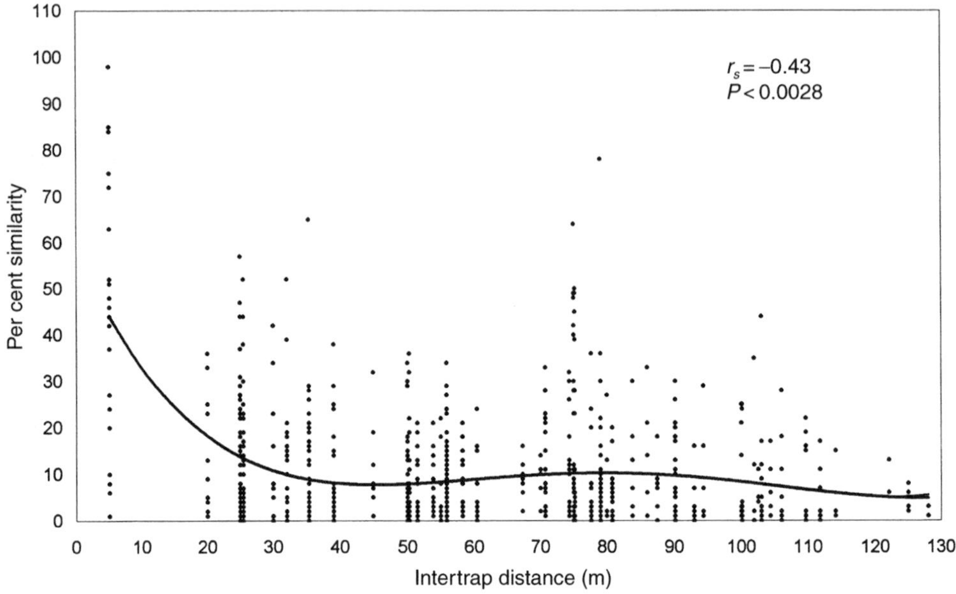

Fig. 1.6. Spatial autocorrelation of composition of seed rain falling into 40 traps arrayed in a grid within the 2.25 ha tree plot at Cocha Cashu, Peru. Spatial correlation drops to background level within approximately 30 m, showing that most seeds fall within 30 m of adults.

seeds, *Sapium ixiamasense* 11 seeds and *Guarea guidonia* (*Meliaceae*) five seeds. These species (all bird-dispersed) contributed to the seed rain, even though no known individuals of any of them occur within 500 m of the plot.

With the exception of the relatively few seeds imported from far outside the plot, the impression given by the results is that most rain-forest trees experience extremely limited dispersal. Similar, but even more extensive measurements made with seed traps on Barro Colorado Island, Panama, convey the same impression (Hubbell *et al.*, 1999). Over a 13-year period, 1.3 million seeds were collected from 200 traps, and yet, on average, seeds of roughly a third of the species with adults in the plot failed to hit any of the traps in a given year (H. Muller-Landau, personal communication). However, all species represented by adults were registered in traps at some time over the 13 years. Results such as these have given rise to the theoretical notion of 'winner by default', which refers to the occupancy of recruitment sites by species other than the best competitor in the community because of dispersal limitation (Hubbell *et al.*, 1999).

But, if dispersal is as limited as seed-trap data seem to imply, many tree species populations should be extremely clumped. It is true that the populations of most tropical tree species are clumped, but not nearly to the degree implied by empirically determined seed shadows. Why not? Undoubtedly, it is because seed traps capture seeds before they are exposed to terrestrial seed predators, whereas saplings represent seeds that escaped seed predators. In addition, so-called secondary dispersal may play a crucial role for some species, and secondary dispersal is invisible to seed traps (Andresen, 1999).

A lot of biology transpires between the moment a seed hits the ground and the time it morphs into a sapling. This is evident when one compares a middle-of-the-road seed-shadow model with the observed distribution of saplings (Clark *et al.*, 1999; Fig. 1.7). If seeds were to fall to the ground in accord with a negative exponential with distance, the expected density of seeds would drop to an undetectable level only a few crown diameters away from a given fruiting adult (Nathan and Muller-Landau, 2000).

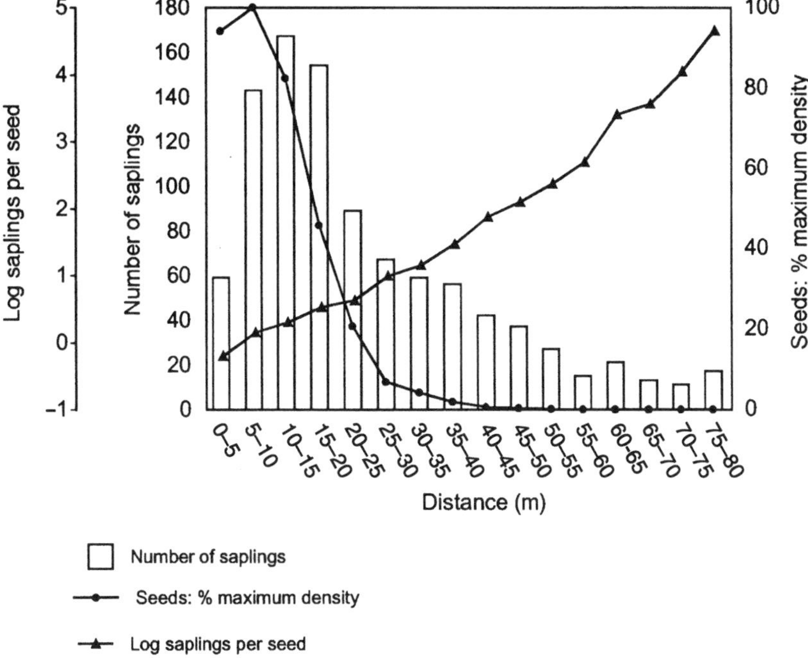

Fig. 1.7. Contrast between a generalized seed shadow that mimics Fig. 1.6 (negative exponential, line marked by solid dots) and the distribution of saplings in respect of the nearest conspecific adults of all 19 'common' and 75 'less common' species combined (white bars). The line marked by triangles represents the hypothetically increasing 'value' of a seed as it is dispersed away from the parent tree. This value is obtained by dividing the distribution of saplings by the distribution of seeds. The result suggests that the 'value' of a seed can increase several orders of magnitude as it is transported away from the parent tree. The scale is arbitrary.

If each seed that fell to the ground had an equal chance of becoming a sapling, the seed and seedling curves would be congruent, but it is obvious that they are dramatically distinct. This can be demonstrated by dividing the distribution of saplings in respect of nearest conspecific adult by the hypothetical seed shadow. The resulting derived curve (Fig. 1.7) gives a crude estimate of the 'value' of a seed dispersed to increasing distances. It is an avowedly rough estimate, because quantitative aspects of seed shadows vary greatly between species and because many years' seed crops may be required to generate a single surviving sapling. Nevertheless, the exercise is informative because it shows that the probability that a given seed will become a sapling increases by perhaps orders of magnitude when that seed is transported away from the parent.

Pattern at the metre scale

Much effort has recently been devoted to searching for density dependence in tropical tree communities (Hubbell and Foster, 1986; Condit *et al.*, 1994; Wills *et al.*, 1997). The fact that density dependence is concentrated at the earliest stages of the life cycle is brought out by what can be termed 'input–output' analysis. Silman (1996) determined that at least 500 seeds fall on to the average square metre of forest floor every year. (Probably the number is much higher, because the mesh used in his seed traps allowed seeds smaller than 1.5 mm in diameter to fall through.) Yet, despite this massive seed rain, the forest floor at Cocha Cashu is occupied by only about 20 plants m^{-2} (tree seedlings, vines and herbs combined, most of which are < 30 cm tall). These plants turn over at a rate of only 20%

year^{-1}, so that roughly four new individuals appear and replace four that have died every year in the average square metre (Terborgh and Wright, 1994). (To put this in perspective, consider the size of a typical starter pot that a gardener would use to raise seedlings prior to transplanting. Such pots typically have an area of *c.* 0.5 dm^2.) Yet, at 20 m^{-2}, the average seedling on the forest floor has 5 dm^2 it can call its own, roughly ten times the area of a seedling pot. Such plants are thus scattered to a degree that suggests interactions between them are relatively weak.

Discussion

Now, let us return briefly to the list of theories of plant diversity to see which ones have withstood the scrutiny of our analyses at multiple spatial scales (see Table 1.1).

Broken stick, niche pre-emption

These are black-box mechanisms that generate species abundance distributions not even remotely approximating the rank abundance curves typical of tropical forest tree communities (May and Stumpf, 2000; Plotkin *et al.*, 2000).

Intermediate disturbance

Connell (1978) proposed that the species diversity of space-limited communities (trees, intertidal organisms, etc.) will be low at high and low rates of disturbance and maximal at some intermediate rate. Phillips *et al.* (1994) have presented evidence suggesting an important role for disturbance in promoting tropical forest tree diversity, but the details of any such relationship are greatly in need of clarification.

Community drift

The community drift or 'non-equilibrium' hypothesis of Hubbell (1979) states that all tree species are adaptively equal, and that community composition will consequently vary over time as described by a random walk. Our findings argue strongly against community drift, which implies uncorrelated species abundances in spatially disjunct forests. Instead, we found that species in a few key families consistently dominate western Amazonian forests located 1400 km apart. Even more notable is the finding that abundance relationships are conserved over this great distance (Fig. 1.1). Such a high degree of spatial coherence of community composition cannot be reconciled by this model (Terborgh *et al.*, 1996). Chance thus fails as a means of accounting for forest homogeneity on such large spatial scales.

Lottery competition

Year-to-year climate fluctuations can have important consequences for tree demography, as abundantly affirmed by the work of Condit *et al.* (1995, 1996) on Barro Colorado Island, Panama. However, the fact that the highest tree diversities occur in regions with the least variable climates suggests that this mechanism makes, at best, only a minor quantitative contribution to the overriding question of how diversity is maintained through time.

Resource limitation

Spatial heterogeneity in the availability of limiting nutrients can generate a corresponding mosaic of species composition. Species turnover occurs on strong edaphic gradients in the tropics as well as elsewhere, but the low beta diversity observed at a landscape scale and the homogeneity of Amazonian forests at medium and large spatial scales seem to preclude an important role for this mechanism as an explanation for alpha diversity.

Dispersal limitation (winner by default)

The notion that dispersal limitation is a powerful force in tropical forest dynamics is partly an illusion resulting from the fact that the seeds captured in seed traps are nearly all (> 99%) destined for failure. If we could magically pick out of the seed rain those seeds that were earmarked for success, an entirely different picture would result because nearly all saplings appear to arise from dispersed seeds (see below).

Spatial heterogeneity

The notion that the forest floor is a complex mosaic of microsites that can enhance or depress the prospects of individual seeds is a powerful one, much in need of further empirical study.

Escape in space (Janzen–Connell), pathogens

These hypotheses are really different aspects of the Janzen–Connell mechanism. Evidence presented here strongly affirms the operation of Janzen–Connell at a broad community level, and further shows that the offspring of rare species are more strongly inhibited by the proximity of conspecific adults than those of common species.

Where are we and where do we go from here?

Perhaps a good place to begin is to point out that, of the ten theories of plant diversity listed in Table 1.1, there is only one that explicitly provides a role for animals. And that, not surprisingly, is Janzen–Connell. And yet how curious, for animals play crucial roles at every step in plant reproduction and recruitment: they pollinate flowers, they prey upon seeds, both before and after dispersal, they disperse both fruits and seeds, they carry out secondary dispersal – a much neglected mechanism – and they destroy or weaken seedlings and saplings through herbivory. So how can one have a theory of plant diversity in which animals play no explicit role? We are at a loss to say, other than to express our astonishment that so many students of plant ecology have overlooked the animals!

Animals are the heart and soul of Janzen–Connell, for they both disperse seeds and destroy them, thereby creating the large disparities we have noted between seed shadows and the distribution of saplings. Our results demonstrate that large numbers of species (in the aggregate) show the expected 'escape with distance' pattern anticipated by Janzen and Connell, providing strong affirmation of the postulated mechanism at a broad community level. These observations can be extended to suggest a new interpretation of 'commonness' and 'rarity' as consequences of interactions underlying the Janzen–Connell mechanism, namely, the events that transpire between dispersal and the successful establishment of seedlings.

On average, the saplings of common species recruit closer to potential parents than those of less common species (see also Condit *et al.*, 1992). Perhaps this is partly a consequence of the fact that common tree species at Cocha Cashu produce larger seeds on average than less common species. Otherwise, we know of no reason why the dispersal biology of the two categories of species should differ in any systematic way. However, other things being equal, it can be presumed that the seed shadows cast by adults of common species would overlap more than those of less common species, thereby elevating the seed rain falling on to the forest floor. Indeed, that is suggested by the fruit-trap data, which show that a small minority of species in the community produce a disproportionately large fraction of seeds captured in traps.

These observations lead us to surmise that what makes a 'common' species is the ability to recruit near an adult, as Schupp (1988) found with *Faramea occidentalis* and Hubbell and Foster (1986) found with *Trichilia tuberculata*, two of the commonest species on Barro Colorado Island, Panama. Conversely, what may make a 'rare' species is inability to recruit near a conspecific adult. Indeed, the median distance of saplings of 75 'less common' species to the nearest conspecific adult was 32 m, whereas the median for 19 'common' species was only 14 m. Nevertheless, there is strong evidence that the rarer a tree species is, the more clumped its distribution (Hubbell, 1979; Condit *et al.*, 2000). Statistical clumping is thus likely to be explained by the scale of the analysis; rare species tend to be clumped at larger scales (e.g. 1 ha), but apparently not at the smaller scales considered here.

Hubbell (1979) was both right and wrong when he maintained that density dependence must be very low, especially in less common species, because few saplings in tropical forests are nearest neighbours of conspecifics. The observation that few saplings of most species are conspecifics of their nearest neighbours is correct, but the inference that consequently there is little or no density dependence is incorrect. Janzen–Connell provides the density dependence in another form – one that is mediated by the actions of seed and seedling predators and pathogens. Comparisons of seed shadows with seedling distributions show density dependence to be very strong, but it operates most stringently before the seedling stage and is manifested as the nearly universal failure of seeds falling near the parent tree.

At Cocha Cashu, an annual input of ≥ 500 seeds m^{-2} results in an output of only four new seedlings. Inescapably, this means that > 99% of all seeds that fall to the ground fail to produce seedlings that survive even 1 year. Here, then, is where the important biology is happening that determines the future of the forest. The biology operates through both abiotic and biotic processes, which determine which seeds succeed and which fail. Clearly, if the forest is to perpetuate itself, seeds of every species succeed somewhere, but mostly they fail. Learning the 'rules' that determine success vs. failure for species having different seed sizes and dispersal modes emerges as a major challenge in tropical plant ecology.

The median adult tree (arbitrarily defined as those ≥ 10 cm dbh) in the forest at Cocha Cashu is 14 m tall. The median crown radius of trees this tall or taller in the flood-plain forest is 4 m (Terborgh and Petren, 1991). If we take this figure to represent the spread of an average adult tree, from Figs 1.4 and 1.5 it can easily be calculated that > 94% of the saplings of 'common' species establish at distances ≥ 4 m from the nearest conspecific adult and for 'rare' species the corresponding figure is > 98%.

Looking at this from another perspective, the median distance at which saplings of 'common' species establish from the nearest conspecific adult is 14 m, which is equivalent to 3.5 adult crown radii. In contrast, the median sapling of the 75 'less common' species is 32 m from the nearest conspecific adult, equivalent to eight adult crown radii. These distances describe an ample space around the adults of even common species in which the recruitment of conspecific saplings appears to be inhibited and in which, consequently, recruitment of heterospecifics is favoured. In this manner, 'rarity' can help promote diversity, as Janzen (1970) and Connell (1971) so cogently pointed out 30 years ago.

The finding that large numbers of saplings are growing at 20 m, 40 m or even further from the nearest conspecific adult is revealing, because it underscores the importance of dispersal away from parent trees. True dispersal distances are undoubtedly greater than our results suggest, because the parent of a given sapling is not always the nearest adult (Nathan and Muller-Landau, 2000). Indeed, our results make it clear that the vast majority of saplings in the forest originate from seeds dispersed well away from the parent tree. Conversely, undispersed seeds appear to have an extremely low success rate. The hugely increased probability of survival of dispersed seeds thus acts strongly to offset dispersal limitation and helps to explain the oft-cited enigma of isolated individual trees located hundreds of metres from the nearest conspecific.

Some implications of this can be explored in a thought experiment. What if the dispersal and/or seed and seedling predation regimes of a forest were to change as, say, a consequence of seed-disperser populations being decimated by hunting? Such perturbations of the animal community could be expected to alter the optimal recruitment distance for many species, as outlined by Janzen in his original 1970 paper. If this were to happen, the relative abundances of species in the next generation of adult trees would be dramatically modified. In particular, if recruitment distances were to decrease (via reduced dispersal with no concurrent change in seed predation and seedling herbivory), density dependence would increase (i.e. fewer seeds would escape) and diversity would be expected to decrease (see Dirzo and Miranda, 1991). Conversely, if recruitment distances were to increase via intensified seed predation and/or herbivory with no concurrent change in dispersal, more rare species could participate in the community and diversity should increase. If these imaginary scenarios could be shown to be true, what it would imply is that animals are fundamentally regulating plant diversity. In retrospect, this conclusion seems so obvious, but why have we ecologists so long avoided it?

Janzen (1970) and Connell (1971) (with recent support from Givnish, 1999) got it right, but fully appreciating this gets us only part way to a deeper understanding of the mechanisms that perpetuate tree species diversity through time. The Janzen–Connell mechanism ensures that a few lucky seeds will escape in distance from the largely biotic and deterministic mortality factors that operate near fruiting adults. But, once a seed has escaped these biotic mortality factors, the abiotic properties of the site will determine whether it succeeds or fails – whether it germinates and, if so, whether the conditions of light, moisture, etc. are adequate for the seedling to become vigorously established. Thus, spatial heterogeneity and abiotic processes have a large role to play too, one in which chance is prominent.

What perpetuates diversity, then, is the precise way animal activities (represented by Janzen–Connell) map on to the mosaic of physical microsites on the forest floor. It is therefore the intersection between Janzen–Connell and the abiotic world that should define the frontier of tropical forest ecology for the next generation of theoreticians and empiricists alike.

Chance vs. determinism as a function of scale

Finally, we return to larger spatial scales. A high degree of determinism is required to explain the observed homogeneity of western Amazonian forests at intermediate and large spatial scales. Exactly what the deterministic forces are that bring about the large-scale patterning of tropical tree communities remains to be elucidated, but we can hardly doubt that the structure of the animal community is an important component of the puzzle. At small scales, more chance enters into the process of tree establishment, as both abiotic and biotic factors (e.g. seed predation) control which seeds among the hundreds that fall in a given microsite survive and prosper. None of the existing theories of plant diversity incorporates this kind of scale-dependent complexity.

Our analysis of tree distribution at the landscape scale comes to a very different conclusion from Gentry's (1988). Beta diversity does not appear to be very high; indeed, it appears to be surprisingly low. A large majority of the tree species in the south-west Amazon, up to 90% or more, appear in two or more edaphically distinct plant communities, albeit at frequently contrasting abundance levels (Pitman *et al.*, 1999). Segregation of species on edaphic gradients does not seem to play a conspicuous role in organizing these communities.

However, this statement should be qualified by placing it in context. The entire landscape of Madre de Dios is an alluvial outwash plain of the Andes, nearly all of it of Pleistocene or Holocene age (Kalliola *et al.*, 1993). None of it is influenced by local bedrock, which, in any case, is buried under thousands of metres of sediment. The 'edaphically distinct' elements of this landscape that define its principal vegetation formations (*terra firme* forest, swamps, successional stands, etc.) are all constructed of the same or similar substrate but differ in such factors as hydrology, time of exposure to weathering and pH (Terborgh *et al.*, 1996). The region's floristic homogeneity thus reflects its geological homogeneity. In contrast, geologically complex tropical regions tend to display much greater floristic heterogeneity (Ashton and Hall, 1992; ter Steege, 2000).

The *terra firme* forests of eastern Ecuador and south-eastern Peru, although 1400 km apart, share hundreds of tree species, many of which occur in the two regions at similar relative abundances. Geographically, what ties the two regions together is that both lie in the Andean foreland region, which is built on recent alluvial sediments originating in the Andes (Salo *et al.*, 1986; Kalliola *et al.*, 1993). Perhaps this is what lies behind the strong floristic similarity of the two regions, notwithstanding the fact that eastern Ecuador is non-seasonal and receives 4000 mm of rain annually, whereas south-eastern Peru is markedly seasonal and receives only half as much precipitation (Pitman, 2000). The fact that 70% of the tree species recorded from south-eastern Peru also occur in eastern Ecuador (Pitman, 2000) argues against a prominent role of climate as a determinant of the floristic composition of the respective regions (see Condit *et al.*, 1995).

Another test of the role of climate can be found by comparing the flora of eastern Ecuador with that of the Iquitos region in north-eastern Peru. The Iquitos flora is markedly distinct from that of eastern Ecuador, even though Iquitos is only half as distant as south-eastern Peru (*c.* 650 km) and lies in the same non-seasonal climate zone (Vásquez, 1997). The difference is that Iquitos is much further from the Andes and lies in a Tertiary basin containing weathered sediments derived from a variety of sources, including the Guiana Shield (Kalliola *et al.*, 1993; Räsänen *et al.*, 1995).

We argue that these patterns call for a sea change in the way ecologists think about tropical forests. The traditional view of small-scale vegetational mosaics must yield to a new picture of very large areas dominated by predictable species associations, not unlike the situation in temperate forests.

Conclusions

- We examined community-level patterns in western Amazonian tree communities at spatial scales ranging from subcontinental to 1 m^2 as a means of evaluating ten prominent theories of plant diversity.
- More than 70% of tree species in south-eastern Peru are also found in eastern Ecuador, 1400 km distant. Abundances of shared species are positively correlated in the two regions.
- Within each region, a set of approximately 150 common tree species predominates in both abundance and frequency. These regional forests are thus dominated by an 'oligarchy' of species, much as temperate forests.
- Within south-eastern Peru, more than 85% of all tree species are found in two or more major habitats (*terra firme*, flood-plain, successional stands, swamps), indicating that beta diversity is very low at the regional scale.
- At the hectare scale, we report that > 95% of saplings appear outside the projected crown radius of the nearest conspecific adult, implying that nearly all saplings originate from dispersed seeds.
- The median sapling of 19 'common' tree species was 14 m from the nearest conspecific adult, whereas the corresponding distance for saplings of 75 'less common' species was 32 m. These distances indicate that saplings typically appear at distances equivalent to several crown radii from potential parents, thereby leaving ample space in the neighbourhood for heterospecific recruitment, as postulated by Janzen (1970) and Connell (1971).
- These results lead to the suggestion that 'common' tree species are those that are able to recruit near conspecific adults, whereas 'less common' species are those that are unable to do so.
- 'Input–output' analysis of the seed rain at Cocha Cashu in Peru indicated that > 500 seeds fall on to each square metre of forest floor every year and yet give rise to only four new plants. We suggest that density dependence operates in this community mainly at the seed and early seedling stages, thereby explaining the difficulty investigators have had in demonstrating strong density dependence at later ontogenetic stages.
- The intersection between the Janzen–Connell mechanism and the abiotic world is what should define the frontier of tropical forest ecology for the next generation of theoreticians and empiricists, alike.

Conservation relevance and avenues for future research

- The tree communities of western Amazonia are composed largely of species with broad geographical distributions, which display a wide range of tolerance of local edaphic gradients.

- The pre-eminence of alpha over beta diversity at the regional scale implies that randomly situated conservation areas will capture most tree species inhabiting the region.
- The importance of animals as dispersal agents in tropical forests is underscored by the observation that >95% of all saplings in the understorey of a Peruvian forest arose from seeds that had been transported away from the nearest potential parent.
- Consequently, the widespread decimation of dispersers by overhunting can be predicted to have devastating long-term consequences for the maintenance of tree species diversity in tropical forests.

Acknowledgements

We gratefully acknowledge financial support from the Pew Charitable Trusts, Andrew W. Mellon Foundation and National Science Foundation. Ran Nathan, Helene Muller-Landau and Carlos Martínez del Rio kindly read an early draft of the manuscript and made many helpful suggestions. Renata Leite is thanked for help with data analysis. We are most grateful to Helene Muller-Landau for calculating the distributions of randomly arrayed saplings of common and less common species.

References

Andresen, E. (1999) Seed dispersal by monkeys and the fate of dispersed seeds in a Peruvian rainforest. *Biotropica* 31, 145–158.

Ashton, P.S. and Hall, P. (1992) Comparisons of structure among mixed dipterocarp forests of north-western Borneo. *Journal of Ecology* 80, 459–481.

Brown, J.H. (1995) *Macroecology*. University of Chicago Press, Chicago, Illinois.

Chesson, P.L. and Warner, R.R. (1981) Environmental variability promotes coexistence in lottery competitive systems. *American Naturalist* 117, 923–943.

Clark, D.A. and Clark, D.B. (1984) Spacing dynamics of a tropical rain forest tree: evaluation of the Janzen–Connell model. *American Naturalist* 124, 769–788.

Clark, J.S., Silman, M., Kern, R., Maclin, E. and HilleRisLambers, J. (1999) Seed dispersal near and far: patterns across temperate and tropical forests. *Ecology* 80, 1475–1494.

Condit, R., Hubbell, S.P. and Foster, R.B. (1992) Recruitment near conspecific adults and the maintenance of tree and shrub diversity in a neotropical forest. *American Naturalist* 140, 261–286.

Condit, R., Hubbell, S.P. and Foster, R.B. (1994) Density dependence in two understory tree species in a neotropical forest. *Ecology* 75, 671–680.

Condit, R., Hubbell, S.P. and Foster, R.B. (1995) Mortality rates of 205 neotropical tree species and the responses to a severe drought. *Ecological Monographs* 65, 419–439.

Condit, R., Hubbell, S.P. and Foster, R.B. (1996) Changes in tree species abundance in a Neotropical forest over eight years: impact of climate change. *Journal of Tropical Ecology* 12, 231–256.

Condit, R., Ashton, P.S., Baker, P., Bunyavejchewin, S., Gunatilleke, S., Bunatilleke, N., Hubbell, S.P., Foster, R.B., Itoh, A., LaFrankie, J.V., Seng, L.H., Losos, E., Manokaran, N., Sukumar, R. and Yamakura, T. (2000) Spatial patterns in the distribution of tropical tree species. *Science* 288, 1414–1418.

Connell, J.H. (1971) On the role of natural enemies in preventing competitive exclusion in some marine animals and in rain forest trees. In: den Boer, P.J. and Gradwell, G.R. (eds) *Dynamics of Numbers in Populations*. Centre for Agricultural Publication and Documentation, Proceedings of the Advanced Study Institute, Osterbeek, Wageningen, The Netherlands, pp. 298–312.

Connell, J.H. (1978) Diversity in tropical rain forests and coral reefs. *Science* 199, 1302–1310.

Dirzo, R. and Miranda, A. (1991) Altered patterns of herbivory and diversity in the forest understory: a case study of the possible consequences of contemporary defaunation. In: Price, P.W., Lewinsohn, P.W., Fernandes, G.W. and Benson, W.W. (eds) *Plant–Animal Interactions: Evolutionary Ecology in Tropical and Temperate regions*. John Wiley & Sons, New York, pp. 273–287.

Forget, P.-M. (1992) Regeneration ecology of *Eperua grandiflora* (*Caesalpiniaceae*), a large-seeded tree in French Guiana. *Biotropica* 24, 146–156.

Gentry, A.H. (1988) Changes in plant community diversity and floristic composition on environmental and geographical gradients. *Annals of the Missouri Botanical Garden* 71, 1–34.

Givnish, T.J. (1999) On the causes of gradients in tropical tree diversity. *Journal of Ecology* 87, 192–210.

Hubbell, S.H. (1979) Tree dispersion, abundance, and diversity in a tropical dry forest. *Science* 203, 1299–1309.

Hubbell, S.H. and Foster, R.B. (1986) Biology, chance, and history and the structure of tropical rain forest tree communities. In: Diamond, J. and Case, T.J. (eds) *Community Ecology*. Harper and Row, New York, pp. 314–329.

Hubbell, S.P., Foster, R.B., O'Brien, S.T., Harms, K.E., Condit, R., Wechsler, B., Wright, S.J. and Loo de Lao, S. (1999) Light-gap disturbances, recruitment limitation, and tree diversity in a Neotropical forest. *Science* 283, 554–557.

Hurtt, G.C. and Pacala, S.W. (1995) The consequences of recruitment limitation: reconciling chance, history, and competitive differences between plants. *Journal of Theoretical Biology* 176, 1–12.

Janzen, D.H. (1970) Herbivores and the number of tree species in tropical forests. *American Naturalist* 104, 501–528.

Kalliola, R., Puhakka, M. and Danjoy, W. (eds) (1993) *Amazonia peruana: vegetación húmeda tropical en el llano subandino*. Gummerus Printing, Jyväskylä, Finland.

MacArthur, R.H. (1957) On the relative abundances of bird species. *Proceedings of the National Academy of Sciences USA* 43, 293–295.

May, R.M. and Stumpf, M.P.H. (2000) Species–area relations in tropical forests. *Science* 290, 2084–2086.

Nathan, R. and Muller-Landau, H.C. (2000) Spatial patterns of seed dispersal, their determinants and consequences for recruitment. *Trends in Ecology and Evolution* 15, 278–285.

Pacala, S.W. and Tilman, D. (1994) Limiting similarity in mechanistic and spatial models of plant competition in heterogeneous environments. *American Naturalist* 143, 222–257.

Phillips, O.L., Hall, P., Gentry, A.H., Sawyer, S.A. and Vásquez, R. (1994) Dynamics and species richness of tropical rain forests. *Proceedings of the National Academy of Sciences USA* 91, 2805–2809.

Pitman, N.C.A. (2000) A large-scale inventory of two Amazonian tree communities. PhD thesis, Duke University, Durham, North Carolina.

Pitman, N.C.A., Terborgh, J., Silman, M.R. and Núñez V., P. (1999) Tree species distributions in an upper Amazonian forest. *Ecology* 80, 2651–2661.

Pitman, N.C.A., Terborgh, J., Silman, M.R., Núñez V., P., Neill, D.A., Cerón, C.E., Palacios, W.A. and Aulestia, M. (2001) Dominance and distribution of tree species in two upper Amazonian *terra firme* forests. *Ecology* 82, 2101–2117.

Plotkin, J.B., Potts, M.D., Yu, D.W., Bunyavejchewin, S., Condit, R., Foster, R., Hubbell, S., LaFranjkie, J., Manokaran, N., Seng, L.H., Sukumar, R., Nowak, M.A. and Ashton, P.S. (2000) A new method of analyzing species diversity in tropical forests. *Proceedings of the National Academy of Sciences USA* 97, 10850–10855.

Räsänen, M., Linna, A.M., Santos, J.C.R. and Negri, F.R. (1995) Late Miocene tidal deposits in the Amazonian foreland basin. *Science* 269, 386–390.

Salo, J., Kalliola, R., Häkkinen, I., Mäkinen, Y., Niemelä, P., Puhakka, M. and Coley, P.D. (1986) River dynamics and the diversity of Amazon lowland forest. *Nature* 322, 254–258.

Schupp, E.W. (1988) Seed and early seedling predation in the forest understory and treefall gaps. *Oikos* 51, 221–227.

Silman, M.R. (1996) Regeneration from seed in a neotropical rain forest. PhD thesis, Duke University, Durham, North Carolina.

Terborgh, J. and Andresen, E. (1998) The composition of Amazonian forests: patterns at local and regional scales. *Journal of Tropical Ecology* 14, 645–664.

Terborgh, J. and Petren, K. (1991) Development of habitat structure through succession in an Amazonian floodplain forest. In: Bell, S.S., McCoy, E.D. and Mushinsky, H.R. (eds) *Habitat Structure: the Physical Arrangement of Objects in Space*. Chapman & Hall, London, pp. 28–46.

Terborgh, J. and Wright, S.J. (1994) Effects of mammalian herbivores on plant recruitment in two Neotropical forests. *Ecology* 75, 1829–1833.

Terborgh, J., Foster, R.B. and Núñez V., P. (1996) Tropical tree communities: a test of the nonequilibrium hypothesis. *Ecology* 77, 561–567.

ter Steege, H. (2000) *Plant Diversity in Guyana with Recommendations for a National Protected Area Strategy*. Tropenbos Series 18, Tropenbos Foundation, Wageningen, The Netherlands.

Tilman, D. (1982) *Resource Competition and Community Structure*. Princeton University Press, Princeton, New Jersey.

Tilman, D. (1988) *Plant Strategies and the Dynamics and Structure of Plant Communities*. Princeton University Press, Princeton, New Jersey.

Tilman, D. (1994) Competition and biodiversity in spatially structured habitats. *Ecology* 75, 2–16.

Tuomisto, H. and Ruokolainen, K. (1994) Distribution of *Pteridophyta* and *Melastomataceae* along an edaphic gradient in an Amazonian rain forest. *Journal of Vegetation Science* 5, 25–34.

Tuomisto, H., Ruokolainen, K., Kalliola, R., Linna, A., Danjoy, W. and Rodriguez, Z. (1995) Dissecting Amazonian biodiversity. *Science* 269, 63–66.

Vásquez M., R. (1997) *Flórula de las reservas biológicas de Iquitos, Perú.* Missouri Botanical Garden, St Louis, Missouri.

Wenny, D.G. (2000) Seed dispersal, seed predation, and seedling recruitment of a Neotropical montane tree. *Ecological Monographs* 70, 331–351.

Whittaker, R.H. (1965) Dominance and diversity in land plant communities. *Science* 147, 250–260.

Wills, C., Condit, R., Foster, R.B. and Hubbell, S.P. (1997) Strong density- and diversity-related effects help to maintain tree species diversity in a neotropical forest. *Proceedings of the National Academy of Sciences USA* 94, 1252–1257.

2 Dissemination Limitation and the Origin and Maintenance of Species-rich Tropical Forests

Eugene W. Schupp[1], Tarek Milleron[1] and Sabrina E. Russo[2]

[1]Department of Rangeland Resources and the Ecology Center, Utah State University, Logan, UT 84322-5230, USA;
[2]Department of Ecology, Ethology and Evolution, University of Illinois, Urbana, IL 61801-3707, USA

Introduction

An enduring question in biogeography is: 'Why are tropical forests so diverse?' (Janzen, 1970; Connell, 1971; Hubbell, 1979; Palmer, 1994; Pitman *et al.*, 1999; Wright, 1999; Colwell and Lees, 2000). Although many hypotheses explain high tropical species richness (see Terborgh *et al.*, this volume), much recent interest has focused on whether processes that limit recruitment of new individuals can determine tropical forest community structure by influencing species richness, spatial structure of populations and dynamics of species composition through time (Hurtt and Pacala, 1995; Schupp and Fuentes, 1995; Clark *et al.*, 1998, 1999a; Hubbell *et al.*, 1999; Ehrlén and Eriksson, 2000; Nathan and Muller-Landau, 2000).

A theoretical test of the hypothesis that recruitment limitation can be an important force structuring communities is provided by Hurtt and Pacala (1995). In their model, recruitment limitation is defined as 'the failure to have any viable juveniles at an available site' (p. 2). As a consequence of this failure, species often 'win' recruitment sites by forfeit – that is, not because they are the superior competitors under the given environmental conditions, but because better competitors never established in that site. The basics of their model are: (i) recruitment sites are opened by random adult deaths; (ii) potential recruits arrive at sites by random dispersal; (iii) the winner of a site is the superior competitor of those potential recruits that arrived; (iv) competitive rankings of potential recruits are determined by environmental conditions; and (v) environmental conditions vary in time and space. Despite highly deterministic outcomes of interspecific competition, their model demonstrated that an 'enormous' number of species can be maintained by recruitment limitation. Furthermore, as recruitment limitation was strengthened by increasing species richness or decreasing mean fecundity, more sites were won by forfeit rather than by absolute dominance. As a result, population and community dynamics slowed and ecological drift in species composition became a more important determinant of community structure.

Results of these models have implications for understanding the structure and dynamics of species-rich communities, such as tropical forests. By definition, rare species are more recruitment-limited than common species because at the population level they are less fecund and therefore less likely to have seeds reach available sites. Because highly diverse communities are comprised of many rare species, recruitment limitation should be an important force structuring such communities. As a consequence, in more diverse communities the pace of population and community dynamics slows and ecological drift becomes more important. In this view, species-rich tropical forests are non-equilibrium communities structured more by chance and history than by competition. Furthermore, both the high diversity and the ecological drift are driven by rarity, which is a direct result of the initial high diversity. Such an interpretation of recruitment limitation may help explain the maintenance of high species richness, but not its origin.

Hurtt and Pacala (1995) focused on rarity's contribution to recruitment limitation. There are, however, other mechanisms that can promote recruitment limitation in tropical forests. These alternative mechanisms are likely to differ in their effects on population and community dynamics, species coexistence and the origin and maintenance of diversity. Consequently, a more thorough exploration of mechanisms driving recruitment limitation will help us understand how recruitment limitation can structure communities.

In this chapter we first present an overview of classes of mechanisms causing recruitment limitation. We then define three mechanisms of dissemination limitation, two of which, distance-restricted seed dispersal and, especially, spatially contagious seed dispersal, we discuss in detail and for which we provide empirical examples. We place particular emphasis on how the behaviours and physiologies of animal seed-dispersers may lead to dissemination limitation. We then examine the consequences of dissemination limitation for patterns of recruitment in plant populations and how these patterns in turn affect plant community structure. Finally, we explore how dissemination limitation can contribute to the origin and maintenance of species diversity in ecological communities. We conclude with an overview of what we believe to be the major gaps in our understanding of the role of dissemination limitation in population and community structure and dynamics.

Mechanisms Causing Recruitment Limitation

Recruitment limitation can result from three broadly defined classes of mechanisms: (i) source limitation, which occurs when recruitment is limited by low population-level seed availability; (ii) dissemination limitation, which occurs when recruitment is limited by a failure to disperse seeds to potential recruitment sites; and (iii) establishment limitation, which occurs when recruitment is limited by inappropriate biotic or abiotic environments (Clark *et al.*, 1998, 1999b). These classes of mechanisms are neither mutually exclusive nor independent. For example, recruitment may be limited by seeds being dispersed to only a very small portion of the landscape (dissemination limitation), coupled with the delivery of many of these dispersed seeds to sites where establishment is unlikely (establishment limitation). In this case, dispersal processes, environmental conditions and their interactions limit recruitment.

This framework helps clarify how different mechanisms limit recruitment. Hurtt and Pacala's (1995) rarity is a form of source limitation that limits recruitment by causing population-level seed production to be too low to reach any but a fraction of available sites. Hypotheses that address how high levels of granivory, herbivory and pathogen attack limit recruitment and promote species coexistence (e.g. Gillett, 1962; Janzen, 1970, 1971; Connell, 1971) imply source limitation if mortality is pre-dispersal (e.g. destruction of flowers and developing seeds) but establishment limitation if mortality is post-dispersal (e.g. post-dispersal seed predation and pathogen attack). Indeed, Hurtt and Pacala (1995) essentially modelled pre-dispersal pest attack when they increased recruitment limitation by decreasing individual fecundity, just as increased pre-dispersal predation decreases fecundity. Such examples demonstrate the wide variety of mechanisms operating at multiple stages of a plant's life to contribute to recruitment limitation; both empirical and theoretical approaches are needed to address this variety of mechanisms. The focus of the remainder of this chapter, however, is on dissemination limitation, because we believe it is a major contributor to recruitment limitation that has received insufficient attention.

Dissemination limitation encompasses three processes: (i) quantitatively restricted seed dispersal; (ii) distance-restricted seed dispersal; and (iii) spatially contagious seed dispersal. Quantitatively restricted seed dispersal means that, independent of seed production, the quantity of seeds dispersed away from parents is limited by disperser activity. Due to low disperser visitation, disperser movement patterns, disperser feeding behaviours and other factors, many seeds fall or are dropped beneath parents undispersed. Thus, seeds reach fewer – in many cases, far fewer – recruitment sites than expected based on population-level seed production. Distance-restricted seed dispersal means that, even in the hypothetical case where the number of seeds dispersed is sufficient to saturate all potential sites, many sites will remain unoccupied because so many seeds are dispersed only short distances. Thus, seed rain may be higher than needed to saturate available sites near fruiting adults and yet be very limited far from adults. The importance of this mechanism of dissemination limitation increases with rarity and clumping, both of which result in less of the landscape being near fruiting trees. Lastly, spatially contagious seed dispersal means that, independent of dispersal distance, seeds are deposited very patchily, so that some sites receive very many seeds and others receive few or none. We focus on these last two forms of dissemination limitation.

Distance-restricted seed dispersal

As recognized by Janzen (1970) and Connell (1971), seed dispersal is frequently spatially restricted, with seed densities being high beneath parents and decreasing rapidly with distance (Fig. 2.1; Fleming and Heithaus, 1981; Howe *et al.*, 1985; Augspurger and Kitajima, 1992; Portnoy and Willson, 1993; Willson, 1993; Wenny, 2000). The peak in seed density beneath parents is largely due to quantitatively restricted seed dispersal, but the shape of the seed dispersal curve beyond the canopy edge can be greatly influenced by distance-restricted dispersal. Distance-restricted seed dispersal contributes to recruitment limitation by limiting the chance of seeds reaching suitable recruitment sites that are distant from parent trees. For example, in the hemiepiphytic *Ficus subtecta* (*Moraceae*) in Borneo, seed traps along transects radiating from individual trees demonstrated a sharp decrease in seed density with distance (Laman, 1996). As a result, a 'safe site' (estimated to be on the scale of 20 cm × 20 cm) 60 m from an adult *Ficus* is expected to be colonized by a seed only once every 100 reproductive events. Thus, it is extremely unlikely that new *F. subtecta* will establish far from existing adults in the population, simply because very few seeds land far from adults in this relatively low-density population.

Spatially contagious seed dispersal

The smooth curves of seed dispersal depicted in most publications do not represent the true shapes of animal-produced seed shadows. Some studies have revealed densities of dispersed seeds along individual transects around focal trees varying by an order of magnitude or more, with the highest densities not always near parent plants (Fleming and Heithaus, 1981; Laman, 1996). Seed fall is extremely heterogeneous in space, and this heterogeneity is not simply a function of distance from an adult. It is largely driven by animal behaviour.

The Janzen–Connell model (*sensu* Clark and Clark, 1984) is remembered for the smooth curves of the seed shadows, but Janzen (1970) stressed that seed fall is highly heterogeneous, with higher densities in habitats preferred by dispersal agents, beneath perches, along travel routes, etc. (Fig. 2.2). Such spatially contagious seed dispersal has been generally ignored in studies of seed dispersal

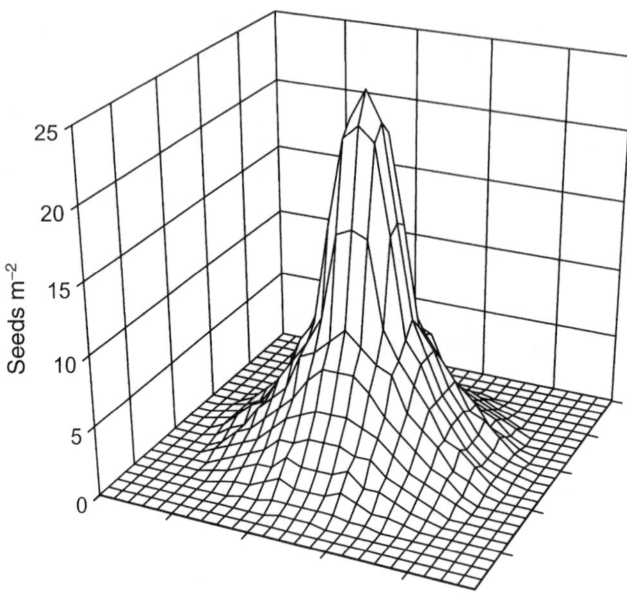

Fig. 2.1. Hypothetical seed shadow showing effects of distance-restricted seed dispersal. The peak in density represents the location of the focal fruiting tree. Quantitatively restricted seed dispersal is largely responsible for the peak of seed dispersal being at the parent tree. The very rapid drop-off in seed density away from the parent crown is due to large numbers of seeds being dispersed only short distances (distance-restricted seed dispersal) combined with a rapidly increasing surface area with increasing distance from the parent. The result is very little chance of seeds reaching suitable recruitment sites far from parents.

(but see Howe, 1989). Most research on spatial patterns of seed dispersal has focused on dispersal as a function of distance from the nearest fruiting conspecific. We contend that spatially contagious seed dispersal is far more widespread than generally acknowledged and has far-reaching implications for plant recruitment.

Many examples in the literature illustrate different forms of spatially contagious seed dispersal and their effects on patterns of seed fall. Contagion is evident at spatial scales ranging from small-scale clumping of seeds in individual defecations (e.g. Fleming and Heithaus, 1981; Loiselle, 1990; Wrangham *et al.*, 1994; Andresen, 1999; Voysey *et al.*, 1999a) to large-scale variation in seed-fall density among habitats, driven by post-feeding habitat use by dispersers (e.g. McDiarmid *et al.*, 1977; Thomas *et al.*, 1988; Thomas, 1991). Here we present a selective review, demonstrating the variety and prevalence of spatially contagious seed dispersal. We focus on five non-mutually exclusive categories. For each category we first present a case-study demonstrating the pattern and then discuss the generality of the pattern.

Dispersal to fruit-processing roosts

Many vertebrate frugivores remove fruits from parents and process them at repeatedly used sites or roosts. For example, in Costa Rica, Janzen *et al.* (1976) found that most *Andira inermis* (*Papilionoideae*) seeds dispersed by *Artibeus* spp. (Phyllostomidae) bats fell beneath a single fruit-processing roost. Consequently, seeds were concentrated in two areas; 48.2% fell beneath the parent tree undispersed, while 34.9% were deposited beneath a *Cocoloba* sp. (*Polygonaceae*) tree > 100 m away. Very few seeds fell beneath several other trees and no seeds were found beneath most trees sampled.

Seed dispersal to fruit-processing roosts appears widespread, especially among bats. Many bats pluck fruits and fly to a roost, where they process the fruit, dropping seeds and fruit pieces, and frequently defecating. This

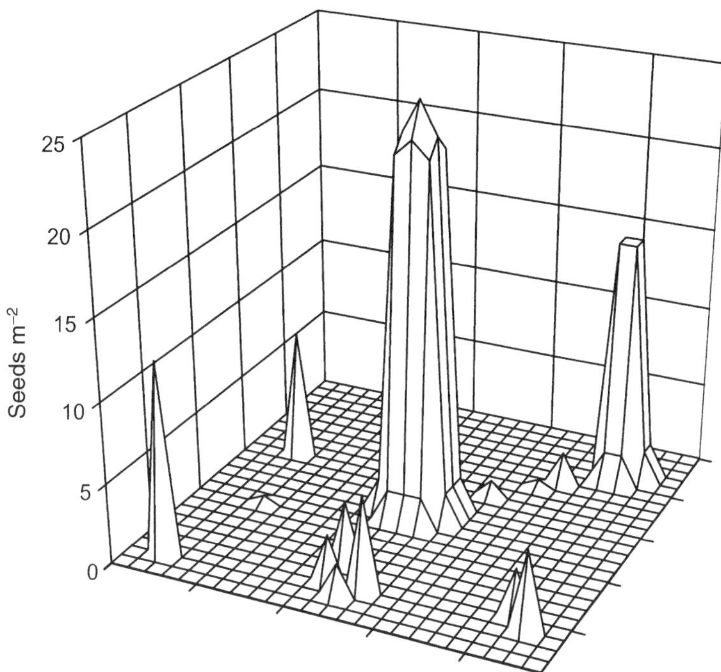

Fig. 2.2. Hypothetical seed shadow showing effects of spatially contagious seed dispersal. The peak in seed density represents the location of the focal fruiting tree. The occurrence of the peak at the parent is due to a combination of quantitatively restricted seed dispersal and dispersal of conspecific seeds to the focal tree by frugivores that had previously fed at other trees (dispersal to fruiting trees). Many processes could produce other peaks. For example, the peak at the back right represents seeds accumulating beneath a habitually used roost where fruits are processed (dispersal to fruit-processing roosts), while the small peaks between this roost and the focal tree represent seeds dropped or defecated *en route*. The peak in the centre foreground represents seed accumulation at a sleeping site comprised of several trees and used by a large troop of monkeys (dispersal to sleeping and resting sites). As with distance-restricted seed dispersal, much of the forest fails to receive any seeds, although the spatial patterns of seed accumulation and, consequently, of recruitment differ greatly.

behaviour results in the accumulation of seeds beneath roosts (Fleming, 1981; Fleming and Heithaus, 1981; Bizerril and Raw, 1998). With the exception of fruit processing at breeding display sites (see below), the habitual use of particular perches after feeding on fruits is poorly documented in birds. None the less, many birds probably use the same roost repeatedly, as is obvious when watching *Turdus viscivorus* (Muscicapidae) regularly flying between the *Prunus mahaleb* (*Rosaceae*) trees, where they feed, and a clump of *Pinus nigra* (*Pinaceae*) trees, where they process the fruits (Jordano and Schupp, 2000).

The degree of spatial contagion resulting from dispersal to fruit-processing roosts depends on several factors. For example, in bats, whether fruits are removed to roosts for processing or are swallowed at the parent plant with seeds defecated in flight depends on fruit size relative to bat size (Bizerril and Raw, 1998). Thus, bat-generated seed shadows range from very contagious distributions, as in *A. inermis* dispersed by *Artibeus* (Janzen *et al.*, 1976), to more homogeneous seed shadows around parents, as in *Piper arboreum* (*Piperaceae*) dispersed by *Carollia perspicillata* (Phyllostomidae) and *Glossophaga soricina* (Phyllostomidae) (Bizerril and Raw, 1998). Additionally, the degree of spatial contagion of seed fall will be influenced by the propensity of a disperser to use a roost repeatedly, which will depend on both intrinsic (e.g. foraging behaviour) and extrinsic (e.g. spatial pattern of fruit resources) factors.

Dispersal to display sites

The reproductive behaviours of some vertebrate frugivores, primarily birds, cause them to spend significant amounts of time at breeding display sites, where they defecate seeds. In French Guiana, Krijger *et al.* (1997) found significantly higher soil seed-bank densities at display sites of male *Corapipo guttaralis* (Pipridae) than in the surrounding forest, whether they considered all bird-dispersed seeds or only seeds in the *Melastomataceae*, a family comprised of many manakin-dispersed species. Consequently, bird behaviour can result in concentrated seed deposition in a central place, which results in reduced deposition in the rest of the forest and, hence, reduced seed availability for widely spread potential recruitment sites.

Disproportionate dispersal of seeds to breeding display sites should characterize dispersal by males of polygynous frugivores. During the breeding season, males spend most of the day at display sites, making only rapid feeding flights to and from fruiting trees. *Pipra erythrocephala* (Pipridae) males in Trinidad are on display sites (leks) about 88% of the day during the breeding season, and most departures last only 30–90 s (Snow, 1962). The degree and spatial scale of seed contagion created by this behaviour will vary with the reproductive behaviour of the species. The contagion of displaying males, and thus of seed deposition, varies widely among species. Species may display in leks containing numerous males, such as some manakins (Pipridae) and cock-of-the-rocks (*Rupicola rupicola*, Rupicolidae), solitarily or in small groups, such as ochre-bellied flycatchers (*Mionectes oleagineus*, Tyrannidae), in exploded leks, such as pihas (*Lipaugus* spp., Cotingidae), or as solitary males, as in bellbirds (*Procnias* spp., Cotingidae) (Snow, 1961, 1970; Trail, 1985; Wheelwright, 1988; Théry and Larpin, 1993; Westcott, 1997; Wenny, 2000). Bellbirds should produce very small-scale clumping beneath isolated dead snags and cock-of-the-rocks should produce larger-scale clumping (> 200 m^2) (Théry and Larpin, 1993). All species, however, produce high densities of seeds at some spatial scale and consequently deny seeds to the remainder of the forest.

Dispersal to latrines

Some vertebrate frugivores use latrines, sometimes with several individuals defecating in the same location. *Maximiliana maripa* (*Palmae*) grows in monodominant clumps in Brazil (Fragoso, 1997). Rodents, peccaries, deer and primates generally disperse seeds < 5 m from clumps. *Tapirus terrestris* (Tapiridae), on the other hand, disperses seeds to widely scattered latrines up to 2 km from palm clumps, where they deposit hundreds of seeds. Rodents secondarily disperse seeds short distances from latrines, eventually producing, it is believed, new palm clumps. Thus, the pattern of dispersal to latrines may explain the spatial pattern of palms, which are present in high densities in some areas but virtually absent from most of the forest.

This form of spatially contagious seed dispersal is probably not widespread, but it does have major consequences for some tree species. For example, seed dispersal to latrines by the greater one-horned rhinoceros (*Rhinoceros unicornis*, Rhinocerotidae) (Dinerstein and Wemmer, 1988; Dinerstein, 1991) may explain the highly clumped spatial pattern of riverine trees invading flood-plain grasslands in Nepal.

Dispersal to resting or sleeping sites

Many vertebrate frugivores defecate seeds at resting and sleeping sites that vary in the degree to which they are reused. In Manú National Park, Peru, *Ateles paniscus* (Cebidae) is the major disperser of *Virola calophylla* (*Myristicaceae*) trees. Many seeds are defecated under trees where the monkeys rest during the day and, especially, under trees where they sleep at night ('rest trees' and 'sleep trees', respectively). Seed densities beneath sleep trees, which are often used repeatedly, did not differ from densities beneath fruiting *V. calophylla* trees during peak fruiting (mean = 11.1 seeds m^{-2}, SD = 19.4, $n = 6$ sleep trees; mean = 8.32 seeds m^{-2}, SD = 6.29, $n = 4$ fruiting trees (S.E. Russo, unpublished data)). Densities beneath rest trees, which are less frequently reused, appear to be intermediate between the high densities beneath sleeping and fruiting trees and the very low background densities throughout the rest of the forest.

Dispersal to resting and sleeping sites appears to be very widespread. Fruits are easy to find and capture and are high in bulk (i.e. indigestible seeds), so frugivores fill their guts quickly and cannot feed again until the meal is processed (Levey and Grajal, 1991; Levey and Duke, 1992). Consequently, many frugivores spend significant time resting while processing fruit in the gut, rumen or mouth, potentially depositing high densities of seeds at resting and sleeping sites. Howler monkeys (*Allouata* spp., Cebidae) spend extensive time in trees eating, resting and defecating (Julliot, 1997; Andresen, 1999). They also defecate *en route* between trees, but most defecations appear to be at sleeping sites. Duikers (*Cephalophus* spp., Bovidae), medium-sized African forest ruminants, spit out hard seeds at resting sites after rumination; consequently, duiker dispersal is characterized by long-distance dispersal to scattered resting sites (Feer, 1995). Similarly, gorillas (*Gorilla gorilla*, Pongidae) (Voysey, *et al.*, 1999b) and dwarf cassowaries (*Casuarius bennetti*, Casuariidae) (Mack, 1995) defecate extensively at sleeping sites. Seeds dispersed to resting and sleeping sites are generally deposited in high-density, multispecies seed clumps, which may or may not be associated with fruiting trees, while much of the forest receives no seeds. The true degree of spatial contagion of seed fall that results will be a complex function of social behaviour, activity and defecation patterns, gut and rumen capacity, passage rates and the propensity to reuse the same resting and sleeping sites.

Dispersal to fruiting trees

Many vertebrate frugivores spend considerable time feeding and resting at fruiting trees. As a result, seeds ingested elsewhere are often deposited beneath fruiting trees. Data from Venezuela (T. Milleron, unpublished data) show that fruiting trees can be foci for the dispersal of seeds of concurrently fruiting species. *Virola* sp. and *Maximiliana* sp. are dispersed primarily by monkeys and large birds. Density and species richness of seeds of other animal-dispersed species were an order of magnitude higher beneath these species than beneath *Clathrotropis* sp. (*Fabaceae*), which has a dry pod that is ignored by monkeys and birds. These findings suggest that animal-dispersed species are disproportionately dispersed to concurrently fruiting heterospecific species that share dispersal agents. Similar patterns have been found in Cameroon (C. Clark, Brazil, 2000, personal communication) and Peru (S.E. Russo, unpublished data).

Few studies have directly addressed dispersal to fruiting trees in intact forests, but indirect evidence indicates that such dispersal may be widespread. For example, in deforested areas in Uganda (Duncan and Chapman, 1999), Costa Rica (Harvey, 2000) and Mexico (Guevara and Laborde, 1993), densities and species diversity of dispersed seeds were significantly greater beneath fruiting remnant trees than in open grassland. Though not conclusive, this suggests that fruiting trees can be powerful attractants for frugivores and the seeds they disperse.

Dispersers moving between fruiting trees rapidly relative to gut passage and then slowing or stopping movement while feeding should deposit a suite of seed species together beneath or in the vicinity of fruiting trees. Such dispersal is probably frequent in many birds (e.g. *Penelope* spp., Cracidae; and *Ramphastos* spp., Rhamphastidae) and monkeys (e.g. *Cebus* spp., Cebidae; and *Ateles* spp.). For example, howler monkeys (*Allouata* spp.) can spend hours at fruiting trees, eating, resting and defecating (see above) (Andresen, 1999). Because of their long gut passage time, many seeds defecated beneath fruiting trees are probably from other trees. The degree of spatial contagion resulting should vary with both disperser and tree species. As an extreme example, dispersers feeding in groups in large fruit-filled trees and then travelling directly to other fruiting trees will create much more contagious patterns of seed fall than will dispersers feeding solitarily in trees with little fruit and feeding on insects between bouts of frugivory.

Dispersal to fruiting trees has similarities with dispersal to rest and sleep trees, and the same disperser may do both. *Ateles belzebuth* (Cebidae) and *A. paniscus* mostly deposit seeds from morning and early afternoon feeding beneath fruiting and rest trees and seeds from late afternoon feeding beneath sleep trees (J.L. Dew, Brazil, 2000, personal communication; S.E. Russo, unpublished data). Seed deposition at fruiting trees and sleep trees should, however, differ in respect of the density and diversity of dispersed seeds. A fruiting tree should receive a relatively lower density of foreign seeds – seeds swallowed somewhere other than at the deposition site – deposited against a background of high-density seed fall from the fruiting tree. In contrast, because of the time spent there, foreign seed rain beneath a sleep tree should be relatively higher in density and perhaps in species richness. Further, because these seeds are deposited against a virtually seed-free background, deposition beneath sleep trees should have greater evenness as well. The consequences of these alternative patterns for tree recruitment are probably very different.

And the mechanisms go on . . .

We have highlighted five major processes that lead to spatially contagious seed dispersal, but many other processes can contribute to contagious patterns of seed deposition. High densities and diversities of seeds can accumulate around the bases of nesting trees where female and young hornbills (Bucerotidae) are sealed in the nest for up to 4 months, being fed and regurgitating through a hole (Gilliard, 1958). Monkeys such as *Allouata* spp. and *Ateles* spp. eat leaves in addition to fruit and may therefore also deposit high densities of seeds beneath non-fruiting trees in which they are feeding (S.E. Russo, unpublished data). Movement patterns of frugivorous birds may be influenced by even subtle topographic variation, resulting in very uneven use of the habitat and thus very uneven distribution of dispersed seeds (Westcott, 1994, 1997). There are surely many other documented processes promoting spatially contagious seed dispersal that we have not mentioned, and probably others that have yet to be documented.

The temporal component of spatially contagious seed dispersal

It can be argued that these contagious patterns of seed fall reflect only what happens over the short term, and that seed fall will eventually become more homogeneous as locations of seed clumps shift through the dispersal season and even more so from year to year. Perhaps seed densities tend to average out over a long time frame. Thus, the temporal component of spatially contagious seed dispersal cannot be ignored. The clumping created by gorillas defecating at nest sites that are changed daily (Voysey *et al.*, 1999b) is quantitatively and qualitatively different from clumping created by lek-breeding birds, which often use the same leks for many years (Théry and Larpin, 1993). But do longer-term patterns negate the impact of spatially contagious seed dispersal that has been documented primarily over very short time frames?

Wright (1999) suggested that the two most abundant tree species on Barro Colorado Island, Panama, have escaped dissemination limitation because both had seeds dispersed to each of 200 0.5 m^2 litter traps over a 13-year period. We believe this is a premature conclusion. Shifting patterns of seed dispersal over long time periods may eventually result in all sites receiving seeds, but dissemination limitation occurs when seeds fail to arrive in a site when it is suitable for recruitment. One to several seeds dispersed to a trap over a 13-year period need not mean that the species has escaped dissemination limitation. Spatial contagion of seed dispersal is important even if the foci of contagion change daily. Recruitment sites are not simply open and waiting for seeds of a given species to arrive; they open and close unpredictably in time. Spatial contagion is important because it deprives sites of seeds during their 'open window of opportunity'.

Consequences of Dissemination Limitation: Recruitment of Individual Species

Dissemination limitation will have both immediate and delayed consequences for recruitment of tree species. Both will influence the degree of recruitment limitation individual species face and the potential for promoting and maintaining high species richness.

Immediate consequences of dissemination limitation: seed shadows

The immediate consequence of dissemination limitation is reflected in the seed shadow.

Firstly, dissemination limitation affects occupancy of potential recruitment sites. Distance-restricted seed dispersal increases the number of sites failing to receive seeds by denying seeds to sites far from adults. Spatially contagious seed dispersal further limits the number of sites receiving seeds by depriving sites of seeds independent of distance from adults. The major effect on recruitment limitation is that sites that do not receive seeds cannot generate recruits.

Secondly, dissemination limitation affects variation in seed density among sites receiving seeds. Distance-restricted seed dispersal yields a negative correlation between distance and seed density, with densities decreasing with increasing distance from adults. Spatially contagious seed dispersal weakens this correlation by generating both low and high seed densities both near and far from adults. The effects of various patterns of seed-density variation on recruitment are not straightforward because they depend on delayed consequences; that is, they depend on how varying seed shadows affect the probability of a seed successfully producing an adult.

Delayed consequences of dissemination limitation: recruitment from seeds

Delayed consequences of distance-restricted seed dispersal have received extensive attention, so we shall only consider them briefly. Janzen (1970) and Connell (1971) predicted a lack of recruitment near parents because of near-total destruction of recruits by density- and distance-dependent enemies. Indeed, numerous empirical studies have shown greater mortality near to than far from conspecific adults (e.g. Augspurger, 1983; Clark and Clark, 1984; Howe *et al.*, 1985; Schupp, 1988; Schupp and Frost, 1989; Condit *et al.*, 1992). Because mortality is concentrated where most seeds fall, recruitment at the population level appears to be limited. Whether this pattern of mortality necessarily limits population growth is uncertain, however (see below). Further, results from experimentally manipulated seed shadows of *Tachigalia versicolor* (*Caesalpinioideae*) in Panama suggest that distance-restricted seed dispersal may actually increase rather than limit overall recruitment to the population (Augspurger and Kitajima, 1992); at larger spatial scales (100 m radius), distance-restricted dispersal increased overall seed survival through predator satiation, even while mortality at smaller spatial scales (1 m^2) was density-dependent.

Delayed consequences of spatially contagious seed dispersal have received far less attention. Several issues are involved, three of which we consider. Firstly, to what extent are seed and seedling fates determined by density rather than by distance from an adult? Because spatially contagious seed dispersal weakens the correlation between seed density and distance from adults, distinguishing effects of density versus distance on seed and seedling fate is critical for understanding the demographic implications of spatial patterns of seed dispersal. Studies distinguishing effects of density independent of those of distance have found evidence that high seed density increases mortality even far from parents (Augspurger and Kelly, 1984; Janzen, 1986; Augspurger and Kitajima, 1992). If density-dependent mortality acting on seed clumps far from conspecific adults is significant, spatially contagious seed dispersal will further limit recruitment. Not only will sites lacking seeds fail to produce recruits (dissemination limitation), but some sites receiving abundant seeds will likewise fail (establishment limitation).

Secondly, what are the effects on plant recruitment of the species composition of codispersed seeds or of seeds already at sites to which new seeds are dispersed? Seeds dispersed to monkey sleep trees are deposited in mixed-species assemblages; seeds of many species accumulate as these trees are reused (S.E. Russo, unpublished data). Seeds dispersed to fruiting trees may be dispersed to conspecific or heterospecific trees. Few data exist on either the effects of interactions among seed/seedling species that have been codispersed or the effects of background seed density and identity on recruitment.

Seeds dispersed under a conspecific crown may suffer the same fate as seeds that were not dispersed at all. Alternatively, survival may be less than expected relative to seeds falling beneath their parent if a propagule is resistant to its parent's pathogens, or greater than expected if a pathogen is locally adapted to the genotype of the parent. Experimental studies of the Janzen–Connell model have typically addressed the effects of distance from any conspecific adult, not distance from the parent. Evidence from *T. versicolor*, however, suggests that seedling survival may be greater in mixed- than single-genotype seed shadows (Augspurger and Kitajima, 1992), suggesting some level of local adaptation to pathogens.

The consequences of being dispersed to a fruiting heterospecific tree or to a sleep tree, where seeds fall into a patch with many seeds of a variety of species, are even less well understood. Fates of seeds and seedlings in these conditions will depend on many factors, including overall seed and seedling densities, relative densities of individual species, relative desirability of species to herbivores, the degree to which species share pathogens and competitive abilities of seedlings. These factors may, in turn, affect the probability of being secondarily dispersed or consumed, the probability of being attacked by herbivores and pathogens and the balance of intra- and interspecific competitive relationships.

Thirdly, does spatially contagious seed dispersal result in directed dispersal (*sensu* Howe and Smallwood, 1982) of seeds to sites that are most suitable for successful recruitment? Or does it result in placement into sites where death is virtually guaranteed? If frugivores dispersing seeds contagiously are qualitatively effective dispersers (i.e. carrying seeds preferentially to suitable sites) (Schupp, 1993), spatially contagious dispersal may greatly increase plant recruitment. Alternatively, if these dispersers are not effective, recruitment will be limited by spatially contagious seed dispersal. Few well-designed studies exist for evaluating the directed dispersal hypothesis, but at least some seed-dispersers preferentially deposit seeds in favourable sites. In fact, this phenomenon may occur more commonly than generally appreciated (Wenny, 2001). Three-wattled bellbirds (*Procnias tricarunculata*, Cotingidae) in Costa Rica use standing dead trees as display sites and thus deposit seeds in high-light environments, thereby promoting seedling establishment (Wenny and Levey, 1998; Wenny, 2000). Other bellbirds have similar behaviours (Snow, 1961, 1970), and both manakin and cock-of-the-rock (*Rupicola* spp.) leks tend to be in sites with greater insolation (Endler and Théry, 1996). Nests of gorillas (Voysey *et al.*, 1999b) and resting sites of nocturnal duikers (Feer, 1995) tend to be located in more disturbed areas, which may be more favourable for establishment. Fruit-induced diarrhoea in *Saguinus fuscicollis* (Callitrichidae) results in fluid defecations that coat seeds of the hemiepiphyte *Asplundia peruviana* (*Cyclanthaceae*) on to trunks, where they grow (Knogge *et al.*, 1998). Of course, not all spatially contagious dispersal will direct seeds to the most favourable sites. For example, large trees often selected by monkeys as sleep trees may provide a poor light environment for emerging seedlings.

Ultimately, the consequences for plant recruitment of distance-restricted and spatially contagious seed dispersal depend on the seed shadow produced (immediate consequences) and on the responses of potential recruits and their enemies and mutualists to the seed shadow (delayed consequences). Although much work has addressed the immediate and delayed consequences of distance-restricted seed dispersal, gaps in our knowledge remain. In contrast, both the immediate and delayed consequences of spatially contagious seed dispersal have been virtually ignored. Consequently, only a preliminary analysis of the effects of dissemination limitation on species richness can be attempted at this time.

Consequences of Dissemination Limitation: Origin and Maintenance of Species-rich Forests

Distance-restricted seed dispersal is one of two key processes at the core of the Janzen–Connell model (*sensu* Clark and Clark, 1984); recruitment is limited near parent trees by distance- and density-dependent seed and seedling enemies, but it is limited far from parents by a lack of seeds. In fact, when Hurtt and Pacala (1995) included distance-restricted seed dispersal in their model, it resulted in increased recruitment limitation, winning by forfeit and drift in species composition. Thus, distance-restricted seed dispersal potentially contributes to the maintenance of high species richness. As with rarity *per se*, however, it cannot contribute to the origin of high species richness because the effect of distance-restricted seed dispersal on recruitment limitation and, ultimately, species richness depends on the prior existence of a diverse community comprised of rare species. All else being equal, the ability of distance-restricted dispersal to limit recruitment should decrease with increasing abundance of a species, simply because more potential recruitment sites would be accessible to seeds.

Mortality from distance- and density-dependent enemies, rather than limitation by distance-restricted dispersal *per se*, provides the potential for the Janzen–Connell model to contribute to the origin of species-rich communities by introducing implicit negative density dependence (Schupp, 1992). Although many studies have addressed the Janzen–Connell model at the individual tree level, however, few have considered population-level effects. In *Faramea occidentalis* (*Rubiaceae*), seed predation at the scale of individual trees conformed to Janzen–Connell predictions (Schupp, 1988). At the scale of populations, though, seed survival increased with increasing adult density, a correlate of population-level seed availability. Seed predators appeared to be satiated by high densities of seed patches, yielding positively density-dependent recruitment, contrary to the model's implicit predictions (Schupp, 1992). Further, the concentration of seeds near parents caused by distance-dependent seed dispersal may satiate predators, resulting in greater overall seedling recruitment at the population level than would occur if seeds were more evenly distributed (Augspurger and Kitajima, 1992). Understanding the contribution of distance-restricted seed dispersal to the maintenance and origin of species-rich communities requires a shift in experimental approaches (see below). Further demonstrations that survival is greater away from than beneath conspecific trees will provide further evidence that dispersal is advantageous, but will not demonstrate that these local effects translate into the larger-scale population and community effects predicted by the model.

In contrast to distance-restricted seed dispersal, spatially contagious seed dispersal can potentially contribute to recruitment limitation when plant species are abundant. Even with a very large number of seeds, as expected with an abundant species, many potential recruitment sites may be forfeited if dispersed seeds are concentrated in relatively few sites rather than being widely dispersed to many sites. Thus, spatially contagious seed dispersal has the potential to severely limit recruitment even when species are abundant. Consequently, spatially contagious seed dispersal may contribute to the origin as well as the maintenance of species richness.

Dissemination Limitation and Gradients in Species Diversity

If dissemination limitation contributes to the origin of latitudinal gradients in species richness, then dissemination limitation should be more prevalent or powerful in species-rich tropical forests than in species-poor temperate forests. Is there any evidence supporting such a geographical pattern in dissemination limitation? Using inverse approaches that assign seeds to probable parents, Clark *et al.* (1998, 1999b) estimated that median dispersal distances were less for animal- than for wind-dispersed species, and suggested that animal-dispersed species are therefore more distance-restricted in their dispersal in both temperate and tropical regions. Many problems exist with such approaches to estimating seed shadows; among others, it is difficult to separate effects of seed size and dispersal mode, spatially contagious dispersal makes it difficult to capture sufficient seeds to reliably estimate curves, dispersal to conspecific feeding trees is indistinguishable from seeds dropped beneath parents and the tails of dispersal curves are unknown. None the less, the major conclusion for distance-restricted dispersal may still be true. Even if animal dispersal produces longer tails, as might be expected, animal-dispersed species may still be more distance-restricted if dispersal results in a greater concentration of seeds near parents; the very low densities of seeds in long tails will miss most potential recruitment sites. Additionally, we would expect spatially contagious seed dispersal to be much more common in animal- than in wind-dispersed species, because spatial contagion is largely driven by animal behaviour. So it is possible that both forms of dissemination limitation better characterize animal-dispersed than wind-dispersed species.

Further, a higher proportion of tropical than of temperate species have fleshy fruits dispersed by birds and mammals (Howe and Smallwood, 1982). As a consequence, recruitment limitation driven by dissemination limitation may indeed be more prevalent in tropical than in temperate forests, independent of rarity, and could in fact contribute to rarity and the origin of species-rich communities. If true, however, a critical question arises. If animal dispersal leads to recruitment limitation, and presumably to reduced population growth and fitness, why is animal dispersal so common in tropical forests? Other advantages to seed dispersal by animals may outweigh these disadvantages. Perhaps longer distance dispersal (longer tails of seed shadows) and thus colonization of new habitats are a key advantage that allows for longer-term survival of species dispersed by animals. Although this suggestion sounds contrary to the results of Clark *et al.* (1998, 1999b), it need not be. As they acknowledge, the method they employ is especially unreliable for estimating tails of distributions. It is not unreasonable in closed-canopy forests to expect longer, fatter tails in animal-generated than in wind-generated seed shadows, even if median dispersal distance is less.

Obviously, more prevalent dissemination limitation in tropical forests due to a greater representation of animal-dispersed species cannot fully explain worldwide patterns of tree species richness. The highly speciose *Dipterocarpaceae*, which contribute so much to South-East Asian tree community diversity, are, in fact, wind-dispersed. We do not believe that dissemination limitation created species-rich tropical forests – or, for that matter, that any single explanation for the origin of latitudinal gradients of species richness was exclusively responsible. None the less, we do believe that dissemination limitation is a powerful force that contributes strongly to the maintenance, and perhaps somewhat to the origin, of species-rich forests.

Avenues for Future Research

This review points to several challenges that need to be overcome if we are to understand the role of seed dispersal in population and community dynamics. Here we briefly present those we believe are most critical.

1. We need to more thoroughly explore through models the consequences of differing seed shadows in respect of successful occupation of available sites. Such an approach should include both distance-restricted and spatially contagious seed dispersal. For example, what intensity of spatially contagious seed dispersal is necessary to significantly alter patterns of site occupation?
2. We need to characterize natural seed shadows more thoroughly and for more dispersal systems, concentrating on the contributions of spatially contagious seed dispersal to the spatial pattern of seed deposition.
3. We need empirical studies designed to address more effectively the population-level consequences of Janzen–Connell spacing mechanisms.
4. We need empirical studies to address the immediate and delayed consequences of spatially contagious and distance-restricted seed dispersal at both individual and population levels. What are the environmental characteristics of sites where seeds are concentrated and what are the site-specific consequences for plant recruitment? How do the responses of enemies and mutualists and the outcomes of competition change under different spatial, qualitative and quantitative patterns of seed contagion?
5. We must understand the phenomenon of codispersal of suites of species and how it affects seed fate, plant recruitment and the spatial structure of populations and communities.

Acknowledgements

The overview presented here of mechanisms causing recruitment limitation is based on ideas that have been jointly developed by P. Jordano and E.W. Schupp, and that will be published in detail at a later date. The paper on which this chapter is based was supported by the Ecology Center (E.W.S., T.M.) and the Utah Agricultural Experiment Station (E.W.S.), Utah State University, the National Science Foundation (T.M., S.E.R.), the American Ornithologist's Union (S.E.R.), the University of Illinois (S.E.R.) and the Organization for Tropical Studies (S.E.R.). We thank Carol Augspurger, Carlos Herrera and Pedro Jordano for fruitful discussions and K. Böhning-Gaese, T. Masaki and Doug Levey for helping improve the presentation of our ideas. We especially thank Wesley Silva, Mauro Galetti, Marco Aurélio Pizo, Doug Levey and Ronda Green for a superb symposium–workshop that we feel honoured to have participated in.

References

Andresen, E. (1999) Seed dispersal by monkeys and the fate of dispersed seeds in a Peruvian rain forest. *Biotropica* 31,145–158.

Augspurger, C.K. (1983) Seed dispersal of the tropical tree, *Platypodium elegans*, and the escape of its seedlings from fungal pathogens. *Journal of Ecology* 71, 759–771.

Augspurger, C.K. and Kelly, C.K. (1984) Pathogen mortality of tropical tree seedlings: experimental studies of the effects of dispersal distance, seedling density, and light conditions. *Oecologia* 61, 211–217.

Augspurger, C.K. and Kitajima, K. (1992) Experimental studies of seedling recruitment from contrasting seed distributions. *Ecology* 73, 1270–1284.

Bizerril, M.X.A. and Raw, A. (1998) Feeding behaviour of bats and the dispersal of *Piper arboreum* seeds in Brazil. *Journal of Tropical Ecology* 14, 109–114.

Clark, D.A. and Clark, D.B. (1984) Spacing dynamics of a tropical rain forest tree: evaluation of the Janzen–Connell model. *American Naturalist* 124, 769–788.

Clark, J.S., Macklin, E. and Wood, L. (1998) Stages and spatial scales of recruitment limitation in southern Appalachian forests. *Ecological Monographs* 68, 213–235.

Clark, J.S., Beckage, B., Camil, P., Cleveland, B., HilleRisLambers, J., Lichter, J., McLachlan, J., Mohan, J. and Wyckoff, P. (1999a) Interpreting recruitment limitation in forests. *American Journal of Botany* 86, 1–16.

Clark, J.S., Silman, M., Kern, R., Macklin, E. and HilleRisLambers, J. (1999b) Seed dispersal near and far: patterns across temperate and tropical forests. *Ecology* 80, 1475–1494.

Colwell, R.K. and Lees, D.C. (2000) The mid-domain effect: geometric constraints on the geography of species richness. *Trends in Ecology and Evolution* 15, 70–76.

Condit, R., Hubbell, S.P. and Foster, R.B. (1992) Recruitment near conspecific adults and the maintenance of tree and shrub diversity in a neotropical forest. *American Naturalist* 140, 261–286.

Connell, J.H. (1971) On the role of natural enemies in preventing competitive exclusion in some marine animals and in rain forest trees. In: den Boer, P.J. and Gradwell, G.R. (eds) *Dynamics of Populations*. Centre for Agricultural Publishing and Documentation, Wageningen, The Netherlands, pp. 298–313.

Dinerstein, E. (1991) Seed dispersal by greater one-horned rhinoceros (*Rhinoceros unicornis*) and the flora of *Rhinoceros* latrines. *Mammalia* 55, 355–362.

Dinerstein, E. and Wemmer, C. (1988) Fruits *Rhinoceros* eat: dispersal of *Trewia nudiflora* (*Euphorbiaceae*) in lowland Nepal. *Ecology* 69, 1768–1774.

Duncan, R.S. and Chapman, C.A. (1999) Seed dispersal and potential forest succession in abandoned agriculture in tropical Africa. *Ecological Applications* 9, 998–1008.

Ehrlén, J. and Eriksson, O. (2000) Dispersal limitation and patch occupancy in forest herbs. *Ecology* 81, 1667–1674.

Endler, J.A. and Théry, M. (1996) Interacting effects of lek placement, display behavior, ambient light, and color patterns in three neotropical forest-dwelling birds. *American Naturalist* 148, 421–452.

Feer, F. (1995) Seed dispersal in African forest ruminants. *Journal of Tropical Ecology* 11, 683–689.

Fleming, T.H. (1981) Fecundity, fruiting pattern, and seed dispersal in *Piper amalgo* (*Piperaceae*), a bat-dispersed tropical shrub. *Oecologia* 51, 42–46.

Fleming, T.H. and Heithaus, E.R. (1981) Frugivorous bats, seed shadows, and the structure of tropical forests. *Biotropica* 13 (suppl.), 45–53.

Fragoso, J.M.V. (1997) Tapir-generated seed shadows: scale-dependent patchiness in the Amazon rain forest. *Journal of Ecology* 85, 519–529.

Gillett, J.B. (1962) Pest pressure, an underestimated factor in evolution. *Systematics Association Publication* 4, 37–46.

Gilliard, E.T. (1958) *Living Birds of the World.* Doubleday, Garden City, New York, 400 pp.

Guevara, S. and Laborde, J. (1993) Monitoring seed dispersal at isolated standing trees in tropical pastures: consequences for local species availability. *Vegetatio* 107/108, 319–338.

Harvey, C.A. (2000) Windbreaks enhance seed dispersal into agricultural landscapes in Monteverde, Costa Rica. *Ecological Applications* 10, 155–173.

Howe, H.F. (1989) Scatter- and clump-dispersal and seedling demography: hypothesis and implications. *Oecologia* 79, 417–426.

Howe, H.F. and Smallwood, J. (1982) Ecology of seed dispersal. *Annual Review of Ecology and Systematics* 13, 201–228.

Howe, H.F., Schupp, E.W. and Westley, L.C. (1985) Early consequences of seed dispersal for a neotropical tree (*Virola surinamensis*). *Ecology* 66, 781–791.

Hubbell, S.P. (1979) Tree dispersion, abundance, and diversity in a tropical dry forest. *Science* 203, 1299–1309.

Hubbell, S.P., Foster, R.B., O'Brien, S.T., Harms, K.E., Condit, R., Wechsler, B., Wright, S.J. and Loo de Lao, S. (1999) Light-gap disturbances, recruitment limitation, and tree diversity in a neotropical forest. *Science* 283, 554–557.

Hurtt, G.C. and Pacala, S.W. (1995) The consequences of recruitment limitation: reconciling chance, history and competitive differences between plants. *Journal of Theoretical Biology* 176, 1–12.

Janzen, D.H. (1970) Herbivores and the number of tree species in tropical forests. *American Naturalist* 104, 501–528.

Janzen, D.H. (1971) Seed predation by animals. *Annual Review of Ecology and Systematics* 2, 465–492.

Janzen, D.H. (1986) Mice, big mammals, and seeds: it matters who defecates what where. In: Estrada, A. and Fleming, T.H. (eds) *Frugivores and Seed Dispersal.* Dr W. Junk, Dordrecht, The Netherlands, pp. 251–271.

Janzen, D.H., Miller, G.A., Hackforth Jones, J., Pond, C.M., Hooper, K. and Janos, D.P. (1976) Two Costa Rican bat-generated seed shadows of *Andira inermis* (*Leguminosae*). *Ecology* 57, 1068–1075.

Jordano, P. and Schupp, E.W. (2000) Seed-disperser effectiveness: the quantity component and patterns of seed rain for*Prunus mahaleb*. *Ecological Monographs* 70, 591–615.

Julliot, C. (1997) Impact of seed dispersal by red howler monkeys (*Alouatta seniculus*) on the seedling population in the understorey of tropical rainforest. *Journal of Ecology* 85, 431–440.

Knogge, C., Heymann, E.W. and Tirado Herrera, E.R. (1998) Seed dispersal of *Asplundia peruviana* (*Cyclanthaceae*) by the primate *Saguinus fuscicollis*. *Journal of Tropical Ecology* 14, 99–102.

Krijger, C.L., Opdam, M., Théry, M. and Bongers, F. (1997) Courtship behaviour of manakins and seed bank composition in a French Guianan rain forest. *Journal of Tropical Ecology* 13, 631–636.

Laman, T.G. (1996) *Ficus* seed shadows in a Bornean rain forest. *Oecologia* 107, 347–355.

Levey, D.A. and Duke, G.E. (1992) How do frugivores process fruit? Gastrointestinal transit and glucose absorption in cedar waxwings (*Bombycilla cedrorum*). *Auk* 109, 722–730.

Levey, D.A. and Grajal, A. (1991) Evolutionary implications of fruit-processing limitations in cedar waxwings. *American Naturalist* 138, 171–189.

Loiselle, B.A. (1990) Seeds in the droppings of tropical fruit-eating birds: the importance of considering seed composition. *Oecologia* 82, 494–500.

McDiarmid, R.W., Ricklefs, R.E. and Foster, M.S. (1977) Dispersal of *Stemmadenia donnell-smithii* (*Apocynaceae*) by birds. *Biotropica* 9, 9–25.

Mack, A.L. (1995) Distance and non-randomness of seed dispersal by the dwarf cassowary *Casuarius bennetti*. *Ecography* 18, 286–295.

Nathan, R. and Muller-Landau, H.C. (2000) Spatial patterns of seed dispersal, their determinants and consequences for recruitment. *Trends in Ecology and Evolution* 15, 278–285.

Palmer, M.W. (1994) Variation in species richness: towards a unification of hypotheses. *Folia Geobotanica and Phytotaxonomica* 29, 511–530.

Pitman, N.C.A., Terborgh, J., Silman, M.R. and Nuñez, V.P. (1999) Tree species distributions in an upper Amazonian forest. *Ecology* 80, 2651–2661.

Portnoy, S. and Willson, M.F. (1993) Seed dispersal curves: behavior of the tail of the distribution. *Evolutionary Ecology* 7, 25–44.

Schupp, E.W. (1988) Seed and early seedling predation in the forest understory and in treefall gaps. *Oikos* 51, 71–78.

Schupp, E.W. (1992) The Janzen–Connell model for tropical tree diversity: population implications and the importance of spatial scale. *American Naturalist* 140, 526–530.

Schupp, E.W. (1993) Quantity, quality and the effectiveness of seed dispersal by animals. *Vegetatio* 107/108, 15–29.

Schupp, E.W. and Frost, E.J. (1989) Differential predation of *Welfia georgii* seeds in treefall gaps and the forest understory. *Biotropica* 21, 200–203.

Schupp, E.W. and Fuentes, M. (1995) Spatial patterns of seed dispersal and the unification of plant population ecology. *EcoScience* 2, 267–275.

Snow, B.K. (1961) Notes on the behavior of three Cotingidae. *Auk* 78, 150–161.

Snow, B.K. (1970) A field study of the bearded bellbird in Trinidad. *Ibis* 112, 299–329.

Snow, D.W. (1962) A field study of the golden-headed manakin, *Pipra erythrocephala*, in Trinidad. *Zoologica* 47, 183–198.

Théry, M. and Larpin, D. (1993) Seed dispersal and vegetation dynamics at a cock-of-the-rock's lek in the tropical forest of French Guiana. *Journal of Tropical Ecology* 9, 109–116.

Thomas, D.W., Cloutier, D., Provencher, M. and Houle, C. (1988) The shape of bird- and bat-generated seed shadows around a tropical tree. *Biotropica* 20, 347–348.

Thomas, S.C. (1991) Population densities and patterns of habitat use among anthropoid primates of the Ituri forest, Zaire. *Biotropica* 23, 68–83.

Trail, P.W. (1985) Territoriality and dominance in the lek-breeding Guianan cock-of-the-rock. *National Geographic Research* 1, 112–123.

Voysey, B.C., McDonald, K.E., Rogers, M.E., Tutin, C.E.G. and Parnell, R.J. (1999a) Gorillas and seed dispersal in the Lopé Reserve, Gabon. I: Gorilla acquisition by trees. *Journal of Tropical Ecology* 15, 23–38.

Voysey, B.C., McDonald, K.E., Rogers, M.E., Tutin, C.E.G. and Parnell, R.J. (1999b) Gorillas and seed dispersal in the Lopé Reserve, Gabon. II: Survival and growth of seedlings. *Journal of Tropical Ecology* 15, 39–60.

Wenny, D.G. (2000) Seed dispersal, seed predation, and seedling recruitment of a neotropical montane tree. *Ecological Monographs* 70, 331–351.

Wenny, D.G. (2001) Advantages of seed dispersal: a re-evaluation of directed dispersal. *Evolutionary Ecology Research* 3, 51–74.

Wenny, D.G. and Levey, D.J. (1998) Directed seed dispersal by bellbirds in a tropical cloud forest. *Proceedings of the National Academy of Sciences USA* 95, 6204–6207.

Westcott, D.A. (1994) Leks of leks: a role for hotspots in lek evolution? *Proceedings of the Royal Society of London B* 258, 281–286.

Westcott, D.A. (1997) Lek locations and patterns of female movement and distribution in a Neotropical frugivorous bird. *Animal Behaviour* 53, 235–247.

Wheelwright, N.T. (1988) Four constraints on coevolution between fruit-eating birds and fruiting plants: a tropical case history. In: Ouellet, H. (ed.) *Acta XIX Congressus Internationalis Ornithologici.* Kluwer, Dordrecht, the Netherlands, pp. 827–845.

Willson, M.F. (1993) Dispersal mode, seed shadows, and colonization patterns. *Vegetatio* 107/108, 261–280.

Wrangham, R.W., Chapman, C.A. and Chapman, L.J. (1994) Seed dispersal by forest chimpanzees in Uganda. *Journal of Tropical Ecology* 10, 355–368.

Wright, S.J. (1999) Plant diversity in tropical forests. In: Pugnaire, F.I. and Valladares, F. (eds) *Handbook of Functional Plant Ecology.* Marcel Dekker, New York, pp. 449–471.

3 Assessing Recruitment Limitation: Concepts, Methods and Case-studies from a Tropical Forest

Helene C. Muller-Landau,[1, 2] S. Joseph Wright,[2] Osvaldo Calderón,[2] Stephen P. Hubbell[2, 3] and Robin B. Foster[2, 4]

[1]*Department of Ecology and Evolutionary Biology, Princeton University, Princeton, NJ 08544-1003, USA;* [2]*Smithsonian Tropical Research Institute, Apartado 2072, Balboa, Ancón, Republic of Panama;* [3]*Botany Department, University of Georgia, Athens, GA 30602-7271, USA;* [4]*The Field Museum, 1400 S. Lake Shore Drive, Chicago, IL 60605-2496, USA*

Introduction

Seed production and seed dispersal are critically important processes in population dynamics, precisely because they are almost never completely successful – that is, because not all sites suitable for a given species are reached by its seeds. The failure of seeds to arrive at all suitable sites limits population growth rates and abundances, a phenomenon called seed limitation (Crawley, 1990; Eriksson and Ehrlen, 1992; Turnbull *et al.*, 2000). Seed limitation has important consequences for population and community dynamics and for species diversity at multiple scales (Tilman, 1994; Hurtt and Pacala, 1995; Pacala and Levin, 1997; Zobel *et al.*, 2000).

Seed limitation can arise from limited seed numbers and/or limited dispersal of available seeds. The total number of seeds available, determined by a species' adult abundance and fecundity, places an upper limit on the number of sites that can possibly be reached by seeds and determines overall mean seed density (Clark *et al.*, 1998). The variance in seed density depends primarily upon the shapes and sizes of seed shadows and the clumping or contagion of the seed rain (Clark *et al.*, 1998). If adult trees are clumped (as they often are, Condit *et al.*, 2000), seed limitation will be further increased (Ribbens *et al.*, 1994).

To assess the consequences and importance of restricted seed rain, we need to quantify seed limitation and compare it with establishment limitation (also known as site limitation) – the limitation of a plant population by the number of sites suitable for establishment or, more generally, by the suitability of sites for establishment (Eriksson and Ehrlen, 1992; Clark *et al.*, 1998; Nathan and Muller-Landau, 2000). Plant ecologists have long recognized the contribution of both these factors, although they have differed on the issue of their relative importance, with profound

 Seed Dispersal and Frugivory: Ecology, Evolution and Conservation (eds D.J. Levey, W.R. Silva and M. Galetti)

implications for how one views plant communities (Clark *et al.*, 1999). In particular, if establishment limitation dominates, then species relative abundances are determined mainly by their regeneration niches and the relative abundances of microhabitats (Grubb, 1977). In contrast, if seed limitation dominates, then fewer sites are won by the best possible competitor in that microhabitat and more by whichever species happens to arrive (Hubbell and Foster, 1986; Cornell and Lawton, 1992).

While the consequences of establishment limitation are well understood, the mechanisms and implications of seed limitation are less widely appreciated. Seed limitation essentially slows rates of change in abundance – species cannot increase as quickly if they do not reach all suitable sites (nor, as a result, can their competitors decrease as quickly). Further, dispersal limitation in particular reduces the frequency of interspecific competition and increases the frequency of intraspecific competition, because propagules of a species are aggregated rather than equally distributed across sites, further reducing potential rates of changes in population size (Pacala and Levin, 1997). These forces will operate even in the face of strong niche differences. For example, even if all species have strict competitive rankings in every available habitat and are able to win only within their preferred habitat, seed limitation will increase the probability that the best competitor is not present at a given site and that the site will be won instead by a lesser competitor that is present. This results in more stochastic dynamics on the community level, despite deterministic dynamics at each site (Hurtt and Pacala, 1995). Whether species are equivalent or differentiated by life-history trade-offs, model communities in which dispersal is localized maintain higher total diversity (albeit lower local diversity) than those in which dispersal is global (Hubbell, 2001; Chave *et al.*, 2002).

Despite its importance to plant populations and communities and its obvious link to seed dispersal, few studies of dispersal explicitly quantify seed limitation and its components (but see Clark *et al.*, 1998). In this chapter, we first outline methods for measuring seed limitation, establishment limitation and their components. These methods are applicable to any study that quantifies seed rain at an unbiased sample of locations in a community or explicitly measures the shapes of seed shadows. Then we apply these methods to several species in a tropical forest to evaluate their usefulness. Finally, we assess the implications of observed seed and establishment limitation for tropical forest diversity and conservation.

General Methods for Quantifying Recruitment Limitation and its Components

Recruitment limitation is the reduction in a species' abundance from the maximum set by the environment that can be attributed to limited numbers of recruits. That is, how much smaller is the abundance than it would be if there were unlimited numbers of recruits? To apply this very general definition, we must specify whose abundance (e.g. adults, juveniles), and which stage recruits (e.g. seeds, seedlings) are of interest. In the theoretical literature, recruitment limitation means limitation of the adult population by arrival of the mobile, dispersing stage; in this case, recruitment limitation can be directly juxtaposed with total establishment limitation, which reflects all post-dispersal processes (Hurtt and Pacala, 1995; Pacala, 1997). This accords with the use of the term in the marine literature, where it was introduced to describe limitation of adult density by the rate of larval arrival (Chesson, 1998). In the terrestrial plant literature, the term has been used in different ways corresponding to different definitions of recruit: in some cases, recruits are seeds and recruitment limitation is simply seed limitation (Tilman, 1997; Hubbell *et al.*, 1999); in others, recruits are an older juvenile class such as seedlings and recruitment limitation thus reflects a combination of seed limitation and early establishment limitation (Ribbens *et al.*, 1994; Clark *et al.*, 1998). Here, we use the more specific terms seed limitation and seedling limitation, respectively, to describe these two cases, and use recruitment limitation in the general sense, to encompass limitation by any stage recruits of a measure of the abundance of any later stage.

We start by defining seed limitation and establishment limitation of adult abundance, where establishment means establishment to adulthood, including all post-arrival processes (e.g. germination, competition, herbivory). We can assess seed and establishment limitation in two distinct ways: (i) seed addition experiments; and (ii) measurement of seed rain and establishment patterns. These provide different but complementary measures of each type of limitation.

Seed addition experiments

The most straightforward way to assess seed limitation of a population is to experimentally add large numbers of seeds of that species, and compare the results with controls in which no seeds are added. Many such experiments have been conducted; of those reviewed by Turnbull *et al.* (2000), approximately half found an increase in density following sowing, evidence that the populations are seed-limited. To determine how much larger population density would be if seed availability were not limiting would require addition experiments using optimal densities of seeds – that is, seed densities at which adult density is maximized (this may mean simply seed densities greater than a threshold value above which further addition of seeds does not result in further increases in adults, or it might mean densities within a limited range above which yield actually decreases due to overcompensating density dependence). From such experiments, we can calculate exactly the reduction in population density due to reduced (or excessive) numbers of seeds alone, in the context of all other limitations imposed during establishment. By analogy with the concept of realized niches (niche size in the context of all other factors), Nathan and Muller-Landau (2000) termed this realized seed limitation:

$$\text{Realized seed limitation} = 1 - \frac{\text{adult density in control plots}}{\text{adult density in seed addition plots}}$$

The fraction essentially gives actual adult density (density under actual seed densities and actual establishment conditions) as a proportion of potential density, given optimal seed densities and actual establishment conditions. Note that experiments that add seeds at non-optimal densities will allow for estimation of a lower bound on realized seed limitation, since adult densities in the addition plots will not be as large as they could be.

Seed addition experiments also provide information on the total limitation of population size by factors other than seed availability. Because seed numbers are not limiting in the addition plots, any difference in adult density from the maximum possible must be due to limitation by other factors acting on early establishment and survival to adulthood. Maximum possible density could be determined from maximum densities observed in the field, experimental monocultures or, potentially, calculations based on organism size and total resource availability. We can then calculate the reduction in population density due to establishment factors from the maximum possible if neither seed availability nor establishment were limiting. By analogy with the concept of fundamental niches, this is termed fundamental establishment limitation (Nathan and Muller-Landau, 2000):

$$\text{Fundamental establishment limitation} = 1 - \frac{\text{adult density in seed addition plots}}{\text{maximum possible adult density}}$$

The numerator of the fraction in this equation is essentially potential density given optimal seed densities and actual establishment conditions, while the denominator is potential density given optimal seed densities and optimal establishment conditions. Experiments in which seeds are added at non-optimal densities provide an upper bound on fundamental establishment limitation.

Measurement of seed rain and establishment patterns

Another way to assess seed limitation is to measure patterns of seed rain in the field or to simulate patterns of seed arrival, to determine the proportion of sites that are reached by seeds (Ribbens *et al.*, 1994; Clark *et al.*, 1998; Hubbell *et al.*, 1999). Where there are no seeds, there can be no subsequent seedlings or adults, regardless of establishment conditions. Thus, the proportion of all sites at which seeds do not arrive is a measure of fundamental seed limitation – seed limitation measured as if no other factors were limiting. To assess fundamental seed limitation of adult density, we define a site as the area occupied by a single adult and observe or estimate seed rain to sites that constitute an unbiased sample of the community. Following Nathan and Muller-Landau (2000), we then define:

$$\text{Fundamental seed limitation} = 1 - \frac{\text{sites reached by seeds}}{\text{total number of sites}}$$

(see also Ribbens *et al.*, 1994; Clark *et al.*, 1998). The numerator above is essentially the potential adult density given actual seed arrival patterns and optimal establishment conditions; the denominator is potential adult density given seeds everywhere and optimal establishment conditions. Optimal establishment conditions would be conditions under which an adult establishes at every site receiving one or more seeds.

If both seed arrival and subsequent establishment are measured, we can also obtain an estimate of establishment limitation. Where seeds arrive but establishment does not occur, establishment must be limiting. Thus, the proportion of sites receiving seeds at which establishment does not occur is a measure of realized establishment limitation – establishment limitation in the context of other limiting factors. To assess realized establishment limitation, we observe or estimate seed rain to sites constituting an unbiased sample of the community, and then observe or estimate establishment at sites sampling the community (preferably, but not necessarily, the same sites). We then define:

$$\text{Realized establishment limitation} = 1 - \frac{\text{sites in which establishment occurs}}{\text{sites reached by seeds}}$$

(Nathan and Muller-Landau, 2000). The numerator above is actual density (under actual seed arrival patterns and actual establishment conditions); the denominator is potential density given actual seed arrival patterns and optimal establishment conditions (see Box 3.1).

Further decomposing limitation

Both seed and establishment limitation reflect a variety of factors and thus can be decomposed into corresponding component limitations. For example, seed limitation arises from both limited numbers of seeds and limited distribution of available seeds. We can separate these two influences quantitatively by considering what would happen if available seeds were distributed uniformly across sites. Clark *et al.* (1998) pioneered such analyses by decomposing what we call fundamental seed limitation into source limitation and dispersal limitation. Their source limitation, which we call fundamental source limitation, is failure of seeds to reach sites due simply to insufficient seed numbers: there are not enough seeds to go around, even if all seeds are uniformly distributed among sites (Clark *et al.*, 1998). Clark *et al.* (1998) calculate this as:

$$\text{Fundamental source limitation} = 1 - \frac{\begin{array}{c}\text{sites that would be reached by seeds}\\ \text{if seeds were uniformly distributed}\end{array}}{\text{total number of sites}}$$

Source limitation is contrasted with dispersal limitation – seed limitation due to non-uniform distribution of seeds among sites (Clark *et al.*, 1998). Non-uniform distribution of seeds is nearly ubiquitous, because seeds are dispersed limited distances from their sources and are often dispersed in clumps (Clark *et al.*, 1998) and because adult trees are themselves clumped, increasing the number of sites that are very far from any sources (Ribbens *et al.*, 1994). Clark *et al.* (1998) quantify dispersal limitation of seed arrival, which we call fundamental dispersal limitation, as:

$$\text{Fundamental dispersal limitation} = 1 - \frac{\text{sites reached by seeds}}{\begin{array}{c}\text{sites that would be reached by seeds}\\ \text{if seeds were uniformly distributed}\end{array}}$$

(see Box 3.1 for a worked example).

Box 3.1. Calculating recruitment limitation from observational data.

Data on seed arrival and subsequent seedling establishment at an unbiased sample of sites allow us to assess fundamental seed limitation and its components, as well as realized seedling establishment limitation. Consider, for simplicity, an example of $n = 5$ sites, of which $a = 3$ receive seeds. (Much larger sample sizes are needed for statistically significant estimates, of course; the small numbers used here are for illustration only.)

 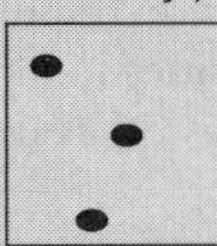 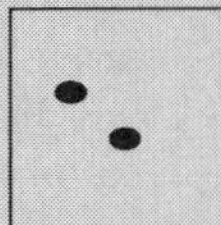 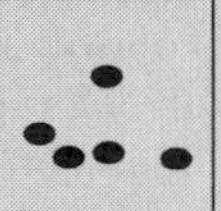 

Fundamental seed limitation – the proportion of sites not receiving seeds – can then be defined as:

$$\text{Seed limitation} = 1 - \frac{a}{n} = 1 - \frac{3}{5} = 0.4$$

at this spatiotemporal scale (see main text for a discussion of how to choose an appropriate scale).

We can calculate the limitation due solely to seed number by considering how many sites would be reached if seeds were distributed uniformly, with an expectation of s/n seeds per site. Following Clark *et al.* (1998), we can define uniform distribution stochastically as a Poisson seed rain with equal expectation everywhere – that is, a random distribution. Then the proportion of sites at which no seeds arrive under such a distribution is simply the Poisson probability of zero events given an expectation of s/n events, that is:

$$\text{Source limitation (stochastic)} = \exp\left(-\frac{s}{n}\right) = \exp\left(-\frac{10}{5}\right) = 0.14$$

Note that under this stochastic definition, source limitation is non-zero even though there are more seeds than sites. An alternative, deterministic interpretation of a uniform distribution would distribute seeds evenly, and thus non-independently, across all sites to the degree possible without producing fractional seeds per site. In this case, we have

$$\text{Source limitation (deterministic)} = \max\left\{1 - \frac{s}{n}, 0\right\} = \max\left\{1 - \frac{10}{5}, 0\right\} = 0$$

Thus, source limitation is zero when there are more seeds than sites, as here, and is the proportion of sites that would not receive a seed if no more than one seed were deposited on each site otherwise. In the results presented in this chapter, we apply the stochastic definition.

By comparing the proportion of sites reached by seeds in reality with the proportion of sites that would be reached by seeds if dispersal were uniform, we can assess the influence of restricted dispersal of seeds among sites as:

$$\text{Dispersal limitation} = 1 - \frac{a/n}{1 - \text{source limitation}}$$

(Clark *et al.*, 1998). For our example, this is 0.3 for the stochastic definition of a uniform distribution and 0.4 for the deterministic definition.

From a study of seedling establishment finding that seedlings recruit in $r = 2$ of $n = 5$ sites,

 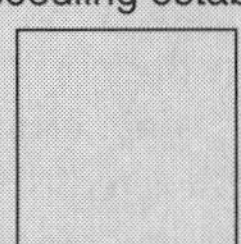  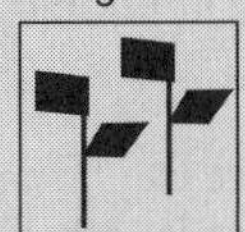 

we can calculate fundamental seedling limitation as:

$$\text{Seedling limitation} = 1 - \frac{r}{n} = 1 - \frac{2}{5} = 0.6$$

Given information on both seed arrival and seedling establishment at the same spatiotemporal scale, the reduction in seedling site occupancy due to failure of establishment in sites where seeds arrive can be quantified as:

$$\text{Establishment limitation} = 1 - \frac{r}{a} = 1 - \frac{2}{3} = \frac{1}{3}$$

Calculations of establishment limitation can be made even if seedling recruitment is not measured in the same sites as seed arrival, as long as site sizes are the same in both studies and both sets of sites are an unbiased sample of the area.

Like fundamental seed limitation, realized seed limitation can also be broken down into contributions due to source and dispersal limitation. This requires seed redistribution experiments, in which all seeds produced in an area are collected and redistributed uniformly across sites. Comparison with controls in which seeds dispersed naturally would allow calculation of realized dispersal limitation – the decrease in population caused by clumping of dispersed seeds. Comparison with seed addition experiments in which optimal seed densities were added would allow calculation of realized source limitation – the decrease in population caused by limited seed numbers alone, in the absence of limited dispersal. Such seed redistribution experiments have rarely been conducted (but see Augspurger and Kitajima, 1992).

Dispersal limitation could be further decomposed into contributions due to clumping of adults, variance in seed production among adults, short dispersal distances of seeds and so forth. For example, we can examine the contribution of clumping of adults by simulating or experimentally manipulating seed rain under actual spatial patterns of adults as well as under uniform spatial patterns, all other things being equal. Establishment limitation can similarly be broken down into contributions at different stages or by different agents, by examining the proportion of sites occupied at different stages or with and without particular agents (e.g. herbivores, pathogens, physical damage).

Generalizing and applying limitation measures

The methods described above can be adapted to examine limitation of and by other stages as well – for example, instead of examining limitation of adult density by seed availability, we could examine limitation of juvenile density by seed availability or limitation of adult biomass by seedling availability. Of course, the relative magnitudes of seed and establishment limitation for juveniles may be very different from that for adults, because factors act differently at different stages (Schupp, 1995). In particular, seed limitation of juveniles will generally be larger than seed limitation of adults because there is more scope for establishment factors to manifest themselves in the longer time to adulthood. It may be possible to extrapolate from limitation of juvenile densities to limitation of adult densities by using information on later survival patterns.

Both sets of methods for calculating seed and establishment limitation essentially divide the number of missed opportunities (sites a species does not capture) between those missed due to failure of seed arrival and those missed due to failure of establishment. They differ in how they attribute failure at sites in which both seed rain and establishment conditions are more than minimally adequate but less than optimal – sites in which one or more seeds arrive but no establishment occurs in nature, and yet establishment does occur when seeds are added at optimal numbers. Realized seed limitation will almost always be greater than fundamental seed limitation because the probability of having an adult establish almost always increases as additional seeds (above one) are added and thus the proportion of sites not receiving seeds is smaller than the proportion of sites in which no adults establish because of limited seed availability overall. Similarly, realized establishment limitation will almost always be greater than fundamental establishment limitation, because the proportion of sites in which establishment is totally impossible is smaller than the proportion of sites receiving any seeds in which establishment does not occur. The relationship between fundamental and realized limitation will also depend upon the correlation, if any, between seed arrival and establishment conditions across sites. Directed dispersal alone produces a strong positive correlation between seed arrival and establishment conditions; negatively density-dependent survival alone results in a negative correlation.

Choosing spatiotemporal scales

Definition of a suitable scale at which to calculate limitation measures depends in large part on the stage whose limitation is being assessed. To assess seed limitation of adult populations, addition experiments or seed-rain measurement should be conducted on a spatial scale similar to the area occupied by an adult, and on a temporal scale similar to the time it takes for an adult to establish (as in Ribbens *et al.*, 1994). On the other hand, for examining limitation of seedling or sapling densities in the same forest, the appropriate scale would be smaller – a seedling might occupy only 0.1 m × 0.1 m, a sapling 1 m × 1 m (as in Clark *et al.*, 1998). Choice of temporal scale can likewise be based on the age of the class whose limitation is being evaluated. It may be useful to examine seed and establishment limitation at spatial or temporal scales smaller than those occupied by an establishing individual of the focal stage, however, in order to take account of heterogeneity of seed rain and establishment conditions at such smaller scales. For any given system or species, there will be a range of spatiotemporal scales at which limitation measures provide useful information; in general, no single scale will be most appropriate for all questions because processes change across scales (Kollmann, 2000).

Limitation measures change both absolute and relative values with scale. The proportion of 1 m^2 sites receiving seeds will obviously be smaller than the proportion of 25 m^2 sites receiving seeds. Further, the scale of the analysis will affect the relative magnitude of different components of limitation. Source limitation is a declining function of the total seed fall expected per sample plot, which will increase with area and time period. Seed limitation will also decline with increasing sample plot size or time period, but much less quickly, because real seed rain is spatiotemporally autocorrelated. Thus, dispersal limitation will become a proportionately larger component of seed limitation as the scale increases – at some point, plots are large enough (or time periods long enough) for uniform distribution of seeds to result in all plots receiving seeds, and thus any failure of seeds to arrive is attributed entirely to limited dispersal. Because limitation measures are scale-dependent, the spatiotemporal scale should be given for every measure reported, measures should be calculated for multiple scales when possible and the utmost care should be taken in comparing limitation measures between species and systems when absolute or relative scales vary.

Case-studies from Barro Colorado Island

Fundamental seed limitation can be calculated from either of two kinds of data routinely collected in studies of seed dispersal.

1. We can make calculations of observed seed limitation in a set of randomly or regularly spaced sites (e.g. seed traps), providing an unbiased sample of the area in respect of distance to and density of source trees (see Box 3.1).
2. We can make calculations of projected limitation, given sufficient information on adult density, spatial pattern and seed shadows to allow projection of seed rain across the area of interest (Ribbens *et al.*, 1994; Clark *et al.*, 1998).

In this section, we use data on seed arrival to calculate both observed and projected seed limitation and its components at the 0.5 m^2 scale and periods of 1–12 years for four tropical tree species varying in life history and abundance. We then combine estimates of 1 m^2 seed limitation (obtained via extrapolation) with data on seedling abundances to calculate seedling limitation and establishment limitation at the 1 m^2 scale and periods of 1–6 years.

The spatial scales examined here were chosen for the seed and seedling censuses because they seemed appropriate for sampling heterogeneity in seed rain and establishment conditions in a tropical forest. They are of the order of scales at which seed rain and establishment conditions vary and over which seedlings compete – that is, small enough for seed rain and establishment conditions within sites to be relatively homogeneous and large enough to encompass competitors for the same regeneration site (although not large enough to encompass all competitors for the same canopy position). The temporal scales used in calculating the limitation measures are of the order of the time it takes for an open regeneration site to be pre-empted in this forest and for seedlings and saplings to establish.

Study site and species

The case-studies we present are from a seasonally moist tropical forest on Barro Colorado Island (BCI), Panama. Annual rainfall averages 2600 mm, with a dry season from late January through mid-April. The geology and hydrology of the 1500 ha island are described in Dietrich *et al.* (1982) and its flora and vegetation in Croat (1978) and Foster and Brokaw (1982). The study was conducted within the 50 ha Forest Dynamics Plot on the central plateau (described in Hubbell and Foster, 1983).

The target species were chosen to provide a diverse sampling of the range of abundances, dispersal characteristics and adult spatial patterns present among tree species on BCI (Table 3.1): *Beilschmiedia pendula* (Sw.) Hemsl. (*Lauraceae*), *Cordia alliodora* (Ruiz & Pav.) Oken (*Boraginaceae*), *Terminalia amazonia* (J. F. Gmel.) Exell (*Combretaceae*) and *Trichilia tuberculata* (Triana & Planch.) C. DC. (*Meliaceae*).

Data collection

Seed-fall data were collected from 200 seed traps placed along trails within the 50 ha Forest Dynamics Plot (Wright *et al.*, 1999). Each seed trap consists of a square, 0.5 m^2 polyvinyl chloride (PVC) frame supporting a shallow, open-topped, 1 mm nylon-mesh bag, suspended 0.8 m above the ground on four PVC posts. Beginning in January 1987 and continuing to the present, seed traps have been emptied weekly and all seeds, fruits and seed-bearing fruit fragments > 1 mm in diameter have been identified to species. The count of mature fruits was multiplied by the total number of seeds per fruit (S.J. Wright, unpublished data) and was added to the count of

Table 3.1. Demographic and life-history characteristics of the focal species. Tree density is the density of trees greater than 10 cm diameter at breast height (dbh) on the Forest Dynamics Plot in 1995. Adult size is the estimated minimal size of first reproduction (R.B. Foster and S.J. Wright, unpublished data). Adult density is the density of trees greater than the threshold adult size on the Forest Dynamics Plot. Seedling density is the average density of first-year seedlings per year as estimated from 7 years of censuses of 600 1 m^2 plots. Seed density is the density of seeds arriving per year as estimated from 13 years of censuses of 200 0.5 m^2 seed traps. Regeneration habitat of *Beilschmiedia* from Hubbell and Foster (1983) and Welden *et al.* (1991), of *Cordia* from Augspurger (1984) and Welden *et al.* (1991), of *Terminalia* from Augspurger (1984) and of *Trichilia* from Welden *et al.* (1991). Seed masses of *Beilschmiedia* from Wenny (2000), of *Cordia* and *Terminalia* from Augspurger (1986) and of *Trichilia* from S.J. Wright (unpublished data). Dispersers from Croat (1978), Leighton and Leighton (1982) and S.J. Wright (unpublished data).

	Beilschmiedia pendula	*Cordia alliodora*	*Terminalia amazonia*	*Trichilia tuberculata*
Tree density (ha^{-1})	5.9	1.3	0.56	34
Adult size (cm dbh)	20	13	20	20
Adult density (ha^{-1})	3.6	1.1	0.40	17
Seedling density (ha^{-1})	1.5×10^3	1.1×10^2	0	6.8×10^3
Seed density (ha^{-1})	6.8×10^3	1.9×10^4	7.1×10^4	2.5×10^5
Dioecious	No	No	No	Yes
Regeneration habitat	Prefers slopes, shade-tolerant	Requires gaps	Requires large gaps	Generalist, very shade-tolerant
Seed mass (g)	12.89	0.0063	0.0041	0.15
Dispersers	Mammals, large birds and a bat	Wind	Wind	Mammals and large birds

simple seeds to obtain the estimated total number of seeds falling into the traps. To avoid having partial data for a fruiting season, only data collected in the 12 complete phenological fruiting years of each species falling between 1 January 1987 and 1 January 2000 were used in the analyses here.

Seedling data were collected from 600 seedling plots, three matched to each seed trap (Harms *et al.*, 2000). Each seedling plot is 1 m × 1 m and located 2 m distant from its associated seed trap on one of the sides away from the nearest trail. Beginning in January–March 1994 and continuing annually until the present, all seedlings < 50 cm tall were measured and, if new, identified and marked.

Observed seed limitation

Observed seed limitation, source limitation and dispersal limitation were calculated directly from the data for the 200 seed traps (see Box 3.1 for methods). Measures were calculated at the 0.5 m^2 scale of the traps and for temporal periods of 1, 2, 3, 4, 6 and 12 years for non-overlapping subsets of the data (Fig. 3.1). We used the stochastic definition of a uniform distribution in calculating source and dispersal limitation (see Box 3.1). We found considerable interannual variation in limitation measures, a reflection of interannual variation in seed production (Wright *et al.*, 1999) and seed dispersal (unpublished analyses).

Comparing seed limitation measures among the species, we observe patterns consistent with their abundances, distributions and life histories (Tables 3.1 and 3.2, Fig. 3.1). *Beilschmiedia* is quite common but concentrated on the slopes of the plot (Hubbell and Foster, 1983). Its seeds fail to reach a large proportion of the plot (high seed limitation), even at the larger temporal scales at which population-level production of its large seeds is adequate to reach most sites (low source limitation). *Cordia* is not very common and only moderately well dispersed (Augspurger, 1986). It also shows high levels of seed limitation, despite high seed production (of small seeds) per adult. *Terminalia*, while rare, produces its tiny seeds in very large numbers and disperses them very well (Augspurger, 1986). It has low seed limitation even at the 1-year scale, and almost no seed limitation at the 6- and 12-year scales. *Trichilia* is abundant throughout the BCI plot but not very well dispersed (Muller-Landau *et al.*, 2001). It also has relatively low seed limitation; what seed limitation it does have is due entirely to dispersal limitation.

Projected seed limitation

We calculated projected seed limitation and its components from simulations of seed rain across the plot, with the simulations based on fitted functions for seed production and seed dispersal.

Data on the locations and numbers of seeds in seed traps and on locations and sizes of adults in the 50 ha plot were used to fit the probability of seed arrival as a function of distance from an adult tree and to fit fecundity as a function of tree size. Starting from a set of parameters specifying these functions, expected seed rain into a given trap was calculated as the sum of contributions from conspecific adult trees on the plot. These contributions were determined by tree sizes and distances to traps, according to the parameter values. We then searched for parameter values that produced the best fit to the observed seed rain, using maximum likelihood methods (Ribbens *et al.*, 1994; Clark *et al.*, 1998). Likelihood ratio tests were used to determine whether the best-fit model was significantly better than a null model that assumed uniform expected seed rain across the plot. Seed dispersal kernels giving expected seed rain as a function of distance from source tree were fitted with negative exponential functions (Turchin, 1998), and fecundity was assumed to be proportional to basal area (as in Ribbens *et al.*, 1994). We used a negative binomial distribution to model variation in annual seed fall into each trap, thus allowing for contagion or clumping of seeds due to factors not included in the model (Clark *et al.*, 1998).

Using the fitted functions for seed production and seed dispersal, we then simulated seed rain to one in every 50 sites of size 0.5 m^2 across the entire 50 ha plot, and calculated seed

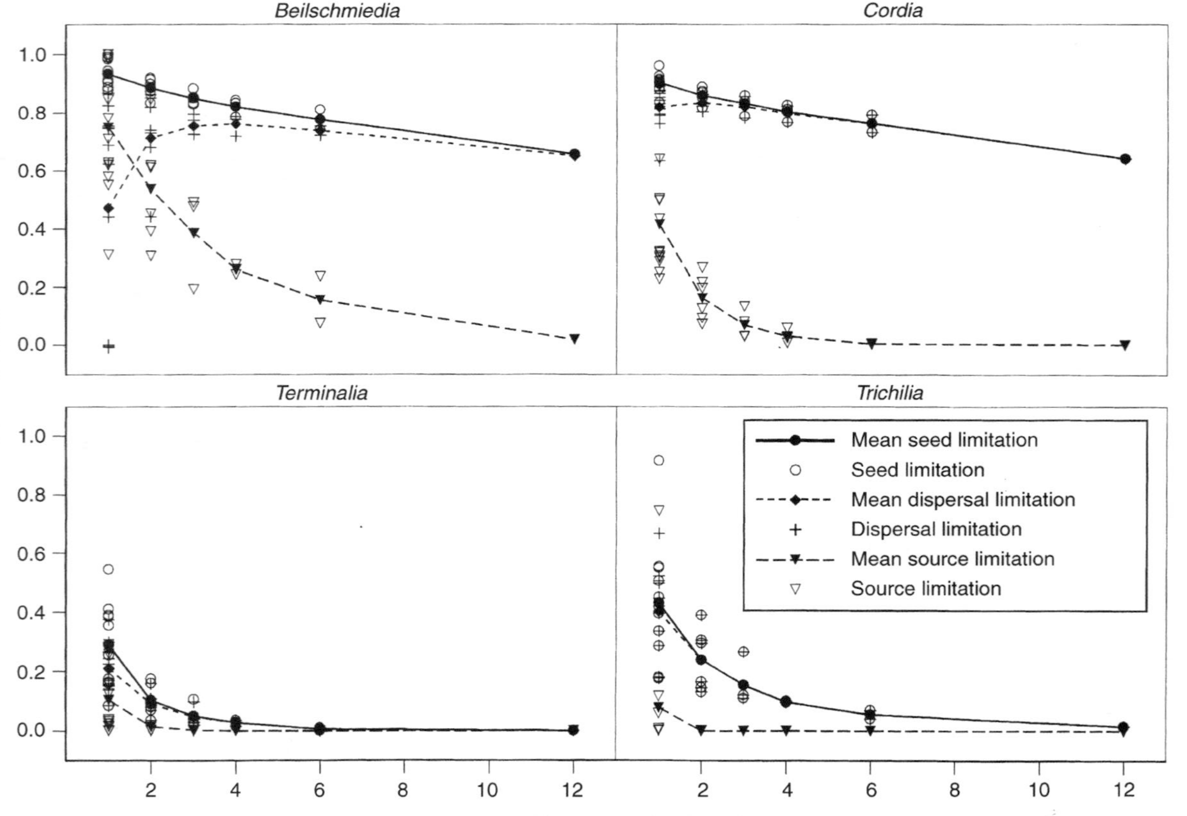

Fig. 3.1. Fundamental seed limitation (the proportion of sites not receiving seeds), source limitation (the proportion of sites that would not receive seeds if all were distributed randomly) and dispersal limitation (the proportion of sites that would have been reached under random dispersal but are not actually reached) at the spatial scale of 0.5 m^2 and temporal scales of 1–12 years, as calculated from the seed-trap data. Values for individual years or non-overlapping blocks of years are shown with open symbols; mean values across periods are shown with filled symbols.

limitation measures from the simulated seed rain. We refer to these as measures of projected seed limitation because they are calculated, not from direct observations, but from estimates of seed rain, themselves based on a model that was fitted to the data. One advantage of projected measures is that they can be calculated across a much larger area than could realistically be sampled; thus, we calculate projected limitation for simulated seed rain across the whole plot, as well as for simulated seed rain only to the 200 traps. There were some small differences between projections to the 200 traps and to the plot as a whole, reflecting the fact that traps happen to sample sites on average closer to *Beilschmiedia* and further from *Cordia* than the plot as a whole (Table 3.2).

Projected limitation measures were always within the range of observed limitation measures for the same species and scales, although sometimes quite different from the mean of observed values (Table 3.2). For all species, projected seed limitation closely matched mean observed seed limitation. Projected source limitation was quite different from mean observed source limitation, except in the case of *Beilschmiedia*. The poor match for *Cordia* was due to much lower projected mean seed arrival per site than was observed among the 200 traps, where a single trap received more seeds than all other traps combined (thus, projected source limitation may actually better represent source limitation for the plot as a whole). For *Terminalia* and *Trichilia*, mean seed densities were well fitted, but mean observed source limitation was much higher than projected source limitation because of huge interannual variation in mean seed densities. The non-linear dependence of source limitation upon seed density makes the mean of source limitation for years with varying seed density much higher than source limitation for a year with mean seed density – mean observed seed limitation reflects the former, while projected seed limitation effectively reflects the latter. *Beilschmiedia* source limitation values matched, despite considerable interannual variation in seed density, because source limitation is a nearly

Table 3.2. Observed and projected seed rain and limitation measures at the 0.5 m^2 spatial scale and 1-year temporal scale. Observed values were calculated separately for 12 different years and are given as mean ± standard deviation [minimum, maximum].

Tree species	Observed (200 traps)	Projected (200 traps)	Projected (whole plot)
Beilschmiedia pendula			
Seeds per site	0.34 ± 0.34 [0.00, 1.17]	0.30	0.23
Seed limitation	0.93 ± 0.05 [0.88, 1.00]	0.92	0.92
Source limitation	0.75 ± 0.22 [0.31, 1.00]	0.74	0.79
Dispersal limitation	0.47 ± 0.37 [–0.01, 0.87]	0.68	0.63
Cordia alliodora			
Seeds per site	0.97 ± 0.41 [0.12, 1.49]	0.48	0.65
Seed limitation	0.90 ± 0.03 [0.84, 0.96]	0.90	0.89
Source limitation	0.41 ± 0.19 [0.23, 0.89]	0.62	0.52
Dispersal limitation	0.82 ± 0.07 [0.63, 0.89]	0.75	0.76
Terminalia amazonica			
Seeds per site	3.57 ± 2.52 [1.23, 9.96]	3.73	3.66
Seed limitation	0.29 ± 0.13 [0.09, 0.55]	0.28	0.27
Source limitation	0.10 ± 0.11 [0.00, 0.29]	0.02	0.03
Dispersal limitation	0.21 ± 0.08 [0.08, 0.37]	0.26	0.25
Trichilia tuberculata			
Seeds per site	12.47 ± 12.36 [0.30, 37.31]	13.86	13.09
Seed limitation	0.43 ± 0.20 [0.18, 0.92]	0.43	0.44
Source limitation	0.08 ± 0.21 [0.00, 0.74]	0.00	0.00
Dispersal limitation	0.40 ± 0.14 [0.18, 0.67]	0.43	0.44

linear function of seed density for seed densities below 1. (*Beilschmiedia* and *Cordia* values were mostly within this range; *Terminalia* and *Trichilia* densities were mostly much higher.) Projected dispersal limitation was relatively close to the mean observed dispersal limitation for all species but *Beilschmiedia*. Mean observed dispersal limitation in *Beilschmiedia* was depressed by 4 years with zero dispersal limitation; in these years, only three or fewer seeds were captured, each in different traps.

Observed establishment limitation

We calculated observed seedling limitation – the proportion of plots in which seedlings did not emerge – at the 1 m^2 scale of the seedling plots and for temporal periods of 1, 2, 3 and 6 years (see Box 3.1 for methods). Because seed data and seedling data were collected at different spatial scales, realized establishment limitation could not be calculated exactly, but only bounded. First, we calculated bounds on seed limitation at the 1 m^2 scale from data at the 0.5 m^2 scale, by considering the probability that seeds arrive in the two halves of the 1 m^2 plots. Seed arrival in the two halves will almost certainly be correlated; in the absence of any data on this correlation, we made calculations based on the extreme assumptions of perfect positive correlation and no correlation. If seed arrival in adjacent 0.5 m^2 plots is perfectly positively correlated, then seed limitation at the 1 m^2 scale is equal to seed limitation at the 0.5 m^2 scale. If seed arrival is entirely uncorrelated, then seed limitation at the 1 m^2 scale is equal to the square of seed limitation at the 0.5 m^2 scale. From these estimates, we then calculated corresponding bounds upon establishment limitation.

Establishment limitation is strong in all four species; correspondingly, seedling limitation is always substantially higher than seed limitation (Fig. 3.2). Differences in establishment limitation across species parallel differences in shade tolerance. No seedlings of the very light-demanding *Terminalia* ever appear in any of the census plots, despite its abundant seed rain. Moderately light-demanding *Cordia* appears in some plots, but still exhibits high establishment limitation, which combines with moderate seed limitation to effect strong seedling limitation. *Trichilia* and *Beilschmiedia*, both shade-tolerant, show lower establishment limitation. Given *Trichilia*'s low seed limitation, establishment limitation is nevertheless the main factor contributing to significant seedling limitation at the 1 m^2 scale – although this limitation is still much lower than for any other species examined. For *Beilschmiedia*, with its very large seeds and high shade tolerance, seed limitation appears to be somewhat more important than establishment limitation at these scales, although better estimates of seed limitation at the 1 m^2 scale are needed to evaluate their relative magnitude with confidence.

Discussion

Interpreting limitation measures

The examples presented here illustrate how we can assess seed and seedling limitation and their components from data routinely collected in studies of seed dispersal. The results of such analyses can provide information about the success of a species in distributing its seeds and seedlings and on the relative importance of factors limiting its ability to do so. However, as is the case with any analysis, the insight that can be obtained is circumscribed by the amount and scale of data collection and the appropriateness of the underlying assumptions to the study system.

As our results illustrate, limitation measures change in both absolute and relative magnitude across scales. The spatial and temporal scales we examined in our case-studies are relevant for understanding seed and establishment limitation of seedling and small sapling abundances in this tropical forest, but they are not the only relevant scales. We anticipate that future research that examines how spatiotemporal correlations change across scales will provide a basis for extrapolation across scales within individual systems. Even then, however, it will be necessary to gather at least some data at all scales at which limitation measures are calculated or extrapolated.

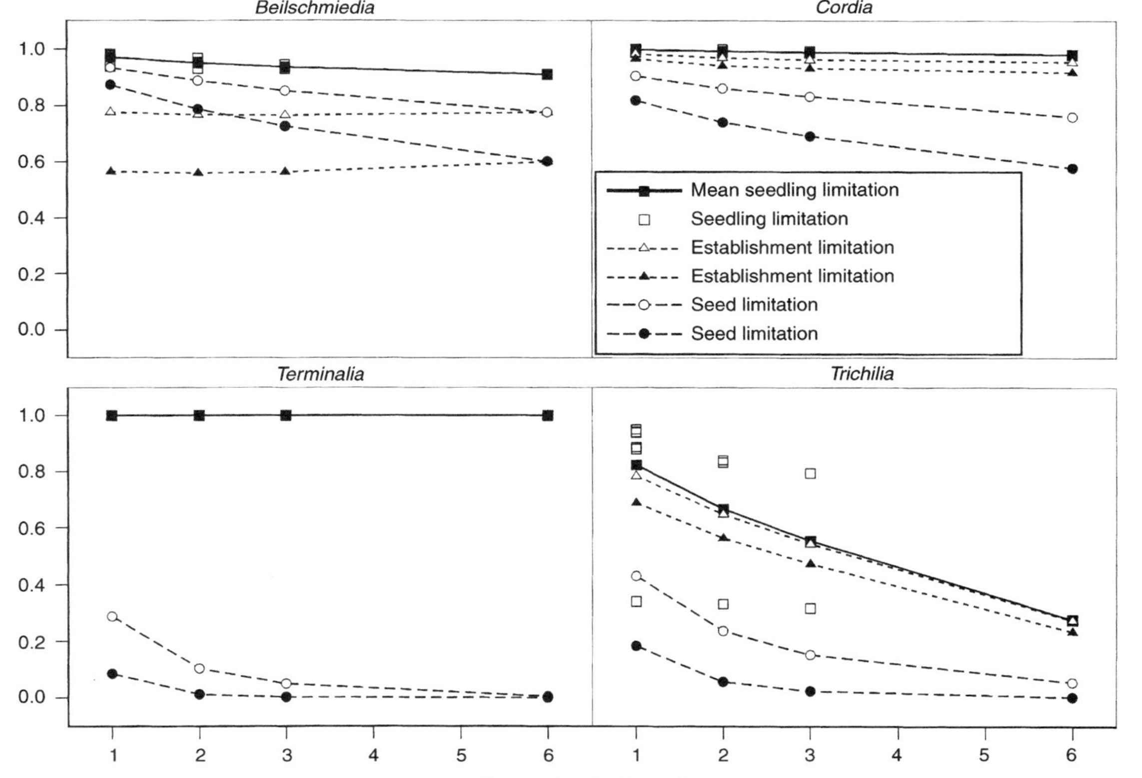

Fig. 3.2. Fundamental seedling limitation (the proportion of sites in which no new seedling recruits emerge in the given time period), establishment limitation (the proportion of sites receiving seeds in which no seedlings emerge) and seed limitation (the proportion of sites not receiving seeds) at the spatial scale of 1 m^2 and temporal scales of 1–6 years, as estimated from the seedling plot and seed-trap data. Values of seed limitation for individual years or non-overlapping blocks of years are shown with open squares; mean values across periods are shown with filled squares. Upper and lower bounds on seed and establishment limitation were estimated using two extreme assumptions on how seed arrival at the 1 m^2 spatial scale relates to seed arrival at the 0.5 m^2 spatial scale: perfect independence of seed rain into adjacent 0.5 m^2 plots (upper bound on seed limitation, lower bound on establishment limitation) and perfect positive correlation (lower bound on seed limitation, upper bound on establishment limitation).

One shortcoming of our study is that rare microhabitats, such as canopy gaps, were inadequately sampled. The 200 sets of seed traps and seedling plots we used provide a representative sample of microhabitats – which necessarily means very few fall within gaps, a rare but important habitat for establishment (Hubbell *et al.*, 1999). For species with restricted regeneration niches, larger samples of their rare but preferred habitats may be needed to reliably estimate establishment limitation. For example, we encountered not a single seedling of *Terminalia*, a large gap specialist, and thus calculated its seedling limitation and establishment limitation to be 1, both obviously overestimates. In contrast, in their 1996 censuses of all the canopy gaps in the plot, Dalling *et al.* (1998) found a total of 19 *Terminalia* seedlings of various sizes in 15 different 1 m × 1 m plots. If these are the only seedlings on the 50 ha plot, then the density of first-year seedlings is less than 0.38 ha^{-1}, seedling limitation at the 1 m^2 1-year scale is at least 0.99997 and establishment limitation is between 0.99996 and 0.99997.

While the examples presented involved only fundamental seed limitation, data such as these can potentially be combined with information on establishment to estimate realized seed limitation as well. Data on the proportions of sites with different establishment conditions and on establishment likelihood in each of these sets of conditions can be combined with information on seed rain to predict likely seedling numbers under current conditions and under conditions of seed augmentation. Density-dependent establishment could also be incorporated in such a framework; indeed, this can be expanded to a full-scale individual-based, spatially explicit model enabling simulations to test population limitation resulting from many different factors (as in Pacala *et al.*, 1996). However, care should be taken that model complexity does not outstrip the quality of the data available to estimate model components.

Ideally, observational studies of seed rain should be combined with experimental studies to develop a complete understanding of seed and establishment limitation. It will soon be possible to compare the results of the observational studies reported here with those of seed addition experiments currently being conducted on BCI by Jens-Christian Svenning. Preliminary results from these experiments with 32 species suggest strong realized seed limitation in this community (Jens-Christian Svenning, November 2000, personal communication), which accords with our results here.

Implications for understanding communities

To understand the importance of seed rain and regeneration niches for community structure and dynamics, we need to consider seed and establishment limitation of all species in a community. Our results suggest that seedling abundances of most tropical tree species are likely to be both strongly seed-limited and strongly establishment-limited, with considerable variation across species. Extrapolation from the species we examined is difficult because all are somewhat atypical for BCI; they have either high abundance (*Beilschmiedia, Trichilia*) or small seeds (*Cordia, Terminalia*). Not coincidentally, these properties ensure that seeds arrive in our seed traps in sufficient numbers to allow detailed analyses. In contrast, 67% of the 305 tree species present in the 50 ha plot are rare (< 1 adult ha^{-1}) and 81% have seeds larger than those of *Cordia* and *Terminalia* – that is, too large to allow production of millions per tree (Grubb, 1998). Thus, most species produce insufficient seeds to reach a majority of potential regeneration sites; they are source-limited at annual 1 m^2 scales. Further, even if we consider larger spatiotemporal scales at which population seed production is in theory sufficient to cover all sites, most sites will not be reached by seeds of any given species because virtually all species are strongly dispersal-limited. Most seeds remain near parent trees and those that do travel further are often deposited in clumps, reducing the number of sites that are colonized. Exceptions include a few pioneer species (e.g. *Terminalia*) that have very high seed production and long dispersal distances, allowing them to reach most sites in most years. To do so requires small seeds, which results in strong establishment limitation. Thus, most species in this forest are strongly seed-limited, with 88% reaching on average fewer than 5% of traps per year. The only species to largely escape both seed and establishment limitation are extremely abundant and widely distributed shade-tolerant species with moderate or large seeds, such as *Trichilia* in this forest and monodominants elsewhere (Hart *et al.*, 1989).

Our results provide some support for the hypothesis of a trade-off among species between colonization ability and establishment ability, mediated by seed size. The four species examined here exhibit a positive relationship between seed mass and establishment probability (the number of seedlings per seed) and a negative relationship between seed mass and seed production per unit basal area, as do species in the BCI assemblage as a whole (S. Joseph Wright, unpublished data) and in other systems (Westoby *et al.*, 1996; Grubb, 1998). Thus, in general, larger-seeded species are expected to suffer more seed limitation (due to more source limitation) and less establishment limitation than smaller-seeded species, consistent with results in temperate forests (Clark *et al.*, 1998) and grasslands (Kiviniemi and Telenius, 1998; Turnbull *et al.*, 1999). However, both seed and establishment limitation also depend upon abundance; higher abundance tends to reduce both types of limitation because it corresponds to increased seed sources (reproductive adults) and increased average seed number per site reached (providing more chances to establish at each site).

Pervasive seed limitation depresses local species diversity (alpha diversity) because not all potential species that might coexist in an area will reach it. At the same time, it enhances larger-scale diversity (beta diversity), because of stochastic variation in which species end up arriving and dominating in different areas (Horn, 1981). Essentially, seed limitation slows competitive dynamics and enhances opportunities for non-equilibrium coexistence on large scales (Hubbell, 1979; Hubbell and Foster, 1986). Experimental studies show that when seed limitation is decreased through addition of seeds of multiple species, local species diversity within seed addition plots increases (Tilman, 1997).

The negative relationship between seed limitation and establishment limitation across species also acts to increase local species diversity, enhancing coexistence of species according to a competition–colonization trade-off (Hastings, 1980; Tilman, 1994). This diversity enhancement is qualitatively different from that of pervasive seed limitation alone because it is equilibrium, rather than non-equilibrium: differences among species in competition–colonization strategies stabilize their coexistence, while pervasive seed limitation merely slows competitive exclusion. This effect has been demonstrated in large-scale, spatially explicit models: model communities in which species differ according to a competition–colonization trade-off have more species than those in which species are equivalent (Chave *et al.*, 2001). In real plant communities, this trade-off appears to be mediated by seed size, as discussed earlier. The best demonstration of the effects of seed size on seed and establishment limitation is a multispecies seed addition experiment by Turnbull *et al.* (1999). When no or few seeds of each species were added, small-seeded species were somewhat more abundant than large-seeded species. When many seeds of each species were added in equal numbers, seed limitation was essentially removed completely; thus, the large-seeded species dominated, nearly excluding the small-seeded species.

Conservation Implications

Consideration of seed and establishment limitation can help us understand the effects of human activity on tropical forests, and thus inform conservation. Anthropogenic habitat modification has a direct impact on establishment limitation and indirectly affects seed limitation. When a habitat is modified to become entirely unsuitable for regeneration (e.g. paved, built upon, farmed), total establishment limitation increases for all species, and seed limitation within the remaining suitable habitat may also increase, due to the loss of seed input from the modified area. Species with longer dispersal distances and those that rely on vectors that cross into the newly unsuitable habitat (wind, some small animals) will be disproportionately affected, because they will deposit more seeds in these areas. At the same time, the creation of an unsuitable habitat is usually accompanied by habitat modification to sites on the edge, which experience greater light availability, reducing the establishment limitation of light-demanding species and giving them an advantage relative to shade-tolerant species (Laurance *et al.*, 1998).

Loss of frugivorous animals to hunting can lead tree species that depend on these animals for seed dispersal to suffer increased dispersal limitation and seed limitation in their absence (e.g. Wright *et al.*, 2000). These tree species may also experience increased establishment limitation, since seeds in high concentrations near parents are more likely to suffer predation and less likely to successfully establish (Janzen, 1970; Connell, 1971; Wright *et al.*, 2000). Large-seeded trees are disproportionately affected, because they are most likely to depend on large animals for dispersal and large animals are the most often hunted (Redford and Robinson, 1987).

The differential effects of these anthropogenic disturbances on different tree species upset the competitive balances that contribute to the equilibrium coexistence of tree species, changing community composition and diversity (Leigh *et al.*, 1993; Laurance *et al.*, 1998). The best hope for conserving and restoring tropical forests is to conserve and restore the processes that maintain these competitive balances, including differential seed and establishment limitation, which depend fundamentally on the disperser assemblage and the relative abundances of establishment conditions.

Conclusions and future directions

Much remains to be learned about the magnitude, causes and consequences of recruitment limitation in tropical forests and elsewhere. The available evidence suggests that most tropical tree species are likely to be strongly seed-limited, with seeds reaching only a small minority of potential regeneration sites because of low adult abundances and limited dispersal. Theoretical studies show that such limitation and observed trade-offs among species between seed and establishment limitation could strongly influence community dynamics and contribute to the maintenance of species diversity. More studies are needed on more species comprising a wider and more representative sample of abundances and life histories and at larger spatial and temporal scales. Integrated studies of seed dispersal and subsequent establishment, especially if they include seed-sowing experiments, will greatly contribute to our understanding of the population dynamics of individual species and, ultimately, of community structure and dynamics.

Acknowledgements

We gratefully acknowledge the work of the many people who have contributed to the seedling census and the 50 ha plot census over the years, especially Andrés Hernández, Steve Paton, Rick Condit, Rolando Peréz and Suzanne Loo de Lao. The Environmental Sciences Program of the Smithsonian Institution provided funding for the seed and seedling censuses, the National Science Foundation (NSF), the Smithsonian Tropical Research Institute and many other institutions provided funding for the 50 ha plot census. H.C.M was supported by a predoctoral fellowship from the Smithsonian Tropical Research Institute. Travel funds to present the paper on which this chapter is based were contributed by the Association of Princeton Graduate Alumni and an NSF grant to D. Levey. We thank Doug Levey, Wesley Silva and Mauro Galetti for organizing the 2000 Frugivory and Seed Dispersal Symposium and editing this volume. This manuscript was greatly improved thanks to comments from Jérôme Chave, Doug Levey, Ran Nathan, Henry Stevens, Jens-Christian Svenning and David Tilman.

References

Augspurger, C.K. (1984) Light requirements of neotropical tree seedlings: a comparative study of growth and survival. *Journal of Ecology* 72, 777–795.

Augspurger, C.K. (1986) Morphology and dispersal potential of wind-dispersed diaspores of neotropical trees. *American Journal of Botany* 73, 353–363.

Augspurger, C.K. and Kitajima, K. (1992) Experimental studies of seedling recruitment from contrasting seed distributions. *Ecology* 73, 1270–1284.

Chave, J., Muller-Landau, H.C. and Levin, S.A. (2002) Comparing classical community models: theoretical consequences for patterns of diversity. *American Naturalist* (in press).

Chesson, P. (1998) Recruitment limitation: a theoretical perspective. *Australian Journal of Ecology* 23, 234–240.

Clark, J.S., Macklin, E. and Wood, L. (1998) Stages and spatial scales of recruitment limitation in Southern Appalachian forests. *Ecological Monographs* 68, 213–235.

Clark, J.S., Beckage, B., Camill, P., Cleveland, B., HilleRisLambers, J., Lichter, J., McLachlan, J., Mohan, J. and Wyckoff, P. (1999) Interpreting recruitment limitation in forests. *American Journal of Botany* 86, 1–16.

Condit, R., Ashton, P.S., Baker, P., Bunyavejchewin, S., Gunatilleke, S., Gunatilleke, N., Hubbell, S.P., Foster, R.B., Itoh, A., LaFrankie, J.V., Lee, H.S., Losos, E., Manokaran, N., Sukumar, R. and Yamakura, T. (2000) Spatial patterns in the distribution of tropical tree species. *Science* 288, 1414–1418.

Connell, J.H. (1971) On the roles of natural enemies in preventing competitive exclusion in some marine animals and in rain forest trees. In: den Boer, P.J. and Gradwell, G.R. (eds) *Dynamics of Populations: Proceedings of the Advanced Study Institue on Dynamics of Numbers in Populations, Oosterbeek, 1970.* Centre for Agricultural Publication and Documentation Wageningen, The Netherlands, pp. 298–312.

Cornell, H.V. and Lawton, J.H. (1992) Species interactions, local and regional processes, and limits to the richness of ecological communities: a theoretical perspective. *Journal of Animal Ecology* 61, 1–12.

Crawley, M.J. (1990) The population dynamics of plants. *Philosophical Transactions of the Royal Society of London B* 330, 125–140.

Croat, T.B. (1978) *Flora of Barro Colorado Island.* Stanford University Press, Stanford, California.

Dalling, J.W., Hubbell, S.P. and Silvera, K. (1998) Seed dispersal, seedling establishment and gap partitioning among pioneer trees. *Journal of Ecology* 86, 674–689.

Dietrich, W.E., Windsor, D.M. and Dunne, T. (1982) Geology, climate and hydrology of Barro Colorado Island. In: *The Ecology of a Tropical Forest: Seasonal Rhythms and Long-term changes.* Smithsonian Institution Press, Washington, DC, pp. 21–46.

Eriksson, O. and Ehrlen, J. (1992) Seed and microsite limitation of recruitment in plant populations. *Oecologia* 91, 360–364.

Foster, R.B. and Brokaw, N.V.L. (1982) Structure and history of the vegetation of Barro Colorado Island. In: *The Ecology of a Tropical Forest: Seasonal Rhythms and Long-term Changes.* Smithsonian Institution Press, Washington, DC, pp. 67–82.

Grubb, P.J. (1977) The maintenance of species-richness in plant communities: the importance of the regeneration niche. *Biological Reviews* 52, 107–145.

Grubb, P.J. (1998) Seeds and fruits of tropical rainforest plants: interpretation of the range in seed size, degree of defence and flesh/seed quotients. In: Newbery, D.M., Prins, H.H.T. and Brown, N.D. (eds) *Dynamics of Tropical Communities.* Blackwell Scientific, Oxford, pp. 1–24.

Harms, K.E., Wright, S.J., Calderon, O., Hernandez, A. and Herre, E.A. (2000) Pervasive density-dependent recruitment enhances seedling diversity in a tropical forest. *Nature* 404, 493–495.

Hart, T.B., Hart, J.A. and Murphy, P.G. (1989) Monodominant and species-rich forests of the humid tropics: causes for their co-occurrence. *American Naturalist* 133, 613–633.

Hastings, A. (1980) Disturbance, coexistence, history, and competition for space. *Theoretical Population Biology* 18, 363–373.

Horn, H.S. (1981) Some causes of variety in patterns of secondary succession. In: West, D.C., Shugart, H.H. and Botkin, D.B. (eds) *Forest Succession: Concepts and Applications.* Springer-Verlag, New York, pp. 24–35.

Hubbell, S.P. (1979) Tree dispersion, abundance, and diversity in a tropical forest. *Science* 203, 1299–1309.

Hubbell, S.P. (2001) *The Unified Neutral Theory of Biodiversity and Biogeography.* Princeton University Press, Princeton, New Jersey.

Hubbell, S.P. and Foster, R.B. (1983) Diversity of canopy trees in a neotropical forest and implications for conservation. In: Sutton, S.L., Whitmore, T.C. and Chadwick, A.C. (eds) *Tropical Rain Forest: Ecology and Management.* Blackwell Scientific Publications, Oxford, pp. 25–41.

Hubbell, S.P. and Foster, R.B. (1986) Biology, chance, and history and the structure of tropical rain forest tree communities. In: Diamond, J.

and Case, T.J. (eds) *Community Ecology*. Harper and Row, New York, pp. 314–329.

Hubbell, S.P., Foster, R.B., O'Brien, S.T., Harms, K.E., Condit, R., Wechsler, B., Wright, S.J. and Loo de Lao, S. (1999) Light-gap disturbances, recruitment limitation, and tree diversity in a neotropical forest. *Science* 283, 554–557.

Hurtt, G.C. and Pacala, S.W. (1995) The consequences of recruitment limitation: reconciling chance, history and competitive differences between plants. *Journal of Theoretical Biology* 176, 1–12.

Janzen, D.H. (1970) Herbivores and the number of tree species in tropical forests. *American Naturalist* 104, 501–528.

Kiviniemi, K. and Telenius, A. (1998) Experiments on adhesive dispersal by wood mouse: seed shadows and dispersal distances of 13 plant species from cultivated areas in southern Sweden. *Ecography* 21, 108–116.

Kollmann, J. (2000) Dispersal of fleshy-fruited species: a matter of spatial scale? *Perspectives in Plant Ecology, Evolution and Systematics* 3, 29–51.

Laurance, W.F., Ferreira, L.V., Rankin-de Merona, J.M. and Laurance, S.G. (1998) Rain forest fragmentation and the dynamics of Amazonian tree communities. *Ecology* 79, 2032–2040.

Leigh, E.G.J., Wright, S.J., Herre, E.A. and Putz, F.E. (1993) The decline of tree diversity on newly isolated tropical islands: a test of a null hypothesis and some implications. *Evolutionary Ecology* 7, 76–102.

Leighton, M. and Leighton, D.R. (1982) The relationship of size of feeding aggregate to size of food patch: howler monkeys (*Alouatta palliata*) feeding in *Trichilia cipo* fruit trees on Barro Colorado Island. *Biotropica* 14, 81–90.

Muller-Landau, H.C., Dalling, J.W., Harms, K.E., Wright, S.J., Condit, R., Hubbell, S.P. and Foster, R.B. (2001) Seed dispersal and density-dependent seed and seedling mortality in *Trichilia tuberculata* and *Miconia argentea*. In: Losos, E.C., Condit, R., LaFrankie, J.V. and Leigh, E.G. (eds) *Tropical Forest Diversity and Dynamism: Findings from a Network of Large-Scale Tropical Forest Plots*. Chicago University Press, Chicago.

Nathan, R. and Muller-Landau, H.C. (2000) Spatial patterns of seed dispersal, their determinants and consequences for recruitment. *Trends in Ecology and Evolution* 15, 278–285.

Pacala, S.W. (1997) Dynamics of plant communities. In: Crawley, M.J. (ed.) *Plant Ecology*. Blackwell Scientific, Oxford, pp. 532–555.

Pacala, S.W. and Levin, S.A. (1997) Biologically generated spatial pattern and the coexistence of competing species. In: Tilman, D. and Kareiva, P. (eds) *Spatial Ecology: the Role of Space in Population Dynamics and Interspecific Interactions*. Princeton University Press, Princeton, New Jersey, pp. 204–232.

Pacala, S.W., Canham, C.D., Saponara, J., Silander, J.A.J., Kobe, R.K. and Ribbens, E. (1996) Forest models defined by field measurements: estimation, error analysis and dynamics. *Ecological Monographs* 66, 1–43.

Redford, K.H. and Robinson, J.G. (1987) The game of choice: patterns of Indian and colonist hunting in the Neotropics. *American Anthropologist* 89, 650–667.

Ribbens, E., Silander, J.A., Jr and Pacala, S.W. (1994) Seedling recruitment in forests: calibrating models to predict patterns of tree seedling dispersion. *Ecology* 75, 1794–1806.

Schupp, E.W. (1995) Seed–seedling conflicts, habitat choice, and patterns of plant recruitment. *American Journal of Botany* 82, 399–409.

Tilman, D. (1994) Competition and biodiversity in spatially structured habitats. *Ecology* 75, 2–16.

Tilman, D. (1997) Community invasibility, recruitment limitation, and grassland biodiversity. *Ecology* 78, 81–92.

Turchin, P. (1998) *Quantitative Analysis of Movement*. Sinauer, Sunderland, Massachusetts,

Turnbull, L.A., Rees, M. and Crawley, M.J. (1999) Seed mass and the competition/colonization trade-off: a sowing experiment. *Journal of Ecology* 87, 899–912.

Turnbull, L.A., Crawley, M.J. and Rees, M. (2000) Are plant populations seed-limited? A review of seed sowing experiments. *Oikos* 88, 225–238.

Welden, C.W., Hewett, S.W., Hubbell, S.P. and Foster, R.B. (1991) Sapling survival, growth, and recruitment: relationship to canopy height in a Neotropical forest. *Ecology* 72, 35–50.

Wenny, D.G. (2000) Seed dispersal of a high quality fruit by specialized frugivores: high quality dispersal? *Biotropica* 32, 327–337.

Westoby, M., Leishman, M. and Lord, J. (1996) Comparative ecology of seed size and dispersal. *Philosophical Transactions of the Royal Society of London B* 351, 1309–1318.

Wright, S.J., Carrasco, C., Calderón, O. and Paton, S. (1999) The El Niño Southern Oscillation, variable fruit production and famine in a tropical forest. *Ecology* 80, 1632–1647.

Wright, S.J., Zeballos, H., Dominguez, I., Gallardo, M.M., Moreno, M.C. and Ibanez, R.

(2000) Poachers alter mammal abundance, seed dispersal, and seed predation in a Neotropical forest. *Conservation Biology* 14, 227–239.

Zobel, M., Otsus, M., Liira, J., Moora, M. and Möls, T. (2000) Is small-scale species richness limited by seed availability or microsite availability? *Ecology* 81, 3274–3282.

4 Have Frugivores Influenced the Evolution of Fruit Traits in New Zealand?

Janice M. Lord, Adrienne S. Markey and Jane Marshall
Department of Botany, University of Otago, PO Box 56, Dunedin, New Zealand

Introduction

Fruit-eating animals tend to consume many species of fruit and, likewise, the fruits of plants tend to be consumed by a wide range of animals (Herrera and Jordano, 1981; Wheelwright *et al.*, 1984; Pratt and Stiles, 1985; Charles-Dominique, 1993; Howe, 1993; Larson, 1996; Corlett, 1998; Herrera, 1998). Because of this, the relationship between fleshy-fruited plants and frugivores is diffuse rather than consisting of tight mutualisms (Janzen, 1980; Howe, 1984; Jordano, 1987; Charles-Dominique, 1993). It is thus not surprising that frugivore characteristics rarely appear to influence the evolution of fruit traits (Hedge *et al.*, 1991; Mazer and Wheelwright, 1993; Rey *et al.*, 1997). There is, however, clear empirical evidence that frugivores could exert selective pressure on fruit traits. Frugivorous birds, for example, show preferences related to fruit colour, presentation, accessibility, level of insect damage and fruit size (Moermond and Denslow, 1983; Gautier-Hion *et al.*, 1985; Jordano, 1987; Hedge *et al.*, 1991; Whelan and Willson, 1994; Puckley *et al.*, 1996; Sanders *et al.*, 1997; Gervais *et al.*, 1999), when selecting among fruits on a single plant, fruits of individuals of the same species or fruits of different species (Levey, 1987; Debussche and Isenmann, 1989). However, even when preferences are clear, they may have no directional effect on the evolution of fruit traits, because the composition and abundance of the frugivore assemblage can vary significantly over time (Herrera, 1998). Instead, if frugivore preference influences the evolution of fruit traits at all, it would most probably be via general characteristics of the frugivore assemblage or of dominant guilds of frugivores.

Differences in feeding ability or preferences between guilds of frugivores (e.g. birds and mammals) have been linked to differences among fleshy-fruited species in fruit colour, size, odour, protein and lipid content and degree of protection. This has lead to the description of fruit syndromes associated with particular frugivore guilds (van der Pijl, 1969; Janson, 1983; Gautier-Hion *et al.*, 1985; Howe, 1986; Debussche and Isenmann, 1989). However, the concept of syndromes is problematic; the traits used to define them vary from study to study, as does the composition of frugivore guilds. Consider mammalian frugivores as an example: the classical mammal dispersal syndrome involves large, husked or protected, brown, green, orange or yellow fruit, which are often odoriferous and low in protein content

 Seed Dispersal and Frugivory: Ecology, Evolution and Conservation (eds D.J. Levey, W.R. Silva and M. Galetti)

(Janson, 1983; Howe, 1986; Debussche and Isenmann, 1989; Herrera, 1989; Fischer and Chapman, 1993; Tamboia *et al.*, 1996). However, mammals differ enormously in their size, visual acuity and ability to access and manipulate fruit (e.g. primates vs. rodents vs. elephants: Gautier-Hion *et al.*, 1985). Thus, the classical mammal dispersal syndrome does not apply, in its entirety, to all mammals, and it is wrong to expect it to do so. Conversely it does not necessarily exclude some mammals (e.g. Fischer and Chapman (1993) erroneously associated the syndrome with diurnal mammals). A more constructive approach is to identify frugivore traits that should be logically associated with certain fruit traits, and vice versa. For example, fruits dispersed by terrestrial frugivores should fall at maturity; large, fleshy, single-seeded fruits should be consumed by large-gaped frugivores; fruits mainly dispersed by nocturnal frugivores are unlikely to be coloured but likely to be odoriferous; and husked fruits are likely to be consumed by a dexterous frugivore. Specifically testing for an association between, for example, green, odoriferous fruits and a nocturnal mammalian frugivore (Tamboia *et al.*, 1996) is a much more productive way to examine the relationship between frugivores and fruit traits.

Another very productive approach to examining the influence of frugivores on the evolution of fruit traits is to make use of biogeographical variation in the composition of frugivore assemblages. In particular, studies of fruit traits on islands that lack, or have lost, particular guilds of frugivores (e.g. Fischer and Chapman, 1993; see also McConkey and Drake, this volume) can provide a wealth of insights. The aim of this study is to examine the New Zealand fleshy-fruited flora for traits associated with the features of the main New Zealand frugivore guilds. New Zealand has a flora with some unusual features and, because of its isolation, it also has, or had, an unusual frugivore assemblage. The next section describes the New Zealand flora and then identifies three frugivore guilds. These are then each examined in detail with reference to specific fruit traits to test whether these frugivores have influenced fruit evolution in New Zealand.

Characteristics of the New Zealand Fleshy-fruited Flora

Approximately 25% of New Zealand's indigenous plant genera and *c.* 12% of its species produce fleshy fruits (we use 'fruit' in the functional, not structural, sense). The distribution of fleshy-fruitedness in the flora has a historical component; fleshy-fruitedness is more common among genera that have a pre-Miocene pollen record in New Zealand than among genera with more recent pollen records, perhaps because the fragmented nature of the New Zealand land mass during the early–mid-Tertiary selected for highly vagile taxa (Lord, 1999). In prehuman New Zealand, the majority of fruits would have been consumed by birds or reptiles, because, with the exception of two species of bat (see below), no mammalian frugivores were present.

The New Zealand fleshy-fruited flora has three interesting features:

1. Fruits tend to be small (Fig. 4.1) ($n = 246$, diameter: $\bar{X} = 6.4 \pm 3.8$ mm (SD), maximum = 18.9 mm; length: $\bar{X} = 7.7 \pm 5.4$ mm, maximum = 33.5 mm). This is significantly less than the mean for 332 tropical and temperate bird-dispersed angiosperm species (diameter: $n = 332$, $\bar{X} = 10.2 \pm 5.4$ mm; length: $n = 266$, $\bar{X} = 11.9 \pm 7.2$ mm (Jordano, 1995); two-sample T-tests: length, $|T| = 4.2$, d.f. = 510, $P < 0.001$; diameter, $|T| = 3.8$, d.f. = 576, $P < 0.001$).

2. Fruits tend to become more elliptic in shape with increasing size (Fig. 4.2); when log (fruit width) is regressed on log (fruit length), following Herrera (1992), the slope is significantly less than 1 (reduced major axis regression, slope = 0.8342, 95% confidence interval = 0.7530–0.9235, $n = 67$; methods on fruit dimensions provided in section on volant birds, below). This relationship between fruit size and shape is not a feature of all floras; fruit length and width scale isometrically in the flora of the Iberian Peninsula (Herrera, 1992) and among bird-dispersed plants in Malawi (Mazer and Wheelwright, 1993). Mazer and Wheelwright (1993) did find that fruit became more elliptic with increasing size within a single species, within the *Lauraceae* and among

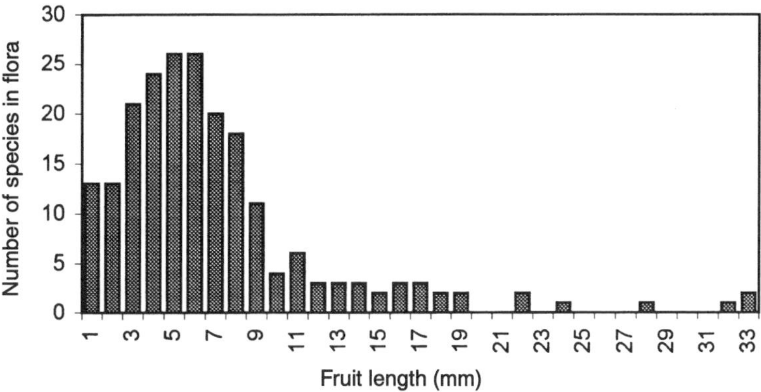

Fig. 4.1. The distribution of fruit length (mm) for 246 indigenous New Zealand fleshy-fruited species.

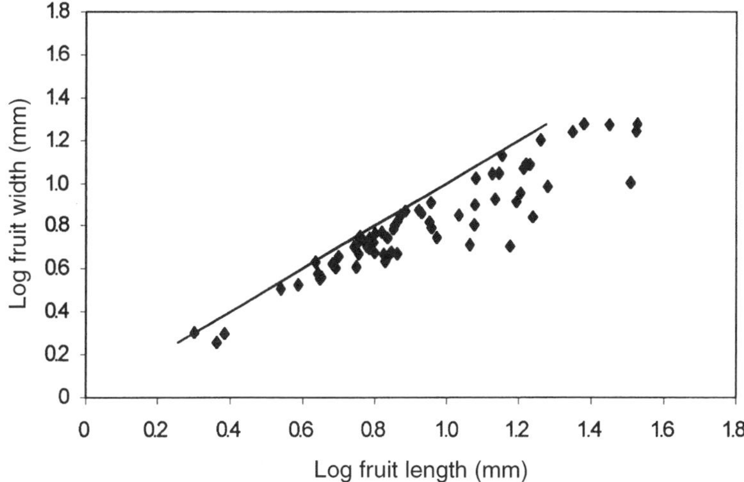

Fig. 4.2. The relationship between log (fruit length) and log (fruit width) for 67 New Zealand fleshy-fruited species, representing 51 of the 94 fleshy-fruited genera in New Zealand. The solid line denotes a 1 : 1 relationship.

bird-dispersed species of lower montane forests in Costa Rica. The relationship between fruit size and shape in the New Zealand flora is not just a product of the taxonomic mix of species present; when the same analysis was applied to 32 temperate and subtropical Australian species in 21 New Zealand genera, the 95% confidence interval for the regression slope encompassed one.

3. The distribution of fruit colours is unusual (all colours mentioned refer to the human visual spectrum). No species have green fruits at maturity and only two species have brown fruits (*Freycinetia baueriana, Pandanaceae* – see discussion below of mammalian frugivory; *Peperomia urvilleana, Piperaceae*, minute brownish drupes). Species with white or pale blue to sky-blue fruits, simultaneous bicoloured fruit in which one colour is white or pale blue, or species polymorphic for fruit colour in which one morph has white fruit, make up 21.2% of the fleshy-fruited flora (Fig. 4.3; 9.8% white, 8.6% polymorphic or bicolour, 2.9% pale blue or sky-blue; henceforth, these colour classes will be referred to collectively as white- or blue-fruited). This is collectively a higher proportion than has been observed for other temperate floras (Iberian Peninsula: 2.7% white, 15.3% blue (Herrera, 1989); Florida: 7.9% white, 3.9% blue; Europe: 5.2% white,

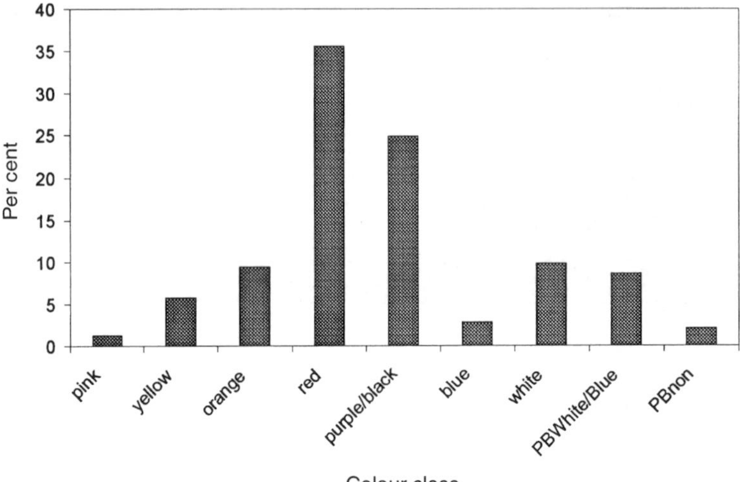

Fig. 4.3. The frequency of different fruit colour classes among 245 indigenous fleshy-fruited species in the New Zealand flora. 'Pink' includes cerise and pink-brown; 'yellow' includes yellow-green; 'orange' includes yellow-orange; 'red' includes red-orange, crimson and brick; 'purple/black' includes lilac, dark purple, brown-black and brown; 'blue' ranges from pale blue to sky-blue; 'white' includes cream; 'PBWhite/Blue' means species with simultaneous bicolour fruit, or species consisting of morphs with different-coloured fruits, and fruit colours include white and/or blue; 'PBnon' means that species are either bicolour or polymorphic for colours not including white or blue.

7.3% blue (Wheelwright and Janson, 1985); note that 'blue' may involve different hues and brightnesses in these floras).

New Zealand Frugivore Guilds

New Zealand frugivores can be classified into three guilds differing in size and ability to access fruits. These guilds are: (i) flightless birds, many of which were large and most of which are extinct; (ii) volant birds, including both extinct and extant species; and (iii) lizards. Mammalian frugivory has probably not been important in New Zealand. The only land mammals native to New Zealand are three species of Microchiroptera. Two of these species, the lesser short-tailed bat (*Mystacina tuberculata*) and the greater short-tailed bat (*Mystacina robusta*), are known or thought to have eaten fruit; however, *M. tuberculata* is now uncommon and *M. robusta* became extinct in the 1960s (Daniel, 1976; King, 1998). *Mystacina tuberculata* is known to consume the fruits of *Collospermum* species, monocot epiphytes with small red or white berries, and *F. baueriana*, a monocot liana with large, cream, succulent bracts surrounding a brown, pulpy, infructescence (Daniel, 1976). In areas where *M. tuberculata* is locally extinct, *F. baueriana* bracts and fruits are apparently being eaten by the introduced marsupial *Trichosurus vulpecula* (Lord, 1991).

Flightless avian frugivores

Humans arrived in New Zealand approximately 1200 years ago. Since then, nearly half of the avifauna, including many flightless birds, have become extinct, either as a direct result of hunting or as a result of introduced predators, such as rats (*Rattus rattus*, *Rattus norvegicus*, *Rattus exulans*), cats (*Felis catus*) and stoats (*Mustela erminea*). Many extinct flightless birds are known (from subfossil gizzard remains) or thought to have eaten fruit (Burrows *et al.*, 1981; Clout and Hay, 1989; Holdaway, 1989; Gill and Martinson, 1991). Among extant flightless birds, weka (*Gallirallus australis*) and kiwi (*Apteryx* species) are minor frugivores (Clout and Hay, 1989). Fruit can be a seasonally important component of the diet of kakapo (*Strigops habroptilus*), but, like most parrots, this species chews and cracks larger seeds and so must be regarded primarily as a seed predator, although small seeds can be voided intact (Best, 1984; Clout and Hay, 1989).

If flightless birds were significant seed-dispersers in prehuman New Zealand, one would expect to find fleshy-fruited species that dropped their fruits at maturity. New Zealand has several large-fruited species that do so (e.g. *Elaeocarpus* spp., *Dysoxylum spectabile*, *Cornyocarpus laevigatus*), but insufficient information is available to determine the frequency of this trait in the flora and whether it is associated with large fruits. However, features of some species that drop their fruit suggest that conspicuousness rather than food value *per se* may play a role in attraction. *Dysoxylum* seeds are surrounded by a conspicuous, thin, orange aril. The drupes of *Elaeocarpus* spp. are purple-brown and have very little pulp (17.3% of fruit dry weight for *E. hookerianus*), but the fruit exocarp has a metallic sheen that is highly visible on the forest floor. *Sophora microphylla* (*Fabaceae*) is not fleshy-fruited, but has conspicuous, tough-coated, bright yellow seeds which drop to the ground when the pods rupture. Some extant flightless birds are known to consume objects with apparently little food value; Australian emus pick up conspicuous objects readily (Eastman, 1969, in McGrath and Bass, 1999), and brown kiwi (*Apertyx australis*) pick up fruit of *Elaeocarpus dentatus*, possibly as substitute gizzard stones (Clout and Hay, 1989).

Moas as frugivores

The group of extinct flightless birds about which we know the most are the moas (Aves: Dinornithiformes), which were ratites most closely related to South American rhea species (*Rhea americana* and *Pterocnemia pennata*) (Cooper *et al.*, 1993). New Zealand had at least 11 moa species prior to the arrival of humans. These species ranged in body mass from approximately 15 kg to 270 kg and were distributed throughout New Zealand (Cooper *et al.*, 1993). Among extant ratites, the Australian cassowary (*Casuarius casuarius*) is the most highly frugivorous. Its diet tends to include species with large fruits that are often black or yellow/orange (Crome, 1976; Willson *et al.*, 1989). Emus (*Dromaius novaehollandiae*) and rhea species also consume fruit, although their diets are composed mainly of grasses and herbs (Davies, 1978; Martella *et al.*, 1996; Quin, 1996).

WERE MOAS MAJOR OR MINOR FRUGIVORES? Gizzard samples retrieved from swamps in South Island, New Zealand, representing 19 individuals in three moa genera (16 *Dinornis*, one *Eurapteryx* and two *Emeus*) (Falla, 1941; Gregg, 1972; Burrows *et al.*, 1981), contain a wide variety of fruit, often in substantial amounts. The most complete gizzard sample described to date contains 851 seeds representing 16 fleshy-fruited species, in 4250 cm^3 (420 g) of plant material, with a further 2200 cm^3 (5.6 kg) of gizzard stones (*Dinornis* 122B, Pyramid Valley) (Burrows *et al.*, 1981). A second, probably complete, gizzard sample from the same location contains 2216 seeds, representing 12 fleshy-fruited species in 1500 cm^3 of plant material (*Dinornis* XA) (Burrows *et al.*, 1981). These samples suggest that fruit was an important component of the diet of some moa species.

DID MOAS EXHIBIT FEEDING SELECTIVITY WITH REGARD TO FRUITING SPECIES OR FRUIT SIZE? The majority of the gizzards described to date (16 of 19) originate from one site, Pyramid Valley Swamp, and those that have been radiocarbon-dated are between 3450 and 3740 years old (Gregg, 1972; Burrows *et al.*, 1981). Collectively, these 16 individuals had ingested the fruits of 20 species of shrubs, small trees and lianas, including all but three fruiting species known to be present at the site from pollen and macrofossil evidence (Moar, 1970; Burrows *et al.*, 1981). The three fruiting species not consumed (two species of canopy tree, *Dacrydium cupressinum* and *Dacrycarpus dacrydioides*, *Podocarpaceae*, and a mistletoe, *Tupeia antarctica*) were probably not accessible to moas, as fruits of these species tend to persist on the plant and are consumed by volant frugivores. Some fruits do seem to have been favoured by moas, as individual birds appear to have concentrated on particular fruits in a feeding bout. In two of the 13 individuals with more than 50 seeds in their gizzard, *Corokia cotoneaster* (shrub with berries 7.2 mm × 6.3 mm) accounted for 60% and 79% of seeds. *Coprosma* species accounted for more than 75% of seeds in seven of these 13 individuals. The berries of *Coprosma* species can be moderately large (12 mm × 7 mm, *C. lucida*), but the only species definitely known to have occurred at the site are small-fruited species (berries 3.5–6.5 mm × 3.5–4.1 mm). Only three gizzard samples contained seeds from moderately large fruits. In two individuals with more than 50 seeds in their gizzard, *Prumnopitys taxifolia* (drupes 9 mm × 5 mm) accounted for 75% and 87% of seeds. The largest fruit found thus far in gizzard samples is *E. hookerianus* (drupes 10.8 mm × 7.1 mm), represented by one seed in each of two samples (Burrows *et al.*, 1981).

DOES THE FLORA CONTAIN 'ANACHRONISMS' THAT COULD BE RELATED TO MOA DISPERSAL? 'Anachronistic' or puzzling characteristics of extant plants have been used as evidence for selection pressures exerted by extinct herbivores and frugivores (Janzen and Martin, 1981; Givnish *et al.*, 1994), although the existence of tight coevolved mutualisms between fruits and frugivores has been questioned on numerous occasions (e.g. Howe, 1984; Jordano, 1987, 1995; Herrera, 1998). Herbivory by moas has been suggested as the selective pressure behind the evolution of heteroblasty and the divaricating growth form (densely interlacing tough stems, with small sparse leaves) in New Zealand (Greenwood and Atkinson, 1977; Cooper *et al.*, 1993; Givnish *et al.*, 1994). The role of moas as frugivores might also be judged by an examination of the New Zealand flora. Are there species with apparently 'anachronistic' fruits? The obvious distinguishing feature of moas is size; the largest moa species would have been capable of swallowing fruits > 5 cm in diameter (Clout and Hay, 1989). However, the New Zealand flora contains no extant species with very large fruits; as mentioned in the previous section, New Zealand fruits are characteristically small and the largest fruits are well below the gape size of the largest moa species (*C. laevigatus, Corynocarpaceae*, drupes 33.3 mm × 17.6 mm; *Beilschmeidia* spp., *Lauraceae*, drupes 33.5 mm × 18.9 mm). If there were species specialized for moa dispersal, it is unlikely that they would have completely disappeared in the short time since moas were extirpated. Moa species were plentiful until about 1000 years ago (Cooper *et al.*, 1993), which is within the lifetime of the longest-lived New Zealand tree species and no more than a few generations for many other fleshy-fruited species. Also, no unusual plant remains, such as very large seeds, have been reported from subfossil deposits or moa gizzard contents.

Conclusion: frugivory by flightless birds

New Zealand's extinct flightless birds may have been important in the dispersal of seeds from fallen fruit. However, information on fruit impersistence is lacking and present-day studies of fruit fate could be regarded as ambiguous – the fruit of extant fruiting species that were once consumed by volant birds may now fall to the ground uneaten, due to the loss of bird species and reduction in numbers of extant frugivorous species. However, it is likely that the extinct flightless avifauna, in general, played a role in seed dispersal simply because of their abundance. Fruit was undoubtedly an important component of moa diets, and *Dinornis*, at least, appears to have selectively consumed fruits of certain shrubby species. There is little evidence that large fruits were selectively consumed. The lack of information on diets of moas in the North Island is a problem, as the three largest-fruited species in the flora occur mainly on the North Island. These species do occur in the northern third of the South Island but not at any of the sites from which gizzard contents have been described. As the flora contains no obvious fruit 'anachronisms' suggestive of specialization to moa dispersal, it seems that moas were not sufficiently selective or important to influence the evolution of, at least, fruit size. This does not mean, however, that moas were unimportant for seed dispersal; like the Australian emu, they would have been capable of carrying large numbers of seeds over large distances (McGrath and Bass, 1999) and probably played an important role in the maintenance of regional plant biodiversity in New Zealand.

Characteristics of volant New Zealand frugivorous birds

Clout and Hay (1989) list 17 volant indigenous birds that consume or consumed fruit. They have gape sizes ranging from < 0.5 cm to 1.5 cm (Fig. 4.4), and are small-bodied compared with Australian volant frugivorous birds (Fig. 4.5) (one-way analysis of variance (ANOVA): $F_{1,\ 70} = 6.18$, $P < 0.05$). Of these 17 species, two are extinct (huia, *Heterolocha acutirostris*, gape size 1.5 cm; piopio, *Turnagra capensis*, gape size 1.1 cm) and many others are severely reduced in numbers and distribution (Clout and Hay, 1989). The largest extant volant frugivore is the New Zealand wood-pigeon (*Hemiphaga novaeseelandiae*, gape size 1.4 cm) (Clout and Hay, 1989). Birds introduced by European settlers have become important frugivores in New Zealand

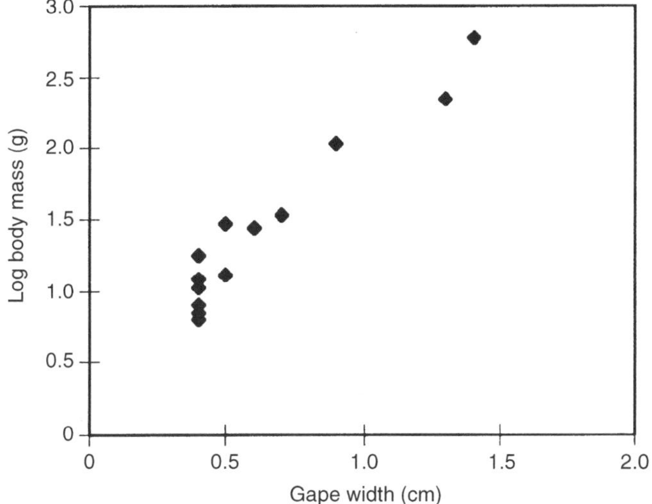

Fig. 4.4. The relationship between gape width (data from Clout and Hay, 1989) and body mass (data from Dunning, 1993) for 13 extant volant indigenous New Zealand frugivorous birds.

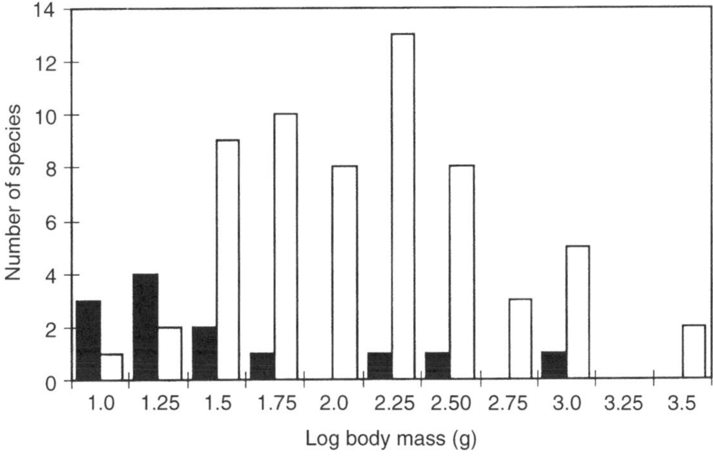

Fig. 4.5. A comparison of body masses of extant volant frugivorous birds in New Zealand (solid bars) and Australia (open bars). Information on bird diet from Clout and Hay (1989) and MacDonald (1973). Body-mass data from Dunning (1993).

(Ferguson and Drake, 1999) and some, such as the blackbird (*Turdus merula*), song thrush (*Turdus philemelos*) and starling (*Sturnus vulgaris*), have gape sizes similar to extinct or rare indigenous passerines (e.g. piopio; saddleback, *Philesturnus carunculatus*) (Clout and Hay, 1989). However, the degree to which introduced frugivores can replace extinct or rare indigenous frugivores is debatable; introduced frugivorous birds in New Zealand may preferentially consume the fruit of introduced plants (Williams and Karl, 1996). All extant volant frugivorous birds in New Zealand are non-migratory generalists that eat a wide range of fruits (Lee *et al.*, 1991).

Birds with broader gapes generally eat larger fruits and, conversely, large fruits are generally consumed by large frugivores (Wheelwright, 1985; Pratt and Stiles, 1985; Debussche and Isenmann, 1989; Williams and Karl, 1996; Corlett, 1998). Birds also select among fruits on the basis of size, even in a single feeding bout; they are more likely to reject or mishandle larger fruits (Levey, 1987; Rey *et al.*, 1997). Theoretically, then, the gape-size distribution of the frugivore assemblage could act as an agent for selection on fruit size. Fruit size appears to be more sensitive to disperser-driven selection than other fruit traits; in a study of phylogenetic conservatism in fruit traits, Jordano (1995) found that fruit diameter showed recent evolutionary divergence correlated with disperser type. However, the probability that a fruit will be swallowed is not simply a function of its size, but also of its shape. For a given volume, spherical shapes will be less easily swallowed than elongated shapes, as the maximum diameter will be larger for the former. This leads to the prediction that, if gape size has influenced the evolution of fruit shape, then, for a flora dispersed by a common set of frugivores, larger-fruited species should be more elongate than smaller-fruited species (Levey, 1987; Herrera, 1992; Mazer and Wheelwright, 1993).

The difference in body-mass distribution and, by implication, gape size between New Zealand and Australian volant frugivorous birds provides an opportunity to test for an effect of frugivore size on the evolution of fruit size. Body mass is certainly correlated with gape size, at least among New Zealand frugivorous birds for which data are available (Pearson's $r = 0.956$, $n = 13$) (Fig. 4.4). New Zealand and Australia share many fleshy-fruited plant genera; 63 out of 93 fleshy-fruited New Zealand genera also occur in Australia (though represented by different species). If frugivore size, particularly gape size, acted as a selective pressure on the evolution of fruit size, one would expect that fruit of New Zealand species in these genera would be smaller than fruit of congeneric Australian species.

Within-genus contrasts between New Zealand and Australian species

For 20 genera, representing 18 families, that occur in both New Zealand and Australia, five to 20 fresh, ripe fruit (depending on accessibility and availability) were collected for New Zealand and temperate–subtropical Australian species (Tasmania and New South Wales). None of the Australian species had brown or green fruit – a trait often associated with dispersal by mammals. Fruit length and width were measured to the nearest 0.1 mm with vernier calipers and averaged for each species. Paired *T*-tests showed that fruits of New Zealand (NZ) species were significantly smaller overall than their Australian congeners (NZ width − Australian width: mean diff. = −2.780, $T = -3.00$, d.f. = 19, $P < 0.01$; NZ length − Australian length: mean diff. = −3.009, $T = -2.94$, d.f. = 19, $P < 0.01$).

Conclusion: frugivory by volant birds

The difference in size between New Zealand and Australian fruits and the general small size of fruits in New Zealand (Fig. 4.1) are consistent with the prediction that the smaller-sized volant avian frugivore assemblage in New Zealand has acted as a selective pressure on the fruit size of New Zealand fleshy-fruited species, both in respect of related species in Australia and across the flora as a whole. Also, the fact that fruit species in the New Zealand flora become significantly more elongated with increasing size (Fig. 4.2) suggests that selection has acted on fruit shape to maintain 'swallowability', despite increased fruit mass.

Frugivory by lizards

Studies of frugivory by lizards are relatively uncommon, but the phenomenon has been noted in many areas, including South Africa (Whiting and Greeff, 1997), the Bahamas (Iverson, 1985), the Canary Islands (Valido and Nogales, 1994; Nogales *et al.*, 1998), the Balearic Islands (Traveset, 1995), New Caledonia (Bauer and Sadlier, 1994) and Central and South America (Traveset, 1990; Figueira *et al.*, 1994; Willson *et al.*, 1996). Whitaker (1987) lists four gecko species (Gekkonidae) and nine skink species (Scincidae) as fruit consumers in New Zealand. Many of these species are restricted to open grassland or shrubland habitats; however, lizards are also relatively common in forest habitats. Lizards would have been more widespread and abundant throughout New Zealand prior to the introduction of mammalian predators. For extinct species, we have no way of reconstructing diet. *Hoplodactylus delcourtii*, at 370 mm snout–vent length with a head width of 87 mm, was the largest gecko in the world. It is known from a single, poorly preserved, specimen in the Museé d'Histoire Naturelle de Marseille, France (Bauer and Russell, 1986). Extant members of the genus, which is restricted to New Zealand, consume fruit regularly (Whitaker, 1987). We shall never know the role of *H. delcourtii* as a frugivore, but it may very well have consumed large fruits and dispersed seeds.

What seed-dispersal services might a lizard be able to provide a fruiting plant that a bird could not? Lizards might be more abundant in certain habitats than birds and able to access fruits in dense growth, where the fruits are inaccessible to birds. Also, lizards might deposit seeds in more suitable microsites than would birds (e.g. crevices) (Valido and Nogales, 1994; Traveset, 1995; Wotton, 2000). Fruit preferences have been documented among frugivorous lizards. *Gallotia galloti* (Lacertidae) on the Canary Islands shows feeding selectivity with regard to fruiting species (Valido and Nogales, 1994). The Cape Flat lizard (*Platysaurus capensis*, Sauria, Cordylidae) has been shown to discriminate between figs on the basis of ripeness and colour/brightness, choosing red- over white-painted figs (Whiting and Greeff, 1997). A different colour preference has recently been demonstrated in two common New Zealand skinks, *Oligosoma macannii* and *Oligosoma nigriplantare* (J. Marshall, unpublished data). In replicated trials under controlled conditions, captive skinks showed a clear preference for white or blue *Coprosma* (*Rubiaceae*) berries (*C. cheesemanii* white morph, *C.* aff. *parviflora* sp.'t', *C. propinqua*, *C. rugosa*) over red berries (*C. cheesemanii* red morph, *C. decurva*, *C. robusta*), when simultaneously offered fruit of each colour. An association between fruit colour and consumption by lizards is also apparent from field observations in New Zealand: eight of the 16 fleshy-fruited species listed by Whitaker (1987) as visited by lizards are blue- or white-fruited. This is more than would be expected based on the prevalence of these fruit colours in the fleshy-fruited flora as a whole (21.5%; $\chi^2 = 7.6496$, d.f. = 1, $P < 0.01$). Chemosensory response is also important to lizard feeding behaviour, and has been shown to shift with corresponding changes in diet (Cooper, 2000). The *Coprosma* fruits used in the colour preference trial described above also appear to differ in volatile compounds, which may underlie lizard preference patterns. Initial trials with gas chromatography–mass spectrometry suggest that blue and, to a lesser extent, white fruits contain resinous compounds (e.g. terpenes), whereas red fruits contain fruity and sulphurous volatiles (J.-P. DuFour and J.M. Lord, unpublished data).

We suggest that lizard dispersal, in general, should be a feature of densely growing shrubs in which fruit is borne towards the interior of the plant and which occur in an arid environment in which seed deposition in a humid microsite is important to germination. The relationship between fruit size and colour and dispersal by lizards is likely to vary between countries, depending on the size and colour sensitivity of the herpetofauna. In New Zealand, we predict that a densely growing shrub growth form and open habitats should also be associated with small, white or blue fruits, given the colour preference described above for two common frugivorous skinks and given that the frugivorous New Zealand lizards listed by Whitaker (1987) are generally small (maximum snout–vent length ranges from 77 to 160 mm). New Zealand has a particular abundance of densely growing shrub species, termed 'divaricates'. Divaricates are shrubs that have a high branching angle, twiggy interlacing growth and small, sparsely distributed leaves (Greenwood and Atkinson, 1977; Kelly, 1994). The fruits are often borne on the undersides of the branches, where they are relatively inaccessible to birds (although the small native passerine *Zosterops lateralis lateralis* has been seen inside divaricate *Coprosma* species (A.S. Markey, 1999, personal observation)). Lizards, on the other hand, can move easily through divaricate shrubs. These shrubs may also provide relatively protected basking sites, as the shrub *Larrea divaricata* (*Zygophyllaceae*) does for a species of Argentinian lizard (Deviana *et al.*, 1994). The divaricating habit is also associated with exposed or dry habitats (McGlone and Webb, 1981). In such environments, lizards may deposit seeds in more sheltered microsites than would birds (Wotton, 2000). Lord and Marshall (2002) tested for an association between fruit colour, fruit size, growth form and distribution, first using the flora as a whole and then using various subsets of the flora, such as divaricate species, shrubs and subshrubs, and the genus *Coprosma*. They found that white or blue fruit colours were more common than expected among divaricate species, and among shrubs and subshrub species. White and blue fruits were also more common in open habitats and were associated with small fruit size, both among all fleshy-fruited species and among fleshy-fruited shrubs.

Many of the divaricating species in New Zealand are in the woody Australasian–Pacific genus *Coprosma* (*Rubiaceae*). *Coprosma* in New Zealand is highly variable for fruit colour; many species have red or orange fruit, but all colours, from white, pink and yellow to deep wine and purple-black, are represented among the approximately 55 species. Thus, the genus provides an opportunity to test for an evolutionary association between fruit colour, growth form, leaf size and habitat. Fruit colour in *Coprosma* is known to be related to both leaf size and habitat. Lee *et al.* (1988) found that red fruits were more common among large-leaved *Coprosma* and at higher altitudes and provided a greater colour contrast in leafy environments (Lee *et al.*, 1994). Small-leaved *Coprosma* species at lower altitudes tended to have blue, white–cream or dark purple-black fruits. A recently constructed molecular phylogeny for *Coprosma*, based on the ITS (internal transcribed spacer) region of nuclear ribosomal DNA (A.S. Markey and S. Wichman, unpublished data), indicates that fruit, leaf and growth form traits are all highly labile. A relationship between plant traits and fruit colour is, however, apparent among closely related groups of species. In two well-supported terminal clades containing both red- and white- or blue-fruited species and with a large-leaved, red-fruited, probable ancestor, the white- and blue-fruited members of the clade have small leaves, smaller fruits and are of smaller stature and most have a divaricate growth form, suggesting that these traits evolved in concert.

Conclusion: frugivory by lizards

White and blue fruit colours are non-randomly associated with divarication, small fruit size, open habitats and a shrub growth form in the New Zealand flora as a whole, in various subsets of the flora and among related *Coprosma* species. This is consistent with the prediction that frugivory and seed dispersal by lizards with a preference for certain fruit colours may have been important in certain environments in New Zealand.

General Summary

Three features of the New Zealand fleshy-fruited flora are of interest; the general small size of fruits, the strong tendency for fruits to become more elongated with increasing size and the number of white- and blue-fruited species. We suggest that these features may have evolved partially in response to frugivory by two of the three main guilds of frugivores. The prevalence of small-gaped volant frugivorous birds is logically related to the small size of fruits and the tendency for fruits to become more elongate with increasing size, as would be expected if selection was acting to keep fruits within a swallowable size range independent of selection on fruit or seed mass. The frequency of white- and blue-fruited species and especially the non-random association of these fruit colours with small fruit size, open habitats and a shrubby growth form imply that frugivory by lizards may have influenced the evolution of fruit traits in New Zealand. There is ample evidence that extinct, large-bodied, flightless birds were important frugivores and showed some selectivity; however, there is no evidence from the fruit size distribution of the extant flora for tight mutualisms between extinct frugivores and New Zealand fruiting species.

Implications for Conservation

Judging from the fruit size distribution of the extant flora, no fruiting species has been left without a disperser, despite the large number of New Zealand frugivorous birds that have been driven to extinction in the last 1000 years. It is, of course, possible that any fruiting species closely dependent on dispersal by, for example, moa species, has already become extinct. However, it seems unlikely that such a species would disappear without a trace in such a short space of time. The extinction of and decline in the numbers of effective dispersers have left New Zealand's largest-fruited species (fruit diameter > 1 cm) very vulnerable, as these species are now virtually totally dependant on one avian frugivore, *H. novaeseelandiae*, for seed dispersal. This species is arguably the most important seed-disperser in New Zealand forests (Clout and Hay, 1989; Lee *et al.*, 1991). The growing realization of the importance of lizards as frugivores in New Zealand also raises conservation concerns. Four of the 16 currently described gecko species, and nine of the 27 currently described skink species are classified as endangered (International Union for the Conservation of Nature (IUCN) definition), and most species have restricted distributions. Many of them are also highly susceptible to environmental disturbance from various forms of land development (Pickard and Towns, 1988; Gill and Whitaker, 1998). The loss or reduction in numbers of frugivorous skinks in open habitats may have adverse effects on the regeneration of many shrub species, especially those with a divaricate growth form that limits avian access to fruit.

Avenues for Future Research

Too little use has been made of biogeographical comparisons among floras in different parts of the world. New Zealand, like many islands, offers a wealth of possibilities in the field of frugivory, because of the limited range of animals naturally present and because of recent extinctions and introductions. For example, the absence of native terrestrial mammals affords an opportunity to test for the loss of protective mechanisms against seed predation, and for a shift in fruit traits within lineages. The reduction in numbers of frugivores and the loss of species are likely to have had an impact on the population structure and regeneration of fruiting species. The introduction of new frugivores could be altering the distribution and abundance of native and introduced fruiting species. The way forward is not, however, in analyses of whole floras for the occurrence of 'syndromes'. Best use can be made of natural 'experiments' like New Zealand by testing predictions based on the characteristics of specific groups of frugivores and fruiting species.

Acknowledgements

We are grateful to Pedro Jordano, Richard Corlett and Doug Levey for helpful comments on earlier drafts. This project has been supported financially by Foundation for Research Science and Technology Grant MQI401 and University of Otago Research Grants to J.M.L. and a University of Otago Postgraduate Scholarship to A.S.M.

References

Bauer, A.M. and Russell, A.P. (1986) *Hoplodactylus delcourtii* n. sp. (Reptilia: Gekkonidae), the largest known gecko. *New Zealand Journal of Zoology* 13, 141–148.

Bauer, A.M. and Sadlier, R.A. (1994) Diet of the New Caledonian gecko *Rhacodactylus auriculatus* (Squamata, Gekkonidae). *Russian Journal of Herpetology* 1, 108–113.

Best, H.A. (1984) The foods of kakapo on Stewart Island as determined from their feeding sign. *New Zealand Journal of Ecology* 7, 71–83.

Burrows, C.J. (1989) Moa browsing: evidence from the Pyramid Valley mire. In: Rudge, M.R. (ed.) Moas, mammals and climate in the ecological history of New Zealand. *New Zealand Journal of Ecology* 12 (suppl.), 51–56.

Burrows, C.J., McCulloch, B. and Trotter, M.M. (1981) The diet of moas based on gizzard contents samples from Pyramid Valley, North Canterbury, and Scaifes Lagoon, Lake Wanaka, Otago. *Records of the Canterbury Museum* 9, 309–336.

Charles-Dominique, P. (1993) Speciation and coevolution: an interpretation of frugivory phenomena. *Vegetatio* 107/108, 75–84.

Clout, M.N. and Hay, J.R. (1989) The importance of birds as browsers, pollinators and seed dispersers in New Zealand forests. In: Rudge, M.R. (ed.) Moas, mammals and climate in the ecological history of New Zealand. *New Zealand Journal of Ecology* 12 (suppl.), 27–34.

Cooper, A., Atkinson, I.A.E., Lee, W.G. and Worthy, T.H. (1993) Evolution of the moa and their effect on the New Zealand flora. *Trends in Ecology and Evolution* 8, 433–437.

Cooper, W.E. (2000) Correspondence between diet and food chemical discriminations by omnivorous geckos (*Rhacodactylus*). *Journal of Chemical Ecology* 26, 755–763.

Corlett, R.T. (1998) Frugivory and seed dispersal by vertebrates in the Oriental (Indo Malayan) region. *Biological Review* 73, 413–448.

Crome, F.H.J. (1976) Some observations on the biology of the Cassowary in northern Queensland. *Emu* 76, 8–14.

Daniel, M.J. (1976) Feeding by the short-tailed bat (*Mystacina tuberculata*) on fruit and possibly nectar. *New Zealand Journal of Zoology* 3, 391–398.

Davies, S.J.J.F. (1978) The food of emus. *Australian Journal of Ecology* 3, 411–422.

Debussche, M. and Isenmann, P. (1989) Fleshy fruit characters and the choices of bird and mammal seed dispersers in a Mediterranean region. *Oikos* 56, 327–338.

Deviana, M.L., Jovanovich, C. and Valdes, P. (1994) Density, sex-ratio and use of space of *Liolaemus darwinii* (Sauria, Iguanidae) in Valle de Tin Tin, Argentina. *Revista de Biologia Tropical* 42, 281–287.

Dunning, J.B. (1993) *CRC Handbook of Avian Body Masses.* CRC Press, Boca Raton, Florida.

Falla, R.A. (1941) Preliminary report on excavations at Pyramid Valley Swamp, Waikari, North Canterbury: the avian remains. *Records of the Canterbury Museum* 4, 339–353.

Ferguson, R.N. and Drake, D.R. (1999) Influence of vegetation structure on spatial patterns of seed deposition by birds. *New Zealand Journal of Botany* 37, 671–678.

Figueira, J.E.C., Vasconcellos-Neto, J., Garcia, M.A. and de Souza, A.L.T. (1994) Saurochory in *Melocactus violaceus* (Cactaceae). *Biotropica* 26, 292–301.

Fischer, K.E. and Chapman, C.A. (1993) Frugivores and fruit syndromes: differences in patterns at the genus and species level. *Oikos* 66, 472–482.

Gautier-Hion, A., Duplantier, J.-M., Quris, R., Feer, F., Sourd, C., Decoux, J.-P., Dubost, G., Emmons, L., Erard, C., Hecketsweiler, P., Moungazi, A., Roussihon, C. and Thiollay, J.-M. (1985) Fruit characters as a basis of fruit choice and seed dispersal in a tropical forest vertebrate community. *Oecologia* 65, 324–337.

Gervais, J.A., Noon, B.R. and Willson, M.F. (1999) Avian selection of the color-dimorphic fruits of salmonberry, *Rubus spectabilis*: a field experiment. *Oikos* 84, 77–86.

Gill, B. and Martinson, P. (1991) *New Zealand's Extinct Birds.* Random Century, Auckland.

Gill, B. and Whitaker, T. (1998) *New Zealand Frogs and Reptiles.* David Bateman, Auckland.

Givnish, T.J., Sytsma, K.J., Smith, J.F. and Hahn, W.J. (1994) Thorn-like prickles and heterophylly in Cyanea – adaptations to extinct avian browsers on Hawaii. *Proceedings of the National Academy of Sciences (USA)* 91, 2810–2814.

Greenwood, R.M. and Atkinson, I.A.E. (1977) Evolution of divaricating plants of New Zealand in relation to moa browsing. *Proceedings of the New Zeland Ecological Society* 24, 21–33.

Gregg, D.R. (1972) Holocene stratigraphy and moas at Pyramid Valley, North Canterbury, New Zealand. *Records of the Canterbury Museum* 9, 151–158.

Hedge, S.G., Ganashaiah, K.N. and Uma Shaanker, R. (1991) Fruit preference criteria by avian frugivores: their implications for the evolution of clutch size in *Solanum pubescens*. *Oikos* 60, 20–26.

Herrera, C.M. (1989) Frugivory and seed dispersal by carnivorous mammals, and associated fruit characteristics, in undisturbed Mediterranean habitats. *Oikos* 55, 250–262.

Herrera, C.M. (1992) Interspecific variation in fruit shape: allometry, phylogeny, and adaptation to dispersal agents. *Ecology* 73, 1832–1841.

Herrera, C.M. (1998) Long-term dynamics of Mediterranean frugivorous birds and fleshy fruits: a 12-year study. *Ecological Monographs* 68, 511–538.

Herrera, C.M. and Jordano, P. (1981) *Prunus mahaleb* and birds: the high-efficiency seed dispersal system of a temperate fruiting tree. *Ecological Monographs* 51, 203–218.

Holdaway, R.N. (1989) New Zealand's pre-human avifauna and its vulnerability. In: Rudge, M.R. (ed.) Moas, mammals and climate in the ecological history of New Zealand. *New Zealand Journal of Ecology* 12 (suppl.), 11–26.

Howe, H.F. (1984) Constraints on the evolution of mutualisms. *American Naturalist* 123, 764–777.

Howe, H.F. (1986) Seed dispersal by fruit-eating birds and mammals. In: Murray, D.R. (ed.) *Seed Dispersal.* Academic Press, New York, pp. 123–189.

Howe, H.F. (1993) Specialized and generalized dispersal systems: where does 'the paradigm' stand? *Vegetatio* 107/108, 3–13.

Iverson, J.B. (1985) Lizards as seed dispersers? *Journal of Herpetology* 19, 292–293.

Janson, C.H. (1983) Adaptation of fruit morphology to dispersal agents in a neotropical forest. *Science* 219, 187–189.

Janzen, D.H. (1980) When is it coevolution? *Evolution* 34, 611–612.

Janzen, D.H. and Martin, P.S. (1981) Neotropical anachronisms: the fruits the Gomphotheres ate. *Science* 215, 19–27.

Jordano, P. (1987) Patterns of mutualistic interactions in pollination and seed dispersal: connectance, dependance asymmetries, and coevolution. *American Naturalist* 129, 657–677.

Jordano, P. (1995) Angiosperm fleshy fruits and seed dispersers: a comparative analysis of adaptation and constraints in plant–animal interactions. *American Naturalist* 145, 163–191.

Kelly, D. (1994) Towards a numerical definition for divaricate (interlaced small-leaved) shrubs. *New Zealand Journal of Botany* 32, 509–518.

King, C. (1998) *The Handbook of New Zealand Mammals.* Oxford University Press, Oxford.

Larson, D.L. (1996) Seed dispersal by specialist versus generalist foragers: the plant's perspective. *Oikos* 76, 113–120.

Lee, W.G., Wilson, J.B. and Johnson, P.N. (1988) Fruit colour in relation to the ecology and habit of *Coprosma* (*Rubiaceae*) species in New Zealand. *Oikos* 53, 325–331.

Lee, W.G., Clout, M.N., Robertson, H.A. and Wilson, J.B. (1991) Avian dispersers and fleshy fruits in New Zealand. In: Clout, M.N. and Paton, D.C. (eds) *Bird–Plant Interactions, Symposium* 28, *Acta XX Congressus Internationalis Ornithologici.* New Zealand Ornithological Trust Board, Wellington.

Lee, W.G., Weatherall, I.L. and Wilson, J.B. (1994) Fruit conspicuousness in some New Zealand *Coprosma* (*Rubiaceae*) species. *Oikos* 69, 87–94.

Levey, D.J. (1987) Seed size and fruit-handling techniques of avian frugivores. *American Naturalist* 129, 471–485.

Lord, J.M. (1991) Pollination and seed dispersal in *Freycinetia baueriana*, a dioecious liane that has lost its bat pollinator. *New Zealand Journal of Botany* 29, 83–86.

Lord, J.M. (1999) Fleshy-fruitedness in the New Zealand flora. *Journal of Biogeography* 26, 1249–1253.

Lord, J.M. and Marshall, J. (2002) Correlations between growth form, habitat and fruit colour in New Zealand flora, with reference to frugivory by lizards. *New Zealand Journal of Botany* (in press).

MacDonald, J.D. (1973) *Birds of Australia: a Summary of Information.* Reed, Sydney.

McGlone, M.S. and Webb, C.J. (1981) Selective forces influencing the evolution of divaricating plants. *New Zealand Journal of Ecology* 4, 20–28.

McGrath, R.J. and Bass, D. (1999) Seed dispersal by emus on the New South Wales north-east coast. *Emu* 99, 248–252.

Martella, M.B., Navarro, J.L., Gonnet, J.M. and Monge, S.A. (1996) Diet of greater Rheas in an agroecosystem of central Argentina. *Journal of Wildlife Management* 60, 586–592.

Mazer, S.J. and Wheelwright, N.T. (1993) Fruit size and shape: allometry at different taxonomic levels in bird-dispersed plants. *Evolutionary Ecology* 7, 556–575.

Moar, N.T. (1970) A new pollen diagram from Pyramid Valley swamp. *Records of the Canterbury Museum* 8, 455–461.

Moermond, T.C. and Denslow, J.S. (1983) Fruit choice in neotropical birds: effects of fruit type and accessibility on selectivity. *Journal of Animal Ecology* 52, 407–420.

Nogales, M., Delgado, J.D. and Medina, F.M. (1998) Shrikes, lizards and *Lycium intricatum* (Solanaceae) fruits: a case of indirect seed dispersal on an oceanic island (Alegranza, Canary Islands). *Journal of Ecology* 86, 866–871.

Pickard, C.R. and Towns, D.R. (1988) *Atlas of the Amphibians and Reptiles of New Zealand.* Conservation Sciences Publication 1, Department of Conservation, Wellington.

Pratt, T.K. and Stiles, E.W. (1985) The influence of fruit size and structure on composition of frugivore assemblages in New Guinea. *Biotropica* 17, 314–321.

Puckley, H.L., Lill, A. and O'Dowd, D.J. (1996) Fruit colour choices of captive silvereyes (*Zosterops lateralis*). *Condor* 98, 780–789.

Quin, B.R. (1996) Diet and habitat of Emus *Dromaius novaehollandiae* in the Grampians Ranges, south-western Victoria. *Emu* 96, 114–122.

Rey, P.J., Gutiérrez, J.E., Alcántara, J. and Valera, F. (1997) Fruit size in wild olives: implications for avian seed dispersal. *Functional Ecology* 11, 611–618.

Sanders, M.J., Owen-Smith, R.N. and Pillay, N. (1997) Fruit selection in the olive thrush: the importance of colour. *South African Journal of Zoology* 32, 21–23.

Tamboia, T., Cipollini, M.L. and Levey, D.J. (1996) An evaluation of vertebrate seed dispersal syndromes in four species of black nightshade (*Solanum* sect. *Solanum*). *Oecologia* 107, 522–532.

Traveset, A. (1990) *Ctenosaurus similis* Gray (Iguanidae) as a seed disperser in a Central American deciduous forest. *American Midland Naturalist* 123, 402–404.

Traveset, A. (1995) Seed dispersal of *Cneorum tricoccon* L. (*Cneoraceae*) by lizards and mammals in the Balearic Islands. *Acta Oecologia* 16, 171–178.

Valido, A. and Nogales, M. (1994) Frugivory and seed dispersal by the lizard *Gallotia galloti* (Lacertidae) in a xeric habitat of the Canary Islands. *Oikos* 70, 403–411.

van der Pijl, L. (1969) *Principles of Dispersal in Higher Plants.* Springer, Berlin.

Wheelwright, N.T. (1985) Fruit size, gape width, and the diets of fruit-eating birds. *Ecology* 66, 808–818.

Wheelwright, N.T. and Janson, C.H. (1985) Colors of fruit displays of bird-dispersed plants in two tropical forests. *American Naturalist* 126, 777–799.

Wheelwright, N.T., Haber, W.A., Murray, K.G. and Guindon, C. (1984) Tropical fruit-eating birds and their food plants: a survey of a Costa Rican lower montane forest. *Biotropica* 16, 173–192.

Whelan, C.J. and Willson, M.F. (1994) Fruit choice in migrating North American birds: field and aviary experiments. *Oikos* 71, 137–151.

Whitaker, A.H. (1987) The roles of lizards in New Zealand plant reproductive strategies. *New Zealand Journal of Botany* 25, 315–328.

Whiting, M.J. and Greeff, J.M. (1997) Facultative frugivory in the Cape Flat Lizard, *Platysaurus capensis* (Sauria: Cordylidae). *Copeia* 1997, 811–818.

Williams, P.A. and Karl, B.J. (1996) Fleshy fruits of indigenous and adventive plants in the diet of birds in forest remnants, Nelson, New Zealand. *New Zealand Journal of Ecology* 20, 127–146.

Willson, M.F., Irvine, A.K. and Walsh, N.G. (1989) Vertebrate dispersal syndromes in some Australian and New Zealand plant communities, with geographic comparisons. *Biotropica* 21, 133–147.

Willson, M.F., Sabag, C., Figueira, J., Armesto, J.J. and Caviedes, M. (1996) Seed dispersal by lizards in Chilean rainforest. *Revista Chilena de Historia Natural* 69, 339–342.

Wotton, D.M. (2000) Frugivory and seed dispersal by the common gecko *Hoplodactylus maculatus.* Msc thesis, Victoria University, Wellington, New Zealand.

5 Mechanistic Models for Tree Seed Dispersal by Wind in Dense Forests and Open Landscapes

Ran Nathan, Henry S. Horn, Jerome Chave and Simon A. Levin
Department of Ecology and Evolutionary Biology, Princeton University, Princeton, NJ 08544-1003, USA

Introduction

Seed dispersal is the main process linking spatial patterns of parent plants and their descendants (Harper, 1977; Schupp and Fuentes, 1995; Nathan and Muller-Landau, 2000). To examine the consequences of seed dispersal for patterns of plant recruitment, the spatial patterns of dispersed seeds must be quantified and their determinants understood (Nathan and Muller-Landau, 2000). A powerful framework towards this end is a combination of theoretical models and fieldwork (Okubo and Levin, 1989; Nathan and Muller-Landau, 2000).

Wind is a common dispersal agent of seeds, especially of temperate and boreal trees (Howe and Smallwood, 1982; van der Pijl, 1982). Early mechanistic models of seed dispersal by wind have recently been extended to a broad array of species and systems (reviewed in Nathan *et al.*, 2001). However, while our understanding of the mechanisms of seed dispersal by wind continues to improve, fundamental issues remain largely unexplored. In particular, it is well known that the roughness structure of the ground surface strongly affects wind conditions at the 'surface layer' (the lower part of the 'atmospheric planetary boundary layer' (Oke, 1987; Stull, 1988)), where seed dispersal mostly takes place. Thus, landscape structure can strongly influence deposition patterns of wind-dispersed seeds. For example, it has long been suggested that tree-fall gaps act as sinks for wind within and over a forest and therefore should receive a disproportionate number of wind-blown seeds (Augspurger and Franson, 1988; Schupp *et al.*, 1989). Overall, the suggested mechanisms and the predictions about spatial distributions of wind-dispersed seeds remain relatively unexplored.

Only two mechanistic wind-dispersal models have explicitly incorporated landscape structure: a model of seed dispersal from a forest into a clearing (Greene and Johnson, 1996), and one of secondary dispersal on snow (Greene and Johnson, 1997). No comparison has yet been made between the predicted distributions of dispersal distances under different landscape structures. Yet such comparisons are important for assessing how dispersal may determine recruitment in different environments.

In this study we compare wind dispersal of tree seeds in two distinct landscape structures:

 Seed Dispersal and Frugivory: Ecology, Evolution and Conservation (eds D.J. Levey, W.R. Silva and M. Galetti)

dense forests and trees scattered in an open landscape. These two environments differ greatly in wind conditions. The characteristic wind profile, which describes mean horizontal wind speed at different heights, is typically logarithmic over low vegetation in open landscapes and exponential within dense forests (Kaimal and Finnigan, 1994; see Methods; Fig. 5.1). Winds in open landscapes are typically stronger than winds within dense forests. Moreover, the different shape of the wind profiles implies that seeds dispersed in open landscapes encounter relatively strong horizontal winds over a much wider vertical range during their flight than seeds released from a similar height within dense forests. This can be illustrated by comparing the length of the arrows indicating the horizontal wind speed in the two profiles in Fig. 5.1. Thus, everything else being equal, the distribution of the dispersal distances is expected to be more limited in dense forest than in open landscapes. We shall examine this prediction with two mathematical models of seed dispersal by wind that differ only in the shape of the wind profile. Then, we shall apply and test the model's predictions against extensive seed-trap data collected in an isolated stand of Aleppo pine (*Pinus halepensis* Mill.) in Israel. In particular, we compare fits of the two models for dispersal data collected within versus outside the stand, expecting the model that incorporates the exponential profile to perform better within the stand, and the one that incorporates the logarithmic profile to perform better outside the stand.

Methods

Wind-dispersal simulations

The logarithmic wind profile has already been implemented in a mechanistic wind-dispersal model (Sharpe and Fields, 1982) and in WINDISPER (Nathan *et al.*, 2001). WINDISPER simulates the temporally and spatially explicit

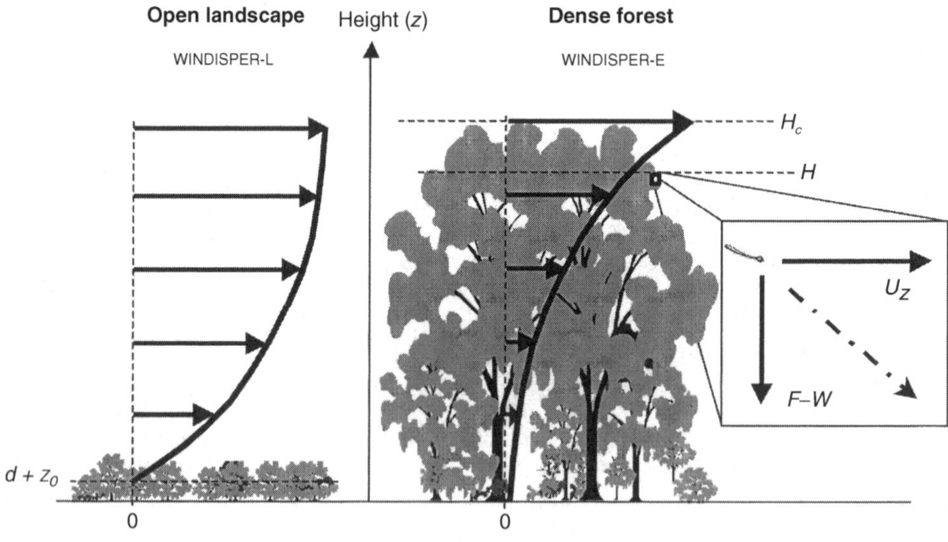

Fig. 5.1. Main factors influencing tree seed dispersal in open landscapes, as modelled in WINDISPER-L, and in dense forests, as modelled in WINDISPER-E. In both landscape types, the horizontal component of seed flight is determined by the horizontal wind speed (U), and the vertical component by the terminal velocity (F), i.e. the constant rate of seed fall in still air, and the vertical wind speed (W). The two landscape types differ in the vertical wind profiles describing the change in U with height (z) above the ground surface. A logarithmic profile is typically observed in open landscapes above short vegetation, characterized by the roughness length (z_0) and the displacement height (d), where $d + z_0$ is the height at which $U = 0$. An exponential profile is typically observed within a forest below the canopy height (H_c).

dynamics of seed dispersal by wind by incorporating stochasticity in all operative factors. Using the same modelling approach, we introduce here a new mechanistic model to describe the dispersal of seeds by wind within dense forest canopy. We call this model WINDISPER-E ('E' indicates the use of an exponential wind profile). For consistency, we refer to the original model, WINDISPER, as WINDISPER-L ('L' indicates the use of a logarithmic wind profile). We emphasize that the general structure of the two models is identical; they differ only in the wind profiles used.

In the following three subsections, we first describe the general modelling approach and the main assumptions. In the second subsection, we briefly summarize the original model (WINDISPER-L); we refer readers to Nathan *et al.* (2001) for a complete description of this model. The third subsection describes the new model (WINDISPER-E) in detail. The main parameters are defined in Table 5.1.

General modelling approach and main assumptions

In both models, dispersal of individual seeds is simulated in a square grid with 1 m^2 cells representing a simulated landscape. Seeds are dispersed from source (tree canopy) cells and can be deposited in any cell depending on the dispersal direction (*R*) and the dispersal distance (*D*). The dispersal direction is randomly selected from the wind directions observed during the simulated period and is assumed to be constant during flight. The dispersal

Table 5.1. Main variables used in WINDISPER-L and WINDISPER-E. Apart from *Q*, which varies only among the simulated periods, all parameters given in capital letters vary among individual dispersal events.

Symbol	Definition (units)	Formulation/distribution [model]*	Standard values (mean (SD))
State parameters			
D	Horizontal dispersal distance (m)	Eqn. 3 [L]; eqn. 10 [E]	
R	Dispersal direction (radians)	Follows meteorological data [L, E]	
Species parameters			
Q	Number of seeds released (seeds·canopy section^{-1} day^{-1})†	Follows meteorological data [L, E]	
F	Terminal falling velocity (m s^{-1})	Normal‡ [L, E]	0.81 (0.14)
H	Height of seed release (m)	*TH*·*PT* [L, E]	
TH	Tree height (m)	Normal§ [L, E]	9.09 (1.94)
PT	Proportion of *TH* from which seeds are released	Normal‡ [L, E]	0.61 (0.07)
Meteorological parameters			
R	Wind direction (radians)	Follows meteorological data [L, E]	
U	Horizontal windspeed (m s^{-1})	Log-normal; ‡ follows meteorological data [L, E]	
U_*	Friction velocity (m s^{-1})	Follows meteorological data [L]	
d	Displacement height (m)	[L]	0.36/0.30‖
z_0	Roughness length (m)	[L]	0.21/0.07‖
α	Attenuation coefficient	[E]	2.0
W	Vertical windspeed (m s^{-1})	Normal‡ [L, E]	0.10 (0.35)

*Capital letters in square brackets indicate the factors used in the two models, incorporating logarithmic [L] or exponential [E] wind profile.
†Each canopy section is 1 × 1 × *TH* m.
‡In agreement with Greene and Johnson (1989, 1996) and with site-specific empirical data (Nathan *et al.*, 2001).
§In agreement with site-specific empirical data (Nathan *et al.*, 2001).
‖The left and right values refer to the study site and its reference meteorological station of the Israeli Meteorological Service at 'En-Karmel, respectively.

distance, defined as the horizontal distance a seed is carried by the wind, can be calculated as

$$D = \frac{H \cdot U}{F - W} \quad (1)$$

where H is the height of seed release, U is the mean horizontal wind speed during seed flight, F is the terminal velocity of a seed falling in still air and W the mean vertical wind speed during flight, which is negative downward and positive upward.

Because wind-dispersed tree seeds typically reach terminal velocity (F) shortly after release, the vertical distance travelled before reaching F (called the relaxation distance) is typically small compared with the vertical distance travelled after F is reached (Guries and Nordheim, 1984; Nathan *et al.*, 1996). Therefore, we assume that the relaxation distance is negligible. We also assume that W is constant during flight and, to ensure finite dispersal distances, we force it to be smaller than F. Thus, events of seed uplifting are precluded, as we assume they play only a minor role in short-distance dispersal of most seeds (although they may be crucial for long-distance dispersal).

Instead of using the mean horizontal wind speed U as in eqn. 1, the two models incorporate the vertical profile of U, assumed to be logarithmic in open landscapes (WINDISPER-L) and exponential in dense forests (WINDISPER-E). In both models, the parameters that determine U (see below), as well as all the other parameters of eqn. 1, are randomly selected from their measured distribution (Table 5.1) for each seed dispersed. The horizontal wind speed deterministically decreases during seed fall, as dictated by the respective wind profile. Thus, the models do incorporate variation between different dispersal events, but do not incorporate random fluctuations in the horizontal wind speed during a seed's flight, assuming that these fluctuations do not have a significant effect during the typically short time most seeds remain airborne. We also assume that seeds are released at random in respect of either W or U. We emphasize that these assumptions do not always hold. Rather, we assert that they generally hold for short-distance dispersal. Models of long-distance dispersal, however, will probably require that they be relaxed.

WINDISPER-L – *the logarithmic wind profile model*

The logarithmic wind profile describes the decline in horizontal wind speed, U, with decreasing height above the surface, due to the surface resistance, as:

$$U_Z = \frac{u_*}{K} \ln\left(\frac{z - d}{z_0}\right) \quad (2)$$

where U_z is the mean U at height z above the ground; u_* is the friction velocity; K is the von Kármán constant (≈ 0.40); and z_0 and d are two roughness parameters, termed roughness length and displacement height, respectively (Stull, 1988). Equation 2 only applies for $z \geq d + z_0$, below which $U = 0$ (Fig. 5.1). The roughness length scales the amount of drag the ground surface exerts on the wind and is closely related to the average height of the roughness elements, i.e. the plants. Tightly packed plants (e.g. those forming a dense canopy) act as though the surface is located at some height above the real surface. The height of this 'displaced surface' is called the displacement height. These two roughness parameters are not directly measurable physical quantities; they are best determined empirically by measuring U at several heights above a surface and fitting eqn. 2. Recall that WINDISPER-L describes seed dispersal from trees scattered in an open landscape of much shorter vegetation. Thus, its wind profile is assumed to be determined only by the dominant short vegetation, not by the scattered trees (Fig. 5.1). Numerous studies have shown that the logarithmic profile works well in describing the horizontal wind speed above various vegetation types (Stull, 1988; Wieringa, 1993; Kaimal and Finnigan, 1994). As detailed in Nathan *et al.* (2001), eqn. 2 can be incorporated into eqn. 1 to provide the following equation for dispersal distance (D):

$$D = \frac{u_*}{K(F - W)}\left((H - d)\ln\left(\frac{H - d}{e \cdot z_0}\right) + z_0\right) \quad (3)$$

WINDISPER-E – *the exponential wind profile model*

The profile of the horizontal wind speed within plant canopies usually follows an exponential relationship (Cionco, 1965):

$$U_z = U_{H_C} \exp\left(\alpha\left(\frac{z}{H_C} - 1\right)\right) \tag{4}$$

where H_C is the height of the canopy top; U_{H_C} is U at H_C; and α is the attenuation coefficient (Cionco, 1965), also called the canopy flow index (Cionco, 1978). The attenuation coefficient tends to increase with increasing canopy density (Cionco, 1978; Raupach, 1988; Kaimal and Finnigan, 1994), i.e. the decline in horizontal wind speed from the canopy top (H_C) downwards is most rapid in forests of high foliage and stem density. The exponential profile generally fits observed data well (Cionco, 1978, Amiro and Davis, 1988; Amiro, 1990; Gardiner, 1994; Kaimal and Finnigan, 1994). One exception occurs when a bare-trunk layer results in a small secondary peak in wind speed at the lower half of the canopy height. This effect does not occur in stands with well-developed understorey vegetation (Amiro, 1990; Gardiner, 1994).

We shall now describe how the exponential wind profile is incorporated in the basic equation of dispersal distance (eqn. 1). As stated above, the vertical wind velocity is assumed to have a constant value, W, during flight. Then D is equal to the distance a seed travels between the time of release ($t_0 = 0$) and t_1, the time it hits the ground. Therefore,

$$D = \int_0^{t_1} U(t) \tag{5}$$

where $U(t)$ is U at time t and, following eqn. 1:

$$t_1 = \frac{H}{F - W} \tag{6}$$

i.e. the time until a seed falling at an average velocity ($F - W$) from height H reaches the ground ($z = 0$). The vertical position, $z(t)$, of a seed during flight at time t is:

$$z(t) = H - (F - W)\cdot t \tag{7}$$

Substituting $z(t)$ from eqn. 7 into eqn. 4 results in:

$$U(t) = U_{H_C} \exp\left(\alpha\left(\frac{H - H_C - (F - W)\cdot t}{H_C}\right)\right) \tag{8}$$

and integration within the limits of eqn. 5 yields:

$$D = \frac{U_{H_C}\cdot H_C}{\alpha(F - W)} \times \exp\left(\frac{\alpha(H - H_C - (F - W)\cdot t)}{H_C}\right)\Bigg|_0^{t_1} \tag{9}$$

Given $t_0 = 0$ and t_1 as in eqn. 6, eqn. 9 can be simplified to:

$$D = \frac{U_{H_C}\cdot H_C}{\alpha(F - W)} \left(\exp\left(\frac{\alpha(H - H_C)}{H_C}\right) - \exp(-\alpha)\right) \tag{10}$$

Model evaluation

Species and site

Pinus halepensis is a native Mediterranean tree (Mirov, 1967; Barbéro *et al.*, 1998) that has been widely introduced throughout the world (Richardson and Higgins, 1998). Adult trees reach relatively low heights (usually < 15 m) for pines (Nathan and Ne'eman, 2000). Seed release is stimulated by fire and by Sharav events, which are dry and hot spells of < 1 week, characteristic of the eastern Mediterranean. These events typically occur during spring and autumn and have stronger positive (upwards) vertical wind speeds than other periods (Nathan *et al.*, 1999). We do not deal here with fire-induced seed release.

Pinus halepensis seeds are samara-like structures typical of wind-dispersed pines, with a single asymmetric wing that generates autorotation during fall. The species is considered a very successful colonizer (Acherar *et al.*, 1984; Lepart and Debussche, 1991; Rejmánek and Richardson, 1996). Isolated individuals have been found several kilometres from a stand (Lepart and Debussche, 1991; R. Nathan, unpublished data). Most seeds, however, do not attain distances > 20 m from the canopy edge (Acherar *et al.*, 1984; Nathan *et al.*, 1999, 2000).

Of the two study sites described in Nathan *et al.* (1999, 2001), we selected the site at Nir-'Ezyon, on the lower western slopes of Mt Carmel (32°41′ N; 34°58′ E, 116 m altitude) for this analysis because of the larger database and the greater isolation of the focal stand

(Fig. 5.2). Seed dispersal was monitored by 94 identical seed traps (0.99 m × 0.84 m × 0.15 m, length × width × height) placed in 62 stations within and around the stand. Fourteen stations, each with a single trap, were placed within the stand, half under tree canopies and half in the gaps between the trees. The remaining 48 stations were arranged along eight transects, radiating from the focal stand in the eight main compass directions. Each transect had six stations, at approximate distances of 5, 10, 15, 25, 50 and 100 m from the focal stand. Two traps were placed in each station at 50 m and four at 100 m, to compensate for the increase in area as the distance from the source increased. This design provided 11–16 traps per 10 m annular interval up to 40 m from the nearest tree, and four to eight traps per 10 m interval from 40 to 110 m from the nearest tree.

Seeds were collected from traps regularly between October 1993 and November 1994 ($n = 36$ collections, mean interval 11 days) and later only during the dispersal seasons (spring and autumn 1995 and spring 1996, $n = 34$, mean interval 7 days). Wind measurements were taken from the Israel Meteorological Service (IMS) station at 'En-Karmel, 1.5 km south-west of the site. This station is located in an open landscape over short vegetation, in which the wind profile is typically logarithmic. A comparison between contemporaneous measurements taken at the IMS reference station and in the site (outside the stand, Fig. 5.2) during 32 days in autumn 1995 did not reveal significant differences in wind direction and horizontal wind velocity (Nathan *et al.*, 1999). We simulate seed dispersal only during the intensive periods of seed release, thus taking into account the particular wind conditions that characterize Sharav events (there were no fires in the study site during the study period). More detailed information on the study site and the procedures for seed trapping and wind measurements are given in Nathan *et al.* (1999, 2000).

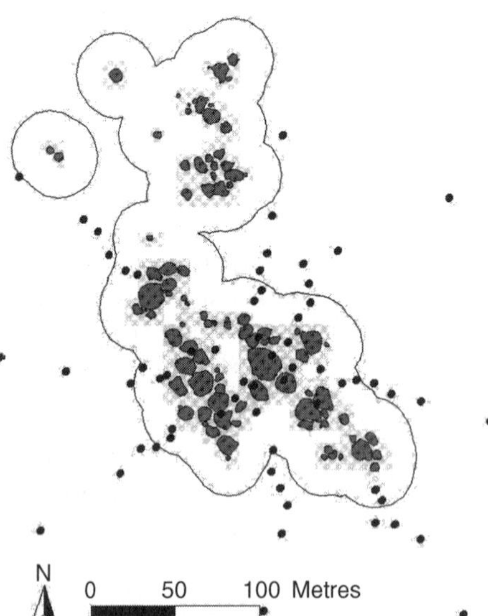

Fig. 5.2. The study site south of Nir-Ezyon on the lower western slopes of Mt Carmel. Shaded blotches show adult (seed-producing) trees, and circles indicate seed-trap stations. The areas 'within' and 'outside' the focal stand are distinguished by a line 20 m from the nearest central location of an adult tree.

Parameter estimation

Terminal velocity (F) was estimated by analysing video photos of falling seeds (Nathan *et al.*, 1996). The height of seed release (H) was calculated as the product of tree height (TH) and the proportional distribution of seeds with tree height (PT), both measured directly in the field (Nathan *et al.*, 2001). Estimates of vertical wind speed (W), displacement height (d) and roughness length (z_0) were based on wind measurements taken in the study site. The values of d and z_0, calculated by Robinson's (1962) iterative method as modified by Haenel (1993), are typical of similar vegetation surfaces (shrubland) surrounding the stand (Wieringa, 1993). Data on wind direction (R) and horizontal windspeed (U) were taken from wind measurements taken in the IMS reference station during the dispersal seasons. As described by Nathan *et al.* (2001), u_* was assumed to be identical for the Nir-'Ezyon and IMS sites. Thus, from eqn. 2:

$$u_* = \frac{K \cdot U_{10_r}}{\ln\left(\dfrac{10 - d_r}{z_{0_r}}\right)} \tag{11}$$

where the subscript r symbolizes values of the reference station, and U_{10} is U measured 10 m above the ground. The mean horizontal wind speed at canopy height (U_{H_C}) was estimated as three times u_*. This relationship is extracted from eqn. 2, given the empirically based approximation that d and z_0 equal roughly two-thirds and one-tenth of the canopy height, respectively (Oke, 1987). Because wind was not measured within the stand, we used the association with stand density (larger values characterizing stands with denser canopy (see Methods: WINDISPER-E)) to select the attenuation coefficient (α). Reported values of α for forest canopies range from 1.7 to 4.8 (Cionco, 1978; Pinker and Moses, 1982; Amiro and Davis, 1988; Amiro, 1990; Mursch-Radlgruber and Kovacic, 1990; Kaimal and Finnigan, 1994; Su *et al.*, 1998). We selected a relatively low value, $\alpha = 2.0$, for our site because our pine stand was relatively small and sparse.

Model validation

We use the proportions of seeds at each seed-trap station relative to the total number of seeds counted in all seed traps (corrected for differences in the number of traps between sampling stations) as a descriptor of the spatial pattern of dispersal. The dispersal data collected in Nir-'Ezyon during autumn 1993 and spring 1994 were used to estimate the temporal pattern of seed release, and the predictions of the two models were tested for the remaining four dispersal seasons. We also test the model's prediction for two partial sets of seed-trap stations within (0–20 m from the nearest central location of a tree) and outside (20–110 m) the focal stand (Fig. 5.2). The areas covered by the two sectors were 28,644 and 48,736 m^2, with 23 and 39 stations (23 and 71 seed traps), respectively.

For statistical validation of predicted vs. observed data, we apply linear regression analysis testing for zero intercept and slope = 1. Square root transformation corrected heteroscedasticity and non-normality of both predicted and observed data. The coefficient of determination (R^2), measuring the proportion of explained variance, was calculated as the squared multiple correlation coefficient (SMCC) between the response and the predictor variables, and also by the method of Kvålseth (1985) for situations in which the variables were transformed (see Sokal and Rohlf, 1995, p. 538). We also report the mean squared error of prediction (MSE) as a measure of predictive accuracy (Wallach and Goffinet, 1989).

Results

Predicted distance distributions

For each model, we randomly selected 1,000,000 dispersal distances calculated during an ordinary model run. Simulations of WINDISPER-E were repeated for values of the attenuation coefficient (α) other than the chosen value ($\alpha = 2.0$), covering the range reported in the literature. All distance distributions are both positively skewed and leptokurtic, with the mode just a few metres away from the point of release (Fig. 5.3, Table 5.2). However, the distributions generated by the two models differ markedly in their shape at very short distances (Fig. 5.3). WINDISPER-L generated markedly larger dispersal distances, as indicated by all summary statistics (Table 5.2). Even very sparse forests ($\alpha = 1.0$) generated distance distributions that were much more shifted towards the source than those generated by isolated trees in open landscapes. Very dense forests ($\alpha = 5.0$) produced very restricted seed shadows, with 99% of the seeds travelling less than 2.7 m (Table 5.2). Note that the kurtosis value does not accurately reflect the fatness of the distribution's tail because it is influenced by the proportions of the dispersal distances, not only at the tail but also near the source, in relation to the shoulders.

Field validation

The model's predictions of the proportions of seeds dispersed to a seed-trap station fit the empirical data sets reasonably well (Table 5.3). For the entire study area (all seed-trap stations), the 95% confidence limits (CL) for the regression slopes are narrowly distributed around unity in all cases. The CL for the intercepts are narrowly distributed around zero (for all WINDISPER-L simulations) or are slightly larger than zero (for some of the WINDISPER-E simulations). The model's predictions

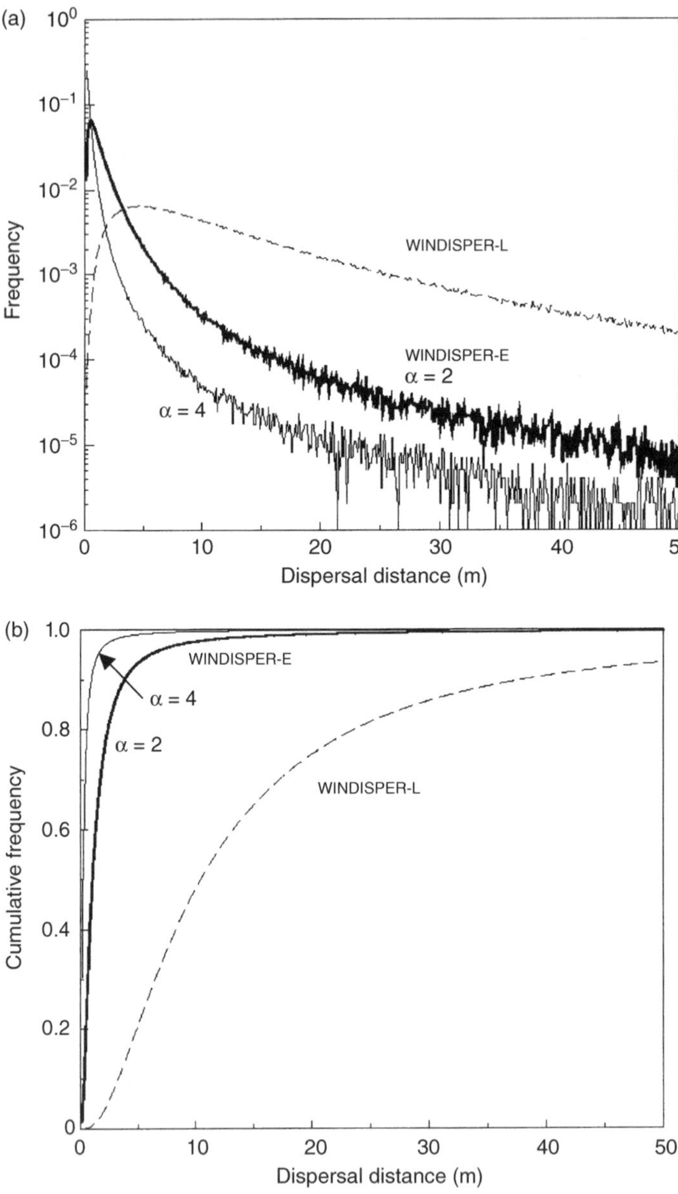

Fig. 5.3. Frequency (a) and cumulative frequency (b) distributions of 1,000,000 randomly selected dispersal distances from a typical run of the logarithmic (dashed line), and the exponential (solid lines) wind profile with attenuation coefficient (α) of 2.0 (thick line) and 4.0 (thin line).

accounted for 83–90% of the variation in the observed data for WINDISPER-L and for slightly higher percentages (87–94%) for WINDISPER-E. The maximum absolute error in the model's predictions for the proportions of seeds in a seed-trap station was low (< 0.2%; Table 5.3).

The model's predictions for the area defined as 'within' the focal stand (see Methods: Model validation) also fit the empirical data set collected between autumn 1994 and spring 1996 (Table 5.3). As for the entire data set, WINDISPER-L exhibited better performance for

Table 5.2. Summary statistics of 1,000,000 randomly selected dispersal distances (in metres) calculated during simulations of the logarithmic (WINDISPER-L) and the exponential (WINDISPER-E) models. Simulations of the exponential model were repeated for values of the attenuation coefficient (α) other than the chosen value (α = 2.0), covering the range reported in the literature.

						Percentile		
Wind profile	Mean	SE	Skewness	Kurtosis	Mode	50	95	99
Logarithmic	32.9	1.8	460	254,987	3.6	10.5	60.1	198.7
Exponential								
α = 1.0	7.5	0.5	382	282,804	1.0	2.2	12.4	40.8
α = 2.0	3.4	0.2	541	341,870	0.5	1.1	6.1	20.2
α = 3.0	1.8	0.1	435	224,186	0.2	0.5	3.0	10.0
α = 4.0	0.9	0.1	903	862,328	0.1	0.3	1.6	5.1
α = 5.0	0.4	0.1	978	968,800	0.1	0.1	0.8	2.7

Table 5.3. Statistical validation of model's predictions (means of 50 independent model runs) of the proportion of seeds dispersed to a seed-trap station against observed data. Both predicted and observed data are square-root-transformed.

		R^2†		Linear regression	
Dispersal season(s)	MSE*	SMCC	Kvålseth	Slope (95% CL)	Intercept (95% CL)
Logarithmic wind profile					
Autumn 1994	0.0010	0.83	0.84	1.05 (0.92, 1.17)	–0.011 (–0.031, 0.008)
Spring 1995	0.0008	0.86	0.87	0.99 (0.88, 1.09)	–0.001 (–0.017, 0.016)
Autumn 1995	0.0008	0.86	0.86	1.03 (0.92, 1.14)	–0.008 (–0.025, 0.009)
Spring 1996	0.0008	0.87	0.84	1.01 (0.91, 1.11)	–0.005 (–0.021, 0.011)
Autumn 1994 – Spring 1996‡	0.0006	0.90	0.90	1.03 (0.94, 1.12)	–0.008 (–0.023, 0.007)
Within the stand‡	0.0017	0.81	0.86	1.11 (0.87, 1.35)	–0.028 (–0.082, 0.026)
Outside the stand‡	0.0002	0.59	0.61	0.82 (0.59, 1.05)	0.017 (–0.007, 0.041)
Exponential wind profile					
Autumn 1994	0.0004	0.93	0.94	0.95 (0.88, 1.02)	0.007 (–0.004, 0018)
Spring 1995	0.0008	0.87	0.87	0.88 (0.79, 0.97)	0.020 (0.005, 0.034)
Autumn 1995	0.0008	0.87	0.90	0.90 (0.81, 0.99)	0.015 (0.000, 0.029)
Spring 1996	0.0006	0.90	0.87	0.91 (0.83, 0.98)	0.015 (0.003, 0.028)
Autumn 1994 – Spring 1996‡	0.0004	0.94	0.94	0.93 (0.87, 0.99)	0.012 (0.002, 0.022)
Within the stand‡	0.0009	0.90	0.91	0.86 (0.73, 1.00)	0.031 (0.000, 0.061)
Outside the stand‡	0.0001	0.57	0.59	1.96 (1.39, 2.53)	–0.097 (–0.155, –0.039)

*Mean squared error of prediction.
†The coefficient of determination (R^2) calculated as squared multiple correlation coefficient (SMCC) or by Kvålseth's (1985) method.
‡Validation of model's predictions for the total data set, and for the sectors within and outside the focal stand, separately.

the hypotheses assuming unity regression slope and zero intercept, while the predictions of WINDISPER-E accounted for a higher percentage (90% vs. 81%) of the variance in the observed data. The performance of both models is considerably lower for the area defined as 'outside' the focal stand, with WINDISPER-L explaining a slightly higher percentage (59% vs. 57%) of the variance in the observed data.

Discussion

How does landscape structure affect seed dispersal?

Variation in seed dispersal has critical implications for the population and community dynamics of plants (Harper, 1977; Howe and Smallwood, 1982; Schupp and Fuentes, 1995;

Nathan and Muller-Landau, 2000). Major sources of variation include internal (parent-controlled) factors, such as seed morphology and height of release, and external factors that influence the performance of the dispersal agent(s). The structure of the landscape affects wind flow and thus can determine the spatial dynamics of wind-dispersed species inhabiting diverse habitats. In particular, the wind profile experienced by a tree seed during flight typically has an exponential shape within dense forests and a logarithmic shape above short vegetation in open landscapes. We found a significant difference between the distributions of dispersal distances generated in the two landscape types – dispersal distances in dense forests are considerably shorter than in open landscapes. A review of seed dispersal in dipterocarps in South-East Asia reached a similar conclusion based on empirical studies (Tamari and Jacalne, 1984).

Short- versus long-distance dispersal

In general, most seeds are dispersed over short distances; only a few travel far from the source (Cain *et al.*, 2000). Although both short- and long-distance dispersal can be generated by stochastic effects, these types of dispersal often result from different mechanisms (Nathan and Muller-Landau, 2000). Moreover, the implications for recruitment after short-distance dispersal could be fundamentally different from those after long-distance dispersal. We therefore need to distinguish between the effects of landscape structure on short- and on long-distance dispersal, although the distinction between the two is not sharp. Using tree height as a qualitative criterion traditionally used by foresters (see Bullock and Clarke, 2000), one can refer to the scale of up to a few tree heights (a few tens of metres for *P. halepensis*), where most seeds are deposited, as short-distance dispersal, and the scale of tens of tree heights (hundreds of metres for *P. halepensis*), where only very few seeds are deposited, as long-distance dispersal. Our results show that such a rough criterion can be misleading, since trees of the same height can generate considerably different seed shadows in different landscapes. For example, the distance travelled by 99% of the seeds can vary between almost 200 m (22 times the measured mean tree height) and less than 3 m (less than a third of the measured tree height), depending on the landscape type (Table 5.2).

We developed the two models to describe short-distance dispersal. Extrapolation to long-distance dispersal is inappropriate because five basic assumptions become unrealistic when larger spatial and temporal scales are considered. First, it is assumed that a seed flies in a straight line in the selected wind direction. Secondly, seed release is assumed to be independent of wind speed; however, seeds may typically be released in higher-than-average wind speeds (Greene and Johnson, 1992). Thirdly, the roughness parameters of the two wind profiles are assumed to be constant in time and space. Fourthly, variation in U and W (the mean vertical wind speed) during individual flights is not considered. Fifthly, and most importantly, for each dispersal event, F (the seed falling velocity in still air), is constrained to be larger than W; thus seed uplifting, a process critical for long-distance dispersal (Greene and Johnson, 1995), is not taken into account. Accordingly, our tests have shown that both models explain short-distance dispersal considerably better than long-distance dispersal.

Seed dispersal under different wind profiles: small-scale implications

Short-distance dispersal of the vast majority of the seeds generates a small-scale pattern that has critical implications for local recruitment (Janzen, 1970; Connell, 1971; Schupp and Fuentes, 1995; Nathan and Muller-Landau, 2000). A basic feature of this pattern, the rapid decline in seed densities with distance from the source (Harper, 1977; Willson, 1993), leads to strong sibling competition and intensive seed predation in the vicinity of adult trees (Janzen, 1970; Connell, 1971). Because dispersal distances are considerably shorter within dense forests than in open landscapes, Janzen–Connell effects are likely to be more powerful in forests. This is amplified by differences in fecundity. But, even without differences in fecundity, overlapping seed shadows in forests are likely to generate much higher seed densities close to adult trees than in open landscapes. Moreover, the high seed densities within forests could reach a level at which predators become satiated, hence increasing seed survival (Janzen, 1971) and favouring selection for reduced dispersal distances. Strong Janzen–Connell effects and especially predator satiation effects are less likely in open landscapes. Overall, evaluation of the potential implications of the restricted dispersal distances within forests requires data on fecundity, dispersal and seed survival. Our models could help to estimate the effects of fecundity and dispersal and to guide experimental studies on seed survival.

Seed dispersal under different wind profiles: large-scale implications

The critical importance of seed dispersal for various post-dispersal processes at the local (small) spatial scale has long been recognized (Harper, 1977; Howe and Smallwood, 1982). Recently, studies have emphasized the disproportionate importance of long-distance dispersal in determining large-scale patterns, such as spatial spread, gene flow and metapopulation dynamics (Kot *et al.*, 1996; Clark *et al.*, 1998; Cain *et al.*, 2000; Nathan, 2001). Although application of our models to long-distance dispersal would require relaxation of some basic assumptions, we can still suggest which landscape type favours long-distance dispersal. Seeds dispersed by wind in open landscapes travel much further than those dispersed within a forest, due to differences in the shape of the wind profile. Accordingly, the exponential profile better fits dispersal data within a forest, while the logarithmic profile does better outside the stand. Furthermore, dense forests and open landscapes differ not only in the shape of the wind profile but also in the absolute wind velocities; winds in open landscapes are typically stronger (Stull, 1988) and produce greater dispersal distances. This difference is further amplified by the typically stronger wind up-draughts above rather than within a plant canopy (Stull, 1988; Kaimal and Finnigan, 1994): stronger horizontal winds generate stronger shear-generated up draughts. Buoyancy-generated up-draughts (thermals) are also more frequent in open landscapes (Stull, 1988).

Because of the predominance of better wind conditions for long-distance dispersal in open landscapes, isolated trees or trees at the forest edge could be important for tree spatial spread, gene flow and metapopulation dynamics. This supports the notion that long-distance dispersal events can generate 'great leaps forward' that determine population spread (Mollison, 1972; Kot *et al.*, 1996; Clark *et al.*, 1998). However, the effect of landscape structure on population spread depends on multiple factors, including the relative fecundity of isolated versus forest trees, on the transition between the logarithmic and exponential models and on the relative probability of seed survival to adulthood in different landscapes.

Prospects for future research

We show that mechanistic understanding of the physical and biological conditions affecting seed dispersal by wind at small scales can be translated to simple tools that reliably predict the dispersal of most seeds at a spatial resolution of 1 m^2. However, questions on how landscape structure affects dispersal will require consideration of wind-dispersal mechanisms acting at large scales, too. In the absence of field data on long-distance dispersal (Cain *et al.*, 2000), models are even more important for speculating about large-scale processes. Indeed, models may help in the design of experiments, which may increase the likelihood of observing intrinsically rare events. Mechanistic models of long-distance dispersal can enhance the recent progress in predicting seed-dispersal patterns over large scales through phenomenological models (Kot *et al.*, 1996; Clark, 1998; Clark *et al.*, 1998, 1999; Higgins and Richardson, 1999). This is because only mechanistic models provide the means for generalization beyond the studied systems and yield insights into the main operative factors (Okubo and Levin, 1989; Nathan and Muller-Landau, 2000). Thus, they allow examination of other puzzles: for example, how winds behave (and carry seeds) in the transition between the forest and open fields and whether the biological parameters of dispersal (e.g. height of release and seed terminal velocity) also vary with landscape type. Answering these questions will necessitate integration of models and empirical studies and will require creative solutions to challenges imposed by the complicated and yet important aspects of long-distance dispersal (see Greene and Johnson, 1995; Bullock and Clarke, 2000). Further, given the success of mechanistic models in predicting wind dispersal (Nathan and Muller-Landau, 2000), we need to develop a comprehensive mechanistic approach for seed dispersal by animals and by other agents.

Acknowledgements

We gratefully acknowledge the support by the National Science Foundation (IBN-9981620), the Andrew Mellon Foundation, the Desertification and Restoration Ecology Research Centre of the Keren Kayemeth Leisrael (Jewish National Fund) and the Israel National Parks and Nature Reserve Authority. For their help with wind measurements, we thank E. Zachs (Minerva Arid Ecosystems Research Centre, the Hebrew University of Jerusalem) and I. Seter (IMS). We also thank D. Roitemberg for his intensive help with fieldwork, and I. Gilderman and J. Ma for their help in C programming. Finally, we are grateful to M. Cain, D. Levey, R. Avissar, G. Katul, H. Muller-Landau, J. Dushoff and K. Kitajima for their helpful comments and discussions.

References

Acherar, M., Lepart, J. and Debussche, M. (1984) La colonisation des friches par le pin d'Alep (*Pinus halepensis* Miller) en Languedoc méditerranéen. *Acta Oecologica* 5, 179–189.

Amiro, B.D. (1990) Comparison of turbulence statistics within three boreal forest canopies. *Boundary-Layer Meteorology* 51, 99–121.

Amiro, B.D. and Davis, P.A. (1988) Statistics of atmospheric turbulence within a natural black spruce forest canopy. *Boundary-Layer Meteorology* 44, 267–283.

Augspurger, C.K. and Franson, S.E. (1988) Input of wind-dispersed seeds into light-gaps and forests sites in a neotropical forest. *Journal of Tropical Ecology* 4, 239–252.

Barbéro, M., Loisel, R., Quézel, P., Richardson, D.M. and Romane, F. (1998) Pines of the Mediterranean basin. In: Richardson, D.M. (ed.) *Ecology and Biogeography of* Pinus. Cambridge University Press, Cambridge, pp. 153–170.

Bullock, J.M. and Clarke, R.T. (2000) Long distance seed dispersal by wind: measuring and modelling the tail of the curve. *Oecologia* 124, 506–521.

Cain, M.L., Milligan, B.G. and Strand, A.E. (2000) Long-distance seed dispersal in plant populations. *American Journal of Botany* 87, 1217–1227.

Cionco, R.M. (1965) A mathematical model for air flow in a vegetative canopy. *Journal of Applied Meteorology* 4, 517–522.

Cionco, R.M. (1978) Analysis of canopy index value for various canopy densities. *Boundary-Layer Meteorology* 15, 81–93.

Clark, J.S. (1998) Why trees migrate so fast: confronting theory with dispersal biology and the paleorecord. *American Naturalist* 152, 204–224.

Clark, J.S., Fastie, C., Hurtt, G., Jackson, S.T., Johnson, C., King, G.A., Lewis, M., Lynch, J., Pacala, S., Prentice, C., Schupp, E.W., Webb, T. III and Wyckoff, P. (1998) Reid's paradox of rapid plant migration: dispersal theory and interpretation of paleoecological records. *BioScience* 48, 13–24.

Clark, J.S., Silman, M., Kern, R., Macklin, E. and HilleRisLambers, J. (1999) Seed dispersal near and far: patterns across temperate and tropical forests. *Ecology* 80, 1475–1494.

Connell, J.H. (1971) On the role of natural enemies in preventing competitive exclusion in some marine animals and in forest trees. In: den Boer, P.J. and Gradwell, G.R. (eds) *Dynamics of Populations.* Centre for Agricultural Publishing and Documentation, Wageningen, pp. 298–312.

Gardiner, B.A. (1994) Wind and wind forces in a plantation spruce forest. *Boundary-Layer Meteorology* 67, 161–186.

Greene, D.F. and Johnson, E.A. (1992) Fruit abscission in *Acer saccharinum* with reference to seed dispersal. *Canadian Journal of Botany* 70, 2277–2283.

Greene, D.F. and Johnson, E.A. (1995) Long-distance wind dispersal of tree seeds. *Canadian Journal of Botany* 73, 1036–1045.

Greene, D.F. and Johnson, E.A. (1996) Wind dispersal of seeds from a forest into a clearing. *Ecology* 77, 595–609.

Greene, D.F. and Johnson, E.A. (1997) Secondary dispersal of tree seeds on snow. *Journal of Ecology* 85(3), 329–340.

Guries, R.P. and Nordheim, E.V. (1984) Flight characteristics and dispersal potential of maple samaras. *Forest Science* 30, 434–440.

Haenel, H.D. (1993) Surface-layer profile evaluation using a generalization of Robinson's method for the determination of d and z_0. *Boundary-Layer Meteorology* 65, 55–67.

Harper, J.L. (1977) *Population Biology of Plants.* Academic Press, London.

Higgins, S.I. and Richardson, D.M. (1999) Predicting plant migration rates in a changing world: the role of long-distance dispersal. *American Naturalist* 153, 464–475.

Howe, H.F. and Smallwood, J. (1982) Ecology of seed dispersal. *Annual Review of Ecology and Systematics* 13, 201–228.

Janzen, D.H. (1970) Herbivores and the number of tree species in tropical forests. *American Naturalist* 104, 501–528.

Janzen, D.H. (1971) Seed predation by animals. *Annual Review of Ecology and Systematics* 2, 465–492.

Kaimal, J.C. and Finnigan, J.J. (1994) *Atmospheric Boundary Layer Flows: Their Structure and Measurement.* Oxford University Press, New York.

Kot, M., Lewis, M.A. and Van Den Driessche, P. (1996) Dispersal data and the spread of invading organisms. *Ecology* 77, 2027–2042.

Kvålseth, T.O. (1985) Cautionary note about R^2. *American Statistician* 39, 279–285.

Lepart, J. and Debussche, M. (1991) Invasion processes as related to succession and disturbance. In: Groves, R.H. and di Castri, F. (eds) *Biogeography of Mediterranean Invasions.* Cambridge University Press, Cambridge, pp. 159–177.

Mirov, N.T. (1967) *The Genus* Pinus. Ronald Press, New York.

Mollison, D. (1972) The rate of spatial propagation of simple epidemics. *Proceedings of the Six Berkeley Symposium on Mathematics, Statistics, and Probability* 3, 579–614.

Mursch-Radlgruber, E. and Kovacic, T. (1990) Mean canopy flow in an oak forest and estimation of the foliage profile by a numerical-model. *Theoretical and Applied Climatology* 41, 129–136.

Nathan, R. (2001) Dispersal biogeography. In: Levin, S.A. (ed.) *Encyclopedia of Biodiversity,* Vol. II. Academic Press, San Diego, pp. 127–152.

Nathan, R. and Muller-Landau, H.C. (2000) Spatial patterns of seed dispersal, their determinants and consequences for recruitment. *Trends in Ecology and Evolution* 15, 278–285.

Nathan, R. and Ne'eman, G. (2000) Serotiny, seed dispersal and seed predation in *Pinus halepensis.* In: Ne'eman, G. and Trabaud, L. (eds) *Ecology, Biogeography and Management of* Pinus halepensis *and* P. brutia *Forest Ecosystems in the Mediterranean Basin.* Backhuys, Leiden, The Netherlands, pp. 105–118.

Nathan, R., Safriel, U.N., Noy-Meir, I. and Schiller, G. (1996) Samara's aerodynamic properties in *Pinus halepensis* Mill., a colonizing tree species, remain constant despite considerable variation in morphology. In: Steinberger, Y. (ed.) *Preservation of Our World in the Wake of Change.* Israel Society for Ecology and Environmental Quality Sciences, Jerusalem, pp. 553–556.

Nathan, R., Safriel, U.N., Noy-Meir, I. and Schiller, G. (1999) Seed release without fire in *Pinus halepensis,* a Mediterranean serotinous wind-dispersed tree. *Journal of Ecology* 87, 659–669.

Nathan, R., Safriel, U.N., Noy-Meir, I. and Schiller, G. (2000) Spatiotemporal variation in seed dispersal and recruitment near and far from adult *Pinus halepensis* trees. *Ecology* 81, 2156–2169.

Nathan, R., Safriel, U.N. and Noy-Meir, I. (2001) Field validation and sensitivity analysis of a mechanistic model for tree seed dispersal by wind. *Ecology* 82, 374–388.

Oke, T.R. (1987) *Boundary Layer Climates,* 2nd edn. Routledge, London.

Okubo, A. and Levin, S.A. (1989) A theoretical framework for data analysis of wind dispersal of seeds and pollen. *Ecology* 70, 329–338.

Pinker, R.T. and Moses, J.F. (1982) On the canopy flow index of a tropical forest. *Boundary-Layer Meteorology* 22, 313–324.

Raupach, M.R. (1988) Canopy transport process. In: Steffan, W.L. and Denmead, O.T. (eds) *Flow and Transport in the Natural Environment: Advances and Applications.* Springer-Verlag, Berlin, pp. 95–127.

Rejmánek, M. and Richardson, D.M. (1996) What attributes make some plant species more invasive? *Ecology* 77, 1655–1661.

Richardson, D.M. and Higgins, S.I. (1998) Pines as invaders in the southern hemisphere. In: Richardson, D.M. (ed.) *Ecology and Biogeography of* Pinus. Cambridge University Press, Cambridge, pp. 450–473.

Robinson, S.M. (1962) Computing wind profile parameters. *Journal of the Atmospheric Sciences* 19, 189–190.

Schupp, E.W. and Fuentes, M. (1995) Spatial patterns of seed dispersal and the unification of plant population ecology. *Ecoscience* 2, 267–275.

Schupp, E.W., Howe, H.F., Augspurger, C.K. and Levey, D.J. (1989) Arrival and survival in tropical treefall gaps. *Ecology* 70, 562–564.

Sharpe, D.M. and Fields, D.E. (1982) Integrating the effects of climate and seed fall velocities on seed dispersal by wind: a model and application. *Ecological Modelling* 17, 297–310.

Sokal, R.R. and Rohlf, F.J. (1995) *Biometry*, 3rd edn. Freeman, New York.

Stull, R.B. (1988) *An Introduction to Boundary-layer Meteorology*. Kluwer Academic Publishers, Dordrecht.

Su, H.B., Shaw, R.H., U, K.T.P., Moeng, C.H. and Sullivan, P.P. (1998) Turbulent statistics of neutrally stratified flow within and above a sparse forest from large-eddy simulation and field observations. *Boundary-Layer Meteorology* 88, 363–397.

Tamari, C. and Jacalne, D.V. (1984) Fruit dispersal of dipterocarps. *Bulletin of Forestry and Forest Products Research Institute* 325, 127–140.

van der Pijl, L. (1982) *Principles of Dispersal in Higher Plants*, 3rd edn. Springer, Berlin, 162 pp.

Wallach, D. and Goffinet, B. (1989) Mean squared error of prediction as a criterion for evaluating and comparing system models. *Ecological Modelling* 44, 299–306.

Wieringa, J. (1993) Representative roughness parameters for homogeneous terrain. *Boundary-Layer Meteorology* 63, 323–363.

Willson, M.F. (1993) Dispersal mode, seed shadows, and colonization patterns. *Vegetatio* 107/108, 261–280.

6 The Role of Vertebrates in the Diversification of New World Mistletoes

Carla Restrepo,[1] Sarah Sargent[2], Douglas J. Levey[3] and David M. Watson[4]

[1]*Department of Biology, University of New Mexico, Albuquerque, NM 87108, USA;* [2]*Department of Environmental Sciences, Allegheny College, Meadville, PA 16335, USA;* [3]*Department of Zoology, University of Florida, Gainesville, FL 32611, USA;* [4]*Environmental Studies Unit, Charles Sturt University, Bathurst, NSW 2795, Australia*

Birds feeding on berries of *Loranthaceae* become quite a nuisance in gardens and citrus plantations, by spreading these pests on to trees.

(Haverschmidt, 1968)

Introduction

Mistletoes are unique among woody plants in that all species are parasites (Kuijt, 1969; Calder and Bernhardt, 1983). Even though being a parasite may seem to constrain possibilities for diversification, mistletoes have successfully radiated into a wide array of species, genera and families with different lifestyles (Barlow, 1964; Polhill and Wiens, 1998). Some mistletoes, for example, are terrestrial and others are aerial root parasites (Hoehne, 1931; Kuijt, 1963; Fineran and Hocking, 1983). Most species, however, are aerial stem parasites that vary in their preferences for hosts (Barlow and Wiens, 1977; Hawksworth and Wiens, 1996), substrates within compatible hosts (Sargent, 1995) and degree of dependence on the host for the acquisition of resources (Fisher, 1983; Lamont, 1983; Ehleringer and Marshall, 1995). This has led to the hypothesis that host–mistletoe interactions may have driven the diversification of mistletoes (Fig. 6.1; Norton and Carpenter, 1998), as postulated for other host–parasite systems (Price, 1980; Brooks, 1988). In fact, variation in parasite virulence and host resistance have been shown to influence host switching and host specificity, two phenomena that can explain diversification among parasites mediated through host–parasite interactions (Page, 1994; Hoberg *et al.*, 1997).

An alternative, but not mutually exclusive, hypothesis is that diversification among mistletoes has been driven by vertebrate–mistletoe interactions and their outcome, seed dispersal (Fig. 6.1). Indeed, most mistletoes are dispersed by vertebrates and patterns of mistletoe distribution within and among hosts are strongly influenced by the behaviour of these dispersers (Restrepo, 1987; Sargent, 1994;

 Seed Dispersal and Frugivory: Ecology, Evolution and Conservation (eds D.J. Levey, W.R. Silva and M. Galetti)

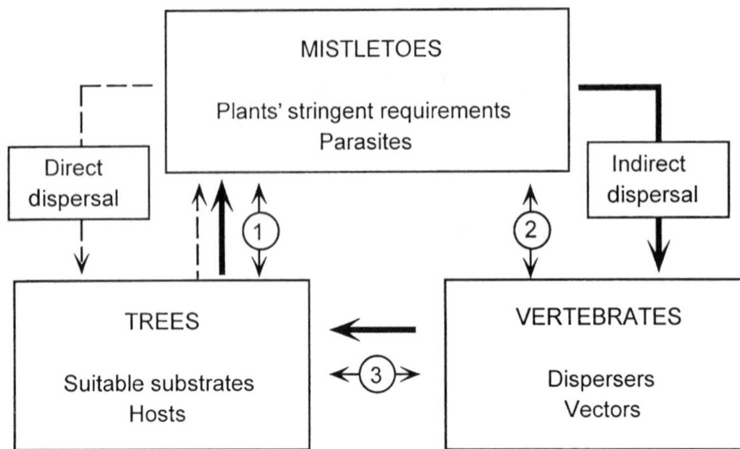

Fig. 6.1. Three hypotheses, alone or in combination, may explain the diversification of mistletoes mediated through interactions under the apparent constraints imposed by a parasitic life: (1) host–parasite, (2) vector–parasite and (3) host–vector interactions.

Martínez del Río *et al.*, 1996). This has led to the suggestion that mistletoes may be more dependent on their vectors than non-parasitic plants for the delivery of seeds to suitable sites (Reid, 1991). Mistletoes, however, exhibit a variety of dispersal modes, which range from direct to indirect or vector-mediated dispersal (Fig. 6.1; Orfila, 1978; Reid, 1991; Hawksworth and Wiens, 1996). We hypothesize that variation in dispersal mode, coupled with specialized requirements for germination and establishment (Frochot and Sallé, 1980; Sallé, 1983; Hoffman *et al.*, 1986; Sargent, 1995), may have influenced the diversification of mistletoes through vector–parasite interactions.

In other parasitic organisms there is strong evidence that vector-mediated dispersal has strongly influenced the diversification of parasites. First, traits that influence dispersal of parasites, such as parasite virulence and compatibility and vector competence and resistance, are under strong selection (Collins *et al.*, 1986; Yan *et al.*, 1997; Failloux *et al.*, 1999). Secondly, vectors contribute to the movement of parasite propagules to 'safe sites' and facilitate the exploration of 'host space' (Kim, 1985; Collins and Besansky, 1994; Azad and Beard, 1998). Thirdly, vector, rather than host, phylogenies explain parasite diversification either through associations by descent–vector specificity or associations by colonization–vector switching (Chiykowski, 1981; Davis, 1992; Carreno *et al.*, 1997; Luke *et al.*, 1997), as postulated for host–parasite systems (Brooks, 1988; Brooks and McLennan, 1993). The signatures for associations by descent and colonization are congruent and incongruent phylogenies, respectively (Hoberg *et al.*, 1997). Alternatively, associations between vectors and parasites that seem to be the result of vector switching may actually be the result of sorting events, such as parasite extinction, parasite absence from the vector founder population at a speciation event or sampling error (Paterson and Gray, 1997).

To evaluate the hypothesis that vector–parasite interactions have played a role in the diversification of mistletoes, we tested three predictions. First, mistletoe taxa that are most diverse are predominantly vertebrate-dispersed. Secondly, vectors associated with mistletoes represent a narrow subset of local vertebrate assemblages. If true, we expect to find little variation among mistletoe species in terms of their vectors. Thirdly, vector–mistletoe associations found within local assemblages are likely to reflect a long-term history of association. If true, we expect to find a high degree of congruence between bird and mistletoe phylogenies. To test these predictions, we focused on New World mistletoes and their dispersers, because both are relatively well known.

Methods

Mistletoe diversity and dispersal mode

Mistletoes are in the *Santalales* and represent the largest assemblage of parasitic woody angiosperms (Kuijt, 1968). In the New World, they are represented by four families (*Eremolepidaceae*, *Loranthaceae*, *Misodendraceae* and *Viscaceae*), 24 genera and approximately 700 species (Table 6.1; J. Kuijt, personal communication). *Loranthaceae* and *Misodendraceae* represent basal clades, whereas *Eremolepidaceae* and *Viscaceae* are derived clades (Nickrent, in press).

Most mistletoes produce fleshy fruits with a single seed associated with viscin, a highly sticky tissue (Kuijt, 1969). Unlike seeds of other angiosperms, those of mistletoes lack a seed-coat; thus, the unit of dispersal consists of an embryo and endosperm, often photosynthetic, and various tissues, including viscin (Bhandari and Vohra, 1983; Bhatnagar and Johri, 1983). Viscin functions in the attachment of seeds to the host (for other functions, see Gedalovich and Kuijt, 1987). The achenes of *Misodendraceae* with their feather-like structures represent the only exception to this pattern (Kuijt, 1969).

Mistletoe species were classified according to dispersal mode and family to test the prediction that dispersal mode has influenced mistletoe diversity. At least three dispersal modes have been reported among New World mistletoe species: anemochory – wind dispersal (Orfila, 1978) – autochory – ballistic dispersal (Hawksworth and Wiens, 1996) – and endozoochory – animal dispersal via fruit ingestion (Walsberg, 1975; Davidar, 1987; Restrepo, 1987; Sargent, 1994). The autochorous genus *Arceuthobium* produces explosive fleshy fruits, capable of short-distance seed dispersal; however, long-distance dispersal may occur when the sticky seeds get attached to vertebrate feathers and fur (Hawksworth and Wiens, 1996). We term this type of dispersal facultative epizoochory. In addition, species in the predominantly endozoochorous genus *Dendrophthora* may have explosive fruits that eject their single seed when manipulated within birds'

Table 6.1. Diversity of New World mistletoes and associated dispersal modes. 'Mixed' mode includes autochory/facultative epizoochory and facultative autochory/endozoochory (see text).

		Dispersal		
		Anemochory	Endozoochory	Mixed
Misodendraceae	*n* (genera)	1	–	–
	n (species)	12	–	–
	Species per genus	12	–	–
Loranthaceae	*n* (genera)	–	17	–
	n (species)	–	278	–
	Species per genus	–	16.3 (18.2)	–
Eremolepidaceae	*n* (genera)	–	3	–
	n (species)	–	12	–
	Species per genus	–	4.0 (3.4)	–
Viscaceae	*n* (genera)	–	3	2
	n (species)	–	359	35
	Species per genus	–	120.0 (119.5)	17.0 (22.6)
Total	*n* (genera)	1	23	2
	n (species)	12	650	35
	Species per genus	12	28.3 (53.7)	17.0

Number of species (in parentheses): *Loranthaceae*: *Desmaria* (1), *Gaiadendron* (1), *Notanthera* (1), *Ligaria* (2), *Tripodanthus* (2), *Tristerix* (12), *Maracanthus* (3), *Oryctanthus* (12), *Oryctina* (8), *Panamanthus* (1), *Phthirusa* (35), *Struthanthus* (*c.* 55), *Cladocolea* (30), *Dendropemon* (30), *Ixocactus* (15), *Aetanthus* (15), *Psittacanthus* (*c.* 55). *Eremolepidaceae*: *Lepidoceras* (2), *Eubrachion* (2), *Antidaphne* (8). *Misodendraceae*: *Misodendrum* (12). *Viscaceae*: *Dendrophthora* (120), *Phoradendron* (240), *Arceuthobium* (34).

bills (Sargent, 1994). We term this type of dispersal facultative autochory. Anemochory and autochory represent direct modes of dispersal, whereas endozoochory and epizoochory represent indirect or vector-mediated modes (Fig. 6.1).

Mistletoe diversity and vector-mediated dispersal

Local-scale assemblages

We conducted observations at two sites rich in mistletoe species, one in Colombia and one in Costa Rica (Table 6.2). These sites are heavily covered by forest and second-growth vegetation and are classified as lower montane wet forest (Restrepo, 1987; Haber, 2000). At the Colombian site, we recorded bird activity at clumps of five mistletoe species (Table 6.2); we spent an average of 25 h month^{-1} observing each species (February 1983–March 1984). A feeding visit was defined as a bird landing on a mistletoe clump to feed and departing afterwards (Restrepo, 1987). At the Costa Rican site, we recorded bird activity at mistletoe clumps belonging to six species (Table 6.2). These clumps were found along a 3 km loop transect, which was surveyed on a weekly basis (September 1989–July 1990). At both sites, we identified and recorded the number of visits and behaviour of the birds feeding on mistletoe fruits. We used bird visits to mistletoe species as the response variable. Bird visits were classified according to mistletoe and bird species to test the prediction that vectors associated with mistletoe species represented a narrow subset of vertebrates.

Regional-scale assemblages

We compiled a database on vertebrates feeding on mistletoe fruits or dispersing their seeds. Each record (542 in total) includes information on mistletoe (family and species) and vertebrate (class, family/subfamily and species) taxonomy, vertebrate fruit- and seed-handling methods, site of the observation (site name, geographical coordinates, elevation) and source. Records were compiled from the literature, an on-line mistletoe database (http://www.rms.nau.edu/mistletoe/mtbib.html), and from unpublished records and theses made available through personal communication with researchers. We used vertebrate species as our response variable. Although simplistic, this was necessary for several reasons. First, few studies report data on the relative importance of vertebrate species to mistletoes. Secondly, most studies report only casual observations of vertebrates feeding on mistletoes. Thirdly, some 'sites' are over-represented in the database, yielding several records for the same species. Fourthly, we could not differentiate between 'legitimate' seed-dispersers and seed predators because the fate of seeds was rarely mentioned. More importantly, however, 'legitimate' dispersers and seed predators are both likely to influence seed-dispersal systems (Herrera, 1984).

Table 6.2. Location and characteristics of study sites in Colombia and Costa Rica. Species that were intensively studied are listed below the table.

Site	Coordinates	Elevation (m)	Families and number of species
Zingara and La Frizia, Colombia	3°20′N, 76°38′W	1950	*Viscaceae* (9) *Loranthaceae* (5) *Eremolepidaceae* (1)
Monteverde, Costa Rica	10°18′N, 84°48′W	1550	*Viscaceae* (5) *Loranthaceae* (6) *Eremolepidaceae* (1)

Colombia: *Phoradendron colombianum*, *Phoradendron inaequidentatum* and *Phoradendron dipterum*, *Viscaceae*; *Cladocolea lenticellata*, *Loranthaceae*; and *Antidaphne viscoidea*, *Eremolepidaceae*. Costa Rica: *Phoradendron robustissimum*, *Phoradendron chrysocladon*, *Phoradendron robaloense*, *Viscaceae*; *Struthanthus oerstedii* and *Oryctanthus spicatus*, *Loranthaceae*; and *A. viscoidea*, *Eremolepidaceae*.

Vertebrate species were classified into taxonomic groups to test the prediction that vector–mistletoe associations found within local assemblages are likely to reflect a long-term history of associations. These groups were: non-passerine (NONP), suboscine passerines (PSOS), oscine passerine birds (POSC) and other (OTHE); the latter includes mammals and fish. We further classified PSOS and POSC species into families and/or subfamilies to explore the occurrence of associations by descent–vector specificity and associations by colonization–vector switching. We made use of recent phylogenies based on molecular data showing that: (i) *Euphonia/Chlorophonia* forms a clade that is sister to the Carduelini (Fringillinae) and not the Emberizinae; (ii) *Tersinia, Cyanerpes, Dacnis, Coereba, Diglossa, Chlorophanes* and *Saltator* belong to the Thraupini; and (iii) *Spiza, Pheuticus, Passerina* and *Piranga* represent a clade within the Cardinalini (Sibley and Ahlquist, 1990; Burns, 1997; Klicka *et al.*, 2000).

In all instances we used chi-square tests and included individual taxa when sample sizes were large enough so that > 80% of the expected cell frequencies were > 5; taxa that did not meet this criterion were pooled (Siegel and Castellan, 1988). We partitioned the $r \times k$ contingency tables into a series of 2×2 subtables to identify the cells contributing to significant results; the 2×2 subtables were analysed as if they were independent from each other by using a modified χ^2 test (Siegel and Castellan, 1988). The groups were arranged a priori to reflect meaningful comparisons between mistletoes and vertebrates. We calculated the standardized residuals for the 2×2 subtables for which the chi-square values were significant. We used Matlab to program the routines used to partition the contingency tables and to calculate the chi-square values.

Results

Mistletoe diversity and dispersal mode

Endozoochory is disproportionately common among New World mistletoes, whether analysed with number of genera or number of species (goodness-of-fit test, $\chi^2 = 28.5$, d.f. = 2, $P \leq 0.0001$ and $\chi^2 = 1324.0$, d.f. = 2, $P \leq 0.0001$, respectively (Table 6.1)). Yet, when we discriminate among mistletoe families, we find that there is a significant association between dispersal mode and family. In the *Loranthaceae* more genera than expected are endozoochorous, whereas in the *Misodendraceae* and *Viscaceae* (*Arceuthobium* and *Dendrophthora*) more genera than expected exhibit the anemochorous and 'mixed' dispersal modes, respectively (chi-square test, $\chi^2 = 14.8$, d.f. = 3, $P \leq 0.002$ (Table 6.1)). The same pattern holds at the species level (chi-square test, $\chi^2 = 193.0$, d.f. = 3, $P \leq 0.0001$, anemochorous and 'mixed' dispersal modes pooled (Table 6.1)).

Within endozoochorous genera and species, *Loranthaceae* are genera-rich and *Viscaceae* are species-rich (goodness-of-fit test, $\chi^2 = 9.8$, d.f. = 2, $P \leq 0.002$ and $\chi^2 = 10.5$, d.f. = 2, $P \leq 0.001$, respectively (Table 6.1)). Three mutually non-exclusive hypotheses may explain this pattern. First, differences in the biogeographical origin and age of the lineages (Barlow, 1983) may have contributed to the differential accumulation of genera among mistletoe families. Secondly, differences among vectors (i.e. vector competence) may have contributed to the diversification of mistletoes in different ways. Thirdly, mistletoe families may differ in terms of their compatibility with vectors: that is, some mistletoe taxa may be more restrictive in terms of the vectors they attract.

Mistletoe diversity and vector-mediated dispersal

Local-scale assemblages

We recorded 33 bird species belonging to eight taxa (Columbidae, Tyrannidae, Pipridae, Vireonidae, Muscicapidae, Fringillinae, Thraupini and Cardinalini) feeding on mistletoe fruits in Colombia and Costa Rica (21 and 12 species, respectively (Fig. 6.2)). Even though mistletoes belonging to different species could be found parasitizing the same or neighbouring trees, they attracted distinctly different subsets of birds. In both sites, the proportion of visits contributed by each bird species differed significantly among mistletoe

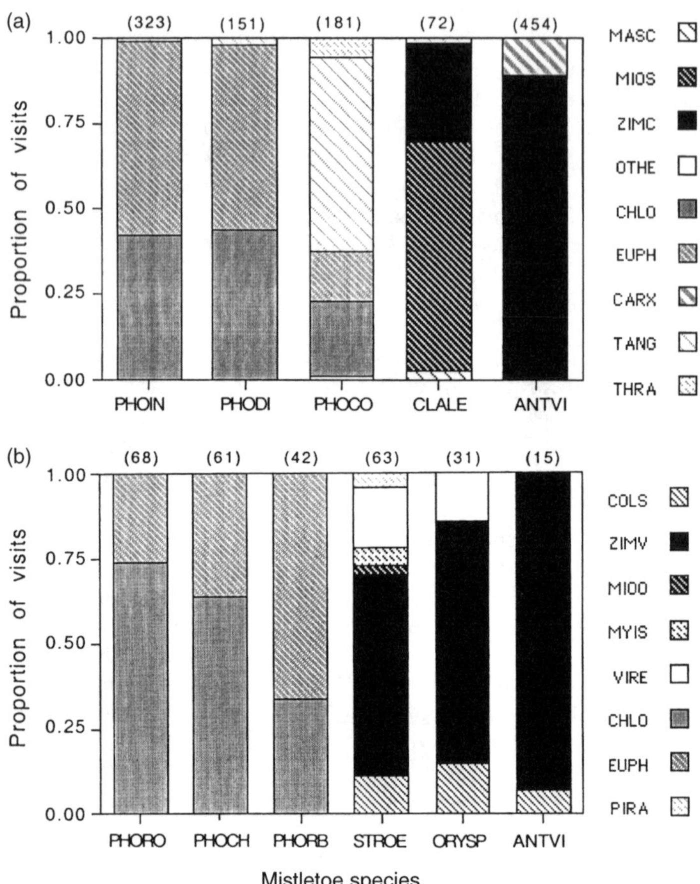

Fig. 6.2. Local assemblages of mistletoes and their avian dispersers. (a) Colombia. EUPH: *Euphonia* spp. (*E. xanthogaster*, *E. laniirostris* and *E. musica*); CHLO: *Chlorophonia* spp. (*C. pyrrophrys* and *C. cyanea*); TANG: *Tangara* spp. (*T. labradorides*, *T. nigroviridis*, *T. ruficervix*, *T. xanthocephala*, *T. arthus*, *T. heinei*, *T. cyanicollis* and *T. vitriolina*); THRA: Thraupini–other (*Chlorochryssa nitidissima* and *Anisognathus flavinucha*); MIOS: *Mionectes striaticollis*; ZIMC: *Zimmerius chrysops*; MASC: *Masius chrysopterus*; CARX: *Carduelis xanthogastra*; OTHE: *Entamodestes coracinnus* and *Cyclarhis nigrirostris*; PHOIN: *Phoradendron inaequidentatum*; PHODI: *Phoradendron dipterum*; PHOCO: *Phoradendron colombianum*; CLALE: *Cladocolea lenticellata*; ANTVI: *Antidaphne viscoidea*. (b) Costa Rica. CHLO: *Chlorophonia callophrys*; EUPH: *Euphonia* spp. (*E. luteicapilla*, *E. hirundinaceae* and *E. musica*); ZIMV: *Zimmerius villisimus*; COLS: *Columba subvinaceae*; VIRE: *Vireo* spp. (*V. flavifrons*, *V. leucophrys* and *V. philadelphicus*); PIRA: *Piranga flava*; MIOO: *Mionectes olivaceus*; MYIS: *Myiozetetes similis*; PHORO: *Phoradendron robustissimum*; PHOCH: *Phoradendron chrysocladon*; PHORB: *Phoradendron robaloense*; STROE: *Struthanthus oerstedii*; ORYSP: *Oryctanthus spicatus*. Number of species in parentheses.

taxa (chi-square test, $\chi^2 = 2551$, d.f. = 24, $P \leq 0.0001$ and chi-square test, $\chi^2 = 339$, d.f. = 25, $P \leq 0.0001$, for Colombia and Costa Rica, respectively (Fig. 6.2)). Partitioning the contingency table revealed large differences within and among mistletoe species in terms of the birds feeding on their fruits (Tables 6.3 and 6.4). First, species in the genus *Phoradendron* (*Viscaceae*) differed in the proportion of visits made by *Euphonia* and *Chlorophonia*. Secondly, species in the genus *Phoradendron* differed in the proportion of visits made by the *Euphonia*/*Chlorophonia* group and *Tangara* spp.; this difference became more noticeable

Table 6.3. Birds feeding on mistletoe fruits in Colombia. The 2 × 2 subtables were used to establish the contribution of bird and mistletoe taxa to the overall significant results. The modified χ^2 for each subtable is in italics; $^*P \leq 0.05$. Abbreviations as in Fig. 6.2. 'Other' includes *Chlorochryssa nitidissima*, *Anisognathus flavinucha* and *Masius chrysopoterus*. Rows and columns represent mistletoe and bird species, respectively. + indicates pooling of species during the generation of the 2 × 2 subtables.

	Euph	*Chlo*	*Euph + Chlo*	*Tang*	*Euph + Chlo + Tang*	*Carx*	*Euph + Chlo + Tang + Carx*	*Mios*	*Euph + Chlo + Tang + Carx + Mios*	*Zimc*	*Euph + Chlo + Tang + Carx + Mios + Zimc*	Other
Phoco	27	39	66	103	169	0	169	0	169	0	169	12
Phoin	185	136	321	2	323	0	323	0	323	0	323	0
	*6.4**		*502.1**		*0.1*		*0.1*		*0.3*		*43.5**	
Phoco +												
Phoin	212	175	387	105	492	0	492	0	492	0	492	12
Phodi	82	67	149	2	151	0	151	0	151	0	151	0
	0.0		*60.5**		*0.0*		*0.01*		*0.0*		*5.6**	
Phoco +												
Phoin +												
Phodi	294	242	536	107	643	0	643	0	643	0	643	12
Clale	0	0	0	1	1	0	1	48	49	21	70	2
	0.0		*0.1*		*8.0*		*803.4**		*24.7**		*0.5*	
Phoco +												
Phoin +												
Phodi +												
Clale	294	242	536	108	644	0	644	48	692	21	713	14
Antvi	0	0	0	0	0	50	50	0	50	404	454	0
	0.0		*0.0*		*196.1**		*1.0*		*890.7**		*0.8**	

Table 6.4. Birds feeding on mistletoe fruits in Costa Rica. The 2 × 2 subtables were used to establish the contribution of bird and mistletoe taxa to the overall significant results. The modified χ^2 for each subtable is in italics; $^*P \leq 0.05$. Abbreviations as in Fig. 6.2. Rows and columns represent mistletoe and bird species, respectively. + indicates pooling of species during the generation of the 2 × 2 subtables.

	Chlo	*Euph*	*Chlo +* *Euph*	*Cols*	*Chlo +* *Euph +* *Cols*	*Vire*	*Euph +* *Chlo +* *Cols +* *Vire*	*Pira +* *Mioo +* *Myis +*	*Euph +* *Chlo +* *Cols +* *Vire +* *Pira,* *Mioo,* *Myis*	*Zimv*
Phoro	50	18	68	0	68	0	68	0	68	0
Phoch	39	22	61	0	61	0	61	0	61	0
	2.0		*0.0*		*0.0*		*0.0*		*0.0*	
Phoro + *Phoch*	89	40	129	0	129	0	129	0	129	0
Phorb	14	28	42	0	42	0	42	0	42	0
	*27.5**		*0.0*		*0.0*		*0.0*		*0.0*	
Phoro + *Phoch +* *Phorb*	103	68	171	0	171	0	171	0	171	0
Stroe	0	0	0	7	7	11	18	8	26	37
	0.0		*32.3**		*48.5**		*37.6**		*82.4**	
Phoro + *Phoch +* *Phorb +* *Stroe*	103	68	171	7	178	11	189	8	197	37
Orysp	0	0	0	5	5	4	9	0	9	22
	0.0		*19.1**		*9.0**		*0.2*		*43.2**	
Phoro + *Phoch +* *Phorb +* *Stroe +* *Orysp*	103	68	171	12	183	15	198	8	206	59
Antvi	0	0	0	1	1	0	1	0	1	14
	0.0		*1.4**		*0.0*		*0.0*		*37.2**	

when the 'other' category (*Anisognathus flavinucha, Chlorochryssa nitidissima, Entamodestes coracinnus* and *Cyclarhis nigrirostris*) was included. These results demonstrate that subsets of species within *Phoradendron* may be associated with different bird taxa. Thirdly, when *Viscaceae* (*Phoradendron* species were pooled) were compared with *Loranthaceae* (*Cladocolea lenticellata, Struthanthus oerstedii* and *Oryctanhus spicatus*), it became clear that *Viscaceae* and *Loranthaceae* differed significantly in terms of the bird taxa associated with them. The addition of *Carduelis xanthogastra* (Fringillinae), Tyrannidae (*Mionectes* spp., *Zimmerius* spp., *Myiozetetes similis*), *Columba subvinaceae* (Columbidae) and Vireonidae (*Vireo* spp.) during the generation of the 2 × 2 subtables resulted in significant χ^2 values (Tables 6.3 and 6.4). This shows that there is little overlap between bird taxa visiting *Viscaceae* and *Loranthaceae.* Fourthly, we found that the *Viscaceae* and *Loranthaceae* differed significantly from *Eremolepidaceae* because of the high proportion of visits made by *Zimmerius* spp. to *Antidaphne viscoidea* at both sites. This suggests a high degree of association between *Zimmerius* spp. and *A. viscoidea.*

Regional-scale assemblages

A total of 221 species of vertebrates have been reported feeding on mistletoe fruits or dispersing their seeds: 95.5% are birds, 4.0% are mammals and 0.5% are fish. *Arceuthobium*, the mistletoe genus with the northernmost distribution, is associated with a small assemblage of vertebrates, which includes 28 bird and five mammal species (Appendix 6.1; appendices cited in this chapter are available from the authors upon request). Because of the prevalence of autochory and epizoochory in *Arceuthobium*, we excluded this genus from the analyses that follow.

We found a significant association between mistletoe families and the vertebrate taxa feeding on their fruits (chi-square test, $\chi^2 = 17.0$, d.f. = 6, $P \leq 0.008$ (Fig. 6.3 and Table 6.5)). Partitioning of the contingency table showed that this pattern was largely due to differences between *Loranthaceae–Eremolepidaceae* and *Viscaceae*. In fact, more NONP/PSOS species feed on *Loranthaceae–Eremolepidaceae* than expected by chance, whereas an equal proportion of POSC species feed on fruits of *Loranthaceae–Eremolepidaceae* and *Viscaceae* (Table 6.5).

Most records of vertebrates feeding on mistletoes are of passerine birds. Within passerines we found a significant association between mistletoes and high-order passerine

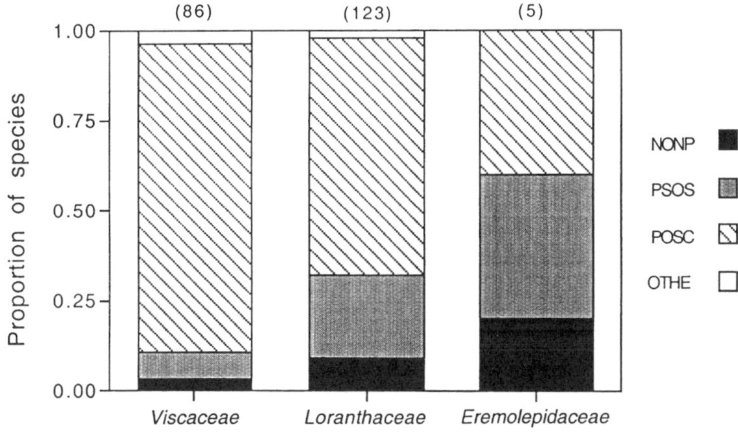

Fig. 6.3. Regional assemblages of mistletoes, *Viscaceae*, *Loranthaceae* and *Eremolepidaceae* (excluding *Arceuthobium*), and the vertebrates feeding on their fruits. NONP: non-passerine; PSOS: passerine suboscines; POSC: passerine oscines; OTHE: mammals and fish. Number of species in parentheses.

Table 6.5. Vertebrates feeding on mistletoe fruits in the New World. The 2 × 2 subtables were used to establish the contribution of bird and mistletoe taxa to the overall significant results. The modified χ^2 for each subtable is in italics. $^*P \leq 0.05$. Abbreviations as in Fig. 6.3. Rows and columns represent mistletoe and bird species, respectively. + indicates pooling of species during the generation of the 2 × 2 subtables.

	NONP	PSOS	NONP + PSOS	POSC	NONP + PSOS + POSC	OTHE
Loranthaceae	12	28	40	81	121	2
Eremolepidaceae	1	2	3	2	5	0
	0.0		*1.9*		*0.0*	
Loranthaceae + *Eremolepidaceae*	13	30	43	83	126	2
Viscaceae	3	6	9	74	83	3
	0.0		*14.4**		*0.8*	

Table 6.6. Passerine species feeding on *Loranthaceae* and *Viscaceae* fruits. Numbers indicate number of bird species. Bird taxa with small samples sizes are indicated by – in the χ^2 column. Taxa (subfamilies/tribes) in grey were pooled for analyses. $^*P \leq 0.05$, $^{**}P \leq 0.01$.

Suborder	Passerine taxa	*Loranthaceae*	*Viscaceae*	χ^2
Suboscines	Cotingidae	5	0	4.6*
	Pipridae	4	2	–
	Tyrannidae	19	4	9.8**
Oscines	Corvidae	3	1	6.2*
	Vireonidae	8	1	–
	Bombycillidae	5	4	0.1
	Turdinae	6	9	0.6
	Mimini	6	5	0.1
	Certhiidae	2	1	–
	Paridae	1	0	–
	Carduelini	3	1	–
	Euphonia/Chlorophonia	5	16	5.8*
	Emberizini	0	2	6.2*
	Cardinalini	5	1	–
	Icterini	3	0	–
	Parulini	5	0	–
	Thraupini	29	33	0.3

taxa (Table 6.6). First, species in the Cotingidae/Pipridae, Tyrannidae, Corvoidea (Corvidae and Vireonidae) and Emberizini/Cardinalini/Icterini/Parulini feed more often than expected on *Loranthaceae* fruits. Secondly, *Euphonia/Chlorophonia* species have been recorded more often than expected feeding on *Viscaceae* fruits. Thirdly, species in the Bombycillidae, Turdinae, Minimi and Thraupini feed with equal frequency on fruits of *Viscaceae* and *Loranthaceae*.

Discussion

Results of this study support the hypothesis that vector–parasite interactions have contributed to the diversification of New World mistletoes. First, we found a strong association between mistletoe diversity and dispersal mode. In particular, species-poor mistletoe taxa are anemochorous or autochorous, whereas species-rich taxa are endozoochorous. Secondly, we found that mistletoe families are associated with a narrow subset of vertebrate taxa; the mistletoe families, however, differ in terms of the vertebrates with whom they are associated. Thirdly, these associations have an important phylogenetic component, which may help to establish the origin of these associations either through colonization and vector switching or through descent and vector specificity.

Mistletoe diversity and dispersal mode

Anemochory is found only in *Misodendraceae*, a small family restricted to the Andes of Argentina and Chile (Orfila, 1978). The low diversity of *Misodendraceae* and its restricted distribution may result from dispersal limitation and a high degree of host specificity. In fact, the only reported hosts are species of *Nothofagus* (Orfila, 1978). Recent work indicates that seeds of *Misodendraceae* disperse over short distances (maximum 10 m from the parent (N. Tercero, personal communication)) and that there are strong preferences for hosts within *Nothofagus* stands (Vidal Russell, 2000). Specifically, it has been shown that *Nothofagus* stands that are genetically diverse are heavily parasitized by *Misodendrum* spp. (Vidal Russell, 2000).

Autochory, in combination with epizoochory and endozoochory – our 'mixed' dispersal mode – has probably been important in the diversification of mistletoes. Epizoochory can result in the movement of seeds over long distances; in fact, a small but important proportion of species on oceanic islands produce fleshy fruits with 'sticky' seeds that are carried on the feathers of birds (Carlquist, 1967). All mistletoe genera exhibiting this dispersal mode belong to the *Viscaceae* (*Arceuthobium*, *Dendrophthora*, *Korthalsella* and *Notothixos*, the latter two being Old World genera) (Leiva and Bisse, 1983; Liddy, 1983; Kuijt, 1987; Sargent, 1994; Hawksworth and Wiens, 1996). This characteristic, in combination with the observation that the *Viscaceae* is the only mistletoe family consistently found on continental and oceanic islands (Kuijt, 1961; Barlow, 1983), supports the idea that the 'mixed' dispersal mode has contributed to the diversification of mistletoes, possibly through 'founder' effects.

The 'mixed' dispersal mode and its long-distance dispersal component are diverse; seeds may get dispersed when they either attach to the feathers of birds or when fruits discharge seeds (*Arceuthobium*) (Hawksworth and Wiens, 1996) or are dispersed when fruits are manipulated in birds' bills (*Arceuthobium bicarinatum*, *Arceuthobium verticilliflorum*, *Dendrophthora corynarthron* and *Dendrophthora cupressoides*) (Etheridge, 1971; Leiva and Bisse, 1983; Sargent, 1994; Hawksworth and Wiens, 1996). Fruits of *D. corynarthron*, for example, are routinely taken by *Chlorophonia callophrys* in Costa Rica and their single seed can be ejected up to 1 m when the bird squeezes the fruit in its bill (S. Sargent and S. Mitra, unpublished data).

Mistletoe diversity and vector-mediated dispersal

Most mistletoe species are endozoochorous, and locally are associated with a narrow subset of birds. Furthermore, vector–mistletoe associations found within local assemblages seem to reflect a long-term history of associations, as suggested by the analyses of regional assemblages of vertebrates and mistletoes. We found a significant association between high-order vertebrate and mistletoe taxa, suggesting that vector–mistletoe associations resulted from colonization–vector switching or from descent–vector specificity. The signatures for associations by colonization and descent are incongruent and congruent phylogenies, respectively.

Loranthaceae–*vector associations: a case of vector switching?*

Two observations suggest that vector switching may explain the origin and maintenance of *Loranthaceae*–vector associations and thus the diversification of these mistletoes. First, New World *Loranthaceae* derive from one of two Gondwana lineages that underwent extensive radiation once South America separated from Antarctica, some time in the middle Cretaceous (Barlow, 1983). Secondly, the *Loranthaceae* are consumed and dispersed by both PSOS (in particular, Tyrannidae and Cotingidae) and POSC (in particular, Thraupini and Vireonidae), two groups of birds that do not share a common history in the New World.

The suboscines (PSOS) also have a Gondwana origin and represent the oldest lineage within the passerines (late Cretaceous to middle Tertiary) (Mayr, 1964; Sibley and Ahlquist, 1990; Boles, 1995; Feduccia, 1999; Raikow and Bledsoe, 2000). The largest radiation of PSOS occurred in South America, after which they dispersed into Central America and southern North America and underwent a secondary radiation during the Pliocene (Mayr, 1964; Sibley and Ahlquist, 1990). The oscine passerines (POSC), on the other hand, have a Laurasian origin and appeared in the American continent much later than the PSOS (Oligocene–middle Miocene) (Sibley and Ahlquist, 1990; Burns, 1997; Feduccia, 1999). They entered America through the Bering Strait and the North Atlantic land bridge. Vireonidae represent an exception to this pattern. They arrived in South America via Antarctica, where they radiated and moved north, experiencing a secondary radiation (*Vireo*) in North America during the Pliocene (Sibley and Ahlquist, 1990).

We postulate that associations between *Loranthaceae* and PSOS have a long history, which originated in South America and remained restricted to that continent until connections between South America and the Antilles (Eocene–Oligocene) and North

America (Pliocene) were established (Iturralde-Vinent and MacPhee, 1999). The connection between South America and the Antilles, however, was brief. During the Pliocene, genera within the *Loranthaceae* became associated with POSC birds. Furthermore, we postulate that associations between *Loranthaceae* and PSOS originated through associations by descent and vector specificity and those between *Loranthaceae* and POSC through association by colonization and then vector switching. The combination of these two processes may explain the generic diversity of *Loranthaceae*.

Although vector switching may explain associations between *Loranthaceae* and their vectors, we cannot exclude the possibility of a spurious effect resulting from unresolved relationships among New World *Loranthaceae*. More precisely, two lines of evidence suggest that New World *Loranthaceae* may have derived from, in addition to the South American Gondwana lineage, a second lineage derived from Laurasian stocks. First, three genera in the *Loranthaceae* (*Gaiadendron*, *Aetanthus* and *Psittacanthus*) are significantly associated with POSC: of 47 bird genera, 33 in the POSC feed on these mistletoes, with only eight in the PSOS and five in the NONP (Appendix 6.3). Secondly, fossil pollen of several *Loranthaceae* (*Loranthus* spp. from North American and European Eocene deposits and *Aetanthus* sp. from a Puerto Rican Oligocene deposit) indicate that *Loranthaceae* were present in North America before the connection between South and North America was established (Graham and Jarzen, 1969; Muller, 1981; Taylor, 1990). More recent records of fossil pollen from Caribbean basin upper Miocene and Pliocene deposits include *Aetanthus*, *Oryctanthus*, cf. *Psittacanthus* and cf. *Sruthanthus* (Graham and Jarzen, 1969; Graham, 1990, 1991; Graham and Dilcher, 1998), substantially predating the central American land bridge, which formed in the Pliocene.

Viscaceae–*vector associations: a case of vector specificity?*

Two observations suggest that vector specificity may explain the origin and maintenance of *Viscaceae*–vector associations in the New World. First, New World *Viscaceae* derive from a common Laurasian ancestor that reached the North American continent through the Bering strait during the early Tertiary (Barlow, 1983; Kuijt, 1988; Nickrent *et al.*, 1998). Secondly, species in the *Viscaceae* (excluding *Arceuthobium*) are associated with a homogeneous assemblage of vertebrate vectors, the vast majority POSC; these include Thraupini and *Euphonia*/ *Chlorophonia*, plus a few records among Bombycillidae, Turdinae and Mimini.

POSC, like the *Viscaceae*, entered the New World from Laurasia sometime in the Oligocene–middle Miocene (Sibley and Ahlquist, 1990; Burns, 1997). The POSC, in particular the nine-primaried oscines, then experienced an explosive radiation (Raikow and Bledsoe, 2000), which mirrors that of the *Viscaceae*. Thus, we postulate that the association between *Viscaceae* and POSC is relatively recent, originating in North America and the Caribbean and reaching its greatest diversity in Central and South America in recent times. The non-monophyly of *Phoradendron* and *Dendrophthora* (Ashworth, 2000), in combination with our data on local mistletoe–bird assemblages, suggests that vector specificity may have contributed to the diversification of *Viscaceae* in terms of species numbers.

Eremolepidaceae–*vector associations*

The associations between *Antidaphne* spp. and their vectors are intriguing and may indicate a case of extreme vector specificity. Our data, although limited to *Antidaphne* (*A. viscoidea*), indicate a high degree of specificity between this family and its vectors. In Colombia and Costa Rica, *A. viscoidea* is mostly associated with *Zimmerius* spp. (*Z. chrysops* and *Z. villisimus*, respectively; PSOS – Tyrannidae). Also, *Antidaphne* is often parasitized by *Ixocactus*, a genus dispersed by *Z. chrysops* (C. Restrepo, unpublished data). The high degree of association between *Zimmerius* spp. and *Antidaphne* may be tied to a South American origin of *Zimmerius* spp. (Ridgely and Tudor, 1994) and *Antidaphne* (Kuijt, 1988). This extreme case of vector specificity may explain why the *Eremolepidaceae* are much less diverse than *Viscaceae* and *Loranthaceae*.

Mistletoe–vertebrate associations: the next step

Our work has generated several hypotheses about vector–parasite-mediated diversification of New World mistletoes. In addition, it has revealed that, at local scales, mistletoe–vertebrate associations appear constrained by the long history of these associations. Two broad questions should be addressed to further explain the diversification of mistletoes through vector–parasite interactions. First, to what degree can vector switching and vector specificity explain the diversification of mistletoes at the generic level? Secondly, what is the potential for evolutionary change among those traits that mediate vector–mistletoe interactions?

To address the first question, we suggest a macroevolutionary approach similar to the one developed here. This will require new data on mistletoe–vertebrate associations and mistletoe fossils, the ordination of mistletoe taxa based on their associated vectors (C. Restrepo and D. Levey, unpublished data), the generation of phylogenetic trees for both mistletoes and vertebrates and information on fruit- and seed-handling methods. Geographical regions for which data on mistletoe–vertebrate associations are badly needed include Mexico, the Antilles, Brazil's *mata Atlantica* and *cerrado* ecosystems and the temperate forests of Chile. For example, are other genera in the *Eremolepidaceae* associated with PSOS? How do vector–mistletoe associations map on to the biogeographical history of the Antilles?

To address the second question, we suggest study of traits that are under selection and that influence the dispersal of parasites. This includes assessing the fitness benefits for both mistletoes and vectors. In vector–parasite systems, the traits include parasite virulence and compatibility and vector competence and resistance (Collins *et al.*, 1986; Yan *et al.*, 1997; Failloux *et al.*, 1999). For example, do mistletoe fruits have compounds, both nutritive and toxic, that affect the behaviour of their vertebrate dispersers? Can mistletoe seeds survive processing in the guts of most vertebrates? Do vector physiology and behaviour affect the chance of mistletoes being dispersed to suitable hosts and suitable substrates within hosts? Do vectors overcome the toxicity of mistletoe fruits? Most of these questions remain relatively unexplored.

Conservation and management implications

Mistletoes are often regarded as pests and, likewise, birds feeding on mistletoe berries are considered a nuisance in anthropogenic ecosystems and landscapes. Yet, in many ecosystems, particularly those in which plants bearing non-fleshy fruits are dominant, mistletoes may not only represent a food resource for vertebrates but may serve as 'hot spots' for the recruitment of other plants bearing fleshy fruits. In this regard, mistletoe–vector associations can contribute significantly to ecosystem diversity and ecosystem function.

Acknowledgements

We are indebted to M. Aizen, G. Amico, S. Arango, M.P. Velasquez, H. Alvarez, A. Cruz, L. Lopez del Buen, C. Martínez del Río, J.F. Ornelas, C. Samper, C. Smith and R. Vidal for kindly sharing unpublished data with us. C. Bosque, J. Lambert, D. Nickrent and especially J. Kuijt provided useful comments on early versions of the manuscript.

References

Ashworth, V.E.T.M. (2000) Phylogenetic relationships in Phoradendrae (*Viscaceae*) inferred from three regions of the nuclear ribosomal cistron. I. Major lineages and paraphyly of *Phoradendron*. *Systematic Botany* 25, 349–370.

Azad, A.F. and Beard, C.B. (1998) Rickettsial pathogens and their arthropod vectors. *Emerging Infectious Diseases* 4, 179–186.

Barlow, B.A. (1964) Classification of the *Loranthaceae* and *Viscaceae*. *Proceedings of the Linnean Society of New South Wales* 89, 268–272.

Barlow, B.A. (1983) Biogeography of *Loranthaceae* and *Viscaceae*. In: Calder, M. and Bernhardt, P. (eds) *The Biology of Mistletoes*. Academic Press, Sydney, Australia, pp. 19–46.

Barlow, B.A. and Wiens, D. (1977) Host–parasite resemblance in Australian mistletoes. *Evolution* 31, 69–84.

Bhandari, N.N. and Vohra, S.C.A. (1983) Embryology and affinities of *Viscaceae*. In: Calder, M. and Bernhardt, P. (eds) *The Biology of Mistletoes*. Academic Press, New York, pp. 69–86.

Bhatnagar, S.P. and Johri, B.M. (1983) Embryology of *Loranthaceae*. In: Calder, M. and Bernhardt, P. (eds) *The Biology of Mistletoes*. Academic Press, New York, pp. 47–67.

Boles, W.W. (1995) A preliminary analysis of the Passeriformes from Riversleigh, northwestern Queensland, Australia, with the description of a new species of lyrebird. *Courier Forschunginstitut Senckenberg* 181, 163–170.

Brooks, D.R. (1988) Macroevolutionary comparisons of host and parasite phylogenies. *Annual Review of Ecology and Systematics* 19, 235–259.

Brooks, D.R. and McLennan, D.A. (1993) Historical ecology: examining phylogenetic components of community evolution. In: Ricklefs, R.E. and Schluter, D. (eds) *Species Diversity in Ecological Communities: Historical and Geographical Perspectives*. University of Chicago Press, Chicago, Illinois, pp. 267–280.

Burns, K.J. (1997) Molecular systematics of Tanagers (Thraupinae): evolution and biogeography of a diverse radiation of Neotropical birds. *Molecular Phylogenetics and Evolution* 8, 334–348.

Calder, M. and Bernhardt, P. (1983) *The Biology of Mistletoes*. Academic Press, New York, 348 pp.

Carlquist, S. (1967) The biota of long-distance dispersal. V. Plant dispersal to Pacific islands. *Bulletin of the Torrey Botanical Club* 94, 129–162.

Carreno, R.A., Kissinger, J.C., McCutchan, T.F. and Barta, J.R. (1997) Phylogenetic analysis of haemosporinid parasites (Apicomplexa: Haemosporina) and their coevolution with vectors and intermediate hosts. *Archiv für Protisten Kunde* 148, 245–252.

Chiykowski, L.N. (1981) Epidemiology of diseases caused by leafhopper-borne pathogens. In: Maramorosch, K. and Harris, K.F. (eds) *Plant Diseases and Vectors: Ecology and Epidemiology*. Academic Press, New York, pp. 106–161.

Collins, F.H. and Besansky, N.J. (1994) Vector biology and the control of malaria in Africa. *Science* 264, 1874–1875.

Collins, F.H., Sakai, R.K., Vernick, K.D., Paskewitz, S., Seeley, D.C., Miller, L.H., Collins, W.E., Campbell, C.C. and Gwadz, R.W. (1986) Genetic selection of a *Plasmodium*-refractory strain of the malaria vector *Anopheles gambiae*. *Science* 234, 607–610.

Davidar, P. (1987) Fruit structure in two neotropical mistletoes and its consequences for seed dispersal. *Biotropica* 19, 137–139.

Davis, G.M. (1992) Evolution of prosobranch snails transmitting Asian *Schistosoma*; coevolution with *Schistosoma*: a review. *Progress in Clinical Parasitology* 3, 145–204.

Ehleringer, J.R. and Marshall, J.D. (1995) Water. In: Press, M.C. and Graves, J.D. (eds) *Parasitic Plants*. Chapman & Hall, London, pp. 125–140.

Etheridge, D.E. (1971) *Inventario y fomento de los recursos forestales, Republica Dominicana*. FO:SF/DOM 8, Informe Técnico 1, FAO, Rome, Italy, 27 pp.

Failloux, A.-B., Vazeille-Falcoz, M., Mousson, L. and Rodhain, F. (1999) Contrôle génétiqué de la compétence vectorielle des moustiques du genre *Aedes*. *Bulletin de la Société de Pathologie Exotique* 92, 266–273.

Feduccia, A. (1999) *The Origin and Evolution of Birds*. Yale University Press, New Haven, Connecticut, 466 pp.

Fineran, B.A. and Hocking, P.J. (1983) Features of parasitism, morphology, and haustorial anatomy in Loranthaceaous root parasites. In: Calder, M. and Bernhardt, P. (eds) *The Biology of Mistletoes*. Academic Press, Sydney, Australia, pp. 205–227.

Fisher, J.T. (1983) Water relations of mistletoes and their hosts. In: Calder, M. and Bernhardt, P. (eds) *The Biology of Mistletoes*. Academic Press, Sydney, Australia, pp. 161–184.

Frochot, H. and Sallé, G. (1980) Modalités de dissémination et d'implantation du gui. *Revue Forestiere Francaise* 32, 505–519.

Gedalovich, E. and Kuijt, J. (1987) An ultrastructural study of the viscin tissue of *Phthirusa pyrifolia* (H. B. K.) Eichler (*Loranthaceae*). *Protoplasma* 137, 145–155.

Graham, A. (1990) Late Tertiary microfossil flora from the Republic of Haiti. *American Journal of Botany* 77, 911–926.

Graham, A. (1991) Studies in neotropical paleobotany. IX. The Pliocene communities of Panama – Angiosperms (dicots). *Annals of the Missouri Botanical Garden* 78, 201–223.

Graham, A. and Dilcher, D.L. (1998) Studies in neotropical paleobotany. XII. A palynoflora from the Pliocene Rio Banano Formation of Costa Rica and the Neogene vegetation of Mesoamerica. *American Journal of Botany* 85, 1426–1438.

Graham, A. and Jarzen, D.M. (1969) Studies in neotropical paleobotany. I. The Oligocene communities of Puerto Rico. *Annals of the Missouri Botanical Garden* 56, 308–357.

Haber, W.A. (2000) Plants and vegetation. In: Nadkarni, N.M. and Wheelwright, N.T. (eds) *Monteverde: Ecology and Conservation of a Tropical Cloud Forest.* Oxford University Press, New York, pp. 39–94.

Haverschmidt, F. (1968) *Birds of Surinam.* Oliver and Boyd, Edinburgh, UK, 445 pp.

Hawksworth, F.G. and Wiens, D. (1996) *Dwarf Mistletoes: Biology, Pathology, and Systematics.* US Department of Agriculture, Forest Service, Washington, DC, 410 pp.

Herrera, C.M. (1984) Avian interference of insect frugivory: an exploration into the plant–bird–fruit pest evolutionary triad. *Oikos* 42, 203–210.

Hoberg, E.P., Brooks, D.R. and Siegel-Causey, D. (1997) Host-parasite co-speciation: history, principles, and prospects. In: Clayton, D.H. and Moore, J. (eds.) *Host–Parasite Evolution: General Principles and Avian Models.* Oxford University Press, New York, pp. 212–235.

Hoehne, F.C. (1931) Algo sobre a ecologia do *Phrygilanthus eugenioides* (H. B. K.) Eichl. *Bolletin Agricultura, Secretaria da Agricultura Industria e Commercio Estado São Paulo* 32, 258–290.

Hoffman, A.J., Fuentes, E.R., Cortes, I., Liberona, F. and Costa, V. (1986) *Tristerix tetrandrus* (*Loranthaceae*) and its host plants in the Chilean matorral: patterns and mechanisms. *Oecologia* 69, 202–206.

Iturralde-Vinent, M.A. and MacPhee, R.D.E. (1999) Paleogeography of the Caribbean region: implications for Cenozoic biogeography. *Bulletin of the American Museum of Natural History* 238, 1–95.

Kim, K.C. (1985) Parasitism and coevolution. In: Kim, K.C. (ed.) *Coevolution of Parasitic Arthropods and Mammals.* John Wiley & Sons, New York, pp. 661–682.

Klicka, J., Johnson, K.P. and Lanyon, S.M. (2000) New World nine-primaried oscine relationships: constructing a mitochondrial DNA framework. *The Auk* 117, 321–336.

Kuijt, J. (1961) A revision of *Dendrophthora* (*Loranthaceae*). *Wentia* 6, 1–145.

Kuijt, J. (1963) On the ecology and parasitism of the Costa Rican tree mistletoe, *Gaiadendron punctatum* (Ruiz & Pavon) G Don. *Canadian Journal of Botany* 41, 927–938.

Kuijt, J. (1968) Mutual affinities of Santalalean families. *Brittonia* 20, 136–147.

Kuijt, J. (1969) *The Biology of Parasitic Flowering Plants.* University of California Press, Berkeley, California, 246 pp.

Kuijt, J. (1987) Miscellaneous mistletoe notes, 10–19. *Brittonia* 39, 447–459.

Kuijt, J. (1988) Monograph of the *Eremolepidaceae. Systematic Botany Monographs* 18, 1–60.

Lamont, B. (1983) Mineral nutrition of mistletoes. In: Calder, C. and Bernhardt, P. (eds) *The Biology of Mistletoes.* Academic Press, Sydney, Australia, pp. 185–204.

Leiva, A. and Bisse, J. (1983) Nuevo género de *Loranthaceae* para la flora de Cuba: *Arceuthobium* M. Bieb. *Revista del Jardín Botánico Nacional de Cuba* 4, 57–67.

Liddy, J. (1983) Dispersal of Australian mistletoes: the Cowiebank study. In: Calder, M. and Bernhardt, P. (eds) *The Biology of Mistletoes.* Academic Press, Sydney, Australia, pp. 101–116.

Luke, J., Jirkú, M., Dolezel, D., Kral'ová, I., Hollar, L. and Maslov, D.A. (1997) Analysis of ribosomal RNA genes suggests that trypanosomes are monophyletic. *Journal of Molecular Evolution* 44, 521–527.

Martínez del Río, C., Silva, A., Medel, R. and Hourdequin, M. (1996) Seed dispersers as disease vectors: bird transmission of mistletoe seeds to plant hosts. *Ecology* 77, 912–921.

Mayr, E. (1964) Inferences concerning the Tertiary American bird faunas. *Proceedings of the National Academy of Sciences USA* 51, 280–288.

Muller, J. (1981) Fossil pollen records of extant Angiosperms. *Botanical Review* 47, 1–142.

Nickrent, D.L. (2001) Mistletoe phylogenetics: current relationships gained from analysis of DNA sequences. In: Geils, B. and Mathiasen, R. (eds) *Western International Forest Disease Work Conference.* USDA Forest Service, Kona, Hawaii.

Nickrent, D.L., Duff, R.J., Colwell, A.E., Wolfe, A.D., Young, N.D., Steiner, K.E. and dePamphilis, C.W. (1998) Molecular phylogenetic and evolutionary studies of parasitic plants. In: Soltis, D., Soltis, P. and Doyle, J. (eds) *Molecular Systematics of Plants II. DNA Sequencing.* Kluwer Academic Publishers, Boston, Massachusetts, pp. 211–241.

Norton, D.A. and Carpenter, M.A. (1998) Mistletoes as parasites: Host specificity and speciation. *Trends in Ecology and Evolution* 13, 101–105.

Orfila, E.N. (1978) *Misodendraceae de la Argentina y Chile.* Fundación Elías y Ethel Malamud, Buenos Aires, Argentina, 73 pp.

Page, R.D.M. (1994) Parallel phylogenies: reconstructing the history of host–parasite assemblages. *Cladistics* 10, 155–173.

Paterson, A.M. and Gray, R.D. (1997) Host–parasite cospeciation, host-switching or the missing boat? In: Clayton, D.H. and Moore, J. (eds) *Host–Parasite Evolution: General Principles and Avian Models.* Oxford University Press, Oxford, pp. 236–250.

Polhill, R.M. and Wiens, D. (1998) *Mistletoes of Africa.* Royal Botanical Gardens, Kew, Richmond, UK, 370 pp.

Price, P. (1980) *Evolutionary Biology of Parasites.* Princeton University Press, Princeton, New Jersey, 237 pp.

Raikow, R.J. and Bledsoe, A.H. (2000) Phylogeny and evolution of the passerine birds. *BioScience* 50, 487–499.

Reid, N. (1991) Coevolution of mistletoes and frugivorous birds. *Australian Journal of Ecology* 16, 457–469.

Restrepo, C. (1987) Aspectos ecológicos de la diseminación de cinco especies de muérdagos por aves. *Humboldtia* 1, 1–116.

Ridgely, R.S. and Tudor, G. (1994) *The Birds of South America.* University of Texas, Austin, Texas, USA, 84 pp.

Sallé, G. (1983) Germination and establishment of *Viscum album* L. In: Calder, M. and Bernhardt, P. (eds) *The Biology of Mistletoes.* Academic Press, New York, pp. 145–159.

Sargent, S. (1994) Seed dispersal of mistletoes by birds in Monteverde, Costa Rica. PhD dissertation, Cornell University, Ithaca, New York, USA, 193 pp.

Sargent, S. (1995) Seed fate in a tropical mistletoe: the importance of host twig size. *Functional Ecology* 9, 197–204.

Sibley, C.G. and Ahlquist, J.E. (1990) *Phylogeny and Classification of Birds.* Yale University Press, New Haven, Connecticut, 976 pp.

Siegel, S. and Castellan, N.J.J. (1988) *Non-parametric Statistics for the Behavioural Sciences.* McGraw-Hill, New York, USA, 399 pp.

Taylor, D.W. (1990) Paleobiogeographic relationships of angiosperms from the Cretaceous and early Tertiary of the North American area. *Botanical Review* 56, 279–417.

Vidal Russell, R. (2000) Evidencias de resistencia en *Nothofagus a Misodendrum*: patrones de infección y consecuencias sobre la estructura genética de la planta parásita. BSc, Universidad Nacional del Comahue, Bariloche, Argentina.

Walsberg, G.E. (1975) Digestive adaptations of *Phainopepla nitens* associated with the eating of mistletoe berries. *The Condor* 77, 169–174.

Yan, G., Severson, D.W. and Christensen, B.M. (1997) Costs and benefits of mosquito refractoriness to malaria parasites: implications for genetic variability of mosquitoes and genetic control of malaria. *Evolution* 51, 441–450.

7 Mistletoes as Parasites and Seed-dispersing Birds as Disease Vectors: Current Understanding, Challenges and Opportunities

Juliann E. Aukema[1] and Carlos Martínez del Rio[2]

[1]Department of Ecology and Evolutionary Biology, University of Arizona, Tucson, AZ 85721-0033, USA; [2]Department of Zoology and Physiology, University of Wyoming, Laramie, WY 82071-3166, USA

Introduction

Most mistletoes are vector-borne parasites whose vectors are their avian seed-dispersers (Martínez del Rio *et al.*, 1996). In most vector-borne parasites and diseases, the vector maintains a parasitic or, at best, a commensal relationship with the parasite (Price, 1980). Mistletoes are unique among vector-borne parasites because they maintain a mutualistic interaction with their vectors (Martínez del Rio *et al.*, 1996; Fig. 7.1). Birds obtain nutrients, energy and, in the desert, water from mistletoes (Walsberg, 1975; Reid, 1991). In turn, mistletoes receive directed movement of their propagules into safe germination sites (Reid, 1991).

Because of the apparently specialized nature of the interaction between mistletoes and birds, the dispersal of mistletoes has received considerable attention (Cowles, 1936; Reid, 1991; Overton, 1994; Martínez del Rio *et al.*, 1995; Sargent, 1995; Larson, 1996). Here we attempt to place the interaction between birds, mistletoes and host plants in a broad context. We argue that mistletoes present unique opportunities to integrate seed-dispersal ecology with several other, seemingly disparate, areas of biology, such as plant physiology, parasitology and metapopulation ecology. We also contend that the biology of mistletoes makes them well suited for developing and testing models of how seed dispersal shapes the spatial and temporal dynamics of plant populations. To emphasize the connection between seed dispersal and parasitism, we use parasitology terminology, such as 'prevalence' and 'intensity' of infection to refer to mistletoe infection frequency and the number of mistletoes per host, respectively (Price, 1980). The ideas presented here were shaped by our research on two desert mistletoes: *Tristerix aphyllus* (*Loranthaceae*) and *Phoradendron californicum* (*Viscaceae*). The natural history of these two species is described in detail in Martínez del Rio *et al.* (1995) and Larson (1996). Briefly, *T. aphyllus* infects several species of columnar cacti in semiarid regions of Chile. Its seeds are dispersed primarily by the Chilean mockingbird (*Mimus thenca*).

 Seed Dispersal and Frugivory: Ecology, Evolution and Conservation (eds D.J. Levey, W.R. Silva and M. Galetti)

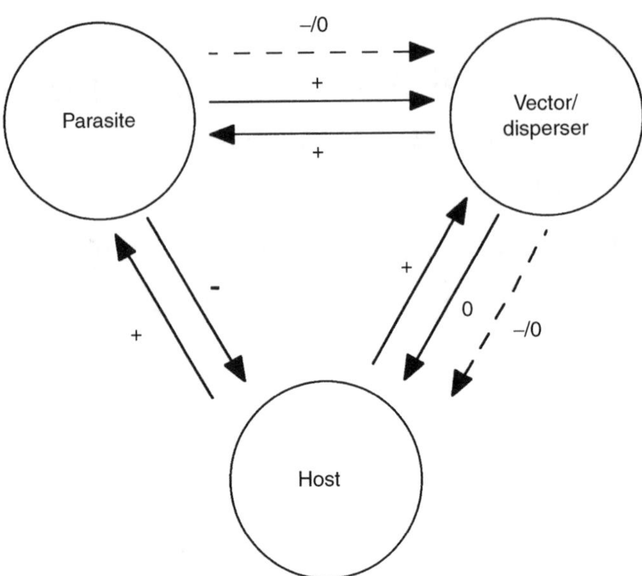

Fig. 7.1. Relationships between mistletoes (parasites), their hosts and the birds (vectors) that disperse mistletoe seeds. Solid arrows indicate the direction of interaction between two participants of the mistletoe system, and the signs above the arrows indicate whether the interaction is beneficial (+), detrimental (–) or has no effect (0) on the participant at the arrow's head. The mistletoe is both a parasite of its host plant and a mutualist of the birds that feed on its berries and disperse its seeds. These birds act both as seed-dispersers and as disease vectors. Dashed arrows indicate interactions in which most other parasites differ from mistletoes. In many parasitic systems, the parasite has a negative or neutral (–/0) effect on its vector. Also, the mistletoe dispersers do not have a direct negative effect on the host, as do many vectors that are themselves parasites, such as mosquitoes or ticks.

Phoradendron californicum infects leguminous trees and shrubs in arid environments from the south-western USA to central Mexico. Its seeds are dispersed primarily by phainopeplas (*Phainopepla nitens*).

First, we outline patterns of variation in mistletoe infection. This section identifies the contrasting scales at which these patterns are found and the mechanisms that shape them. In a second section, we propose a unified framework for the study of mistletoe populations. This framework emphasizes the role of mistletoes as plant parasites, recognizes that they have a patchy metapopulation-like structure and accentuates the fact that their seeds are dispersed by mutualistic birds. The two final sections identify areas that we believe can increase our understanding of mistletoe–host–vector systems, and summarize our primary conclusions.

Mistletoes at Different Scales: Patterns and Mechanisms

Like many other parasites, the distribution of mistletoes among individual hosts is often heavily clumped (Overton, 1996). Why are some individual hosts more intensely parasitized than others? If the site contains several potential host species, the frequency of parasitism may differ among host species (Lamont, 1982; J. Aukema and C. Martínez del Rio, unpublished data). Why does the prevalence of mistletoes differ among host species?

At a larger spatial scale, the overall prevalence and intensity of infection may vary among sites (Overton, 1996). Why are mistletoes more abundant in some sites than in others? Providing mechanistic answers to these questions requires consideration of all the steps in the life history of mistletoes: seed rain, seed establishment, mistletoe persistence and mistletoe reproductive success.

Seed rain, seedling establishment, mistletoe persistence and reproductive success

The actions of seed-dispersers can lead to differential deposition of seeds among hosts or sites. Birds may choose to perch on individual hosts and may move preferentially among sites with predictable characteristics. Seedling establishment is the next step that can lead to variation in mistletoe prevalence and intensity among hosts and sites. Once a seed is deposited on to a host and germinates, the probability that the seedling will become established is dependent on the match between the characteristics of the host and those of the mistletoe. After germination, successful mistletoe seeds establish an intimate haustorial connection with their hosts (Yan, 1993, and references therein). The growth, survival and reproductive output of mistletoes depend, to a large extent, on their success at using this connection to tap their host's resources.

Mistletoes, parasitism and metapopulations

The natural history of mistletoes reveals potential mechanisms that can lead to differences in mistletoe prevalence and intensity among hosts and sites. Here we attempt to incorporate these mechanisms into an integrative, potentially predictive framework that emerges directly from recognizing that mistletoes are parasites with a metapopulation structure. This section poses several predictions that spring from this view and uses data on *P. californicum* and *T. aphyllus* to evaluate these predictions.

What are the consequences of the mutualism between mistletoes and birds for the population biology of mistletoes? Like other consumers, mistletoe-feeding birds tend to concentrate their activity at sites with relatively high resource densities (Martin, 1985; Sargent, 1990). The response of birds to mistletoes is likely to take place at two scales: birds should perch in fruit-bearing parasitized trees more frequently than in non-parasitized trees (Martínez del Rio *et al.*, 1995), and birds should be more abundant and spend more time at sites with higher mistletoe prevalences (Martínez del Rio *et al.*, 1996). Furthermore, prevalence and intensity of infection are often correlated in host–parasite systems (Fig. 7.2, for example). Fruit abundance at a site is a multiplicative function of the number of fruits per parasite, infection intensity and prevalence. Thus, fruit abundance should increase in an accelerating fashion with prevalence.

The mutualism between birds and mistletoes leads to two simple predictions:

1. Because birds should preferentially visit hosts that are infected by mistletoes, already parasitized hosts should receive seeds more frequently than non-parasitized hosts.
2. Because birds should show higher densities and/or spend more time at sites with higher mistletoe infection frequencies, seed deposition by birds should increase with mistletoe prevalence.

Several studies have provided support for the first prediction (see Figs 7.3 and 7.4). In general, seed rain is higher on parasitized than on non-parasitized hosts (Martínez del Rio *et al.*, 1996) and experimental removal of mistletoes from hosts leads to reduced seed deposition (J. Aukema, unpublished data). Unless infection by mistletoes induces host resistance (Hoffmann *et al.*, 1986), increased seed deposition on to already infected hosts should lead to reinfection and increased parasite loads. The extremely clumped distribution of mistletoes among host individuals exhibited by many mistletoe populations is probably a result of the disproportionate number of seeds deposited by birds on to already parasitized hosts (Overton, 1996). Preferential seed dispersal on to already parasitized hosts is a special case of 'conspecific attraction' (i.e. preferential dispersal to occupied patches over suitable empty ones (Smith and Peacock, 1990)), a phenomenon that can lead to a lower frequency of occupied patches and hence to lower mistletoe prevalences (Ray *et al.*, 1991; see also below, Overton's model).

The response of birds to mistletoe-infected hosts can lead to increased seed rain into

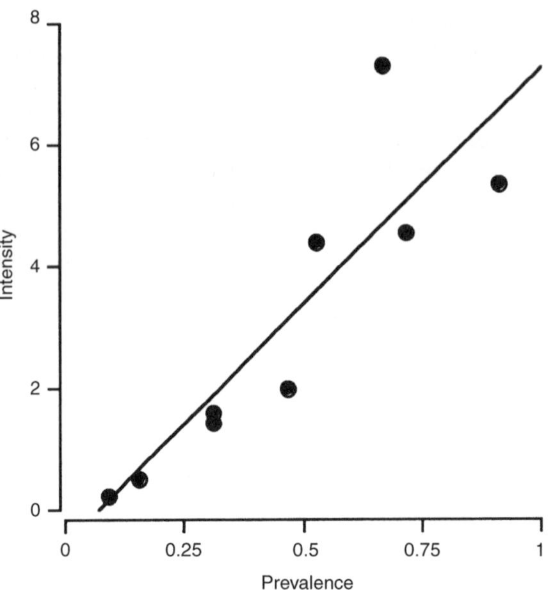

Fig. 7.2. At ten sites in the Santa Rita Experimental Range (Arizona, USA), mean infection intensity (mistletoes per host) of *P. velutina* increased significantly with the prevalence of infection (frequency of infected hosts) by *P. californicum* at each site ($r^2 = 0.77$, $P = 0.0018$; intensity = −0.584 + 7.93 (prevalence)).

already infected patches. Can the response of birds to mistletoes be extended to a larger spatial scale? Do sites with higher prevalences also show higher seed rain? Both *T. aphyllus* and *P. californicum* show a positive, accelerating relationship between seed deposition on to non-infected hosts and prevalence (Fig. 7.4; Martínez del Rio *et al.*, 1996). This relationship could lead to a positive correlation between the rate at which new hosts are infected at a site and prevalence. Thus, the response of birds would lead to a positive feedback in infection that could lead to spatial aggregation of parasitism in a landscape and to a positive autocorrelation in prevalence across it (Martínez del Rio *et al.*, 1996). The spatial scale at which this aggregation can be detected, however, depends on the scale at which individual birds and bird populations respond to the density of mistletoes and generate spatial patterns of seed deposition.

Mistletoes and metapopulations

Overton (1994) treated mistletoes as metapopulations, although it may be more appropriate to call them spatially structured patchy populations. Hosts can be identified as 'patches', infection and loss of infection can be characterized as patch occupancy and patch extinction, respectively, and seed dispersal and establishment can be equated with patch colonization. Mistletoe hosts can be viewed as living patches (hosts) inhabited by mistletoe subpopulations. Strictly, a subpopulation is a set of individuals that interact with each other with high probability (Hanski and Simberloff, 1997). Many mistletoes have animal pollinators that can travel and hence can move pollen and genes among patches, one consequence of which is to homogenize the spatial structure of mistletoe subpopulations (Reid *et al.*, 1995).

Mistletoes form discrete groups that inhabit distinct patches separated by unsuitable habitat. Populations in these patches can become extinct, either when all mistletoes in a patch die or when the host dies. Patches/hosts can be colonized only when propagules from other patches immigrate into them. Mistletoe subpopulations rarely inhabit all patches/hosts available, and hosts show turnover, both because mistletoe populations become extinct and because hosts die. Patch turnover is a

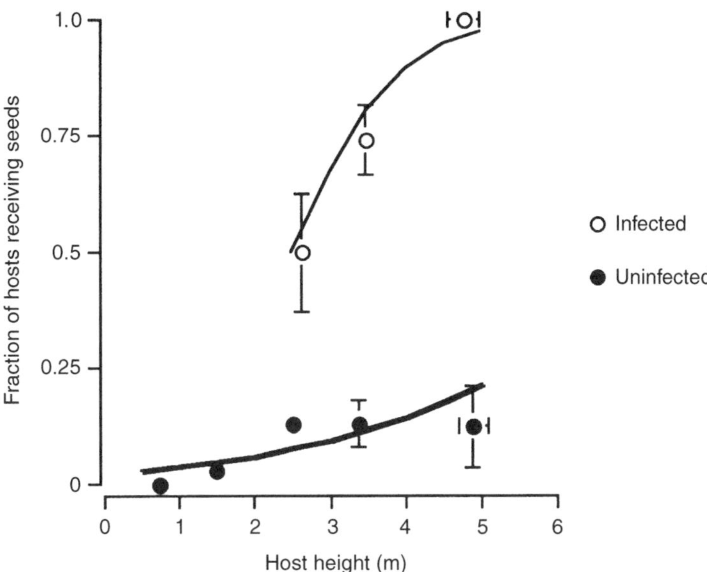

Fig. 7.3. The probability that a *P. velutina* tree host received seeds of the parasite *P. californicum* increased significantly with both height and previous infection (logistic regression $P < 0.001$; open circles are infected hosts and closed circles are uninfected hosts). Points are average values for size classes (bars are standard errors). Data were divided into size classes for visual clarity. Curves were fitted using a logistic regression procedure (logit (π) = −2.56 + 0.66 (height) + 1.46 (infection status)).

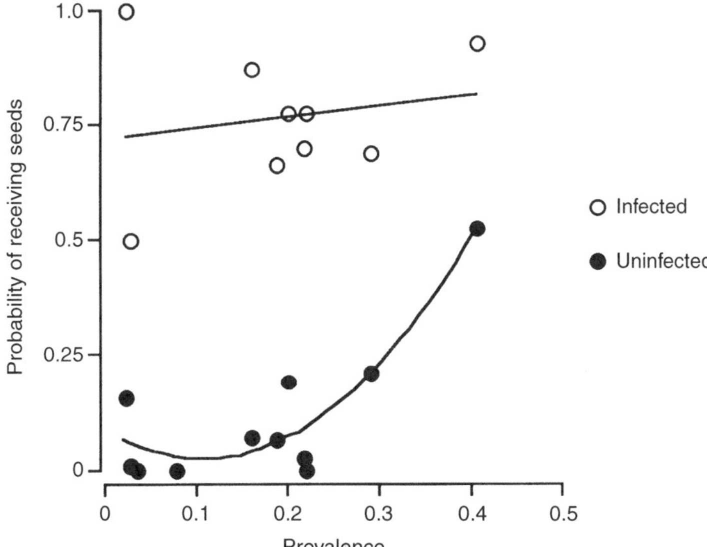

Fig. 7.4. At ten sites in the Santa Rita Experimental Range (Arizona, USA), the probability of receiving *P. californicum* seeds increased with infection prevalence in uninfected *P. velutina* hosts (closed circles). The relationship between the fraction of infected hosts receiving seeds and prevalence, however, was non-linear. For descriptive purposes, we fitted a second-degree polynomial to the data. The quadratic coefficient for this polynomial was significantly positive ($P < 0.03$ indicating an accelerating relationship, $y = 0.05 - 0.91x + 5.10x^2$, $r^2 = 0.81$, $P < 0.007$). For already infected hosts (open circles), there was no significant relationship between the fraction of hosts receiving seeds and prevalence at the site ($r = 0.06$, $P = 0.62$).

key element of the colonization–extinction dynamics that characterize metapopulations (Hanski and Simberloff, 1997).

Overton's model

A variety of host traits can influence the fraction of each host/patch that is occupied and the number of mistletoes inhabiting a host/patch. Overton (1996) modified Levins' (1969, 1970) classical metapopulation model to explore the role of host age on host occupancy. Patch turnover confers an age structure to the patch population, and thus Overton's (1996) model can be used to predict the relationship between host/patch occupancy and age. Overton's model depicts an array of hosts/patches that are either occupied or unoccupied at any point in time. Empty patches are equally likely to receive seeds and occupied hosts are equally likely to produce them. Overton (1996) modified Levins's (1970) model to explore the relationship between host age and probability of infection occupancy by assuming that population occupancy was at equilibrium. Overton's model yields two predictions: (i) occupancy should increase as a function of host age; and (ii) occupancy should increase with host age at an increasing rate at sites with higher among-host/patch dispersal rates.

A positive relationship between host size and infection prevalence is commonly found in mistletoe populations (Figs 7.5–7.7; Donohue, 1995; Overton, 1996; Kelly, 1998; Lei, 1999; J. Aukema and C. Martínez del Rio, unpublished data). Thus, prediction (i) appears to hold true, assuming that size is a good proxy for age. However, the positive relationship between infection prevalence and host age that is commonly found in mistletoes can be attributed to two non-exclusive hypotheses: (i) accumulation of mistletoes with age; or (ii) preference of avian seed-dispersers for taller, and probably older, host individuals. Explanations based on larger hosts receiving more seeds should result in a strongly positive size–intensity relationship. Although many mistletoes exhibit a significant and positive host size–infection intensity relationship (Overton, 1996), these relationships are often weak (Fig. 7.5). Two factors may account for the weakness of this relationship: (i) not only

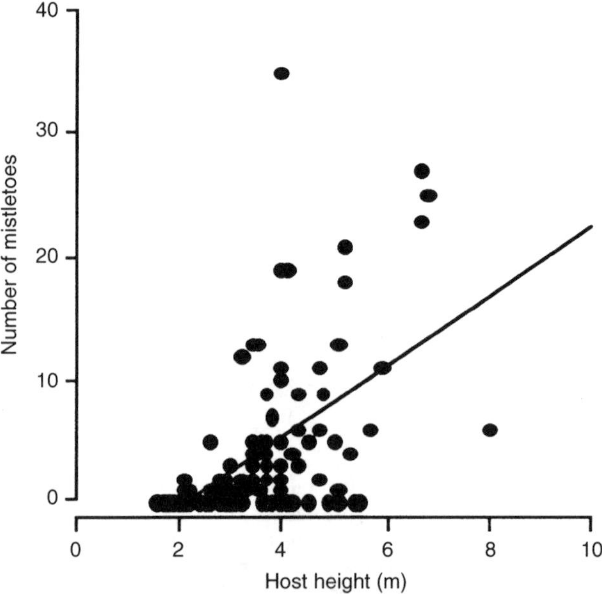

Fig. 7.5. The number of *P. californicum* individuals increased with host (*Prosopis velutina*) height ($r = 0.07$, $P < 0.0001$, $n = 115$), but very little variation was explained by the regression line ($r^2 = 0.269$). Although host height is a good predictor of infection frequency, it is a poor predictor of infection intensity.

are already parasitized hosts more likely to receive seeds and hence become reinfected; but (ii) older/taller trees are also more likely to receive seeds (Fig. 7.3). Age/size-specific differences in seed deposition on to hosts may exacerbate the age–occupancy relationship, but they are not required to generate it (Overton, 1996).

According to Overton's (1996) model, occupancy should increase with host age at an

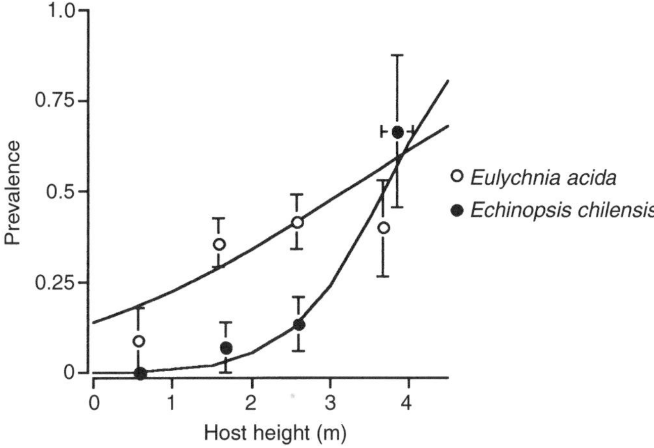

Fig. 7.6. At the Reserva Nacional Las Chinchillas, Chile, the frequency with which hosts were infected by *T. aphyllus* increased with height for both *Echinopsis chilensis* (open circles; logit (π) = −1.779 + 0.5632 (height), $P < 0.01$, $n = 122$) and *Eulychnia acida* hosts (closed circles, logit (π) = −6.244 + 1.7039 (height), $P < 0.00063$, $n = 52$). This positive relationship is in accord with that predicted by Overton's (1996) model of mistletoe metapopulations.

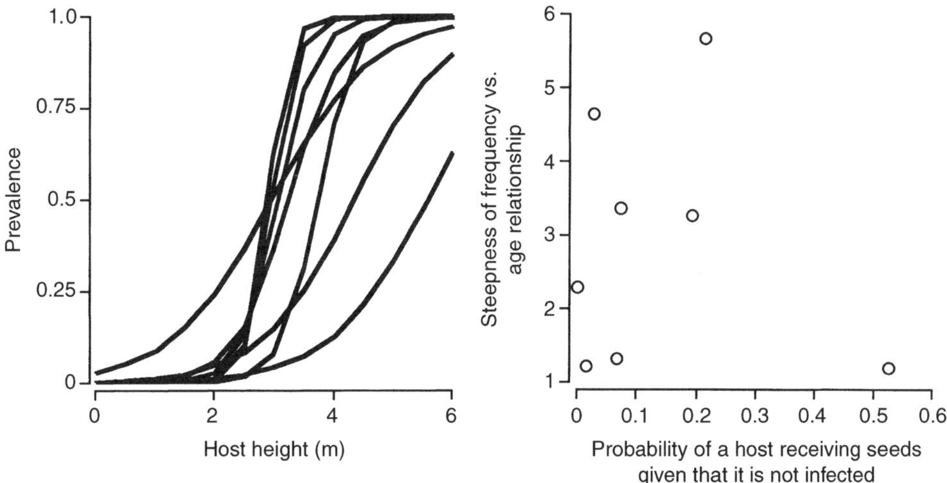

Fig. 7.7. The first prediction of Overton's (1996) model was met by *P. californicum* infecting *P. velutina* at ten sites in the Santa Rita Experimental Range (Arizona, USA). Frequency of infection increased with height at all sites ($P < 0.05$, after logistic regression). Curves in the left panel are relationships between frequency of infection and height at each site constructed using logistic regression. However, the second prediction of Overton's model (1996), namely that the steepness of the relationship between infection frequency and host age/height would increase with the probability of unparasitized trees receiving seeds (dispersal efficiency) was not met ($r = 0.015$, $P = 0.772$, right panel).

increasing rate at sites with higher among host/patch dispersal. For *P. californicum* infecting velvet mesquites (*Prosopis velutina*) at ten sites in Arizona, we found positive relationships between host size and occupancy (Fig. 7.7a), but no relationship between the steepness of these relationships (as measured by the slope of the log odds function (Ramsey and Schafer, 1997)) and seed dispersal into non-parasitized hosts (Fig. 7.7b).

Overton's model assumes that seed dispersal is random. However, most data suggest that dispersal is not random in mistletoes. A disproportionate number of seeds fall on to already infected hosts. Therefore, mistletoes exhibit conspecific attraction (Stamps, 1991). Ray *et al.* (1991) modified Levins's model to incorporate conspecific attraction by assuming that a fraction of all propagules is deposited in already occupied patches. Altering Overton's (1996) model to incorporate conspecific attraction makes mistletoe invasion more difficult, but leaves all other qualitative predictions unchanged. Occupancy still increases as a function of host age, and occupancy increases with host age at an increasing rate at sites with higher among-host/patch dispersal (J. Aukema and C. Martínez del Rio, unpublished data). Conspecific attraction does not explain the lack of a significant positive relationship between the steepness of the occupancy and age curve and dispersal efficiency.

Beyond Overton's model: structured and spatially explicit models

Although conspecific attraction does not change the predictions of Overton's model qualitatively, it is likely to have a significant effect on more realistic, and hence more complex, models. Exploration of the effects of host age and parasite status on infection intensity may be facilitated by construction of structured mistletoe metapopulation models. The goal of these models is to determine how different processes produce a distribution of local population sizes (Gyllenberg *et al.*, 1997, and references therein).

So far, all the models that we have described for mistletoe populations are spatially implicit. They ignore the spatial location of hosts/patches and hence assume that all subpopulations are equally connected (Hanski and Simberloff, 1997). Although we believe that a large number of questions can be explored with these models, other questions require explicit consideration of space. For example, we have argued that the positive, accelerating relationship between seed deposition into non-infected hosts and prevalence (Fig. 7.4) may lead to spatial aggregation of parasitism in a landscape and to a positive autocorrelation in prevalence across it (Martínez del Rio *et al.*, 1996). Testing the logical merit of this hypothesis requires a spatially explicit model. Overton (1996) and Lavorel *et al.* (1999) offer two elegant examples of spatially structured models for mistletoes. Mechanistic models of seed dispersion patterns can be constructed directly from the characteristics of mistletoes and hosts (density, size and distribution) and from the behaviour of the birds that disperse their seeds (Overton, 1996; Lavorell *et al.*, 1999). Spatial patterns of seed dispersal are the key to understanding plant population dynamics in a spatial context (Nathan and Muller-Landau, 2000). Mistletoes offer an unparalleled opportunity to document the consequences of seed dispersal by animals on the spatial dynamics of plant populations.

Future Directions

In this chapter we have emphasized the consequences of the interaction between seed-dispersers and mistletoes for the metapopulation ecology of these plant parasites. We have placed less emphasis on the fact that the patches occupied by mistletoes are exploited living organisms, and we have ignored the roles they play in their biotic communities. We believe that a more complete understanding of mistletoe population ecology must recognize the intricacies of their interactions with hosts and their roles in biotic communities.

Hosts and mistletoes

The biological characteristics of hosts and the physical environment they occupy can determine their quality as patches for mistletoe occupancy. In particular, knowledge about the physiology of hosts/patches can allow one to predict which patches are sources that contribute mistletoe colonists to other patches, and which are sinks, where mistletoe subpopulations would go extinct in the absence of immigration (Pulliam, 1988; Hanski and Simberloff, 1997). Because at least some mistletoes can be accurately aged (Dawson *et al.*, 1990), the effect of the host's physiology on the demography of subpopulations can be studied. Mistletoes and their hosts offer a unique opportunity to integrate the physiological details of a plant–plant interaction with their demographic and even metapopulational consequences.

Mistletoes are not only influenced by the physiological status of their hosts, but have important impacts on it. Mistletoes probably degrade the quality of the patches that they occupy and increase their turnover by killing their hosts. Using Burdon's (1991) colourful classification, mistletoes can be castrators, killers or debilitators. Because some mistletoes can infect and damage economically important plants, there is some information on their effects on host growth, reproduction and survival (reviewed by Reid *et al.*, 1995). Little is known, however, about their effects on host populations (Silva and Martínez del Rio, 1996; Medel, 2000). Most models that explore mistletoe population and metapopulation dynamics assume that mistletoe infection has no effect on host survival and reproduction and that host populations are at equilibrium. Addressing how mistletoes affect host populations and how this interaction affects the temporal and spatial dynamics of mistletoes requires that we obtain better empirical data on the effects of mistletoes on hosts and that we incorporate these effects into models of mistletoe–host interactions.

Mistletoes as community members

Mistletoes are intriguing elements of biotic communities because they play the dual role of host scourges and bird mutualist benefactors (Martínez del Rio *et al.*, 1995; Fig. 7.1). In addition to the direct effects that mistletoes can have on the fecundity and viability of their hosts and mutualists, they may indirectly affect the host's competitors, herbivores, pollinators and seed-dispersers. Although community-level effects of plant pathogens have received significant attention (Dobson and Crawley, 1994, and references therein), we know little about the contribution of mistletoes to community composition and function.

Mistletoes can have significant effects on bird communities (Turner, 1991; Bennets *et al.*, 1996). Because many mistletoes are pollinated and dispersed by birds, their direct influence on birds is through the abundance of nectar and fruit. For example, *T. aphyllus* and *P. californicum* bloom and produce abundant fruit during the winter, when food resources are scarce. Areas heavily infected by these species are hot spots of activity for nectar- and fruit-eating animals (Martínez del Rio *et al.*, 1995; J. Aukema and C. Martínez del Rio, unpublished data). Mistletoes can also have significant indirect effects on bird communities. For example, prevalence of the dwarf mistletoe *Arceuthobium vaginatum* was positively correlated with bird abundance and species richness (Bennets *et al.*, 1996). Because the fruit of dwarf mistletoes are not extensively used by birds, Bennets *et al.* (1996) concluded that mistletoe infections increase bird abundances by enhancing insects that feed on and pollinate mistletoes or that take advantage of the weakened condition of tree hosts.

Dwarf mistletoes promote bird diversity because they create a mosaic of habitat structures within a forest stand through their effect on tree growth and mortality. They also increase nesting habitat. Several forest bird species use the dense clumps ('witches' brooms') that are formed by branches of the host tree for roosts and nest sites (Bennets, 1991, and references therein). Mistletoes are often considered insidious forest pests that reduce the economic value of timber stands (Wicker, 1984). As such, mistletoe removal is practised in managed forests with the objective of increasing timber production (Hawksworth and Wiens, 1995; Kelly *et al.*, 1997). In areas where management goals are not strictly focused on timber production, the value of mistletoes for biodiversity may make their control unjustified, impractical or undesirable (Bennets *et al.*, 1996).

Conclusions: a Few Relatively Solid Patterns and Much Work Ahead

The main messages of this chapter can be summarized in the following sentence: 'Mistletoes are *parasitic plants* that exhibit a *metapopulation* structure, and whose seeds are dispersed by *mutualistic* avian seed-dispersers.' The elements emphasized in this sentence are responsible for several patterns that may characterize many, if not most, mistletoe populations. In this final section, we list these patterns and reiterate the mechanisms that probably shape them. Because the patterns listed here have been well documented in just a few mistletoe–host systems, and primarily in desert mistletoes, their generality is uncertain. These patterns should be viewed as testable hypotheses, rather than as general and firmly established results.

Mutualistic avian seed-dispersers seem to respond to the abundance of mistletoes at two scales: individual hosts and sites. The consequence of this response is that seed dispersal is not random among hosts and across landscapes. We hypothesize the following:

1. Seeds fall disproportionately more frequently on already parasitized than on non-parasitized hosts.
2. Seed rain increases with mistletoe prevalence across sites.

Hypotheses 1 and 2 yield two ancillary hypotheses:

1a. Disproportionate seed deposition on to already parasitized hosts leads to super-infection and to a highly aggregated distribution of parasite individuals among hosts.
2a. Disproportionate seed deposition at sites with higher prevalence leads to spatial autocorrelation in parasitism prevalence across a landscape.

Mistletoe populations can be perceived as metapopulations in which hosts are patches. Under fairly general conditions, a simple metapopulation model suggests the following:

3. The frequency of occupied hosts/patches increases with host age.

The logical and empirical validity of hypotheses 1a and 2a and the generality of hypothesis 3 must be tested by the complementary use of structured (1a and 3) and spatially explicit (2a and 3) metapopulation models and, of course, by field research.

Mistletoes provide theoreticians and empiricists with unique opportunities and peculiar challenges. Many of the processes that are difficult to investigate in other species are relatively straightforward to study in mistletoes because they are sessile and hence relatively easy to count. Because in some cases they can be aged (Dawson *et al.*, 1990), their demography can be studied. Their seeds are large and visible and are dispersed by birds whose movements are relatively easy to follow (e.g. relative to vectors such as mosquitoes and tsetse flies (Kitron, 1998)). Wheelwright and Orians (1982) have characterized the task of distinguishing safe germination sites as 'nearly impossible', but this task is relatively straightforward in mistletoes because seeds only establish on suitable hosts (Sargent, 1995). Mistletoes are ideal systems to integrate the ecology of seed dispersal into the larger framework of the temporal and spatial dynamics of plant metapopulations. Because the fine points of the interaction between hosts and mistletoes probably have significant consequences for the population biology of mistletoes, these plants provide a unique opportunity to determine the ecological penetrance of physiological processes. Finally, because data can be generated relatively rapidly in mistletoe systems, they provide an ideal arena for the testing and refinement of plant metapopulation models. We hope that the themes developed here will stimulate empiricists to explore other mistletoe systems and challenge theoreticians to model them.

Acknowledgements

Carla Restrepo and Anna Traveset commented on the manuscript. This work was partially funded by a National Science Foundation Graduate Research Fellowship. Many of our ideas were shaped while tramping in the field.

Our field companions, Arturo Silva, John King and Emma Aukema, served as intellectual catalysts.

References

Bennets, R.E. (1991) The influence of dwarf mistletoe infestation on bird communities in Colorado ponderosa pine forests. PhD thesis, Colorado State University, Fort Collins, Colorado.

Bennets, R.E., White, G.C., Hawksworth, F.G. and Severs, S.E. (1996) The influence of dwarf mistletoe on bird communities in Colorado ponderosa pine forests. *Ecological Applications* 6, 899–909.

Burdon, J.J. (1991) Fungal pathogens as selective forces in plant populations and communities. *Australian Journal of Ecology* 16, 423–432.

Cowles, R.B. (1936) The relation of birds to seed dispersal of the desert mistletoe. *Madroño* 3, 352–356.

Dawson, T., King, E. and Ehleringer, J. (1990) Age structure of *Phoradendron juniperinum* (*Viscaceae*), a xylem-tapping mistletoe: inferences from a non-destructive morphological index of age. *American Journal of Botany* 77, 573–583.

Dobson, A. and Crawley, M. (1994) Pathogens and the structure of plant communities. *Trends in Ecology and Evolution* 9, 393–398.

Donohue, K. (1995) The spatial demography of mistletoe parasitism on a Yemeni acacia. *International Journal of Plant Science* 156, 816–823.

Gyllenberg, M., Hanski, I. and Hastings, A. (1997) Structured metapopulation models. In: Hanski, I.A. and Gilpin, M.E. (eds) *Metapopulation Biology: Ecology Genetics and Evolution.* Academic Press, New York, pp. 93–122.

Hanski, I.A. and Simberloff, D. (1997) The metapopulation approach, its history, conceptual domain, and application to conservation. In: Hanski, I.A. and Gilpin, M.E. (eds) *Metapopulation Biology: Ecology Genetics and Evolution.* Academic Press, New York, pp. 5–26.

Hawksworth, F.G. and Wiens, D. (1995) *Dwarf Mistletoes: Biology, Pathology, and Systematics.* Forest Service Agriculture Handbook Number 450, USDA, Washington, DC.

Hoffmann, A.J., Fuentes, E.R., Cortes, I., Liberona, F. and Costa, V. (1986) *Tristerix tetrandrus* (*Loranthaceae*) and its host-plants in the Chilean matorral: patterns and mechanisms. *Oecologia* 69, 202–206.

Kelly, D. (1998) Spatial clumping of *Tupeia antarctica* at Wainui. *Canterbury Botanical Society Journal* 32, 62–65.

Kelly, P., Reid, N. and Davies, I. (1997) Effects of experimental burning, defoliation, and pruning on survival and vegetative resprouting in mistletoes (*Amyema miquelii* and *Amyema pendula*). *International Journal of Plant Science* 158, 856–861.

Kitron, U. (1998) Landscape ecology and epidemiology of vector-borne diseases: tools for spatial analysis. *Journal of Medical Entomology* 35, 435–445.

Lamont, B. (1982) Host range and germination requirements of some South African mistletoes. *South African Journal of Science* 78, 41–42.

Larson, D. (1996) Seed dispersal by specialist versus generalist foragers: the plant's perspective. *Oikos* 76, 113–120.

Lavorell, S., Stafford Smith, M. and Reid, N. (1999) Spread of mistletoes (*Amyema preisii*) in fragmented Australian woodlands. *Landscape Ecology* 14, 147–160.

Lei, S. (1999) Age, size and water status of *Acacia greggii* influencing the infection and reproductive success of *Phoradendron californicum. American Midland Naturalist* 141, 358–365.

Levins, R. (1969) Some demographic and genetic consequences of environmental heterogeneity for biological control. *Bulletin of the Entomological Society of America* 15, 237–240.

Levins, R. (1970) Extinction. In: Gestenhaber, M. (ed.) *Some Mathematical Problems in Biology.* American Mathematical Society, Providence, Rhode Island, pp. 77–107.

Martin, T.E. (1985) Resource selection by tropical frugivorous birds: integrating multiple interactions. *Oecologia* 66, 563–573.

Martínez del Rio, C., Hourdequin, M., Silva, A. and Medel, R. (1995) The influence of cactus size and previous infection on bird deposition of mistletoe seeds. *Australian Journal of Ecology* 20, 571–576.

Martínez del Rio, C., Silva, A., Medel, R. and Hourdequin, M. (1996) Seed dispersers as disease vectors: bird transmission of mistletoe seeds to plant hosts. *Ecology* 77, 912–921.

Medel, R. (2000) Assessment of parasite-mediated selection in a host–parasite system in plants. *Ecology* 81, 1554–1564.

Nathan, R. and Muller-Landau, H.C. (2000) Spatial patterns of seed dispersal, their determinants and consequences for recruitment. *Trends in Ecology and Evolution* 15, 278–285.

Overton, J.M. (1994) Dispersal and infection in mistletoe metapopulations. *Journal of Ecology* 82, 711–723.

Overton, J.M. (1996) Spatial autocorrelation and dispersal in mistletoes: field and simulation results. *Vegetatio* 125, 83–98.

Price, P.W. (1980) *The Evolutionary Biology of Parasites.* Princeton University Press, Princeton, New Jersey.

Pulliam, R.H. (1988) Sources, sinks, and population regulation. *American Naturalist* 132, 652–661.

Ramsey, F.L. and Schafer, D.W. (1997) *The Statistical Sleuth: a Course in Methods of Data Analysis.* Duxbury Press, Boston, Massachusetts.

Ray, C., Gilpin, M. and Smith, A.T. (1991) The effect of conspecific attraction on metapopulation dynamics. In: Gilpin, M. and Hanski, I. (eds) *Metapopulation Dynamics: Empirical and Theoretical Investigations.* Academic Press, New York, pp. 123–134.

Reid, N. (1991) Coevolution of mistletoes and frugivorous birds. *Australian Journal of Ecology* 16, 457–469.

Reid, N., Smith, N.M. and Yan, Z. (1995) Ecology and population biology of mistletoes. In: Lowman, M.D. and Nadkarni, N.M. (eds) *Forest Canopies.* Academic Press, San Diego, California, pp. 285–310.

Sargent, S. (1990) Neighborhood effects on fruit removal by birds: a field experiment with *Viburnum dentatum* (Caprifoliaceae). *Ecology* 71, 1289–1298.

Sargent, S. (1995) Seed fate in a tropical mistletoe: the importance of host twig size. *Functional Ecology* 9, 197–204.

Silva, A. and Martínez del Rio, C. (1996) Effects of mistletoe parasitism on the reproduction of cacti hosts. *Oikos* 75, 437–442.

Smith, A.T. and Peacock, M.M. (1990) Conspecific attraction and the determination of metapopulation colonization rates. *Conservation Biology* 4, 320–323.

Stamps, J.A. (1991) The effects of conspecifics on habitat selection in territorial species. *Behavioural Ecology and Sociobiology* 28, 29–36.

Turner, R.J. (1991) Mistletoe in eucalypt forests – a resource for birds. *Australian Forestry* 54, 226–235.

Walsberg, G.E. (1975) Digestive adaptations of *Phainopepla nitens* associated with the eating of mistletoe berries. *Condor* 77, 169–174.

Wheelright, N.T. and Orians, G.H. (1982) Seed dispersal by animals: contrasts with pollen dispersal, problems of terminology, and constraints on coevolution. *American Naturalist* 119, 402–413.

Wicker, E.F. (1984) Dwarf mistletoe insidious pest of North American conifers. In: Hawksworth F.G. and Scharpf (eds) *Biology of the Dwarf Mistletoes: Proceedings of the Symposium.* Forest Service General Technical Report RM 111, USDA, p. 1.

Yan, Z. (1993) Resistance to haustorial development of two mistletoes, *Amyema preissii* (Miq.) and *Lysiana exocarpi* (Behr) *tieghem* ssp. *exocarpi* (*Loranthaceae*) on host and non-host species. *International Journal of Plant Science* 154, 386–394.

8 Secondary Metabolites of Ripe Fleshy Fruits: Ecology and Phylogeny in the Genus *Solanum*

Martin L. Cipollini,[1] Lynn A. Bohs,[2] Kim Mink,[1] Eric Paulk[1] and Katrin Böhning-Gaese[3]

[1]*Department of Biology, PO Box 430, Berry College, Mount Berry, GA 30149, USA;* [2]*Department of Biology, University of Utah, 257 South 1400 East, Salt Lake City, UT 84112, USA;* [3]*Institut für Biologie II, RWTH Aachen, Kopernikusstr. 16, 52074 Aachen, Germany*

Introduction and Overview

Why do ripe fleshy fruits contain secondary metabolites, sometimes in concentrations that make the fruits toxic to vertebrates? And how do the chemical and physical characteristics of fruits influence choice by frugivores? These questions have drawn the attention of those interested in understanding the adaptive roles of fruit traits (see Cipollini and Levey, 1997a, b,c; Cipollini, 2000). However, a criticism of such studies is that few have addressed the possible influence of physiological and phylogenetic constraints on fruit traits (see Cipollini and Levey, 1998; Eriksson and Ehrlen, 1998). We are addressing this concern by studying fruit traits of the genus *Solanum*, while assessing and accounting for phylogenetic effects. Here we provide a preliminary analysis of our data, which we hope will stimulate others to conduct similar analyses.

Theoretically, predictable differences in the quality of seed dispersal among frugivores should influence the evolution of fruit traits (see Janson, 1983; Wheelwright, 1985; Jordano 1987a; Debussche and Isenmann, 1989; Willson *et al.*, 1989; Gautier-Hion, 1990; Stiles and Rosselli, 1992). Nevertheless, studies using phylogenetic null models have not supported this theory (see Bremer and Eriksson, 1992; Herrera, 1992; Jordano, 1995). This lack of support is assumed to result, in part, because of extensive decoupling of interactions between particular frugivores and the plants whose seeds they disperse (see Howe, 1984; Jordano, 1987b; Herrera, 1998). However, most studies on the evolution of fruit traits have overlooked the effects of secondary metabolites, which are probably important mediators of fruit–frugivore interactions (Cipollini and Levey, 1997c, 1998).

We have presented several adaptive hypotheses on the adaptive significance of secondary metabolites in ripe fleshy fruits (Cipollini, 2000). In general, these hypotheses assume a selective advantage to plants bearing fruits containing secondary metabolites and thus predict that patterns of secondary chemistry in fruits can be explained, at least in part, by interactions with frugivores. Likewise, the hypotheses predict that fruit-use patterns by frugivores can be explained, at least in part, by the presence of secondary metabolites. The putatively adaptive pattern upon which we focus is an apparent correlation of ripe-fruit glycoalkaloid (GA) (potentially toxic metabolite) content with other fruit traits, a pattern that could result from diffuse coevolution leading to broadly defined seed-dispersal syndromes (suites of species showing similar, but independently evolved, fruit characteristics (*sensu* van der Pjil, 1969)). In particular, ripe fruits of *Solanum* species that are dispersed by birds have very low GA levels; those dispersed by bats have variable levels; and those dispersed by terrestrial mammals have high levels (Cipollini and Levey, 1997c). Such patterns may have resulted from coadaptive fruit–frugivore interactions. Or they might simply reflect physiological constraints related to leaf and unripe-fruit defence and/or the effects of shared ancestry. For example, Ehrlen and Eriksson (1993) postulated that 'toxic' fruits are more common in related plant taxa and that ripe-fruit toxicity primarily results from physiological constraints associated with selection for defences in leaves and/or unripe fruits (see also Eriksson and Ehrlen, 1998). Considering these alternatives, we focus on two questions.

1. Are ripe *Solanum* fruits that are high in GAs primarily found in species:
(a) whose leaves are high in GAs?
(b) whose unripe fruits are high in GAs?
(c) whose closest relatives produce fruits high in GAs?
2. Are patterns of fruit-trait covariation consistent with independent selection pressures, or are patterns associated strongly with phylogeny?

General Phylogenetic Approach

To answer our two questions, we are collecting molecular, physical and chemical data for a large group of *Solanum* species whose fruit traits contrast markedly. By using techniques that account for phylogeny, we hope to provide rigorous, phylogenetically corrected statistical tests of the relationships among traits (e.g. the correlation between leaf and fruit chemical traits). This is, to our knowledge, the first use of molecular data as a basis for a comparative study of fruit chemistry in wild plants (see Bremer and Eriksson (1992) for an analysis of fruit morphology).

A primary requirement of our approach is a rigorous phylogenetic hypothesis for our species. To avoid circularity, this phylogeny should be derived independently of the fruit traits we are examining (Givnish, 1997; but see de Queiroz, 1996). Our approach is thus to determine a phylogeny for our species using gene sequence data and to measure a suite of fruit traits considered to be relevant to frugivores. We then use the methods of independent contrasts (IC), phylogenetic autocorrelation (PA), and signed Mantel (MAN) tests to control for species relatedness in statistical tests and to examine whether predicted patterns exist among the traits after the effects of phylogeny have been removed or minimized.

Study Species and Taxonomic Background

The genus *Solanum* was selected primarily because traits relating to seed dispersal vary tremendously in this group. It is one of the largest plant genera (*c.* 1400 species) and encompasses remarkable diversity in morphology, habit and distribution. The most widely used infrageneric classification divides *Solanum* into seven subgenera and about 70 sections (D'Arcy, 1972, 1991). Some sections have been the subject of intensive study because of their economic importance (e.g. the potatoes (section *Petota*)) and others have been the focus of recent taxonomic revisions and/or phylogenetic studies (see Olmstead and Palmer, 1997; Olmstead *et al.*, 1999; Spooner *et al.*, 1999). While many groups are poorly known, a consistent picture is emerging of phylogenetic relationships in the genus as a whole, and progress has been made in defining monophyletic sections (see Bohs and Olmstead 1997, 1999). Analyses of sequence data from the chloroplast gene *ndhF* and the nuclear ITS (internal transcribed spacer) region identify about 11 major clades in the genus (Bohs, 2000). Some of these clades are congruent with traditional taxonomic subdivisions, whereas others are not. Our preliminary approach has been to sample species from disparate clades as a means of obtaining a large sample that is diverse in fruit traits.

Solanum GAs: Ecological and Phylogenetic Context

Total glycolalkaloid (TGA) content

We focus on TGA, because studies suggest that quantitative variation probably overrides most differences in the deterrent effects of specific GAs toward consumers. For example, Cipollini and Levey (1997a,b,c) found little difference in the deterrence of the two most prevalent GAs (solamargine and solasonine) towards a wide variety of organisms. The potato GAs (solanine and chaconine) show similar general deterrent effects (see van Gelder, 1990), although some variation in toxicity among organisms has been noted, depending on the specific compound tested. Moreover, the patterns that we are examining are principally quantitative, especially regarding correlations of putatively physiologically constrained traits within plant species (e.g. leaf vs. ripe-fruit GA).

Ecological patterns of Solanum *fruit GAs*

Among temperate North American *Solanum*, only the large, low-nutrient, yellow, odorous, winter-dispersed, mammal-syndrome fruit, *S. carolinense*, is known to contain high concentrations of GAs when ripe (Cipollini and Levey, 1997a,b,c). This suggests that ripe-fruit toxins might defend against pests when dispersal is rare or unpredictable, and thus may represent a trade-off between defence against pests and palatability for dispersers. In a study of other *Solanum*, including tropical species, this finding was corroborated; high levels of GAs were commonly found in fruits having traits suggesting dispersal by terrestrial mammals (Cipollini and Levey, 1997c). All 'bird' fruits (small, high-nutrient, red, black, odourless) showed little or no detectable GA, whereas 'bat' fruits (variable-sized, high-nutrient, dull green or yellow-green, strong odours) had variable levels. Nee (1991) likewise suggested an association of high GA content with some mammal-dispersed species. High GA content in terrestrial mammal-dispersed fruits may be possible because the large body size of such mammals may confer some tolerance to GAs (van Gelder, 1990). Plants dispersed by such animals might more easily afford to protect their fruits from pests via high levels of GAs. Consumption of such fruits might even reduce parasitic infection in some species (e.g. the maned wolf, *Chrysocyon brachyurus*, which feeds on *Solanum lycocarpon* (Courtenay, 1994)).

Known phylogenetic patterns

While much is known about *Solanum* GAs in leaves and fruits (see Schreiber, 1968; Ripperger and Schreiber, 1981), little use has been made of these data from an evolutionary perspective. Based upon a few studies, we note the following patterns.

1. In a reanalysis of data collected by Bradley *et al.* (1979) on 47 *Solanum* species, we found a significant correlation between leaf and unripe-fruit GA concentration ($r = 0.58$, $P \leq 0.0001$). However, 16 species with no GAs in their leaves had significant levels (0.1–1.5% dry weight) in the fruits. Because few ripe fruits were analysed, however, these data are inadequate to rigorously test for a correlation between leaf and ripe-fruit GA content.

2. While many *Solanum* fruits lose GAs with ripening (e.g. Schreiber, 1963; Bradley *et al.*, 1979), plants with 'toxic' ripe fruits are found in at least three subgenera and at least ten groups or sections of the genus (Schreiber, 1963; Kingsbury, 1964; Bradley *et al.*, 1979; M.L. Cipollini, D. Levey, E. Paulk, K. Mink and L.A. Bohs, unpublished data). These results suggest that fruit toxicity is relatively unconstrained by phylogeny.

3. Using published data, we conducted nested analysis of variance of unripe- and ripe-fruit TGA based upon a widely accepted taxonomy of the genus (D'Arcy, 1991). These analyses indicated that a significant amount of variation in TGA content resides among species within sections and hence could be adaptive (77.6% and 35.2% of variation for unripe and ripe fruits, respectively). This suggests only weak effects of phylogeny on fruit chemistry, provided that D'Arcy's (1991) taxonomic scheme accurately represents phylogenetic relationships within the genus.

Methods

Species selection, growth and sample collection

To avoid biases found in the literature, we collect all data from greenhouse-grown plants using standard molecular and phytochemical methods. We obtained seeds of about 90 *Solanum* species and are growing these plants in common garden conditions (voucher data for species used in this chapter are given in Table 8.1). Seeds were collected from field sites and from the Botanical Garden at Nijmegen, The Netherlands. After growing plants to maturity, we collect about 100 g of both unripe and ripe fruits and 20–30 leaves.

Fruit separation and morphological analyses

For ripe-fruit samples, we measure the following morphological traits: whole-fruit wet mass, pulp and seed wet and dry masses, seed and pulp dry-matter content, seed number per fruit and mean individual dry seed mass. Ripe-fruit colour is recorded by classifying fruits as black/purple, red, orange, yellow, white and/or green.

Chemical analyses

Immature and mature fruits

We analyse freeze-dried pulp samples for total protein using the Bradford assay (Jones *et al.*, 1989), for total soluble sugars using the anthrone technique (Smith, 1981) and for total phenolics using the Prussian blue method (Budini *et al.*, 1980). Based upon preliminary data, total lipids are assumed to be low and unimportant.

GA of fruits and leaves

Freeze-dried leaf and fruit-pulp samples are analysed for GA (TGA analysis) using the Birner (1969) technique.

Molecular analyses

DNA was extracted from fresh or silica-dried leaves using the modified CTAB procedure of Doyle and Doyle (1987) or by a mini-extraction protocol (available upon request from Lynn Bohs). The phylogenies we use are based on sequence data from the nuclear ITS region (ITS 1, ITS 2 and the intervening 5.8S ribosomal DNA (rDNA) subunit) (Baldwin *et al.*, 1995). Polymerase chain reaction (PCR) amplification, clean-up of DNA and PCR products, sequencing and sequence editing and alignment followed the techniques described in Bohs and Olmstead (2001). The data matrix containing the aligned sequences is available from Lynn Bohs upon request.

Phylogenetic analyses were performed using PAUP* 4.0b3a (Swofford, 2000). The parsimony analyses reported here used the heuristic search algorithm with the TBR and MulTrees options, equal weights for all nucleotide positions, gaps treated as missing data and 100 random-order entry replicates. Trees were rooted using *Lycianthes heteroclita* as the outgroup. This species was subsequently deleted from the tree file used in the PA, IC and MAN tests (see below). When multiple, most-parsimonious trees resulted from the searches, a single representative tree was used in the PA, IC and MAN analyses.

Statistical analyses involving fruit traits

TIPS parametric analyses

We first conducted the following parametric analyses using raw data for all taxa under the assumption that the taxa are independent (TIPS analyses). TIPS analyses allow comparisons with the results of phylogenetically corrected analyses using the same data (PA, IC and MAN tests, below); patterns that disappear upon phylogenetic correction via these

Table 8.1. Names of species included in phylogenetically corrected (PA, IC and MAN) analyses, including raw data for ripe-fruit colour, mean whole-fruit wet mass (g) and mean seed number per fruit.

Species	Voucher*†	Colour	Mass	Seed
Lycianthes heteroclita (Sendtn.) Bitter	Bohs 2376	n/a	n/a	n/a
Solanum abutiloides (Griseb.) Bitter & Lillo	RGO S-73 (DNA)/Cipollini 94 (fruits)	O	1.5	199
Solanum acerifolium Dunal	Bohs 2714	G/W	0.7	41
Solanum aculeatissimum Jacq.	Cipollini 60 (DNA)/Cipollini SK (fruits)	Y	4.6	91
Solanum adhaerens Roem. & Schult.	Bohs 2473 (DNA)/Cipollini SL (fruits)	O	–	–
Solanum aphyodendron Knapp	RGO S-92 (DNA)/Cipollini SAP (fruits)	G/Y	–	–
Solanum capsicoides All.	Bohs 2451 (DNA)/Cipollini 37 (fruits)	O	15.3	117
Solanum carolinense L.	Cipollini SC	Y	2.1	103
Solanum cordovense Sessé & Moc.	Bohs 2693	B	0.4	–
Solanum dasyphyllum Schum. & Thonn	Cipollini 7	Y/G	15.0	50
Solanum diflorum Vell.	Cipollini 11	R/O	2.5	85
Solanum dulcamara L.	No voucher (DNA)/Cipollini SD (fruits)	R	0.4	25
Solanum glaucophyllum Desf.	Cipollini 125	B	2.4	17
Solanum jamaicense Mill.	RGO S-85 (DNA)/Bohs 2481 (fruits)	O	0.3	77
Solanum laciniatum Ait.	Bohs 2528	O/Y	1.5	91
Solanum macrocarpon L.‡	Cipollini 101	Y	28.8	405
Solanum mammosum L.	RGO S-89 (DNA)/Cipollini 40 (fruits)	O/Y	22.5	33
Solanum melongena L.	RGO S-91 (DNA)/Cipollini 85 (fruits)	B/Y	18.1	348
Solanum myriacanthum Dunal	Cipollini 83	Y/G	12.7	118
Solanum nigrum L.	Bohs 2534	B	0.2	20
Solanum opacum A. Braun & Bouché	Bohs 2459	G	0.2	35
Solanum physalifolium Rusby	Bohs 2467	G	0.2	20
Solanum prinophyllum Dunal	Bohs 2725	G	–	–
Solanum pseudocapsicum L.	Cipollini 95	R/O	2.6	6
Solanum ptychanthum Dunal	RGO S-94 (DNA)/Cipollini SP (fruits)	B	0.3	55
Solanum rudepannum Dunal	Bohs 2712 (DNA)/Cipollini SRD (fruits)	Y	–	–
Solanum rugosum Dunal	Bohs 3011 (DNA)/Cipollini SRG (fruits)	Y/G	–	–
Solanum scabrum Mill.	Bohs 2729	B	1.3	89
Solanum sciadostylis (Sendtn.) Bohs	Bohs 2453	Y/G	2.5	91
Solanum sessilistellatum Bitter‡	Cipollini 54	–	–	–
Solanum terminale Forssk.	Cipollini 134	R	0.3	13
Solanum torvum Swartz‡	Cipollini 64	Y	–	–
Solanum tucumanense Griseb.	Cipollini 25	R/O	0.8	60
Solanum umbellatum Mill.	Bohs 2560	Y/G	–	–
Solanum variabile Mart.‡	Cipollini 84	O	0.7	21
Solanum viarum Dunal	Cipollini 67	Y	4.9	30
Solanum villosum Mill.	Bohs 2553	O	0.3	43
Solanum virginianum L.	Cipollini 17	Y	4.1	268

‡Determination provisional.
*Bohs vouchers deposited at University of Utah; Cipollini vouchers and Berry College; RGO vouchers at University of Washington.
†If one voucher listed, DNA and fruit analyses were done on same accession. If two vouchers are listed, DNA and fruit analyses were performed on different accessions.
O, orange; G, green; W, white; Y, yellow; B, black/purple; R, red.

methods are assumed to be associated with phylogeny.

REGRESSIONS To provide ahistorical tests of hypotheses relating to physiological constraints, we calculated the linear regressions of fruit TGA content on leaf TGA content. We likewise examined linear regressions of ripe-fruit traits on unripe-fruit traits, and among ripe-fruit traits. Regression parameters were estimated using SPSS for Windows (SPSS, Inc., 1999).

PRINCIPAL COMPONENTS ANALYSIS (PCA) To describe overall relationships among fruit traits and thus test for fruit-trait covariation consistent with fruit-dispersal syndromes, we conducted PCAs using fruit-trait data using SPSS for Windows (SPSS, Inc., 1999). We conducted three TIPS-based PCAs using the following groups of species: (i) 26 species with both fruit chemical and morphological data; (ii) 30 species with both fruit chemical and DNA data; and (iii) 26 species with both fruit morphological and DNA data. Factor loadings were used to determine the strength of association of each fruit trait with each factor. Results using raw and orthogonally rotated matrices were similar and we thus report results only for unrotated matrices.

Phylogenetically corrected analyses

Following TIPS analyses, we applied three types of phylogenetic correction to our data (PA, IC and MAN) and then repeated regression and PCAs. Analyses were performed on data sets for which DNA data were available (sets ii and iii, above). We used all three methods because all are designed to control for inflated degrees of freedom in statistical analyses resulting from the non-independence of related species, and because none is universally accepted as the best approach to phylogenetic correction (Harvey and Pagel, 1991).

PHYLOGENETIC AUTOCORRELATION (PA) ANALYSIS Using one of the most parsimonious molecular phylogenies, we used COMPARE 4.3 (Martins, 2000) to quantify the phylogenetic component of each fruit trait in a PA analysis (Cheverud *et al.*, 1985; Gittleman and Kot, 1990). We then used the specific residual (putatively adaptive) component of each trait value in regression and PCA analyses.

INDEPENDENT CONTRASTS (IC) The IC method controls for phylogeny by splitting trait variation among related species into independent parts (Felsenstein, 1985). This is done via the estimation of nodal values from the phenotypes of descendants below nodes in a phylogeny. Comparisons are made using $(n - 1)$ independent standardized contrasts from a data set of n species. So, using COMPARE 4.3 (Martins, 2000) and the same molecular phylogeny used in PA analysis, we generated IC contrasts for each fruit trait and used these contrasts in regression and PCAs. Because the sign of IC contrasts is arbitrary, we computed regressions through the origin when evaluating relationships among traits.

SIGNED MANTEL (MAN) TESTS In signed MAN tests the dissimilarity between pairs of species in a dependent variable (Y variable) is compared with their phylogenetic distance and with their dissimilarity in the other independent variables (X variables) (Legendre *et al.*, 1994; Böhning-Gaese and Oberrath, 1999; Böhning-Gaese *et al.*, 2000). The Y matrix is regressed on the X matrices and tested for significance using MAN tests (Mantel, 1967; Smouse *et al.*, 1986). MAN tests use Monte Carlo randomizations, whereby the X matrices are held constant and the species in the Y matrix are randomly permuted (Smouse *et al.*, 1986; Legendre *et al.*, 1994). This method is a statistical approach to testing the effect of phylogeny on the Y variable and on the relationship between the Y and the X variables. MAN does not assume any particular microevolutionary process except that the mean dissimilarity of any pair of species in a Y variable is a linear function of their phylogenetic distance. To be consistent with TIPS-, IC- and PA-based PCAs, we conducted MAN-based PCAs using the chemical or morphological dissimilarity variables along with the genetic distance values. PCA factor loadings thus correspond to the chemical and morphological variables as well as genetic distance.

Results

Phylogenetic analyses

Fruit chemistry study

The data matrix for the fruit chemistry study included 30 *Solanum* species plus the outgroup *L. heteroclita*. The total aligned length of the sequences was 675 characters, including gaps; of these, 451 were invariant, 224 were variable and 152 were parsimony-informative. Parsimony analysis resulted in one most parsimonious tree of 528 steps (Fig. 8.1), with a

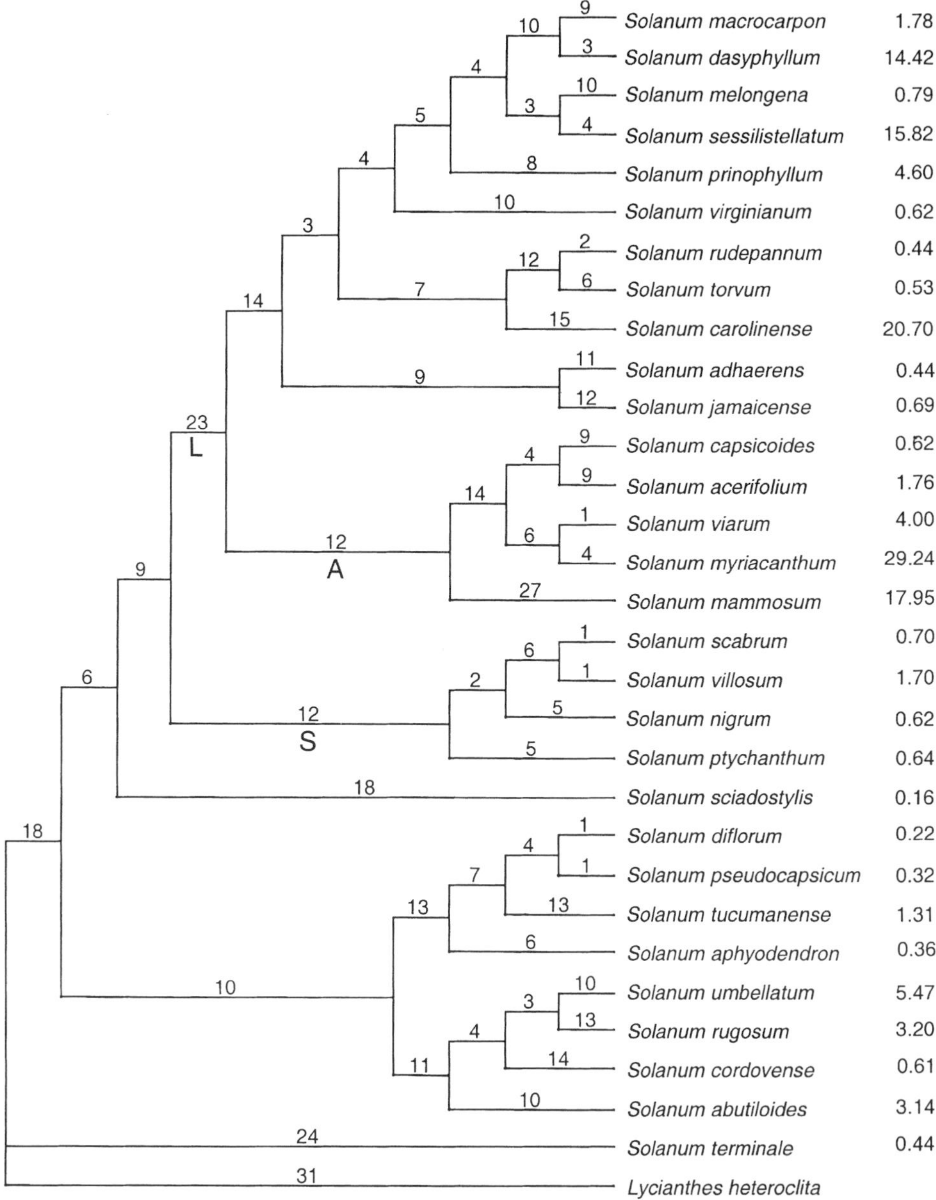

Fig. 8.1. Single most parsimonious tree resulting from analysis of nuclear ITS sequence data for 30 taxa of *Solanum* plus outgroup *Lycianthes heteroclita*. Length = 528 steps, consistency index (CI) excluding uninformative characters = 0.514, retention index (RI) = 0.743. Branch lengths are numbers of nucleotide substitutions; all characters weighted equally and gaps treated as missing data. Branches marked with letters delimit the following monophyletic groups: L = *Solanum* subgenus *Leptostemonum*; A = *Solanum* section *Acanthophora*; S = *Solanum* section *Solanum*. Data following species names are mean values for TGA of ripe-fruit pulp (mg g^{-1} dry mass).

consistency index (CI) (excluding uninformative characters) of 0.514 and a retention index (RI) of 0.743.

Looking across the phylogenetic tree in Fig. 8.1, fruit GA levels vary widely within some clades but are relatively similar in others. For instance, GA levels among species in the subgenus *Leptostemonum* (clade L in Fig. 8.1) vary by over an order of magnitude. This variation is especially notable in section *Acanthophora* (clade A), where fruits of *Solanum myriacanthum* contained nearly 50 times the concentration of those of *Solanum capsicoides*. On the other hand, GA levels among the sampled species of section *Solanum* (clade S) appear to be evolutionarily conservative. This pattern requires further investigation by including data from more species within these clades, and by confirming that GA content does not vary significantly within species.

Fruit morphology study

The fruit morphology data set contained ITS sequence data for 26 *Solanum* species plus the outgroup, *L. heteroclita*. The aligned sequence matrix included 682 characters per taxon (including gaps), of which 442 were invariant and 150 were parsimony-informative. Parsimony analysis found 60 equally parsimonious trees of 534 steps, with a CI (excluding uninformative characters) of 0.521 and an RI of 0.696. One of the 60 most parsimonious trees (not shown) was randomly chosen for input into the PA, IC and MAN tests.

Regression analyses

TIPS regressions indicated no relationship between leaf TGA and either unripe (R^2 = 0.001, P > 0.05, n = 18) or ripe fruit TGA (R^2 = 0.064, P > 0.05, n = 18). In contrast to our reanalysis of Bradley *et al.*'s (1979) data, these results show no evidence for physiological constraints of leaf chemistry on fruit chemistry, albeit for a smaller data set.

TIPS regressions for 30 species with fruit chemical data indicate significant relationships between unripe- and ripe- fruit TGA and between other unripe- and ripe-fruit variables (e.g. total phenolics and proteins (Table 8.2A)). Reanalysis using IC, PA and MAN methods left these results relatively unchanged

Table 8.2. Summary of TIPS-, IC-, PA- and MAN-based regressions for fruit chemical and morphological traits. TIPS results are the R^2 of regressions using raw data for species for which DNA data were available, IC results are the R^2 of regressions through the origin for standardized contrasts for each trait, PA results are the R^2 of regressions of specific residuals derived from phylogenetic autocorrelation analysis and MAN results are the R^2 values for whole-model regressions incorporating the genetic distance matrix as a covariate. All chemical data were in mg g^{-1} dry mass.

Regression	TIP	IC	PA	MAN[a]
A. Fruit chemical traits (n = 30 species)				
TGA ripe on unripe	0.263***	0.411***	0.265***	0.164**; ns
Phenolics ripe on unripe	0.148*	0.088ns	0.149*	0.068***; ns
Proteins ripe on unripe	0.408***	0.605***	0.413***	0.220***; ns
TGA ripe on phenolics ripe	0.256***	0.538***	0.257***	0.174***; ns
B. Ripe fruit morphological traits (n = 26 species)				
Pulp wet mass (g) on seed wet mass (g)	0.644***	0.751***	0.620***	0.497***; ns
Pulp dry mass (g) on seed dry mass (g)	0.794***	0.638***	0.792***	0.717**; ns
Pulp dry mass (g) on seed number	0.413***	0.347**	0.414***	0.334ns; ns
Seed dry mass (g) on yellow	0.199*	0.020ns	0.145*	0.059**; ns
Seed number on yellow	0.177*	0.028ns	0.172*	0.039*; ns

[a] P value for the partial regression coefficient of the dependent variable on the independent variable, followed by the P value for the partial regression coefficient of the dependent variable on the genetic distance matrix.

ns, $P > 0.05$; *, $P \leq 0.05$; **, $P \leq 0.01$; ***, $P \leq 0.001$.

(only occasionally was the significance of a relationship lost). This suggests that, for fruit chemical variables, phylogenetic effects are negligible and hence a correlation between unripe- and ripe-fruit secondary chemistry is supported. The magnitude of these effects is weak, however, as evidenced by relatively low R^2 values (Table 8.2A). Results of all regression analyses also suggest that, independent of phylogeny, fruits high in TGAs tend also to be high in phenolics (Table 8.2A).

TIPS regressions for fruit morphological traits were also relatively unaffected by phylogenetic correction (Table 8.2B). In all analyses, positive relationships were found between seed mass, pulp mass and seed number, indicative of allometric relationships among the traits. Seed number and seed mass were also positively related with yellow fruit colour (Table 8.2B; the relationship between yellow colour and whole-fruit mass was marginally significant (results not shown)). These and the regression of ripe-fruit phenolics on unripe-fruit phenolics were the only analyses where the IC method gave results incongruent with TIPS, PA or MAN analyses (being non-significant). Apart from those regressions in Table 8.2, we found no other significant regressions among fruit traits.

PCA analyses

Fruit morphology and chemistry

TIPS-based PCA using 26 species with fruit chemical and morphological traits shows a pattern consistent with preliminary expectations about overall fruit-trait variation (Fig. 8.2; Table 8.3A). In particular, the first PCA axis (which accounted for 28.85% of total variation) differentiates high-GA, high-phenolic, low-protein, low-sugar, large, dense (high dry-matter content), yellow fruits (mammal syndrome?) from low-GA, low-phenolic, high-protein, high-sugar, small, watery (low dry-matter content), red and black fruits (bird syndrome?). The second axis (14.68% of total variation) further differentiates yellow fruits with high TGA, high phenolics and large seed mass from larger orange and red fruits.

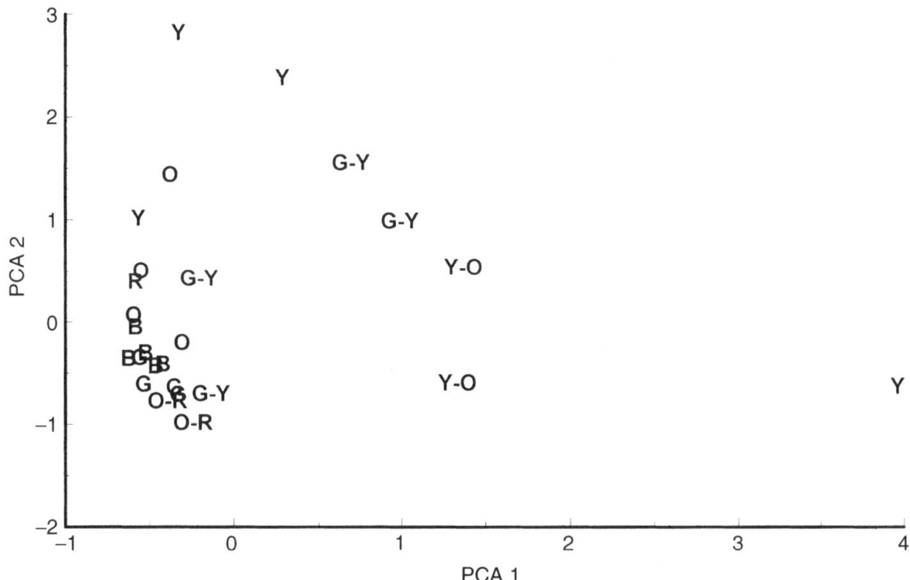

Fig. 8.2. Graph of PCA scores for the first two PCA axes, for the analysis using uncorrected fruit chemical and morphological traits (TIPS data). Factor loadings for the axes correspond to those listed in Table 8.3A. Symbols indicate ripe fruit colours of each species: R = red, B = black/purple, O = orange, Y = yellow, G = green.

Table 8.3. Results of principal component analysis. Data are factor loadings exceeding a value of 0.20 for the first two factors, as well as the per cent of total variation (%VAR) explained by each factor.

A. Fruit chemical and morphological traits (n = 26 species)

Factor	%VAR	RTGA	RPHN	RPRO	RSUG	UTGA	UPHN	UPRO	USUG	BLCK	GREN	YELL	ORNG	RED	WMS	PMS	SMS	PDM	DMP	DMS	SDM	ASM	SDN
TIPS		Factor loadings																					
1	28.98	0.54	0.41	–0.6	–0.35	0.40	0.40	–0.52	0.29	–0.38	–	0.82	–	–0.27	0.90	0.88	0.84	0.84	0.84	0.23	–	–	0.9
2	13.85	–0.50	–0.77	–	–	–0.73	–0.57	0.48	–	–	–	–0.20	0.34	0.29	0.33	0.28	0.42	0.47	0.37	–0.2	0.25	–0.31	0.24

B. Fruit chemical traits (n = 30 species)

Factor	%VAR	RTGA	RPHN	RPRO	RSUG	UTGA	UPHN	UPRO	USUG	GEND
TIPS		Factor loadings								
1	35.69	0.66	0.80	–	–0.62	0.70	0.68	–	–	n/a
2	21.15	0.22	0.31	–0.82	0.35	0.45	–	–0.84	0.51	n/a
IC										
1	35.30	0.87	0.92	–	–	0.85	0.39	–	0.28	n/a
2	30.15	0.22	–	0.92	–0.77	–0.41	–	0.83	–	n/a
PA										
1	36.16	0.67	0.81	–	–0.59	0.72	0.70	–	–	n/a
2	21.49	0.21	0.29	–0.82	0.40	0.43	–	–0.84	0.54	n/a
MAN										
1	20.91	0.66	0.82	–	–	0.86	–	–	–	–
2	16.65	–	–	0.71	0.57	–	–	0.80	–	–

C. Fruit morphological traits (n = 26 species)

Factor	%VAR	BLCK	GREN	YELL	ORNG	RED	WMS	PMS	SMS	PDM	DMP	DMS	SDM	ASM	SDN	GEND
TIPS		Factor loadings														
1	36.81	–0.28	–	–	–	–	0.95	0.90	0.96	0.95	–	–	0.96	–	0.74	n/a
2	17.61	–0.35	0.98	0.98	–0.41	–0.20	–	–	–	–	–	–	–	–0.24	–	n/a
IC																
1	36.01	–	–	–	–	–	0.97	0.94	0.96	0.89	–	–	0.96	–	0.70	n/a
2	19.87	–	–0.55	–	0.83	–	–	–	–	–	–0.75	0.91	–	–	–	n/a
PA																
1	38.72	–0.20	–	0.49	–	–	0.94	0.89	0.95	0.95	–	–	0.96	–	0.75	n/a
2	13.01	–0.22	–0.90	–0.42	0.62	0.27	–	–	–	–	0.31	–	–	–	–	n/a
MAN																
1	32.89	–	–	–	–	–	0.94	0.87	0.93	0.95	–	–	0.94	–	0.71	–
2	9.72	–0.55	–	–	0.41	0.62	–	–	–	–	0.37	–0.25	–	–0.26	–	0.57

Variable definitions:%VAR, percentage of total trait variation explained by each factor; RTGA, ripe-fruit TGA; RPHN, ripe-fruit total phenolics; RPRO, ripe-fruit total protein; RSUG, ripe-fruit total sugar; UTGA, unripe-fruit TGA; UPHN, unripe-fruit total phenolics; UPRO, unripe-fruit total protein; USUG, unripe-fruit total sugars; BLCK, ripe fruits dark blue/purple/black; GREN, ripe fruits green; YELL, ripe fruits yellow; ORNG, ripe fruits orange; RED, ripe fruits red; WMS, whole wet fruit mass (g); PMS, pulp wet mass (g); SMS, seed wet mass (g); PDM, pulp dry mass (g); DMP, dry-matter pulp; DMS, dry-matter seeds; SDM, seed dry mass (g); ASM, average seed dry mass (g); SDN, seed number per fruit; GEND, genetic distance. All chemical data are in mg g^{-1} dry mass.

Fruit chemistry

TIPS-based PCA of fruit chemical traits of 30 species (Fig. 8.3A; Table 8.3B) likewise shows a differentiation among species that may correspond to a gradient from mammal to avian dispersal. In this analysis, species are differentiated along the first PCA axis (35.69% of total variation) from fruits low in sugar and high in GAs and phenolics to those high in sugar and low in GAs and phenolics. As with regression analyses, PA-, IC- and MAN-based PCA analyses using the same data set suggest little effect of phylogeny (Fig. 8.3B; Table 8.3B). In all cases, species are distinguished primarily by fruit secondary chemistry (PCA axis 1) and secondarily by protein and sugar concentrations (PCA axis 2).

Fruit morphology

As for chemical traits, PCA analyses of morphological traits were relatively unaffected by phylogenetic correction (Fig. 8.3C, D; Table 8.3C). The results of these PCAs were consistent with those of the other PCAs; in this case, the difference is between species with black, orange and/or red fruits with low mass and low seed number and those with yellow or green fruits with high mass and high seed number. TIPS- and PA-based PCAs were most similar, whereas IC- and MAN-based PCAs differed from the other PCAs in not showing differentiation based upon fruit colour on the first axis, suggesting that fruit colour was somewhat related to phylogeny. This was supported by high factor loadings for genetic distance and for black, orange and red fruit colour on the second MAN-based PCA axis.

Conclusions and Significance of Our Study

Physiological and phylogenetic constraints

If secondary metabolites of ripe fruits serve some adaptive purposes, then our results should show substantial variation among species that is unexplained by phylogeny or by physiological constraints. Our preliminary analyses provide some evidence for physiological constraints on chemical traits of ripe fruits (i.e. some association between unripe- and ripe-fruit chemistry), but do not suggest that phylogeny has an important influence on fruit chemical or morphological trait variation within the species studied. Regarding the positive relationship between unripe- and ripe-fruit chemical characteristics, one might ask, 'Which is the chicken and which is the egg?' For example, is a positive relationship between GA content in ripe and unripe fruits more likely to be a consequence of the fruit's inability to remove all GAs during ripening of immature fruits (where the GAs presumably have some functional role) or a consequence of the need to build GA levels over a long growth period (perhaps coupled with functional roles in both fruit stages)? A constraint that 'accidentally' produces fruits toxic to dispersers would seem to be very costly in terms of fitness; meanwhile, some species (i.e. those with high TGA levels in unripe fruits and virtually none in ripe fruits) reduce TGA levels with apparent ease.

Our preliminary analyses also support the existence of independently evolved fruit-dispersal syndromes. Nevertheless, we cannot yet draw strong conclusions from these data because our phylogenetically corrected analyses are currently based on too few species (40 is generally considered a minimum (Martins, 2000)) and because phylogenetically corrected analyses based on fruit chemical traits were done separately from those using fruit morphological traits. We also cannot yet report phylogenetically corrected results examining the relationship of leaf chemistry to fruit chemistry (although our preliminary TIPS analysis suggest no relationship). In future work, we hope to assess all traits (genetic relatedness, leaf chemistry, fruit chemistry and fruit morphology) using a larger, single set of species. Related to the problem of low sample size is an incomplete sampling of fruit types and lack of replication at the lower levels of the phylogeny. As more (presumably similar) closely related species are added to the analysis, the strength of the phylogenetic effect may increase. Increased sampling within monophyletic clades will also strengthen our ability to estimate ancestral states, which is necessary for IC analysis.

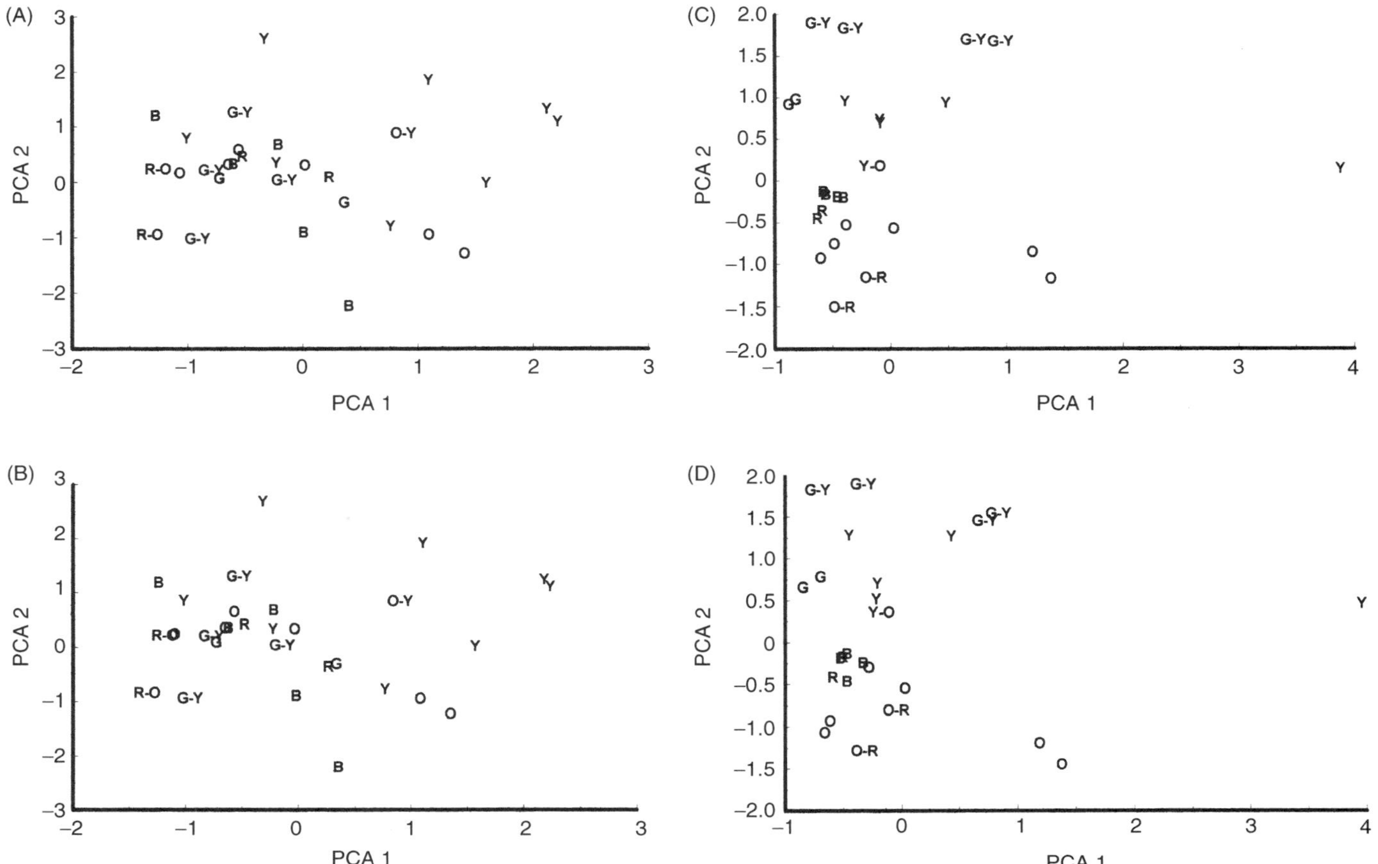

Fig. 8.3. Graphs of PCA scores for the first two PCAs using (A) uncorrected fruit chemical traits (TIPS data), (B) PA-corrected fruit chemical traits, (C) uncorrected fruit morphological traits (TIPS data), and (D) PA-corrected fruit morphological traits. Factor loadings correspond to those in Table 8.3B and C. Colour symbols correspond to those in Fig. 8.2.

Absent from our current analyses are estimates of variance due to within-species error and estimates of variance due to incorrect assumptions about the evolutionary model and/or incorrect specification of the phylogeny. So, in most cases (except MAN), we used standard parametric methods to determine the significance of statistical results. Martins and Hansen (1997) and Garland and Ives (2000) describe general approaches to incorporating within-species and phylogenetic sources of error into such analyses, which we plan to incorporate into our final analyses. Because our current analyses show little effect of phylogeny, we can only say that results for all methods of phylogenetic correction were basically similar.

'Phylogeny vs. ecology' vs. 'phylogeny and ecology'

Our study is focused mainly on the issue of phylogenetic effects and yet the distinction between ecology and phylogeny is not all that clear. In fact, some ecologists (e.g. Westoby *et al.*, 1995a,b,c) argue that trait variation across taxa is so inextricably intertwined with history that 'phylogeny' cannot be separated from 'ecology'. This is because related species may inherit similar habitats and selective regimes from their ancestors (the 'phylogenetic niche conservation' of Harvey and Pagel (1991)). Our inclination is that the IC and MAN methods might be more acceptable than PA, because variation attributed to phylogeny is not entirely removed from consideration when using IC or MAN. IC has lately become the method of choice for studies of plant evolution (Silvertown and Dodd, 1997), and progress has been made in reconciling the PA and IC methods into a general approach (see Martins and Hansen, 1997). On the other hand, since ancestral states must be estimated for IC, this method seems to require exhaustive sampling within clades to ensure valid ancestral state reconstructions. Considering all factors, the MAN technique appears to be a very promising approach to phylogenetic correction.

Westoby *et al.* (1995 a,b,c) and Givnish (1997) argue that all statistical 'phylogenetically correct' methods can be biased under certain circumstances: i.e. when the evolutionary model is incorrect (as in the case of strong directional selection), or when a trait or combination of traits of interest arises only a few times and it is assumed a priori that their persistence within a clade is a 'constraint'. In the latter case, the association of a trait (e.g. high ripe-fruit GA content) with an ecological correlate (e.g. dispersal by mammals) may seem to be explained by phylogeny and yet may have been adaptive when it first appeared and could have been maintained by selection thereafter. It may be equally plausible for correlations among traits to be maintained within clades by selective forces as it is for them to result from time-lags or from genetic, physiological and/or developmental constraints. Thus, neither adaptive nor non-adaptive explanations are necessarily more parsimonious. Adaptive explanations for traits and for covariation among traits are thus warranted in the absence of direct evidence for mechanisms of constraint, coupled with the presence of evidence of their current functions and/or fitness effects. This is a difficult issue to address and bolsters our belief that examinations of the functions and fitness values of fruit traits must remain an important approach to determining adaptive significance.

Future Work

How variable are closely related plant species in ripe-fruit secondary chemistry, and are differences among species the result of selection pressures related to frugivory? The rationale and study design needed to answer this question depend strongly on the results of a phylogenetically based study. The general approach is to study frugivory in the field, focusing on plants identified as key species for comparison. If highly contrasting taxa are common within clades (weak phylogenetic effects), one might ask questions concerning differences in the dispersers of these species. Associations of certain frugivore types with certain fruit chemical types would provide the strongest evidence of coadaptation. If one concludes, on the other hand, that variation in fruit chemical traits primarily reflects variation among clades (strong phylogenetic effects), one might question whether contrasting clades differ in the dispersers associated with each. One might find that species within clades are, indeed, dispersed by similar disperser types (e.g. some clades by passerine birds, others by bats). In either case, extant frugivore–fruit-type associations could be interpreted either as 'old' coadaptive relationships or as the result of similar extant frugivores selecting related fruit species because of phylogenetically constrained physiological and ecological similarities – the 'ecological fitting' of Janzen (1985). Regardless, evidence would exist for the importance of secondary chemistry in explaining current fruit-use patterns.

The general approach described here differs from that taken in most comparative studies of fruit traits, which tend to focus on the traits of many species occupying one habitat whose fruit vary strongly in secondary chemistry. Although more difficult to conduct, examination of evolutionary divergence within a group of phylogenetically related species occupying different habitats or selective regimes could be more revealing than studies focusing on such habitat-defined species assemblages, and analyses of fruit secondary chemical profiles are technically facilitated. Regardless, good estimates of phylogenetic relationships are essential for our approach.

We recommend continued exploration of the ecological functions of ripe-fruit secondary metabolites as a means of better understanding possible consequences for frugivory, seed dispersal and plant fitness. Regardless of conclusions about the evolution of fruit traits, a focus on secondary metabolites should continue to enlighten the understanding of the ecology of fruit–frugivore interactions. In addition to their importance for addressing evolutionary and ecological questions, studies of fruit secondary metabolites have significance for medicinal phytochemistry and for conservation. Using our study as an example, fruit GAs are of potential use as: precursors for steroid synthesis, anticancer agents, fungicides, molluscicides, pesticides, herbicides, antiparasitic agents, neurologically active agents and cholesterol-lowering agents (Cipollini, 2000). Focused study of such chemicals could result in the identification of new sources of known compounds or sources of novel compounds. This possibility provides an important argument for conservation efforts directed towards such taxa of potential medical importance (Tewksbury *et al.*, 1999).

Acknowledgements

We thank K. Patrick-Cipollini, I. Cipollini, M. Cipollini, R. Cipollini, S. King-Jones, D. Graybosch, A. Moore, K. Vaughn, A. Sanders and C. Manous for laboratory and field assistance, R.G. Olmstead for providing DNA and protocols, M. Nee for species determinations and the Nijmegen Botanical Garden for providing seed accessions. This work was supported by a National Science Foundation grant (DBI-9601505), Berry College School of Mathematics and Natural Sciences Environmental Science, Development of Undergraduate Research and Summer Travel grants to M.L.C., by Berry College Professional Development grants to M.L.C. and by National Science Foundation grants (DEB-9207359 and DEB-9996199) to L.A.B. We thank I. Izhaki and D. Levey for comments on the manuscript.

References

Baldwin, B.G., Sanderson, M.J., Porter, J.M., Wojciechowski, M.F., Campbell, D.F. and Donoghue, M.J. (1995) The ITS region of nuclear ribosomal DNA: a valuable source of evidence on angiosperm phylogeny. *Annals of the Missouri Botanical Garden* 82, 247–277.

Birner, J. (1969) Determination of total steroid bases in *Solanum* species. *Journal of Pharmaceutical Science* 58, 258–259.

Böhning-Gaese, K. and Oberrath, R. (1999) Phylogenetic effects on morphological, life-history, behavioural, and ecological traits of birds. *Evolutionary Ecology Research* 1, 347–364.

Böhning-Gaese, K., Halbe, B., Lemoine, N. and Oberrath, R. (2000) Factors influencing the clutch size, number of broods and annual fecundity of North American and European land birds. *Evolutionary Ecology Research* 2, 823–839.

Bohs, L. (2000) Slicing up the Solanums: major lineages and morphological synapomorphies. *American Journal of Botany* 87(6), 115 (Abstract).

Bohs, L. and Olmstead, R.G. (1997) Phylogenetic relationships in *Solanum* (*Solanaceae*) based on ndhF sequences. *Systematic Botany* 22, 5–17.

Bohs, L. and Olmstead, R.G. (1999) *Solanum* phylogeny inferred from chloroplast DNA sequence data. In: Nee, M., Symon, D.E., Lester, R.N. and Jessop, J.P. (eds) *Solanaceae IV: Advances in Biology and Utilization.* Royal Botanic Gardens, Kew, UK, pp. 97–110.

Bohs, L. and Olmstead, R.G. (2001) A reassessment of *Normania* and *Triguera* (*Solanaceae*). *Plant Systematics and Evolution* (in press).

Bradley, V., Collins, D.J., Eastwood, F.W., Irvine, M.C. and Swan, J.M. (1979) Distribution of steroidal alkaloid in Australian species of *Solanum.* In: Hawkes, J.G., Lester, R.N. and Skelding, A.D. (eds) *The Biology and Taxonomy of the Solanaceae.* Academic Press, London, UK, pp. 203–209.

Bremer, B. and Eriksson, O. (1992) Evolution of fruit characters and dispersal modes in the tropical family *Rubiaceae. Biological Journal of the Linnean Society* 47, 79–95.

Budini, R., Tonelli, D. and Girotti, S. (1980) Analysis of total phenols using the Prussian blue method. *Journal of Agricultural and Food Chemistry* 28, 1236–1238.

Cheverud, J.M., Dow, M.M. and Leutenegger, W. (1985) The quantitative assessment of phylogenetic constraints in comparative analyses: sexual dimorphism in body weights among primates. *Evolution* 39, 1335–1351.

Cipollini, M.L. (2000) Secondary compounds in fleshy fruits: evidence for adaptive functions. *Revista Chilena de Historia Natural* 73, 243–252.

Cipollini, M.L. and Levey, D.J. (1997a) Why are some fruits toxic? Glycoalkaloids in *Solanum* and fruit choice by vertebrates. *Ecology* 78, 782–798.

Cipollini, M.L. and Levey, D.J. (1997b) Antifungal activity of *Solanum* fruit glycoalkaloids: implications for frugivory and seed dispersal. *Ecology* 78, 799–809.

Cipollini, M.L. and Levey, D.J. (1997c) Secondary metabolites of fleshy vertebrate-dispersed fruits: adaptive hypotheses and implications for seed dispersal. *American Naturalist* 150, 346–372.

Cipollini, M.L. and Levey, D.J. (1998) Secondary metabolites as traits of ripe fleshy fruits: a response to Eriksson and Ehrlen. *American Naturalist* 152, 908–911.

Courtenay, O. (1994) Conservation of the maned wolf: fruitful relationships in a changing environment. *Canid News* 2, 1–5.

D'Arcy, W.G. (1972) Solanaceae studies II: typification of subdivisions of *Solanum. Annals of the Missouri Botanical Garden* 59, 262–278.

D'Arcy, W.G. (1991) The *Solanaceae* since 1976, with a review of its biogeography. In: Hawkes, J.G., Lester, R.N. and Skelding, A.D. (eds) *Solanaceae III: Taxonomy, Chemistry, Evolution.* Royal Botanic Gardens, Kew, UK, pp. 75–137.

Debussche, M. and Isenmann, P. (1989) Fleshy fruit characters and the choices of bird and mammal seed-dispersers in a Mediterranean region. *Oikos* 56, 327–338.

de Queiroz, K. (1996) Including characters of interest during tree reconstruction and the problems of circularity and bias in studies of character evolution. *American Naturalist* 148, 700–708.

Doyle, J.J. and Doyle, J.L. (1987) A rapid DNA isolation procedure for small quantities of fresh leaf tissue. *Phytochemical Bulletin* 19, 11–15.

Ehrlen, J. and Eriksson, O. (1993) Toxicity in fleshy fruits – a non-adaptive trait? *Oikos* 66, 107–113.

Eriksson, O. and Ehrlen, J. (1998) Secondary metabolites in fleshy fruits: are adaptive explanations needed? *American Naturalist* 152, 905–907.

Felsenstein, J. (1985) Phylogenies and the comparative method. *American Naturalist* 125, 1–15.

Garland, T., Jr and Ives, A.R. (2000) Using the past to predict the present: confidence intervals for regression equations in phylogenetic comparative methods. *American Naturalist* 155, 346–364.

Gautier-Hion, A. (1990) Interactions among fruit and vertebrate fruit-eaters in an African tropical rain forest. In: Bawa, K. and Hadley, M. (eds) *Reproductive Ecology of Tropical Forest Plants.* Man and the Biosphere Series, Vol. 7, Parthenon Press, Paris, France, pp. 219–232.

Gittleman, J.L. and Kot, M. (1990) Adaptation: statistics and a null model for estimating phylogenetic effects. *Systematic Zoology* 39, 227–241.

Givnish, T.J. (1997) Adaptive radiation and molecular systematics: issues and approaches. In: Givnish, T. and Sytsma, K. (eds) *Molecular Evolution and Adaptive Radiation.* Cambridge University Press, Cambridge, UK, pp. 1–54.

Harvey, P.H. and Pagel, M.D. (1991) *The Comparative Method in Evolutionary Biology.* Oxford University Press, Oxford, UK.

Herrera, C.M. (1992) Interspecific variation in fruit shape: allometry, phylogeny, and adaptation to dispersal agents. *Ecology* 73, 1832–1841.

Herrera, C.M. (1998) Long-term dynamics of Mediterranean frugivorous birds and fleshy fruits: a 12-year study. *Ecological Monographs* 68, 511–538.

Howe, H.F. (1984) Constraints on the evolution of mutualisms. *American Naturalist* 123, 764–777.

Janson, C.H. (1983) Adaptation of fruit morphology to dispersal agents in a neotropical forest. *Science* 219, 187–188.

Janzen, D.H. (1985) On ecological fitting. *Oikos* 45, 308–310.

Jones, C.G., Hare, J.D. and Compton, S.J. (1989) Measuring plant protein with the Bradford assay. 1. Evaluation and standard methodology. *Journal of Chemical Ecology* 15, 979–992.

Jordano, P. (1987a) Diet, fruit choice and variation in body condition of frugivorous warblers in Mediterranean scrubland. *Ardea* 76, 193–209.

Jordano, P. (1987b) Patterns of mutualistic interactions in pollination and seed dispersal: connectance, dependence asymmetries, and coevolution. *American Naturalist* 129, 657–677.

Jordano, P. (1995) Angiosperm fleshy fruits and seed dispersers: a comparative analysis of adaptation and constraints in plant–animal interactions. *American Naturalist* 145, 163–191.

Kingsbury, J.M. (1964) *Poisonous Plants of the United States and Canada*. Prentice-Hall, Englewood Cliffs, New Jersey.

Legendre, P., Lapointe, F.-J. and Casgrain, P. (1994) Modeling brain evolution from behaviour: a permutational regression approach. *Evolution* 48, 1487–1499.

Mantel, N. (1967) The detection of disease clustering and a generalized regression approach. *Cancer Research* 27, 209–220.

Martins, E.P. (2000) *COMPARE: Computer Programs for the Statistical Analysis of Comparative Data*, Version 4.3. Department of Biology, University of Oregon, Eugene, Oregon. http://darkwing.uoregon.edu/~compare4/

Martins, E.P. and Hansen, T.F. (1997) Phylogenies and the comparative method: a general approach to incorporating phylogenetic information into the analysis of interspecific data. *American Naturalist* 149, 646–667.

Nee, M. (1991) 17. Synopsis of *Solanum* Section Acanthophora: a group of interest for glycoalkaloids. In: Hawkes, J.G., Lester, R.N., Nee, M. and Estrada, J. (eds) *Solanaceae III: Taxonomy, Chemistry, Evolution*. Royal Botanical Gardens Kew and Linnean Society of London, London, pp. 257–266.

Olmstead, R.G. and Palmer, J.D. (1997) Implications for the phylogeny, classification, and biogeography of *Solanum* from cpDNA restriction site variation. *Systematic Botany* 22, 19–29.

Olmstead, R.G., Sweere, J.A., Spangler, R.E., Bohs, L. and Palmer, J.D. (1999) Phylogeny and provisional classification of the *Solanaceae* based on chloroplast DNA. In: Nee, M., Symon, D.E., Lester, R.N. and Jessop, J.P. (eds) *Solanaceae IV: Advances in Biology and Utilization*. Royal Botanic Gardens, Kew, pp. 111–137.

Ripperger, H. and Schreiber, K. (1981) Solanum steroid alkaloids. In: Manske, R.H.F. and Rodrigo, R.G.A. (eds) *The Alkaloids: Chemistry and Physiology*, Vol. XIX. Academic Press, New York, pp. 81–192.

Schreiber, K. (1963) Isolierung von Solasodinglykosiden aus Pflanzen der Gattung *Solanum* L. *Solanum*-Alkaloide. XXVIII. *Mitteilung. Kulturpflanze* 11, 451–501.

Schreiber, K. (1968) Steroid alkaloids: the *Solanum* group. In: Manske, R.H.F. (ed.) *The Alkaloids, Chemistry and Physiology*, Vol. X. Academic Press, New York, pp. 1–192.

Silvertown, J. and Dodd, M. (1997) Comparing plants and connecting traits. In: Silvertown, J., Franco, M. and Harper, J.L. (eds) *Plant Life Histories: Ecology, Phylogeny and Evolution*. Cambridge University Press, Cambridge, UK, pp. 3–35.

Smith, D. (1981) *Removing and Analyzing Total Nonstructural Carbohydrates from Plant Tissue*. Publication R2107, Wisconsin University Extension Service, Madison, Wisconsin, USA.

Smouse, P.E., Long, J.C. and Sokal, R.R. (1986) Multiple regression and correlation extensions of the Mantel test of matrix correspondence. *Systematic Zoology* 35, 627–632.

Spooner, D.M., Olmstead, R.G. and Bohs, L. (1999) Current data on the systematics of the *Solanaceae*, with a focus on potatoes and tomatoes. In: *Plant and Animal Genome VII. Abstracts*, San Diego, CA. p.67.

SPSS, Inc. (1999) *SPSS for Windows*, Version 10.0. SPSS, Inc., Chicago, Illinois.

Stiles, F.G. and Rosselli, L. (1992) Consumption of fruits of the *Melastomataceae* by birds: how diffuse is coevolution? *Vegetatio* 107, 57–74.

Swofford, D. (2000) *PAUP* 4.0*. Beta version 4.0b3a. Sinauer Associates, Sunderland, Massachusetts.

Tewksbury, J.J., Nabhan, G.P., Norman, D., Suzan, H., Tuxhill, J. and Donovan, J. (1999) 'In situ' conservation of wild chiles and their biotic associates. *Conservation Biology* 13, 98–107.

van der Pijl, L. (1969) *Principles of Dispersal in Higher Plants*. Springer-Verlag, Berlin, Germany.

van Gelder, W.M. (1990) Chemistry, toxicology, and occurrence of steroidal glycoalkaloids: potential contaminants of the potato (*Solanum tuberosum* L.). In: Rizk, A.-F.M. (ed.) *Poisonous Plant Contamination of Edible Plants*. CRC Press, Boca Raton, Florida, pp. 117–156.

Westoby, M., Leishman, M. and Lord, J. (1995a) On misinterpreting the phylogenetic correction. *Journal of Ecology* 83, 531–534.

Westoby, M., Leishman, M. and Lord, J. (1995b) Further remarks on phylogenetic correction. *Journal of Ecology* 83, 727–729.

Westoby, M., Leishman, M. and Lord, J. (1995c) Issues of interpretation after relating comparative datasets to phylogeny. *Journal of Ecology* 83, 892–893.

Wheelwright, N.T. (1985) Fruit size, gape width, and the diets of fruit-eating birds. *Ecology* 66, 808–818.

Willson, M.F., Irvine, A.K. and Walsh, N.G. (1989) Vertebrate dispersal syndromes in some Australian and New Zealand plant communities, with geographic comparisons. *Biotropica* 21, 133–147.

9 The Seed-dispersers and Fruit Syndromes of *Myrtaceae* in the Brazilian Atlantic Forest

Marco A. Pizo

Departamento de Botânica – IB, Universidade Estadual Paulista, Caixa Postal 199, 13506-900 Rio Claro, São Paulo State, Brazil

Introduction

Historically, fruit syndromes have been interpreted as adaptations of plants to their seed-dispersers (Ridley, 1930; van der Pijl, 1982). According to this view, fruit characteristics have been moulded by coevolution between groups of seed-dispersers and plants to facilitate the location of fruits and dispersal of seeds. Recent studies, however, have questioned this exclusively adaptationist interpretation of fruit syndromes, pointing to historical (Herrera, 1992a) and phylogenetic constraints (Herrera, 1992b; Jordano, 1995). For example, considering fruit morphological and chemical traits in 910 angiosperm species and using statistical procedures that control for taxonomic relatedness, Jordano (1995) concluded that only fruit size (mainly fruit diameter) was associated with type of seed-disperser. However, the possibility exists that patterns of correlated evolution in particular clades are obscured by more general trends through the whole angiosperm clade. Thus, Jordano (1995) suggested that local analyses with narrower taxonomic scope (e.g. within-family analyses) are needed to refine our understanding of fruit syndromes in particular and plant–seed-disperser mutualisms in general.

To study dispersal syndromes in a narrow taxonomic scope, one must pick a taxon with wide variation in fruit traits and disperser taxa. The *Myrtaceae* species occurring in the Brazilian Atlantic Forest are particularly suited for this. Besides being a dominant plant family in the Atlantic Forest (Mori *et al.*, 1983), all *Myrtaceae* species in this region produce berry-like, fleshy fruits with a wide diversity of characteristics (Landrum and Kawasaki, 1997). Therefore, a diverse assemblage of vertebrate frugivores is expected. Exactly which frugivores disperse seeds is largely unknown because only anecdotal observations are available for a few species (Kuhlmann and Kühn, 1947). Moreover, as a consequence of the severe degradation of the Atlantic Forest, some *Myrtaceae* species endemic to this biome are disappearing before we can gather even a basic knowledge of their biology. This is particularly regrettable because even though several *Myrtaceae* fruits are economically exploited in Brazil and worldwide (e.g. *Eugenia uniflora, Psidium guajava*), these species represent only a fraction of the economic potential of the family. In particular, many non-commercial species produce fruits edible by humans (Kawasaki and Landrum, 1997).

I identified the seed-dispersers and investigated fruit syndromes among the *Myrtaceae* species in the Brazilian Atlantic Forest, one of the most threatened biomes in the world, and rich in *Myrtaceae* species. I observed frugivore visits to 20 species of *Myrtaceae* and examined removal of fallen fruits of an additional five species in a well-preserved lowland Atlantic Forest site in south-east Brazil. Additional information was gathered from the literature to provide a general overview of the seed-dispersers of a total of 68 *Myrtaceae* species in the Atlantic Forest. These data were combined with data on associated variation in fruit morphology and chemical composition to determine if suites of fruit traits are associated with distinct sets of frugivores. Specifically, I asked:

1. Are some *Myrtaceae* genera closely associated with particular groups of seed-dispersers?
2. Do genera with a diverse assemblage of dispersers display greater variation in fruit morphology than genera dispersed by only one or a few groups of frugivores?
3. Do fruits dispersed by birds and monkeys, the most common seed-dispersers of Atlantic forest *Myrtaceae*, differ in their morphological and chemical attributes?

Study Site and Methods

Study site

Data were collected at the Saibadela Research Station (24° 14′ S, 48° 04′ W; 70 m a.s.l.) in the Parque Estadual Intervales, a 490 km^2 reserve located in São Paulo state, south-east Brazil. Parque Intervales forms, with adjacent reserves, one of the largest blocks of Atlantic Forest remaining in Brazil, totalling 1200 km^2. The reserve contains a complete suite of potential seed-dispersers, including large frugivorous birds, monkeys and terrestrial mammals (Aleixo and Galetti, 1997). Old-growth forest (*sensu* Clark, 1996) predominates at Saibadela. *Myrtaceae* are the dominant plant family in number of species and individuals with diameter at breast height (dbh) > 5 cm (Almeida-Scabbia, 1996). To date, 40 *Myrtaceae* species representing ten genera have been collected at Saibadela.

Seed-dispersers

Between January 1999 and June 2000, to detect bird and monkey activity in fruiting trees at Saibadela, I observed one to five individual trees for each species during 2 h observation sessions set from 0600 to 1000 h (mean observation time per species = 4.8 h, $n = 20$). For bats, observations were conducted on the same trees with the aid of a torch from 1900 to 2300 h during 1 h observation sessions (mean observation time per species = 1.5 h, $n = 20$). Trees were observed on non-rainy days during their peak period of fruit production. Although I identified birds and monkeys to species, I made no attempt to identify bats. The fruit-handling behaviour of animals visiting the trees allowed me to classify them as seed-dispersers, fruit-pulp consumers and/or seed predators.

Because rodent-dispersed seeds are usually large-sized (Forget, 1990), *Myrtaceae* species with seeds ≥ 2 cm diameter (*Eugenia melanogyna, Eugenia mosenii, Eugenia cambucarana, Eugenia multicostata* and *Eugenia neoverrucosa*) were tested for dispersal by rodents. Seeds were marked by attaching a coloured flag (3 cm × 10 cm) to the end of a 20–30 cm long piece of monofilament line that passed through a 3 mm hole drilled in the seed (see Forget, 1990). The holes made in the seeds did not preclude germination and apparently did not make them more vulnerable to insect and fungal infestation. Depending on the availability of seeds, groups of three to five marked seeds were placed on the forest floor beneath two to three parent trees during the peak fruiting period of each species, totalling 30–76 seeds per species. No more than 25 marked seeds were tested simultaneously under a given tree. Marked seeds were monitored monthly for 5 months. The following fates were recorded: (i) eaten, if the seed was gnawed by rodents or removed from the monofilament line; (ii) cached, if the seed was buried below the soil surface or the leaf litter; and (iii) unknown fate, if the seed was removed from the immediate vicinity of the tree still attached to its line and not found. If a seed was cached and then recovered and gnawed, it was considered eaten. I searched for marked seeds within 10 m of their original positions. Seeds were considered dispersed if they were cached and not recovered by rodents by the end of the 5-month monitoring period. Based on previous experience at the site, most, if not all, seeds are cached in the first month and the recovery of cached seeds, if any, takes place in the next two months (M.A. Pizo, personal observation).

I compiled published records of frugivores eating *Myrtaceae* fruits in the Atlantic Forest. Most often these records came from studies focusing on the diet of a particular frugivore species. I paid special attention to information on fruit-handling behaviour, which I used to infer whether a given frugivore actually disperses seeds. When such information was not provided, the frugivore's role as a seed-disperser was inferred from comparisons with related taxa or from the morphology of its mouth and digestive system. Records from outside the Atlantic Forest were only used if the species also occurred in this region. I did not include records of immature fruit consumption.

Morphology and chemical composition of fruits

Fruit colour and size (maximum width (W) and length (L)), seed size (seed length, which is the most common seed measure provided in the literature) and mean seed number per fruit were recorded for 32 species (one to five individual trees were sampled per species) at Saibadela. In addition, the ratio between length and width (L/W) was used as an index of fruit shape (globose fruits, L/W = 1; elongated fruits, L/W > 1; depressed fruits, L/W < 1). Black (including dark red) and red fruits were pooled for analysis, as were yellow and orange fruits. I pooled these colours because they are not always easily distinguishable (e.g. dark red fruits may appear black) and usually form a continuum that reflects fruit maturation. The length and width (diameter) of 10 to 15 fruits and seeds of each species collected at the study site were measured with callipers to the nearest 0.1 mm. For species not occurring at Saibadela, published data were used. When a range of measures was provided, I took the mid-point of the distribution.

Major nutritional components (water, lipids, proteins, water-soluble carbohydrates, fibre and ash) were analysed from fruits collected in the field. Pulp samples from fruits collected directly from the canopy or from recently fallen fruits were frozen until they were chemically analysed. An attempt was made to collect fruits from as many individual trees as possible. Insufficient material, however, precluded a complete analysis for some species. Lipids were analysed according to Bligh and Dyer (1959). Total nitrogen (N) was analysed by the micro-Kjeldahl method (AOAC, 1990) and converted into crude protein by multiplying N by 6.25. It should be noted, however, that a conversion factor of 6.25 overestimates the protein actually available for frugivores in some fruits (Levey *et al.*, 2000). Water-soluble carbohydrates and acid-detergent fibre were analysed using methods no. 325.35B and no. 973.18, respectively, of the Association of Official Analytical Chemists (AOAC, 1990). Ash content was calculated after incineration of the sample to constant mass in a muffle furnace at 550°C.

Statistical procedures

Correspondence analysis was performed on the presence–absence matrix of species × type of disperser to investigate the 'radiation' of *Myrtaceae* genera across the different seed-disperser groups. Two principal component analyses (PCAs) were performed on the matrix of *Myrtaceae* species × fruit traits, one with morphological variables and the other with chemical variables. Soluble carbohydrates were not considered in the latter analysis, due to small sample size.

If fruit traits are evolutionarily determined by seed-dispersers, as the concept of fruit syndromes implies, genera with many disperser types should exhibit greater variation in morphological traits than genera dispersed by only one or few disperser types. To test this prediction, I carried out across-genera Pearson correlations between the mean number of distinct disperser types per species and the coefficient of variation of morphological fruit traits. Insufficient data precluded such analysis for chemical traits.

Given that birds and monkeys were by far the most frequent seed-dispersers (see below) and that these two groups of seed-dispersers are often reported to be associated with different groups of fruits (bird fruits tend to be smaller and more elongated than monkey fruits (Janson, 1983; Mazer and Wheelwright, 1993; but see Gautier-Hion *et al.*, 1985)), I looked for differences in morphological traits between the sets of fruits eaten by birds and monkeys. I randomly selected congeneric pairs of species, one member of the pair bird-dispersed and one monkey-dispersed. I then tallied the number of these pairs in which fruit width and length were lower and L/W higher for bird-dispersed than for monkey-dispersed species (i.e. the number of pairs with fruit morphology consistent with an 'adaptive' hypothesis of decreasing fruit size and increasing elongation with bird syndrome). Statistical significance was assessed with a binomial test. Insufficient data precluded the use of such pairwise congeneric comparisons for seed number and size and for all chemical components. For these traits, Student's *t*-tests were applied to log-transformed (seed traits) and arcsin-transformed (chemical traits) data to test for differences between bird- and monkey-dispersed fruits. For the *t*-tests I used a Bonferroni-corrected *P* level of 0.008. None the less, it should be noted that the statistical significance of the *t*-tests is inflated, because of lack of phylogenetic independence among species (Felsenstein, 1985). However, given that the majority of the tests did not reach significance in spite of type I error, I believe my conclusions are robust.

Mean ± SD is reported throughout the text. A significance level of 0.05 was used unless otherwise stated.

Results

Seed-dispersers

Seed-dispersers of 68 species representing 14 genera of Atlantic Forest *Myrtaceae* were identified (Table 9.1). Birds and monkeys were the most frequent seed-dispersers, recorded for 44 and 38 species, respectively (many species were dispersed by both birds and monkeys). They were followed by carnivorous mammals (for eight *Myrtaceae* species), ungulates (four species), bats (three species), rodents (three species) and marsupials (one species). Each species was dispersed on average by 1.6 different disperser types. Some genera were dispersed by several taxa (e.g. *Eugenia, Psidium*), whereas others were dispersed by only one or a few groups (e.g. *Gomidesia, Marlierea*) (Table 9.1).

Birds as small as piprids (e.g. the swallow-tailed manakin, *Chiroxiphia caudata*, 24 g) and as large as cracids (e.g. the jacutinga, *Pipile jacutinga*, ~1.5 kg) eat *Myrtaceae* fruits and disperse their seeds. Monkeys also displayed great variation in body size, from 500 g (tamarins, *Saguinus* spp.) to approximately 9 kg (muriqui monkeys, *Brachyteles arachnoids*).

The South American coati (*Nasua nasua*) and several canids (*Cerdocyon thous, Lycalopex vetulus, Chrysocyon brachyurus*) were the carnivorous mammals for which effective dispersal of *Myrtaceae* has been documented in the Atlantic Forest. These mammals eat fallen fruits underneath parent plants. The same is true for the ungulates, tapir (*Tapirus terrestris*) and deer (*Mazama* spp.).

Bats were recorded eating only the fruits of *Eugenia stictosepala, Psidium catleyanum* and *P. guajava*. Evidence of effective dispersal by rodents at Saibadela was found for *E. cambucarana, E. multicostata* and *E. neoverrucosa*. For these species, 5–10% of the seeds were found cached (Table 9.2). These percentages probably underestimate the actual frequency of caching because some of the seeds removed from their lines were included in the 'eaten' category, but may actually have been cached. Seeds were found cached either 1–2 cm below the soil surface or under the leaf litter, from 1.6 to 6.1 m away from parent trees. Agoutis (*Dasyprocta agouti*) and spiny rats (*Proechymis iheringi*), two common rodents at the study site, were the most likely dispersers of the seeds tested.

One opossum (*Didelphis marsupialis*) is the only marsupial recorded eating a *Myrtaceae* fruit (*P. guajava*) in the Atlantic Forest (Cordero and Nicolas, 1987).

The first two dimensions of the correspondence analysis accounted for 62% of the total variation associated with the *Myrtaceae*

Table 9.1. Frequency of each seed-disperser group recorded of *Myrtaceae* genera in the Brazilian Atlantic forest.

Genus	Number of species	Per cent of species dispersed by							Mean no. of distinct disperser types/species
		Birds	Bats	Carnivorous mammals	Monkeys	Rodents	Marsupials	Ungulates	
Calycorectes	1	100	0	0	0	0	0	0	1.0
Calyptranthes	1	100	0	0	0	0	0	0	1.0
Campomanesia	7	14.3	0	14.3	100	0	0	28.6	1.6
Eugenia	22	54.5	4.5	13.6	50.0	13.6	0	0	1.4
Gomidesia	6	100	0	0	0	0	0	0	1.0
Marlierea	6	66.7	0	0	66.7	0	0	0	1.3
Myrceugenia	1	100	0	0	100	0	0	0	2.0
Myrcia	12	83.3	0	16.7	50.0	0	0	0	1.5
Myrcianthes	1	100	0	0	100	0	0	0	2.0
Myrciaria	3	33.3	0	33.3	100	0	0	0	1.7
Neomitranthes	2	50.0	0	50.0	0	0	0	0	1.0
Plinia	2	100	0	0	0	0	0	0	1.0
Psidium	3	66.7	66.7	33.3	100	0	33.3	66.7	3.6
Siphoneugena	1	100	0	0	100	0	0	0	2.0

species × seed-disperser matrix. The first dimension (eigenvalue = 1.0000; 39% of total variation) clearly separated the three *Eugenia* species that are rodent-dispersed, while the second dimension (eigenvalue = 0.5919; 23% of total variation) distinguished between the genera and species that are exclusively dispersed by birds (e.g. *Gomidesia*) and those that are either predominantly dispersed by monkeys (e.g. *Campomanesia*) or have a mix of disperser types (Fig. 9.1).

Fruit morphology, chemistry and syndromes

Among the 68 *Myrtaceae* species whose seed-dispersers were identified, fruit size varies widely: 5–47 mm in width, 6–50 mm in length. Fruit shape varies from elongated (e.g. *E. stictosepala*, L/W ratio = 1.70) to markedly depressed (e.g. *Eugenia riedeliana*, L/W ratio = 0.70). Most genera have few-seeded fruits (one to three seeds per fruit), but

Table 9.2. Fate of marked *Myrtaceae* seeds exposed to rodents at Parque Intervales, Saibadela Research Station. Seeds were monitored for 5 months or until they disappeared.

Species	Number of trees	Number of seeds	Not removed	Removed* Eaten	Removed* Cached	Removed* Unknown fate
Eugenia cambucarana	3	48	25	14	3	6
Eugenia melanogyna	3	46	43	0	0	3
Eugenia multicostata	3	76	18	50	4	4
Eugenia neoverrucosa	2	47	31	7	5	4
Eugenia mosenii	2	30	25	5	0	0

*'Eaten' refers to seeds gnawed by rodents on the spot or removed from the attached lines; 'Cached' refers to seeds buried below the soil surface or leaf litter; 'Unknown fate' refers to seeds removed from the immediate vicinity of the tree with line still attached but not found.

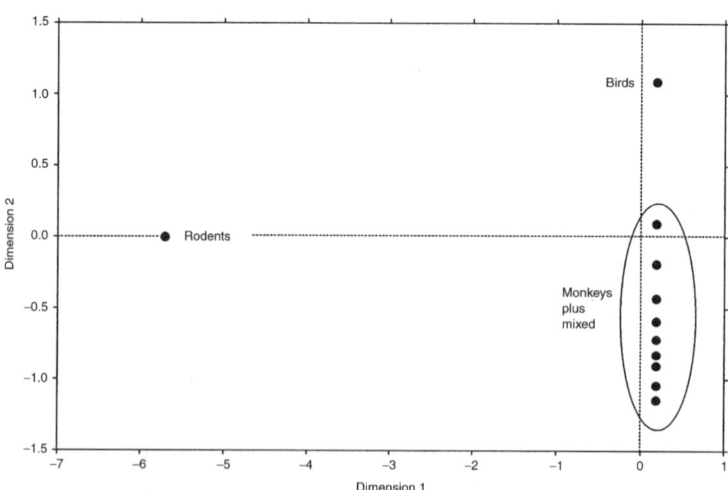

Fig. 9.1. Ordination from the correspondence analysis performed on the presence–absence matrix of *Myrtaceae* species × type of seed-disperser. The first dimension separates the three rodent-dispersed *Eugenia* species (*E. cambucarana, E. multicostata* and *E. neoverrucosa*). The second dimension separates species exclusively dispersed by birds (e.g. *Gomidesia* spp.) from species that are exclusively dispersed by monkeys (e.g. most *Campomanesia*) or have a mix of seed-dispersers (inside the ellipse).

Campomanesia (four to 18 seeds per fruit) and, especially, *Psidium* (up to 250 seeds per fruit) have many small seeds. There was also considerable variation in colour. Black (including purple) was the most common colour, followed by yellow, red, orange, grey and green (Table 9.3).

The first two factors of the PCA conducted with the fruit morphological traits accounted for 79.5% of total variation (Table 9.4). The first factor is related to fruit size, with fruit width, fruit length and seed size (which is positively correlated with fruit width, $r = 0.54$, $n = 43$, $P \leq 0.001$) having large positive loadings on it. This axis separates the three rodent-dispersed *Eugenia* species from the others (Fig. 9.2A). The second factor is basically explained by a gradient of fruit form (species with elongated fruit having positive scores on it) and fruit seediness (species with seedy fruits having negative scores on it). Fruit width has a marginal positive loading on this axis, probably due to the marginal positive correlation between fruit diameter and fruit seediness ($r = 0.27$, $n = 45$, $P = 0.07$). The second axis roughly separates the predominantly monkey-dispersed *Campomanesia* species from species that are bird-dispersed or have a mix of seed-dispersers (Fig. 9.2A).

The predicted increase in within-genera morphological variation with the increase in number of disperser types did not hold for any of the traits considered (width: $r = 0.59$, $P = 0.09$, $n = 9$; length: $r = 0.67$, $P = 0.09$, $n = 7$; L/W ratio: $r = 0.42$, $P = 0.41$, $n = 6$; number of seeds per fruit: $r = 0.64$, $P = 0.08$, $n = 8$; seed size: $r = -0.23$, $P = 0.62$, $n = 7$).

Birds and monkeys were not associated with any fruit colour (chi-square test with Yates correction: all P values > 0.20). Pairwise congeneric comparisons between bird- and monkey-dispersed fruits revealed that the former have significantly smaller widths than the latter (binomial test: $P = 0.03$), but do not differ in length ($P = 0.09$) or L/W ratio ($P = 0.31$). No significant differences were found when t-tests were applied to the whole pool of morphological traits considered for bird- and monkey-dispersed fruits (Table 9.5).

The three species dispersed by rodents are among the largest-fruited neotropical *Myrtaceae* (width = 39.4 ± 8.5 mm, range 30–47 mm). They are all green or yellow, muricate fruits (i.e. with the surface covered by small tubercles) with a few, large, well-protected seeds and fibrous pulp.

In contrast to the wide interspecific variation in morphological traits, species were fairly similar in the chemical composition of the fruit pulp (Table 9.6). *Myrtaceae* fruits are typically high in water and carbohydrates and low in lipids and proteins. Lipid is the most variable component, but does not exceed 19% dry mass for any species (Table 9.6).

The PCA conducted with fruit chemical traits revealed that two factors accounted for 64.7% of total variation but failed to clearly group species according to their seed-dispersers, even for the morphologically distinctive rodent-dispersed *Eugenia* species (Table 9.4; Fig. 9.1B). The first factor reflects a gradient in fruit succulence; species with watery fruits, rich in protein and ash, have positive loadings on it. The second factor is simply a gradient between species with fibrous fruits, which have negative scores on this axis, and species with oily, energy-rich fruits, which have positive scores on it (Fig. 9.2B).

No significant differences were found between fruit species eaten by birds and monkeys in any of the chemical components analysed (Table 9.5).

Discussion

Seed-dispersers

My survey of plant–seed-disperser interactions conducted over a large geographical area (the Atlantic Forest extends from 8° to 28°S) and focused on a diverse plant family such as *Myrtaceae* is obviously incomplete. Published reports on diet were scarce, especially for mammals other than monkeys. I tried to overcome such limitations by focusing on fieldwork at a *Myrtaceae*-rich site possessing a complete suite of potential seed-dispersers. However, even this strategy has drawbacks. In particular, focal tree observations tend to overlook rare visitors. Such constraints call for caution in the interpretation of the patterns that emerged from my data.

Table 9.3. Fruit morphological traits and colours of the genera of *Myrtaceae* sampled in the Brazilian Atlantic Forest. Values refer to means (sample sizes in parentheses) for the species whose seed-dispersers were identified.

Genus	Fruit width (W) (mm)	Fruit length (L) (mm)	L/W ratio	Mean no. of seeds per fruit	Seed length (mm)	Fruit colour (per cent of species)						
						n	Grey	Black	Green	Orange	Red	Yellow
Calycorectes	17.5 (1)	14.0 (1)	0.8 (1)	1.1 (1)	11.5 (1)	1	0	0	0	0	100	0
Calyptranthes	12.9 (1)	11.4 (1)	0.9 (1)	1.1 (1)	9.4 (1)	1	0	100	0	0	0	0
Campomanesia	23.9 (7)	23.4 (6)	0.9 (6)	8.8 (5)	7.5 (7)	7	0	0	0	0	0	100
Eugenia	19.2 (22)	22.2 (16)	1.1 (16)	1.1 (13)	18.7 (14)	22	4.5	50.0	9.1	13.6	13.6	9.1
Gomidesia	15.3 (6)	13.2 (5)	0.9 (5)	1.4 (5)	10.2 (5)	6	16.7	83.3	0	0	0	0
Marlierea	17.6 (6)	16.3 (6)	0.9 (6)	1.3 (4)	12.7 (4)	6	0	100	0	0	0	0
Myrceugenia	16.3 (1)	18.7 (1)	1.1 (1)	2.6 (1)	8.0 (1)	1	0	0	0	100	0	0
Myrcia	8.3 (11)	9.5 (5)	1.3 (4)	1.8 (6)	5.4 (4)	10	10.0	90.0	0	0	0	0
Myrcianthes	12.5 (1)	12.5 (1)	1.0 (1)	1.5 (1)	12.5 (1)	1	0	100	0	0	0	0
Myrciaria	14.7 (3)	–	–	1.2 (2)	10.0 (1)	3	0	100	0	0	0	0
Neomithrantes	15.9 (2)	15.6 (1)	0.8 (1)	1.8 (2)	13.6 (1)	2	0	100	0	0	0	0
Plinia	11.3 (2)	15.0 (2)	1.3 (2)	1.3 (2)	12.0 (2)	2	0	0	0	0	100	0
Psidium	21.7 (2)	37.5 (2)	0.8 (1)	6.0 (1)	3.0 (2)	3	0	0	0	0	33.3	66.7
Siphoneugena	10.5 (1)	–	–	–	–	1	0	100	0	0	0	0
All	16.5 (66)	18.8 (47)	1.0 (45)	2.3 (44)	12.1 (44)	66	4.5	59.1	3.0	6.1	10.6	16.7

Table 9.4. Pattern of rotated factors (factor loadings, extracted by principal components) for separate analyses of fruit morphological and nutrient content traits of *Myrtaceae* fruits. Varimax rotation used.

Morphology	Factor 1	Factor 2	Nutrient content	Factor 1	Factor 2
Fruit width	0.89	0.41	Water	0.82	0.08
Fruit length	0.97	0.07	Lipids	0.01	0.88
No. of seeds per fruit	0.09	0.78	Protein	0.82	0.23
Seed length	0.88	–0.39	Fibre	–0.24	–0.61
Length/width ratio	0.03	–0.72	Ash	0.78	0.11
Eigenvalue	2.5400	1.4376		2.2550	0.9805
Cumulative variance %	50.8	79.5		45.1	64.7

As the most speciose genus among American *Myrtaceae* (Landrum and Kawasaki, 1997), *Eugenia* is also the most variable in fruit morphology and the taxa of its seed-dispersers. With the exception of marsupials, it attracts all the disperser types recorded in this study. Most of the other genera are much more conservative in both these aspects. For instance, *Campomanesia* species have medium- to large-sized, yellow fruits chiefly dispersed by mammals, especially monkeys, while *Gomidesia* fruits are small, red or black and exclusively dispersed by birds. The primary exception is *Psidium*, which, although morphologically homogeneous (medium- to large-sized fruits, coloured green or yellow), is dispersed by a variety of animals. The key feature of *Psidium* that permits such a varied assemblage of seed-dispersers is the presence of small seeds scattered through the pulp; any animal feeding on its fruits will ingest and disperse its seeds.

The largest *Myrtaceae* fruit eaten whole by birds is *E. mosenii* (2 cm diameter). It is eaten by medium to large frugivorous birds (e.g. trogons) and is probably close to the upper size limit for ingestion by birds at Saibadela. Larger fruits are eaten piecemeal by birds, with effective seed dispersal depending on the size of the seeds. Although many bird-dispersed fruits are also eaten by monkeys, some *Myrtaceae* species appear to be exclusively dispersed by birds. These species are typically small trees of the forest understorey with small (< 17 mm diameter), red or black fruits, usually produced in small numbers (< 20 per plant) (e.g. *Calyptranthes lanceolata, Eugenia cuprea, Gomidesia flagelaris*).

Monkeys appear to rely heavily upon *Myrtaceae* fruits at Saibadela; it is the dominant plant family in the diet of capuchin (*Cebus apella*) and muriqui monkeys (*B. arachnoids*) (Izar, 1999). Indeed, all seven species of *Campomanesia* I examined were chiefly dispersed by monkeys. Seed dispersal by birds was only recorded for *C. xanthocarpa*, which has the smallest fruit (20.7 mm diameter) among the *Campomanesia* examined. Birds eat large *Campomanesia* fruits piecemeal but usually avoid ingesting the seeds (Landrum, 1986). In a different genus, *Marlierea regeliana* is possibly dispersed exclusively by monkeys. Its fruits, albeit small, are protected by a hard husk, greatly differing from other *Marlierea* species (see Janson, 1983).

Neotropical bats eat and disperse the seeds of a few *Myrtaceae* species. Although a large number of frugivorous bats and *Myrtaceae* species occur at Saibadela (25 and 40 species, respectively), only the fruits of *E. stictosepala* were recorded in the diet of bats there. In fact, neotropical bats in general (Heithaus *et al.*, 1975; Vazquez-Yanes *et al.*, 1975) and Atlantic Forest bats in particular (Marinho-Filho, 1990) eat and disperse the seeds of few *Myrtaceae* species.

As far as I know, species dispersed exclusively by rodents are unique among neotropical *Myrtaceae*. The fruits of *Eugenia cambucarana, E. multicostata* and *E. neoverrucosa* are apparently not eaten by birds or monkeys at Saibadela (this study; Izar, 1999), although I cannot rule out the possibility of rare events (e.g. seed dispersal by tapirs). These fruits are among the largest of any *Myrtaceae*. The green-coloured mature fruits of *E. multicostata* and *E. neoverrucosa* are unique among the species studied because they are green, a trait that may reflect photosynthetic capacity (Cipollini and Levey, 1991).

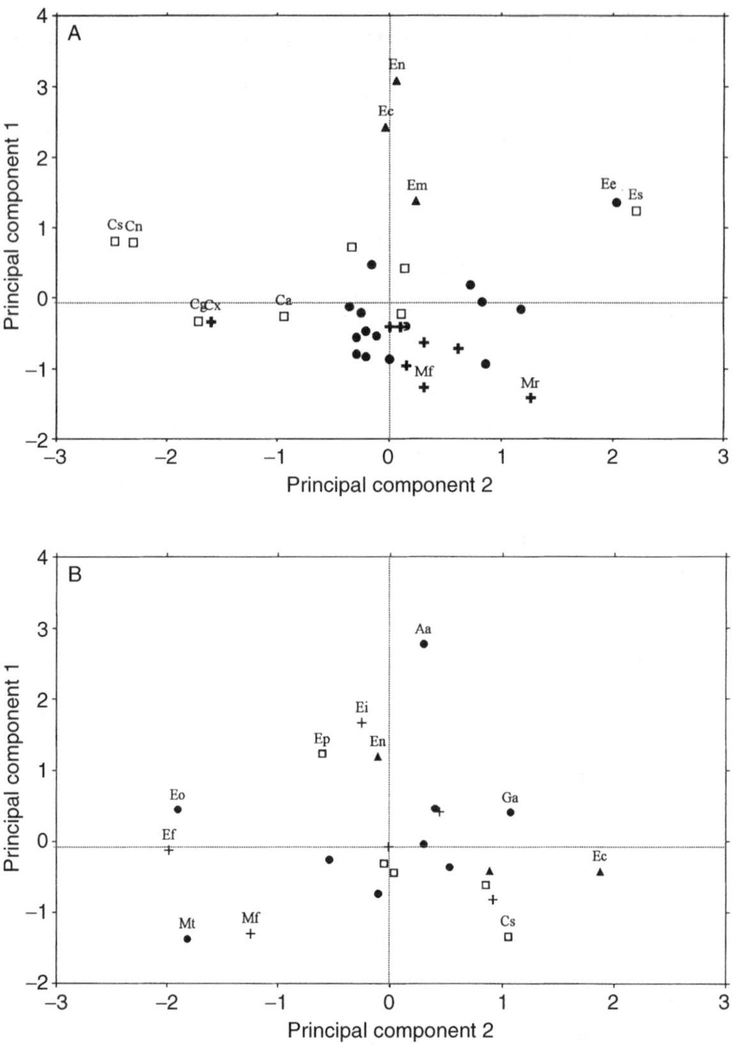

Fig. 9.2. Locations of *Myrtaceae* species on the first two principal component axes of (A) morphological and (B) chemical fruit traits. Species are divided according to their seed-dispersers: monkeys (squares), birds (filled circles), rodents (filled triangles) and mixed (monkeys plus birds) (crosses). For sake of clarity, only peripheral species are identified, as follows: Aa, *Calycorectes australis*; Ca, *Campomanesia adamantium*; Cg, *C. guaviroba*; Cn, *C. neriiflora*; Cs, *C. schlechtendaliana*; Cx, *C. xanthocarpa*; Ec, *Eugenia cambucarana*; Ee, *E. mosenii*; Ef, *E. florida*; Ei, *E. involucrata*; Em, *E. multicostata*; En, *E. neoverrucosa*; Eo, *E. oblongata*; Ep, *E. pyriformis*; Es, *E. stictosepala*; Ga, *Gomidesia anacardifolia*; Mf, *Myrcia fallax*; Mr, *M. rostrata*; Mt, *Marlierea tomentosa*.

Fruit syndromes

My results agree in several ways with fruit syndrome studies on more taxonomically diverse assemblages of plants (Janson, 1983; Knight and Siegfried, 1983; Gautier-Hion *et al.*, 1985; Jordano, 1995) and on single plant taxa (Tamboia *et al.*, 1996). In particular, fruit size separates bird from mammal fruits in *Myrtaceae*, as also reported by Janson (1983) and Knight and Siegfried (1983). In all cases, mammal fruits tend to be larger than bird

Table 9.5. Mean values ± SD for morphological and chemical traits of *Myrtaceae* fruits eaten by birds and monkeys in the Brazilian Atlantic Forest. Sample sizes are indicated in parentheses.

Traits	Birds	Monkeys	P^*
Morphological			
Width (mm)	14.4 ± 6.1 (40)	16.6 ± 7.3 (37)	0.18
Length (mm)	16.5 ± 8.6 (32)	20.7 ± 9.4 (22)	0.06
Length/width ratio	1.0 ± 0.2 (30)	1.0 ± 0.8 (21)	0.79
No. seeds per fruit	1.6 ± 1.3 (29)	3.3 ± 5.5 (22)	0.03
Seed length (mm)	11.3 ± 5.9 (28)	9.8 ± 2.5 (24)	0.41
Chemical (proportions)			
Water	0.84 ± 0.06 (19)	0.82 ± 0.06 (15)	0.52
Lipid	0.06 ± 0.05 (19)	0.05 ± 0.04 (15)	0.51
Protein	0.08 ± 0.04 (19)	0.09 ± 0.04 (15)	0.74
Soluble carbohydrates	0.36 ± 0.12 (13)	0.33 ± 0.15 (12)	0.45
Fibre	0.42 ± 0.13 (19)	0.40 ± 0.19 (13)	0.72
Ash	0.04 ± 0.01 (19)	0.04 ± 0.01 (15)	0.15

*Student's *t*-tests applied on log-transformed data for morphological traits and arcsin-transformed data for chemical traits. Bonferroni-corrected *P* level = 0.008.

Table 9.6. Fruit chemical traits of the *Myrtaceae* genera from the Brazilian Atlantic Forest. Values are mean proportions of water and of the major chemical constituents (based on dry mass) of the pulp of *Myrtaceae* fruits whose seed-dispersers were identified. Number of species analysed indicated in parentheses.

Species	Water	Lipids	Protein	Soluble carb.*	Fibre	Ash
Calycorectes	0.94 (1)	0.07 (1)	0.19 (1)	–	0.49 (1)	0.07 (1)
Calyptranthes	–	–	–	–	0.33 (1)	–
Campomanesia	0.80 (4)	0.03 (4)	0.07 (4)	0.20 (3)	0.51 (4)	0.03 (4)
Eugenia	0.84 (14)	0.07 (14)	0.10 (14)	0.35 (7)	0.40 (10)	0.04 (14)
Gomidesia	0.82 (4)	0.03 (4)	0.07 (4)	0.33 (3)	0.47 (3)	0.04 (4)
Marlierea	0.77 (1)	0.15 (1)	0.06 (1)	0.21 (1)	0.47 (3)	0.03 (1)
Myrceugenia	0.86 (1)	0.06 (1)	0.09 (1)	–	–	0.05 (1)
Myrcia	0.81 (3)	0.08 (3)	0.07 (3)	0.38 (3)	0.36 (3)	0.03 (3)
Neomithranthes	0.87 (1)	0.05 (1)	0.06 (1)	0.57 (1)	0.19 (1)	0.03 (1)
Psidium	0.80 (1)	0.02 (1)	0.04 (1)	0.34 (1)	0.57 (1)	0.03 (1)
All	0.83 (30)	0.06 (30)	0.09 (30)	0.33 (19)	0.42 (27)	0.04 (30)
CV (%)	6.7	79.7	43.8	41.7	40.8	29.3

*Water-soluble carbohydrates.
CV, coefficient of variation.

fruits. Fruit size is likely to be important in distinguishing between bird- and monkey-dispersed fruits, because most birds cannot swallow large fruits (> 2 cm diameter (Wheelwright, 1985)), not because small fruits are avoided by monkeys. Indeed, monkeys exploit a wide variety of fruits (Knight and Siegfried, 1983). However, as noted by Debussche and Isenmann (1989), the fruit size distinction between bird and monkey fruits is weakened, because birds may disperse seeds from large, small-seeded fruits (e.g. *Psidium* spp.). Consequently, bird and monkey fruits may not form a clear dichotomy (Gautier-Hion *et al.*, 1985). This is true not only for fruit size but also for fruit colour. Some authors have found birds associated with black, purple and red fruits, and monkeys associated with yellow, orange, green and brown fruits (Janson, 1983; Knight and Siegfried, 1983), whereas other have not found such a distinction (Gautier-Hion *et al.*, 1985; this study).

Rodent-dispersed species, in contrast, form the most easily recognized syndrome among the *Myrtaceae*, characterized by fruits that are large, few-seeded, muricate and yellow or green in colour with fibrous pulp. These characteristics closely match those described for 'squirrel' and 'large-rodent' fruits by Gautier-Hion *et al.* (1985) and for other rodent-dispersed fruits (Forget, 1990), suggesting that rodents exert a consistent and pervasive evolutionary pressure on the morphology of fruits whose seeds they disperse.

The major chemical components of *Myrtaceae* fruits vary little among species and do not differ between bird- and monkey-dispersed species. It has been shown that mammals in general tend to avoid lipid-rich fruits (Debussche and Isenmann, 1989), whereas birds may favour them (Stiles, 1993). None the less, the chemical characteristics of fleshy fruits seem to be less determined by preferences of seed-dispersers than by common ancestry (Jordano, 1995). Likewise, traits strongly related to the morphology of the ovary and embryogenesis (e.g. seed number and size) (Hodgson and Mackey, 1986; Lord *et al.*, 1995) appear to be most easily explained by phylogeny.

In summary, size is the fruit trait most clearly tied to differences in dispersal agents of Atlantic Forest *Myrtaceae*, in agreement with what other authors have found for other plant families (Mack, 1993; Jordano, 1995). The possibility exists, however, that, within genera, other traits may be important. In *Campomanesia*, for example, fruit size determines whether seeds will be dispersed exclusively by monkeys or by monkeys and birds, whereas in *Marlierea* fruit protection determines a species' dispersers (e.g. *M. regeliana*). Thus, although Jordano (1995) concluded that evolutionary responses to dispersal agents were rare below the genus level among angiosperms, it is possible that evolutionary responses below genus level do indeed occur, at least for one unusually diverse family.

Conservation Considerations and Avenues for Future Research

After a long history of destruction (Dean, 1995), the Brazilian Atlantic Forest is one of the most threatened biomes in the world, represented by only 5–12% of its precolonization distribution (Fonseca, 1985). Most of what remains is highly fragmented (Ranta *et al.*, 1998). Home of many endemic plant and animal species, the Atlantic Forest is expected to lose many species in the near future because of deforestation and habitat alteration (Mori *et al.*, 1981; Brooks and Balmford, 1996). Of great concern for those preoccupied with the integrity of the Atlantic Forest is the disruption of plant–animal interactions through local and global extinction of pollinators and seed-dispersers. These extinctions are feared to affect the long-term sustainability of plant populations (Howe, 1984). Silva and Tabarelli (2000) estimated that approximately 32% of the tree flora occurring in the highly fragmented Atlantic Forest of north-east Brazil (Ranta *et al.*, 1998) are threatened due to the disappearance of large, wide-gaped frugivorous birds, which act as seed-dispersers for these species. My study adds some *Myrtaceae* species to Silva and Tabarelli's (2000) list of threatened species. For instance, the rodent-dispersed species of *Myrtaceae* depend on the agouti, whereas *E. mosenii* and possibly *E. melanogyna* seem to depend heavily on large birds for seed dispersal. These are exactly the organisms that disappear from forest fragments due to hunting (Redford, 1992) or habitat alteration (Willis, 1979). One can thus predict alterations in the recruitment dynamics of plant species in forest fragments, with unknown consequences for their long-term existence. Plant recruitment studies are needed to frame and test such predictions (e.g. Chapman and Chapman, 1995).

Advancements in the understanding of fruit syndromes will be achieved by incorporation of phylogenetic frameworks, as demonstrated by Jordano (1995). Detailed chemical analyses of fruits, careful observations of frugivores in fruiting plants and within-family approaches are also needed.

Chemical analyses that take into account not only major nutrients but also minor constituents of fruit pulp (e.g. secondary metabolites, minerals) should be included in future studies to properly evaluate the role of fruit chemistry in delineating fruit syndromes. Within-family studies, such as the one reported here, represent a means of fine-tuning, because patterns verified in diversified clades may not be valid for subsets of them (Tamboia *et al.*, 1996). Two approaches are particularly promising for within-family studies. First, intrageneric analyses may reveal different evolutionary responses to similar communities of seed-dispersers, as suggested here for *Campomanesia* and *Marlierea*. Secondly, the study of plant communities that differ widely in their assemblages of dispersers will help tease apart environmental and disperser- and phylogenetic-related influences on fruit traits (Keeler-Wolf, 1988; Mack, 1993). These approaches will be superficial if they are not based upon rigorous field studies that document not only the identity of seed-dispersers but also the relative contribution of each disperser to seed removal. Such data will allow refinement of the broad seed-disperser categories traditionally used in fruit syndrome studies to permit, for example, the separation of small from large birds and bats – groups that may differ substantially in their preferences.

Acknowledgements

I am grateful to the Fundação Florestal do Estado de São Paulo for its continuous support of my work at Intervales. I am in debt to Wesley R. Silva, Mauro Galetti, William Zaca and Flávio H.G. Rodrigues for sharing their unpublished data. Comments by Douglas Levey, Pedro Jordano and Colin Chapman greatly improved the chapter. Pedro Jordano also helped decisively with the statistical analyses. Valesca B. Zipparro helped in many ways during the fieldwork. Special thanks to M. Lúcia Kawasaki for plant identification and for providing me with morphological data and fruit samples. Helena T. Godoy (FEA (Faculdade de Engenhania de Alimentos)/ UNICAMP (Universidade Estadual de Campinas)) patiently explained the chemical analyses. This work was possible due to the facilities provided by the Departamento de Botânica of the Universidad Estadual Paulista (UNESP) at Rio Claro and the financial support by FAPESP (Fundação de Amparo à Pesquisa do Estado de São Paulo) (procs. 95/9626 and 98/11185-0). Additional support came from the International Foundation for Science (D/2953-1), British Ecological Society (SEPG 1625) and Idea Wild.

References

Aleixo, A. and Galetti, M. (1997) The conservation of the avifauna in a lowland Atlantic Forest in south-east Brazil. *Bird Conservation International* 7, 235–261.

Almeida-Scabbia, R. (1996) Fitossociologia de um trecho de Mata Atlântica no sudeste do Brasil. MSc thesis, Universidade Estadual Paulista, Rio Claro, Brazil.

AOAC (1990) *Official Methods of Analysis*, 15th edn. Association of Official Analytical Chemists, Washington.

Bligh, E.G. and Dyer, W.J. (1959) A rapid method of total lipid extraction and purification. *Canadian Journal of Biochemical Physiology* 37, 911–917.

Brooks, T. and Balmford, A. (1996) Atlantic Forest extinctions. *Nature* 380, 115.

Chapman, C.A. and Chapman, L.J. (1995) Survival without dispersers: seedling recruitment under parents. *Conservation Biology* 9, 675–678.

Cipollini, M.L. and Levey, D.J. (1991) Why some fruits are green when they are ripe: carbon balance in fleshy fruits. *Oecologia* 88, 371–377.

Clark, D.B. (1996) Abolishing virginity. *Journal of Tropical Ecology* 12, 735–739.

Cordero, R. and Nicolas B., R.A. (1987) Feeding habits of the opossum (*Didelphis marsupialis*) in northern Venezuela. *Fieldiana: Zoology* 39, 125–131.

Dean, W. (1995) *With Broadax and Firebrand: the Destruction of the Brazilian Atlantic Forest.* University of California Press, Berkeley.

Debussche, M. and Isenmann, P. (1989) Fleshy fruit characters and the choices of birds and mammal seed dispersers in a Mediterranean region. *Oikos* 56, 327–338.

Felsenstein, J. (1985) Phylogenies and the comparative method. *American Naturalist* 125, 1–15.

Fonseca, G.A.B. (1985) The vanishing Brazilian Atlantic Forest. *Biological Conservation* 34, 17–34.

Forget, P.M. (1990) Seed dispersal of *Vouacapoua americana* (*Caesalpiniaceae*) by caviomorph rodents in French Guiana. *Journal of Tropical Ecology* 6, 459–468.

Gautier-Hion, A., Duplantier, J.-M., Quris, R., Feer, F., Sourd, C., Decoux, J.-P., Dubost, G., Emmons, L., Erard, C., Hecketsweiler, P., Moungazi, A., Roussilhon, C. and Thiollay, J.-M. (1985) Fruit characters as a basis of fruit choice and seed dispersal in a tropical forest vertebrate community. *Oecologia* 65, 324–327.

Heithaus, E.R., Fleming, T.H. and Opler, P.A. (1975) Foraging patterns and resource utilization in seven species of bats in a seasonal tropical forest. *Ecology* 56, 841–854.

Herrera, C.M. (1992a) Historical effects and sorting processes as explanations for contemporary ecological patterns: character syndromes in Mediterranean woody plants. *American Naturalist* 140, 421–446.

Herrera, C.M. (1992b) Interspecific variation in fruit shape: allometry, phylogeny, and adaptation to dispersal agents. *Ecology* 73, 1832–1841.

Hodgson, J.G. and Mackey, J.M.L. (1986) The ecological specialization of dicotyledoneous families within a local flora: some factors constraining optimization of seed size and their possible evolutionary significance. *New Phytologist* 104, 497–515.

Howe, H.F. (1984) Implications of seed dispersal by animals for tropical reserve management. *Biological Conservation* 30, 261–281.

Izar, P. (1999) Aspectos de ecologia e comportamento de um grupo de macacos-prego (*Cebus apella*) em área de Mata Atlântica, São Paulo. PhD thesis, Universidade de São Paulo, São Paulo, Brazil.

Janson, C.H. (1983) Adaptation of fruit morphology to dispersal agents in a neotropical forest. *Science* 219, 187–189.

Jordano, P. (1995) Angiosperm fleshy fruits and seed dispersers: a comparative analysis of adaptation and constraints in plant–animal interactions. *American Naturalist* 145, 163–191.

Kawasaki, M.L. and Landrum, L.R. (1997) A rare and potentially economic fruit of Brazil: cambuci, *Campomanesia phaea* (*Myrtaceae*). *Economic Botany* 51, 403–407.

Keeler-Wolf, T. (1988) Fruit and consumer differences in three species of trees shared by Trinidad and Tobago. *Biotropica* 20, 38–48.

Knight, R.S. and Siegfried, W.R. (1983) Inter-relationships between type, size and colour of fruits and dispersal in Southern African trees. *Oecologia* 56, 405–412.

Kuhlmann, M. and Kühn, E. (1947) *A flora do Distrito de Ibiti.* Publicações do Instituto de Botânica, Secretaria da Agricultura, São Paulo.

Landrum, L.R. (1986) *Campomanesia, Pimenta, Blepharocalyx, Legrandia, Acca, Myrrhinium,* and *Luma. Flora Neotropica* 45, 1–179.

Landrum, L.R. and Kawasaki, M.L. (1997) The genera of *Myrtaceae* in Brazil: an illustrated synoptic treatment and identification keys. *Brittonia* 49, 508–536.

Levey, D.J., Bissell, H.A. and O'Keefe, S.F. (2000) Conversion of nitrogen to protein and amino acids in wild fruits. *Journal of Chemical Ecology* 26, 1749–1763.

Lord, J., Westoby, M. and Leishman, M. (1995) Seed size and phylogeny in six temperate floras: constraints, niche conservatism, and adaptation. *American Naturalist* 146, 349–364.

Mack, A.L. (1993) The sizes of vertebrate-dispersed fruits? A neotropical–paleotropical comparison. *American Naturalist* 142, 840–852.

Marinho-Filho, J.S. (1990) The coexistence of two frugivorous bat species and the phenology of their food plants. *Journal of Tropical Ecology* 7, 59–67.

Mazer, S.J. and Wheelwright, N.T. (1993) Fruit size and shape? Allometry at different taxonomic levels in bird-dispersed plants. *Evolutionary Ecology* 7, 556–575.

Mori, S.A., Boom, B.M. and Prance, G.T. (1981) Distribution patterns and conservation of eastern Brazilian coastal forest tree species. *Brittonia* 33, 233–245.

Mori, S.A., Boom, B.M., Carvalino, A.M. and Santos, T.S. (1983) Ecological importance of *Myrtaceae* in an eastern Brazilian wet forest. *Biotropica* 15, 68–70.

Ranta, P., Blom, T., Niemelä, J., Joensuu, E. and Siitinen, M. (1998) The fragmented Atlantic rain forest of Brazil: size, shape and distribution of forest fragments. *Biodiversity Conservation* 7, 385–403.

Redford, K.H. (1992) The empty forest. *Bioscience* 42, 412–422.

Ridley, H.N. (1930) *The Dispersal of Plants Throughout the World.* Reeve, Ashford.

Silva, J.M.C. and Tabarelli, M. (2000) Tree species impoverishment and the future flora of the Atlantic Forest of northeast Brazil. *Nature* 404, 72–74.

Stiles, E.W. (1993) The influence of pulp lipids on fruit preference by birds. *Vegetatio* 107/108, 227–235.

Tamboia, T., Cipollini, M.L. and Levey, D.J. (1996) An evaluation of vertebrate seed dispersal syndromes in four species of black nightshade (*Solanum* sect. *Solanum*). *Oecologia* 107, 522–532.

van der Pijl, L. (1982) *Principles of Seed Dispersal in Higher Plants,* 3rd edn. Springer-Verlag, Berlin.

Vazquez-Yanes, C., Orozco, A., François, G. and Trejo, L. (1975) Observations on seed dispersal by bats in a tropical humid region in Veracruz, Mexico. *Biotropica* 7, 73–76.

Wheelwright, N.T. (1985) Fruit size, gape width, and the diets of fruit-eating birds. *Ecology* 66, 808–818.

Willis, E.O. (1979) The composition of avian communities in remanescent woodlots in southern Brazil. *Papéis Avulsos de Zoologia* 33, 1–25.

10 Are Plant Species that Need Gaps for Recruitment More Attractive to Seed-dispersing Birds and Ants than Other Species?

Carol C. Horvitz,[1] Marco A. Pizo,[2] Bernardina Bello y Bello,[3] Josiane LeCorff[4] and Rodolfo Dirzo[3]

[1]*Department of Biology, PO Box 249118, University of Miami, Coral Gables, FL 33124, USA;* [2]*Departamento de Botânica, Universidade Estadual Paulista, 13506-900 Rio Claro, São Paulo State, Brazil;* [3]*Instituto de Ecología, Universidad Nacional Autonoma de México, Mexico DF, Mexico;* [4]*Institut d'Horticulture, 2 rue le Nôtre, 49045 Angers, France*

Introduction

Despite empirical evidence to the contrary (e.g. Herrera, 1998), many seed, fruit and fruit-display traits are assumed to be adaptations for seed dispersal (Van der Pijl, 1982). These traits should thus be associated with dispersal success (Howe, 1977; Howe and Vande Kerckhove, 1979; Van der Pijl, 1982; Mazer and Wheelwright, 1993). Ultimately, dispersal success should be measured by its effects on fitness (Horvitz and Schemske, 1986a). Because it is very rare to have the complete set of demographic data needed to evaluate the effect of any particular stage-specific event on fitness, most studies measure the demographic effect of dispersal by its effects on one component of fitness, plant recruitment (Schupp, 1993; Horvitz and Schemske, 1994). If effects on recruitment translate into effects on fitness and if the movement patterns of seed-dispersers correspond to the spatiotemporal distribution of plant recruitment sites (Wenny and Levey, 1998; Wenny, 2000), then plant traits that non-randomly attract those animals should be selected.

Closely related taxa sometimes have contrasting suites of traits associated with different modes of biotic dispersal (e.g. ant vs. bird dispersal) (O'Dowd and Gill, 1986; Westoby *et al.*, 1991), providing the opportunity to investigate whether the respective advantages of different dispersal modes correspond to differing recruitment requirements. Different animals probably produce different kinds of seed shadows. In particular, they may drop seeds beneath the parent plant, remove seeds from the parent-plant canopy a short distance or remove seeds a long distance, depositing them at random or into sites especially suitable for seedling establishment. Alternatively, they may destroy seeds. The relative importance to plants of these animal behaviours with respect to seeds may vary among plant species. If there is sufficient variation in plant traits that mediate the interaction, then each plant species should adapt to attract the type of animal that provides the 'best' dispersal. 'Best', however, will not be defined in the same way for all plant taxa. For example, short-distance dispersal may be best for some plants and long-distance dispersal best for others. Short-distance dispersal may be sufficient if appropriate recruitment sites are located close to the parent plant. Long-distance seed dispersal may be costly because it is often inefficient, requiring the production of numerous seeds and more expensive structures to attract and reward dispersers. Despite these costs and the uncertainty associated with long-distance dispersal, it is the best way to ensure recruitment for some plants.

In the *Marantaceae*, a family of tropical-forest understorey herbs, ornithochory (seed dispersal by birds) is thought to be the primitive state and myrmecochory (seed dispersal by ants) the derived state (Horvitz, 1991). Ornithochorous (i.e. bird-dispersed) species have fruits that partially dehisce to expose seeds that are displayed with bright colours (e.g. orange, pink, blue, yellow), 1–3 m above the forest floor. Myrmecochorous (i.e. ant-dispersed) species, in contrast, have fruits that are borne near the ground and that split open completely at maturity, dropping their black and white or grey and white seeds on the forest floor. Both types of species are widespread and may occur in disparate habitats, from swamp areas and river edges to shaded understorey to tree-fall gaps and second growth. These species differ in their habitat requirements; some are more shade-tolerant than others (Horvitz, 1991; Horvitz and Schemske, 1994; LeCorff, 1996). The habitat of many species, the tropical-forest understorey, is mostly quite dark. New light gaps are rare. Building phases of forest succession, characterized by intermediate light conditions, are more common than new gaps.

Two predictions about the association of dispersal syndromes with the gap dependency of recruitment are tested in this chapter. First, myrmecochorous species should be more shade-tolerant, because dispersal by ants results in only short-distance movement of seeds. Such an association of ant dispersal with shade tolerance was predicted by a model of demography and dispersal based on data from a Mexican ant-dispersed *Calathea* (Horvitz and Schemske, 1986a). This model found selection against long-distance dispersal for shade-tolerant species in the context of tree-fall gap-driven forest dynamics. Further investigation of the model suggested that the only circumstance in which long-distance dispersal would be favoured was for a hypothesized shade-intolerant species with an order-of-magnitude higher seed production than that observed for ant-dispersed species (Horvitz, 1991), such as might occur for bird-dispersed species. The second hypothesis tested in this chapter is that, among bird-dispersed species, those species that have traits most attractive to dispersers should be those that have the greatest dependency on gaps for recruitment. Two components of recruitment were considered: germination and seedling survivorship. Movement to light gaps is not the only possible advantage of dispersal. We focus on it because it was the hypothesized advantage in the model that was the conceptual foundation of this study (Horvitz, 1991).

To address these hypotheses, data on the dispersal and recruitment of seven plant species at two sites were collected. We also investigated whether the diaspores of ornithochorous species would be attractive to ants if they were removed from plants and placed in the leaf litter. Such data provide insight into the steps needed to evolve from ornithochory to myrmecochory. In our study system, differences among plant species are used to examine the effects of plant traits on dispersal, whereas differences between our two study sites are used to examine the effects of different disperser assemblages on dispersal. Although the two study sites were originally chosen to represent forests of differing light regimes (one evergreen and the other partially seasonally deciduous), the most interesting site differences for the current analysis were the differences in the taxonomic composition of the disperser assemblages. Four kinds of data were compared across plant species and sites: (i) the attractiveness of seeds to ants and birds; (ii) the average distance of seed dispersal; (iii) differences in disperser assemblages; and (iv) the effects of gaps on seedling emergence and survivorship.

Methods

Study sites

The two study sites are lowland tropical rain forests in Costa Rica. The La Selva Biological Station is on the Atlantic slope (Heredia Province) and the Sirena Station at Corcovado National Park is on the Pacific slope (Osa Peninsula, Puntarenas Province). *Marantaceae* comprise a conspicuous portion of the community of understorey herbaceous plants at both sites (Herwitz, 1981; Hartshorn, 1983). La Selva is located at the base of the central chain of volcanic mountains at 10° 26′N and 83° 59′W, at the confluence of the Saripiqui and Puerto Viejo Rivers. It is a 1536 ha preserve and field station, owned and operated by the Organization for Tropical Studies. The climate is tropical wet forest (Holdridge, 1967). Altitude ranges from 35 to 150 m, a marked dry season is lacking and mean annual precipitation is 3962 mm (Hartshorn, 1983). Corcovado Park is located at 8° 27–39′N and 83° 25–45′W and the Sirena Station is situated along the coast, between the mouths of the Sirena and Claro Rivers. Corcovado is a 53,735 ha National Park of Costa Rica, operated by the Servicio de Parques Nacionales. Altitude ranges from 0 to 750 m and a marked dry season occurs from January to March. Annual rainfall and climate type vary within the park, but our study took place within the tropical wet forest zone, which receives > 4000 mm of rainfall annually (Herwitz, 1981; Hartshorn, 1983).

Study species

Seven species of *Marantaceae*, six ornithochores and one myrmecochore, were studied. Additional myrmechocorous species were initially included, but unexpectedly did not fruit during the 3.5 years of the study. Comparative data on ant-dispersed species from other studies will be presented in the Discussion section. *Marantaceae* produce hard seeds with an oily aril, borne in fruit capsules that dehisce at maturity. They include species that appear to be adapted for ant dispersal (Horvitz and Beattie, 1980; Horvitz and Schemske, 1986b, 1994; LeCorff and Horvitz, 1995), bird dispersal (O'Daniel, 1987) or bat dispersal (Horvitz, 1991). Bird dispersal is expected in species characterized by a brightly coloured display (orange, pink, bright blue and yellow), with mature seeds exhibited on the fruiting stalks above the foliage (1.5–3 m above the forest floor) (Horvitz, 1991). Bird-dispersed species in this study are: *Calathea lutea* (Aubl.) Schult., *Calathea lasiostachya* Donn. Sm, *Calathea inocephala* (Kuntze) H.A. Kenn. & Nicolson, *Calathea gymnocarpa* H.A. Kenn., *Calathea marantifolia* Standl. and *Pleiostachya pruinosa* (Regel) K. Schum. Ant dispersal is expected in species characterized by mature seeds with a light aril (white) and a dark seed body (grey or black), displayed in the leaf litter after falling from capsules that are borne near ground level and split open completely. The ant-dispersed species in this study is *Calathea cleistantha* Standl.

Among the bird-dispersed species, three groups were previously postulated (Horvitz, 1991), based on morphology: the 'typical bird-type' species (represented by *C. lutea*, *C. lasiostachya* and *Pleiostachya*), the species with 'large seeds' (represented by *C. gymnocarpa* and *C. inocephala*) and the species 'with clonal propagules' (represented by *C. marantifolia*) (Table 10.1). These groups were often considered separately in analyses, as it was expected that large seed size and clonal reproduction might confer some shade tolerance, thereby reducing their dependency on light gaps and, consequently, long-distance dispersal. Further details on the study species can be found in Horvitz (1991) and Horvitz and LeCorff (1993).

Bird–seed and ant–seed interactions

To determine: (i) the attractiveness of seeds to ants and birds, (ii) the distances seeds were dispersed and (iii) differences in disperser assemblages, diaspores were observed as follows. Bird–seed interactions were recorded during 495 trials (total of 2279 diaspores observed) in 1991 and ant–seed interactions were recorded during 524 trials (total of 1440 diaspores observed) in 1992. Each 'trial' was a 90 min observation period, in which fresh

Table 10.1. Summary of dispersal types of some Costa Rican *Marantaceae* (reprinted with permission from Horvitz, 1991).

Type	Fruit stalk Height (m)	Fruit stalk Branched?	Display	Seed mass (mg)	Reward (aril/aril + seed), (% dry mass)
Bird	2–3	Yes	Diaspores against flattened bracts	500–1000	7–17
Bird w/large seeds	1–2.5	No	Diaspores in waxy fruit capsule	2500–3000	8
Bird w/clonal propagules	1–1.5	No	Diaspores embedded in infructescence	500–1000	5–6
Ant	0.1–0.3	No	White arils against dark seeds, forest floor	< 500	5

diaspores were observed. Fresh diaspores were defined as seeds with arils from capsules that dehisced on the same day as the observations. In *Marantaceae*, there are one to three seeds per capsule. Few capsules mature each day on a plant. For bird–seed observations, diaspores were observed in their natural positions on the infructescences of plants. The number of seeds in a trial was determined by the number of mature fruits that could be seen by an observer positioned nearby (about 5 m from plants) with binoculars. This number was constrained by availability in the habitat and varied among species and sites (mean number per species per site ranged from 2.5 to 15.8 seeds per trial). For ant–seed observations, diaspores were removed from capsules and placed on the leaf litter. The number of seeds in a trial varied by species (one seed per trial for *Pleiostachya*, three seeds per trial for other species), according to the typical number of seeds per capsule. In addition to observing ant–seed interactions of the myrmecochore, we assayed the diaspores of ornithochores for their attractiveness to ants. Previously, it was known that myrmecochores in *Marantaceae* differ from ornithochores in display height, coloration and capsule dehiscence (Horvitz, 1991), but it was not known if they differed in traits of the diaspores as perceived by ants.

In both studies, identities of animals interacting with seeds were recorded and the type of interaction was classified. Bird–seed interactions were classified as: drop beneath plant, take to a perch nearby (< 7 m from the parent plant) and either drop or regurgitate the seed, or take seeds away 'a long distance' (defined as the bird ingesting a seed and flying out of sight). These categories classified the kinds of interactions we were able to observe. The data do not distinguish between seed predation and seed dispersal, nor do they address ultimate seed fates outside our range of vision (about 7 m). Ant–seed interactions were classified as: ignore, examine, recruit (many worker ants recruited to a seed) or remove seed. Seeds removed by ants were tracked and dispersal distances recorded.

These data were used in two different kinds of analyses. The first focused on the trial as the unit of observation. Two components of dispersal are analysed for each seed trial: (i) the probability per seed of being found and manipulated by animals; and (ii) the distance of dispersal. For bird–seed interactions, the distance component was measured as the proportion of all removed seeds that were taken 'a long distance' (as defined above). For ant–seed interactions, the distance component was measured directly as the mean distance moved of all the seeds that ants removed during a trial. These analyses addressed whether plant species and/or site had significant effects on dispersal parameters.

The second kind of analysis focused on the animal–seed interaction as the unit of analysis. These data are used to analyse whether there were differences in disperser assemblages across sites. We asked what proportion of bird–seed or ant–seed interactions ('ignores' were not counted as interactions) at each site were due to each bird or ant species and whether these differed significantly between sites. Differences in diversity (H') of each disperser assemblage at each site were also analysed:

$$H' = -\Sigma p_i \ln(p_i)$$

The summation is over all i and p_i is the proportion of all disperser–seed interactions that were with each disperser species, i.

Gap dependence of seedling emergence and survival

To determine the effects of gaps on seeds and seedlings, seeds were planted in 204 wire-mesh germination boxes (20 cm × 20 cm × 5 cm), nine seeds to a box (a total of 1936 seeds) in 1990. Boxes were assigned randomly to three different light environments: gap, understorey and intermediate. Mesh size varied with plant species and was chosen to allow germination and seedling growth, while excluding large-seed predators (Horvitz and Schemske, 1994; LeCorff, 1996). Boxes were planted in natural habitats in the vicinity of fruiting plants and filled with soil from the planting site. At Corcovado, all study species co-occurred in the forest and thus all species shared box sites. At La Selva, study species were found in widely different parts of the forest and more box sites were needed. Seeds were planted on the same day they were harvested from newly dehisced fruits and distributed regularly among treatments. We placed at least five boxes per species in each of the three light environments at each site. Seeds were not passed through bird guts before planting, although gut passage may alter response to gaps.

Gap, intermediate and understorey sites were chosen in August 1990. Hemispherical fish-eye lens photographs were taken over each site and analysed for indirect reflected skylight (ISF) and direct sunlight (DSF) with the program CANOPY (Rich, 1990). At the time of planting, gap boxes received twice as much light as intermediate boxes and up to three times the amount of light received by understorey boxes. In gaps, ISF was 12.6% ± 1.3% (mean ± SE; $n = 6$ box sites) and 14.3% ± 0.9% ($n = 22$) at Corcovado and La Selva, respectively. DSF was 21.2% ± 2.4% ($n = 6$) and 23.0% ± 1.5% ($n = 22$) at Corcovado and La Selva, respectively (C. Horvitz and J. LeCorff, unpublished data).

Here we report analyses of seedling emergence (analysed per box) and seedling survivorship (analysed per emerged seedling) from 2 years of censusing these boxes. We tested whether plant species, site and light environment interacted in their effects on seedling emergence. We then asked whether seedling survival time (as estimated by the Kaplan–Meier product-limit method for survivorship data with right-censored observations: Statistical Analysis Software Lifetest Procedure (SAS, 1999)) per species within a site was significantly affected by light environment.

Results and Discussion

Bird–seed and ant–seed interactions

Attractiveness of diaspores

Both plant species and site (i.e. disperser assemblage) significantly affected the probability of seed removal by birds (Friedman block tests for the four species studied at both sites; for the species effect: $F = 37.3$, d.f. = 3, $P < 0.0001$; for the site effect, $F = 12.6$, d.f. = 1, $P < 0.001$ (C. Horvitz and M.A. Pizo, unpublished data)). Of those species studied at two sites, *C. lutea* was the most attractive to birds, followed by *C. inocephala*, *Pleiostachya* and *C. marantifolia*, in decreasing order. The species' order of attractiveness was the same at both sites. All species were more successful in attracting birds at La Selva than they were at Corcovado (1.4–3.8-fold more attractive). For the two species studied only at La Selva, one was quite attractive to birds (*C. gymnocarpa* was as attractive as *C. inocephala*) and the other was much less attractive (*C. lasiostachya*) (Fig. 10.1A; note that the order in which species are presented in this figure is retained throughout the chapter to facilitate comparisons).

The order of attractiveness to birds is related to the conspicuousness of the display. Of all the species, *C. lutea* has the largest display. It is highly branched and often above 3 m. In addition, its pink seed capsules bear orange-arillate seeds, which are quite striking among the brown bracts. The next two most attractive species, *C. inocephala* and *C. gymnocarpa*, are the two large-seeded species and also have quite showy displays. Although their infructescences are solitary and not branched, fruiting heads are composed of

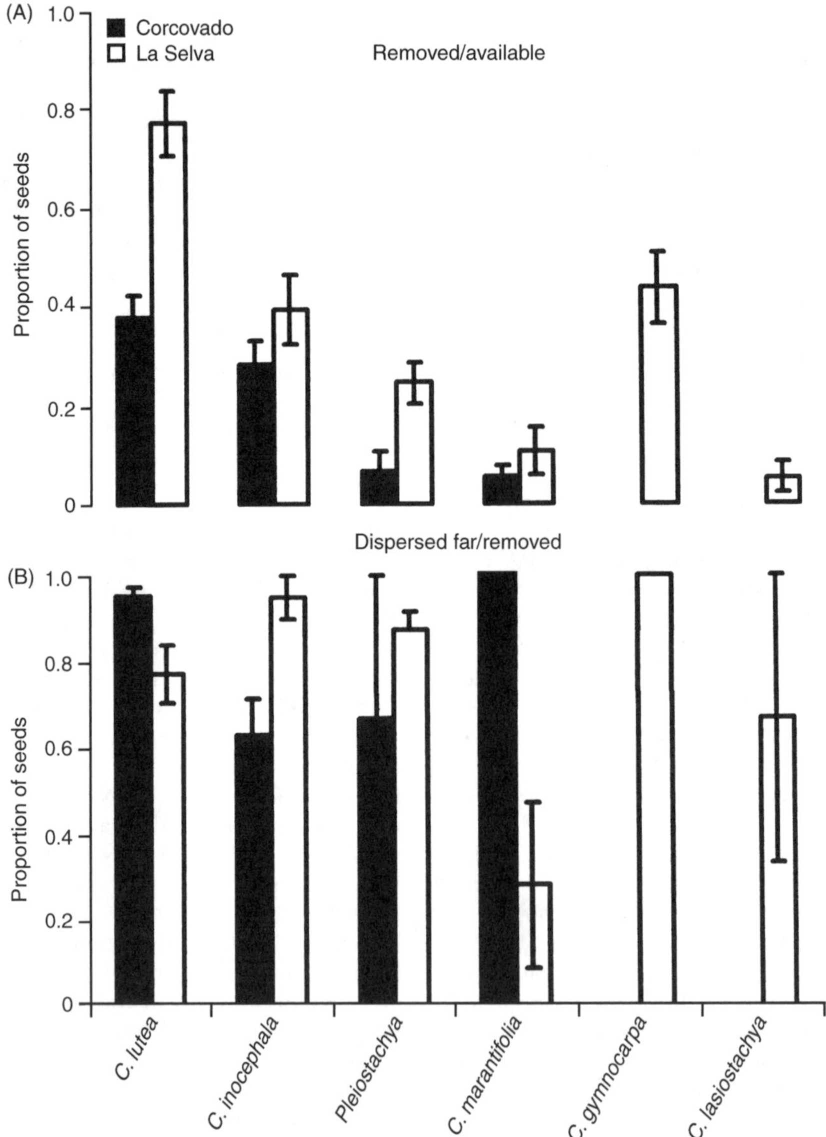

Fig. 10.1. Attractiveness of seeds to birds. (A) Proportion of available seeds per seed trial that were removed by birds (± SE) (*n* = 87, 30, 74, 37, 26, 65, 66, 30, 0, 40, 0, 40 trials of each plant species at each site, from left to right). (B) Proportion of removed seeds per seed trial that were taken 'a long distance' by birds (± SE) (*n* = 48, 26, 28, 20, 3, 36, 6, 5, 0, 21, 0, 3 trials of each plant species at each site, from left to right; only trials in which seeds were removed could be counted). Two species (*C. gymnocarpa* and *C. lasiostachya*) only occurred at La Selva. Plant species that occur at both sites are listed first (from left to right), in the order of their attractiveness to birds; these are followed by species that occurred only in one site.

large, persistent, waxy capsules, which are bright orange or red and which partially dehisce to reveal bright blue seeds (that are about three to five times larger than the seeds of the other species in the study (Table 10.1)). It was surprising to observe such large seeds being taken readily by small birds, including flycatchers (e.g. *Mionectes*) and manakins (e.g. *Pipra*). An unexpected result was the very low attractiveness of *C. lasiostachya*, a species characteristic of small gaps in primary forest. It has branched infructescences and a display of blue seeds against yellow, flattened bracts, but its display is not as large as that of *C. lutea*. The low attractiveness of the remaining two species is not surprising. *Pleiostachya* infructescences are not colourful and, although branched, are compact. *Calathea marantifolia* infructescences are colourful (bright yellow), but are solitary, relatively small and low in stature (about 1–1.5 m above the ground).

The attractiveness of diaspores to ants was similar in three ways to the attractiveness of diaspores to birds: (i) both plant species and site (i.e. disperser assemblage) were important; (ii) La Selva had a higher probability of seed removal than Corcovado; and (iii) *C. lutea* was the most attractive species. Similar to the bird–seed study, both plant species and site significantly affected the probability of seed removal. In the ant–seed study, however, there was also a significant interaction of species and site (two-way analysis of variance for the three species studied at both sites; for species effect, $F = 8.8$, d.f. = 2, $P < 0.001$; for site effect, $F = 19.7$, d.f. = 1, $P < 0.0001$; for the interaction effect, $F = 3.8$, d.f. = 2, $P < 0.05$ (C. Horvitz, D. Bello y Bello and R. Dirzo, unpublished data). For those species studied at both sites, the order of attractiveness differed. *Calathea lutea* was the most attractive to ants at both sites. However, at La Selva, *C. lutea* was followed by *C. marantifolia* and *Pleiostachya* (the two species that were least attractive to birds at both sites), while, at Corcovado, these latter two species were nearly equal in attractiveness. All species were more successful in attracting ants at La Selva than they were at Corcovado (from 1.1- to 3.3-fold more attractive). For the two species studied only at La Selva, one (*C. cleistantha*, the myrmecochore) was quite attractive to ants (as attractive as *C. lutea*) and the other was much less attractive (*C. gymnocarpa*, the least attractive to ants of all the species studied at La Selva). *Calathea inocephala* was assayed for ant attractiveness only at Corcovado; it was the species least attractive to ants at that site (Fig. 10.2A). *Calathea gymnocarpa* and *C. inocephala* had large seeds (Table 10.1), which may be the reason they were unattractive to ants. The species most attractive to ants in this study had removal probabilities quite similar to those of previously studied myrmecochores (*Calathea micans*, studied at La Selva, and *Calathea ovandensis* studied in Mexico), which had probabilities of removal of 0.87 and 0.86 (Horvitz and Schemske, 1986b; LeCorff and Horvitz, 1995).

In considering how ant dispersal might arise, an interesting comparison is the relative attractiveness of diaspores to ants vs. birds. This comparison can be made for eight cases: *C. lutea*, *C. marantifolia* and *Pleiostachya* at both sites; *C. gymnocarpa* at La Selva; and *C. inocephala* at Corcovado. Except for the large-seeded *C. inocephala* and *C. gymnocarpa*, which were unattractive to ants, diaspores were more readily discovered and removed by ants than by birds. Ants may be more 'reliable' as dispersers, because they are ubiquitous in the leaf litter. In contrast, fruit-eating birds are especially variable in occurrence over space and time (Levey, 1988). A similar observation was made for interactions of Australian *Acacia* species with ants and birds (O'Dowd and Gill, 1986). These results suggest possible evolutionary steps from ornithochory to myrmecochory. In particular, they indicate that arillate seeds of ornithochorous species (when removed from fruits and placed into the leaf litter) are attractive to litter-foraging ants that are typically involved in myrmecochory (including ponerines, such as *Odontomachus*, *Pachycondyla* and *Ectatomma*, myrmicines, such as *Aphaenogaster*, and others). A similar result has been found for a variety of ornithochorous arillate species of several plant families in the Brazilian Atlantic Forest (Pizo and Oliveira, 2000). If interactions with ants positively influence recruitment and if seeds come into contact with litter-foraging ants (by occasionally falling from plants or being dropped beneath plants by birds who have not consumed the entire aril), then ants may begin to select on plant traits that will favour more ant–seed interactions.

Site differences will be interpreted in the context of disperser assemblages (below).

Distance of dispersal

Plant species did not significantly affect the probability that removed seeds were taken far by birds, but site (i.e. disperser assemblage)

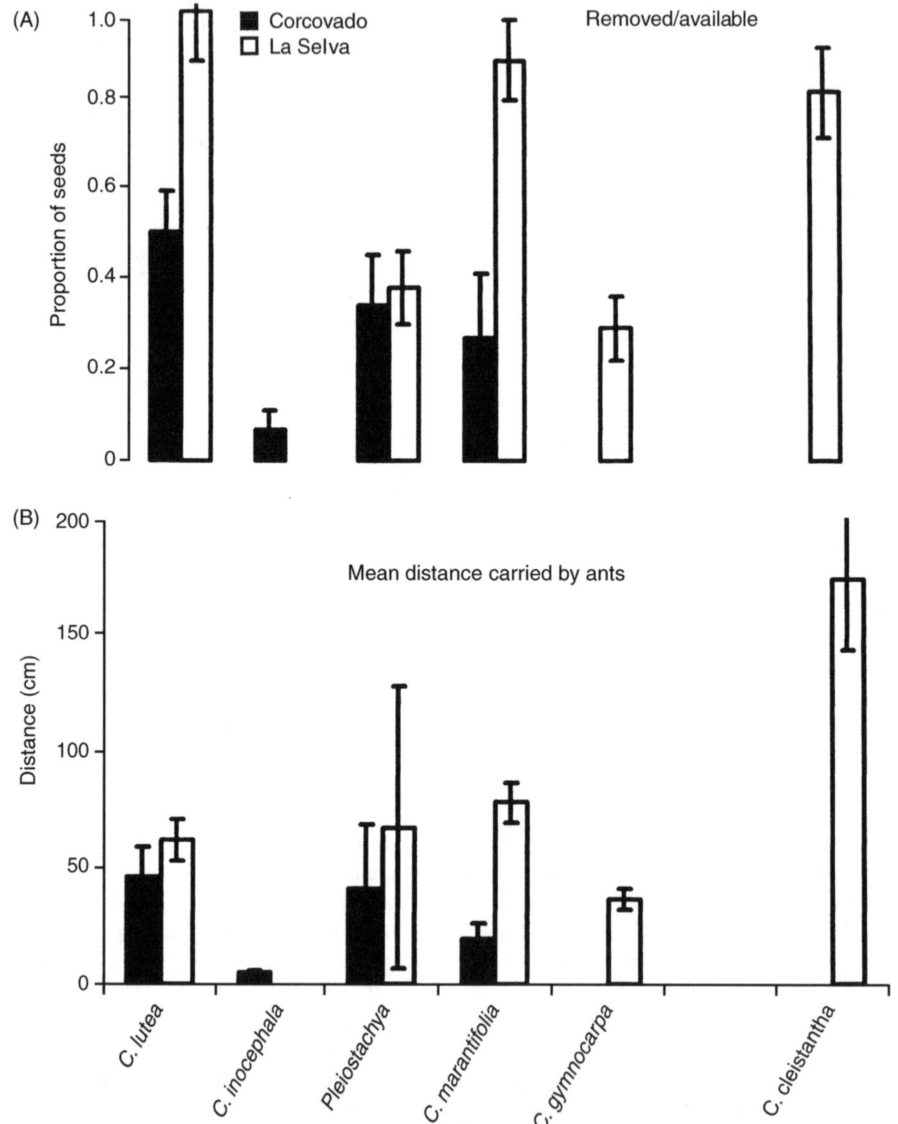

Fig. 10.2. Attractiveness of seeds to ants. (A) Proportion of available seeds per seed trial that were removed by ants (± SE) (*n* = 32, 38, 32, 0, 32, 33, 11, 32, 0, 32, 0, 40 trials of each plant species at each site, from left to right). (B) Distance that seeds were moved per trial (± SE) (*n* = 19, 35, 4, 0, 9, 18, 4, 31, 0, 14, 0, 33 trials of each plant species at each site, from left to right; only trials in which seeds were removed could be counted). Two species (*C. gymnocarpa* and *C. cleistantha*) only occurred at La Selva. One species was studied only at Corcovado (*C. inocephala*), because fruiting was extremely rare in this species at La Selva during the year of this study. Only *C. cleistantha* has the 'ant-dispersal syndrome'. Plant species are listed in the same order as in Fig. 10.1.

did (Friedman block tests; for species effect, $F = 0.94$, d.f. = 3, NS; for site effect, $F = 6.3$ d.f. = 1, $P < 0.01$ (Horvitz and Pizo, 2001; Fig. 10.1B)). Seeds that were taken by birds had a slightly, but significantly, higher probability of being taken a long distance at Corcovado than at La Selva. The means were only slightly different, 0.84 vs. 0.83, respectively, but the non-parametric analysis is based upon ranks, which differed considerably, 30.7 vs. 24.6, respectively, for Corcovado and La Selva (C. Horvitz and M.A. Pizo, unpublished data). The site effect was not consistent across all plant species. Site differences probably resulted from differences in dispersal assemblages (discussed below).

The proportion 'seeds moved far/seeds found by birds' was not dependent upon plant species, but the probability of being found by birds (seeds found by birds/seeds available) was dependent upon plant species and was the rate-limiting component (C. Horvitz and M.A. Pizo, unpublished data). In other words, to be found plants had to be very attractive, but they thus attracted many birds of various behaviours (discussed below), such that further specialization by plants on particular bird species (e.g. those that carry seeds a long distance) appears to be constrained. It was not possible for species to attract long-distance avian dispersers without also attracting short-distance avian dispersers and even predators (C. Horvitz and M.A. Pizo, unpublished data).

The distance component of dispersal for ants was different from the distance component of dispersal for birds. Only site (i.e. disperser assemblage), but not plant species, significantly affected the distance seeds were carried by ants (two-way analysis of variance for the three species studied at both sites; for species effect, $F = 0.01$, d.f. = 2, NS; for site effect, $F = 6.7$, d.f. = 1, $P < 0.05$; for the interaction effect, $F = 0.9$, d.f. = 2, NS (C. Horvitz, D. Bello y Bello and R. Dirzo, unpublished data)). This result at first seems similar to that found for birds. However, all plant species were dispersed longer distances by ants at La Selva than at Corcovado, a site effect opposite to the distance results for bird dispersal. More importantly, if one compares all the plant species (rather than only the three species studied at both sites), there was a very large plant-species effect on dispersal distance. *Calathea cleistantha* (the myrmecochore), a species studied only at La Selva, was carried nearly threefold further than all the others (Fig. 10.2B).

This analysis shows that, even though ornithochorous seeds were attractive to ants, they were abandoned after very short distances. This contrasts markedly with the way ants behave toward the myrmecochorous species. The bird-adapted diaspores did not elicit from ants the kind of carrying behaviour seen with true myrmecochores; further specialization of diaspore traits, perhaps chemical or morphological, seem to be necessary to convert an ornithochore into a myrmecochore.

Site differences will be interpreted in the context of disperser assemblages (see below).

Disperser assemblages

The bird assemblages differed significantly between sites ($G^2 = 564.2$, d.f. = 10, $P < 0.001$ (Fig. 10.3)), although *Mionectes oleagineus*, the ochre-bellied flycatcher (Tyrannidae), was responsible for the majority of bird–seed interactions at both sites. La Selva's disperser assemblage had a higher H', due to a more even representation of species, and included many interactions with birds in the Emberizidae (e.g. *Cyanocompsa*, *Arremonops* and *Ramphocelus* (Fig. 10.3)). This group of birds had a much lower probability of carrying seeds a long distance (38.0% of 213 bird interactions) than did the Pipridae or the Tyrannidae (91.9% and 96.8% of 185 and 347 interactions, respectively (C. Horvitz and M.A. Pizo, unpublished data)). In addition, bird–seed interactions were more evenly distributed across the forest (among the various study locations) at La Selva than at Corcovado (C. Horvitz and M.A. Pizo, unpublished data). These features of La Selva's bird-disperser assemblage probably accounted for both the higher probability that diaspores were found by birds and the lower probability that such diaspores were moved long distances, compared with Corcovado. However, at Corcovado, the second most abundant interactions were with a species of dove, *Claravis pretiosa* (Columbidae), which is more probably a seed predator than a disperser (C. Herrera, D. Levey and B. Loiselle, personal

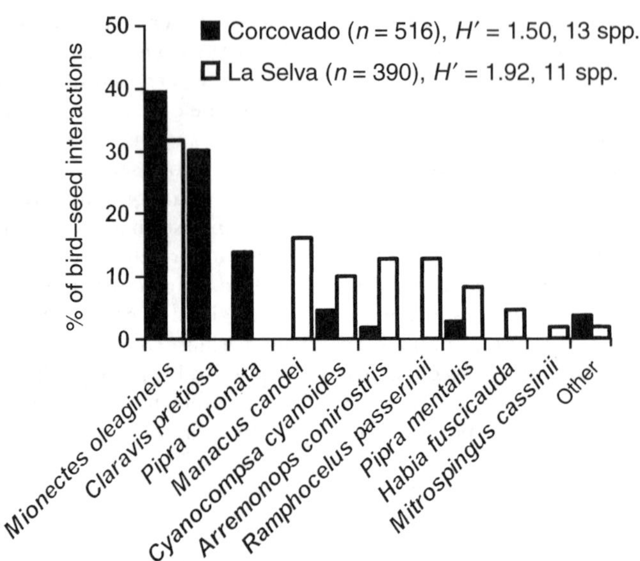

Fig. 10.3. Disperser assemblages: per cent of bird–seed interactions by each bird species at each site (n is the total number of interactions at each site). Species are sorted in order of abundance, where abundance is indicated by the sum of the number of interactions across both sites. Rare species are lumped into the category 'Other' to enable statistical comparison of the sites.

communication). If long-distance seed removal by Columbidae is not counted, Corcovado still has a higher probability of long-distance dispersal; 86.3% of the 359 interactions with other birds at Corcovado were long-distance events, compared with 72% of 390 such interactions at La Selva. More data on the fates of seeds after being ingested by birds are needed in future studies.

Ant assemblages also differed significantly between sites ($G^2 = 373.0$, d.f. = 20, $P < 0.0001$ (Fig. 10.4)). Unlike the bird communities, there was not a species of ant that dominated at both sites (Fig. 10.4). At Corcovado, the ant species responsible for most interactions belonged to the genus *Solenopsis*, species of which recruit large numbers of workers to seeds and bury them in soil tunnels, where the workers remove the arils bit by bit (Horvitz and Schemske, 1986b). In contrast, at La Selva, the ant species responsible for most interactions belonged to the genus *Aphaenogaster* (Fig. 10.4), species of which carry seeds away (LeCorff and Horvitz, 1995). These differences probably account for the higher probability of seed removal and the longer dispersal distances at La Selva than at Corcovado. At both sites, there were also frequent interactions involving ponerine ants (*Ectatomma*, *Odontomachus* and *Pachycondyla*), which also carry seeds away (Horvitz and Beattie, 1980; Horvitz and Schemske, 1986b; LeCorff and Horvitz, 1995; C. Horvitz, D. Bello y Bello and R. Dirzo, unpublished data). Ant assemblages of two myrmecochores studied at La Selva, *C. cleistantha* (C. Horvitz, D. Bello y Bello and R. Dirzo, unpublished data) and *C. micans* (LeCorff and Horvitz, 1995), were very similar.

Gap dependence of seedling emergence and survival

We now return to the behaviour of the seeds and seedlings and to the main theme of the chapter: to relate dispersal to gap dependency of recruitment. Plant species and light environment significantly affected seedling emergence. The effects of light differed among species and species differed between sites in seedling emergence. A three-way analysis of variance indicated the following significant results: main effects of plant species, $F = 17.0$, d.f. = 3, $P < 0.0001$, and light environment,

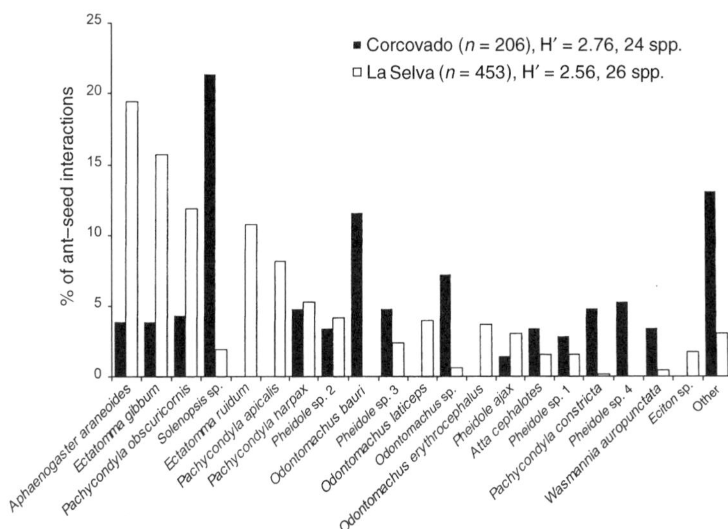

Fig. 10.4. Disperser assemblages: per cent of ant–seed interactions by each ant species at each site (*n* is the total number of interactions at each site). Species are sorted in order of abundance, where abundance is indicated by the sum of the number of interactions across both sites. Rare species are lumped into the category 'Other' to enable statistical comparison of the sites.

$F = 6.3$, d.f. = 2, $P < 0.0025$; interaction of species with light, $F = 3.4$, d.f. = 6, $P < 0.0041$, and species with site, $F = 13.3$, d.f. = 3, $P < 0.0001$ (C. Horvitz and J. LeCorff, unpublished data). For example, *C. lutea* (the species with the most conspicuous display that was most attractive to birds at both sites) germinated much better at La Selva, where it showed very strong gap dependency (Fig. 10.5B), than it did at Corcovado (Fig. 10.5A). In contrast, *C. inocephala* (a large-seeded species that ranked second in its attractiveness to birds at both sites) germinated much better at Corcovado than it did at La Selva; at both sites its germination was strong in both gaps and understorey, but weak in intermediate light conditions (Fig. 10.5A, B). Except for this species and *C. gymnocarpa* (the other large-seeded species, which was also quite attractive to birds), all other species at La Selva had higher seedling emergence in gaps (Fig. 10.5B). This pattern was not seen at Corcovado (Fig. 10.5A).

Was gap dependency of seedling emergence associated with dispersal syndrome or attractiveness to birds? Our data do not support the hypothesized associations between gap dependency and dispersal. Although there was a strong gap advantage for *C. lutea* (very attractive to birds) as hypothesized, there were also strong gap advantages for *C. lasiostachya* (least attractive to birds) and *C. cleistantha* (the myrmecochore), which were not predicted. Comparing other myrmecochorous *Marantaceae*, we note that *C. micans*, another widespread myrmecochore at La Selva, was able to germinate readily in shady conditions (LeCorff, 1996), but a Mexican species, *C. ovandensis*, germinated much more readily in gaps than in shaded understorey (Horvitz and Schemske, 1994). The pattern shown by *C. lasiostachya* remains perplexing: it looks as though it should be attractive to dispersers and its germination is gap-dependent and yet the probability of avian dispersal was quite low.

Calathea gymnocarpa and *C. inocephala*, the large-seeded species, did not depend upon gaps for germination and yet they were very attractive to birds. Large seed size is often associated with shade tolerance (Garwood, 1989). The question remains, however, why these species invest in such showy displays and how avian dispersal may benefit them. They may use dispersal to escape enemies under the parent plant rather than to colonize light gaps or to reach some other preferred habitat. *Calathea*

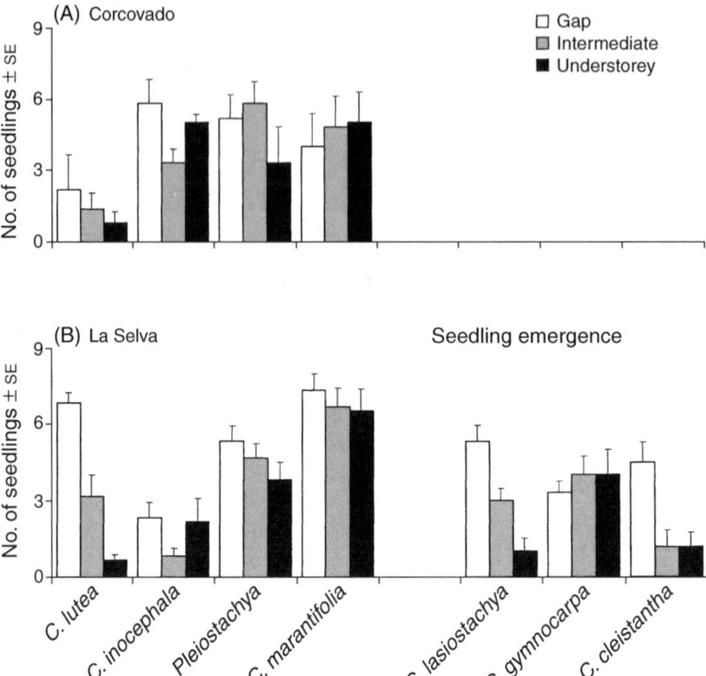

Fig. 10.5. Seedling emergence (number of seedlings emerged after 1 year) per box (± SE) in each light environment. The number of boxes per light environment was the same for a given species within a given site. (A) Corcovado (n = 5, 6, 6 and 5 boxes per species per light environment; species listed from left to right). (B) La Selva (n = 6, 6, 6, 6, 10, 6 and 6 boxes per species per light environment; species listed from left to right). Plant species are listed in the same order as in Fig. 10.1.

inocephala is very common throughout the Corcovado forest, but it occurs only patchily in the La Selva forest. Although it had a higher probability of seed removal by birds at La Selva, it had higher germination at Corcovado; these intersite comparisons suggest that its populations may be germination-limited in some way.

In summary, in contradiction to the hypothesized relationships, gap dependency of seedling emergence was strong for the myrmecochore and variation in gap dependency of the ornithochores was unrelated to their attractivenes to birds. We next examine whether dispersal was associated with gap dependency of seedling survival, the second component of recruitment.

Survival time of seedlings was significantly enhanced by gaps in general, as indicated by analysis of the pooled data (test of homogeneity of survival across light environments, log-rank analysis, $\chi^2 = 25.7$, d.f. = 2, $P < 0.0001$ (C. Horvitz and J. LeCorff, unpublished data)). However, analyses of each species at each site revealed variation among species and between sites in survival responses to gaps. For example, at both sites, the large-seeded *C. inocephala* survived significantly longer in understorey than in gaps (Fig. 10.6). Thus, this species can germinate in the shade and actually prefers shade for seedling survival. There was no significant effect of light environment on survival time for *C. lutea* at Corcovado (but there was for his species at La Selva), for *C. gymnocarpa* or for the myrmecochore *C. cleistantha* (Fig. 10.6). The two species with the strongest gap advantage for seedling survival time were *C. marantifolia* and *Pleiostachya* (ornithochores with relatively low attractiveness to birds), but germination of these species was relatively independent of gaps.

Was gap dependency of seedling survival associated with gap dependency of germination, with dispersal syndrome or with attractiveness to birds? First, a comparison of Figs 10.5

and 10.6 reveals no clear association between germination and survival responses of species to light environments. This result is counter to a general intuition that a species germination response is 'adaptive' – that seeds germinate in environments that are most favourable for seedlings. Seed dormancy responses also varied among the species, were complex and do not resolve all the issues brought up here (C. Horvitz and J. LeCorff, unpublished data). Secondly, the myrmecochore's seedling survival time was independent of light environment – which may argue for relative shade tolerance – but overall it had the lowest survival times in the study. Low survival was unexpected as this species is very common throughout the La Selva forest. Thirdly, attractiveness of species to avian dispersers was not associated with gap dependency of survival. The order of attractiveness to avian dispersers was *C. lutea*, *C. inocephala*, *Pleiostachya* and *C. marantifolia*. In contrast, the order of gap dependency of survival for these same species was *C. marantifolia*, *Pleiostachya*, *C. lutea* and *C. inocephala* at Corcovado (Fig. 10.6A). The order was similar at La Selva (Fig. 10.6B).

Conclusions

It is difficult to reach a simple, synthetic conclusion from this rich complexity of results, although the data do suggest some steps in the pathway from ornithochory to myrmecochory. The initial hypothesis – that ant-dispersed species should have more shade-tolerant recruitment than bird-dispersed species – is not supported by the data. It is true, though, that

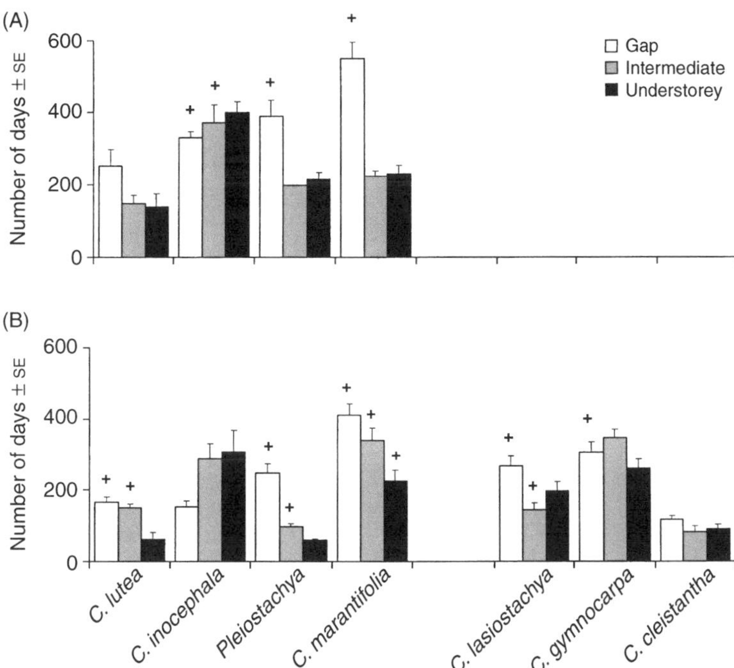

Fig. 10.6. Survival time (number of days, estimated by the product-limit method) per emerged seedling (± SE) in each light environment. A 'plus' sign above the bar indicates that the survival time estimate may be an underestimate due to censored observations. The number of emerged seedlings per light environment per species per site was not fixed, but was itself a result of the experiment. (A) Corcovado (n = 11, 8, 5, 35, 20, 30, 31, 35, 21, 20, 24 and 25 seedlings per species per light environment as listed from left to right). (B) La Selva (n = 41, 19, 4, 14, 8, 14, 32, 28, 23, 44, 40, 40, 53, 30, 10, 21, 25, 24, 27, 7 and 10 seedlings per species per light environment as listed from left to right). Plant species are listed in the same order as in Fig. 10.1.

the bird-dispersed species with the largest display is a species that strongly prefers gaps for germination. However, the array of responses of the other species makes it difficult to generalize about the relationship between gap dependency and dispersal traits. From a demographic viewpoint, it is intriguing that two different components of recruitment success (i.e. seed germination and seedling survival) behave quite differently. Further analyses of the demography of the complete life cycle of these species are in progress and may provide more clues as to the selective advantage of one type of dispersal over another.

The initial hypothesis was derived from a megamatrix analysis, a particular kind of analysis of population dynamics in a heterogeneous landscape (Horvitz and Schemske, 1986a). Recent work on the demography of understorey plants in heterogeneous landscapes has indicated that a more appropriate analysis for dispersal questions might be a stochastic sequence analysis (S. Tuljapurkar, C. Horvitz and J. Pascarella, unpublished data). These two analyses address different kinds of average population growth rates: the aggregate average over space at a given time in a very large area comprised of many patches (megamatrix) and the long-term temporal average in a given patch (stochastic sequence). Both are of interest in heterogeneous landscapes. It remains to be seen whether the predictions about shade tolerance and dispersal will be altered by framing dispersal questions in terms of the stochastic sequence analysis.

Improving our understanding of the match between plant recruitment requirements, plant–disperser interactions and the evolution of seed and fruit traits may be significant for conservation of species that depend upon mutualisms at key parts of their life cycle (Jordano and Schupp, 2000). The extent to which variation across sites in plant–disperser ecology may or may not result in functionally equivalent mutualisms (e.g. Jordano, 1994) is of special interest in this context.

Acknowledgements

We thank Marcos Molina and Gustavo Vargas for extensive assistance in the field, and the staff of La Selva Biological Station Organization for Tropical Studies, especially Station Directors David and Deborah Clark, and the staff of Sirena Station in Corcovado National Park, especially Paulino Valverde (Servicio de Parques Nacionales), for logistical support. We also thank Rick Seavey, Jean Seavey, Devon Graham and Jeannette Paniagua, and the 1991 Universidad de Costa Rica field ornithology class taught by Gilbert Barrantes, for additional field assistance. We thank Bette Loiselle, Pedro Jordano, Carlos Herrera, Doug Levey, Ted Fleming and Martin Cipollini for discussions and critiques. The field work was supported by National Science Foundation (NSF) grant BSR-8906637 to C.C. Horvitz; analysis and writing were supported by grants from FAPESP (Fundaçao de Amparo à Pesquisa do Estado de São Paulo) (97/3105-3) to M.A. Pizo and from the PRIA (Pole de Recherches et Innovations d'Angers) to C.C. Horvitz.

References

Garwood, N. (1989) Tropical soil seed banks: a review. In: Leck, M.A., Palmer, V.T. and Simpson, R.L. (eds) *Ecology of Soil Seed Banks.* Academic Press, San Diego, California, pp. 149–209.

Hartshorn, G.S. (1983) Plants. In: Janzen, D.H. (ed.) *Costa Rican Natural History.* University of Chicago Press, Chicago, Illinois, pp. 118–157.

Herrera, C.M. (1998) Long-term dynamics of Mediterranean frugivorous birds and fleshy fruits: a 12-yr study. *Ecological Monographs* 68, 511–538.

Herwitz, S.R. (1981) *Regeneration of Selected Tropical Tree Species in Corcovado National Park, Costa Rica.* University of California Press, Berkeley, 111 pp.

Holdridge, L.R. (1967) *Life Zone Ecology.* Tropical Science Center, San José, Costa Rica.

Horvitz, C.C. (1991) Light environments, stage structures and dispersal syndromes of Costa Rican *Marantaceae.* In: Huxley, C.R. and Cutler, D.F. (eds) *Ant–Plant Interactions.* Oxford University Press, Oxford, pp. 463–485.

Horvitz, C.C. and Beattie, A.J. (1980) Ant dispersal of *Calathea* (*Marantaceae*) seeds by carnivorous ponerines (Formicidae) in a tropical rain forest. *American Journal of Botany* 67, 321–326.

Horvitz, C.C. and LeCorff, J. (1993) Spatial scale and dispersion pattern of ant- and bird-dispersed herbs in two tropical lowland rain forests. *Vegetatio* 107/108, 351–362.

Horvitz, C.C. and Schemske, D.W. (1986a) Seed dispersal and environmental heterogeneity in a neotropical herb: a model of population and patch dynamics. In: Estrada, A. and Fleming, T.H. (eds) *Frugivores and Seed Dipsersal.* Dr W. Junk Publishers, Dordrecht, The Netherlands, pp. 169–186.

Horvitz, C.C. and Schemske, D.W. (1986b) Seed dispersal of a neotropical myrmecochore: variation in removal rates and dispersal distances. *Biotropica* 18, 319–323.

Horvitz, C.C. and Schemske, D.W. (1994) Effects of dispersers, gaps, and predators on dormancy and seedling emergence in a tropical herb. *Ecology* 75, 1949–1958.

Howe, H.F. (1977) Bird activity and seed dispersal of a tropical wet forest tree. *Ecology* 58, 539–550.

Howe, H.F. and Vande Kerckhove, G.A. (1979) Fecundity and seed dispersal of a tropical tree. *Ecology* 60, 180–189.

Jordano, P. (1994) Spatial and temporal variation in avian-frugivore assemblage of *Prunus mahaleb*: patterns and consequences. *Oikos* 71, 479–491.

Jordano, P. and Schupp, E.W. (2000) The quantity component and patterns of seed rain for *Prunus mahaleb. Ecological Monographs* 70, 591–615.

LeCorff, J. (1996) Establishment of chasmogamous and cleistogamous seedlings of an ant-dispersed understorey herb, *Calathea micans* (*Marantaceae*). *American Journal of Botany* 83, 155–161.

LeCorff, J. and Horvitz, C.C. (1995) Dispersal of seeds from chasmogamous and cleistogamous flowers in an ant-dispersed neotropical herb. *Oikos* 73, 59–64.

Levey, D.J. (1988) Spatial and temporal variation in Costa Rican fruit and fruit-eating bird abundance. *Ecological Monographs* 58, 251–269.

Mazer, S.J. and Wheelwright, N. (1993) Fruit size and shape: allometry at different taxonomic levels in bird-dispersed plants. *Evolutionary Ecology* 7, 556–575.

O'Daniel, D. (1987) Seed dispersal and seed predation in two species of *Calathea* in a Costa Rican rainforest. Master's dissertation, University of Texas at Austin, Austin, Texas.

O'Dowd, D.J. and Gill, A.M. (1986) Seed dispersal syndromes in Australian *Acacia.* In: Murray, D.R. (ed.) *Seed Dispersal.* Academic Press, North Ryde, NSW, pp. 87–121.

Pizo, M.A. and Oliveira, P.S. (2000) The use of fruits and seeds by ants in the Atlantic forest of southeast Brazil. *Biotropica.*

Rich, P.M. (1990) Characterizing plant canopies with hemispherical photographs. *Remote Sensing Reviews* 5, 13–29.

SAS Institute Inc. (1999) *SAS System®*, Version 8 (TSMO). SAS Institute Inc., Cary.

Schupp, E.W. (1993) Quantity, quality and the effectiveness of seed dispersal by animals. *Vegetatio* 107/108, 15–29. In: Fleming, T.H. and Estrada, A. (eds) *Frugivory and Seed Dispersal: Evolutionary and Ecological Aspects.* Kluwer Academic Publishers, Belgium 32, 851–861.

Van der Pijl, L. (1982) *Principles of Seed Dispersal in Higher Plants*, 3rd edn. Springer-Verlag, Berlin, Germany.

Wenny, D.G. (2000) Seed dispersal, seed predation and seedling recruitment of a neotropical tree. *Ecological Monographs* 70, 331–351.

Wenny, D.G. and Levey, D.J. (1998) Directed seed dispersal by bellbirds in a tropical cloud forest. *Proceedings of the National Academy of Sciences USA* 95, 6204–6207.

Westoby, M., French, K., Hughes, L., Rice, B. and Rodgerson, L. (1991) Why do more plant species use ants for dispersal on infertile compared with fertile soils? *Australian Journal of Ecology* 16, 445–455.

11 The Role of Fruit Traits in Determining Fruit Removal in East Mediterranean Ecosystems

Ido Izhaki

Department of Biology, University of Haifa at Oranim, Tivon 36006, Israel

Introduction

The removal of seeds from parent plants by dispersers has been studied for several decades (Ridley, 1930; Van der Pijl, 1982; Janzen, 1983). The reliance of many fleshy-fruited plants, especially in the tropics, on birds and mammals for seed dispersal gave rise to the theory of a coevolved mutualism between fruiting plants and seed-dispersing birds (Snow, 1971; McKey, 1975; Janzen, 1983). On the basis of historical and phylogenetic effects, the current belief is that seed-dispersers are neither strongly selecting any fruit traits nor promoting evolutionary radiation (Herrera, 1985). It is still plausible, however, that selection pressures of fleshy-fruited plants on the seed-dispersers shape physiological, digestive and behavioural adaptations of the dispersers (Herrera, 1987, 1992, 1995).

Because fruits must be removed before seeds can be dispersed, fruit removal is probably an important component of plant fitness. The degree to which fruits are removed by seed-dispersers is likely to influence the strength of the plant–frugivore interactions. Substantial intraspecific variation in fruit removal has been demonstrated among individual plants within sites in the same year (Courtney and Manzur, 1985; Denslow, 1987; Sargent, 1990; Laska and Stiles, 1994; Izhaki, 1998a), within sites between years (Jordano, 1987; Sallabanks, 1992) and between sites (Alcántara *et al.*, 1997). Interspecific variation in fruit removal has also been found among species in the same site (Herrera, 1984; Izhaki and Safriel, 1985). Furthermore, fruit removal is highly habitat-dependent, as shown by Herrera (1995), who found that fruit-removal success is much higher in lowland habitats (mean = 90.2%) than in highlands (mean = 62.1%) in the western Mediterranean region. These studies reveal that many factors contribute simultaneously to variation in fruit-removal rates, that different factors operate in different habitats and that complex multilevel interactions make it difficult to distinguish general patterns.

I test three hypotheses that account for within- and among-species variation in fruit removal:

1. Fruit morphology (size, pulp–seed ratio, number of seeds, etc.) is an important factor in birds' ability to discriminate between fruits (e.g. Mazer and Wheelwright, 1993). Fruit size is thought to be especially important. According to the 'fruit-size hypothesis', frugivore body size is the predominant factor that determines

 Seed Dispersal and Frugivory: Ecology, Evolution and Conservation (eds D.J. Levey, W.R. Silva and M. Galetti)

the frugivore's ability to feed upon fruits of a given size (Herrera, 1984; Wheelwright, 1985; but see Levey, 1987). In ecosystems where most avian frugivores are relatively small (< 50 g), selection against large-fruited plants may occur. Indeed, the close agreement between fruit size and disperser gape width suggests that birds have constrained the evolution of large fruits (Herrera, 1984; Jordano, 1987).

2. The assumption that frugivores should forage for fruits that best satisfy their nutritional demands is the basis for the 'nutritional-content hypothesis' (Manasse and Howe, 1983; Denslow, 1987; Levey, 1987). However, nutritional demands may differ among and within species. For example, transient individuals may forage for maximal energy gain, whereas reproductive individuals may forage for protein-rich items. Indeed, the nutritional profile of fruits appears to be related to the season of ripening and to the seasonally shifting nutritional demands of birds. For example, fat-rich fruits appear mainly in autumn and winter, whereas water- and carbohydrate-rich fruits appear in the summer (Herrera, 1982). Birds may selectively remove fat-rich fruit during migration, when energy demands are great (Stiles, 1980). But nutrient content of fruit may simultaneously attract frugivores that do not disperse seeds and are therefore detrimental to plant fitness. Thus, the nutritional content of fruits should represent a trade-off between competition for avian dispersers and avoidance of fruit destruction by non-dispersers (Stiles, 1980; Borowicz and Stephenson, 1985).

3. Optimal foraging theory predicts that, within a population, individual plants that offer high relative abundance of fruit should attract disproportionately more attention from frugivores – the 'fruit crop hypothesis' (Snow, 1971; McKey, 1975; Howe and Estabrook, 1977). However, as higher-order interactions may be involved, the relationship between crop size and fruit removal may be much more complex (Jordano, 1987). For example, large-fruit crops may attract more insects, thus increasing infestation level and reducing attractiveness of fruits to dispersers (Jordano, 1987).

I present a case-study at the interspecific level on the relationship between fruit traits and fruit removal in eastern Mediterranean scrublands. My goal is to determine which plant and fruit traits, if any, govern fruit-removal success within a plant community (interspecific level) and among individuals within a population (intraspecific level).

Study Site: Beit Jimal

The study site is a 50,000 m^2 plot near Beit Jimal monastery on the northern slope of a hill, 280–310 m above sea level (34° 58′ E, 31° 40′ N). Mean annual precipitation is 493 mm. Regionally, precipitation drops rapidly in a southward direction; 50 km away is the desert edge, where mean annual precipitation is 200 mm. The vegetation is Mediterranean evergreen scrub, dominated by *Quercus calliprinos* and *Pistacia palaestina* (family names provided in legend of Fig. 11.2). This plant formation is found above 200 m altitude (Zohary, 1962) and reaches its southern limit in the Beit Jimal region. The scrub has trees of *Q. calliprinos* and *P. palaestina* 2–4 m high, shrubs of *Rhamnus lycioides* and *Pistacia lentiscus* 0.8–2 m high and of *Cistus* spp. and *Sarcopoterium spinosum* 0.3–0.6 m high, and vines, *Clematis cirrhosa, Smilax aspera, Rubia tenuifolia* and *Lonicera etrusca.*

The study plot is structurally simple, with only eight fleshy-fruited species (Izhaki and Safriel, 1985; Izhaki, 1986): one tree (*P. palaestina*), two shrubs (*R. lycioides* and *P. lentiscus*), a semiparasitic shrub (*Osyris alba*), a dwarf shrub (*Asparagus aphyllus*) and three vines (*L. etrusca, R. tenuifolia* and *S. aspera*). These species represent 11% of the fruit-producing species in Israel (Table 11.1).

Methods

Fruit removal

On each focal plant I prevented fruit removal by covering five branches with netting and I counted the number of ripe fruits in the bags

Table 11.1. The total number of fleshy-fruited species in the Mediterranean region in Israel and the number included in this study, according to their life-forms.

	Total number of species	Number of species included in the present study	
		Fruit removal	Fruit traits
Trees	18	1	9
Shrubs	17	3	8
Vines	16	3	6
Parasites	5	1	2
Geophytes	7		1
Perennials	5		0
Annuals	3		1
Total	71	8 (11%)	27 (38%)

every 5 days. By comparing this number with the number of fruits on the exposed branches, I could estimate the proportion and absolute number of fruits that had been removed by frugivores. For a full description of this technique see Izhaki *et al.* (1991).

Three statistics estimate the fruit-removal success of a plant:

1. Removal efficiency (after Willson and Whelan, 1993) is the per cent of total crop removed from an individual plant. This value estimates seed-dispersal success relative to the number of seeds produced by the plant. In effect, it is a benefit : cost ratio (Herrera, 1988, 1991; Jordano, 1991) that is not directly correlated to the absolute number of dispersed seeds.

2. Absolute removal is the absolute number of fruits removed from an individual plant by avian frugivores. This index is probably more directly related to plant fitness than the proportion of fruits removed (Murray, 1987; Sargent, 1990; Herrera, 1991).

3. Removal success is the proportion of fruit removed from each individual (on the intraspecific level) or species (on the interspecific level) relative to the total number of fruits removed in the population (on the intraspecific level) or the community (on the interspecific level) (after Willson and Whelan, 1993).

Fruit and plant traits

Do the three hypotheses account for observed variation in fruit removal? Each hypothesis required me to measure a different set of traits. I measured morphological traits to test the fruit-size hypothesis, nutritional traits to test the nutritional-content hypothesis and crop size to test the fruit-crop hypothesis.

Morphological traits

Fruit traits of the eight plant species that produce fleshy fruits at the study site and 19 other species from sites up to 120 km north were quantified. These species belong to 17 families and may be categorized into several life-forms (Table 11.1, and see Fig. 11.2 for a list of plant species). Most of these species are bird-dispersed plants (Izhaki and Safriel, 1985; Izhaki, 1986; Barnea *et al.*, 1991; Izhaki *et al.*, 1991), but at least two of them (*Ziziphus spina-christi* and *Styrax officinalis*) are mainly mammal-dispersed (e.g. Korine *et al.*, 1998).

Fruits were collected from at least ten individuals per species, brought to the laboratory and measured. Morphological measurements, based on 50–120 fruits of each species, included fruit wet mass, single and total wet seed mass and the number of seeds. Fresh pulp and seeds were then oven-dried at 60°C to constant mass. Pulp and seed water content, pulp dry mass, and single and total dry seed mass were calculated. From these morphological measurements I calculated three ratios that reflect the profitability of the fruits for frugivores in terms of the ratio between the nutritious and non-nutritious material (pulp and seeds, respectively) (Herrera, 1987):

1. The relative yield (RY) of a fresh fruit is (dry mass of pulp)/(fresh mass of whole fruit). This ratio indicates how much dry material is gained by the frugivore per unit mass of whole fruit ingested and processed.

2. PTS is (dry mass of pulp)/(total dry mass of seeds per fruit). This ratio indicates how much dry material is gained by the frugivore relative to the total seed load of a whole fruit ingested.
3. PS is pulp dry mass per seed. This ratio indicates how much dry material is gained by the frugivore per load of one seed.

I also calculated the ratio between wet mass of pulp and the mass of total seeds.

Nutritional and mineral fruit content

Pulp of the 27 fruit species (38% of the fruit-producing species in Israel) (Table 11.1) was analysed for lipids, nitrogen (Kjeldahl method), reducing sugars, ash and several minerals (sodium (Na), potassium (K), calcium (Ca), phosphorus (P)), according to the Association of Official Analytical Chemists (AOAC, 1984) procedures (see also Izhaki, 1992). The amino acid content in the pulp of each fruit species was analysed using an amino acid analyser (Izhaki, 1993, 1998b). Total protein was calculated as the sum of all amino acids. This method contrasts with most other studies on fruit nutritional content (e.g. Herrera, 1987), in which total protein is calculated by multiplying total nitrogen by 6.25 to yield crude proteins. As demonstrated by Izhaki (1993), the traditional method of estimating protein does not reflect the true protein content of fruits. However, the amino acid method is limited to only 17 amino acids and cannot distinguish between toxic and non-toxic amino acids (Izhaki, 1993; Levey *et al.*, 2000). I also calculated the caloric content of pulp using the average gross-energy equivalents of protein (17.2 kJ g^{-1}), fat (38.9 kJ g^{-1}) and carbohydrates (17.2 kJ g^{-1}) (see Karlson, 1972). To estimate the maximum energetic benefit of each species for the consumer, I calculated the ratios of total energy to pulp wet mass, to wet fruit mass and to wet seed mass. I emphasize that this approach to calculating the nutritional value of fruits to frugivores is simplistic because it ignores the complexities of frugivore digestive physiology (Martínez del Rio and Restrepo, 1993).

Crop size and plant abundance

The crop size of ten to 16 focal plants belonging to each of the eight fruit-bearing plant species was estimated during two fruiting seasons (May–February). I counted the number of ripe fruits on five branches of each plant every 5 days and recorded the number of branches with fruits per plant.

The presence/absence of the eight focal plants was recorded in 17 100 m × 100 m plots. The frequency of occurrence of each species was calculated by dividing the number of plots in which it was found by the total number of plots. This index was used for estimating the abundance of the eight focal plants in the study area.

Statistics

Principal component analysis (PCA) was performed to evaluate the relationships among fruit traits. The objective of PCA is to reduce the number of independent variables and thereby explain variation in a few dimensions. Following PCA, Varimax orthogonal rotation was applied to construct a new, more easily interpretable pattern of component loadings. All morphological traits were log-transformed and the nutritional and mineral proportions were arcsin-square-root-transformed prior to analysis. Correlations between fruit traits among the eight focal species and between fruit traits and fruit-removal indices were analysed by Spearman rank correlation. Statistical analyses were done with SYSTAT (1998).

Results

Fruit removal

Most or all of the fruit crop produced by the eight species was removed by birds; average dispersal efficiency (± SD) was 88.5% ± 11.6%. Fruit-removal efficiency of four plant species was > 90%, with three species having complete removal. Fruit-removal efficiency of another three species was 82–84%, and one species (*O. alba*) had relatively low removal efficiency (67%) (Table 11.2).

Nearly half of the fruits removed in Beit Jimal belonged to *P. lentiscus*, as indicated by its removal success (49.2%) (Table 11.2). *Pistacia palaestina* and *R. lycioides* contributed 20.3% and 25.9%, respectively, to the annual amount of removed fruits. Thus, the remaining five species had minor removal success (range = 0.1–1.5%). Fruit-removal success was largely dependent on absolute removal ($r_s = 0.88$, $P < 0.01$), and absolute removal positively correlated with plant abundance (Table 11.3). Thus, more fruits were removed from those plants that were more common in the site, regardless of their removal efficiency.

Morphological fruit traits

To evaluate the fruit-size hypothesis I measured morphological traits of 27 fruit species. Frequency distributions of all morphological parameters were right-skewed

Table 11.2. Crop size and fruit-removal parameters in Beit Jimal scrubland.

Plant species	*n*	Crop size (SD)	Absolute removal*	Removal efficiency(%)†	Removal success(%)‡	Plant abundance(%)§
Pistacia palaestina	10	8,722 (8,654)‖	8722	100	20.3	65
Pistacia lentiscus	12	12,595 (10,501)	11461	91	49.2	82
Rhamnus lycioides	16	6,405 (11,866)	5381	84	25.9	88
Osyris alba	13	65 (42)	44	67	0.4	41
Rubia tenuifolia	10	310 (215)	254	82	1.5	59
Lonicera etrusca	10	196 (134)	165	84	0.1	6
Asparagus aphyllus	10	1,315 (616)	1315	100	1.3	27
Smilax aspera	10	542 (466)	542	100	1.2	24

*Number of fruits removed from an individual plant (species means).
†Number of fruits removed/number of fruits produced by an individual plant (species means).
‡Number of fruits removed from plant species in the study site/total fruits removed from all plant species in the study site.
§Number of plots in which a plant species was recorded/total number of plots sampled × 100.
‖Only developed fruits (see Izhaki, 1998a).

Table 11.3. Spearman's correlation coefficients between fruit removal (expressed by three parameters) and nutritional pulp content, morphological traits, energetic content, crop size and presence in Beit Jimal, Israel (*n* = eight plant species). Only significant correlations are shown.

		Fruit removal		
		Removal efficiency	Absolute removal	Removal success
Plant abundance			0.69*	0.88**
Nutritional content	NSC	−0.66*		
	Cellulose	0.69*		
	Ash	0.84**		
	Water	−0.80**	−0.86**	
Morphological traits	Fruit wet mass		−0.74*	−0.81*
Energetic content	Energy/pulp wet mass		0.81*	0.86**
	Energy/fruit wet mass	0.63*	0.86**	
	Energy/wet seed mass			
Crop size			1.0**	0.88**

*0.01 < *P* < 0.05.
***P* < 0.01.
NSC, non-structural carbohydrates.

(Fig. 11.1a). The coefficient of variation (CV) was > 100% for all parameters. Most fruits were small; only a few were > 0.5 g (Fig. 11.1a). Number of seeds per fruit was very right-skewed; most fruits (80%) had only one seed and five had more than ten seeds (Fig. 11.1a). Seed wet mass was extremely right-skewed, with > 70% of the species having seeds < 0.1 g (Fig. 11.1a). This is in contrast to the situation in the western Mediterranean, where a bimodal distribution was detected for seed size and most seeds were > 0.1 g (Herrera, 1987). A right-skewed distribution was also apparent for the calculated morphological ratios, such as dry pulp per seed (PTS) (Fig. 11.1a).

The first three principal components accounted for 85.0% of the total variation in the morphological traits of the fleshy fruits (Table 11.4). The first axis of the PCA included parameters associated with fruit size (Fig. 11.2a) and explained 37.1% of the variance (Table 11.4). The second axis was mainly determined by dry pulp per total dry mass of seeds (PTS) and explained 26.1% of the variation (Table 11.4). The distribution of species on the ordination diagram (Fig. 11.2a) demonstrated much higher variation on the second axis among small-fruited species than among large-fruited species. The small-fruited species were divided into those with high PTS ratios (e.g. *L. etrusca*) and those with low ratios (e.g. *Whithania somnifera* and *Phillyrea latifolia*). This clear dichotomy was much less pronounced among the large-fruited species (except for *Arbutus andrachne*, which had a high PTS value).

No significant correlation was detected between any morphological fruit traits and fruit-removal efficiency. Absolute removal was negatively correlated with fruit wet mass and a similar correlation trend was observed in removal success (Table 11.3). I also correlated fruit removal of the eight focal species and

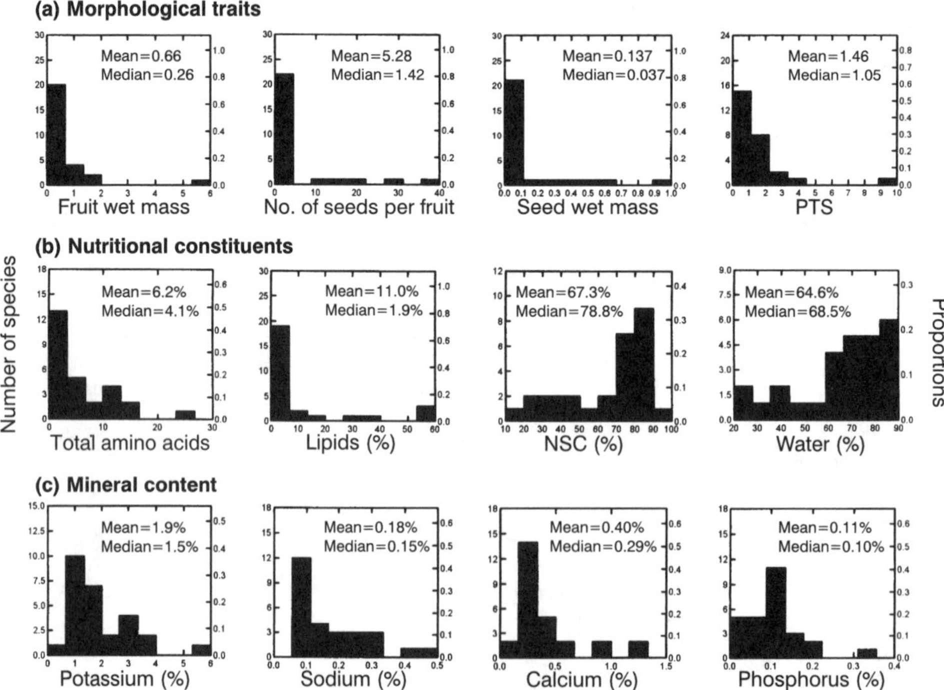

Fig. 11.1. Frequency distribution of (a) four morphological fruit traits, (b) four nutritional constituents and (c) four fruit minerals in 27 plant species from east Mediterranean habitats. Mean and median of each trait are shown.

Table 11.4. Principal component analyses for morphological traits and nutritional and mineral content among 27 fruit species in eastern Mediterranean ecosystems. Each factor represents an ordination axis. Only loadings > 0.5 are shown. See text for definitions.

	Factors		
Variable	F1	F2	F3
Morphological traits			
Fruit wet mass	0.966		
Number of seeds			–0.944
Seeds wet mass	0.934		
Seed dry mass	0.516		0.784
Seeds dry mass	0.918		
Pulp wet mass	0.913		
Pulp dry mass	0.871		
Seed wet mass	0.520		
Wet pulp per seed		0.821	
PS	0.589		0.765
PTS		0.983	
RY		0.558	
Eigenvalue	5.19	3.66	3.05
% variance explained	37.1	26.1	21.8
Cumulative variance	37.1	63.2	85.0
Nutritional content			
Protein (total amino acids)	0.835		
Fat		–0.896	
NSC	–0.508	0.689	
Ash	0.947		
Water		0.833	
Cellulose			0.961
Eigenvalue	2.85	1.36	1.05
% variance explained	47.4	22.7	17.5
Cumulative variance	47.4	70.1	87.6
Mineral content			
N	0.737		
P	0.872		
K	0.930		
Na	0.810		
Ca		0.937	
Eigenvalue	2.83	1.14	
% variance explained	56.6	22.9	
Cumulative variance	56.6	79.5	

their factor scores from the PCA analyses of fruit morphology. No significant correlation was found between removal efficiency, absolute fruit removal and fruit-removal success and the loadings on the first two morphological axes. Thus, the fruit-size hypothesis was only weakly supported.

Nutritional and mineral fruit content

To evaluate the nutritional-content hypothesis, I measured nutritional and mineral contents of 27 fruit species. Pulp averaged ~65% water. Seven species contained < 60% water (Fig. 11.1b). For most species, the main

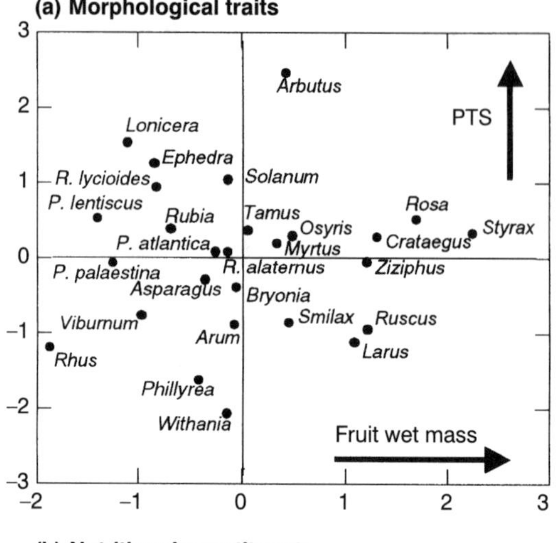
(a) Morphological traits
PTS
Fruit wet mass
Arbutus
Lonicera
Ephedra
R. lycioides
Solanum
P. lentiscus
Rubia
Tamus
Rosa
Osyris
Styrax
P. atlantica
Myrtus
Crataegus
P. palaestina
R. alaternus
Ziziphus
Asparagus
Bryonia
Viburnum
Smilax
Ruscus
Arum
Larus
Rhus
Phillyrea
Withania

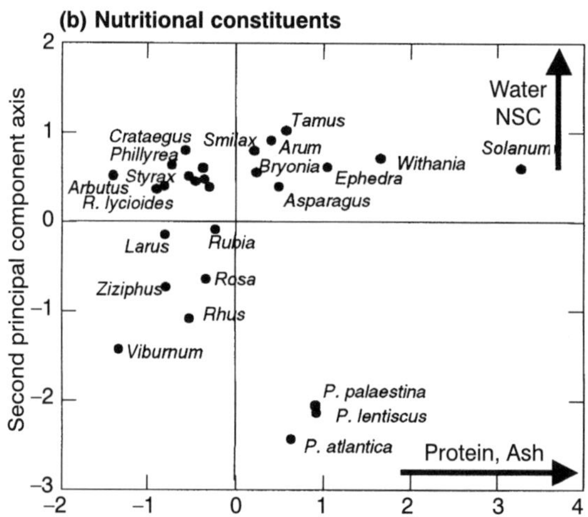
(b) Nutritional constituents
Second principal component axis
Water
NSC
Tamus
Crataegus
Smilax
Arum
Phillyrea
Solanum
Withania
Bryonia
Ephedra
Styrax
Arbutus
R. lycioides
Asparagus
Larus
Rubia
Rosa
Ziziphus
Rhus
Viburnum
P. palaestina
P. lentiscus
P. atlantica
Protein, Ash

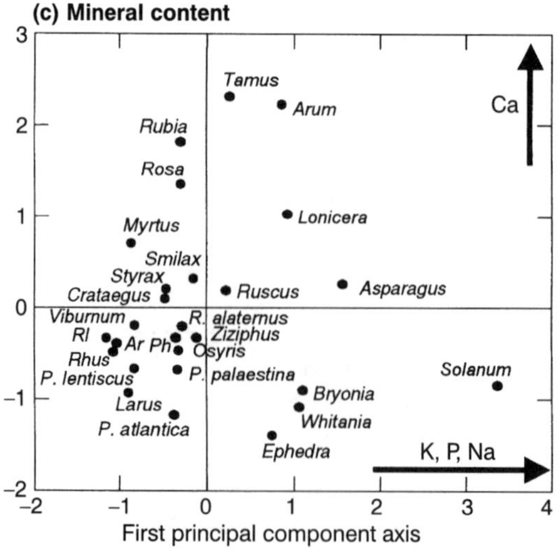
(c) Mineral content
Tamus
Arum
Ca
Rubia
Rosa
Lonicera
Myrtus
Smilax
Styrax
Ruscus
Asparagus
Crataegus
Viburnum
R. alaternus
Rl
Ar Ph
Ziziphus
Osyris
Rhus
Solanum
P. lentiscus
P. palaestina
Bryonia
Larus
Whitania
P. atlantica
Ephedra
K, P, Na
First principal component axis

constituent of dry pulp was non-structural carbohydrates (NSC), averaging 67.3% ± 22.6% (Fig. 11.1b). Structural carbohydrates (cellulose) averaged 9.5% ± 6.9% of dry pulp. The distributions of fats and proteins (total amino acids) were right-skewed, with averages of 11.0% ± 18.4% and 6.4% ± 5.2%, respectively (Fig. 11.1b). Negative correlations were found between fats and NSC ($r = -0.88$, $P < 0.001$) and between protein and NSC ($r = -0.70$, $P < 0.01$). A positive correlation was found between protein and ash ($r = 0.58$, $P < 0.05$).

The first three principal components accounted for 87.6% of the total variation in the nutritional content of the fleshy fruits (Table 11.4). The first axis represents an axis of increasing protein and ash; the second axis represents increasing NSC and water and decreasing fat (Fig. 11.2b). The three *Pistacia* species were clumped and isolated from the other species, due to their extreme fat content. The other 24 species displayed decreased variation on the second axis, with increased values on the first axis. Thus, protein- (and ash-) poor fruits were markedly varied in their water, NSC and fat contents. For example, *P. latifolia* and *Crataegus monogyna* contain high amounts of water and NSC but are low in fats (upper left quarter of PCA diagram), whereas *Viburnum tinus* contains low amounts of water and NSC but has high fat values. However, the protein-rich fruits, such as *Solanum nigrum*, *W. somnifera* and *Ephedra aphylla*, are much less varied on the second axis and characterized by high water and carbohydrate content and low fat content.

A PCA on the essential amino acid profile of the 27 species indicated two distinct groups of fruits (Izhaki, 1998b). The main group consisted of 23 species that were similar in their relatively low total amino acid content. The second group included four species (*E. aphylla*, *Bryonia* spp., *W. somnifera* and *S. nigrum*) that had relatively high total amino acid content. These species and three other species (*Arum dioscoridis*, *Tamus orientalis* and *A. aphyllus*) were distinctive in their protein level, as shown in Fig. 11.2b (upper right quarter). However, these seven species are known for their toxicity and any of them could be rich in secondary metabolites that have chemical structures similar to the 17 amino acids measured by the amino acid analyser (Izhaki, 1998b). Thus, their high amino acid level might be an artefact.

Potassium (1.88% ± 1.22%) was the primary constituent among the minerals, and phosphorus (0.11% ± 0.07%) was the least abundant (Fig. 11.1c). Positive correlations were found between potassium and sodium ($r = 0.75$, $P < 0.001$), phosphorus ($r = 0.72$, $P < 0.01$) and nitrogen ($r = 0.63$, $P < 0.05$). The first two principal components accounted for 79.5% of the total variation in the mineral content of the fleshy fruits (Table 11.4). Loadings on the first axis are positively correlated with levels of all minerals, except calcium, which is represented on the second axis (Fig. 11.2c). Four mineral fruit types are apparent from the ordination diagram, particularly for those species positioned far from the diagram's centre

Fig. 11.2. Ordination of 27 fruit species on the first two axes of a principal component analysis of (a) fruit morphological traits, (b) fruit nutritional constituents, and (c) fruit minerals. Fruit wet mass = wet mass of fruit, pulp and seed, PTS = (dry mass of pulp)/(total dry mass of seeds per fruit), proportion of water in fresh pulp, proportion of protein, ash, non-structural carbohydrates (NSC), K, P, Na and Ca in the dry pulp. Full species (and family) names are: *Arbutus andrachne* (*Ericaceae*), *Arum dioscoridis* (*Araceae*), *Asparagus aphyllus* (*Liliaceae*), *Bryonia* spp. (*Cucurbitaceae*), *Crataegus monogyna* (*Rosaceae*), *Ephedra aphylla* (*Ephedraceae*), *Larus nobilis* (*Lauraceae*), *Lonicera etrusca* (*Caprifoliaceae*), *Myrtus communis* (*Myrtaceae*), *Osyris alba* (*Santalaceae*), *Phillyrea latifolia* (*Oleaceae*), *Pistacia atlantica* (*Anacardiaceae*), *Pistacia lentiscus* (*Anacardiaceae*), *Pistacia palaestina* (*Anacardiaceae*), *Rhamnus alaternus* (*Rhamnaceae*), *Rhamnus lycioides* (*Rhamnaceae*), *Rhus coriaria* (*Anacardiaceae*), *Rosa canina* (*Rosaceae*), *Rubia tenuifolia* (*Rubiaceae*), *Ruscus aculeatus* (*Liliaceae*), *Smilax aspera* (*Liliaceae*), *Solanum nigrum* (*Solanaceae*), *Styrax officinalis* (*Styracaceae*), *Tamus orientalis* (*Dioscoreaceae*), *Viburnum tinus* (*Caprifoliaceae*), *Withania somnifera* (*Solanaceae*), *Ziziphus spina-christi* (*Rhamnaceae*). The five points without names on the upper left quarter of the middle figure represent *Lonicera etrusca, Myrtus communis, Osyris alba, Rhamnus alaternus* and *Ruscus aculeatus.* Rl = *Rhamnus lycioides,* Ar = *Arum dioscoridis,* Ph = *Phillyrea latifolia.*

(Fig. 11.2c). A large group of fruits is mineral-poor (lower left quarter), while a second group is mineral-poor but with relatively high amounts of calcium (upper left quarter). Fruits of the other two groups contain more minerals (right quarters); one is calcium-poor (lower quarter) and the other is calcium-rich (upper quarter).

Fruit-removal efficiency was associated with nutritional and energetic traits of the fruits. It was positively correlated with cellulose, ash and energy/fruit-mass ratio, and negatively correlated with NSC and water (Table 11.3). Absolute removal was positively correlated with two energetic traits and negatively correlated with pulp water content and fruit wet mass (Table 11.3). Similar correlations were observed in removal success, except that no correlations were significant with water content and the energy/fruit wet-mass ratio (Table 11.3). I also correlated fruit removal of the eight focal species and their factor scores from the PCA analyses of nutritional and mineral content. No significant correlation was found between the first two nutritional and mineral factors and removal efficiency. However, absolute removal was positively correlated with loadings on the first nutritional axis ($r = 0.8$, $P < 0.05$) and negatively correlated with loadings on the second nutritional axis ($r = -0.73$, $P < 0.05$). Removal success was negatively correlated with loadings on the second nutritional axis ($r = -0.72$, $P < 0.05$). These results indicate that plants that produce protein- and fat-rich fruits achieved higher absolute removal and higher fruit-removal success. These analyses suggest that the nutritional profile of fruits is an important component governing fruit-removal success in east Mediterranean scrublands, thus supporting the nutritional-content hypothesis.

Fruit crop

To evaluate the fruit-crop hypothesis, I measured the size of the fruit crop of the eight focal species. Huge among-species variability was observed in fruit-crop size, which ranged from < 100 fruits per individual in *Osyris* to more than 10,000 fruits in *P. lentiscus* (Table 11.2). Within-species variability was also high; CV values ranged between 47% in *Asparagus* to 99% in *P. palaestina* (Table 11.2).

No significant correlation was detected between fruit-crop size and fruit-removal efficiency (Table 11.3). However, absolute removal was positively correlated with crop size and a similar correlation trend was observed in removal success (Table 11.3). Thus the crop-size hypothesis was partially supported.

To summarize the results on the validity of the three fruit-removal hypotheses in this eastern Mediterranean site, fruit-removal efficiency was primarily dependent on fruit nutritional traits, whereas absolute removal and removal success were primarily dependent on crop size, plant abundance and fruit energy.

Are fruit morphology, nutrient, mineral contents and crop size related?

A synthesis based on correlations among each of the first two factors of each of the three PCA sets (a total of six factors) was used to explore the relationships between fruit nutritional and mineral content and morphological attributes. Three linear correlations (out of the 12 possible combinations) were significant. The first nutritional factor was negatively correlated with the first morphological factor ($r = -0.42$, $P < 0.05$) and with the second mineral factor ($r = -0.45$, $P < 0.05$). The overall pattern revealed from these analyses is that larger fruits tend to be poor in protein and rich in calcium, while smaller fruits tend to be relatively rich in protein but poor in calcium. The second nutritional factor was positively correlated with the first mineral factor ($r = 0.66$, $P < 0.001$), demonstrating that fruits that are rich in NSC are also rich in K, Na and P. A negative correlation was detected between crop size and fruit wet mass among the eight focal species ($r_s = -0.79$, $P < 0.05$).

Discussion

Is there a general source for the variation in fruit removal?

The fruit-size hypothesis predicts that small fruits will be most attractive to small and medium-sized frugivores. Because birds in this size range are especially common, plants that produce small fruits will be rewarded with high fruit removal. Evidence in support of this prediction has emerged from several studies in southern Spain, where a negative correlation exists between fruit-removal efficiency and fruit size, on both the intra- and interspecific levels (Herrera, 1984, 1988; Jordano, 1987, 1989). The study presented here suggests that fruit-size traits do not affect fruit-removal efficiency among the eight most common species at my study site. However, plants that produce small fruits are rewarded with higher absolute removal and removal success. Thus, it seems that fruit size has a general role in determining seed-dispersal success in the entire Mediterranean region. However, since fruit size was negatively correlated with crop size, it is difficult to isolate the effects of fruit size and crop size.

Leaving all other factors constant, plant species producing fleshy fruits that better match the nutritional demands of birds should be rewarded with higher fruit removal. Studies have shown that this nutritional-content hypothesis is false, as fruit-removal efficiency is not necessarily related to fruit quality (Herrera and Jordano, 1981; Sorensen, 1981; Herrera, 1984). At Beit Jimal, removal efficiency was positively correlated with energy/fruit wet-mass ratio, as expected. But, unexpectedly, it was negatively correlated with carbohydrates and water and positively correlated with cellulose. This result may be partially explained by the intercorrelations among the nutritional attributes of the fruits. For example, *Pistacia* species had high fruit-removal efficiency, while their fruits were poor in carbohydrates and water and rich in fats and have a high energy/fruit wet-mass ratio. However, their average cellulose content is similar to the average for all fruit species. This fits the prediction that, where most fruit removal is by migrating birds, as is the case in the eastern Mediterranean (but not the western Mediterranean), energy-rich fruits should be highly preferred.

The prediction emerging from the fruit-crop hypothesis is that both absolute removal and fruit-removal efficiency should be greater in plants with large crop sizes. Supportive evidence for this hypothesis emerged from several studies that demonstrated higher absolute removal from plants with large crop sizes in the tropics (e.g. Howe and De Steven, 1979; Howe and Vande Kerckhove, 1979; Murray, 1987), in North America (e.g. Moore and Willson, 1982; Stapanian, 1982; Sallabanks, 1992) and in the Mediterranean basin (Alcántara *et al.*, 1997; Izhaki, 1998a). However, the empirical evidence for the role of crop size in fruit-removal efficiency is rather limited (Stapanian, 1982; Davidar and Morton, 1986; Denslow, 1987; Sargent, 1990; Sallabanks, 1992). Furthermore, several studies failed to show such dependency in the tropics (Howe and Vande Kerckhove, 1979; Murray, 1987), in the temperate region (Moore and Willson, 1982; Courtney and Manzur, 1985; Davidar and Morton, 1986; Laska and Stiles, 1994) and in the Mediterranean region (Alcántara *et al.*, 1997; Izhaki, 1998a).

At my study site and on the interspecific level, fruit-crop size does not affect fruit-removal efficiency. However, crop size is positively correlated with absolute removal and removal success. Species that produce small fruits are the most successful species in terms of their abundance in the habitat. The combination of their high abundance and large crop sizes makes their fruiting display dominant in the study area. This is relevant if birds respond to the general fruit display in the area rather than to the fruit crop of an individual plant. This is probably the case for transient birds, which often forage during short stopovers in unfamiliar sites, such as Beit Jimal. Under these circumstances it is difficult to isolate the effects of crop size from the total fruit display of a species (see also Sargent, 1990).

Synthesis

None of the above hypotheses is mutually exclusive. As described above, fruit removal may be dependent upon intrinsic plant and fruit traits, such as fecundity, fruit size and nutritional content, which are under the direct genetic control of the plant. However, many extrinsic factors not under genetic control also influence fruit removal (e.g. presence of fruit-bearing neighbours) (Sargent, 1990; Thébaud and Debussche, 1992). In fact, environmental conditions may play an overriding role in the dispersal efficiency of fleshy-fruited plants (Wheelwright and Orians, 1982). Abiotic conditions, for example, may govern the presence of predators, disperser availability in space and time, availability of alternative food sources for potential dispersers, etc. The pattern of diffuse coevolution between plants and their dispersers may be partially explained by these extrinsic factors, as they may overwhelm any selection pressures generated by seed-dispersers on plant traits (Herrera, 1985; Jordano, 1987; Sargent, 1990).

The lack of tight plant–disperser coevolution may be explained by other factors as well. Plant–disperser interactions do not take place in a vacuum; many other interactions and constraints (e.g. pollination, physiology) may simultaneously shape the evolution of both partners' traits (Herrera, 1986). Furthermore, it is particularly difficult to separate the effect of consistent selection from marked phylogenetic inertia (Jordano, 1995). A concrete example for this is the three *Pistacia* species, which appeared in isolated clumps in all PCA analyses, even though they have evolved separately since the Early Tertiary (Zohary, 1952). Since the Tertiary, each of them has had a different geographical distribution and a different assemblage of dispersal agents, but they have retained similar fruit traits (Herrera, 1985; I. Izhaki, this study).

Although the 'diffuse coevolution' concept also applies for the weak plant–disperser interactions in the Mediterranean region, whether coevolution processes may be occurring to some extent at present remains under debate (Blondel and Aronson, 1995). Currently, however, there are several differences in the composition of the frugivorous assemblages between the eastern and western Mediterranean regions. These may result in different selective pressures on plant and fruit attributes. The western Mediterranean is an overwintering site for numerous frugivorous birds (Herrera, 1982, 1984, 1995); the eastern Mediterranean is mainly a stopover site for frugivorous birds (Izhaki, 1986). For example, many blackcaps (*Sylvia atricapilla*), one of the most abundant frugivores in the Mediterranean region, overwinter in southern Europe and consume huge amounts of fruit during winter (Jordano and Herrera, 1981; Herrera, 1995). This species rarely overwinters in Israel (Maitav and Izhaki, 1994), but it is none the less an important seed-disperser when it migrates through the region (Izhaki, 1986). One of the most important frugivorous warblers in Israel, the lesser whitethroat (*Sylvia curruca*), which migrates through Israel, is a much less pronounced frugivore in the Iberian Peninsula. On the other hand, the garden warbler, *Sylvia borin*, is very abundant in the western Mediterranean, where it is a major seed-disperser (Herrera, 1995), but is relatively rare and has a marginal role in seed dispersal in Israel (Izhaki, 1986).

It is expected that an assemblage of frugivorous species dominated by transients, such as in the eastern Mediterranean, will generate different selective pressures on fruit traits from those of an otherwise similar assemblage dominated by overwintering birds, such as in the western Mediterranean. Although the migration pattern that we see today is only about 10,000 years old (Moreau, 1972), a short evolutionary time for speciation in plant-fruit attributes, the massive flux of small-sized birds during a short season (mainly September–October) with very short stopover periods in east Mediterranean habitats should generate a strong selective pressure for small, conspicuous, energy-rich fruits.

To summarize, the primary constraint on the development of coevolved plant–disperser complexes has probably been the absence of sufficiently intense directional selection. Most plant–disperser systems are more complex (multiple fleshy-fruited plants and frugivore species) than the one studied here. Such complexity may swamp the effects of selection on heritable traits. My results, however, indicate that, in a relatively simple ecosystem, selective pressures currently generated by frugivorous birds upon fleshy-fruited plants are much more consistent than previously suggested for western Mediterranean ecosystems.

Future Research Needs

Many fruit traits that were not studied here may affect fruit removal. For example, recent studies have evaluated how secondary metabolites in fleshy fruits may influence plant–disperser interactions (reviewed in Cipollini and Levey, 1997a). Secondary metabolites in fruits can have many functions that are not related to seed dispersal and probably evolved under different, sometimes conflicting, constraints (Eriksson and Ehrlén, 1998). A definitive answer to the question of the adaptive role of secondary metabolites for plant reproductive success requires both experimental work and field studies. This research avenue was recently taken in a study on the effect of secondary metabolites in *Solanum*. Indeed, this study showed that secondary metabolites have relevant functions for legitimate and non-legitimate frugivores (i.e. seed-dispersers and seed or pulp predators, respectively) (Cipollini and Levey, 1997b; Levey and Cipollini, 1998; Wahaj *et al.*, 1998). A similar study on *Rhamnus alaternus* in the eastern Mediterranean is under way. Preliminary results indicate a correlation between fruit-removal efficiency and the concentrations of the anthraquinone emodin in fruit pulp (I. Izhaki, E. Tsahar and J. Friedman, unpublished data). However, the effect of secondary metabolites on plant reproductive success remains unknown. Because of potential strong and immediate effects on fruit choice, the significance of secondary metabolites in fleshy fruits may override many of the fruit and plant traits discussed earlier in this chapter, a hypothesis that begs to be tested.

Acknowledgements

I thank Carlos Peres, Martin Cipollini and Doug Levey for their helpful comments. Nutritional analyses were performed in Miloda Laboratories under D. Izhar's supervision, in AminoLab Ltd. under Z. Harduf's supervision and in the Plant Extract Chemistry Laboratory in Migal under E. Kvitnitsky's supervision.

References

Alcántara, J.M., Rey, P.J., Valera, F., Sánchezlafuente, A.M. and Gutiérrez, J.E. (1997) Habitat alteration and plant intra-specific competition for seed dispersers – an example with *Olea europaea* var. *sylvestris*. *Oikos* 79, 291–300.

AOAC (1984) *Official Methods of Analysis of the Association of Official Analytical Chemists*. AOAC, Washington, DC.

Barnea, A., Yomtov, Y. and Friedman, J. (1991) Does ingestion by birds affect seed germination. *Functional Ecology* 5, 394–402.

Blondel, J. and Aronson, J. (1995) Biodiversity and ecosystem function in the Mediterranean basin: human and non-human determinants. In: Davis, G.W. and Richardson, D.M. (eds) *Mediterranean Type Ecosystems: the Function of Biodiversity*. Springer-Verlag, Berlin, pp. 43–119.

Borowicz, V.A. and Stephenson, A.G. (1985) Fruit composition and patterns of fruit dispersal of two *Cornus* spp. *Oecologia* 67, 435–441.

Cipollini, M.L. and Levey, D.J. (1997a) Secondary metabolites of fleshy vertebrate-dispersed fruits – adaptive hypotheses and implications for seed dispersal. *American Naturalist* 150, 346–372.

Cipollini, M.L. and Levey, D.J. (1997b) Why are some fruits toxic? Glycoalkaloids in *Solanum* and fruit choice by vertebrates. *Ecology* 78, 782–798.

Courtney, S.P. and Manzur, M.I. (1985) Fruiting and fitness in *Crataegus monogyna*: the effects of frugivores and seed predators. *Oikos* 44, 398–406.

Davidar, P. and Morton, E.S. (1986) The relationship between fruit crop sizes and fruit removal rates by birds. *Ecology* 67, 262–265.

Denslow, J.S. (1987) Fruit removal rates from aggregation and isolated bushes of the red elderberry, *Sambucus pubens*. *Canadian Journal of Botany* 65, 1229–1235.

Eriksson, O. and Ehrlén, J. (1998) Secondary metabolites in fleshy fruits: are adaptive explanations needed? *American Naturalist* 152, 905–907.

Herrera, C.M. (1982) Seasonal variation in the quality of fruits and diffuse coevolution between plants and avian dispersers. *Ecology* 63, 773–785.

Herrera, C.M. (1984) A study of avian frugivores bird-dispersed plants and their interaction in

Mediterranean scrublands. *Ecological Monographs* 54, 1–23.

Herrera, C.M. (1985) Determinants of plant–animal coevolution: the case of mutualistic dispersal of seeds by vertebrates. *Oikos* 44, 132–141.

Herrera, C.M. (1986) Vertebrate-dispersed plants: why they don't behave the way they should. In: Estrada, A. and Fleming, T.H. (eds) *Frugivores and Seed Dispersal.* Dr W. Junk Publishers, Dordrecht, pp. 5–18.

Herrera, C.M. (1987) Vertebrate-dispersed plants of the Iberian Peninsula: a study of fruit characteristics. *Ecological Monographs* 57, 305–331.

Herrera, C.M. (1988) Habitat-shaping host plant use by a hemiparasitic shrub, and the importance of gut fellows. *Oikos* 51, 383–386.

Herrera, C.M. (1991) Dissecting factors responsible for individual variation in plant fecundity. *Ecology* 72, 1436–1448.

Herrera, C.M. (1992) Interspecific variation in fruit shape – allometry, phylogeny, and adaptation to dispersal agents. *Ecology* 73, 1832–1841.

Herrera, C.M. (1995) Plant–vertebrate seed dispersal systems in the Mediterranean: ecological, evolutionary and historical determinants. *Annual Review of Ecology and Systematics* 26, 705–727.

Herrera, C.M. and Jordano, P. (1981) *Prunus mahaleb* and birds: the high-efficiency seed dispersal system of a temperate fruiting tree. *Ecological Monographs* 51, 203–218.

Howe, H.F. and De Steven, D. (1979) Fruit production, migrant bird visitation, and seed dispersal of *Guarea glabra* in Panama. *Oecologia* 39, 185–196.

Howe, H.F. and Estabrook, G.F. (1977) On intraspecific competition for avian dispersers in tropical trees. *American Naturalist* 111, 817–832.

Howe, H.F. and Vande Kerckhove, G.A. (1979) Fecundity and seed dispersal of a tropical tree. *Ecology* 60, 180–189.

Izhaki, I. (1986) Seed dispersal by birds in an east Mediterranean scrubland. PhD dissertation, Hebrew University of Jerusalem, Israel (in Hebrew).

Izhaki, I. (1992) A comparative analysis of the nutritional quality of mixed and exclusive fruit diets for yellow-vented bulbuls. *Condor* 94, 912–923.

Izhaki, I. (1993) Influence of nonprotein nitrogen on estimation of protein from total nitrogen in fleshy fruits. *Journal of Chemical Ecology* 19, 2605–2615.

Izhaki, I. (1998a) The relationships between fruit ripeness, wasp seed predation, and avian fruit removal in *Pistacia palaestina. Israel Journal of Plant Sciences* 46, 273–278.

Izhaki, I. (1998b) Essential amino acid composition of fleshy fruits versus maintenance requirements of passerine birds. *Journal of Chemical Ecology* 24, 1333–1345.

Izhaki, I. and Safriel, U.N. (1985) Why do fleshy-fruit plants of the Mediterranean scrub intercept fall- but not spring-passage of seed-dispersing migratory birds? *Oecologia* 67, 40–43.

Izhaki, I., Walton, P.B. and Safriel, U.N. (1991) Seed shadows generated by frugivorous birds in an eastern Mediterranean scrub. *Journal of Ecology* 79, 575–590.

Janzen, D.H. (1983) Insects. In: Janzen, D.H. (ed.) *Costa Rican Natural History.* University of Chicago Press, Chicago, Illinois, pp. 619–645.

Jordano, P. (1987) Avian fruit removal: effects of fruit variation crop size and insect damage. *Ecology* 68, 1711–1723.

Jordano, P. (1989) Pre-dispersal biology of *Pistacia lentiscus* (Anacardiaceae): cumulative effects on seed removal by birds. *Oikos* 55, 375–386.

Jordano, P. (1991) Gender variation and expression of monoecy in *Juniperus phoenicea* (L) (*Cupressaceae*). *Botanical Gazet* 152, 476–485.

Jordano, P. (1995) Angiosperm fleshy fruits and seed dispersers: a comparative analysis of adaptation and constraints in plant–animal interactions. *American Naturalist* 145, 163–191.

Jordano, P. and Herrera, C.M. (1981) The frugivorous diet of blackcap populations *Sylvia atricapilla* wintering in southern Spain. *Ibis* 123, 502–507.

Karlson, P. (1972) *Biochemie,* 8th edn. Thieme, Stuttgart, Germany.

Korine, C., Izhaki, I. and Arad, Z. (1998) Comparison of fruit syndromes between the Egyptian fruit-bat (*Rousettus aegyptiacus*) and birds in East Mediterranean habitats. *Acta Oecologica* 19, 147–153.

Laska, M.S. and Stiles, E.W. (1994) Effects of fruit crop size on intensity of fruit removal in *Viburnum prunifolium* (*Caprifoliaceae*). *Oikos* 69, 199–202.

Levey, D.J. (1987) Seed size and fruit-handling techniques of avian frugivores. *American Naturalist* 129, 471–485.

Levey, D.J. and Cipollini, M.L. (1998) A glycoalkaloid in ripe fruit deters consumption by cedar waxwings. *Auk* 115, 359–367.

Levey, D.J., Bissell, H.A. and O'Keefe, S.F. (2000) Conversion of nitrogen to protein and amino acids in wild fruits. *Journal of Chemical Ecology* 26, 1749–1763.

McKey, D. (1975) The ecology of coevolved seed dispersal systems. In: Gilbert, L.E. and Raven, P.H. (eds) *Co-evolution of Animals and Plants.* University of Texas Press, Austin, pp. 159–191.

Maitav, A. and Izhaki, I. (1994) Stopover and fat deposition by blackcaps *Sylvia atricapilla* following spring migration over the Sahara. *Ostrich* 65, 160–166.

Manasse, R.S. and Howe, H.F. (1983) Competition for dispersal agents among tropical trees: influences of neighbors. *Oecologia* 59, 185–190.

Martínez del Rio, C. and Restrepo, C. (1993) Ecological and behavioral consequences of digestion in frugivorous animals. *Vegetatio* 107/108, 205–216.

Mazer, S.J. and Wheelwright, N.T. (1993) Fruit size and shape: allometry at different taxonomic levels in bird-dispersed plants. *Evolutionary Ecology* 7, 556–575.

Moore, L.A. and Willson, M.F. (1982) The effect of microhabitat, spatial distribution, and display size on dispersal of *Lindera benzoin* by avian frugivores. *Canadian Journal of Botany* 60, 557–560.

Moreau, R.E. (1972) *The Palaearctic–African Bird Migration System.* Academic Press, London, 384 pp.

Murray, K.G. (1987) Selection for optimal fruit-crop size in bird-dispersed plants. *American Naturalist* 129, 18–31.

Ridley, H.N. (1930) *The Dispersal of Plants Throughout the World.* L. Reeve, Ashford, UK.

Sallabanks, R. (1992) Fruit fate, frugivory, and fruit characteristics: a study of hawthorn, *Crataegus monogyna* (*Rosaceae*). *Oecologia* 91, 296–304.

Sargent, S. (1990) Neighborhood effects on fruit removal by birds: a field experiment with *Viburnum dentatum* (*Caprifoliaceae*). *Ecology* 71, 1289–1298.

Snow, D.W. (1971) Evolution aspects of fruit-eating by birds. *Ibis* 113, 194–202.

Sorensen, A.E. (1981) Interaction between birds and fruit in temperate woodland. *Oecologia* 50, 242–249.

Stapanian, M.A. (1982) A model for fruiting display: seed dispersal by birds for mulberry trees. *Ecology* 63, 1432–1443.

Stiles, E.W. (1980) Patterns of fruit presentation and seed dispersal in bird-disseminated woody plants in the eastern deciduous forest. *American Naturalist* 116, 670–688.

SYSTAT (1998) *SYSTAT for Windows: Statistics*, Version 8.0 edition. SYSTAT, Evanston, Illinois.

Thébaud, C. and Debussche, M. (1992) A field test of the effects of infructescence size on fruit removal by birds in *Viburnum tinus*. *Oikos* 65, 391–394.

Van der Pijl, L. (1982) *Principles of Dispersal in Higher Plants*, 3rd edn. Springer, Berlin, 215 pp.

Wahaj, S.A., Levey, D.J., Sanders, A.K. and Cipollini, M.L. (1998) Control of gut retention time by secondary metabolites in ripe *Solanum* fruits. *Ecology* 79, 2309–2319.

Wheelwright, N.T. (1985) Fruit size, gape width, and the diets of fruit-eating birds. *Ecology* 66, 808–818.

Wheelwright, N.T. and Orians, G.H. (1982) Seed dispersal by animals: contrasts with pollen dispersal, problems of terminology and constraints on coevolution. *American Naturalist* 119, 402–413.

Willson, M.F. and Whelan, C.J. (1993) Variation of dispersal phenology in a bird-dispersed shrub, *Cornus drummondii*. *Ecological Monographs* 63, 151–172.

Zohary, M. (1952) A monographical study of the genus *Pistacia*. *Palestine Journal of Botany Jerusalem Series* 5, 187–238.

Zohary, M. (1962) *Plant Life of Palestine.* Ronald Press, New York. 262 pp.

12 Seed Dispersal of Mimetic Fruits: Parasitism, Mutualism, Aposematism or Exaptation?

Mauro Galetti

Plant Phenology and Seed Dispersal Research Group, Departamento de Ecologia, Universidade Estadual Paulista (UNESP), CP 199, 13506-900 Rio Claro, São Paulo, Brazil

Introduction

Fruits and their seed-dispersers are a classic example of a mutualistic relationship. Seed-dispersers benefit from consuming the nutritious tissues surrounding the seeds, whereas plants benefit from the dispersal of their seeds away from the competition of the parent to newly opened habitat and to places away from the zone of high mortality near the parent plant (Snow, 1971; Howe and Smallwood, 1982).

Several attributes of fruits, including colour, morphology, seed size, phenology and pulp chemistry, are considered plant adaptations to enhance the chances of being eaten by seed-dispersers (van der Pijl, 1982; Janson, 1983; Gautier-Hion *et al.*, 1985). The energy allocated to produce the pulp or aril represents a cost to the plant that probably has no purpose other than to attract seed-dispersers and to protect seeds (Howe, 1993; Mack, 2000). seed-dispersers, on the other hand, also incur costs. Transportation of non-digestible material (seeds) in their guts and exposure to predators while feeding ultimately reduce the rate at which fruit pulp can be processed and nutrients obtained (McKey, 1975; Levey and Grajal, 1991; Murray *et al.*, 1993; but see Witmer, 1998).

As in most mutualistic relationships, some frugivores and fruiting plants have evolved strategies to overcome these costs without losing the benefits. For instance, some bird species eat fruit pulp and discard the seeds below the parent tree, thereby avoiding the costs of seed ingestion (Levey, 1987). A few plant species have also evolved fruits with no nutritive rewards, presumably deceiving frugivores into swallowing their seeds. These plants enjoy the benefit of dispersal without the cost of pulp production. One such strategy is to hide small fruits and seeds among leaves that are ingested by large herbivores ('foliage is the fruit' hypothesis (Janzen, 1984)). Another strategy is to display colourful seeds resembling fleshy ornithochoric (i.e. bird-dispersed) fruits, so-called 'mimetic fruits' (Ridley, 1930; van der Pijl, 1982). The first strategy is relatively well documented (Janzen, 1984; Quinn *et al.*, 1994; Malo and Suaréz, 1998; Ortmann *et al.*, 1998). The second strategy, on the other hand, has been debated since the first monographs on seed dispersal (Ridley, 1930; van der Pijl, 1982), but

 Seed Dispersal and Frugivory: Ecology, Evolution and Conservation (eds D.J. Levey, W.R. Silva and M. Galetti)

very few studies have tested the effectiveness of mimetic fruits in deceiving seed-dispersers (Peres and van Roosmalen, 1996; Foster and Delay, 1998).

In a classic sense, mimicry involves matching colours of edible (in Batesian mimicry) or non-edible species (Müllerian mimicry), to avoid being eaten by predators (Endler, 1981). In the case of mimetic fruits, plants produce seeds that mimic fleshy fruits, thereby facilitating consumption and dispersal of the seed, which provides no nutritive reward to the bird. Fruit mimicry systems apparently share many elements with deceit pollination systems. These include exploitation of naïve consumers, taking advantage of the exploratory behaviour of the animals and use of the signal ordinarily used by truly rewarding systems (in this case, nutritious endocarp) (Dafni, 1984).

Taxonomic Affinities of Mimetic Fruits

Mimetic fruits have been described and discussed by several authors (Ridley, 1930; Corner, 1949; McKey, 1975; van der Pijl, 1982; Williamson, 1982). All of these studies except van der Pijl (1982) focused particularly on *Leguminosae*, primarily because its mimetic species are well known due to their use as jewellery (Armstrong, 1992). We define a 'mimetic fruit' as a brightly coloured fruit or seed with no associated pulp or aril: it consequently does not provide a nutritional reward for potential seed-dispersers. Although such seeds are often termed 'mimetic seeds', I prefer 'mimetic fruits' because the latter terminology makes explicit the model and, hence, the ecological function of the mimicry. My classification is ecological rather than botanical, since different species of plants make use of different structures to mimic fruits. Moreover, we are not concerned with trying to find the probable models for the mimetic fruit, because all species match colour and display patterns common among fleshy-fruited bird-dispersed species.

Mimetic fruits that have a bright visual display are widespread in several non-related plant families (Colour Plate 1 – see Frontispiece; Table 12.1). At least 21 genera have mimetic fruits with colourful displays, but this phenomenon is particularly common in *Leguminosae* and especially in *Mimosoidea*. Mimetic fruits can be found in herbs (*Gahnia*, *Paeonia*) and vines (*Rhyncosia*, *Abrus*), but are most common in trees (*Abarema*, *Pithecellobium*, *Ormosia*, *Erythrina*, *Harpullia*) (characteristics and family names provided in Table 12.1).

The visual display of most mimetic fruits is a colourful (usually red or red and black or even blue) or black seed against a contrasting colourful background such as yellow, red or orange pods (e.g. *Adenanthera pavonina*, *Abarema* spp., *Pithecellobium* spp., *Pararchidendron pruinosum* or *Archidendron grandiflorum*) or other structures (such as in *Paeonia broteroi* (Colour Plate 1)). In the case of *P. broteroi* and other peonies, the non-fertilized ovules are red and contrast with black fertilized ovules (seeds with sarcotesta) immersed in a red carpellary wall (van der Pijl, 1982; C. Herrera, Seville, 2000, personal communication).

Most species with mimetic fruits have long fruiting seasons, which means that fruits are available to seed-dispersers for an unusually long time. Also, the seeds typically have long dormancy periods and several species are rich in secondary compounds that may deter pathogens and other seed predators (Table 12.1). Alkaloids are the main secondary compound found in mimetic fruits, but saponins and flavonoids also occur (Table 12.1).

One of the most well-known and widespread mimetic species is *A. pavonina* (*Leguminosae*, *Mimosoidea*), which has been introduced worldwide. Van der Pijl (1982) fed captive barbets (Capitonidae) *A. pavonina* seeds, but did not mention whether the seeds germinated after gut passage. In addition, Steadman and Freifeld (1999) found six seeds of *A. pavonina* in the crop of the purple-capped fruit-dove *Ptilinopus porphyraceus*, a likely seed-disperser, in Samoa. Ridley (1930) mentioned parrots and pigeons eating *A. pavonina*, but these birds are probably seed predators, not dispersers, of medium or large seeds (see Lambert, 1989). Several long-term studies on diets of neotropical fruit-eating birds have not

Table 12.1. Plant families and genera containing mimetic fruits.

Family	Genus	Species	Life-form	Distribution	Secondary compound*	Colour display
Cyperaceae	*Gahnia*	*sieberana*	Herb	Mal.–Austr.	?	Red seed
Euphorbiaceae	*Glochidion*	*sumatranum*	Tree	Austr.	Tannin, terpenoids	Orange-red seed
Euphorbiaceae	*Margaritaria*	spp.	Tree	Pan Tropical	Alkaloids	Metallic blue capsule
Leguminosae	*Abarema*	spp.	Tree	NTA	?	Blue-white seeds, orange pod
Leguminosae	*Abrus*	*precatorius*	Vine	Pan Tropical	Alkaloids	Red and black seeds
Leguminosae	*Acacia*	*auriculaeformis*	Tree	Austr.	?	Black seeds, brown pod, yellow funicles
Leguminosae	*Adenanthera*	spp.	Tree	Austr./Asia /Pacific	Alkaloids	Red or black-red seeds, orange pod
Leguminosae	*Archidendron*	spp.	Tree	Indomalay, Austr.	Saponin	Black seeds, orange to red pods
Leguminosae	*Batesia*	*floribunda*	Tree	NTA	?	Red seeds
Leguminosae	*Erythrina*†	spp.	Tree	Trop./ Subtrop.	Alkaloids	Red or red/black seeds
Leguminosae	*Ormosia*‡	spp.	Tree	Pan Tropical	Alkaloids	Red or black/red seeds
Leguminosae	*Pararchidendron*	*pruinosum*	Tree	Austr.–Mal.	?	Black seeds, red or orange pod
Leguminosae	*Pithecellobium*	spp.§	Tree	Neotrop.	Saponin	Black seeds, pink pod
Leguminosae	*Rhynchosia*	spp.	Vine	Pan Tropical	?	Black-red seeds
Leguminosae	*Sophora*	*secundiflora*	Tree	Pan Tropical	Alkaloids	Red seeds
Liliaceae	*Allium*	*tricoccum*	Herb	North America	Alkaloids	Black seeds
Ochnaceae	*Brackenridgea*	*nitida*	Herb	Pan Tropical	Terpenoids, flavonoids	Black seeds, red sepals
Ochnaceae	*Campylospermum*	*elongatum*	Herb	African	?	Black seeds, red sepals
Ochnaceae	*Ochna*	*atropurpurea*	Herb	Pan Tropical	Isoflavonid	Black seeds, red calyx
Paeoniaceae	*Paeonia*	spp.	Herb	M. Europe	Alkaloids	Black seeds, red carpel
Sapindaceae	*Harpullia*	*arborea*	Tree	Indomalay	Saponin	Black seeds, yellow capsule

*Secondary compounds found in fruits (reference list sent upon request to author).
†Several species of *Erythrina* have black or brownish seeds (see Bruneau, 1996).
‡Some species of *Ormosia* have black seeds or indehiscent fruits (Rudd, 1965).
§Not all *Pithecellobium* species have mimetic fruits. Also includes *Cojoba*.
Mal.–Austr., Malesia–Australasian; NTA, neotropical Americas; M. Europe, middle Europe.

reported consumption of mimetic fruits (Wheelwright *et al.*, 1984; Loiselle and Blake, 1990; Blake and Loiselle, 1992; Galetti and Pizo, 1996). In fact, very few published studies have reported evidence of avian frugivores consuming seeds of mimetic fruits in the wild (French, 1990; Quin, 1996; Foster and Delay, 1998).

Mammals may eat mimetic fruits, but are probably seed predators. For instance, pig-tailed macaques (*Macaca nemestrina*) were recorded preying upon *Ormosia venosa* seeds on the forest floor in Malaysia (Miura *et al.*, 1997). Peres and van Roosmalen (1996) also recorded spider monkeys preying upon *Ormosia* seeds in an Amazonian forest.

How Did Mimetic Fruits Evolve? Three Adaptive Hypotheses

The parasitism hypothesis

Batesian mimicry involves three agents: the selective agent (the bird), the model (an ornithochoric fruit) and the mimic (the mimetic fruit). All agents affect each other, but only the mimic benefits from this relationship (Endler, 1981). Early studies on seed dispersal hypothesized that mimetic fruits deceive seed-dispersers. Because they provide no benefits for the dispersers and, in fact, take advantage of them, the relationship is parasitic.

The mutualism hypothesis

Peres and van Roosmalen (1996) proposed a hypothesis that mimetic fruits of some species are ingested by terrestrial granivorous birds (tinamous, guans and trumpeters) because the hard-stoned seeds are used as grit to break down other food in the bird's gizzard ('hard seed for grit' hypothesis). The abrasive treatment of the mimetic fruits is hypothesized as essential for their germination. Peres and van Roosmalen (1996) did not provide any evidence that seeds ingested by terrestrial granivorous birds germinate better than seeds ingested by other birds.

The aposematism hypothesis

Aposematism refers to a warning signal (e.g. colours) of animals to advertise unpleasant attributes to avoid predation (Edmunds, 1974). Foster and Delay (1998) proposed that the colour of mimetic fruits is a warning signal of toxicity to seed predators, especially parrots. Some mimetic species (*Ormosia, Abrus, Sophora*) are well known to contain alkaloids that can kill domestic animals (Tokarnia and Dobereiner, 1997). Alkaloids in seeds, however, are not restricted to mimetic fruits (Herrera, 1982; Cipollini and Levey, 1997).

The seed dispersal system of mimetic fruits is still controversial and poorly tested. We present experimental tests of three hypotheses, using *Ormosia arborea* (*Leguminosae, Papilionoidea*) as a focal species.

Material and Methods

Plant species

Ormosia spp. are commonly used in studies on mimetic fruits (McKey, 1975; Peres and van Roosmalen, 1996; Foster and Delay, 1998). The genus contains approximately 100 species that have entirely red, red with a black spot (bicoloured) or entirely black seeds (Rudd, 1965). It is widely distributed and its taxonomy is relatively well studied (Rudd, 1965). Apparently all *Ormosia* species have long fruiting periods (Peres and van Roosmalen, 1996; Foster and Delay, 1998). Our study species, *O. arborea*, occurs in south-east Brazil in forest and *restingas* (Rudd, 1965). In some areas, fruits persist for as long as 36 months (M. Galetti, unpublished data). The fruit is a dehiscent pod, exposing a single red and black seed. Mean seed diameter is 12.85 ± 1.18 mm ($n = 20$). Because accurate information on seed-dispersers of *O. arborea* is lacking, we combined field and captive-animal experiments; we assume that experiments with captive animals can provide important clues to the dispersal of *O. arborea* seeds in the field.

Study Areas

All captive experiments were carried out at Bosque dos Jequitibás Zoo, Campinas, southeast Brazil, and at CRAX, a bird breeding facility specializing in Galliformes, at Contagem, Minas Gerais.

Field experiments were carried out in three areas:

1. Bosque dos Jequitibás, a 10 ha forest fragment in Campinas with a high density of agoutis (*Dasyprocta leporina*).
2. Parque Estadual Intervales, a 50,000 ha Atlantic Forest site near Sete Barras, São Paulo. It hosts one of the highest diversities of birds in the entire Atlantic Forest of Brazil (Aleixo and Galetti, 1997).
3. Estação Ecológica (EE) de Caetetus in Gália west of São Paulo. It has 2100 ha of semideciduous forest, with large populations of large frugivorous birds and mammals.

What Eats *Ormosia* Seeds?

Aviary experiments

We first asked: 'What eats and disperses *O. arborea* seeds?' We offered seeds *ad libitum* to several species of captive birds we thought were likely to consume *O. arborea* seeds in the field. These included large frugivores–granivores (jacutinga, *Pipile jacutinga*, $n = 2$ birds; black-legged guan, *Penelope obscura*, $n = 3$; trumpeter, *Psophia viridis*, $n = 2$; curassows, *Crax fasciolata* and *Crax blumenbackii*, $n = 8$ of each; and solitary tinamous, *Tinamous solitarius*, $n = 8$) and large, canopy frugivores (toco toucan, *Ramphastos toco*, $n = 6$; and bellbird, *Procnias nudicollis*, $n = 1$).

All birds were housed in large cages and supplied food (fruits and a synthetic diet) *ad libitum* and were in good health. We also offered *O. arborea* seeds to several mammal species (spider monkey, *Ateles paniscus*, $n = 2$; capuchin monkey, *Cebus apella*, $n = 5$; and tapir, *Tapirus terrestris*, $n = 2$) and reptiles (tegu lizard, *Tupinambis meriane*, $n = 5$; and tortoise, *Geochelone carbonaria*, $n = 2$) to observe how the seeds were treated (swallowed, chewed up, spat out, etc.). These experiments provided qualitative data; quantitative data are not presented because some animals became satiated during the experiments.

Field experiments

To determine *O. arborea* consumers in the field, we watched three fruiting *O. arborea* trees for 60 h (30 h during the dry season and 30 h during the wet season) at EE Caetetus. Because we did not observe any visitors to these trees, we set up five camera traps (Camtrack®) to detect frugivores eating *O. arborea* seeds on the forest floor. One camera was set up below a fruiting *O. arborea* for 4 months at Caetetus, while the others were set up at Parque Estadual Intervales. In total, the cameras were able to detect visits to the seeds of *O. arborea* during 4080 h. All animals that ate *O. arborea* seeds in captivity and that were photographed were assumed to eat *O. arborea* in the wild, even though we did not directly or indirectly (via photography) observe them ingesting seeds.

Testing the Parasitism Hypothesis: Do *O. arborea* Seeds Deceive Avian Frugivores in the Presence of a Putative Model?

If mimetic fruits deceive seed-dispersers, we would expect that birds would not distinguish between the model (rewarding) fruit and the mimetic fruit. In neotropical forests, several plant species produce arillate fruits that may be models of mimetic fruits. These include *Copaifera langsdorffii*, *Copaifera trapezifolia* (*Leguminosae*), *Sloanea* spp. (*Elaeocarpaceae*), *Cupania* spp. (*Sapindaceae*), *Alchornea triplinervia* (*Euphorbiaceae*) and many others (see Galetti, 1996). All experiments were carried out using *C. langsdorffii* (*Leguminosae, Caesalpinoidea*) as the model. This bird-dispersed species has black seeds partially covered by an orange aril and a diaspore similar in size to that of *O. arborea* and is eaten by several bird species (Galetti and Pizo, 1996).

Experiment using adult birds

Three pet wild-caught toucans (*R. toco*) and three guans (*Penelope superciliaris*) were housed in separate enclosures (2.5 m × 1.5 m × 1.5 m). Because none was captive-bred and all came from undetermined origins, it is possible that these birds had previous contact with *O. arborea* and its models before capture. The three toucans were chosen from a group of 15 that were living in a large enclosure. All experiments started 1 week after the birds were housed in our small enclosures, which appeared to be sufficient time for acclimatization. All birds were fed bananas and papaya and seemed to be in good health. Feeding trials with birds were conducted in the morning, when the birds were most active. Two experiments were carried out on adult birds. In the first, we offered a dish containing 20 *O. arborea* and 20 *C. langsdorffii* seeds to the three guans and to the three toucans. After 24 h we counted the number of seeds of each type ingested and defecated. After 48 h we repeated the experiment using the same birds. In the second experiment, we used only toucans. We offered each toucan (not all three together, as in the first experiment) 20 *O. arborea* seeds, 20 *C. langsdorffii* fruits (i.e. seed plus aril) and 20 *O. arborea* seeds with transplanted *C. langsdorffii* arils. After 24 h we counted the number of seeds of each type ingested and defecated. The experiment was repeated twice (i.e. 60 seeds of each type were offered to each toucan).

Experiment using naïve birds

Naïve birds are less selective in their diet than adult birds (Barrows *et al.*, 1980), so we predicted that naïve birds would be more easily 'deceived' by mimetic fruits than adult birds. We tested whether *O. arborea* seeds would deceive naïve, captive-born toucans (i.e. if *O. arborea* seeds would be ingested by them). Three captive-bred *R. toco* were housed in separate enclosures. They were fed their normal diet and offered 30 *O. arborea* seeds. We counted the number of seeds swallowed. After 24 h and 48 h we repeated the experiment. After another 2 days, without offering any *O. arborea* we offered *Ormosia* seeds with attached artificial arils made of red plasticine (without smell or taste). The number and type of seeds swallowed were recorded after 3 h. The trials were replicated after 24 and 48 h. After another 2 days, we offered *O. arborea* seeds without arils to determine if the behaviour of the birds had changed.

Field experiments

At EE Caetetus, we placed five stations below the crown of a fruiting *C. langsdorffii* tree. The tree bore thousands of ripe arillate fruits, which were frequently dropped to the forest floor by monkeys and birds. Fallen fruits were avidly consumed by three species of thrushes (*Turdus albicollis*, *Turdus rufiventris* and *Turdus amaurocalinus*) and by white-lipped (*Tajacu pecari*) and collared peccaries (*Tajacu tajacu*). Both peccary species destroy *C. langsdorffii* seeds and thus were considered seed predators.

Stations were spaced 3 m apart. At each station we placed one *O. arborea* seed, three *C. langsdorffii* seeds with attached arils and one *O. arborea* seed with a transplanted *C. langsdorffii* aril. *C. langsdorffii* arils can be easily fitted on to *O. arborea* seeds in such a way that even humans cannot distinguish the difference between *O. arborea* with *C. langsdorffii* arils and *C. langsdorffii* seeds with *C. langsdorffii* arils. The ratio of three model seeds to one mimetic fruit was arbitrary. The stations were checked every 15 min, the number and type of seeds eaten recorded and removed seeds replaced. Sixty-nine replacements were made during 5 h of observations.

Testing the Mutualism Hypothesis ('Hard Seed for Grit' Hypothesis): Do *O. arborea* Seeds Require Abrasive Treatment to Germinate?

The 'hard seed for grit' hypothesis (Peres and van Roosmalen, 1996) predicts that seeds defecated or regurgitated by birds with non-muscular gizzards (e.g. toucans) would have lower germination rates than those defecated or regurgitated by birds with muscular gizzards (e.g. Galliformes). To test this prediction, all seeds defecated or regurgitated by guans (*Penelope* and *Pipile*) and toucans (*Ramphastos*) in the aviary experiments were collected and tested for germination in the greenhouse. Each seed was sown in vermiculite and watered daily. Control seeds were not defecated or regurgitated but were otherwise treated identically.

Are Red and Black Seeds More Prone to be Eaten by Animals than Black Seeds?

If mimetic fruits of *O. arborea* evolved their conspicuously contrasting colours to attract frugivores, we would predict that red and black seeds are more likely to be eaten than totally black seeds.

Aviary experiments

We offered three captive toucans (*R. toco*) 30 *O. arborea* seeds and 30 *O. arborea* seeds painted completely black with non-toxic black, as described in the first aviary experiment. The painted seeds were left to dry for 24 h before being offered to the birds. We then counted the number of each seed type ingested and defecated, 24 h after offering them.

Field experiment

At Parque Estadual Intervales we set up 29 plots for each treatment. One treatment had 20 *O. arborea* seeds that were painted all black, the second treatment had 20 *O. arborea* seeds in which only the black part of the seed was painted black (to test whether the ink would affect seed predation) and the third had 20 *O. arborea* seeds that were untreated, as controls. Plots were spaced 100 m apart. The number of removed seeds was counted after 40 days. One camera trap was set up to record frugivore visits to each type of seed ($n = 3$ cameras).

Testing the Aposematism Hypothesis: Do the Colours of *O. arborea* Seeds Reduce Seed Predation by Rodents?

Aposematism is a well-studied phenomenon in insects (e.g. Gamberale and Tullberg, 1998), in which bright colours have evolved as a warning display to predators (usually birds) to reduce the probability of attack (Ritland, 1991). If mimetic fruits evolved colourful seeds as a warning signal of toxicity to seed predators, we would expect that black seeds would be more prone to predation by rodents than would red and black seeds. Note that this prediction is the opposite of that generated by the mutualism hypothesis; rather than focusing on the tendency of brightly coloured seeds to attract dispersers, it focuses on the tendency of these seeds to repel seed predators.

Field experiments

We carried out the experiment in Bosque dos Jequitibás, where agoutis (*Dasyprocta leporina*) are tame and can be observed closely. We set up 30 stations, spaced 100 m apart. At each station we placed one *O. arborea* seed (control) and one *O. arborea* painted completely black. We recorded the number and type of seeds eaten by agoutis after 120 h.

Statistical Analysis

Prior to statistical analyses, all variables were tested for normality and homogeneity of variance. When assumptions of normality and equal variance were not met, the variables were transformed (log transformation for mass and linear dimensions; angular transformations for proportions). If assumptions were still not met, we used non-parametric tests. Categorical data were analysed with χ^2 tests or, when expected values were < 5, with *G* tests (Sokal and Rohlf, 1995).

Results

What eats O. arborea *seeds?*

Aviary experiments

In trials with captive animals, *O. arborea* seeds were occasionally swallowed by a small assemblage of birds, including toucans (*R. toco*), guans (*P. jacutinga* and *P. obscura*) and trumpeters (*P. viridis*), but were ignored by bellbirds (*P. nudicollis*), mammals (*T. terrestris, C. apella, A. paniscus*) and reptiles (*G. carbonaria* and *T. meriane*). Curassows (*C. fasciolata* and *C. blumenbackii*) mandibulated the seeds but did not swallow them.

Field observations

In the wild, we observed white-lipped (*T. pecari*) and collared peccaries (*T. tajacu*) preying upon the seeds of *O. arborea* at EE Caetetus. Both peccaries chewed seeds and spat out the seed-coat. We saw no bird consume *O. arborea* seeds during 60 h of observation.

At Intervales, we recorded in the photographs only agoutis (*D. leporina*, $n = 15$ pictures), tinamous (*T. solitarius*, $n = 5$) and a dove (*Geotrygon montana*, $n = 3$) visiting the fallen seeds of *O. arborea.* Both agoutis and tinamous were classified as seed predators based on captive studies.

Testing the parasitism hypothesis: do O. arborea *seeds deceive avian frugivores in the presence of its putative model?*

Experiment using adult birds

In the first experiment, both toucans and guans preferred *C. langsdorffii* fruits to *O. arborea* seeds (Mann–Whitney *U* test, $U = 21$, $P = 0.002$). There was no difference between their degree of preference ($U = 52$, $P = 0.23$). Only eight seeds of *O. arborea* were eaten by toucans and five by guans, compared with 79 and 25 *C. langsdorffii* fruits consumed by toucans and guans, respectively.

In the second experiment (in which we offered *O. arborea* seeds, *O. arborea* seeds with aril and *C. langsdorffii* fruits), there was a statistically significant difference among the three seed types in the number ingested by toucans (Kruskal–Wallis test, $H = 9.02$, $P = 0.009$). Of 120 presentations, adult toucans ingested 13.2 ± 7.5 *C. langsdorffii* fruits, 10.7 ± 3.6 *O. arborea* seeds with aril of *Copaifera* and only 1.3 ± 3.3 *O. arborea* seeds.

Experiment using naïve birds

Naïve toucans consumed significantly more *O. arborea* seeds when only *O. arborea* seeds were offered than did adult, non-naïve toucans (two-way repeated-measures analysis of variance (ANOVA), $F = 19.90$, $P < 0.0001$). After ingesting *O. arborea* seeds with transplanted arils, the toucans rejected bare *O. arborea* seeds and even *O. arborea* seeds with transplanted arils (Fig. 12.1).

Field experiments

In our field experiment, wild birds (*Turdus* spp.) did not ingest any *O. arborea* seeds. They ingested only *C. langsdorffii* fruit (53% of fruits offered, $n = 207$ seeds) and *O. arborea* seeds with transplanted arils (14% of fruits offered, $n = 69$ seeds). The birds strongly preferred *C. langsdorffii* fruits to *O. arborea* seeds with transplanted aril (*G* test, $G = 16.1$, d.f. = 1, $P < 0.0001$).

Testing the mutualism hypothesis ('hard seed for grit' hypothesis): do O. arborea *seeds require abrasive treatment to germinate?*

We collected 19 *O. arborea* seeds from the faeces of *P. jacutinga*, 20 from *P. obscura* and four from *P. viridis. Tinamous solitarius* ingested a few seeds, but we did not find any intact in their faeces. Seeds defecated or regurgitated by captive guans and toucans and the control were planted in a greenhouse and compared (Fig. 12.2).

Eight months after sowing, seeds of *O. arborea* started to germinate. After 13 months, there was a significant difference in per cent germination between treatments and control ($G = 6.4$, d.f. = 2, $P = 0.04$). However, this difference was not the one predicted.

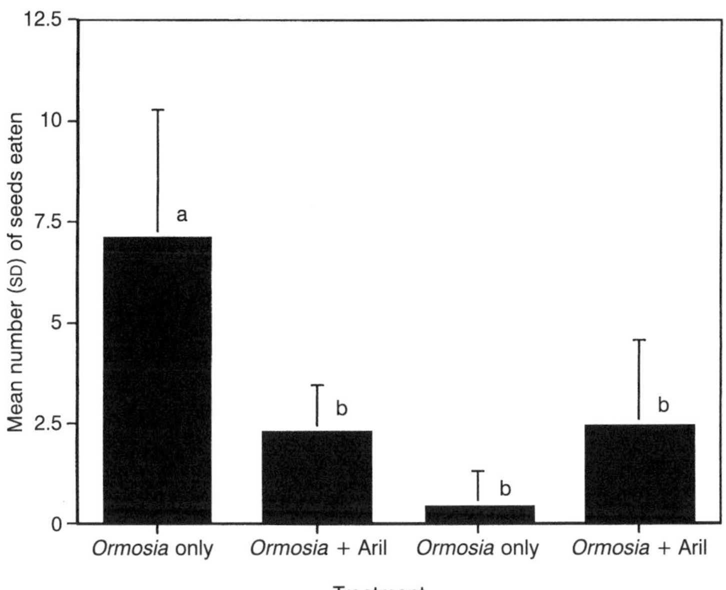

Fig. 12.1. Mean (and standard deviation) of number of *Ormosia arborea* seeds eaten by three naïve toucans (*Ramphastos toco*). Different letters signify statistical differences ($P < 0.05$).

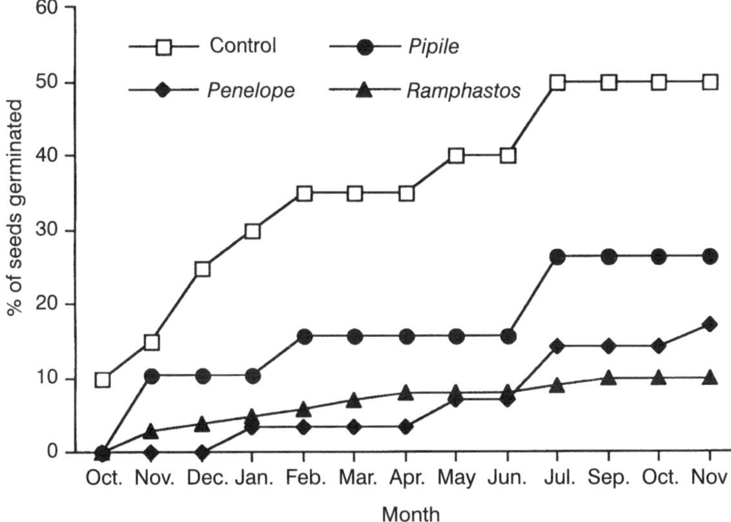

Fig. 12.2. Germination rates of *Ormosia arborea* seeds defecated or regurgitated by three species of birds and the control seeds, which were not ingested by birds. No seeds germinated for the first 7 months after sowing; these months are not shown on the *x* axis.

In particular, there was no difference in per cent germination among seeds ingested by guans and toucans and, in fact, all ingested seeds had substantially lower per cent germination than control seeds (Fig. 12.2). These results do not support the prediction that germination of *O. arborea* seeds is higher when passed through the gut of granivorous birds with muscular gizzards (e.g. guans) than when passed through frugivorous birds with non-muscular gizzards (e.g. toucans).

Are red and black seeds more prone to be eaten by animals than black seeds?

Aviary experiments

There was no difference in the number of natural (red and black) and all-black *O. arborea* seeds consumed by toucans (25 all-black seeds versus 26 red/black seeds). This suggests that the colour of *O. arborea* seeds does not enhance seed ingestion by toucans in these circumstances. The influence of seed colour on probability of ingestion may be different in nature, however.

Field experiments

We found no statistical difference in removal of natural *O. arborea* seeds, *O. arborea* seeds painted all black and *O. arborea* seeds with the naturally black part painted black (ANOVA, $F = 0.6$, d.f. = 2, $P = 0.5$). *Ormosia arborea* seeds that were naturally red and black were consumed slightly more often (2.3 ± 2.6 seeds) than *O. arborea* seeds with only the black part painted (1.9 ± 2.7) and *O. arborea* seeds painted completely black (1.4 ± 2.3, $n = 190$ seeds of each seed type offered).

Testing the aposematism hypothesis: do the colours of O. arborea *seeds reduce seed predation by rodents?*

We found no difference in seed predation levels by agoutis between black-plus-red and all black seeds ($\chi^2 = 12.4$, $P = 0.002$). Sixty-nine per cent of natural *O. arborea* seeds and 79% of *O. arborea* seeds painted all black were preyed upon after 120 h.

Discussion

The experiments described here comprise the first attempt to test alternative hypotheses about mimetic fruits. We caution that the results of experiments using captive birds are tentative because we do not know whether the birds had previous experience with mimetic fruits in the wild. Overall, our results best support the parasitism hypothesis – mimetic fruits deceive avian frugivores – although such deception is extremely rare in the wild (Foster and Delay, 1998).

The mutualism hypothesis

The mutualism hypothesis ('hard seed for grit' hypothesis) posits that the hard seeds of mimetic fruits are dispersed by terrestrial galliform birds that use the seeds as grit. The abrasive treatment in the birds' gut is presumably essential for seed germination (Peres and van Roosmalen, 1996). Our experiments led us to reject this hypothesis; we did not find a difference in seed germination of *O. arborea* seeds defecated by birds that would or would not be likely to use the seeds as grit. In fact, all seeds from bird defecations had lower germination rates than uningested control seeds. Moreover, we would expect that the contrasting colours of *O. arborea* seeds on the forest floor would attract more dispersers than if they were only black (as proposed by Peres and van Roosmalen, 1996). Again, we did not find any difference in seed removal between black and red/black seeds. Studies on the diet of neotropical Galliformes (Cracidae and Tinamidae) have not reported any species of mimetic fruit in their diet (Erard *et al.*, 1991; Galetti *et al.*, 1996; Yumoto, 1999; Santamaria and Franco, 2000). All large terrestrial cracids (*Crax*, *Mitu*) and tinamids (*Tinamous*) are primarily seed predators of large seeds (Bockerman, 1991; Yumoto, 1999; Santamaria and Franco, 2000). Jacutingas (*Pipile*) and guans (*Penelope*) are more arboreal and are mainly seed-dispersers, but they were never recorded visiting areas with *O. arborea* seeds by our cameras.

The aposematism hypothesis

The aposematism hypothesis was also rejected, at least for large-bodied diurnal rodents (e.g. agoutis), but more experiments using other vertebrates, such as parrots, are necessary. *Ormosia* spp. are rich in quinolizinide alkaloids (Ricker *et al.*, 1999), which may be important in deterring seed predation by rodents and insects (P. Guimarães, M. Galetti and J. Trigo, unpublished data). Other taxa with mimetic fruits (e.g. *Erythrina*, *Abrus* and *Sophora*) are well known to be toxic to vertebrates and insects (Ramos *et al.*, 1999), but alkaloids are not exclusive to *Leguminosae* with brightly coloured seeds (A. Tozzi, 2000, unpublished data).

The parasitism hypothesis

The only hypothesis supported by our experiments was the parasitism hypothesis. Naïve birds were more frequently deceived by *O. arborea* seeds than were non-naïve birds, as predicted by the hypothesis. In fact, adult toucans rejected *O. arborea* seeds only when we presented an arillated fruit with a mimetic fruit. Because mimetic fruits have a long fruiting season, it is likely that mimetic fruits are more prone to be dispersed during periods of low fruit availability. Several species considered mimetic have long dormancy and their seeds can be attached to the pod for up to 3 years (e.g. *O. arborea*). All of these characteristics are probably adaptations enabling the plant to maximize the period when seeds are available to dispersers (Peres and van Roosmalen, 1996; Foster and Delay, 1998).

Why are *Ormosia* Seeds Colourful?

Although our experiments support the parasitism hypothesis, we suggest that analysis of fruit morphology of the entire genus may provide clues to the evolution of colourful seeds. Fifty species of *Ormosia* occur in the neotropics and 50 others in the Old World (Rudd, 1965). A puzzling aspect of this genus is that even indehiscent species have colourful seeds and one dehiscent species has totally black seeds (Table 12.2). Assuming that species with abiotic dispersal mechanisms and indehiscent fruits represent the plesiomorphic (ancestral) state of *Papilionoidea* (Janson, 1992), we suggest that seed colour cannot be interpreted as an adaptation to present-day seed-dispersers.

Furthermore, several species of *Ormosia* that occur in flooded forests and have indehiscent fruits are dispersed by water (hydrochory) (Ziburski, 1991; Janson, 1992). Therefore, an evolutionary transition between two passive dispersal modes would be a more parsimonious interpretation than a transition from hydrochoric to endozoochoric syndromes (Janson, 1992; Jordano, 1995). In fact, autochory and hydrochory seem to be the main seed-dispersal syndromes in several mimetic species. Several species with mimetic fruits occur along rivers and have been recorded in studies on water dispersal. Murray (1986) listed two mimetic species (*Abrus precatorius* and *Erythrina variegata*) as capable of long-distance dispersal by ocean currents. It seems astonishing that there is no record of birds dispersing *A. precatorius*, one of the most invasive plants in Florida, USA. *Adenanthera pavonina* is also considered invasive in several islands in Oceania, but we do not have any unambiguous records of birds dispersing viable seeds of this species (only records of gut contents) (Steadman and Freifeld, 1999).

The current worldwide distribution of *Sophora* (which belongs to the same tribe as *Ormosia*), another genus with mimetic fruits, is due to ocean dispersal (Hurr *et al.*, 1999), although *Sophora macrocarpa* is also dispersed by cattle in Chile (R. Bustamante, 2000, personal communication). Most species of *Archidendron* and *P. pruinosum*, also mimetic species, occur along watercourses or in coastal areas (Cowan, 1998a, b). Several species of *Erythrina* occur in flooded areas and are able to float (Bruneau, 1996) and there are scant observations of

Table 12.2. Seed colour, habitat and occurrence of dehiscent fruits in neotropical *Ormosia* (following Rudd, 1965).

	Seed colour			Habitat	
	Red	Black	Red/black	*Terra firme*	Riverine
Indehiscent	5	0	2	0	5
Dehiscent	8	1	34	18	14

Erythrina being dispersed by birds (C.T. Downs, 2000, unpublished data).

Ecological patterns not linked to adaptive processes from current selection but to phylogenetic constraints may explain the seed colour of some mimetic species. For instance, seed colour in *Erythrina* is highly constrained by phylogeny (Bruneau, 1996). Phylogenetic constraints in several traits of fruit morphology, in fact, has been found to be more common than traits moulded by ongoing effects of natural selection (Herrera, 1987, 1992; Fischer and Chapman, 1993; Jordano, 1995).

We propose that the seed-dispersal system of *O. arborea*, and probably of most *Leguminosae* with mimetic fruits, is a typical case of exaptation. Exaptation represents the secondary use of a trait already present for other (generally historical) reasons (i.e. traits fit for their current role but not designed for it) (Gould and Vrba, 1982). Endozoochorous dispersal of mimetic fruits may certainly occur, but it is an extremely rare event. Despite being a rare event, sporadic dispersal by vertebrates might greatly contribute to fitness of rare species with mimetic fruits.

But what constitutes a rare event in vertebrate seed dispersal? There are few long-term studies on fruit fall to evaluate this question for mimetic fruits. Data for 13 years (January 1987 to January 2000) of seed fall on Barro Colorado Island (Panama) revealed only two seeds of *Ormosia* (one *O. coratti*, one *O. macrocalyx*) away from conspecifics in 200 0.5 m^2 traps (100 m^2) in a 50 ha plot. One seed was found more than 400 m from any *Ormosia* adult and the other 95 m away (S.J. Wright and R. Condit, 2000, unpublished data). The long-distance movements of both these *Ormosia* seeds are probably the result of arboreal seed dispersal (perhaps by toucans). In Cameroon, seed-rain samples from 12 months of trapping (totalling 77.9 m^2) below endozoochoric trees contained only one *Erythrina* seed (C. Clark, 2000, unpublished data). The same pattern was found for *P. broteroi* in Sierra de Cazorla, Spain. Three years of data on seed fall (1200 traps) contained only one record of a dispersed seed, 13 m away from the nearest *Paeonia* adult (J.L. Garcia-Castaño and P. Jordano, 2000, unpublished data). Likewise, Foster and Delay (1998) reported that, in 85 h of watching *Ormosia* trees (three species), only 19 seeds were dispersed away from the trees and only one was swallowed by birds. Peres and van Roosmalen (1996) did not record any arboreal frugivores ingesting *Ormosia lignivalvis* in 185 h of focal tree observations.

However, our findings should not be taken out of the context of the plants' demography. In particular, we measured only the number of seeds dispersed, not seedling establishment. The low frequency of seed dispersal means either low selection or low reproduction. Several additional issues should be pursued to understand the mimetic fruit-dispersal system. Would results have been different if the seeds were black in colour? Would a tree actually have higher fitness if it put a nutritive reward on its seeds? To what extent does the maintenance of scarce populations of mimetic-fruited species depend on rare events of seed dispersal by frugivores?

Although mimetic fruits are extremely difficult to study in the wild, study of their seed-dispersal systems may be as informative as the study of more typical fruits and is paramount for understanding the evolution of plant–frugivore interactions. The balance between fruit attraction and chemical defence may be better understood when both extremes of the deception–reward gradient are considered.

Acknowledgements

I am deeply grateful to all students of the Plant Phenology and Seed Dispersal Research Group, particularly Liliane Zumstein, Inez Morato and Paulo Guimarães, Jr, for their help in the field and aviary experiments and to Ana Tozzi for enlightening discussions on *Leguminosae* phylogeny. I also thank the administration of Bosque dos Jequitibás in Campinas and Mr Roberto Azeredo from CRAX for allowing our trials with their captive birds. I am grateful to Instituto Florestal and Fundação Florestal de São Paulo for providing all facilities at EE Caetetus and Parque Intervales. My thanks also go to C. Clark, J.S. Wright and J.L. Garcia-Castaño for allowing me to use their data on seed fall. Daniel Janzen, Pedro Jordano, Carlos Herrera, Douglas Levey and

Marco A. Pizo provided helpful discussion and helped improve the manuscript. Finally, I thank FAPESP (Fundação de Ampara a Pesquisa do Estado de São Paulo) (96/10464-7) and CNPq (Conselho Nacional de Desenvolrimento Científico e Tecnológico) (300025/97-1) for financial support.

References

Aleixo, A. and Galetti, M. (1997) The conservation of the avifauna in a lowland Atlantic forest in southeast Brazil. *Bird Conservation International* 7, 235–261.

Armstrong, W.P. (1992) Jewels of the tropics. *Terra* 30, 26–33. (http://waynesword.palomar.edu/ww0901.htm)

Barrows, E.M., Acquavella, A.P., Wesinstein, P.J.S. and Nosal, R.E. (1980) Response to novel food in captive, juvenile mockingbirds. *Wilson Bulletin* 92, 399–402.

Blake, J.G. and Loiselle, B.A. (1992) Fruits in the diets of neotropical migrant birds in Costa Rica. *Biotropica* 24, 200–210.

Bockerman, W.C.A. (1991) Observações sobre a biologia do macuco *Tinamous solitarius*. PhD thesis, Universidade de São Paulo, Brazil.

Bruneau, A. (1996) Phylogenetic and biogeographical patterns in *Erythrina* (*Leguminosae: Phaseoleae*) as inferred from morphological and chloroplast DNA characters. *Systematic Botany* 21, 587–605.

Cipollini, M.L. and Levey, D.J. (1997) Secondary metabolites of fleshy vertebrate-dispersed fruits: adaptive hypothesis and implications for seed dispersal. *American Naturalist* 150, 346–372.

Corner, E.J.H. (1949) The Durian theory or the origin of the modern tree. *Annals of Botany* 52, 367–414.

Cowan, R.S. (1998a) *Pararchidendron*. In: Orchard, A. (ed.) *Flora of Australia, Mimosaceae (excl. Acacia), Caesalpiniaceae*. CSIRO, Melbourne, Australia, pp. 39–40.

Cowan, R.S. (1998b) *Archidendron*. In: Orchard, A. (ed.) *Flora of Australia, Mimosaceae (excl. Acacia), Caesalpiniaceae*. CSIRO, Melbourne, Australia, pp. 40–48.

Dafni, A. (1984) Mimicry and deception in pollination. *Annual Review of Ecology and Systematics* 15, 259–278.

Edmunds, M. (1974) *Defence in Animals*. Longman, Harlow.

Endler, J.A. (1981) An overview of the relationships between mimicry and crypsis. *Biological Journal of the Linnean Society* 16, 25–31.

Erard, C., Théry, M. and Sabatier, D. (1991) Régime alimentaire de *Tinamous major* (Tinamidae), *Crax alector* (Cracidae) et *Psophia crepitans* (Psophiidae) en fôret guyanaise. *Gibier Faune Sauvage* 8, 183–210.

Fischer, K.E. and Chapman, C.A. (1993) Frugivores and fruit syndromes: differences in patterns at the genus and species level. *Oikos* 66, 472–482.

Foster, M.S. and Delay, L.S. (1998) Dispersal of mimetic seeds of three species of *Ormosia* (*Leguminosae*). *Journal of Tropical Ecology* 14, 389–411.

French, K. (1990) Evidence for frugivory by birds in montane and lowland forests in southeast Australia. *Emu* 90, 185–189.

Galetti, M. (1996) Fruits and frugivores in a Brazilian Atlantic forest. PhD thesis, University of Cambridge.

Galetti, M. and Pizo, M.A. (1996) Fruit-eating birds in a forest fragment in southeastern Brazil. *Ararajuba* 4, 71–79.

Galetti, M., Martuscelli, P., Olmos, F. and Aleixo, A. (1996) Ecology and conservation of the jacutinga *Pipile jacutinga* in the Atlantic forest of Brazil. *Biological Conservation* 82, 31–39.

Gamberale, G. and Tullberg, B.S. (1998) Aposematism and gregariousness: the combined effect of group size and coloration on signal repellence. *Proceedings Royal Society of London B* 265, 889–894.

Gautier-Hion, A., Duplantier, J.M., Quris, R., Feer, F., Sourd, C., Decoux, J.P., Dubost, G., Emmons, L., Erard, C. and Hecketsweiler, P. (1985) Fruit characters as a basis of fruit choice and seed dispersal in a tropical forest vertebrate community. *Oecologia* 65, 324–337.

Gould, S.J. and Vrba, E.S. (1982) Exaptation – a missing term in the science of form. *Paleobiology* 8, 4–15.

Herrera, C.M. (1982) Defense of ripe fruits from pests: its significance in relation to plant–disperser interactions. *American Naturalist* 120, 218–247.

Herrera, C.M. (1987) Vertebrate-dispersed plants of the Iberian peninsula: a study of fruit characteristics. *Ecological Monographs* 57, 305–331.

Herrera, C.M. (1992) Interspecific variation in fruit shape: allometry, phylogeny, and adaptation to dispersal agents. *Ecology* 73, 1832–1841.

Howe, H.F. (1993) Specialized and generalized dispersal systems: where does 'the paradigm' stand? *Vegetatio* 107/108, 3–13.

Howe, H.F. and Smallwood, J. (1982) Ecology of seed dispersal. *Annual Review of Ecology and Systematics* 13, 201–228.

Hurr, K.A., Lockhart, P.J., Heenan, P.B. and Penny, D. (1999) Evidence for the recent dispersal of *Sophora* (*Leguminosae*) around the Southern Oceans: molecular data. *Journal of Biogeography* 26, 565–577.

Janson, C.H. (1983) Adaptation of fruit morphology to dispersal agents in a neotropical forest. *Science* 219, 187–189.

Janson, C.H. (1992) Measuring evolutionary constraints: a Markov model for phylogenetic transitions among seed dispersal syndromes. *Evolution* 46, 136–158.

Janzen, D.H. (1984) Dispersal of small seeds by big herbivores: foliage is the fruit. *American Naturalist* 123, 338–353.

Jordano, P. (1995) Angiosperm fleshy fruits and seed dispersers: a comparative analysis of adaptation and constraints in plant–animal interactions. *American Naturalist* 145, 163–191.

Lambert, F. (1989) Pigeons as seed predators and dispersers of figs in a Malaysian lowland forest. *Ibis* 131, 521–527.

Levey, D.J. (1987) Seed size and fruit-handling techniques of avian frugivores. *American Naturalist* 129, 471–485.

Levey, D.J. and Grajal, A. (1991) Evolutionary implications of fruit-processing limitations by cedar waxwings. *American Naturalist* 138, 171–189.

Loiselle, B.A. and Blake, J.G. (1990) Diets of understory fruit-eating birds in Costa Rica: seasonality and resource abundance. *Studies in Avian Biology* 13, 91–103.

Mack, A.L. (2000) Did fleshy fruit pulp evolve as a defence against seed loss rather than as a dispersal mechanism? *Journal of Bioscience* 25, 93–97.

McKey, D. (1975) The ecology of coevolved seed dispersal systems. In: Gilbert, L.E. and Raven, P.H. (eds) *Coevolution of Animals and Plants.* University of Texas Press, Austin, Texas, pp. 159–209.

Malo, J.E. and Suárez, F. (1998) The dispersal of a dry-fruited shrub by red deer in a Mediterranean ecosystem. *Ecography* 21, 204–211.

Miura, S., Masatoshi, Y. and Ratnam, L.C. (1997) Who steals the fruits? Monitoring frugivory of mammals in a tropical rain forest. *Malayan Nature Journal* 50, 183–193.

Murray, D.R. (1986) Seed dispersal by water. In: Murray, D.R. (ed.) *Seed Dispersal.* Academic Press, London, pp. 49–85.

Murray, K.G., Winnett-Murray, K., Cromie, E.A., Minor, M. and Meyers, E. (1993) The influence of seed packaging and fruit color on feeding preferences of American robins. *Vegetatio* 107/108, 217–226.

Ortmann, J., Schacht, W.H., Stubbendieck, J. and Brink, D.R. (1998) The 'foliage is the fruit' hypothesis: complex adaptations in buffalograss (*Buchloe dactyloides*). *American Midland Naturalist* 140, 252–263.

Peres, C.A. and van Roosmalen, M.G.M. (1996) Avian dispersal of mimetic seeds of *Ormosia lignivalvis* by terrestrial granivores: deception or mutualism? *Oikos* 75, 249–258.

Quin, B.R. (1996) Diet and habitat of emus, *Dromaius novaehollandiae,* in the Grampians ranges, southwestern Victoria. *Emu* 96, 114–122.

Quinn, J.A., Mowrey, D.P., Emanuele, S.M. and Whalley, R.D.B. (1994) The 'foliage is the fruit' hypothesis: *Buchloe dactyloides* (Poaceae) and the shortgrass prairie of North America. *American Journal of Botany* 81, 1545–1554.

Ramos, M.V., Teixeira, C.R., Bomfim, L.R., Madeira, S.V.F. and Moreira, R.A. (1999) The carbohyrate-binding specificity of a highly toxic protein from *Abrus pulchellus* seeds. *Memórias do Instituto Oswaldo Cruz* 94, 185–188.

Ricker, M., Daly, D.C., Veen, G., Robbins, E.F., Sinta, M., Chota, J., Czygan, F. and Kinghorn, D. (1999) Distribution of quinolizidine alkaloid in nine *Ormosia* species (*Leguminosae–Papilionoideae*). *Brittonia* 51, 34–43.

Ridley, H.N. (1930) *The Dispersal of Plants Throughout the World.* L. Reeve, London.

Ritland, D.B. (1991) Revising a classic butterfly mimicry scenario: demonstration of Müllerian mimicry between Florida viceroys (*Limenitis archippus floridensis*) and queens (*Danaus gilippus berenice*). *Evolution* 45, 918–934.

Rudd, V.E. (1965) The American species of *Ormosia* (*Leguminosae*). *Contributions of United States National Herbarium* 32, 279–284.

Santamaria, M. and Franco, A.M. (2000) Diet of the curassow *Mitu salvini* and the fate of ingested seeds in Colombian Amazon. *Wilson Bulletin* 112, 473–481.

Snow, D.W. (1971) Evolutionary aspects of fruit-eating by birds. *Ibis* 113, 194–202.

Sokal, R.R. and Rohlf, F.J. (1995) *Biometry.* W.H. Freeman, New York.

Steadman, D. and Freifeld, H.B. (1999) The food habits of Polynesian pigeons and doves: a systematic and biogeographic review. *Ecotropica* 5, 13–33.

Tokarnia, C.H. and Dobereiner, J. (1997) Cross-immunity by the seeds of *Abrus precatorius* and *Ricinus communis* in cattle. *Pesquisa Veterinária Brasileira* 17, 25–35.

van der Pijl, L. (1982) *Principles of Dispersal in Higher Plants,* 3rd edn. Springer-Verlag, New York.

Wheelwright, N.T., Haber, W.A., Murray, K.G. and Guindon, C. (1984) Tropical fruit-eating birds and their food plants: a survey of a Costa Rican lower montane forest. *Biotropica* 16, 173–192.

Williamson, G.B. (1982) Plant mimicry: evolutionary constraints. *Biological Journal of the Linnean Society* 18, 49–58.

Witmer, M.C. (1998) Do seeds hinder digestive processing of fruit pulp? Implications for plant/frugivore mutualisms. *Auk* 115, 319–326.

Yumoto, T. (1999) Seed dispersal by Salvin's curassow, *Mitu salvini* (Cracidae), in a tropical forest of Colombia: direct measurements of dispersal distance. *Biotropica* 31, 654–660.

Ziburski, A. (1991) *Dissemination, Keimung und Etablierung eininger Baumarten der Uberschwemmungswälder Amazoniens.* 77 Tropische und subtropische Pflanzenwelt. Akademie der Wissenchaften und der Literatur, Mainz.

13 Secondary Dispersal of Jeffrey Pine Seeds by Rodent Scatter-hoarders: the Roles of Pilfering, Recaching and a Variable Environment

Stephen B. Vander Wall
[1]Department of Biology and the Program in Ecology, Evolution and Conservation Biology, University of Nevada, Reno, NV 89557, USA

Introduction

It is increasingly apparent that dispersal of seeds from a plant to safe establishment sites is often a complicated, multistage process (Westoby, 1981; Forget and Milleron, 1991; Levey and Byrne, 1993; Chambers and MacMahon, 1994; Bohning-Gaese *et al.*, 1999; Hoshizaki *et al.*, 1999). At each stage, the vehicle or mode of dispersal can change. The initial stage of dispersal, movement from the parent plant to a substrate, is usually referred to as primary dispersal; all subsequent movements of the seed are termed secondary dispersal. The steps that a seed takes during dispersal and the ways it is treated have important influences on its ultimate fate. Consequently, a detailed knowledge of seed movements is necessary if we are to bridge the gap between studies of seed dispersal and plant demography.

Food-hoarding animals can affect secondary dispersal of seeds when they gather and scatter-hoard seeds that were initially dispersed by some other means, such as by wind or gravity or in vertebrate faeces (Wicklow *et al.*, 1984; Matlack, 1989; Chambers *et al.*, 1991). Additionally, food-hoarding animals can excavate already scatter-hoarded seeds and move them to new storage sites (DeGange *et al.*, 1989; Vander Wall and Joyner, 1998a; Sone and Kohno, 1999). This recaching of seeds can occur repeatedly. Consequently, the term 'secondary dispersal' is somewhat misleading, as propagules can sometimes move three, four or more times before they reach their ultimate fate. In this chapter, I reanalyse and re-evaluate the results of several previously published studies (Vander Wall, 1992b, 1993, 2000; Vander Wall and Joyner, 1998a, b) and present some new data to determine how the recaching of seeds by yellow pine chipmunks (*Tamias amoenus*) and other rodents contributes to secondary dispersal of Jeffrey pine (*Pinus jeffreyi*) seeds. I address the following questions: Why do chipmunks repeatedly excavate and recache seeds? And what effect does this behaviour have on regeneration by the pine?

Study Area

Data were collected in the Whittell Forest and Wildlife Area, a research and teaching reserve owned by the University of Nevada, located

 Seed Dispersal and Frugivory: Ecology, Evolution and Conservation (eds D.J. Levey, W.R. Silva and M. Galetti)

about 30 km south of Reno, Nevada, USA. The area is an isolated, elevated valley on the east slope of the Carson Range in extreme western Nevada (elevation 1960–2680 m). The lower slopes of the valley are dominated by open stands of Jeffrey pine (Fig. 13.1) with an understorey of antelope bitterbrush (*Purshia tridentata*), green-leaf manzanita (*Arctostaphylos patula*), tobacco bush (*Ceanothus velutinus*) and Sierra chinquapin (*Castanopsis sempervirens*). Soils consist of decomposed granite with sparse to thick plant litter under trees and shrubs. Studies were conducted between 1989 and 2000.

Jeffrey Pine

Jeffrey pine, a close relative of ponderosa pine (*Pinus ponderosa*), produces large cone crops at intervals of 2–3 years, with small to moderate cone crops in the interim. Cone production is loosely synchronized. During fairly heavy seed years, most large trees produce from 70 to 100 cones, each containing from 50 to 125 filled seeds. Seeds weigh 132 ± 17 mg and have a large, functional wing (≈3 cm^2) (Vander Wall, 1994). Cones at the study site open in mid–late September. As seeds fall from the cones, they accelerate until they reach terminal velocity and begin to autorotate (Greene and Johnson, 1989). The relatively slow rate of descent allows wind to move the seeds laterally, but because the seeds are relatively massive they are not blown far. Most seeds land within 30 m of the base of the tree. Once the seeds land on the ground, the wind generally does not move them further (Vander Wall and Joyner, 1998b). Consequently, thousands of seeds litter the ground under and around productive trees during September and early October.

Rodent Scatter-hoarders

Several species of rodents that live in the Jeffrey pine–bitterbrush association scatter-hoard seeds and are potential seed dispersers. The yellow pine chipmunk is the most abundant species and the most efficient disperser of seeds. These chipmunks are small (40–50 g), diurnal rodents that live in underground burrows and spend most of their time foraging on the ground surface. They are omnivorous, but focus their foraging on seeds of trees, shrubs and forbs (Broadbooks, 1958). They are generally solitary outside the breeding season. They are not territorial (except around their dens) and occupy fairly large (0.9–1.6 ha), overlapping home ranges; ten to 20 chipmunks may share portions of one another's home ranges (Broadbooks, 1970; S.B. Vander Wall, personal observation). Yellow pine chipmunks are unable to harvest seeds directly from closed cones, as do some of the larger squirrels in the area. However, they are very efficient foragers on wind-dispersed seeds on the ground. They also climb into trees and remove seeds from open cones and gather seeds that get lodged in foliage. Yellow pine chipmunks are avid scatter-hoarders of seeds during summer and autumn, placing one to ten Jeffrey pine seeds in each cache. In late autumn (late October to mid-November) they transfer seeds from scattered caches into a winter larder in preparation for hibernation. Lodgepole chipmunks (*Tamias speciosus*) appear to behave similarly but are much less common at this study site.

Other rodents that store pine seeds in the study area include deer-mice (*Peromyscus maniculatus*), Douglas squirrels (*Tamiasciurus douglasi*), California ground squirrels (*Spermophilus beecheyi*) and golden-mantled ground-squirrels (*Spermophilus lateralis*). Deer mice scatter-hoard seeds in plant litter (Vander Wall *et al.*, 2001). However, they are not very effective dispersers of Jeffrey pine seeds because they are not as abundant and make relatively few caches. Douglas squirrels, important seed predators of Jeffrey pine, store intact, closed cones in cavities in the ground. The ground squirrels harvest many seeds, mostly from closed cones, which they gnaw open. California ground squirrels appear to larder-hoard most of the seeds they obtain, but golden-mantled ground squirrels scatter-hoard many seeds (Saigo, 1969). They are less effective as dispersers than chipmunks, because they bury seeds deeply and place many (ten to 40) seeds in each cache.

To measure the rate at which yellow pine chipmunks and other animals remove

Fig. 13.1. Open Jeffrey pine forest with bitterbrush understorey in the Whittell Forest and Wildlife Area, western Nevada, USA.

wind-dispersed seeds of Jeffrey pine, we established transects through open forests (bare mineral soil with bitterbrush shrubs) and closed forests (heavy deposits of needle litter without shrubs) and placed single, winged Jeffrey pine seeds at ~5 m intervals along the transects. We used twigs, pine cones, pebbles and other natural objects to mark the location of seeds. To prevent wind from displacing seeds, we glued brown strings to the seeds with superglue (we did this several weeks earlier in the laboratory so that the odours of the glue would dissipate) and tethered the seeds to small twigs. Then we walked the transect each day and recorded when seeds disappeared (see Vander Wall (1994) for details on methodology). The results from numerous such transects indicate that Jeffrey pine seeds have a mean (± 1 SD) half-life of 2.5 ± 0.9 days ($n = 7$ transects) in open forests and 6.0 ± 3.3 days ($n = 3$ transects) in closed forests (Vander Wall, 1994; Fig. 13.2; S.B. Vander Wall, unpublished data). These half-lives are equivalent to 24.5% removal per day in open forests and 10.9% removal per day in closed forests. Rates of removal during the 1–2 week long observation period appear to be nearly constant. The greater rate of removal in open forests is probably because of the higher density of rodents in that habitat compared with the much less shrubby, closed forest. Chipmunks appear to prefer the open, shrubby forest habitat, because of increased food resources and increased cover, which presumably lowers predation risk. Seeds continue to fall for about a month after cones open. Although a few seeds gradually get buried by abiotic processes, during the ~2 months between cone opening and the first winter snows there is time for over ten half-lives, suggesting that more than 99% of the wind-dispersed seeds will be gathered by animals.

Secondary Dispersal

Seed removal transects are informative, but they tell us little about which animals remove the seeds or what these animals do with the seeds. To answer these questions, we followed the fates of seeds using radioactive scandium-46 as a tracer. Scandium-46 is a gamma-emitting radionuclide with a half-life of 84.5 days. To label seeds, we removed the wings, numbered the seeds with indelible ink, soaked them in a solution of scandium in distilled water (~1 ml 100 per seeds) for 1 h and then let them dry. The isotope is absorbed by the woody seed hull. Seeds typically receive ~1 μCi of activity. We can detect individual

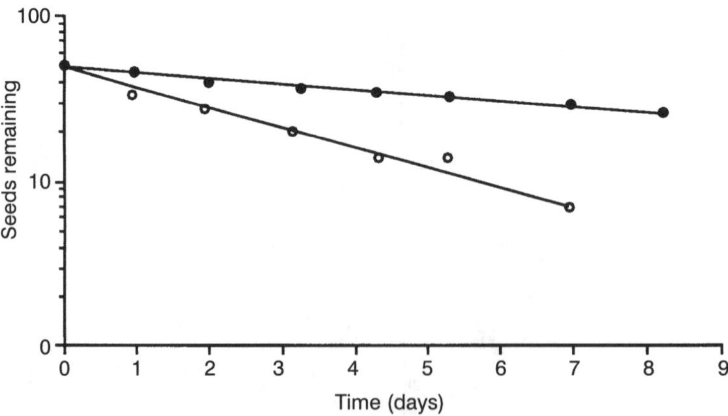

Fig. 13.2. The effect of habitat type on Jeffrey pine seed removal rate. Data from two representative seed transects run concurrently in autumn 1999 are illustrated. Each transect started with 50 seeds. Closed circles, closed Jeffrey pine forest with thick deposits of pine-needle litter and few shrubs (seed half-life = 6.6 days); open circles, open Jeffrey pine forests with exposed mineral soils and dense shrub understorey (seed half-life = 2.4 days). The difference in the rates of removal is significant (proportional hazards model $\chi^2 = 31.9$, $P < 0.0001$; $Z = 5.33$, $P < 0.0001$).

cached seeds from 20–30 cm with a Geiger counter. When rodents eat a seed, they discard the seed hull, which we can find. Thus, we not only locate cached seeds, but we can determine which seeds have been eaten. Scandium appears to be biologically inert; it is not sequestered in the tissues of animals, and the isotope on ingested fragments of the seed hull passes through animals in 1–2 days.

In 1990 and 1991 we conducted two studies on the secondary dispersal of Jeffrey pine seeds from arrays of seeds that simulated wind-generated seed shadows (Vander Wall, 1992b, 1993). We arranged 1064 radioactive and numbered Jeffrey pine seeds around each of four mature Jeffrey pine trees in 12 1 m wide annuli. These arrays simulated those that might be generated under a slight breeze. We did not scatter the seeds, as the wind would have, but rather placed them at 133 stations (eight seeds per station). Consistent with experimental and theoretical studies, the highest densities of seeds in these arrays were just inside the tree canopy (3 m radius) and the greatest number of seeds per annulus was just beyond the tree canopy. A typical wind-generated seed shadow might extend for many metres (at very low densities) but we truncated these distributions at 12 m for practical reasons and because we assumed that the long tail of the wind-dispersal curve, although potentially important in regard to primary dispersal, would have little effect on the study of secondary dispersal. After all, in this nearly pure stand of Jeffrey pine, seeds that travel more than 5 m from their parent are likely to land near another seed-producing tree.

We established these distributions of seeds in the morning and then observed them from a distance of about 50 m for about 1 h. At three of the four source trees, we observed several yellow pine chipmunks and a few lodgepole chipmunks harvest and cache some of the seeds. When we returned the next day, we first counted the number of unharvested seeds and then surveyed the area within 40 m of the source tree with a Geiger counter to find caches and eaten seeds. If we found caches near the periphery of this area, we continued the survey in that direction until we searched more than 10 m beyond the most distant cache (up to 80 m). Of the 1064 seeds deployed at each source tree, 1016 ± 42 seeds were harvested (95.4 ± 4.0%), and only 36 ± 24 of these (3.6 ± 2.6%) were eaten within a few days of seed harvest. For the seeds carried away, chipmunks stored 639 ± 214 seeds (60.0 ± 20.1%) in 171 ± 60 caches ranging up to 69 m from the base of the tree (Fig. 13.3). The area under the source tree (0–3 m) had few caches; there was a peak at 9–12 m from the source tree, followed by a gradual decline in the number of caches to near zero at ~69 m. Most caches contained from one to eight seeds, but golden-mantled ground-squirrels made a few very large caches (maximum = 35 seeds). We did not find 341 ± 245 seeds, and we suspect that rodents carried them outside the search area.

Clearly, secondary dispersal increased the mean dispersal distance and changed the shape of the dispersal curve. Seeds in the wind-dispersed arrays were 4.6 m from the source trees, whereas rodent (primarily chipmunk) caches were 24.3 ± 8.2 m from the trees, an increase in mean dispersal distance of ~20 m. This is likely to benefit seedling establishment by reducing density-dependent seed mortality around the source tree. Many animals that do not store food (e.g. quail, sparrows, bears, deer) are attracted to the concentrations of seeds under trees. These animals can be effective seed predators. Rapid removal of seeds by chipmunks and other scatter-hoarding rodents reduces seed mortality at this time. But several other aspects of secondary dispersal also potentially benefit the seeds. First, chipmunks bury most seeds between ~5 and 25 mm below the surface (Fig. 13.4a). Germination and emergence tests on Jeffrey pine seeds buried over a range of depths have demonstrated that chipmunk caches broadly overlap the range of depths that is the most favourable for seedling establishment (Fig. 13.4b). Secondly, many seeds fall in the closed-canopy forest, which is generally a poor habitat for seedling establishment. Closed Jeffrey pine forests typically have thick accumulations of pine-needle litter, which make poor beds for seedlings. These substrates have low moisture-holding capacity and low nutrient availability relative to mineral soils found in forest openings. Further, Jeffrey pine seedlings do not perform well in the deep shade characteristic of closed forests. Chipmunks and other rodents gather seeds from

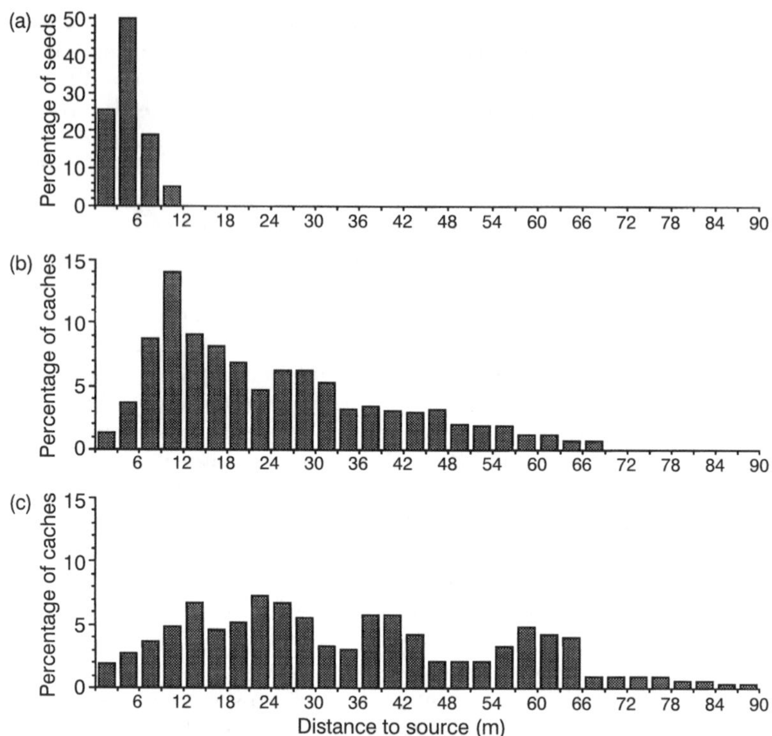

Fig. 13.3. The distribution of seeds and caches around source areas. (a) The array of 'wind-dispersed' seeds used to study secondary dispersal, (b) the composite distribution of primary caches around four simulated source trees (data from Vander Wall, 1992b, 1993), and (c) the distribution of secondary through quintic caches around an array of primary caches. (Data from Vander Wall and Joyner, 1998a.)

closed forests and transport a disproportionate number to open forests or shrubby sites, where the probability of establishment is much greater (Vander Wall, 1993). Thirdly, chipmunks make about a third of their caches under bitterbrush shrubs. These shrubs serve as nurse plants for pine seedlings (Sherman and Chilcote, 1972; Vander Wall, 1992a), increasing establishment by shading the seedlings and protecting them from browsing and physical damage.

In summary, Jeffrey pine appears to be effectively dispersed by a two-stage process: initially by wind dispersal to the ground surface (relatively poor establishment sites) and subsequently by chipmunks and other rodents to cache sites in the soil (relatively good establishment sites). However, this portrayal of a fairly complicated, two-stage seed-dispersal system for Jeffrey pine is actually greatly oversimplified. This is because, after chipmunks initially cache seeds, the same or other rodents dig up the seeds and move them to another site. This may happen repeatedly. A full understanding of seed dispersal in Jeffrey pine must include these subsequent movements from cache site to cache site.

Recaching of Seeds by Chipmunks

We studied the movement of seeds between cache sites by establishing 100 'primary' caches in a 10×10 array with 2 m spacing. Each of the caches contained ten radioactively-labelled and uniquely numbered seeds. We (Vander Wall and Joyner, 1998a) established this array on 1 September 1997 and then returned 1 week later and checked the caches. Of the 100 caches, 91 had disappeared by the end of 1 week (eventually 99 primary caches disappeared). We searched the area on and

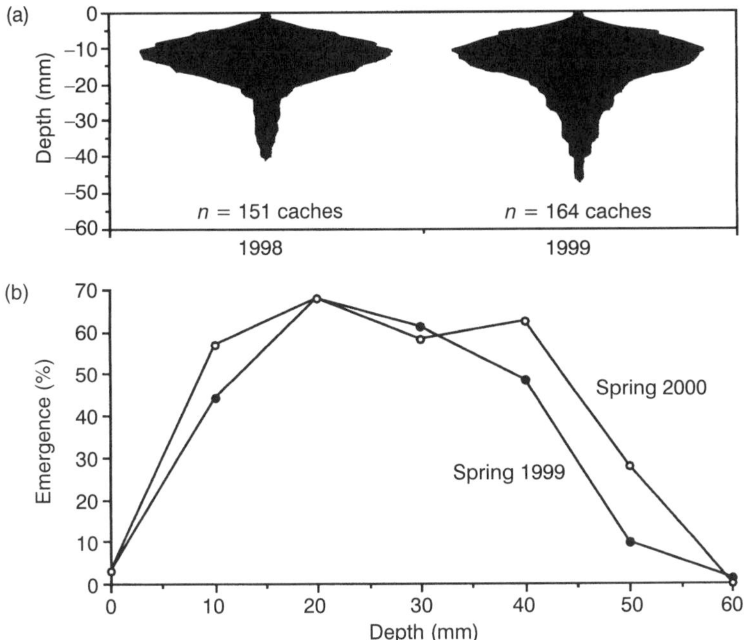

Fig. 13.4. (a) Cache depth profiles of yellow pine chipmunks hoarding Jeffrey pine seeds in autumn 1998 and 1999. The width of the profiles is proportional to the number of seeds buried at that depth. (b) Emergence (%) of Jeffrey pine seedlings from seeds experimentally buried from 0 to 60 mm deep.

around the 'primary' cache grid with a Geiger counter, looking for secondary caches. Over the next 2 months, we searched an area of about 2 ha on three separate occasions, spaced 2–3 weeks apart. We excavated each new cache we found, checked the number on the seeds, replaced the seeds and covered them carefully, leaving no obvious trace of our digging. During the experiment, we accounted for 83% of the seeds taken from primary caches. By comparing numbers on seeds at the end of the experiment (early November), we could trace the routes of individual seeds from one cache site to another.

From the seeds taken from the 99 primary caches, we found a total of 722 new caches. Of these, 377 were secondary caches, 213 were tertiary caches (i.e. seeds in these caches came from secondary caches), 75 were quaternary caches and 13 were quintic (fifth-order) caches (Fig. 13.5). An additional 44 caches were not assigned an order because they contained seeds of mixed order or contained seeds with illegible numbers. Figure 13.6 illustrates a generalized seed-fate pathway for the seeds.

Taken as a whole, the movements of seeds from cache site to cache site were very complicated. The history of primary cache 10 was especially well documented, and we offer it as an example of the ephemeral and dynamic nature of cache sites (see Vander Wall and Joyner (1998a) for another example). The ten seeds in cache 10 were gone by 8 September 1995, but we found them in five secondary caches: 125, 211, 217, 221 and 537 (Fig. 13.7). The two seeds in secondary cache 125 disappeared shortly after we first located it, but we found them cached separately in tertiary caches 389 and 420 on 30 September. The one seed in secondary cache 211 disappeared later in the autumn and was never seen again. Chipmunks cached three seeds in secondary cache 217, but they were moved within a few days. One of these seeds was found in each of two tertiary caches (297 and 299) on 13 September; the third seed disappeared. Chipmunks also placed three seeds in secondary cache 221. This cache had disappeared by 17 September. One seed from this cache was found in tertiary cache 346, one seed was found in tertiary cache

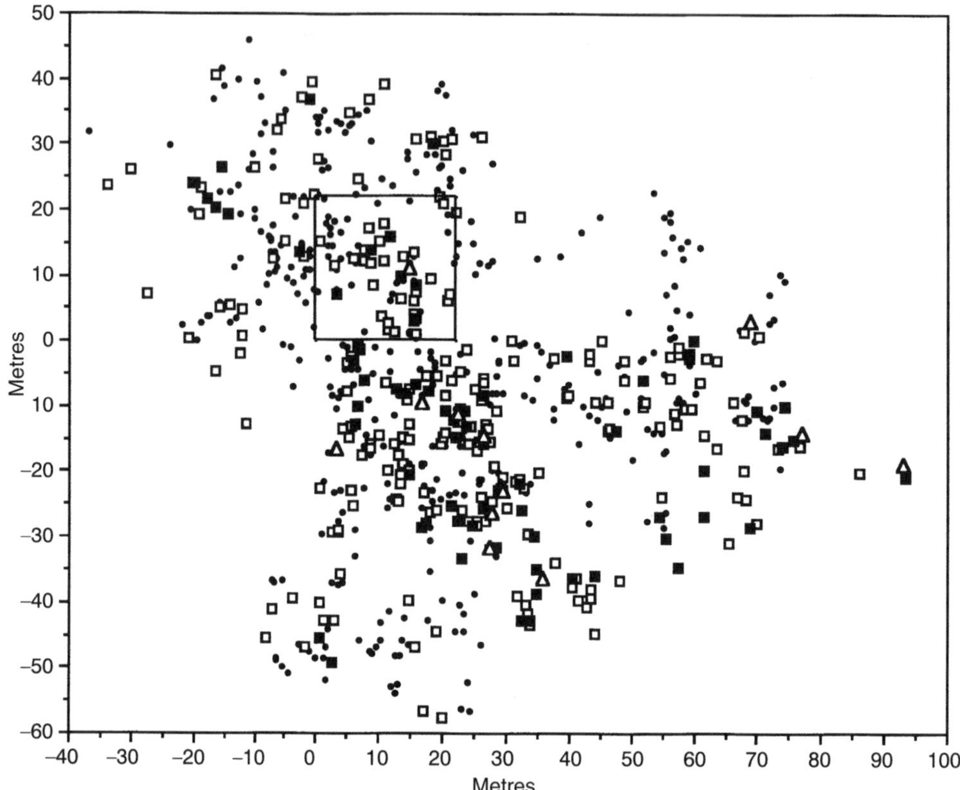

Fig. 13.5. The distribution of 678 caches around a 10 × 10 array of primary caches (large square). Closed circles are secondary caches ($n = 377$), open squares are tertiary caches ($n = 213$), closed squares are quaternary caches ($n = 75$) and open triangles are quintic caches ($n = 13$). (Data from Vander Wall and Joyner, 1998a.)

783 and the third seed disappeared. The last seed from cache 10 was not found until 13 October, when it was discovered in cache 537, along with another seed from primary cache 40. Both of these seeds disappeared from cache 537 and were found in tertiary cache 729 on 3 November. There were three quaternary caches. The single seed in cache 297 was moved to quaternary cache 682 sometime between 13 September and 28 October. The single seed in cache 299 was combined with another seed (from secondary cache 298 ~1 m away) and placed in quaternary cache 355 by 17 September. The two seeds in cache 729 were moved to quaternary cache 806 by 10 November. The only quintic cache (470) was made between 18 September and 6 November from a seed taken from quaternary cache 355 (the seed from cache 298 disappeared). This seed and the seed in quaternary cache 682 eventually germinated in the spring of 1996. The other eight seeds disappeared – seven during the autumn and one during the winter. The fate of seeds from this primary cache illustrates two important points. First, there were at least 16 caches made from the seeds in one fairly large primary cache. (We say 'at least' because some of the seeds that disappeared may have been recached outside the search area.) Secondly, there is a general tendency for seeds to move away from the source area as they get transported from cache to cache.

The number of seeds per cache was inversely correlated with cache order. Because the primary caches in this study were artificial, we obtained data on the size of naturally occurring primary caches from other studies on Jeffrey pine at the same study site (Vander Wall,

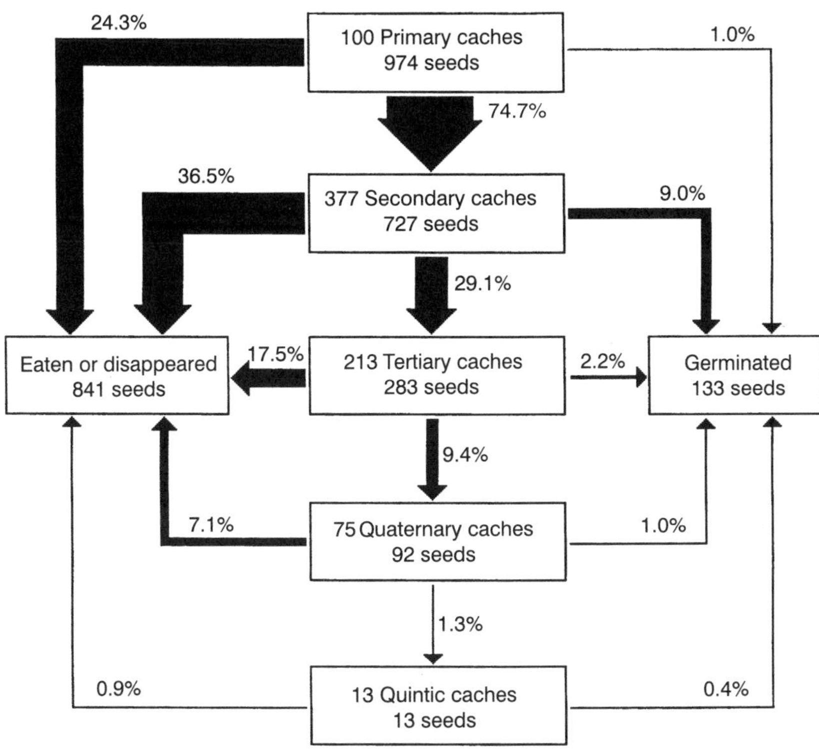

Fig. 13.6. The fate of 974 seeds from 100 primary caches. The width of the arrows is proportional to the number of seeds following a particular pathway. (From Vander Wall and Joyner, 1998a.)

1992b, 1993, 1995). In those studies, primary caches contained 3.7 ± 1.9 seeds ($n = 945$). In the study of secondary dispersal (Vander Wall and Joyner, 1998a), secondary caches contained 1.9 ± 1.1 seeds ($n = 377$), tertiary caches contained 1.3 ± 0.7 seeds ($n = 213$), quaternary caches contained 1.4 ± 0.7 seeds ($n = 75$), and quintic caches contained 1.6 ± 0.7 seeds ($n = 13$). The slight increase in the size of quintic caches may have been a consequence of a small sample size. Secondary caches contained significantly more seeds per cache than did the combined sample of tertiary, quaternary and quintic caches (unpaired *t*-test, $t = 8.174$, d.f. = 676, $P < 0.001$). This decrease in cache size as winter approached apparently occurred because of the splitting of caches and the eating of some seeds when chipmunks recovered them. For example, we found 66 caches (not including the artificial primary caches) where chipmunks excavated seeds and reburied them in two or more smaller caches. Three such cases are illustrated in Fig. 13.7. There were many other cases where a chipmunk retrieved a large cache and made a smaller cache and the other seeds were either eaten or had disappeared. When chipmunks split a cache into two or more caches, the nearest-neighbour distances between daughter caches was 10.0 ± 15.2 m (range 0.3–66.7 m; $n = 84$). Splitting of caches occurred most frequently early in the autumn. Chipmunks also took the seeds from two small caches and combined them to make larger caches, but this was less common ($n = 25$, not counting seeds from primary caches). Two cases of combining seeds in a cache are shown in Fig. 13.7.

The purpose of scatter-hoarding for yellow pine chipmunks is to set aside a reliable food source over the short term and to provide a reserve from which individuals can draw to form a winter larder in late autumn. Consequently, a certain amount of seed consumption is expected whenever chipmunks handle seeds. Table 13.1 summarizes seed consumption and disappearance as a function of cache order.

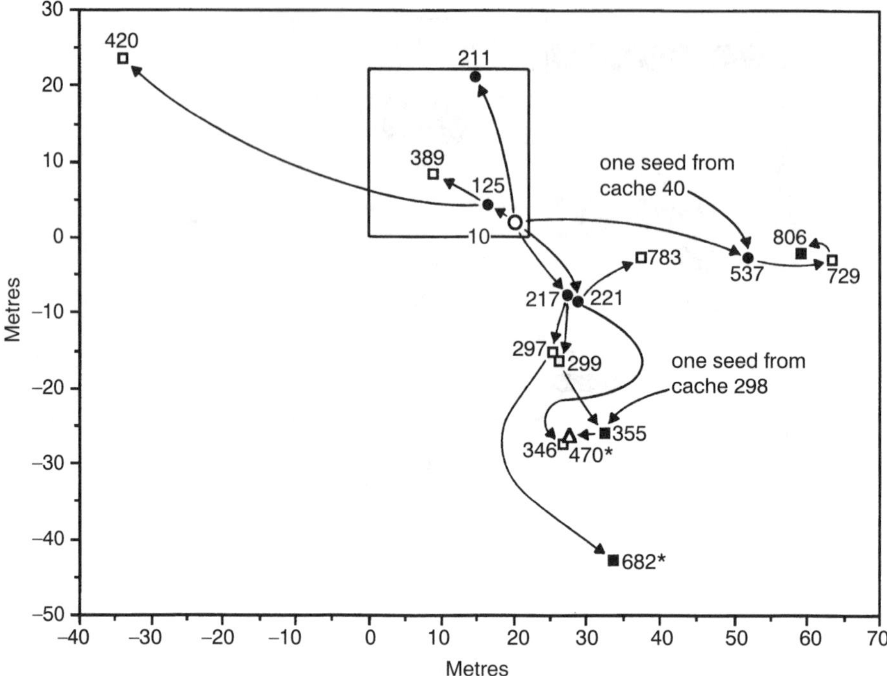

Fig. 13.7. A detailed history of primary cache 10. The large square represents the 10 x 10 array of primary caches. Symbols are the same as those defined in Fig.13.5. Arrows indicate the path of seeds from one cache site (numbers) to another. See text for further details. (Data from Vander Wall and Joyner, 1998a.)

Table 13.1. The percentage of Jeffrey pine seeds eaten or disappearing as a function of cache order.

	Percentage		
Seed origin	Eaten	Disappearing	Total
From primary caches	7.1	17.2	24.3
From secondary caches	5.2	43.7	48.9
From tertiary caches	10.2	50.2	60.4
From quaternary and quintic caches	12.4	61.9	74.3
From all caches	15.3	71.1	86.4

There was a general increase in the proportion of seeds eaten with cache order (i.e. later in the autumn). This increase may reflect a greater need for food as the mean ambient temperature decreases in the autumn. There was also a very definite increase in the proportion of seeds that disappeared with cache order. This increase probably reflects seeds leaving the survey area and chipmunks taking scatter-hoarded seeds and moving them to their winter larders (which may be too deep for us to detect with the Geiger counter).

Approximately 14% of the marked seeds survived to the time of seed germination in the spring (some seedlings actually emerged, but our survey on 19 May interrupted the process of seedling establishment, so we report only survival to the time of germination). This is probably an underestimate because seeds that rodents cached outside the surveyed area may also have germinated. Note that only 1% (of 13.6%) of seed survival was from primary caches (Fig. 13.6). If we had terminated the study of seed dispersal after initial hoarding (i.e. after formation of primary caches) and if we had assumed that cache removal is equivalent to seed mortality (as is commonly done), we would have missed 92% of the seed survival.

Why do Animals Recache Seeds?

There are several possible reasons why chipmunks and other animals might repeatedly excavate and recache seeds. Here we discuss four of these, three concerning the recaching of their own cached seeds and one addressing cache pilferage. One explanation for recaching is that it is a strategy for reducing competition at rich, ephemeral food sources. Animals could rapidly scatter-hoard food around a food source to prevent other animals from harvesting the seeds. This is called rapid sequestering (Jenkins and Peters, 1992). These dense concentrations of caches might be vulnerable to pilferers. But the cacher quickly redistributes the food to safer cache sites spaced further apart and further from the source plant. The resultant lower cache density would, theoretically, reduce cache loss to pilferers (Stapanian and Smith, 1978, 1984; Clarkson *et al.*, 1986). Rapid-sequestering behaviour predicts relatively short initial transport distances and longer distances to secondary caches. This explanation is consistent with the behaviour of some kangaroo rats and grey jays (Jenkins and Peters, 1992; Waite and Reeve, 1995), but it does not seem to explain the hoarding behaviour of yellow pine chipmunks. The reason is that the pattern of transport distances exhibited does not fit the expected pattern; the distances to subsequent cache sites are relatively short compared with the initial distance. In the secondary-hoarding experiment, for example, mean transport distance from caches in the 10×10 array (a relatively rich source of seeds) was 27.8 m and subsequent transport distance to tertiary caches was 12.5 m, to quaternary caches it was 8.8 m and to quintic caches it was 5.5 m. These data suggest that yellow pine chipmunks transport seeds to relatively safe cache sites immediately after they gather them.

Perhaps animals move seeds to monitor their condition. Many food-hoarding animals depend on stored food for sustenance during the winter. Feedback on the condition of the food may be important in surviving the winter (Bock, 1970; Koenig and Mumme, 1987; DeGange *et al.*, 1989). However, this does not appear to be the reason for chipmunks moving caches. First, seeds, in general, do not deteriorate rapidly when they are stored in the soil. Secondly, chipmunks dig up seeds and eat them at a steady rate during the late summer and autumn (Table 13.1). The ease with which they find cached seeds and the condition of the seeds they eat should provide sufficient feedback on the reserve food supply. And, thirdly, it is not necessary to move the seeds to monitor their quality.

Another possibility is that rodents and other food-hoarding animals move stored food to update their memories of storage locations. Spatial memory has been identified as an important means of cache recovery in rodents (Jacobs and Liman, 1991; Vander Wall, 1991; Jacobs, 1992; Macdonald, 1997) and birds (Vander Wall, 1982; Kamil and Balda, 1985; Sherry *et al.*, 1992; Clayton and Krebs, 1994; Bednekoff *et al.*, 1997a). Some studies have indicated that memory of cache locations appears to decay over time in birds (Hitchcock and Sherry, 1990; but see Bednekoff *et al.*, 1997b), although this has not been demonstrated for rodents. However, most studies of rodent memory have been short-term (< 3 days) and have been conducted in relatively small experimental enclosures (e.g. Jacobs and Liman, 1991; Vander Wall, 1991; Jacobs, 1992). Assuming that the memory of chipmunks and other rodents does decay over time, it may be advantageous for individuals to update their memories by revisiting cache sites. But, even if this is true, it is not clear why chipmunks have to move and recache seeds to preserve those memories. Simply visiting cache locations should be sufficient.

A fourth and, in our opinion, far more convincing explanation for recaching is that the movement of seeds from one cache site to another represents the pilfering of caches of one individual by another. All rodents can use olfaction and random digging to find cached seeds, but only the cacher has the added benefit of precise spatial memory. The individual that makes a cache has more information and, consequently, the highest probability of recovering that food item. This should lead to a situation where, whenever a foraging chipmunk encounters a cache that it did not make (or does not remember making), it should dig it up and move those seeds to a new site. By moving the caches of other rodents, the new cacher gains a recovery advantage. If this is true, it could result in a situation where rodents repeatedly pilfer cached seeds from each other. A by-product of the behaviour is multistage seed dispersal.

There is considerable evidence that pilferage of caches does occur. The most informative studies are those that follow the fates of artificial caches because, for these caches, all removal can be attributed to pilferage. Removal of Jeffrey pine seeds from artificial caches by chipmunks and other rodents ranges from 0.3 to 8.8% day^{-1} (Vander Wall and Peterson, 1996; Vander Wall, 1998). Stapanian and Smith (1978) reported removal rates of 8.5–9.4% day^{-1} for walnuts spaced 2.4 m apart, 4.8–5.5% day^{-1} for walnuts 4.6 m apart and 0.4–2.5% day^{-1} for walnuts 9.2 m apart. To put these numbers in perspective, pilferage rates of 3% day^{-1} will eliminate half the caches of an individual in less than 1 month. These and other studies demonstrate that pilferage of caches is widespread and that food-hoarding animals behave in such a way as to minimize pilferage (i.e. hiding food, spacing caches, etc.).

To test the hypothesis that the maker of a cache has the recovery advantage, we conducted an enclosure experiment in which we could control the identity of foragers (Vander Wall, 2000). We constructed five 10 m × 10 m rodent-proof enclosures with wire-mesh walls that extended ~45 cm below ground and ~75 cm above ground. We placed aluminium flashing 20 cm wide around the inside and outside of the top of the enclosure to prevent rodents from leaving or entering. We constructed these enclosures in bitterbrush habitat in Jeffrey pine-forest openings, and used them to test the hoarding and pilfering behaviour of yellow pine chipmunks and one of their close competitors, deer mice.

Experimental trials had four phases. In phase one, we placed 150 radioactively labelled pine seeds in a feeder and allowed chipmunks 1 day or deer mice up to 3 days to cache the seeds. Following hoarding, we located all caches, examined the contents, mapped the locations and replaced the seeds. In phase two, we permitted the cacher, who had knowledge or spatial memories of cache sites, to search for the caches it had made. At the end of this 3-day period, we surveyed the enclosure to determine the number of emptied, intact and new caches. At the conclusion of this phase, we re-established all caches to their positions at the end of phase one. In phases three and four, we repeated the design of phase two but with two naïve individuals. One of these trials used an individual of the same species as the original cacher, and the other trial used an individual of the other species.

We conducted 20 trials, half with yellow pine chipmunks as the initial cacher and half with deer mice as the initial cacher. Previous work (Vander Wall, 1993, 1995, 1998) has shown that the ability of rodents to pilfer caches using olfaction is dependent on the moisture content of the soil: olfaction works well under moist conditions and poorly under dry conditions. To incorporate this environmental variable, we conducted half of the trials under dry conditions (five with chipmunks and five with deer mice as the initial cacher) and half under wet conditions. To create wet conditions, we sprayed the enclosures with ~350 l of water in the afternoon before each recovery phase.

Some of the results of this experiment are illustrated in Fig. 13.8. Under wet conditions, the cacher does not have any apparent advantage in finding seeds it has stored compared with naïve foragers of either species. Olfaction works so well under moist conditions that naïve foragers are as successful in finding caches as knowledgeable foragers. However, under dry conditions naïve foragers find very few caches, while knowledgeable foragers find many caches. This suggests that under dry conditions olfaction works so poorly that memory gives the cacher a huge advantage when searching for seeds it has hidden. A three-way analysis of variance (ANOVA) found a significant forager (knowledgeable versus naïve) effect ($F_{2, 48} = 5.41$, $P = 0.008$), a significant condition (wet versus dry) effect ($F_{1, 48} = 43.11$, $P < 0.001$), and a significant forager × condition interaction ($F_{2, 48} = 16.44$, $P < 0.001$), reflecting the fact that naïve foragers improve their foraging success under wet conditions.

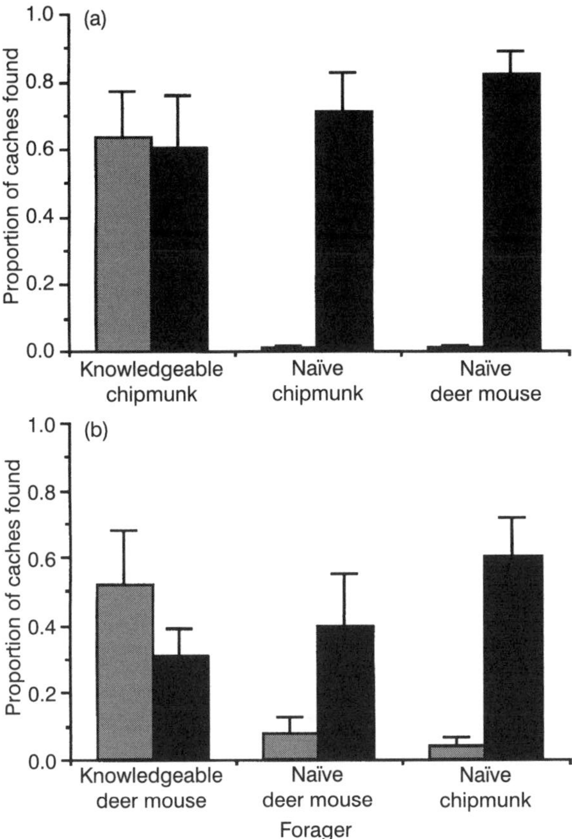

Fig. 13.8. The proportion of caches found during 3-day foraging sessions by yellow pine chipmunks and deer mice that were either knowledgeable of the cache sites or naïve to their presence. Shaded bars represent dry conditions and solid bars represent wet conditions. Vertical lines are standard errors. (a) Original caches made by yellow pine chipmunks. (b) Original caches made by deer-mice. For each treatment, $n = 5$. (Data from Vander Wall, 2000.)

This experiment demonstrates that the environment (i.e. moisture content of the soil) influences the recovery advantage of the hoarder. It appears that under dry conditions chipmunks and deer mice living in arid or semiarid environments can cache seeds with little threat of cache pilfering by other animals. Olfaction works so poorly under dry conditions that naïve foragers cannot find buried seeds effectively. This means that, under dry conditions, animals have nearly exclusive use of their cached seeds, which should promote active foraging and hoarding. But, when it rains and the seeds absorb moisture, they immediately become more apparent to naïve foragers that use olfaction. This permits those foragers to pilfer seeds stored by other individuals and recache them. As long as conditions remain moist, foragers should continue to pilfer the caches of others whenever possible. When the soil is dry again, the pilferage of cached seeds decreases to lower levels. Individuals are compelled to participate in this pilfering free-for-all, because any individual that does not do so risks losing control of its share of the stored food reserve when dry conditions return.

When the items being recached are seeds, this behaviour has potentially important implications for plant dispersal. It may be detrimental for the plants, because every time seeds are handled by animals there is a chance the animals will eat some of the seeds or move them into a larder to be eaten later (Table 13.1). But there are also potential advantages.

Scatter-hoarding animals hide food in widely spaced caches to reduce pilferage (Stapanian and Smith, 1978; Clarkson *et al.*, 1986). Repeated hoarding results in a more even distribution of caches throughout the environment (Fig. 13.3). This is because scatter-hoarding rodents and birds appear to be very sensitive to the local distribution of cached food (Brodin, 1994; Waite and Reeve, 1994, 1995; Jokinen and Suhonen, 1995; Vander Wall, 1995) and move seeds away from dense concentrations to areas more sparsely populated with caches. The distance distribution of secondary and subsequent caches (Fig. 13.3c) is one measure of this. A more even distribution may increase the probability that seeds will be buried in safe sites. This tendency is in marked contrast to the clumped distributions often created by frugivores (Wenny and Levey, 1998; see also Schupp *et al.*, this volume). In addition, recaching may move some seeds further away from the source tree (Fig. 13.7).

In summary, Jeffrey pine seeds are dispersed by a complicated multistage process that begins with wind dispersal. Seeds are then rapidly gathered and cached by rodents, especially yellow pine chipmunks, and later shuffled from place to place by chipmunks that are trying to gain the upper hand in eventual cache recovery from each other. In the process, many seeds are eaten or carried into larders deep below ground, where they will eventually be eaten or die. But a significant number of seeds gets moved to highly favourable establishment sites, where they survive and produce saplings. This 'model' of seed dispersal is probably much more common than we currently realize. Most nuts, acorns and large seeds that are scatter-hoarded by animals are probably moved several times before they encounter their ultimate fate (i.e. germination, consumption or death). It is important to understand these seed movements if we hope to comprehend fully the inter- and intraspecific interactions of granivorous rodents and birds and how such seed-caching systems might have evolved. It is equally important to understand the details of these seed-fate pathways if we are to appreciate how competition for scatter-hoarded seeds influences plant establishment.

Acknowledgements

Ted Thayer, Jenny Hodge, Jamie Joyner, Maurie Beck, Julie Roth, Lisa Crampton, Jennifer Hollander, Jennifer Armstrong, Joe Veech, Liza Bitton, Mark Boon, Janene Auger, Allison Jones and Erin Vander Wall provided valuable assistance in the field. Funding was provided by National Science Foundation grants DEB-9306369, DEB-9707098 and DEB-9708155. We thank the Whittell Board of Control for permission to conduct these studies in the Whittell Forest and Wildlife Area.

References

Bednekoff, P.A., Balda, R.P., Kamil, A.C. and Hile, A.G. (1997a) Long-term spatial memory in four seed-hoarding corvid species. *Animal Behaviour* 53, 335–341.

Bednekoff, P.A., Kamil, A.C. and Balda, R.P. (1997b) Clark's nutcracker (Aves: Corvidae) spatial memory: interference effects on cache recovery performance. *Ethology* 103, 554–565.

Bock, C.E. (1970) The ecology and behavior of the Lewis' woodpecker (*Asyndesmus lewis*). *University of California Publications in Zoology* 92, 1–91.

Bohning-Gaese, K., Gaese, B.H. and Rabemanantsoa, S.B. (1999) Importance of primary and secondary seed dispersal in the Malagasy tree *Commiphora guillaumini*. *Ecology* 80, 821–832.

Broadbooks, H.E. (1958) Life history and ecology of the chipmunk, *Eutamias amoenus*, in eastern Washington. *University of Michigan, Museum of Zoology, Miscellaneous Publications* 103, 1–48.

Broadbooks, H.E. (1970) Home ranges and territorial behavior of the yellow-pine chipmunk, *Eutamias amoenus*. *Journal of Mammalogy* 51, 310–326.

Brodin, A. (1994) The disappearance of caches that have been stored by naturally foraging willow tits. *Animal Behaviour* 47, 730–732.

Chambers, J.C. and MacMahon, J.A. (1994) A day in the life of a seed: movements and fates of seeds and their implications for natural and managed systems. *Annual Review of Ecology and Systematics* 25, 263–292.

Chambers, J.C., MacMahon, J.A. and Haefner, J.H. (1991) Seed entrapment in alpine ecosystems: effects of soil particle size and diaspore morphology. *Ecology* 72, 1668–1677.

Clarkson, K., Eden, S.F., Sutherland, W.J. and Houston, A.I. (1986) Density dependence and magpie food hoarding. *Journal of Animal Ecology* 55, 111–121.

Clayton, N.S. and Krebs, J.R. (1994) Memory for spatial and object-specific cues in food-storing and non-storing birds. *Journal of Comparative Physiology A* 174, 371–379.

DeGange, A.R., Fitzpatrick, J.W., Layne, J.N. and Woolfenden, G.E. (1989) Acorn harvesting by Florida scrub jays. *Ecology* 70, 348–356.

Forget, P.-M. and Milleron, T. (1991) Evidence for secondary seed dispersal by rodents in Panama. *Oecologia* 87, 596–599.

Greene, D.F. and Johnson, E.A. (1989) A model of wind dispersal of winged or plumed seeds. *Ecology* 70, 339–347.

Hitchcock, C.L. and Sherry, D.F. (1990) Long-term memory for cache sites in the black-capped chickadee. *Animal Behaviour* 40, 701–712.

Hoshizaki, K., Suzuki, W. and Nakashizuka, T. (1999) Evaluation of secondary dispersal in a large-seeded tree *Aesculus turbinata*: a test of directed dispersal. *Plant Ecology* 144, 167–176.

Jacobs, L.F. (1992) Memory for cache locations in Merriam's kangaroo rats. *Animal Behaviour* 43, 585–593.

Jacobs, L.F. and Liman, E.R. (1991) Grey squirrels remember the locations of buried nuts. *Animal Behaviour* 41, 103–110.

Jenkins, S.H. and Peters, R.A. (1992) Spatial patterns of food storage by Merriam's kangaroo rats. *Behavioral Ecology* 3, 60–65.

Jokinen, S. and Suhonen, J. (1995) Food hoarding by willow and crested tits: a test of scatter hoarding models. *Ecology* 76, 892–898.

Kamil, A.C. and Balda, R.P. (1985) Cache recovery and spatial memory in Clark's nutcracker (*Nucifraga columbiana*). *Journal of Experimental Psychology: Animal Behavior Processes* 11, 95–111.

Koenig, W.D. and Mumme, R.L. (1987) *Population Ecology of the Cooperatively Breeding Acorn Woodpecker.* Princeton University Press, Princeton, New Jersey.

Levey, D.J. and Byrne, M.M. (1993) Complex ant–plant interactions: rain forest ants as secondary dispersers and post-dispersal seed predators. *Ecology* 74, 1802–1812.

Macdonald, I.M.V. (1997) Field experiments on duration and precision of grey and red squirrel spatial memory. *Animal Behaviour* 54, 879–891.

Matlack, G.R. (1989) Secondary dispersal of seed across snow in *Betula lenta*, a gap-colonizing tree species. *Journal of Ecology* 77, 853–869.

Saigo, B.W. (1969) The relationship of non-recovered rodent caches to the natural regeneration of ponderosa pine. Masters thesis, Oregon State University, Corvallis.

Sherman, R.J. and Chilcote, W.W. (1972) Spatial and chronological patterns of *Purshia tridentata* as influenced by *Pinus ponderosa*. *Ecology* 53, 294–298.

Sherry, D.F., Jacobs, L.F. and Gaulin, S.J.C. (1992) Spatial memory and adaptive specialization of the hippocampus. *Trends in Neurosciences* 15, 298–303.

Sone, K. and Kohno, A. (1999) Acorn hoarding by the field mouse, *Apodemus speciosus* Temminck (Rodentia: Muridae). *Journal of Forest Research* 4, 167–175.

Stapanian, M.A. and Smith, C.C. (1978) A model for seed scatterhoarding: coevolution of fox squirrels and black walnuts. *Ecology* 59, 884–896.

Stapanian, M.A. and Smith, C.C. (1984) Density-dependent survival of scatterhoarded nuts: an experimental approach. *Ecology* 65, 1387–1396.

Vander Wall, S.B. (1982) An experimental analysis of cache recovery in Clark's nutcracker. *Animal Behaviour* 30, 84–94.

Vander Wall, S.B. (1991) Mechanisms of cache recovery in yellow pine chipmunks. *Animal Behaviour* 41, 851–863.

Vander Wall, S.B. (1992a) Establishment of Jeffrey pine seedlings from animal caches. *Western Journal of Applied Forestry* 7, 14–20.

Vander Wall, S.B. (1992b) The role of animals in dispersing a 'wind-dispersed' pine. *Ecology* 73, 614–621.

Vander Wall, S.B. (1993) Cache site selection by chipmunks (*Tamias* spp.) and its influence on the effectiveness of seed dispersal in Jeffrey pine (*Pinus jeffreyi*). *Oecologia* 96, 246–252.

Vander Wall, S.B. (1994) Removal of wind-dispersed pine seeds by ground-foraging vertebrates. *Oikos* 69, 125–132.

Vander Wall, S.B. (1995) Sequential patterns of scatter hoarding in yellow pine chipmunks. *American Midland Naturalist* 133, 312–321.

Vander Wall, S.B. (1998) Foraging success of granivorous rodents: effects of variation in seed and soil water on olfaction. *Ecology* 79, 233–241.

Vander Wall, S.B. (2000) The influence of environmental conditions on cache recovery and cache pilferage by yellow pine chipmunks (*Tamias amoenus*) and deer mice (*Peromyscus maniculatus*). *Behavioral Ecology* 11, 544–549.

Vander Wall, S.B. and Joyner, J.W. (1998a) Recaching of Jeffrey pine (*Pinus jeffreyi*) seeds by yellow pine chipmunks (*Tamias amoenus*): potential effects on plant reproductive success. *Canadian Journal of Zoology* 76, 154–162.

Vander Wall, S.B. and Joyner, J.W. (1998b) Secondary dispersal by the wind of winged pine seeds across the ground surface. *American Midland Naturalist* 139, 365–373.

Vander Wall, S.B. and Peterson, E. (1996) Associative learning and the use of cache markers by yellow pine chipmunks (*Tamias amoenus*). *Southwestern Naturalist* 41, 88–90.

Vander Wall, S.B., Thayer, T.C., Hodge, J.S., Beck, M.J. and Roth, J.R. (2001) Scatter-hoarding behavior of deer mice (*Peromyscus maniculatus*). *Western North American Naturalist* 61, 109–113.

Waite, T.A. and Reeve, J.D. (1994) Cache dispersion by gray jays sequentially exploiting adjacent sources. *Animal Behaviour* 47, 1232–1234.

Waite, T.A. and Reeve, J.D. (1995) Source-use decisions by hoarding gray jays: effects of local cache density and food value. *Journal of Avian Biology* 26, 59–66.

Wenny, D.G. and Levey, D.J. (1998) Directed seed dispersal by bellbirds in a tropical cloud forest. *Proceeding of the National Academy of Sciences USA* 95, 6204–6207.

Westoby, M. (1981) A note on combining two methods of dispersal-for-distance. *Australian Journal of Ecology* 6, 189–192.

Wicklow, D.T., Kumar, R. and Lloyd, J.E. (1984) Germination of blue grama seeds buried by dung beetles (Coleoptera: Scarabaeidae). *Environmental Entomology* 13, 878–881.

14 The Role of Seed Size in Dispersal by a Scatter-hoarding Rodent

Patrick A. Jansen,[1] Martijn Bartholomeus,[2] Frans Bongers,[1] Jelmer A. Elzinga,[2,*] Jan den Ouden[1] and Sipke E. Van Wieren[2]

[1]*Silviculture and Forest Ecology Group, Department of Environmental Sciences, Wageningen University, PO Box 342, NL-6700 AH Wageningen, The Netherlands;* [2]*Tropical Nature Conservation and Vertebrate Ecology Group, Department of Environmental Sciences, Wageningen University, Bornsesteeg 69, NL-6708 PD Wageningen, The Netherlands*

Introduction

Many species of birds and mammals hoard food from locally abundant, ephemeral sources to conserve it for future use (Vander Wall, 1990). The hoarding strategies they follow cover a range bounded by two extremes. 'Larder-hoarders' store their food in one or few caches, each containing much food. 'Scatter-hoarders', in contrast, store food by dispersing it in small amounts among many spatially separated caches. Where larder-hoarding requires active defence or some other mechanism to prevent robbery of the food by competitors, scatter-hoarding relies on spreading the risk of robbery (Vander Wall, 1990). This chapter is about the long-term effects of scatter-hoarding seeds in the upper layer of soil, as is practised by a variety of granivorous birds and mammals.

Scatter-hoarding of seeds has potential advantages to plants. First, it involves transport away from the parent plant, an area that is not only occupied already by the species, but is also often hostile for seeds and seedlings because of pathogens and herbivores associated with adult conspecifics (Janzen, 1970; Hammond and Brown, 1998). This transport is the dispersal needed for the colonization of new sites. Secondly, scatter-hoarding takes seeds away from an area in which they are concentrated and scatters them throughout the surrounding area, isolating individual seeds or small groups of seeds from their siblings. Scattering increases the independence of fates among individual seeds. It reduces the risk of density-dependent mortality, such as consumption by wild pigs and oviposition by granivorous insects (Wilson and Janzen, 1972). It also increases the probability of seeds hitting suitable sites if such sites are patchily distributed in the environment. Thirdly, burial of seeds in the topsoil reduces the probability of other seed-eaters finding and killing the seed

*Present address: Centre for Terrestrial Ecology, Netherlands Institute of Ecology, PO Box 40, NL-6666 ZG Heteren, The Netherlands.

 Seed Dispersal and Frugivory: Ecology, Evolution and Conservation (eds D.J. Levey, W.R. Silva and M. Galetti)

(Stapanian and Smith, 1984; Vander Wall, 1993) and often preserves seeds in better condition for germination and establishment (see Vander Wall, this volume).

The main reason that the advantages of scatter-hoarding are considered potential is that scatter-hoarding has a price: scatter-hoarding animals often recover most hoarded seeds for consumption. Thus, seeds must escape the hoarder in order to profit from scatter-hoarding. For the parent plant, scatter-hoarding is advantageous if the benefits of some seeds surviving outweigh the costs of all others being eaten, i.e. if scatter-hoarding increases the total number of seeds establishing.

Many plant species seem to depend on scatter-hoarding animals for seed dispersal (Vander Wall, 1990). These species, mostly trees, produce larger, more nutritious seeds and in smaller numbers than most plants with other dispersal modes (Smith and Reichman, 1984; Vander Wall, 1990; Leishman *et al.*, 1995; Hammond *et al.*, 1996). Also, large-seeded plants tend to be more seasonal in fruit production than smaller-seeded species and synchronously mature fruit when scatter-hoarding peaks (e.g. Smythe, 1970; Jackson, 1981). These observations have led to the hypothesis that these plant species are adapted to scatter-hoarding: their reproductive strategy has been shaped by scatter-hoarding animals over evolutionary time (Smythe, 1970; Smith and Reichman, 1984; Hallwachs, 1994).

The adaptive explanation of the scatter-hoarding dispersal syndrome is appealing. Many experiments have shown that seed mass, a central characteristic of plant reproductive strategy, is indeed heritable and sensitive to selection (e.g. Cober *et al.*, 1997; Gjuric and Smith, 1997; Malhotra *et al.*, 1997) and so is seed nutrient composition (e.g. Brandle *et al.*, 1993; Rebetzke *et al.*, 1997). Yet the hypothesis of adaptation is difficult to test. Fossil records can be used to determine whether seed size has increased over evolutionary time (e.g. Eriksson *et al.*, 2000). Vander Wall (2001), for example, concludes from fossil records that tree genera currently dispersed by scatter-hoarding animals in the temperate zone have indeed increased seed size since the Palaeocene (~60 million years ago). Still, whether such increases occurred in response to scatter-hoarding animals cannot be determined. What we can do, however, is test the underlying assumption that scatter-hoarding animals impose selective pressure on reproductive traits (in particular, seed size) through their behaviour. Such pressure is a prerequisite for selection towards the production of larger seeds.

Selection towards large seeds

There are several ways in which scatter-hoarding animals could select for larger, more nutritious seeds, all of which assume it is more economical for scatter-hoarders to create and manage a smaller supply of large, nutritious (i.e. high-value) seeds rather than a large supply of small, less nutritious seeds.

First, the likelihood of scatter-hoarders encountering a seed may increase with seed value, because animals may discover high-value seeds more easily or because they may focus their foraging on known sources of high-value seeds. Secondly, high-value seeds may have higher removal rates, because rodents prefer them. Such a preference could give high-value seeds a greater probability of being harvested at all (see Waite and Reeve, 1995). Thirdly, the decision of what to do with a seed after it is harvested may depend on seed value. Scatter-hoarding animals usually eat a few seeds during the process of hoarding many others (see Forget *et al.*, this volume). The seeds they store may preferentially be high-value seeds, while low-value seeds are eaten (Hallwachs, 1994). Fourthly, scatter-hoarders may vary how they cache a seed, depending on its value (Stapanian and Smith, 1978, 1984; Clarkson *et al.*, 1986). High-value seeds could thus get stored in more favourable conditions than low-value seeds (i.e. further away, in lower densities, deeper or at better sites). Finally, the storage life of caches could depend on seed value. Scatter-hoarders could use their high-value seed caches at slower rates than low-value caches, saving highest-value seeds for last. Alternatively, high-value seeds could simply be used at lower rates because they have been cached at lower densities and greater depths and are therefore found less easily by both the cache owner and cache thieves (Stapanian and

Smith, 1984; Vander Wall, 1993). In both cases, the result would be higher survival for high-value seeds.

This study

We conducted a field experiment to test possible selective pressures towards higher seed value by scatter-hoarders. We measured how seed value influenced the fate (death or establishment) of seeds offered to the red acouchy (*Myoprocta exilis*, Wayler, 1831), a neotropical rodent that scatter-hoards seeds by burying them singly. A second aim was to quantify survival probabilities of seeds harvested and cached by acouchies, because direct proof for scatter-hoarding being beneficial to plants is still remarkably scarce (Jansen and Forget, 2001).

We tested the following hypotheses:

1. Large seeds are more likely to be harvested by acouchies than small seeds.
2. Large seeds are harvested by acouchies more quickly than small seeds.
3. Large seeds harvested by acouchies have a higher probability than small seeds to be cached rather than eaten.
4. Large seeds are cached further away and in lower densities than small seeds.
5. Large seeds are recovered from caches and consumed at lower rates than small seeds.
6. Large seeds have higher survival probabilities than small seeds.

Methods

Site and species

We worked at the Nouragues biological station in the Nouragues Reserve, an undisturbed lowland rain-forest site in French Guiana, 100 km south of Cayenne, at 4° 02′ N and 52° 42′ W and 100–150 m above sea level. Annual precipitation averages 2900 mm, with peaks in December–January and April–July. The main fruiting season is from February to May (Sabatier, 1985). Bongers *et al.* (2001) give an extensive description of the site.

To isolate the effect of seed value from other seed characteristics that may influence preferences of animals, such as nutrient composition, secondary compounds, digestibility and odour (Hurly and Robertson, 1986), we took advantage of intraspecific variation in fresh seed mass. Seed wet mass is a good measure of seed value. The nutrient content of seeds is not proportional to seed mass, but larger seeds do contain a larger absolute amount of nutrients and have proportionally less (inedible) seed-coat (Grubb and Burslem, 1998).

We used seeds of *Carapa procera* (*Meliaceae*), henceforth '*Carapa*', a canopy tree species reaching up to 25 m height, occurring throughout the neotropics. *Carapa* produces up to 100 large (*c.* 10 cm diameter), five-valved fruits that contain up to 20 large, fatty seeds (Forget, 1996; Jansen and Forget, 2001). Seeds are shed gradually in February–July, but mostly in May, a period of intense seed-hoarding (see Forget *et al.*, this volume). Fruits burst open upon hitting the ground, scattering seeds under the parent tree. Fresh masses of ripe seeds at Nouragues span a more than 20-fold range, from 3 to 65 g (mean = ~21 g).

Carapa seeds are sought after by acouchies, especially in lean years (Forget, 1996; Jansen and Forget, 2001). The red acouchy, *Myoprocta acouchy* (Erxleben, 1777), is the most common scatter-hoarding animal in French Guiana. It is a caviomorph rodent, 33–39 cm length, and weighs 1.0–1.5 kg (Emmons and Feer, 1990). Acouchies store seeds, their main food, by burying them in shallow caches in the topsoil. They are perfect scatter-hoarders, because they harvest seeds one by one and store them in single-seeded caches. This behaviour led Morris (1962) to introduce the term scatter-hoarding. Acouchies, like agoutis (*Dasyprocta*), scatter-hoard seeds and fruits from many plant species and are therefore considered important seed-dispersers. Acouchies are diurnal, with peak activity at dawn and dusk. They have territories of about 1 ha (Dubost, 1988).

Seed removal

Between 19 April and 24 May 1999, during the peak hoarding season, we established 11 feeding plots of 60 cm × 60 cm below or near reproductive *Carapa* trees. Plots were separated by > 100 m, the average acouchy territory radius according to Dubost (1988), to ensure replication with different individuals. On each plot, we placed 49 *Carapa* seeds that varied widely in fresh mass. Seed masses did not differ between plot samples (Kruskal–Wallis test: χ^2_{10} = 14.0, P = 0.18). Our samples had a higher average seed mass than a random sample taken from the same population in 1995 (analysis of variance (ANOVA): $F_{1,420}$ = 31.4, P < 0.001). While this random sample was skewed to larger seed values (skewness g_1 = 0.68 with SE = 0.08; kurtosis g_2 = 1.02), our sample was approximately normally distributed (g_1 = −0.17 with SE = 1.11; g_2 = 0.14). Seeds were collected a few days before placement from the local population of reproductive *Carapa* (~25 trees). Seeds were weighed, thread-marked (see below), given a number for identification and randomly assigned to positions in a 7 × 7 grid in each plot. Varying seed value within plots enabled us to control for differences in detectability associated with seed size and to account for differences between animals and sites. An animal visiting a plot would simultaneously encounter the entire range of seed values and make its choice among them.

We used an automatic video system to observe selection and removal of seeds without the influence of our presence. We monitored visitation, seed selection and seed removal at 3 frames s^{-1}, using a surveillance camera (Philips VCM 6250/00T) and a time-lapse video recorder (Panasonic AG-1070 DC), mounted on a tree at ~1.5 m. Plots and video equipment were set up at night. Automatic recording took place in daylight during the following 1.5 days, during which time practically all seeds were removed. From the videotapes we recorded the identity of the animal taking each seed and the seeds' order of removal. Recordings were not always complete, due to power problems. One plot completely lacked recordings, due to a defect.

Seed fate

Acouchies carry seeds over large distances and bury them with practically no trace. To be able to retrieve seeds, we attached 1 m of fluorescent green fishing-line with 8 cm of fluorescent pink flagging tape at the end. Acouchies buried these marked seeds but not the line or flagging. Flagged lines protruding from the soil made cached seeds visible up to 10 m. Numbers on the thread marks allowed us to identify seeds without disturbing the cache. Marking seeds in this way is not thought to influence caching behaviour (Forget, 1990), but we cannot rule out the possibility that it influenced our estimate of survival. We believe that any such influence would probably decrease survival, thereby generating conservative estimates of survival.

We searched for seeds immediately after plot depletion, 1.5–2 days after establishment of the plots. We attempted to retrieve all seeds, but stopped after 12 h of searching. We mapped all seed locations, using coordinates of labelled trees (Poncy *et al.*, 2001). Sites of cached seeds were marked with small tags at eye-level on nearby saplings or palms. Flags of caches were covered with leaves to avoid their use by other animals to find caches.

We checked caches at 2, 4, 8, 16, 32, 64 days and 4 months after installation of the plot. We resurveyed the entire area at 32 and 64 days to relocate seeds that had disappeared from caches, because such seeds are often recached rather than eaten (Jansen and Forget, 2001; see also Hoshizaki and Hulme, this volume; Vander Wall, this volume). Whether recaching was done by the cache owner or by a competing acouchy or agouti could not be determined, but this is of no importance from the perspective of the seed. Likewise, we could not determine whether caches were depleted by scatter-hoarding rodents or by peccaries (*Tayassu* spp.), unless seeds had been recached.

The seed fates we distinguished were: still cached, eaten, (re)cached, mark lost (i.e. thread mark separated from seed) and not found. We also noted germination and looked for remains of seedlings around depleted caches. Both acouchies and agoutis sever the epicotyl (including the meristem) when digging up germinated seeds (Jansen and Forget, 2001). Seeds treated in this way cannot form a new seedling, but they do form scar tissue and stay alive for many months as 'zombies'. In this way, acouchies transform rapidly germinating *Carapa* seeds into a non-perishable food suitable for long-term storage. Since *Carapa* seeds normally germinate within a few weeks, we considered seeds that disappeared from caches between day 64 and 4 months as dead.

Analyses

Effects of seed mass were tested using regression techniques and ANCOVA, in which we treated distance, masses and survival as continuous (co)variables and seed fate as a categorical variable. Because the area surrounding plots was unlimited, we could not calculate cache densities. Instead, we calculated 'cache isolation': the median distance to caches from the plot of the seed's origin. This measure behaves as the inverse of cache density and is less sensitive to neighbouring caches being overlooked (for instance, because seeds lost their thread mark) than nearest-neighbour distance.

Distances were $\log_{10}$-transformed to attain normality, except in quantile regressions (see below). Survival time was $\log_2$-transformed to obtain time steps of uniform size. Seed fresh mass was $\log_{10}$-transformed, unless stated otherwise, because we assumed that the importance for rodents of a given increase in mass would be greater in light seeds than in heavy seeds.

Fates of seeds within plots were not independent, especially in the harvesting phase, when one animal handled all seeds in a short time span; what happened to one seed had consequences for what happened to others in the same plot. We used plots, not seeds, as experimental units for all situations in which strong dependence occurred. We used regression techniques to deduce one value per plot describing the trend within each plot: the regression coefficient β (or B) weighted by the inverse square of its standard error (SE),

$$\text{weighted } \beta_i = \beta_i \times \text{SE}_i^{-2} \times \frac{1}{n}\sum_{i=1}^{n} \text{SE}_i^{-2}$$

where i is the plot number and n is the number of plots. These values were tested against $\beta = 0$ using the t-statistic at $\alpha = 0.05$. Overall trends were calculated as the mean of all plot values (G. Gort, Wageningen, 2000, personal communication).

Analyses of post-dispersal survival were done using logistic and Cox proportional-hazard regression models, mostly on pooled data, because sample sizes within plots were too small for plot-level analysis. There may have been some dependence of post-dispersal seed fate within replicates, but we feel that this should not be too problematic for post-dispersal trends, because the foraging decisions concerned are separated in time and were probably made by more than one animal.

We used quantile regression (Koenker and Bassett, 1978) to investigate heteroscedastic variation in errors. Quantile regression gives a more complete picture of the data distribution than common (least-squares) regression and is more robust against the influence of outliers. It can help recognize limiting factors and estimate their effect (Scharf *et al.*, 1998; Cade *et al.*, 1999). We calculated regression quantiles using the least absolute deviation (LAD) quantile regression procedure in BLOSSOM (Midcontinent Ecological Science Center, 1998). The significance of quantile regression factors was tested using rank-score tests, in which the statistic $T_{observed}$ was tested against a distribution of T obtained from 5000 permutations (Slauson *et al.*, 1994). All other analyses were performed with the SPSS 10.0.5 statistical package (SPSS, 1999).

Results

Seed removal

All seeds were removed within 1.5 days, except for three seeds, which were removed 1–2 days later (Table 14.1). Video recordings showed that 50 seeds were taken by white-lipped peccaries (*Tayassu pecari*, 25 seeds), collared peccaries (*Tayassu tajacu*, 13 seeds) and red-rumped agouti (12 seeds). These seeds were excluded from further analyses. We treated all of the remaining 489 seeds as removed by acouchies, including 113 without video proof.

Because all seeds were removed, hypothesis 1 was not confirmed: large seeds did

Table 14.1. Fates of *Carapa procera* seeds harvested by red acouchies (*Myoprocta exilis*) in French Guiana.

	Plot											
	1	2	3	4	5	6	7	8	9	10	11	Total
Initial seed fate												
Cached	40	23*	37	23	17	4	39	44	20	23	34	304
Eaten	1	7	8	1	11	5	4	1	7	6	7	58
Mark lost	1	1	2	1	12	11	1	2	0	0	0	31
Not found	7	18	2	24	9	4	5	2	10	4	8	93
Total (*n*)	49	49	49	49	49	24†	49	49	37†	33†‡	49	486
Initial fate of seeds recovered from primary caches												
Recached	3	3	8	7	4		6	6	3		5	45
Eaten	4	3	11	7	6	2	18	25	5	17	26	124
Not found	32	16*	16	8	3	2	8	7	7	5	3	107
Mark lost	1		2	1	4		6	6	5			25
Not recovered		1§					1			1		3
Total (*n*)	40	23	37	23	17	4	39	44	20	23	34	304
Cache half-life (days)	5	3	6	5	21	23	6	18	4	22	23	9
Ultimate seed fate												
'Survived'		1§	1	1	3		1			1		8
Eaten	5	13	23	16	21	7	27	27	14	27	42	222
Incomplete record	36	18*	21	16	8	2	17	19	16	5	5	163
No record	8	17	4	16	17	15	4	3	7	3	2	96
Total (*n*)	49	49	49	49	49	24†	49	49	37†	36†	49	489

*Includes one seed of unknown weight.
†Not including 50 seeds that were taken by peccaries or agoutis.
‡Not including three seeds that were removed later than the first census.
§The seed had established a seedling at 4 months.

not have a greater probability of being taken by acouchies. The exact sequence of removal (hypothesis 2) was also independent of seed size (*t*-test on weighted linear regression coefficients for ten plots: $t_9 = 0.55$, $P = 0.593$; does not include one plot lacking video data). The exact sequence of removal was strongly related to the position of seeds within the plot: edge, second row or centre (ANOVA: $n = 376$, $F_{2,373} = 37.0$, $P < 0.001$). Even when we pooled data and controlled for position, there was no effect of seed size on sequence of removal (ANCOVA: $n = 375$, $F_{1,372} = 0.15$, $P = 0.70$). The animals spent little time selecting seeds (8 s on average), usually taking the first seed they encountered.

Hoarding versus consumption

We located 362 seeds the day after dispersal. The vast majority (84%) was cached; relatively few seeds (16%) were eaten (Table 14.1). The probability of being cached versus being eaten increased with seed mass at the population level, which confirms hypothesis 3 (Fig. 14.1; logistic regression: $n = 303$ cached, 58 eaten, Wald = 17.0, $P < 0.001$). The probability of being cached also depended on the order of removal: the later seeds were taken from a plot, the lower the probability of being cached ($n = 226$ cached, 46 eaten, Wald = 3.9, $P = 0.049$). Both effects, however, were not significant at the plot level, due to small numbers of eaten seeds.

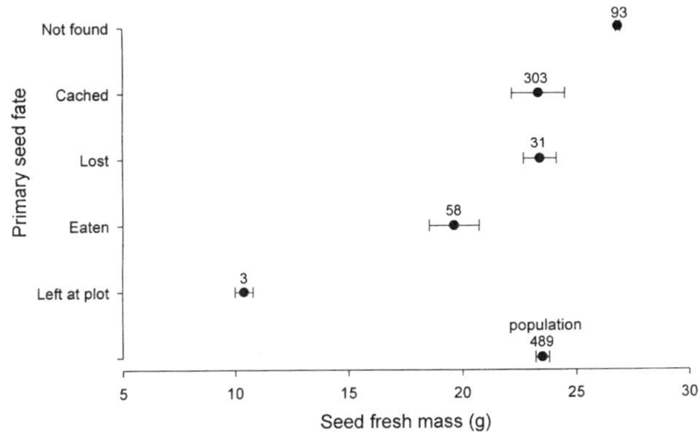

Fig. 14.1. Effect of seed fresh mass on treatment of 488 *Carapa procera* seeds by red acouchies (*Myoprocta exilis*). Data from 11 plots are pooled. Mean values (± SE) of seed fresh mass for different fates at 1 day after dispersal, and sample sizes. Seeds found eaten had lower masses (*t*-test: $n = 58$, $t_{57} = -3.5$, $P = 0.001$), while seeds that were not found at all were heavier (*t*-test: $n = 93$, $t_{92} = 4.7$, $P < 0.001$) than the population mean.

Cache spacing

The distance at which we found acouchy caches immediately after dispersal was highly variable, both within and among plots. The nearest cache was found < 1 m from the plot, the furthest was 124 m away. Cache distance increased with seed size, as predicted by hypothesis 4 (*t*-test on weighted linear regression coefficients for ten plots: $t_9 = 7.7$, $P < 0.001$; does not include a plot that was largely depleted by peccaries and hence had few caches). Most variation in cache distance, however, was not explained by seed mass (Fig. 14.2).

There was also great variation in spatial isolation of acouchy caches, both within and among plots. Larger seeds were more widely spaced than smaller seeds, in support of hypothesis 4. Both cache isolation (*t*-test on weighted linear regression coefficients for ten plots: $t_9 = 5.5$, $P < 0.001$; does not include plot depleted by peccaries) and nearest-neighbour distance ($t_9 = 5.0$, $P < 0.001$) increased with seed mass. However, there was no positive effect of seed mass on cache isolation when we took into account cache distance ($t_9 = -1.9$, $P < 0.096$), not even in the pooled data (multiple regression: $\beta_{distance} = 0.81$, $F_{1,300} = 883$, $P < 0.001$; $\beta_{mass} = -0.25$, $F_{1,300} = 10.9$, $P = 0.001$; model $F_{2,300} = 271$, $P < 0.001$, $R^2 = 0.75$). Apparently, wider spacing was a direct result of further dispersal of larger seeds without an additive effect of seed mass.

Cache exploitation

The lifetime of caches was highly variable. While many caches were depleted within a week, some were still intact 4 months after dispersal. Cache lifetime increased with isolation (Fig. 14.3; Cox regression: $n = 295$ depleted, nine censored, Wald = 9.7, $P = 0.002$). Cache lifetime also increased with cache distance from the source (Wald = 6.0, $P = 0.015$), but the distance to source did not explain variation additional to cache isolation. Seed mass did not affect cache lifetime (Wald = 1.0, $P = 0.32$), contradicting hypothesis 5.

Recovery of seeds, however, did not necessarily lead to seed consumption. Many seeds were recached or were not found again (Table 14.1). Larger seeds had a higher probability of being recached rather than eaten (logistic regression: $n = 124$ eaten, 45 recached, Wald = 4.4, $P = 0.035$). Consumed seeds were found much closer to the cache site than recached seeds (means = 12 m and 31 m, respectively; ANOVA: $F_{1,167} = 25$, $P < 0.001$). This and the fact that the probability of losing seeds increased with distance, due to our inability to search thoroughly at greater distances, suggest that most of the seeds not found were actually recached at greater distances. If we assume that lost seeds were indeed recached, the probability of escaping immediate consumption increased with seed mass more strongly (logistic regression: $n = 124$ eaten, 151 recached or lost, Wald = 11.3, $P = 0.001$). This result supports hypothesis 5.

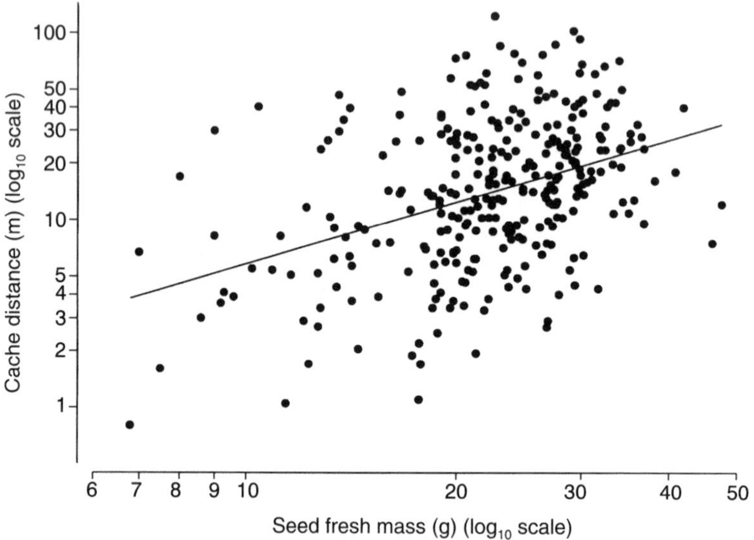

Fig. 14.2. Effect of seed fresh mass on the distance from the source of acouchy caches found 1 day after dispersal ($\log_{10}$ scales). Pooled data ($n = 303$) for 11 plots. Cache distance increased with seed fresh mass (ANCOVA: mass $F_{1,\,291} = 79$, $P < 0.001$; plot $F_{10,\,291} = 16$, $P < 0.001$; model $F_{11,\,291} = 22$, $R^2 = 0.45$). The increase is significant at the plot level (see text).

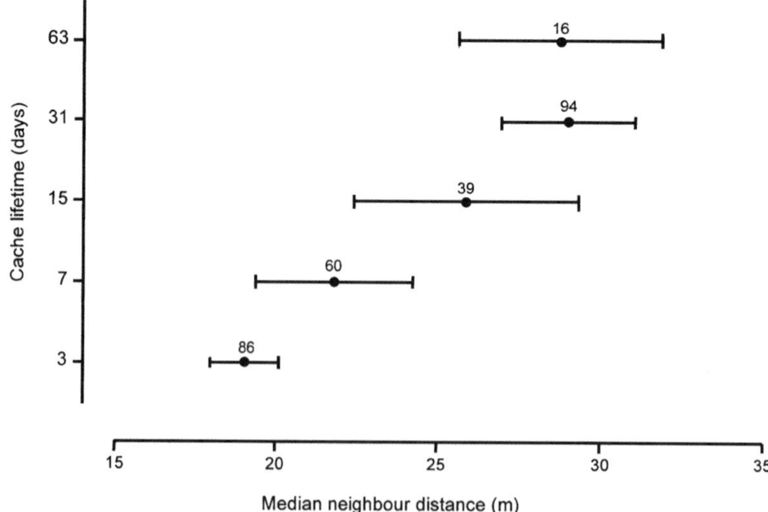

Fig. 14.3. Effect of cache isolation (median distance to other caches) on cache lifetime, i.e. the time until a cache was found depleted (approximately on $\log_2$ scale). Pooled data ($n = 304$) from 11 plots. Mean values (± SE) of isolation for different cache lifetimes, and sample sizes. Cache isolation positively affected the survival probability of caches (see text).

Ultimate seed fate

During subsequent searches and monitoring, we found additional seeds, mostly in caches at great distances. In total, we obtained complete or partial records of seed fate for 393 seeds (80%). The remaining seeds lost their thread mark (35 seeds or 8%) or were never found (61 seeds or 12%). We were able to determine the ultimate fate of 235 seeds (Table 14.1). Of these, only one (0.5%) established a seedling. The remaining seeds were eaten (95%), had been displaced after germination and probably became 'zombies' (1%) or were still cached without an epicotyl (4%). Probably, the latter seeds had germinated and become 'zombies' without our noticing. Fates of all seeds are summarized in a seed-fate pathway diagram (Fig. 14.4), following Price and Jenkins (1986).

Acouchies and other granivorous mammals gradually ate cached seeds. Seeds eaten tended to be lighter than those kept in stock (i.e. cached plus recached), but the difference was significant only during the first 2 weeks after dispersal (Fig. 14.5). The more rapid consumption of small seeds caused a gradual increase of the stock's mean seed mass. The total proportion of seeds with unknown fate also increased over time, because some seeds were lost every time they were handled. Lost seeds tended to be heavier than the stock mean mass. If those seeds were (re-)cached at great distances (see above), the increase of the stock mean mass was even greater than our estimate.

Overall, seed mass strongly affected the fate pathways of seeds and how long seeds were kept in stock (Fig. 14.6; *t*-test on weighted Cox regression coefficients from 11 plots: $t_{10} = -4.1$, $P = 0.002$). The ultimate probability of survival increased with seed mass, as posited by hypothesis 6 (logistic regression: $n = 8$ survived, 480 eaten or lost, Wald = 4.1, $P = 0.043$).

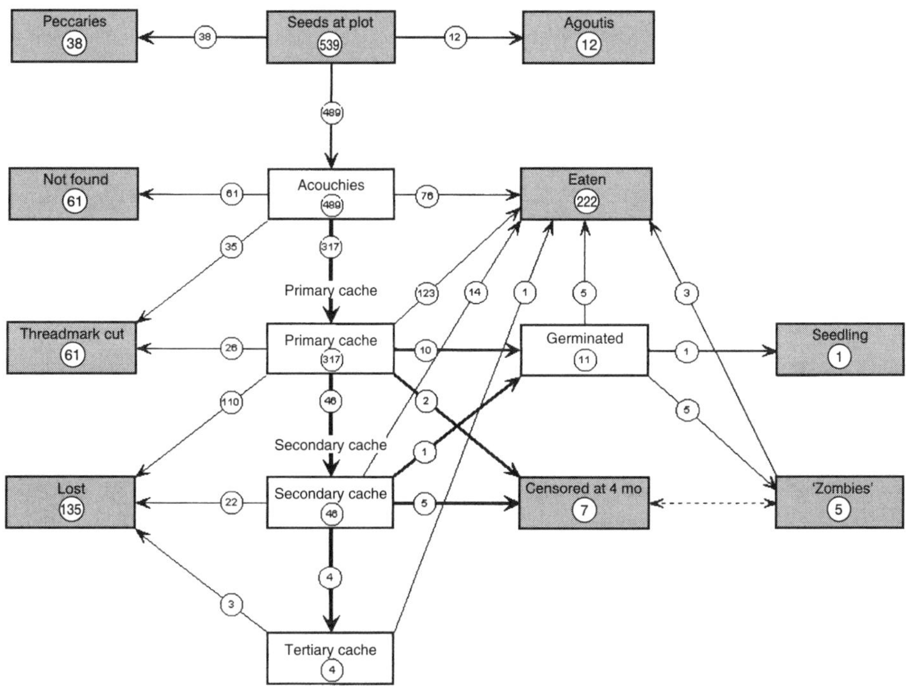

Fig. 14.4. Fate pathways of 539 *Carapa procera* seeds placed in 11 plots in territories of different red acouchies.

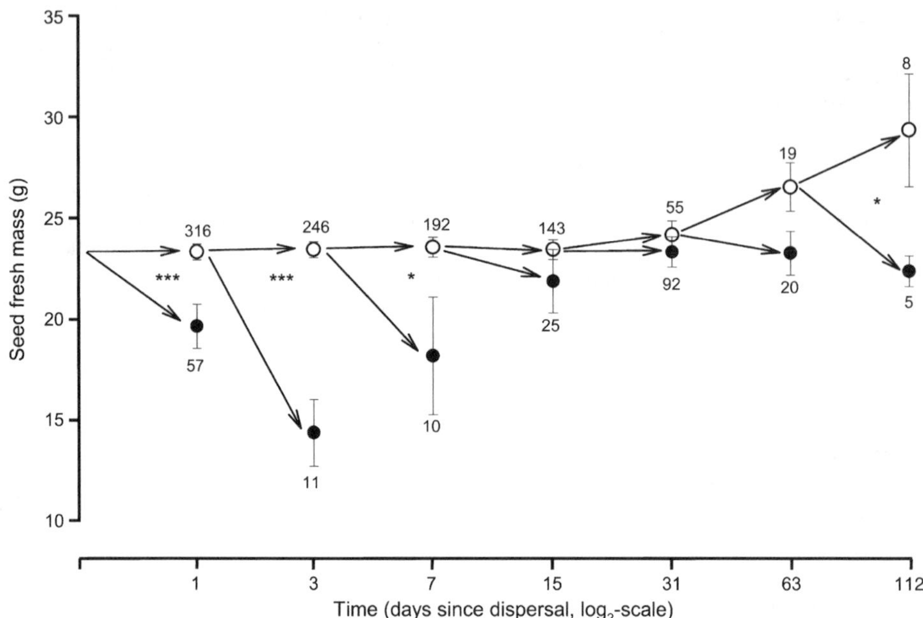

Fig. 14.5. Change of mean mass of cached seeds over time (approximately on $\log_2$ scale). Pooled data from 11 plots. Filled circles are seeds that were found eaten at time *t* but still cached at time *t* – 1, hollow circles are seeds that were still cached. Seeds that were taken from caches but not found are not shown. Numbers are sample sizes. Significance levels are of *t*-tests of differences in mean mass within pairs: ***, $P < 0.001$; *, $P < 0.05$; Seed mass increased the probability of seeds remaining cached rather than being eaten.

Discussion

The belief that many large-seeded tree species depend on scatter-hoarding for regeneration has been largely based on the observation that scatter-hoarding animals were the only vectors for movement of seeds away from the parent tree, where all seeds were killed by seed predators (Jansen and Forget, 2001). There are few studies that measure fate pathways of individual seeds (Chambers and MacMahon, 1994; but see Vander Wall, 1994, 1995a,c; see also Vander Wall, this volume; Hoshizaki and Hulme, this volume). Our study is the first to determine ultimate seed fates of scatter-hoarded seeds in a tropical rain forest, with a known dispersal vector and over distances far beyond 50 m.

We found direct evidence that scatter-hoarding granivores such as acouchies can effectively disperse large-seeded trees. Almost 500 *Carapa* seeds removed by acouchies resulted in one established seedling. This may seem unsubstantial, but 0.2% success is not trivial, given the large numbers of seed produced during a tree's reproductive lifespan (Janzen, 1971; Hallwachs, 1994). Furthermore, caviomorph rodents have high rates of mortality (Dubost, 1988; Hallwachs, 1994) and the death of an acouchy could greatly increase the survivorship of its cached seeds. This prediction could be verified by experimental removal of acouchies from their territories after scatter-hoarding. A more general idea of dispersal effectiveness requires estimations for more years, including years both with high and with low food availability.

Selection towards larger seeds

Our results indicate that scatter-hoarding can result in selection towards larger seeds: the chance of small seeds surviving 4 months was negligible compared with that of large seeds.

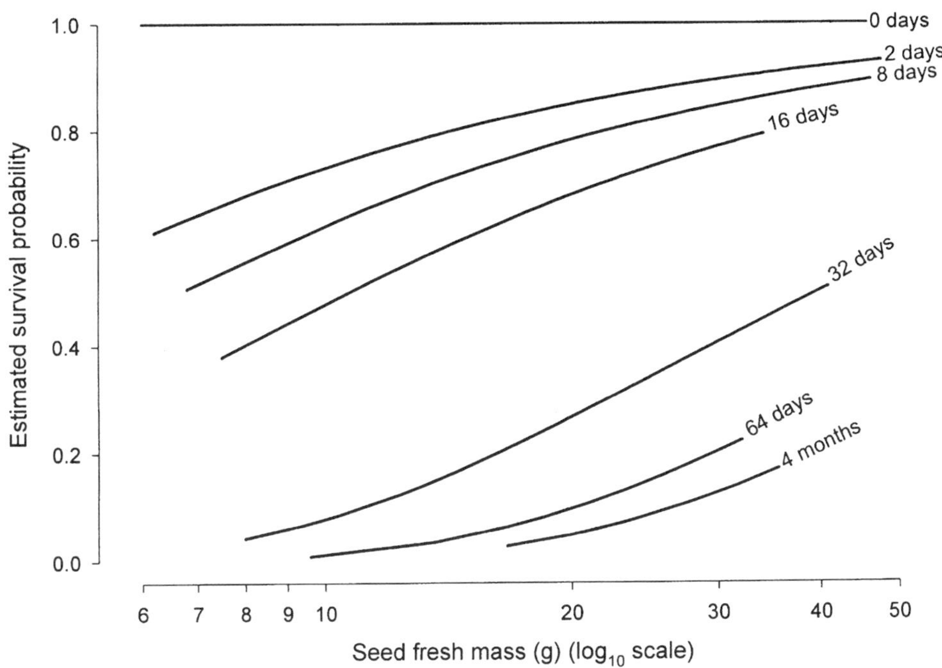

Fig. 14.6. Effect of seed mass (log_{10} scale) on seed lifetime. Survival probability curves as a function of seed mass according to the Cox regression model (see text for further explanation).

Larger seeds were more likely to be cached by acouchies rather than eaten (Fig. 14.1) and were cached further away (Fig. 14.2). Isolation of caches increased with cache distance and translated into longer lifetimes of caches (Fig. 14.3). If removed from their cache, larger seeds were also more likely to be recached than eaten. Overall, larger seeds were consumed at a lower rate than small seeds (Fig. 14.5), which resulted in a higher net probability of survival and seedling establishment for large seeds (Fig. 14.6). Seed mass, however, did not affect the speed and probability of harvest by acouchies.

As was hypothesized by Smith and Reichman (1984), preference for large seeds by scatter-hoarding animals could result in the evolution of larger-seeded crops by providing better dispersal for large-seeded individuals than for small-seeded individuals of the same species. Because we varied seed size within crops, our experiments provide no evidence for acouchies discriminating among seed crops of different seed size. Experiments by Hallwachs (1994) with agoutis in Costa Rica, however, did. Hallwachs found that the proportion and absolute numbers of acorns cached (rather than eaten) by agoutis was higher at large-seeded trees than at small-seeded trees, although the latter had greater numbers of seeds. She also found that artificial seeds (pieces of coconut) from large-seeded crops were cached further away than artificial seeds from small-seeded crops. Acouchies are likely to behave similarly, because they are closely related to agoutis and remarkably alike in behaviour (Smythe, 1978; Dubost, 1988). Another example comes from Waite and Reeve (1995), who found that scatter-hoarding grey jays discriminated among sources of different quality; the birds cached substantially more raisins from a large-item source than from a small-item source when source types were made available on different days.

Selectivity in removal

There are many examples of selectivity by scatter-hoarding animals based on food value (e.g. Smith and Follmer, 1972; Bossema, 1979; Reichman, 1988; Jacobs, 1992; Lucas *et al.*, 1993). The best support for the idea that acouchies should be selective comes from Hallwachs (1994), who found that agoutis in Costa Rica preferred heavier seeds if a range of seed masses was available. Why didn't the acouchies in our study discriminate against small seeds during removal?

One possible explanation comes from models of optimal foraging, in which animals maximize the net rate (Stephens and Krebs, 1986) or efficiency (Waite and Ydenberg, 1994a,b) of hoarding, with time as the primary limiting factor. The risk of food competitors claiming the food source, for example, increases with the time spent on source depletion. According to these models, non-selectivity should occur if the difference in rate of energy gain between large and small seeds is too small for selectivity to pay off. In our study, the amount of time spent per unit food value could have been almost constant because acouchies' investment in caching increased with seed value. However, we think that time was not the primary limiting factor. Acouchies used approximately 1 day to deplete a plot, usually interrupting their work for several hours. The number of seeds removed by competitors was nevertheless quite modest. Competition appeared unimportant at the time-scale of plot depletion.

A more plausible explanation of non-selectivity is that food availability is the limiting factor. Having sufficient food stored is crucial for survival during the period of food scarcity. Selectivity would limit the amount of stored food. Acouchies can afford the luxury of being selective only once a sufficiently large food supply is cached. Perhaps the animals in our study had to cache every *Carapa* seed they found to achieve an adequate supply. We predict that acouchies will show selectivity when food is abundant. Because speed of removal determines the probability of escaping seed predation by peccaries and insects and the probability of being cached rather than eaten, such selectivity would intensify selection towards larger seeds.

Hoarding versus consumption

Scatter-hoarding rodents often eat some seeds while storing many others (e.g. Hallwachs, 1994; Peres and Baider, 1997). We found that larger seeds were more likely to be stored by acouchies than small seeds and were more likely to be recached after recovery from their original cache. Hallwachs (1994) also observed that the proportion of acorns and of pieces of coconut cached by Costa Rican agoutis increased with their size. All small acorns were eaten. Apparently, large seeds are more suitable for storage than smaller ones.

Large seeds may be more frequently cached than small seeds because they have a longer storage life. Large seeds, for example, could be more persistent to drought, which causes *Carapa* seeds to decay (Ferraz-Kossmann and De Tarso Barbosa Sampaio, 1996), due to lower relative permeability to water (lower ratio of surface to volume). Likewise, lower relative water absorption could delay germination of larger seeds. Large seeds might also be more easily managed, as larger reserves take longer to be depleted by a seedling, giving animals more time to intervene and turn seeds into 'zombies', which can be conserved for several months. An alternative explanation is that a preference for larger seeds reduces the number of caches that acouchies must remember and manage for a given mass of food. This explanation, however, immediately begs the question of why acouchies never put more than one seed in a cache.

Cache distance

Our finding that larger seeds were cached at greater distances than small seeds and further away from other seeds agrees with models of optimal cache spacing (Stapanian and Smith, 1978; Clarkson *et al.*, 1986). These models predict that scatter-hoarders hide higher-value food in lower densities (i.e. further away) to compensate for the greater risk of such food being stolen by competitors. Several field studies have confirmed that larger seeds are cached at greater distances (Stapanian and Smith, 1984; Hurly and Robertson, 1986; Hallwachs, 1994; Jokinen and Suhonen, 1995; Vander Wall, 1995b; Forget *et al.*, 1998). Most of these studies, however, were based on interspecific variation in seed size. Food value was thus confounded with other differences among species, including nutrient composition, secondary compounds, digestibility, taste and odour (Hurly and Robertson, 1986). Our study controlled for extraneous variables by varying seed value within species, as did two earlier studies: Hurly and Robertson (1986) observed that whole groundnuts were cached by red squirrels further away than half groundnuts, and Hallwachs (1994) found that the proportion of seeds cached beyond her sight by agoutis increased with seed size.

Constraints on cache distance?

The relationship between cache distance and seed mass was much weaker than expected from cache optimization models. A large proportion of variation in cache distance remained unexplained (Fig. 14.2). What might explain this variation? The untransformed data showed that the range of cache distances varied with seed mass: the maximum distance increased with seed mass, while the minimum remained almost constant. Moreover, the maximum distance seemed to have an optimum at ~29 g, beyond which it dropped off again. The resulting polygonal shape of the scatter diagram could indicate limiting factors (Scharf *et al.*, 1998; Cade *et al.*, 1999).

We investigated this so-called 'envelope effect' (Goldberg and Scheiner, 1993) using quantile regression on pooled data for eight quantiles (Scharf *et al.*, 1998). First, we tested whether regression quantiles had an optimum by calculating the contribution of the cubic and the quadratic factor to the model. Both were significant for none of the quantiles, implying that the apparent 'optimum' could simply be an artefact of low numbers of extremely large seeds (stepwise regression with

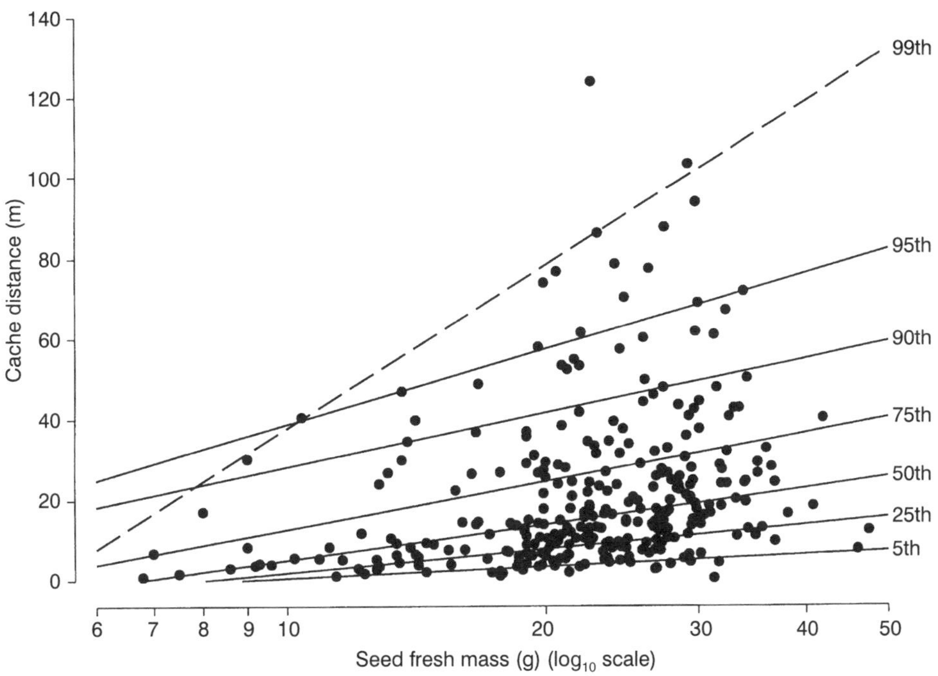

Fig. 14.7. Effect of seed fresh mass ($\log_{10}$ scale) on the distance from the source of acouchy caches found 1 day after dispersal. Pooled data (n = 303) for 11 plots. Plotted lines are significant quantile regressions for seven values of τ (see Table 14.2).

Table 14.2. Quantile regressions of cache distance on fresh mass of *Carapa procera* seeds scatter-hoarded by red acouchies (*Myoprocta exilis*). Data from 11 replicate plots were pooled ($n = 303$). Estimates of β_0 and β_1 are given for models $y = \beta_0 + \beta_1 \cdot x$ (A) and $y = \beta_0 + \beta_1 \cdot \log_{10} x$ (B) with y for cache distance (m) and x for seed fresh mass (g). P values for H_0: $\beta_1 = 0$ were obtained from 5000 permutations and rank-score tests (Slauson *et al.*, 1994). Model B quantile regressions gave a better fit and are shown in Fig. 14.7. A third model, $y = \beta_0 + \beta_1 \cdot x + \beta_2 \cdot x^2$, did not produce a better fit for any of the quantiles.

Model	Quantile	β_0	β_1	$T_{observed}$	$P(\beta_1)$
A	5th	−0.5	0.17	0.045	< 0.001
	10th	−1.5	0.29	0.049	< 0.001
	25th	−1.9	0.45	0.071	< 0.001
	50th	−3.4	0.81	0.076	< 0.001
	75th	10.8	0.70	0.021	< 0.01
	90th	21.9	0.91	0.025	0.052
	95th	26.6	1.33	0.055	< 0.05
	99th	12.9	3.10	0.095	0.091
B	5th	−9.1	9.6	0.043	< 0.001
	10th	−12.1	12.6	0.045	< 0.001
	25th	−17.7	19.6	0.067	< 0.001
	50th	−24.4	29.4	0.072	< 0.001
	75th	−26.8	39.5	0.024	< 0.005
	90th	−16.3	44.5	0.028	< 0.05
	95th	−23.5	62.2	0.062	< 0.05
	99th	−97.3	135.0	0.114	0.06

backward elimination of factors). We then calculated quantile regressions of cache distance on seed mass and on $\log_{10}$-transformed seed mass (Table 14.2). Regression coefficients (β_1) were significant at $\alpha \leq 0.10$ for all quantiles in both models, and increased with τ, implying that most of the variation in distance occurs in the upper quantiles. Quantile regressions with seed mass $\log_{10}$-transformed gave the best fit (Table 14.2, Fig. 14.7). Potential cache distance was clearly far greater for large seeds than for small seeds.

These results suggest that cache distance is limited by some constraint related to seed value, described by the upper regression quantiles. This constraint could correspond to the investment at which the net benefit of seed caching is zero, as suggested by Hurly and Robertson (1987). Cache spacing being governed by a constraint is in disagreement with the models of Stapanian and Smith (1978), Clarkson *et al.* (1986) and Tamura *et al.* (1999), in which cache distance is distributed around an optimum investment, depending on seed size, at which the net energy gain is maximized. A similar 'envelope effect' appears in Hallwachs's (1994) data of agouti cache distance versus acorn size.

Although our data do not prove that potential cache distance had an optimum at 25–30 g of fresh mass rather than continuously increasing with seed mass, the idea of an optimum seed mass makes sense. There must be a seed mass beyond which handling and transport become increasingly difficult and expensive for acouchies. Selection by acouchies will not be directional towards ever-larger seeds but, instead, should stabilize at an optimum. Studies including very large seeds in test samples are needed to test this possibility.

Cache isolation

That large seeds were indeed more likely to be cached far away from other seeds agrees with optimal cache-spacing models (Stapanian and Smith, 1978; Clarkson *et al.*, 1986). However, we found no additive effect of seed mass on dispersal distance. Thus, large seeds were not spaced out more widely than small seeds with the same dispersal distance. We conclude that the two-dimensional cache-spacing pattern is a result of variation in one-dimensional dispersal distance and dispersal direction. Cache isolation and cache density, in other words, are by-products of dispersal, rather than characteristics directly manipulated by acouchies.

Escape from a classic trade-off

The fact that acouchies carried larger seeds further than small seeds is logical from the point of view of foraging theory. However, the result of larger seeds having better dispersal than small seeds is in disagreement with the classic theory of a size–number trade-off between dispersability and vigour (Smith and Fretwell, 1974). This theory assumes that the allocation of nutrients to reproduction is limited and that plants must find an optimal balance between producing large seeds and producing many seeds. Reasoning that small seeds have a higher probability of being effectively dispersed, a need for effective dispersal would select against producing large seeds. Our findings suggest that dispersability in species dispersed by scatter-hoarding animals selects towards the production of large seeds. Scatter-hoarding enables these plant species to produce larger seeds than species with other dispersal modes.

Acknowledgements

We are grateful to Pierre-Michel Forget, Philip Hulme, Han Olff and Herbert Prins for their critical comments and to Gerrit Gort for statistical advice. We thank Pierre Charles-Dominique for his permission to work at the Nouragues Biological Station. This study was funded by the Dutch Foundation for the Advancement of Tropical Research with grant W84-408 (Jansen).

References

Bongers, F., Charles-Dominique, P., Forget, P.M. and Théry, M. (2001) *Nouragues: Dynamics and Plant–Animal Interactions in a Neotropical Rainforest.* Kluwer Academic Publisher, Dordrecht, The Netherlands (in press).

Bossema, I. (1979) Jays and oaks: an eco-ethological study of a symbiosis. *Behaviour* 70, 1–118.

Brandle, J.E., Court, W.A. and Roy, R.C. (1993) Heritability of seed yield, oil concentration and oil quality among wild biotypes of Ontario evening primrose. *Canadian Journal of Plant Science* 73, 1067–1070.

Cade, B.S., Terrell, J.W. and Schroeder, R.L. (1999) Estimating effects of limiting factors with regression quantiles. *Ecology* 80, 311–323.

Chambers, J.C. and MacMahon, J.A. (1994) A day in the life of a seed: movement and fates of seeds and their implications for natural and managed systems. *Annual Review of Ecology and Systematics* 25, 263–292.

Clarkson, K.S., Eden, S.F., Sutherland, W.J. and Houston, A.I. (1986) Density dependence and magpie food hoarding. *Journal of Animal Ecology* 55, 111–121.

Cober, E.R., Voldeng, H.D. and Fregeau-Reid, J.A. (1997) Heritability of seed shape and seed size in soybean. *Crop Science* 37, 1767–1769.

Dubost, G. (1988) Ecology and social life of the red acouchy, *Myoprocta exilis*; comparisons with the orange-rumped agouti, *Dasyprocta leporina*. *Journal of Zoology* 214, 107–113.

Emmons, L.H. and Feer, F. (1990) *Neotropical Rainforest Mammals: a Field Guide.* University of Chicago Press, Chicago.

Eriksson, O., Friis, E.M. and Löfgren, P. (2000) Seed size, fruit size, and dispersal systems in Angiosperms from the early Cretacious to the late Tertiary. *American Naturalist* 156, 47–58.

Ferraz-Kossmann, I.D. and De Tarso Barbosa Sampaio, P. (1996) Simple storage methods for andiroba seeds (*Carapa guianensis* Aubl. e *Carapa procera* D.C.– Meliaceae). *Acta Amazonica* 26, 137–144.

Forget, P.-M. (1990) Seed-dispersal of *Vouacapoua americana* (*Caesalpiniaceae*) by cavimorph rodents in French Guiana. *Journal of Tropical Ecology* 6, 459–468.

Forget, P.-M. (1996) Removal of seeds of *Carapa procera* (*Meliaceae*) by rodents and their fate in rainforest in French Guiana. *Journal of Tropical Ecology* 12, 751–761.

Forget, P.-M., Mileron, T. and Feer, F. (1998) Patterns in post-dispersal seed removal by neotropical rodents and seed fate in relation to seed size. In: Newbery, D.M., Prins, H.H.T. and Brown, N.D. (eds) *Dynamics of Tropical Communities.* Blackwell Science, Oxford, pp. 25–49.

Gjuric, R. and Smith, S.R., Jr (1997) Inheritance in seed size of alfalfa: quantitative analysis and response to selection. *Plant Breeding* 116, 337–340.

Goldberg, D.E. and Scheiner, S.M. (1993) ANOVA and ANCOVA: field competition experiments. In: Scheiner, S.M. and Gurevitch, J. (eds) *Design and Analysis of Ecological Experiments.* Chapman & Hall, London, pp. 69–93.

Grubb, P.J. and Burslem, D.F.R.P. (1998) Mineral nutrient concentrations as a function of seed size within seed crops: implications for competition among seedlings and defence against herbivory. *Journal of Tropical Ecology* 14, 177–185.

Hallwachs, W. (1994) The clumsy dance between agoutis and plants: scatterhoarding by Costa Rican dry forest agoutis (*Dasyprocta punctata*: Dasyproctidae: Rodentia). PhD thesis, Cornell University, Ithaca, New York.

Hammond, D.S. and Brown, V.K. (1998) Disturbance, phenology and life-history characteristics: factors influencing frequency-dependent attack on tropical seeds and seedlings. In: Newbery, D.M., Brown, N. and Prins, H.H.T. (eds) *Dynamics of Tropical Communities.* Blackwell Science, Oxford, pp. 51–78.

Hammond, D.S., Gourlet-Fleury, S., Van der Hout, P., Ter Steege, H. and Brown, V.K. (1996) A compilation of known Guianan timber trees and the significance of their dispersal mode, seed size and taxonomic affinity to tropical rain forest management. *Forest Ecology and Management* 83, 99–116.

Hurly, T.A. and Robertson, R.J. (1987) Scatter hoarding by territorial red squirrels: a test of the optimal density model. *Canadian Journal of Zoology* 65, 1247–1252.

Jackson, J.F. (1981) Seed size as a correlate of temporal and spatial patterns of seed fall in a neotropical forest. *Biotropica* 13, 121–130.

Jacobs, L.F. (1992) The effect of handling time on the decision to cache by grey squirrels. *Animal Behaviour* 43, 522–524.

Jansen, P.A. and Forget, P.-M. (2001) Scatter-hoarding rodents and tree regeneration. In: Bongers, F., Charles-Dominique, P., Forget, P.M. and Théry, M. (eds) *Nouragues: Dynamics and Plant–Animal Interactions in a Neotropical Rainforest.* Kluwer Academic Publisher, Dordrecht, The Netherlands, pp. 275–288.

Janzen, D.H. (1970) Herbivores and the number of tree species in tropical forests. *American Naturalist* 104, 501–528.

Janzen, D.H. (1971) Seed predation by animals. *Annual Review of Ecology and Systematics* 2, 465–492.

Jokinen, S. and Suhonen, J. (1995) Food caching by willow and crested tits: a test of scatterhoarding models. *Ecology* 76, 892–898.

Koenker, R. and Bassett, G. (1978) Regression quantiles. *Econometrica* 46, 33–50.

Leishman, M.R., Westoby, M. and Jurado, E. (1995) Correlates of seed size variation: a comparison among five temperate floras. *Journal of Ecology* 83, 517–530.

Lucas, J.R., Peterson, L.J. and Boudinier, R.L. (1993) The effects of time constraints and changes in body mass and satiation on the simultaneous expression of caching and diet choices. *Animal Behaviour* 45, 639–658.

Malhotra, R.S., Bejiga-Geletu and Singh, K.B. (1997) Inheritance of seed size in chickpea. *Journal of Genetics and Breeding* 51, 45–50.

Midcontinent Ecological Science Center (1998) *BLOSSOM Statistical Software,* version EMVB 0898.L4.5D.S3. Midcontinent Ecological Science Center, Fort Collins, Colorado.

Morris, D. (1962) The behaviour of the green acouchy (*Myoprocta pratti*) with special reference to scatter hoarding. *Proceedings of the Zoological Society of London* 139, 701–733.

Peres, C.A. and Baider, C. (1997) Seed dispersal, spatial distribution and population structure of Brazilnut trees (*Bertholletia excelsa*) in south-eastern amazonia. *Journal of Tropical Ecology* 13, 595–616.

Poncy, O., Sabatier, D., Prevost, F. and Hardy, I. (2001) The lowland rain forest: tree structure and species diversity. In: Bongers, F., Charles-Dominique, P., Forget, P.M. and Théry, M. (eds) *Nouragues: Dynamics and Plant–Animal Interactions in a Neotropical Rainforest.* Kluwer Academic Publisher, Dordrecht, The Netherlands.

Price, M.V. and Jenkins, S.H. (1986) Rodents as seed consumers and dispersers. In: Murray, D.R. (ed.) *Seed Dispersal.* Academic Press, Sydney, pp. 191–235.

Rebetzke, G.J., Pantalone, V.R., Burton, J.W., Carter, T.E., Jr and Wilson, R.F. (1997) Genotypic variation for fatty acid content in selected *Glycine max* × *Glycine soja* populations. *Crop Science* 37, 1636–1640.

Reichman, O.J. (1988) Caching behavior by eastern woodrats, *Neotoma floraidana,* in relation to food perishability. *Animal Behaviour* 36, 1525–1532.

Sabatier, D. (1985) Saisonnalité et déterminisme du pic de fructification en forêt guyanaise. *Revue d'Ecologie (Terre et Vie)* 40, 289–320.

Scharf, F.S., Juanes, F. and Sutherland, M. (1998) Inferring ecological relationships from the edges of scatter diagrams: comparison of regression techniques. *Ecology* 79, 448–460.

Slauson, W.L., Cade, B.S. and Richards, J.D. (1994) *User Manual for BLOSSOM Statistical Software.*

Midcontinent Ecological Science Center, Fort Collins.

Smith, C.C. and Follmer, D. (1972) Food preferences of squirrels. *Ecology* 53, 82–91.

Smith, C.C. and Fretwell, S.D. (1974) The optimal balance between size and number of offspring. *American Naturalist* 108, 499–506.

Smith, C.C. and Reichman, O.J. (1984) The evolution of food caching by birds and mammals. *Annual Review of Ecology and Systematics* 15, 329–351.

Smythe, N. (1970) Relationships between fruiting seasons and seed dispersal methods in a neotropical forest. *American Naturalist* 104, 25–35.

Smythe, N. (1978) The natural history of the Central American agouti (*Dasyprocta punctata*). *Smithsonian Contributions to Zoology* 257, 1–52.

SPSS Inc. (1999) *SPSS for Windows, Release* 10.0.5. SPSS Inc., Chicago.

Stapanian, M.A. and Smith, C.C. (1978) A model for seed scatterhoarding: coevolution of fox squirrels and black walnut. *Ecology* 59, 884–896.

Stapanian, M.A. and Smith, C.C. (1984) Density-dependent survival of scatterhoarded nuts: an experimental approach. *Ecology* 65, 1387–1396.

Stephens, D.W. and Krebs, J.R. (1986) *Foraging Theory*. Princeton University Press, Princeton, New Jersey, 247 pp.

Tamura, N., Hashimoto, Y. and Hayashi, F. (1999) Optimal distances for squirrels to transport and hoard walnuts. *Animal Behaviour* 58, 635–642.

Vander Wall, S.B. (1990) *Food Hoarding in Animals*. University of Chicago Press, Chicago, 445 pp.

Vander Wall, S.B. (1993) Cache site selection by chipmunks (*Tamias* spp.) and its influence on the effectiveness of seed dispersal in Jeffrey pine (*Pinus jeffreyi*). *Oecologia* 96, 246–252.

Vander Wall, S.B. (1994) Seed fate pathways of antelope bitterbrush: dispersal by seed-caching yellow pine chipmunks. *Ecology* 75, 1911–1926.

Vander Wall, S.B. (1995a) Dynamics of yellow pine chipmunk (*Tamias amoenus*) seed caches: underground traffic in bitterbrush seeds. *Ecoscience* 2, 261–266.

Vander Wall, S.B. (1995b) The effects of seed value on the caching behaviour of yellow pine chipmunks. *Oikos* 74, 533–537.

Vander Wall, S.B. (1995c) Sequential patterns of scatter hoarding by yellow pine chipmunks (*Tamias amoenus*). *American Midland Naturalist* 133, 312–321.

Vander Wall, S.B. (2001) The evolutionary ecology of nut dispersal. *Botanical Review* 67, 74–117.

Waite, T.A. and Reeve, J.D. (1995) Source-use decisions by hoarding gray jays: effects of local cache density and food value. *Journal of Avian Biology* 26, 59–66.

Waite, T.A. and Ydenberg, R.C. (1994a) What currency do scatter-hoarding gray jays maximize? *Behavioral Ecology and Sociobiology* 34, 43–49.

Waite, T.A. and Ydenberg, R.C. (1994b) Foraging currencies and the load-size decision of scatter-hoarding grey jays. *Animal Behaviour* 51, 903–916.

Wilson, M.F. and Janzen, D.H. (1972) Predation on *Scheelea* palm seeds by bruchid beetles: seed density and distance from the parent palm. *Ecology* 53, 954–959.

15 Mast Seeding and Predator-mediated Indirect Interactions in a Forest Community: Evidence from Post-dispersal Fate of Rodent-generated Caches

Kazuhiko Hoshizaki[1] and Philip E. Hulme[2]

[1]*Department of Biological Environment, Akita Prefectural University, Akita 010-0195, Japan;* [2]*NERC Centre for Ecology and Hydrology, Hill of Brathens, Banchory AB31 4BY, Scotland, UK*

Introduction

When and how do seeds die? For most temperate trees the answer is soon after seed fall and as a result of seed predation. In temperate deciduous woodland a considerable proportion of seeds is consumed by granivores (Hulme, 1993, 1998; Hulme and Benkman, 2001). Not surprisingly, granivores (in particular, small mammals) have been found to play a pivotal role in the regeneration (Hulme, 1996b, 1997), colonization ability (Myster and Pickett, 1993), spatial distribution (Kollmann, 1995) and reproductive ecology (Sork, 1993) of temperate trees. To date, research on granivory in deciduous woodlands has tended to focus on the relative importance of habitat, microhabitat, plant species and seed density for seed predation rates (Willson and Whelan, 1990; Whelan *et al.*, 1991; Myster and Pickett, 1993; Hulme and Borelli, 1999). However, the frequent absence of consistent directional trends (e.g. Willson and Whelan, 1990) has made it difficult to construct accurate predictive models of post-dispersal seed predation. One factor that might lead to inconsistencies in seed-predation rates may be the temporal and spatial variation in the abundance of alternative food resources.

Polyphagous seed predators, such as small mammals, are likely to respond not only to the absolute abundance of a particular seed species but also to its relative abundance in relation to other seed species (Greenwood, 1985). For example, the rate of seed predation on a particular plant species can decrease dramatically following an increase in availability of a more preferred species. Such frequency-dependent foraging by granivores has considerable potential to structure plant communities (Hulme, 1996a). Furthermore, frequency dependence may explain why inconsistent trends in removal rates are sometimes found among habitats, seasons and years (Schupp, 1990; Willson and Whelan, 1990; Kollmann *et al.*, 1998). These inconsistencies may simply reflect spatial and/or temporal changes in the relative abundance of different seed species. Although several studies have examined seed predation in a community context (Schupp, 1990; Kollmann *et al.*, 1998), only Hulme and Hunt (1999) have analysed how a seed

 Seed Dispersal and Frugivory: Ecology, Evolution and Conservation (eds D.J. Levey, W.R. Silva and M. Galetti)

predator's preference depends on the relative abundance of two species.

In addition to functional changes in rodent foraging behaviour as a result of variation in seed frequency, granivores may also respond numerically to variation in seed abundance (Ostfeld *et al.*, 1996; Ostfeld and Keesing, 2000). A build-up in granivore abundance may occur following an increase in the density of the seeds of one plant species. If granivores forage in a frequency-independent manner, an increase in rodent abundance may lead to an increase in absolute rates of seed predation on the seeds of all plant species in the community but should not alter the relative removal of different species. However, the effect on the less abundant plant species may be disproportionate if they are more vulnerable to the increased impacts of granivores (e.g. recruitment more seed-limited) (see Hulme, this volume).

The impact of variation in seed abundance on seed predation is most likely to be found in plant communities dominated by species that produce large, synchronous seed crops at irregular intervals (e.g. masting species). Mast seeding in tree species and the consequent satiation of seed predators have long been a phenomenon of particular interest (e.g. Janzen, 1971). However, few studies have examined the consequence of masting on the seed/seedling survivorship of co-occurring non-masting trees (but see Curran and Webb, 2000). To what extent, then, does annual variation in seed production among species influence populations of generalist predators and, consequently, their pattern of seed predation?

Indirect interactions often play significant roles in community dynamics (Wootton, 1994; Morin, 1999). Indirect interactions may apply to tree species and their seed predators, because many seed predators are generalists (Crawley, 1992) and respond numerically to seed abundance (Ostfeld *et al.*, 1996; Ostfeld and Keesing, 2000). Temporal variation in seed production by one tree species is expected to have an impact on seed predation on co-occurring species. This type of indirect interaction remains unexplored in seed-predation studies.

Within this context we examined the impact of temporal variations in seed supply on rates of seed removal by rodents. Our goal is to demonstrate that seed predation on *Aesculus turbinata* can depend on the seed abundance of other dominant tree species. In particular, we explore the impact of seasonal variation in *Fagus crenata* seed fall. This is a particularly suitable system to study, since *F. crenata* produces large, irregular mast crops of seeds that are highly preferred by rodents. To examine how masting indirectly affects seed predation in *A. turbinata*, we addressed the following questions:

1. Are rates of *A. turbinata* seed predation indirectly affected by the seed fall of *F. crenata*?
2. Within a mast year, is the effect of *F. crenata* on *A. turbinata* consistent among seasons?
3. What most determines the overall success of *A. turbinata* seeds – its own seed fall or that of *F. crenata*?
4. Do seed predators respond equally to annual variation in seed production of *A. turbinata* and *F. crenata*?

Study Species and Sites

The natural history of *A. turbinata* (*Hippocastanaceae*), a deciduous canopy tree, is described in detail in Hoshizaki *et al.* (1997, 1999). Its seeds are particularly large (2–4 cm in diameter) and contain more seed reserves (6.2 g per seed, dry mass (Hoshizaki *et al.*, 1997)) than any other seeds in the temperate forests of Japan and perhaps in all northern temperate forests. Seed fall begins in mid-September and continues until the beginning of October. The principal granivores of *A. turbinata* are small mammals (*Apodemus speciosus* and *Eothenomys andersoni*). They readily take fallen seeds and either consume or cache them (Hoshizaki, 1999). Spatial variation in seed removal by rodents is low (Hoshizaki *et al.*, 1999), whereas temporal variation is high (Hoshizaki *et al.*, 1997). Caching increases the probability of seeds reaching sites suitable for establishment, because it enlarges the seed shadow (Hoshizaki *et al.*, 1999).

At our study site in the Kanumazawa Riparian Research Forest (39° N, 140° E) (see Hoshizaki *et al.* (1997) and Suzuki *et al.* (2001) for a detailed description), *A. turbinata* occurs with two other common tree species, which also display large interannual variation in seed production. *Fagus crenata* (*Fagaceae*) is overwhelmingly dominant and exhibits strict masting cycles (*sensu* Kelly, 1994; see also Miguchi, 1988, 1996). Annual variation of seed production in a second species, *Quercus mongolica* var. *grosseserrata* (*Fagaceae*), is less synchronized and more regular (Miguchi, 1996; see also Sork, 1993). The seeds of the three species differ in both the quality and quantity of resource they present to rodents. Seeds of *A. turbinata* contain saponins and tannins, seeds of *Q. mongolica* (seed mass, 1.7 g) contain tannins and no saponins and seeds of *F. crenata* (0.13 g) contain no tannins or saponins (K. Hoshizaki, unpublished data).

Research was undertaken in two sites within the Kanumazawa Riparian Research Forest. The long-term monitoring plot represented a 1 ha site in which the seed fall and germination of *A. turbinata* and *F. crenata* have been studied for 7 years (1992–1999). To interpret the long-term trends, a 0.4 ha short-term experimental plot, 150 m from the long-term monitoring plot was established in 1995. The short-term experimental plot was used to examine rates of *A. turbinata* and *F. crenata* seed removal.

Methods

Long-term annual variation in seed predation and seedling establishment

Seed production of the three tree species (*A. turbinata, F. crenata* and *Q. mongolica*) was recorded annually from 1992 to 1998 in the long-term monitoring plot. Between 99 and 121 seed traps were established at regular 10 m intervals throughout the plot in each year. Each seed trap comprised a nylon-mesh bag (1 mm mesh) with a receiving area of 0.5 m^2, set 1.2 m above the ground. Seeds were collected from all traps every 2 weeks. In early June each year, all new *A. turbinata* seedlings within the long-term plot were tagged and their locations mapped. Because *A. turbinata* has no dormancy, overall seed survival can be estimated from the ratio of newly germinated seedlings to seed fall in the previous autumn. We measured seed success as both the proportion and the total number of *A. turbinata* seeds that successfully produced seedlings.

To test whether the annual variation in seed predation in *A. turbinata* is related to seed abundance, we examined *A. turbinata* seed success in relation to the abundance of seeds in the previous autumn. These analyses were also undertaken in relation to variations in seed production of *Fagus* and *Quercus* in order to examine the effects of seed production by other tree species on *A. turbinata* seed success. These data were also interpreted in relation to community-wide food abundance for rodents. The community-wide food abundance, measured as seed-fall energy per hectare, was estimated as the summed value of total seed-fall biomass multiplied by the per-gram energy for each species (K. Hoshizaki, unpublished data).

Short-term assessments of A. turbinata and F. crenata seed predation

In each of 3 years (1995–1997) we followed the fate of *c.* 100 seeds of *A. turbinata* placed under each of five mature trees in the short-term plot. In 1995 and 1996 all seeds came from under these same trees; however, in 1997 seed production was especially low and seeds from other trees also had to be used. Each September, we attached one end of a 40 cm length of wire to a seed and the other end to coloured tape. The number of seeds to which a wire was attached on any one day was determined by the daily seed fall. Every 2–7 days until seedling emergence the following spring (except for a period with snow cover), we searched for marked seeds, as described in Hoshizaki *et al.* (1999). We recorded seed location and status (cached, germinated or eaten). In most cases we could find the seeds because the coloured flag was rarely buried and because the wire was never severed. The mass of the wire (*c.* 2 g) was small compared with the mass of the seeds (*c.* 19 g).

Ideally, predation on *F. crenata* seeds should be evaluated by tracing the fate of seeds as we did for *A. turbinata*. However, due to the relatively small seed size of *F. crenata*, attaching a length of wire to seeds would more than likely influence rodent foraging behaviour. Therefore, *F. crenata* seed predation was assessed indirectly. Under each of two *F. crenata* trees, we established three sample plots. Each plot consisted of four 50 cm × 50 cm quadrats and an adjacent seed trap (see above for description). From October until mid-November in 1995 (a mast year for *F. crenata*), we collected *F. crenata* seeds in a litter sample from one of the four quadrats in each plot every 2 weeks. During the same period and at fortnightly intervals, all *F. crenata* seeds in seed traps were also collected and counted. Daily removal rate (*R*) of *F. crenata* seeds was calculated as:

$$R = \frac{(G_i + F_{i-j}) - G_j}{t_j - t_i} \quad (1)$$

where G_i is the number of *F. crenata* seeds in the litter at time t_i, G_j is the number of seeds in the subsequent litter sample at time t_j, and F_{i-j} is the number of seeds that were collected in the seed trap during the period t_i to t_j. Seed numbers were standardized to seeds per m^2 before calculation. We assume that this removal rate is equivalent to seed consumption rate.

In September 1995, we established an array of seed traps ($n = 69$) to monitor the rate of seed fall for *Fagus* in a 10 m spacing grid in and around the short-term study plot. We collected the contents of traps approximately every 2 weeks and counted sound seeds. The phenology of *Aesculus* seed fall was estimated from the number of seeds tagged per day during the fruiting period. Seedling emergence of both *F. crenata* and *A. turbinata* was assessed following the methodology described for the long-term monitoring plot.

Seed-predator census

We monitored populations of *A. speciosus* and *E. andersoni* from the autumn of 1994 to the autumn of 1999. Rodents were censused between two and five times each year, using Sherman live traps placed 10 m apart in a 4 × 6 or 5 × 8 grid within the Kanumazawa Riparian Research Forest and close to the long- and short-term plots. Each census consisted of three consecutive nights of trapping. Traps were baited with sweet potato and sunflower seeds and checked daily. Rodents were toe-clipped for individual identification and released at the site of capture after species and sex had been recorded. Population density was conservatively estimated as the minimum number alive (MNA) for each species during each trapping session.

Data Analysis

Frequency-dependent seed predation

To test the idea that removal of *A. turbinata* seeds was reduced by *F. crenata* seed fall due to rodents shifting their food resources to the preferred seed, we examined the degree of frequency dependence in seed removal (Hulme and Hunt, 1999). Analysis of frequency dependence followed the method proposed by Greenwood and Elton (1979). Suppose that A_1 and A_2 are the densities of prey1 (*F. crenata*) and pray2 (*A. turbinata*) available and E_1 and E_2 are the densities consumed, a relationship may be fitted between the ratios of frequencies available and consumed, such that:

$$\frac{E_1}{E_2} = \left(v \frac{A_1}{A_2} \right)^b \quad (2)$$

The constant b is a measure of frequency dependence: if it is greater than one, granivores consume proportionally more of the common prey; if it is less than one, granivores consume proportionally less of the common prey; if it is one, granivore feeding is frequency-independent. The constant v may be considered as a measure of frequency-independent selection (Greenwood, 1985). Values of b and log v can be estimated by linear regression on the log-transformed ratios of eqn 2.

Survivorship of **A. turbinata** *seeds*

If *F. crenata* seed fall significantly affects *A. turbinata* survivorship, we would expect greater survivorship during the mast year (1995) than during the non-mast years (1996,

1997). However, direct comparison among years is inappropriate, since the abundance of seed predators also varied among years. To overcome this, we calculated two estimates of seed survivorship in each year. The first estimate was derived from the period of *A. turbinata* seed fall prior to *F. crenata* seed fall (no influence of *F. crenata*) and the second period was when both tree species were seeding (effect of *F. crenata*). We estimated the change in survivorship between these two periods using a proportional-hazards model (StatSoft, 1995) and compared the extent of the change among the 3 years.

The proportional-hazards model analysis allows us to incorporate 'censored data' and estimates conditional probabilities of mortality at any given time from a survivorship curve. It is composed of two elements of the survivorship statistics: hazard and hazard ratio. Suppose S is a survivorship curve function. The hazard h is a function of time and defined as:

$$h(t) \equiv -\frac{dS}{dt} / S(t) \qquad (3)$$

The hazard shows the 'instantaneous risk' of mortality (Hamajima, 1993; StatSoft, 1995; see also Schupp, 1990). Relative risk of mortality among two survivorship curves can be obtained by the ratio of the two hazards, $h_2(t)/h_1(t)$, where $h_2(t)$ is the hazard function for the period with *F. crenata* seed fall and $h_1(t)$ is the period without. We then compared values of the hazard ratio for each of the 3 years. Only data on seeds removed by rodents were included in the analyses.

Results

Long-term trends in A. turbinata *seed success in relation to seed production*

The seed success of *A. turbinata* varied greatly among the 7 years of the study (coefficient of variation (CV) = 83.8% for seed survival, 88.5% for seedling emergence). However, seed success was not correlated with *A. turbinata* seed production (for seed survival, $r = 0.19$, d.f. = 5, $P = 0.67$; seedling emergence, $r = 0.49$, d.f. = 5, $P = 0.29$) (Fig. 15.1). This suggests that there is little evidence for predator satiation or economies of scale for *A. turbinata* (Janzen, 1971; Norton and Kelly, 1988).

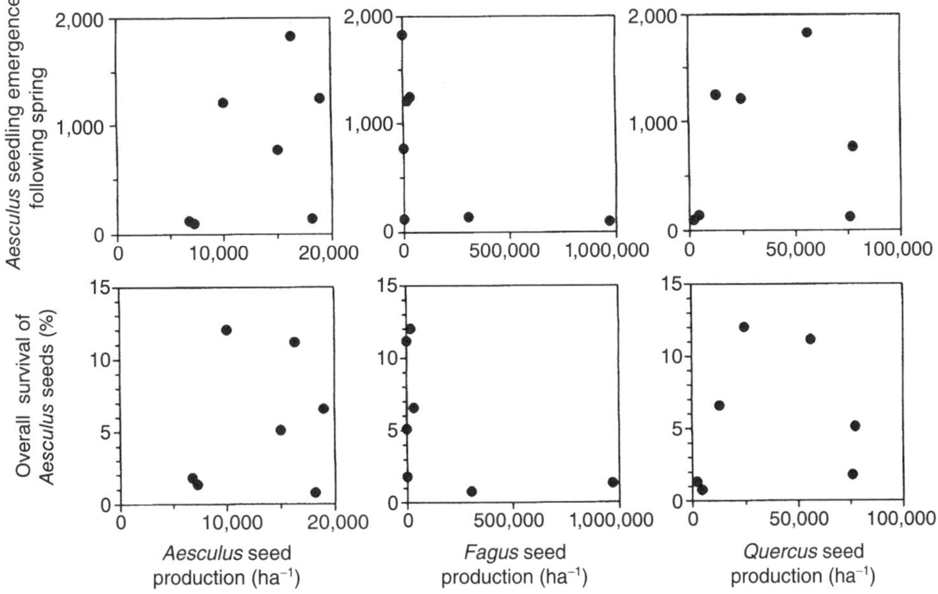

Fig. 15.1. Relationships between seed production of three large-seeded species and overall success of *A. turbinata* seeds. Note that *A. turbinata* seed success was not correlated with its own seed production and was reduced in years of *F. crenata* masting.

Consideration of the availability of the seed of the other tree species provided some support for the indirect effect of *Fagus* masting on *Aesculus* seed survival. *Fagus* exhibited distinct mast years, fruiting only in 1993 (small crop) and 1995 (large crop) during the 7-year monitoring period, while *Q. mongolica* showed an 'alternate bearing' pattern (*sensu* Crawley and Long, 1995) (Fig. 15.2a). Both *A. turbinata* seed survival (< 1.5%) and seedling emergence (< 150 seedlings ha^{-1}) were particularly low during the *F. crenata* mast year and among the lowest in the 7-year observation period (Fig. 15.1). No such trend was observed when *Q. mongolica* massively fruited (Fig. 15.1).

Short-term analysis of seed fate of A. turbinata

Before the seeds of *A. turbinata* established seedlings or were consumed, they experienced various fate pathways through secondary dispersal (Fig. 15.3). Over the 3 years, the initial fate of most seeds was to be placed in a primary cache. Only in 1996 were more seeds

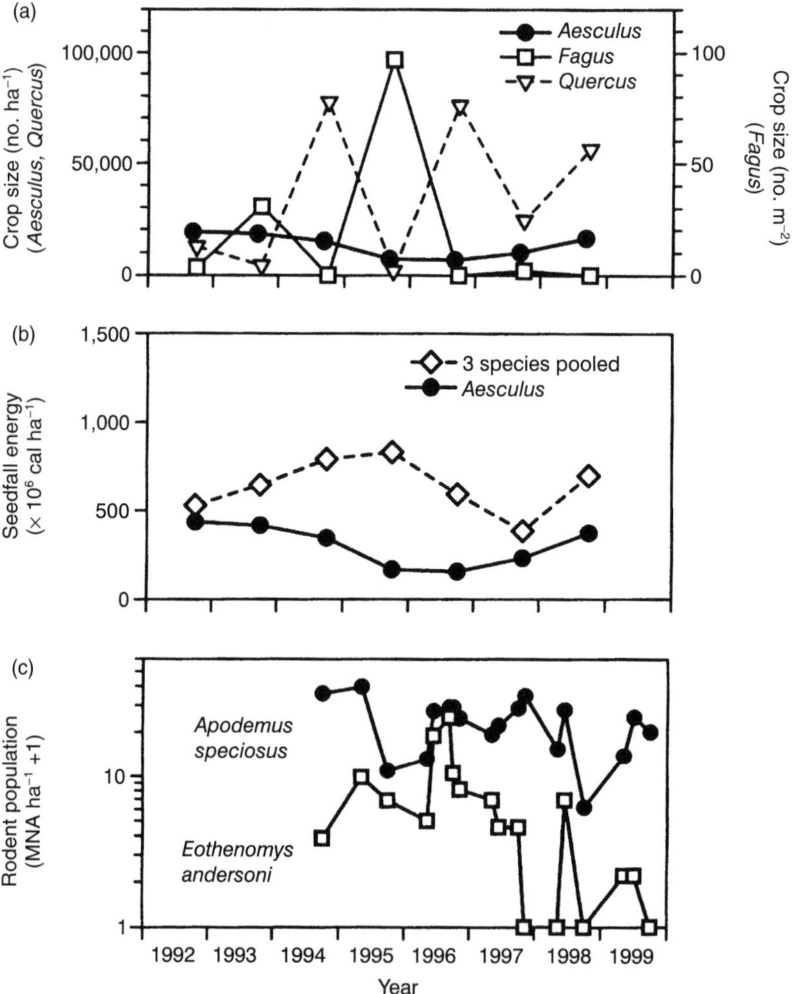

Fig. 15.2. Annual variations in (a) community-wide production of large seeds and (b) their energy values as food resources for consumers, and (c) corresponding numerical responses of rodents.

initially destroyed (60%) than cached (values for 1995 and 1997 were 34% and 17%, respectively). Seed predation was the predominant source of seed mortality, accounting for between 93 and 100% of seed death.

Caches usually contained one seed, though observation in other years indicates that caches with more than ten seeds sometimes occur (Hoshizaki *et al.*, 1997). For most seeds, caching simply delayed subsequent seed predation, as most seeds in primary caches were usually consumed. The ultimate fate of the few seeds that were removed from primary caches and placed into secondary caches was also seed consumption. Although the fate of all seeds was not known, the proportion of seeds that germinated was low, ranging from only 1 to 3%, with over 80% of seeds suffering predation in most years.

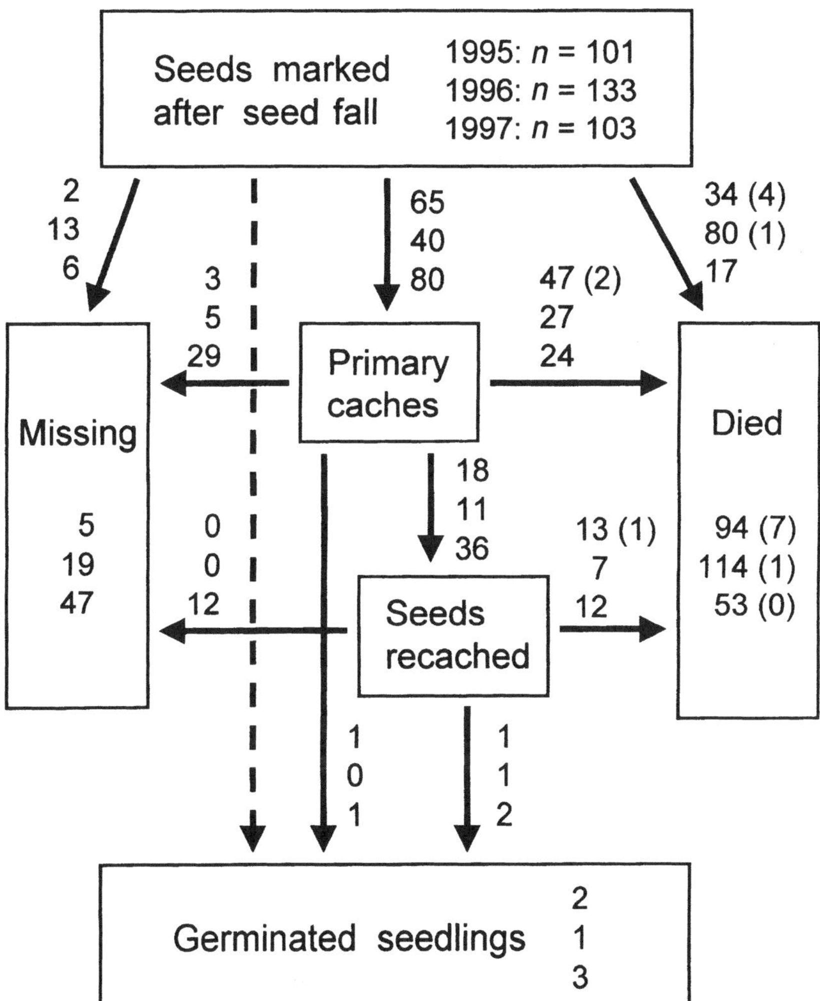

Fig. 15.3. Seed-fate pathways for *Aesculus turbinata*. The three numbers on the side of each arrow indicate transition frequencies for 1995 (top), 1996 (middle) and 1997 (bottom). Because some seeds were moved several times from 'primary cache' to 'seeds recached', summed values for arrows coming into the 'recache' category do not equal the sum of the values for arrows leaving this category. Numbers in parentheses represent the number of seeds that died for reasons other than predation. The broken arrow indicates a pathway which can potentially occur but was never observed during the study.

Seasonal mortality of A. turbinata *seeds*

Levels of seed removal by rodents varied both within the autumn fruiting period and also between years. In 1995, the rate of *A. turbinata* seed removal suddenly decreased in October, despite an abundance of seeds (Fig. 15.4). This coincided with a rapid increase in the density of *F. crenata* seeds and a corresponding increase in *F. crenata* seed removal. This pattern in *A. turbinata* seed predation was not evident in 1996 or 1997 when there was no *F. crenata* masting. Removal of *F. crenata* seed was significantly correlated with its density on the

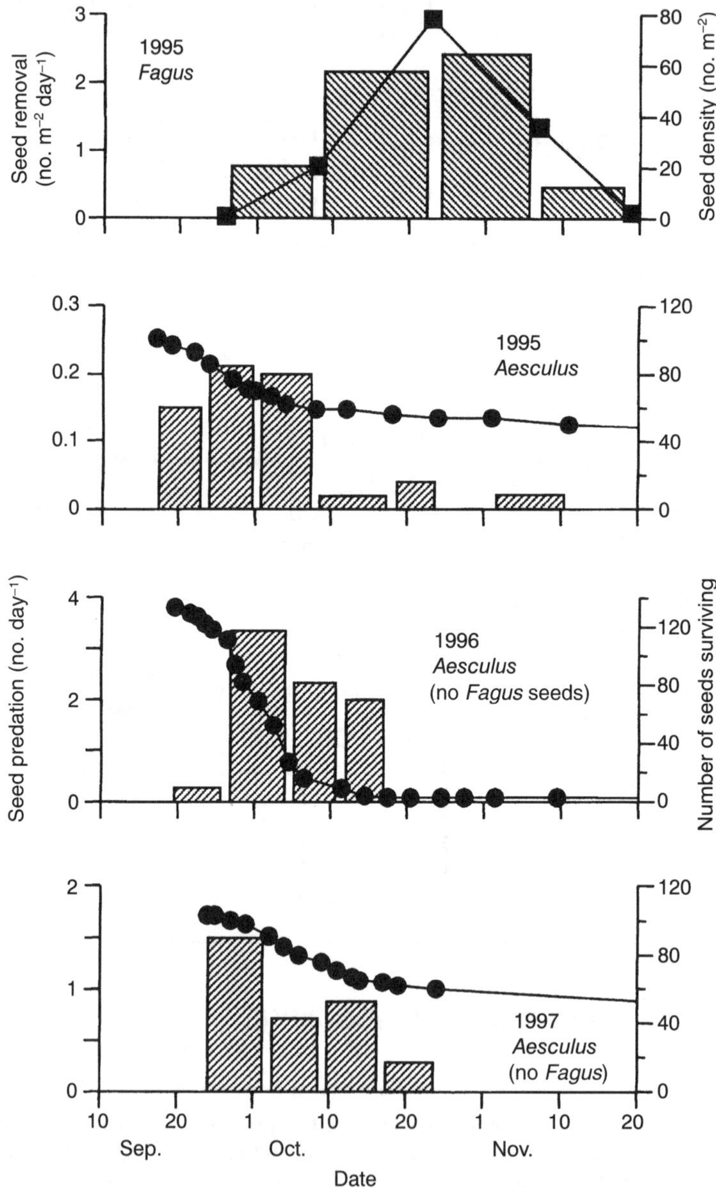

Fig. 15.4. Seasonal changes in seed predation on *Fagus crenata* (top) and on *Aesculus turbinata* (remainder). The rates of predation (bars) and numbers (*A. turbinata*) and density (*F. crenata*) of remaining seeds (lines) are shown. No *F. crenata* seeds were produced in 1996 or 1997.

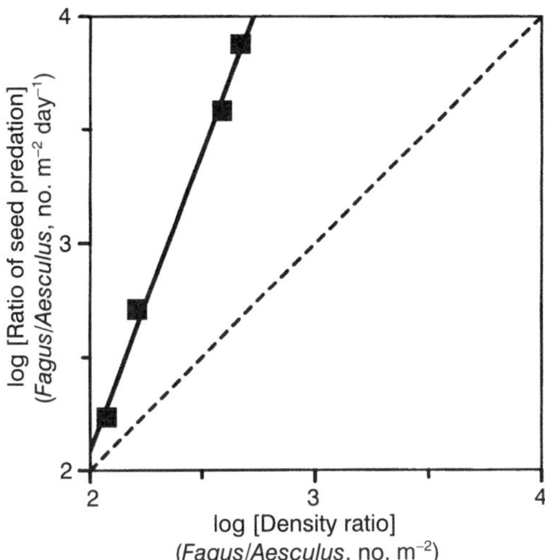

Fig. 15.5. Analysis of frequency dependence in seed predation on *A. turbinata* versus *F. crenata*. A slope of >1 in the regression line indicates stronger preference for *F. crenata* seeds over *A. turbinata* by the predators. Dashed line (slope = 1) shows frequency-independent predation. We obtained a significant regression line with the parameter $b = 3.17$; $r = 0.997$, d.f. = 2, $P = 0.003$.

forest floor ($r^2 = 0.92$, d.f. = 2, $P = 0.04$). Rodents exhibited a strong, consistent preference to *F. crenata* seeds over *A. turbinata* and this preference increased as *F. crenata* seed became more abundant (Fig. 15.5). The results of the proportional-hazards model support the finding that the seasonal change of the seed predation was due to selective consumption of *F. crenata*. The hazard ratio for 1995 was significantly smaller (0.0473, 95% confidence interval 0.0134–0.167) when compared with the ratio in either 1996 (1.426, $CI_{95\%}$ 0.947–2.14, $P < 0.05$) or 1997 (0.483, $CI_{95\%}$ 0.202–1.15, $P < 0.05$). This means that the extent of change in the shape of survivorship curve was greatest in 1995 and was associated with the masting period. In spring, patterns of seed survival were reversed and the mortality of *A. turbinata* seeds was higher in the mast year than in 1996 or 1997 (Fig. 15.6). As a result, the overall survivorship of seeds from the mast year was lower than that from non-mast years.

Annual variation in food resources and the numerical response of rodents

Although the species-pooled seed fall abundance varied greatly among years (CV = 142.0%), seed fall energy showed a very small among-year variation (CV = 24.0%), varying by 2.2-fold for max./min. (Fig. 15.2a, b). This was attributable to annual, relatively constant (CV = 38.7%) fruiting of large seeds of *A. turbinata*. However, densities of the two rodent species varied by an order of magnitude. There was no clear relationship between rodent abundance and any of the seed-fall measures, although their populations increased rapidly in the summer following *F. crenata* masting (Fig. 15.2c).

Discussion

Studies of seed dynamics in the Kanumazawa Riparian Research Forest provide considerable insights into the role of community processes influencing plant–granivore interactions. In each of the 3 years studied, the vast majority of *A. turbinata* seeds were removed by rodents. For those seeds whose fate we were able to record, it was evident that rodents destroyed between 90 and 99% of seeds in any 1 year. Nevertheless, although seed predation was particularly intense, between 1% and 5% of seeds successfully germinated from caches to produce seedlings. If seedlings can only

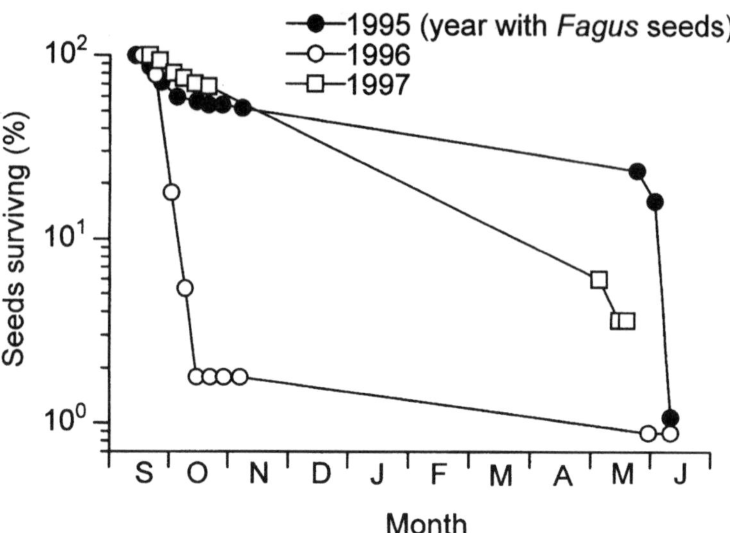

Fig. 15.6. Survivorship curves of *A. turbinata* seeds.

arise from cached seeds, e.g. because seed burial is necessary for successful germination or seeds are transported to suitable microsites, then the net effect of rodent seed removal on plant regeneration will be positive (see Hulme, this volume). Previous research on *A. turbinata* in the Kanumazawa Riparian Research Forest indicates that burial may not be necessary for germination and that seeds are not dispersed to particularly suitable microsites (Hoshizaki *et al.*, 1997, 1999). This suggests that seed dispersal by rodents is not an essential element in the regeneration of *A. turbinata*. Furthermore, there is no guarantee that the small proportion of seeds that actually germinate from caches will subsequently recruit to the adult population. Therefore, the high levels of seed destruction indicate that rodents were acting primarily as seed predators, rather than seed-dispersers.

Although the three dominant tree species show markedly different and uncorrelated fruiting patterns, the resources available to polyphagous rodent seed predators remain, in comparison, relatively constant over the 7 years. We would therefore expect the numerical responses of granivorous rodents to be buffered by the community-wide seed fall. While the temporal dynamics of the rodent populations are variable, there is no clear evidence of them being driven by changes in the seed-fall patterns at either the community or the species level. In such a system, satiation of polyphagous granivores via masting is unlikely to succeed in greater seedling success. Thus, for *A. turbinata* there is no evidence of predator satiation or economies of scale. This situation contrasts with tropical dipterocarp forests, where community-wide synchrony facilitates predator satiation (Curran and Leighton, 2000; Curran and Webb, 2000).

The absence of any clear association between rodent abundance and seed-fall patterns at either the community or the individual species level contrasts with findings where generalist seed consumers respond numerically to increased seed production (Wolff, 1996; Ostfeld and Keesing, 2000). Again, contrary to previous findings (Sork, 1983; Jensen, 1985; Manson *et al.*, 1998), increased seed fall did not increase *A. turbinata* seedling regeneration. These results suggest that, at least in the Kanumazawa Riparian Research Forest, the granivore–plant interaction is not a simple function of the overall abundance of seed resources.

An intuitive assumption of community-wide asynchrony in seed fall is that less-preferred species may benefit from reduced predation during periods when the abundance of a more preferred species increases. The frequency-dependence analysis indicates that rodents strongly prefer smaller *F. crenata* seeds over larger *A. turbinata* seeds. This result is consistent with a field experiment where a choice test was undertaken by providing rodents with both seeds simultaneously (Miguchi, 1996). Since *F. crenata* seeds are rich in per-gram energy and do not contain any saponins or tannins (K. Hoshizaki, unpublished data), this species is expected to be the most palatable among the large-seeded species in our forest. We would therefore expect that there would be less *A. turbinata* seed predation in an *F. crenata* mast year than in a non-mast year. For those seeds whose ultimate fate was known (i.e. not missing), rodents destroyed 91% in the mast year, whereas 99% and 95% were destroyed in the two subsequent non-mast years. This trend is consistent with our expectation, but the significance of the differences among years is questionable. This finding is supported by comparison of the seedling numbers in each of the 3 years. We might expect lower rates of seed predation in the *F. crenata* mast year to facilitate a higher number of *A. turbinata* seedlings. Surprisingly, *A. turbinata* seed survival and seedling emergence were especially low during the *F. crenata* mast period, in complete contrast to our expectations. The highest recorded seedling survival was found in the 1997 non-mast year (K. Hoshizaki, unpublished data). This would suggest that, although *F. crenata* is a highly preferred species, the switching in rodent foraging from *A. turbinata* to *F. crenata* had only a minor effect on subsequent *A. turbinata* seedling regeneration.

Comparison of the daily predation rates (Fig. 15.4) may shed light on this apparent paradox. Rates of *A. turbinata* seed predation in 1995 were an order of magnitude less than in either 1996 or 1997. Although predation rates may have been influenced by the *F. crenata* seed fall (Fig. 15.5), seed predation rates were already low prior to the mast period. These low rates of seed predation appear to reflect both the relatively low levels of *A. turbinata* seed production and the abundance of both rodent species (Fig. 15.2). In 1996, *A. turbinata* seed production was similar to that in 1995 and yet daily rates of *A. turbinata* removal were the highest recorded during the 3 years of study. This coincides with a dramatic increase in the densities of both rodent species in 1996. This may be an indirect effect of the *F. crenata* mast in 1995. By 1997, rodent numbers (especially *E. andersoni*) had declined from the peak in 1996. The result was moderately high rates of seed predation but insufficient to reduce seed numbers as substantially as in 1996. It therefore appears that the patterns of seed predation more or less reflect the change in rodent densities. The abundance of *A. speciosus* in monodominant beech forest of Japan clearly follows the masting cycles of *F. crenata* (Miguchi, 1988). It would therefore be reasonable to expect that the fluctuation in rodent populations reflect, to some extent, the availability of the more preferred species (*F. crenata*) rather than those of the less preferred *A. turbinata*. However, the role of *F. crenata* mast years in the rodent population dynamics cannot be discerned from the data available (Fig. 15.2c). This seems to suggest that there is little evidence for a direct effect of *F. crenata* on these patterns, but the indirect effect through a delayed numerical response in rodent densities cannot be discounted. The considerable annual variation in *A. turbinata* regeneration success contrasts with the relatively constant annual seed fall in this species. This would suggest that other drivers, external to *A. turbinata*, play a role in the regeneration of this species. It is therefore possible that other tree species indirectly affect the regeneration of *A. turbinata* in our multispecies system, though caution must be applied when making generalizations based on only one masting event.

In summary, the value of this research is that it highlights the complexity of the granivore–plant interaction. Clearly, studies based on only 1 year tell us little about the interaction. Even the present 3-year study can only speculate about some of the mechanisms underlying seedling regeneration. Unfortunately, it is evident that the impact of masting needs to be studied over much longer periods. Nevertheless, future studies must assess changes in the focal species as well as other species in the community, not only the granivores but also co-dominant plant species.

Acknowledgements

This chapter greatly benefited from critical comments and helpful suggestions from Doug

Levey and Richard Ostfeld. We also appreciate Takuya Kubo for helping with data analysis; Takashi Masaki for statistical advice and field assistance; and Tohru Nakashizuka and Wajirou Suzuki for continuous encouragement during the course of this study. K.H. was funded by Research Fellowships of the Japan Society for the Promotion of Science for Young Scientists.

References

Crawley, M.J. (1992) Seed predators and plant population dynamics. In: Fenner, M. (ed.) *Seeds: The Ecology of Regeneration in Plant Communities.* CAB International, Wallingford, UK, pp. 157–191.

Crawley, M.J. and Long, C.R. (1995) Alternate bearing, predator satiation and seedling recruitment in *Quercus robur* L. *Journal of Ecology* 83, 683–696.

Curran, L.M. and Leighton, M. (2000) Vertebrate responses to spatiotemporal variation in seed production of mast-fruiting *Dipterocarpaceae. Ecological Monographs* 70, 101–128.

Curran, L.M. and Webb, C.O. (2000) Experimental tests of the spatiotemporal scale of seed predation in mast-fruiting *Dipterocarpaceae. Ecological Monographs* 70, 129–148.

Greenwood, J.J.D. (1985) Frequency-dependent selection by seed predators. *Oikos* 44, 195–210.

Greenwood, J.J.D. and Elton, R.A. (1979) Analysing experiments on frequency-dependent selection by predators. *Journal of Animal Ecology* 48, 721–737.

Hamajima, N. (1993) *Clinical Studies by Multivariate Analysis: Introduction to Proportional Hazards Model and Logistic Model with Application Programs of SAS*, 2nd edn. University of Nagoya Press, Nagoya, 214 pp. (in Japanese).

Hoshizaki, K. (1999) Regeneration dynamics of a sub-dominant tree *Aesculus turbinata* in a beech-dominated forest: interactions between large-seeded tree guild and seed/seedling consumer guild. Doctoral dissertation, Kyoto University, Kyoto, Japan.

Hoshizaki, K., Suzuki, W. and Sasaki, S. (1997) Impacts of secondary seed dispersal and herbivory on seedling survival in *Aesculus turbinata. Journal of Vegetation Science* 8, 735–742.

Hoshizaki, K., Suzuki, W. and Nakashizuka, T. (1999) Evaluation of secondary dispersal in a large-seeded tree *Aesculus turbinata*: a test of directed dispersal. *Plant Ecology* 144, 167–176.

Hulme, P.E. (1993) Post-dispersal seed predation by small mammals. *Symposium of the Zoological Society of London* 65, 269–287.

Hulme, P.E. (1996a) Herbivory, plant regeneration, and species coexistence. *Journal of Ecology* 84, 609–615.

Hulme, P.E. (1996b) Natural regeneration of yew (*Taxus baccata* L.): microsite, seed or herbivore limitation? *Journal of Ecology* 84, 853–861.

Hulme, P.E. (1997) Post-dispersal seed predation and the establishment of vertebrate dispersed plants in Mediterranean scrublands. *Oecologia* 111, 91–98.

Hulme, P.E. (1998) Post-dispersal seed predation: consequences for plant demography and evolution. *Perspectives in Plant Ecology, Evolution and Systematics* 1, 32–46.

Hulme, P.E. and Benkman, C.W. (2001) Granivory. In: Herrera, C.M. and Pellmyr, O. (eds) *Plant–Animal Interactions.* Blackwell Science, Oxford.

Hulme, P.E. and Borelli, T. (1999) Variability in post-dispersal seed predation in deciduous woodland: relative importance of location, seed species, burial and density. *Plant Ecology* 145, 149–156.

Hulme, P.E. and Hunt, M.K. (1999) Rodent post-dispersal seed predation in deciduous woodland: predator response to absolute and relative abundance of prey. *Journal of Animal Ecology* 68, 417–428.

Janzen, D.H. (1971) Seed predation by animals. *Annual Review of Ecology and Systematics* 2, 465–492.

Jensen, T.S. (1985) Seed–seed predator interactions of European beech, *Fagus sylvatica* and forest rodents, *Clethrionomys glareolus* and *Apodemus flavicollis. Oikos* 44, 149–156.

Kelly, D. (1994) The evolutionary ecology of mast seeding. *Trends in Ecology and Evolution* 9, 465–470.

Kollmann, J. (1995) Regeneration window for fleshy-fruited plants during scrub development on abandoned grassland. *Ecoscience* 2, 213–222.

Kollmann, J., Coomes, D.A. and White, S.M. (1998) Consistencies in post-dispersal seed predation of temperate fleshy-fruited species among seasons, years and sites. *Functional Ecology* 12, 683–690.

Manson, R.H., Ostfeld, R.S. and Canham, C.D. (1998) The effects of tree seed and seedling density on predation by rodents in old fields. *Ecoscience* 5, 183–190.

Miguchi, H. (1988) Two years of community dynamics of murid rodents after a beechnut mastyear. *Journal of Japanese Forestry Society* 70, 472–480 (in Japanese with English summary).

Miguchi, H. (1996) Study on the ecological interactions of the regeneration characteristics of

Fagaceae and the mode of life of wood mice and voles. Doctoral dissertation, Niigata University, Niigata, Japan (in Japanese).

Morin, P.J. (1999) *Community Ecology*. Blackwell Science, Malden, Massachusetts, 424 pp.

Myster, R.W. and Pickett, S.T.A. (1993) Effect of litter, distance, density and vegetation patch type on post-dispersal tree seed predation in old fields. *Oikos* 66, 381–388.

Norton, D.A. and Kelly, D. (1988) Mast seeding over 33 years by *Dacrydium cupressinum* Lamb. (rimu) (Podocarpaceae) in New Zealand: the importance of economies of scale. *Functional Ecology* 2, 399–408.

Ostfeld, R.S. and Keesing, F. (2000) Pulsed resources and community dynamics of consumers in terrestrial ecosystems. *Trends in Ecology and Evolution* 15, 232–237.

Ostfeld, R.S., Jones, C.G. and Wolff, J.O. (1996) Of mice and mast: ecological connections in eastern deciduous forests. *BioScience* 46, 323–330.

Schupp, E.W. (1990) Annual variation in seedfall, postdispersal predation, and recruitment of a neotropical tree. *Ecology* 71, 504–515.

Sork, V.L. (1983) Mammalian seed dispersal of pignut hickory during three fruiting seasons. *Ecology* 64, 1049–1056.

Sork, V.L. (1993) Evolutionary ecology of mast-seeding in temperate and tropical oaks (*Quercus* spp.). *Vegetatio* 107/108, 133–147.

StatSoft, Inc. (1995) *STATISTICA Users' Manual*. StatSoft, Tulsa, Oklahoma, USA.

Suzuki, W., Osumi, K., Masaki, T., Takahashi, K., Daimaru, H. and Hoshizaki, K. (2001) Disturbance regimes and community structures of a riparian and an adjacent terrace stand in the Kanumazawa Riparian Research Forest, northern Japan. *Forest Ecology and Management* (in press).

Whelan, C.J., Willson, M.F., Tuma, C.A. and Souza-Pinto, I. (1991) Spatial and temporal patterns of post-dispersal seed predation. *Canadian Journal of Botany* 69, 428–436.

Willson, M.F. and Whelan, C.J. (1990) Variation in postdispersal survival of vertebrate-dispersed seeds: effects of density, habitat, location, season, and species. *Oikos* 57, 191–198.

Wolff, J.O. (1996) Population fluctuations of mast-eating rodents are correlated with production of acorns. *Journal of Mammalogy* 77, 850–856.

Wootton, J.T. (1994) The nature and consequences of indirect effects in ecological communities. *Annual Review of Ecology and Systematics* 25, 443–466.

16 Seasonality of Fruiting and Food Hoarding by Rodents in Neotropical Forests: Consequences for Seed Dispersal and Seedling Recruitment

Pierre-Michel Forget,[1] David S. Hammond,[2] Tarek Milleron[3] and Raquel Thomas[2]

[1]*Muséum National d'Histoire Naturelle, Laboratoire d'Ecologie Générale, CNRS-MNHN UMR 8571, 4 avenue du petit Château, F-91800 Brunoy, France;* [2]*Iwokrama International Centre for Rain Forest Conservation and Development, PO BOX 10630, 67 Bel Air, Georgetown, Guyana;* [3]*Department of Rangeland Resources and the Ecology Center, Utah State University, Logan, UT 84322-5230, USA*

Introduction

> Climatic variation causes a season of scarcity and the agoutis hoard seeds to overcome this.
> (Smythe, 1978, p. 48)

Fruit resource availability is highly seasonal in many tropical forests (Leigh, 1999; Wright *et al.*, 1999). Within-year variation in the diversity and abundance of food resources imposes dietary restrictions on animals by requiring them to consume what food is available in time and accessible in space. During periods of peak fruit abundance, many frugivores forage on the same species of fruiting trees (e.g. Guillotin *et al.*, 1994). When fruit is scarce, diets diverge as vertebrates with different body masses, home-range sizes and foraging tactics seek alternative resources in different ways. For example, peccaries (*Tayassu tajacu*) switch from fruit to worms, and spiny rats (*Proechimys cuvieri* and *Proechimys guianensis*) switch from fruit to insects (Henry, 1994a). The agouti (*Dasyprocta* spp.) and the acouchi (*Myoprocta* spp.), however, are exceptions. They remain frugivorous throughout the year, but alternate from pulp-centred to seed-centred diets during periods of high and low fruit production, respectively (Henry, 1999). In both South and Central American forests, these caviomorph rodents survive during lean periods by eating the seeds they buried when fruit was abundant (Smythe, 1970, 1978; Henry, 1999).

Seed hoarding is crucial to the survival and reproduction of many rodent species (Smythe, 1978; Henry, 1994b; Stapanian, 1986; Vander Wall, 1990). Seeds consumed during lean periods are assumed to be taken from caches or from seedlings with long-lasting seed reserves. In general, food caching takes place in anticipation of a decline in resource availability (e.g. winter in the temperate regions, the dry season in the tropics) by animals that cannot meet their energetic requirements through body fat reserves alone. Seed removal and scatter-hoarding rates most probably depend on seasonal patterns of fruit availability, animal foraging and reproductive behaviour (Smythe, 1978; Dubost, 1988; Henry, 1994b). In temperate regions, it is well known that seed caching peaks when large-seeded plants fruit (Stapanian, 1986; Vander Wall, 1990). In these regions, rodent foraging behaviour may also depend on photoperiod or on interactions between external signals (e.g. day length, food availability) and internal states (e.g. hormone and hunger levels), both of which affect agonistic, reproductive and hoarding behaviour (Vander Wall, 1990; Henry, 1997).

Smythe (1978, p. 25) observed in Panama 'that during times of food abundance, all agoutis hoard considerably more seeds than they eat' and 'when falling fruit becomes scarce, agoutis live almost entirely on scatterhoarded seeds'. Despite recurrent references to the seasonality of rodent seed caching in neotropical forests (Morris 1962; Rankin, 1978; Smythe, 1970, 1978; Vandermeer *et al.*, 1979; Hallwachs, 1986), we lack a theoretical framework for determining the effect of seasonal fruit availability on caching rates and, further, for investigating the implications of this seasonal dependence for plant and animal fitness.

How is seed-hoarding behaviour regulated or triggered in relation to seasonal fluctuations in food abundance in neotropical rain forests? How does seed caching affect pre- and post-dispersal insect seed predators? How does food scarcity affect the residence time of cached seeds? When food is abundant, does caching decline or stop, leaving many seeds vulnerable to other vertebrate and invertebrate seed predators (Smythe, 1970)? Such questions remain largely unexplored.

The aim of this chapter is to present a general, annually based model for hoarding in neotropical forests inhabited by large caviomorph rodents. We first summarize data on the range of seed removal and predation–hoarding ratios for four large-seeded tree species in one neotropical rain forest. Based in part on data collected for these four species, we then present a model of seed hoarding in relation to fruit resource abundance. Finally, we discuss the model's implications for seed fate and seedling recruitment in light of the recent literature on seed dispersal by neotropical rodents.

Patterns of Seed Removal and Seed Fate: Data from Barro Colorado Island

The impact of rodents as seed predators and dispersers was analysed during a 1-year fruiting cycle in 1990 on Barro Colorado Island (BCI), Panama. On BCI, fruit is generally abundant during a 6-month period, from February to July, with peak fruit diversity roughly in the middle of this period (Smythe, 1970; Foster, 1982; Wright *et al.*, 1999). According to Smythe (1978), this is a period of intense seed caching. We expect that, as fruit abundance increases, animals probably become satiated and destroy a decreasing proportion of all seeds and fruits available. Before and after peak fruit fall, rodent removal and consumption of fallen fruits and seeds is proportionately greater. Based on preliminary observations of seasonal seed removal and hoarding in Panama, Costa Rica and French Guiana (Smythe, 1978; Hallwachs, 1986; Forget, 1990), we predict that the proportion of seeds hoarded will peak shortly after fruit production peaks in a community. To assess this prediction, we report seed-removal and seed-fate experiments carried out between January and October 1990, using *Dipteryx panamensis* (*Fabaceae*), *Astrocaryum standleyanum* (*Arecaceae*), *Gustavia superba* (*Lecythidaceae*) and *Attalea butyraceae* (*Arecaceae*). Seeds of each species have largest diameter > 1.5 cm and consist of single-seeded, nutlike fruits (*D. panamensis, A. standleyanum* and *A. butyraceae*) or of large, multiple-seeded capsules (*G. superba*) (Smythe, 1970). Because of minimal overlap in fruiting periods of these species over an annual cycle (D. Windsor, Terrestrial Environmental Sciences Program, personal communication), these four species are key resources for scatter-hoarding rodents, such as agoutis, squirrels and spiny rats; together they provide food throughout the year on BCI (Heaney and Thorington, 1978; Glanz *et al.*, 1982; Smythe *et al.*, 1982; De Steven and Putz, 1984; Sork, 1987; Smythe, 1989; Wright, 1990; Adler,

1995). For these tree species, we present 1 year of seed-fate data, upon which we then build a general model of seed hoarding.

Experiments with *D. panamensis* seeds were conducted in late January and replicated in late February and March 1990, using thread-marked seeds to locate and determine the fate of removed seeds (Forget, 1993). Most seeds were removed by agoutis (*Dasyprocta punctata*). Pooled data show that the per cent of seeds cached (from initial locations > 50 m from fruiting trees, simulating primary seed dispersal by bats) was lowest in February and April and highest in March (Fig. 16.1a), when seeds accumulated beneath parent trees and were eaten by rodents and peccaries (see De Steven and Putz, 1984). Caching of *D. panamensis* seeds therefore reached a maximum just after peak fruit fall of this species.

By April–May, fruit biomass and diversity are at their maximum on BCI. The abundant fruit appears to saturate rodents, leading to accumulation of uneaten seeds (Wright, 1990). Assuming that caching is triggered by food satiation (e.g. Vander Wall, 1990), we expect higher seed removal and caching rates by rodents at this time. Using a population of seven *A. standleyanum* palms, which occupied the intersection of several agouti home ranges, Milleron and Forget (1997) placed two groups of 25 thread-marked seeds under each of five palms, first on 30 April and then on 19 May 1990. We checked seed removal 12 h after the 6:00 a.m. seed placement, and searched for seeds within a 20 m radius of each palm. A large proportion of seeds (91%) were taken within 12 h; all were removed within a week. Most seeds in each trial were not found within 20 m. A smaller proportion was cached and eaten (Fig. 16.1b). We also observed mammal activity under individual palms in two continuous sessions of 3 h and 12.5 h from 9 to 13 June. During these observations, a majority of *A. standleyanum* seeds handled ($n = 105$) by *D. punctata* were cached (61%). The rest were eaten (19%) or removed at least 20 m (20%). If seeds removed > 20 m were cached, as direct observations of foraging animals suggest (Milleron and

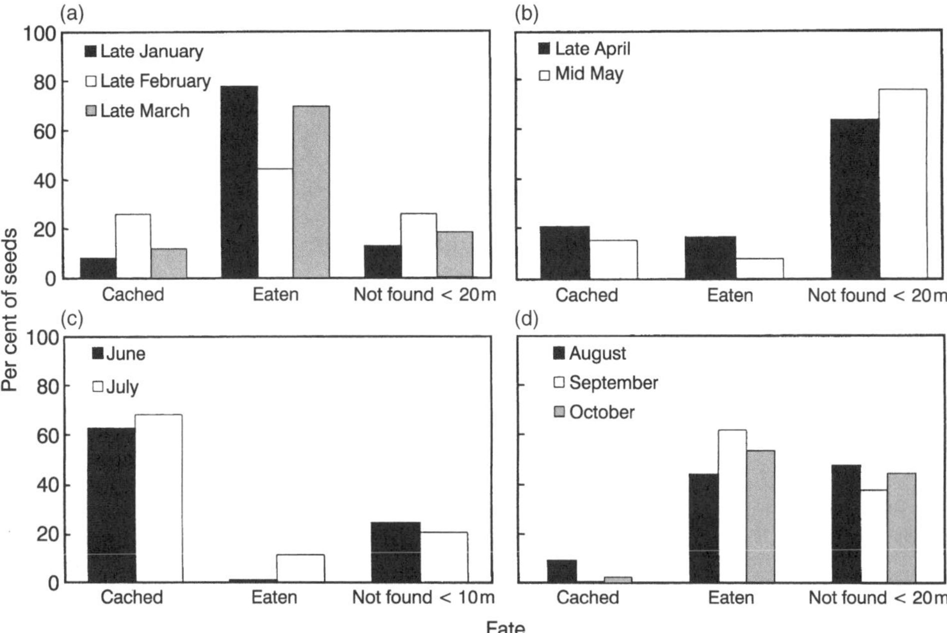

Fig. 16.1. Seed fate in (a) *Dipteryx panamensis* (out of $n = 320$ seeds per month), (b) *Astrocaryum standleyanum* ($n = 250$ seeds per month), (c) *Gustavia superba* ($n = 200$ seeds per month) and (d) *Attalea butyraceae* ($n = 400$ seeds per month) in 1990 at Barro Colorado Island, Panama. (Adapted from Forget, 1993; Forget *et al.*, 1994, 1998; Milleron and Forget, 1997).

Forget, 1997), then our results are consistent with Smythe's (1978) assertion that, when food is abundant, agoutis cache considerably more than they eat – in this case, approximately four seeds were cached for every seed eaten in May–June. Fates of the cached seeds were undetermined. They were rapidly removed, probably taken by rodents to secondary caches (see Jansen and Forget, 2001). Indeed, we occasionally retrieved cached seeds > 100 m from their origin. These seeds had been marked as cached within 20 m of their origin (T. Milleron and P.-M. Forget, personal observations).

Experiments with *G. superba* showed that seed removal was significantly higher in July (100%), when fruit was less abundant, than in June (88%), when fruit was more abundant. Unremoved seeds germinated but all were later eaten by vertebrates. The majority of seeds retrieved within 10 m of their deposition site were cached and few seeds were eaten; the remainder were assumed to be cached further away (Forget, 1992; Forget *et al.*, 1998) (Fig. 16.1c). (Of cached seeds, approximately 13% established and survived in the short term, despite high post-dispersal predation upon germination, especially by rodents (Forget *et al.*, 1998; Dalling and Harms, 1999).)

As food diversity declines on BCI, the spatial distribution of resources becomes concentrated around the few scattered tree species that drop their fruit in July–August (e.g. *A. butyraceae* (Wright, 1990)). All seeds placed beneath four fruiting *Attalea* palms during this period were removed. The majority were eaten and a few were cached < 20 m away (Fig. 16.1d). As in *A. standleyanum*, all of the dispersed seeds were removed by rodents in the following weeks, to be either consumed or recached.

A Model for Seed Fate: Hoarding versus Predation

> When an agouti encounters a concentrated food source, it usually eats a few of the fruits or seeds and then starts to bury them.
>
> (Smythe, 1978, p. 25)

Tropical rodents rely principally on caches – rather than stored body fat – to survive lean periods and, because the daily amount of food they consume is relatively constant throughout the year (e.g. Henry, 1999), seed-caching behaviour probably mirrors the seasonality of fruit and seed resources. On BCI, a 1-year fruiting cycle can be roughly divided in half: a 6-month period of relatively abundant fruit in February–July and a 6-month lean period in June–January (Wright *et al.*, 1999) (Fig. 16.2a). Using the large data set from BCI (1987–1999) and performing a Fourier analysis in the time-series procedure of Systat 9 (SPSS, 1999) on the number of species bearing fruit between January 1987 and December 1999 (O. Calderon and S.J. Wright, personal communication), we fitted data from three sets of four consecutive years (1987–1990, 1991–1994 and 1995–1998) to a sine-wave function. Using the entire 12-year data set, only a 1-year (annual) cycle was evident (data not shown), strongly demonstrating the seasonal periodicity of fruiting on BCI. We expect the hoarding behaviour of rodents to be more tightly related to three 4-month periods (high, declining and low) than to two 6-month periods (peak versus lean) of fruiting. Accordingly, we followed Guillotin *et al.* (1994) and delimited three 4-month periods of fruit diversity summing to the annual cycle for BCI (Table 16.1). Periods were defined as follows: (i) a high fruit season (HFS), centred on the 2 months with the highest fruit diversity; (ii) a declining fruit season (DFS) period, corresponding to a decrease in fruit diversity; and (iii) a low fruit season (LFS), characterized by an increase in fruit diversity, starting from the annual low of fruit diversity.

Data on the percentage of seeds (out of the number of seeds placed) known to be cached and the percentage assumed to be cached (i.e. not found at < 10 or 20 m) were pooled. These 'hoarded seeds' are not necessarily scatter-hoarded seeds, but include those taken to other types of caches in burrows or trees; this is equivalent to seed burial from the animal's perspective. Seed removal is therefore considered equivalent to hoarding, as is often reported in seed-predation studies. However, our use of seed-removal data excludes de facto the fraction of fallen seeds that are immediately consumed by invertebrates and vertebrates, defined *sensu stricto* as pre- and

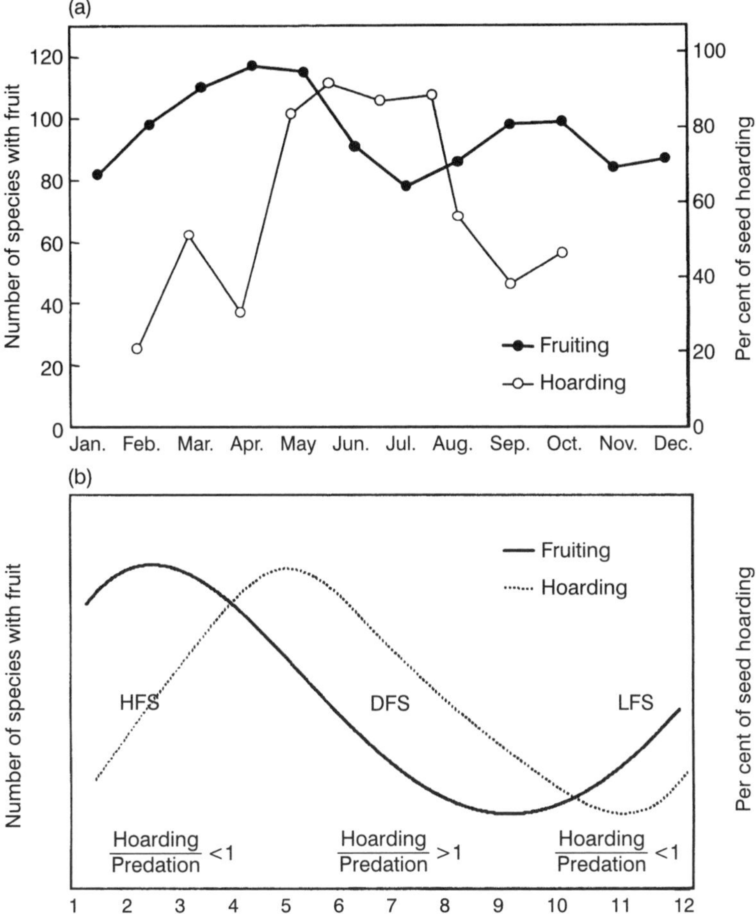

Fig. 16.2. (a) Seasonal fruit-fall pattern and level of hoarding activity for the four study species in the study year (1990) (O. Calderon and S.J. Wright, personal communication). (b) Model of fruit-fall pattern and intensity of hoarding activity during a 1-year cycle, and expected ratio of % seed hoarding vs. % seed predation in the three fruiting seasons. HFS, high fruit season; DFS, declining fruit season; LFS, low fruit season. See text for explanation. The fruiting curve is the polynomial function ($y = -0.0051x^5 + 0.1186x^4 - 0.2607x^3 - 8.1973x^2 + 35.444x + 87.788$; $R^2 = 0.96$), which best fits the data at Paracou, and the hoarding curve was drawn based on results from BCI (Fig. 16.1 and 16.2a) and Paracou (Forget, 1996).

post-dispersal seed predation. The data reveal a temporal pattern, with the maximum proportion of hoards made in the transition between HFS and DFS (Fig. 16.2a).

A graphical model of seed hoarding (as defined above) (Fig. 16.2b) follows the pattern of fruiting with a time-lag of approximately 2 months. First, when the number of fruiting species and fruit biomass is peaking (HFS), seed hoarding should be low, with a hoarding/predation ratio < 1, because rodents are satiated. When the diversity of fruit is highest, resources are widely distributed and rodents probably have access to fruit over their entire home range. Due to both low caching and low predation by rodents, a large fraction of seeds may lie where they fall. Hoarding increases toward the end of the fruit peak, when fruit diversity starts to decline.

Secondly, when fruit diversity is declining (DFS), fruit resources are often concentrated beneath a few, often large-seeded species. Although rodents at this time are not yet food-limited, the difference in spatial distribution of

Table 16.1. Mean number (± SD) of fruiting species during three fruiting seasons and corresponding months in five neotropical rain forests.

Field station	Country	Period	Method	Location	Fruiting season		
					HFS	DFS	LFS
BCI*	Panama	1987–1999 (13 years)	Fruit traps	9° 9′ N, 79° 51′ W	82.8 ± 12.9 (Mar.–Jun.)	56.7 ± 2.0 (Jul.–Oct.)	51.1 ± 9.3 (Nov.–Feb.)
Mabura Hill†	Guyana	1997–1998 (2 years)	Fruit traps	5° 13′ N, 58° 48′ W	12.8 ± 3.1 (Feb.–May)	5.8 ± 0.5 (Jun.–Sep.)	7.8 ± 1.2 (Oct.–Jan.)
Paracou‡	French Guiana	1989–1991 (2 years)	Raked trail censuses	5° 16′ N, 52° 55′ W	29.0 ± 4.0 (Jan.–Apr.)	14.5 ± 6.7 (May–Aug.)	11.8 ± 4.9 (Sep.–Dec.)
Pinkaití§	Brazil	1996–1998 (2 years)	Tree censuses	7° 46′ S, 51° 57′ W	24.9 ± 2.9 (Aug.–Nov.)	18.4 ± 4.8 (Dec.–Mar.)	15.9 ± 5.6 (Apr.–Jul.)
Cocha Cashu‖	Peru	1976–1977 (12 months)	Fruit traps	11° 54′ S, 71° 22′ W	12 ± 1.2 (Sep.–Dec.)	6.5 ± 5 (Jan.–Apr.)	3.8 ± 2.2 (May–Aug.)

*Wright *et al.* (1999) and O. Calderon and S.J. Wright (personal communication).
†Thomas (2000).
‡Henry (1994a) and O. Henry (personal communication).
§Baider (2000).
‖Terborgh (1983) and J. Terborgh (personal communication).
HFS, high fruit season; DFS, declining fruit season; LFS, low fruit season.

resources over their home range may trigger them to hoard seeds before the season of scarcity, just as temperate rodents do in the autumn for consumption in the winter (Stapanian, 1986; Vander Wall, 1990). During this second period, seed-caching rates will steadily increase and predation will remain low, leading to a hoarding/predation ratio > 1. The greater the rodents' reliance on their hoards, the less they will cache seeds they encounter during foraging and the greater their predation on any seeds encountered.

Finally, during a period of low fruit diversity (LFS), predation will probably surpass hoarding, with, for example, two to five seeds cached versus eight seeds eaten (ratio hoarding/predation < 1), as observed in *D. panamensis* in February, when community-wide fruit production was on the upswing (Forget, 1993) (Fig. 16.1a). Evidence of a similar hoarding/predation ratio was observed during the mid-LFS for a large-seeded liana, *Gnetum loyboldii* (*Gnetaceae*), which fruits in August–November on BCI (P.M. Forget, personal observation). Despite heavy predation on freshly fallen seeds, some seeds were obviously cached, as remains of *G. loyboldii* seeds were retrieved that had been dug up and gnawed as late as late February. As fruit diversity increases, rodents progressively hoard seeds in greater proportion to those they consume. Among small-seeded species, however, fruiting when fruit diversity is increasing may lead to significant seed predation and low, if not nil, secondary seed dispersal (Hammond, 1995; Forget *et al.*, 1998).

Supporting Evidence from Other Neotropical Sites

Approximately 20 years after Vandermeer *et al.* (1979) emphasized the lack of information on seed fate, an increasing number of studies are documenting seed fate across the neotropics (Table 16.2). Here we assess seed fates in relation to fruit-fall seasonality to evaluate our model. The methodology used to document fruiting seasonality is inconsistent among sites. The number of species in fruit per month is sometimes based on traps (Cocha Cashu: Terborgh, 1983; Mabura Hill: Thomas, 2000), on ground samples along transects (Paracou: Henry, 1994a) or on direct counts of mature fruit in tree crowns (Pinkaití: Baider, 2000) (Table 16.1). We assume that seasonal oscillation in fruit production is well defined and that the accuracy of each method in detecting seasonal peaks and troughs is similar. Several studies quantify both the mean number of species and the overall mass of fruit produced, showing a clear correlation between the diversity and abundance of fruit (Smythe, 1970; Sabatier, 1985; Terborgh, 1986; Wright *et al.*, 1999). We use the average number of species in fruit per month because it is the metric common to all studies. For Cocha Cashu and Paracou, censuses were biweekly but we used only the mid-month censuses.

The high fruiting season (HFS)

In French Guiana, agoutis (*Dasyprocta leporina*) rely mostly on pulp (48% of diet in fruit-item dry mass) (data in Henry, 1999), not on seeds (21%) in HFS. As a consequence, their impact as seed predators as well as secondary seed-dispersers should be minimal during this season. This is supported in Mabura Hill, Guyana, where 43% of unprotected *Chlorocardium rodiei* (*Lauraceae*) seeds monitored in HFS had been handled by vertebrates. Most (74%) were removed (some probably cached) and only 3% were eaten; the remainder were infested by bruchids (80%) or attacked by ants (10%) (Hammond *et al.*, 1999). In Paracou, French Guiana, *Virola michelii* seeds are rarely taken and secondarily dispersed by rodents in HFS (Forget *et al.*, 2000). Similarly, 77% of marked *Carapa procera* seeds remained after 4 weeks in HFS in Paracou, approximately one seed being hoarded for each seed eaten by rodents (assuming unrecovered seeds were cached further than 10 m) (Forget, 1996). In Paracou, during the HFS, all *Astrocaryum paramaca* seeds were removed by spiny rats and acouchies (*Myoprocta exilis*): 68% were cached, 7% were eaten beneath fruiting palms and the remainder had unknown fates (Forget, 1991).

In Pinkaití, Brazil, Peres and Baider (1997) observed that agoutis (*D. leporina*) rapidly removed and dispersed *Bertholletia excelsa*

Table 16.2. Field studies on post-dispersal seed removal by vertebrates (mostly rodents), observed and expected seed fates in neotropical rain forests during three fruiting periods.

Country	Site	Study species	Study period	Fruit period	Removal (%)	Observed fate	Expected fate	Author
Mexico	Chiapas	*Bursera simaruba*	May	HFS	94	P > H		Hammond (1995)
		Erythrina goldmanii	May		25	H = P		
		Swietenia humilis	May		37	P > H		
		Spondias mombin	July	DFS	76	H > P		
Belize	Bladen Nature Reserve	*Astrocaryum mexicanum*	February–March	HFS	73	P > H		Brewer and Rejmanek (1999)
		Ampelocera hottlei	May	HFS	96	P > H		
		Pouteria sapota	May	HFS	34 (site A)	P > H		
			July–September	DFS	62 (site B)	H ≈ P		
	Bladen Nature Reserve	*Astrocaryum mexicanum*	September–October	LFS	90–100	P > H		Brewer (2001)
Costa Rica	Santa Rosa	*Hymenaea courbaril*	January	LFS	95 (pod)	No data	P > H	Hallwachs (1986)
			March	HFS	12 (pod)	No data	P = H	
			August	DFS	100 (pod)	No data	H > P	
	La Selva	*Welfia regia*	March	HFS	100	P > H		Guariguata *et al.* (2000)
		Minquartia guianensis	April	HFS	100	P > H		
		Virola koschnyi	April	HFS	100	P > H		
		Otoba novogranatensis	April	HFS	97	P > H		
		Lecythis ampla	May	HFS	100	P > H		
		Carapa nicaraguensis	July	DFS	100	P > H		
	La Selva	*Carapa nicaraguensis*	July–September	DFS	54–96	No data	H > P	MacHargue and Harsthorn (1981)
	La Selva	*Welfia regia*	July	DFS	80	No data	H > P	Schupp and Frost (1989)

	La Selva	*Welfia regia*	July–August	DFS	9	No data	H > P	Vandermeer *et al.* (1979)
			September	DFS	15			
			October		49			
	Monteverde	*Ocotea endresiana*	May	HFS	100	P > H		Wenny (2000)
	Monteverde	*Guarea glabra*	June–August	HFS–DFS	95	H > P		Wenny (1999)
		Guarea kunthiana	June–August		100	H > P		
Panama	BCI	*Faramea occidentalis*	January–July	LFS–HFS	*c.* 10–70	No data	P > H H = P	Schupp (1990)
	BCI – Gigante	*Dipteryx panamensis*	January–February	LFS	73–100	No data	P > H	De Steven and Putz (1984)
	BCI	*Dipteryx panamensis*	February–April	LFS–HFS	96–100	P > H		Forget (1993)
	BCI – Gigante – Gatun Lake	*Dipteryx panamensis*	March	HFS	99–100	No data	P > H	Asquith *et al.* (1997)
	BCI	*Astrocaryum standleyanum*	April–May	HFS	100	H > P		Milleron and Forget (1997)
	BCI	*Virola nobilis*	June	HFS	52	H > P		Forget *et al.* (1998)
		Brosimum alicastrum	June		86	P > H		
	Gatun Lake	*Attalea butyraceae*	June	HFS	16	H > P		Adler and Kestell (1998)
	Gatun Lake	*Callophyllum longifolium*	June	HFS	54	P > H		
	BCI	*Gustavia superba*	June–July	HFS–DFS	88–100	H > P		Forget *et al.* (1998)
	BCI	*Gustavia superba*	July	HFS	98	P > H		Asquith *et al.* (1997)
	Gigante				100	H > P		
	Gatun Lake				84–100	P > H		
	BCI	*Licania platypus*	July–August	DFS	99–100	P > H		Forget *et al.* (1998)
	BCI	*Attalea butyracea*	July–September	DFS	100	P > H		Forget *et al.* (1994)
	BCI	*Virola nobilis*	August	DFS	100	P > H		Asquith *et al.* (1997)
	Gigante				100	H > P		
	Gatun Lake				100	P > H		
	BCI	*Cupania latifolia*	August	DFS	100	P > H		Forget *et al.* (1998)
		Doliocarpus olivaceus	August		91	P > H		
		Eugenia coloradensis	August		96	P > H		

continued

Table 16.2. *Continued.*

Country	Site	Study species	Study period	Fruit period	Removal (%)	Observed fate	Expected fate	Author
Trinidad		*Carapa guianensis*	November–February	LFS	60–100	No data	P > H	Rankin (1978)
Guyana	Mabura Hill	*Chlorocardium rodiei*	February–March	HFS	32	H > P		Hammond *et al.* (1999)
French Guiana	Paracou	*Virola michelii*	January	HFS	64	P > H		Forget *et al.* (2000)
	Paracou	*Carapa procera*	March	HFS	23	P > H		Forget (1996)
			April–May	HFS–DFS	75–96	H > P		
	Nouragues	*Vouacapoua americana*	May	DFS	94–100	H > P		Forget (1990)
	Paracou	*Astrocaryum paramaca*	March	HFS	100	H > P		Forget (1991)
Brazil	Pinkaïti	*Bertholletia exelsa*	February	DFS	90	No data	H > P	Peres *et al.* (1997)
			October	HFS	82.5		H = P	
	Pinkaïti	*Bertholletia exelsa*	October	HFS	88	H > P		Peres and Baider (1997)
Peru	Cocha Cashu	*Astrocaryum macrocalyx*	October	HFS	91–97	No data	H = P	Terborgh *et al.* (1993)
		Dipteryx micrantha	October		97			
		Hymenaea courbaril	October		69–81			
		Bertholletia exelsa	October		100			
	Cocha Cashu	*Dipteryx micrantha*	June	LFS	98	P > H		Cintra (1997)
	Jenaro Herrera	*Macoubea guianensis*	July	LFS	95	No data	P > H	Notman *et al.* (1996)
		Pouteria sp.	July		25			

H, hoarding; P, predation; HFS, high fruit season; DFS, declining fruit season; LFS, low fruit season.

(*Lecythidaceae*) seeds in the late HFS. Of 709 marked seeds, 14% were found intact, 44% buried, 2% eaten and 40% not found in October–November (after Table 1 in Peres and Baider (1997)). Peres and Baider (1997) observed a hoarding/predation ratio > 1 for *B. excelsa* with up to seven seeds cached per seed eaten. This result is similar to that of *C. procera* in French Guiana with roughly three seeds cached per seed eaten in late HFS (Forget, 1996).

The declining fruiting season (DFS)

Agouti diets change during the DFS in French Guiana (Henry, 1999), becoming more focused on seeds than pulp. Overall, seeds and pulp accounted for 44.6% and 24.6% of fruit items in DFS, respectively, in reverse proportion to HFS values (after Henry, 1999). In both Guyana and French Guiana, many large-seeded species, including *Astrocaryum sciophilum*, *Attalea maripa* (*Areacaceae*), *Dipteryx punctata* (*Fabaceae*), *Licania alba* (*Chrysobalanaceae*), *Carapa procera* (*Meliaceae*), *Moronobea coccinea* (*Clusiaceae*), *Platonia insignis* (*Clusiaceae*), *Andira inermis* (*Fabaceae*), *Vouacapoua americana* and *Vouacapoua macropetala* (*Caesalpiniaceae*) fruit in the DFS (Sabatier, 1985; Sist, 1989; Forget, 1990, 1991, 1996; Ter Steege and Persaud, 1991; Jansen and Forget, 2001; P.-M. Forget, personal observation), and are clearly rodent-dispersed species (Hammond *et al.*, 1996; Forget and Hammond, 2001; D.S. Hammond, personal observation). In Paracou, seed removal in *C. procera* increases during the DFS, representing seed dispersal by rodents with an average of ten seeds hoarded for each one eaten (Forget, 1996).

The low fruit season (LFS)

Rodent reliance on seeds probably reaches a maximum during the LFS, when few fruit and seed resources are available above ground. In a pilot experiment carried out in LFS at Cocha Cashu, Peru, Cintra (1997) studied the fate of magnet-labelled seeds of *Dipteryx micrantha* and *Astrocaryum murumuru* and found that most removed seeds were preyed upon by rodents. Additional studies are needed in which seeds are stored during peak periods of community-level fruit production for use in seed-fate studies during the LFS.

Variation in seed fate among sites

While studying the effect of local faunal extinction and mammalian exclusion, Guariguata *et al.* (2000) showed that removal rates of medium-sized (2–4 g) and large (6–20 g) seed species were high in HFS at La Selva, which has a fruiting pattern comparable to that of BCI (M. Guariguata, personal communication; see also Loiselle *et al.*, 1996). However, Guariguata *et al.* (2000) observed that only a small proportion (*c.* 8%) of medium-sized seeds was scatter-hoarded within 10 m, suggesting that they were taken mostly by spiny rats (*Heteromys desmarestianus*). A similar result was obtained by Brewer and Rejmanek (1999) by Brewer (2001), and by Brewer and Webb (2001) in Belize, where high removal of *Astrocaryum mexicanum*, *Ampelocera hottlei* (*Ulmaceae*) and *Pouteria zapota* (*Sapotaceae*) in HFS (March–June) and the rest of the year (S. Brewer, personal communication) is explained by seed predation by *H. desmarestianus.* In Monteverde, Costa Rica, Wenny (1999, 2000) showed that seed size influenced whether seeds were eaten or scatter-hoarded, despite overall high seed-removal rates in HFS and DFS. Adler and Kestell (1998) studied *Callophyllum longifolium* (*Clusiaceae*) and *A. butyraceae* seeds during the HFS in Panama on small islands with only one seed predator, spiny rats (*Proechimys semispinosus*), and found low seed removal and contrasting seed fates between species on islands lacking agoutis, the main seed disperser of such large seeds on BCI. Without agoutis, removal of *A. butyraceae* was low (16%) and a majority of seeds found had been cached by spiny rats. In contrast, removal of *C. longifolium* seeds was greater but mainly explained by predation. Seed-removal and hoarding rates in different fruiting seasons probably depend on the relative densities of small and large rodents (Adler, 1998). For example, Asquith *et al.* (1997) observed that 100% removal of *Virola nobilis* and *G. superba* seeds in DFS results in different seed fates between BCI, the mainland and islands in the Panama Canal Zone. Our model implicitly assumes that large rodents, such as agoutis and acouchies, whose scatter-hoarding behaviour contrasts with that of other seed-eaters, are present in sufficient numbers to compete for seeds with smaller rodents, peccaries and other vertebrates.

Implications for Plant Recruitment

In neotropical rain forests, the role of rodents as seed-dispersers or seed predators will probably fluctuate, depending on fruit morphology, seed size and fruiting phenologies. Whether rodents prey upon or disperse seeds has strong implications for seed survival and seedling recruitment. For example, on BCI the peak diversity in wind-dispersed diaspores of lianas and canopy-tree species (which comprise 25% of all woody species) occurs in March–May, the dry season (Foster, 1982) or the HFS (Table 16.1). In contrast, there are two peaks in fruit diversity among animal-dispersed tree species, one in May–June (late HFS) and another in September–October (LFS). In French Guiana, wind-dispersed species (6.8% of the woody species) fruit in the wet season at the same time as animal-dispersed species (Sabatier, 1985). A major consequence of such contrasting fruiting patterns between French Guiana and Panama may be that, when food is scarce during the dry season, large rodents will depend on different food sources at the two sites. Rodents in Panama will rely more upon wind-dispersed and large-seeded animal-dispersed species (Smythe, 1970, 1978; Smythe *et al.*, 1982; Forget, 1993; Forget *et al.*, 1999), while rodents in French Guiana will depend on cached seeds and cotyledons from germinating seedlings (Forget, 1996; Jansen and Forget, 2001). Seedling recruitment in all species, regardless of seed size or dispersal mode, during such periods of low food availability may thus depend on the capacity of local seed predators to become satiated (Janzen, 1971). This capacity will be determined by key resources, such as *Dipteryx* spp. seeds in both Central America and Peru (De Steven and Putz, 1984; Terborgh, 1986; Forget, 1993; Terborgh and Wright, 1994) or large seed crops, such as those of palms, which contribute greatly to hoards (e.g. Forget, 1992; Henry, 1999).

Smythe (1970, p. 33) proposed that 'seeds that are too large to be swallowed and are dispersed by hoarding rodents benefit from fruiting as nearly synchronously as possible'. This argument was based on the observation that large-seeded species, such as *A. standleyanum* or *G. superba*, fruit in May–June, the late HFS. In neotropical forests, large seeds dispersed by hoarding rodents often fruit during the same period, in the late fruiting season when rainfall is high, as in Guyana and French Guiana (Sabatier, 1985; Ter Steege and Persaud, 1991; Forget, 1996; Hammond *et al.*, 1999). Based on several studies of seed removal and fate, this period seems to be strongly characterized by agouti and acouchi caching behaviour (Table 16.3).

Studies on *C. procera* in French Guiana have shown that seeds not removed by rodents in February or March are destroyed by either peccaries or moths. The same patterns hold for several other seed species: *V. americana* seeds dry out or are killed by bruchids (Forget, 1990); *Attalea* spp. and *Astrocaryum* spp. seeds are killed by bruchids or peccaries (Janzen, 1971; Wright, 1983, 1990; Sist, 1989; Smythe, 1989; Terborgh *et al.*, 1993; Forget *et al.*, 1994; Silvius, 1999; Harms and Dalling, 2000); *B. excelsa* seeds rot due to fungal pathogens (Terborgh *et al.*, 1993, Peres and Baider, 1997); *Hymenaea courbaril* seeds are infested by weevils or crushed by peccaries (Janzen, 1983; Hallwachs, 1986); *Guarea kunthiana* seeds are eaten by peccaries and are insect-infested post-dispersal (Wenny, 1999). *Gustavia superba* may be an exception, as unremoved seeds are free from post-dispersal parasitism and germinate and survive on BCI when seed removal is low during a good crop year (Sork, 1987). However, this could be an artefact due to the low populations of large terrestrial vertebrates, especially peccaries, on BCI as observed for large accumulations of *C. procera* seeds at Paracou. There, peccary densities are very low due to hunting. Fruiting of large-seeded species may thus predominate as overall fruit diversity decreases, and caching by rodents is most likely.

Table 16.3 General climate of the three fruiting seasons and predictions drawn from the model on animal energetic state, diet diversity, the ratio between the percentage of seeds hoarded (H) by rodents and suffering predation (P) by all vertebrates and invertebrates, and the rodent seed-dispersal phase that dominates during each fruiting season.

	Fruiting season		
	HFS	DFS	LFS
Rainfall regime	Early to mid wet season	Late wet season	Dry season
Animal energetic state	Satiated	Triggered to hoard	Depleted
Diet diversity	High	Medium	Low
Ratio H/P	< 1	> 1	< 1
	% H < % P	% H > % P	% H < % P
Dispersal phase	Secondary	Primary	Secondary

Studies carried out on islands where environmental factors have been modified by forest fragmentation reveal the range of responses to seeds in habitats lacking large rodents, peccaries and other large mammals (Leigh *et al.*, 1993; Asquith *et al.*, 1997; Adler and Kestell, 1998). When not protected from vertebrates and not cached, seeds are eaten by spiny rats on the mainland and BCI, but seeds survive for at least 1 year on the ground surface on small islands without granivores or at other sites when totally protected from vertebrates.

Large seeds (> 10 g) are less often dispersed by rodents during peak fruiting than subsequently, but this pattern may not hold for smaller seeds (< 5 g). This is because rodents are often secondary dispersers of small- to medium-sized seeds (e.g. Forget and Milleron, 1991; Forget, 1993; Asquith *et al.*, 1997; Forget *et al.*, 1998; Wenny, 1999, 2000; Guariguata *et al.*, 2000). The rate of secondary seed dispersal is highly variable within and between forests. Rodent population density and food availability may interact to create areas of low or high secondary seed dispersal or predation (Notman *et al.*, 1996; Forget *et al.*, 2000, 2001).

The proportion of small- to medium-sized seeds secondarily dispersed by rodents probably depends on the relative rewards of eating versus caching them. Seed species of medium size with a long period of dormancy will be good candidates for caching and, later, establishment. In contrast, edible seeds that germinate rapidly are more likely to be eaten immediately.

In summary, we propose that caviomorph rodent caching behaviour is regulated by a subtle ratio between fruit diversity (greater diversity indicating a broader spatial distribution of fruits) and abundance during the peak fruiting period; caching is greatest at the HFS–DFS transition. The satiation capacity of caviomorph rodents probably depends on the spatiotemporal availability of fruit resources over the rodent's home range and may determine areas favourable (or unfavourable) to primary or secondary seed dispersal or predation.

Acknowledgements

We are very grateful to the editors for their invitation to the symposium and workshop. We thank Doug Levey and the students in his Frugivory and Seed Dispersal course, Patrick Jansen, Tad Theimer and Steve Vander Wall for their constructive comments and suggestions for improving the clarity and presentation of the manuscript. P.-M. Forget would particularly like to thank Martine Perret and Fabienne Aujard (CNRS, Centre National de la Recherche Scientifique, Brunoy) for their help with the time-series statistics. Claudia Baider, Steven Brewer, Manuel Guariguata, Olivier Henry, Evan Notman, Kirsten Silvius, John Terborgh and Joe Wright contributed fruitful discussions during the preparation of this chapter. Thanks also to Joe Wright and Steve Paton for sharing the long-term fruit data set from BCI.

References

Adler, G.H. (1995) Fruit and seed exploitation by central american spiny rats, *Proechimys semispinosus*. *Studies on Neotropical Fauna and Environment* 30, 237–244.

Adler, G.H. (1998) Impacts of resource abundance on populations of a tropical forest rodent. *Ecology* 79, 242–254.

Adler, G.H. and Kestell, D.W. (1998) Fates of Neotropical tree seeds influenced by spiny rats (*Proechimys semispinosus*). *Biotropica* 30, 677–681.

Asquith, N.M., Wright, S.J. and Clauss, M.J. (1997) Does mammal community composition control recruitment in Neotropical forests? Evidence from Panama. *Ecology* 78, 941–946.

Baider, C. (2000) Demografia e ecologia de dispersão de frutos de *Bertholletia excelsa* Humb. & Bonpl. (*Lecythidaceae*) em castanhais silvestres da Amazônia Oriental. Tese de doutorado, IB, Universidade de São Paulo, São Paulo, São Paulo State.

Brewer, S.W. (2001) Predation and dispersal of large and small seeds of a tropical palm. *Oikos* 92, 245–255.

Brewer, S.W. and Webb, M.A.H. (2001) Ignorant seed predators and factors affecting the seed survival of a tropical palm. *Oikos* 92, 245–255.

Brewer, S.W. and Rejmanek, M. (1999) Small rodents as significant dispersers of tree seeds in a Neotropical forest. *Journal of Vegetation Science* 10, 165–174.

Cintra, R. (1997) A test of the Janzen–Connell model with two common tree species in Amazonian forest. *Journal of Tropical Ecology* 13, 641–658.

Dalling, J.W. and Harms, K.E. (1999) Damage tolerance and cotyledonary resource use in the tropical tree *Gustavia superba*. *Oikos* 85, 257–264.

De Steven, D. and Putz, F.E. (1984) Impact of mammals on early recruitment of a tropical canopy tree, *Dipteryx panamensis*. *Oikos* 43, 207–216.

Dubost, G. (1988) Ecology and social life of the red acouchy, *Myoprocta exilis*; comparisons with the orange-rumped agouti, *Dasyprocta leporina*. *Journal of Zoology of London* 214, 107–123.

Forget, P.-M. (1990) Seed-dispersal of *Vouacapoua americana* (*Caesalpiniaceae*) by caviomorph rodents in French Guiana. *Journal of Tropical Ecology* 6, 459–468.

Forget, P.-M. (1991) Scatterhoarding of *Astrocaryum paramaca* by *Proechimys* in French Guiana: comparison with *Myoprocta exilis*. *Tropical Ecology* 32, 155–167.

Forget, P.-M. (1992) Seed removal and seed fate in *Gustavia superba* (*Lecythidaceae*). *Biotropica* 24, 408–414.

Forget, P.-M. (1993) Post-dispersal predation and scatterhoarding of *Dipteryx panamensis* (*Papilionaceae*) seeds by rodents in Panama. *Oecologia* 94, 255–261.

Forget, P.-M. (1996) Removal of seeds of *Carapa procera* (*Meliaceae*) by rodents and their fate in rainforest in French Guiana. *Journal of Tropical Ecology* 12, 751–761.

Forget, P.-M. and Hammond, D.S. (2001) Vertebrates and food plant diversity in Guianan rainforests. In: Hammond, D.S. (ed.) *Tropical Rainforest of the Guianas*. CAB International, Wallingford, UK.

Forget, P.-M. and Milleron, T. (1991) Evidence for secondary seed dispersal in Panama. *Oecologia* 87, 596–599.

Forget, P.M., Munoz, E. and Leigh, E.G. Jr (1994) Predation by rodents and bruchid beetles on seeds of *Scheelea* palms on Barro Colorado Island, Panama. *Biotropica* 26, 420–426.

Forget, P.-M., Milleron, T. and Feer, F. (1998) Patterns in post-dispersal seed removal by Neotropical rodents and seed fate in relation to seed size. In: Newbery, D.M., Prins, H.T.T. and Brown, N.D. (eds) *Dynamics of Tropical Communities*. Blackwell Science, Oxford, UK, pp. 25–49.

Forget, P.-M., Kitajima, K. and Foster, R.B. (1999) Pre- and post-dispersal seed predation in *Tachigali versicolor* (*Caesalpiniaceae*): effects of timing of fruiting and variation among trees. *Journal of Tropical Ecology* 15, 61–81.

Forget, P.-M., Milleron, T., Feer, F., Henry, O. and Dubost, G. (2000) Effects of dispersal pattern and mammalian herbivores on seedling recruitment for *Virola michelii* (*Myristicaceae*) in French Guiana. *Biotropica* 32, 452–462.

Forget, P.-M., Feer, F., Chauvet, S., Julliot, C., Simmen, B., Bayart, F. and Pages-Feuillade, E. (2001) Post-dispersal seed survival in frugivores-dispersed tree species. In: Bongers, F., Charles-Dominique, P., Forget, P.-M. and Théry, M. (eds) *Nouragues: Dynamics and Plant–Animal Interactions in a Neotropical Rainforest*. Kluwer, Dordrecht, The Netherlands, pp. 265–274.

Foster, R.B. (1982) Seasonal rhythm of fruitfall on Barro Colorado island. In: Leigh, E.G., Jr, Rand, A.S. and Windsor, D.M. (eds) *The Ecology of a Tropical Forest: Seasonal Rhythms and Long-term Changes*. Smithsonian Institution Press, Washington, DC, pp. 151–172.

Glanz, W.E., Thorington, R.W., Jr, Giacalone-Madden, J. and Heaney, L.R. (1982) Seasonal food use and demographic trends in *Sciurus granatensis*. In: Leigh, E.G., Jr, Rand, A.S. and Windsor, D.M. (eds) *The Ecology of a Tropical Forest: Seasonal Rhythms and Long-term Changes*.

Smithsonian Institution Press, Washington, DC, pp. 239–252.

Guariguata, M.R., Rosales Adame, J.J. and Finegan, B. (2000) Seed removal and seed fate in two selectively-logged forests with contrasting protection levels in Costa Rica. *Conservation Biology* 14, 1046–1054.

Guillotin, M., Dubost, G. and Sabatier, D. (1994) Food choice and food composition among three major primate species of French Guiana. *Journal of Zoology of London* 233, 551–579.

Hallwachs, W. (1986) Agoutis *Dasyprocta punctata*: the inheritors of guapinol *Hymenaea courbaril* (*Leguminosae*). In: Estrada, R. and Fleming, T.H. (eds) *Frugivores and Seed Dispersal.* Dr. W. Junk, The Hague, The Netherlands, pp. 119–135.

Hammond, D.S. (1995) Post-dispersal seed and seedling mortality of tropical dry forest trees after shifting agriculture, Chiapas, Mexico. *Journal of Tropical Ecology* 11, 295–313.

Hammond, D.S., Gourlet-Fleury, S., Van Der Hout, P., Ter Steege, H. and Brown, V.K. (1996) A compilation of known Guianan timber trees and the significance of their dispersal mode, seed size and taxonomic affinity to tropical rain forest management. *Forest Ecology and Management* 83, 99–116.

Hammond, D.S., Brown, V.K. and Zagt, R. (1999) Spatial and temporal patterns of seed attack and germination in a large-seeded Neotropical tree species. *Oecologia* 119, 208–218.

Harms, K.E. and Dalling, J.W. (2000) A bruchid beetle and a viable seedling from a single diaspore of *Attalea butyraceae. Journal of Tropical Ecology* 16, 319–325.

Heaney, L.R. and Thorington, R.W., Jr (1978) Ecology of Neotropical red-tailed squirrels, *Sciurus granatensis*, in the Panama Canal Zone. *Journal of Mammalogy* 59, 846–851.

Henry, O. (1994a) Caractéristiques et variations saisonnières de la reproduction de quatre mammifères forestiers terrestres de Guyane Française: *Oryzomys capito velutinus* (Rodentia, Cricetidae), *Proechimys cuvieri* (Rodentia, Echimyidae), *Dasyprocta leporina* (Rodentia, Dasyproctidae) and *Tayassu tajacu* (Artiodactyle, Tayassuidae). Influence de l'âge, des facteurs environnementaux et de l'alimentation. Thèse de doctorat de l'Université Paris VII, Paris, France.

Henry, O. (1994b) Saisons de reproduction chez trois rongeurs et un artiodactyle en Guyane française, en fonction des facteurs du milieu et de l'alimentation. *Mammalia* 58, 183–200.

Henry, O. (1997) The influence of sex and reproductive state on diet preference in four terrestrial mammals of the French Guianan rainforest. *Canadian Journal of Zoology* 75, 929–935.

Henry, O. (1999) Dietary choice of the orange-rumped agouti (*Dasyprocta leporina*) in French Guiana. *Journal of Tropical Ecology* 13, 291–300.

Jansen, P.A. and Forget, P.-M. (2001) Scatter-hoarding by rodents and tree regeneration in French Guiana. In: Bongers, F., Charles-Dominique, P., Forget, P.-M. and Théry, M. (eds) *Nouragues: Dynamics and Plant–Animal Interactions in a Neotropical Rainforest.* Kluwer Academic Publisher, Dordrecht, The Netherlands, pp. 275–288.

Janzen, D.H. (1971) Seed predation by animals. *Annual Review of Ecology and Systematics* 2, 465–492.

Janzen, D.H. (1983) Larval biology of *Ectomyelois muriscis* (Pyralidae: Phycitinae), a Costa Rican fruit parasite of *Hymenaea courbaril* (*Leguminosae: Caesalpiniodeae*). *Brenesia* 21, 387–393.

Leigh, E.G., Jr (1999) *Tropical Forest Ecology.* Oxford University Press, New York, 245 pp.

Leigh, E.G., Jr, Wright, S.J., Putz, F.E. and Herre, E.A. (1993) The decline of tree diversity on newly isolated tropical islands: a test of a null hypothesis and some implications. *Evolutionary Ecology* 7, 76–102.

Loiselle, B.A., Ribbens, E. and Vargas, O. (1996) Spatial and temporal variation of seed rain in a tropical lowland wet forest. *Biotropica* 28, 82–95.

MacHargue, L.A. and Harsthorn, G.S. (1983) Seed and seedling ecology of *Carapa guianensis. Turrialba* 33, 399–404.

Milleron, T.E. and Forget, P.-M. (1997) Dispersal and consumption of *Astrocaryum standleyanum* palm seeds on Barro Colorado Island, Panama. *Bulletin of the Ecological Society of America* 78, 285.

Morris, D. (1962) The behavior of the green acouchi (*Myoprocta pratti*) with special reference to scatter hoarding. *Proceedings of the Zoological Society, London* 139, 701–731.

Notman, E., Gorchov, D.L. and Cornejo, F. (1996) Effect of distance, aggregation, and habitat on levels of seed predation for two mammal-dispersed Neotropical rain forest tree species. *Oecologia* 106, 221–227.

Peres, C.A. and Baider, C. (1997) Seed dispersal, spatial distribution and population structure of brazilnut trees (*Bertholletia excelsa*) in south-eastern Amazonia. *Journal of Tropical Ecology* 13, 595–616.

Rankin, J.M. (1978) The influence of seed predation and plant competition on tree species abundances in two adjacent tropical rain forests. PhD thesis, University of Michigan, Ann Arbor, Michigan.

Sabatier, D. (1985) Saisonnalité et déterminisme du pic de fructification en forêt guyanaise. *Revue d'Ecologie (Terre et Vie)* 40, 89–320.

Schupp, E.W. (1990) Annual variation in seedfall, postdispersal predation, and recruitment of a neotropical tree. *Ecology* 71, 504–515.

Schupp, E.W. (1993) Quality, quantity and the effectiveness of seed dispersal by animals. *Vegetatio* 107/108, 15–29.

Schupp, E.W. and Frost, E.J. (1989) Differential predation on *Welfia georgii* seeds in treefall gaps and the forest understory. *Biotropica* 21, 200–203.

Silvius, K. (1999) Interactions among *Attalea* palms, bruchid beetles, and Neotropical terrestrial fruit-eating mammals: implications for the evolution of frugivory. PhD thesis, University of Florida, Gainesville.

Sist, P. (1989) Demography of *Astrocaryum sciophilum*, an understorey palm of French Guiana. *Principes* 33, 142–151.

Smythe, N. (1970) Relationships between fruiting seasons and seed dispersal methods in a Neotropical forest. *American Naturalist* 104, 25–35.

Smythe, N. (1978) The natural history of the Central American agouti (*Dasyprocta punctata*). *Smithsonian Contribution to Zoology* 257, 1–52.

Smythe, N. (1989) Seed survival in the palm *Astrocaryum standleyanum*: evidence for dependence upon its seed-dispersers. *Biotropica* 21, 50–56.

Smythe, N., Glanz, W.E. and Leigh, E.G.J. (1982) Population regulation in some terrestrial frugivores. In: Leigh, E.G., Jr, Rand, A.S. and Windsor, D.M. (eds) *The Ecology of a Tropical Forest: Seasonal Rhythms and Long-term Changes.* Smithsonian Institution Press, Washington, DC, pp. 227–238.

Sork, V.L. (1987) Effects of predation and light on seedling establishment in *Gustavia superba*. *Ecology* 68, 1341–1350.

SPSS (1999) *Systat® 9 Statistics II.* SPSS Inc., Chicago.

Stapanian, M.A. (1986) Seed dispersal by birds and squirrels in the deciduous forests of the United States. In: Estrada, A. and Fleming, T.H. (eds) *Frugivores and Seed Dispersal.* Dr W. Junk Publishers, The Hague, pp. 225–236.

Terborgh, J. (1983) *Five New World Primates: a Study in Comparative Ecology.* Princeton University Press, Princeton, New Jersey, 260 pp.

Terborgh, J. (1986) Community aspects of frugivory in tropical forests. In: Estrada, R. and Fleming, T.H. (eds) *Frugivores and Seed Dispersal.* Dr W. Junk Publishers, The Hague, The Netherlands, pp. 371–384.

Terborgh, J.W. and Wright, S.J. (1994) Effects of mammalian herbivores on plant recruitment in two Neotropical forests. *Ecology* 75, 1829–1833.

Terborgh, J.W., Losos, E., Riley, M.P. and Bolanos Riley, M. (1993) Predation by vertebrates and invertebrates on the seeds of five canopy tree species of an Amazonian forest. *Vegetatio* 107/108, 375–386.

Ter Steege, H. and Persaud, C.A. (1991) The phenology of Guyanese timber species: a compilation of a century of observations. *Vegetatio* 95, 177–198.

Thomas, R.S. (2000) Forest productivity and resource availability in lowland tropical forests of Guyana. PhD thesis, Imperial College, University of London.

Vandermeer, J.H., Stout, J. and Risch, S. (1979) Seed dispersal of a common Costa Rican rain forest palm. *Tropical Ecology* 20, 17–26.

Vander Wall, S.B. (1990) *Food Hoarding in Animals.* Chicago University Press, Chicago, 445 pp.

Wenny, D.G. (1999) Two-stage dispersal of *Guarea glabra* and *G. kunthiana* (*Meliaceae*) in Monteverde, Costa Rica. *Journal of Tropical Ecology* 15, 481–496.

Wenny, D.G. (2000) Seed dispersal, seed predation, and seedling recruitment of a Neotropical montane tree. *Ecological Monographs* 70, 331–351.

Wright, S.J. (1983) The dispersion of eggs by a bruchid beetle among *Scheelea* palm seeds and the effect of distance to the parent palm. *Ecology* 64, 1016–1021.

Wright, S.J. (1990) Cumulative satiation of a seed predator over the fruiting season of its host. *Oikos* 58, 272–276.

Wright, S.J., Carrasco, C., Calderon, O. and Paton, S. (1999) The El Nino Southern Oscillation, variable fruit production, and famine in a tropical forest. *Ecology* 80, 1632–1647.

17 Seed-eaters: Seed Dispersal, Destruction and Demography

Philip E. Hulme
NERC Centre for Ecology and Hydrology, Hill of Brathens, Banchory AB31 4BY, Scotland, UK

A Deceptive Dichotomy

Textbook discussions of plant–animal interactions usually present quite separate treatments of frugivores and granivores (e.g. Howe and Westley, 1988; Abrahamson, 1989; Fenner, 2000; Herrera and Pellmyr, 2001). In general, frugivory has been examined primarily as a mutualistic interaction between animals and plants, whilst granivory has been viewed as an antagonistic interaction within the broader field of herbivory. This dichotomous view is probably inappropriate because of the following.

- Although differences in the taxonomic composition of particular frugivore and granivore guilds can be quite marked, many vertebrates subsist on a diet of both fruit and seeds (e.g. Snow and Snow, 1988; Palacios *et al.*, 1997; see also Chapman and Chapman, this volume). For these species 'frugivore–granivore' is a more accurate description of their trophic niche.
- Some frugivores, although viewed as legitimate seed-dispersers, nevertheless digest or damage a substantial fraction of the seeds they consume (Fig. 17.1a) (see Traveset and Verdú, this volume). Such 'digestive seed predation', when intense, will shift the consequence of frugivory from mutualism towards antagonism. Similarly, some 'seed predators' (e.g. mice) may actually disperse a proportion of seeds ingested (Cipollini and Levey, 1997). In fact, some seed predators may actually be more effective dispersers of seeds than 'legitimate' dispersers, because they exhibit longer retention times and deposit viable seeds from a fruit in more defecations (Levey, 1986).
- A variety of vertebrate granivores store seeds for later consumption. When recovery of their seed stores (caches) is not complete, seeds may survive to germinate. Thus, the overall effect on plant populations of these granivores may in fact be positive, resulting in seed dispersal (see chapters by Vander Wall, Forget *et al.* and Hoshizaki and Hulme, this volume).
- Sublethal consumption of seed endosperm by granivorous curculionid and bruchid weevil larvae may increase germination rates of hard-coated leguminous seeds, while not significantly reducing seedling survivorship (Karban and Lowenberg, 1992; Ollerton and Lack, 1996). Such invertebrate seed-feeders may have a positive impact on plants.

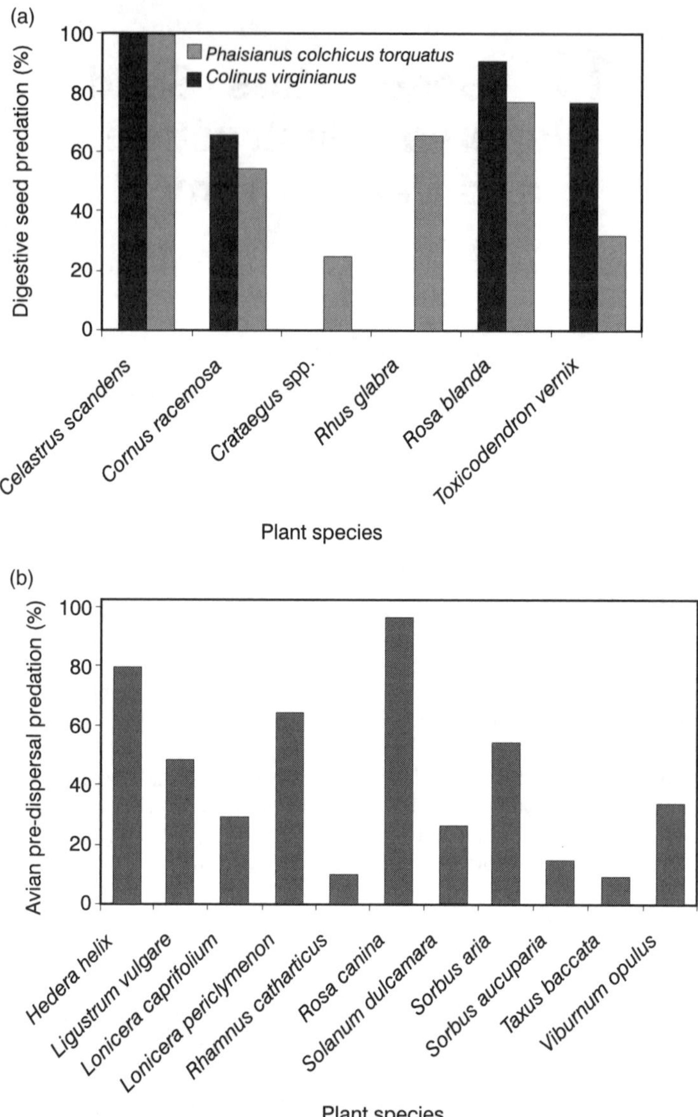

Fig. 17.1. Two forms of pre-dispersal seed predation by birds on the seeds of fleshy fruits: (a) proportion of seeds digested by ring-necked pheasants (*Phaisianus colchicus torquatus*) and bob-white quail (*Colinus virginianus*), indicating that the probability of seed digestion is a function of both the animal and the plant species (data from Krefting and Roe, 1949), and (b) the proportion of seeds destroyed on the parent plant by a diverse assemblage of seed-feeding birds in southern England (data from Snow and Snow, 1988).

Therefore, while legitimate seed-dispersing frugivores and non-caching granivores generally have different impacts on plant population dynamics, they should be viewed as two extremes along a continuum of mutualistic to antagonistic interactions. Furthermore, the impact of frugivores and granivores on plant demography and evolution cannot be viewed independently of each other for the following reasons:

- Granivores frequently consume a sizeable proportion of fruit crops and may decrease the resources available to frugivores (Fig. 17.1b). Granivores often gain the upper hand via their ability to feed on the seeds of unripe fruit. They can therefore strongly shape the dynamics of frugivore–plant interactions (Sallabanks and Courtney, 1992).
- Vertebrate frugivores and granivores harvest both fruits and seeds as individual items. Thus, certain granivores and frugivores may use similar cues when foraging (e.g. crop size, number and size of seeds, secondary metabolites) (Jordano, 2000; Hulme and Benkman, 2001). Hence, the evolution of fruit, seed and crop characteristics is strongly influenced by the selection pressures of both frugivores and granivores.

In this chapter, I aim to integrate current literature with original analyses of both published and unpublished data to highlight key areas in the interface between granivory and frugivory. In particular, I shall assess the relative importance of pre- and post-dispersal granivory on seed-dispersal mutualisms and examine how characteristics of seed shadows influence rates of seed predation. The consequences of seed predation on plant demography will be explored both in relation to different dispersal strategies and from the perspective that granivores may sometimes act as seed-dispersers. I hope to show that our understanding of granivory is far from complete and that its impact on plants is intimately linked to mechanisms of seed dispersal.

Pattern and Process in Pre- and Post-dispersal Predation

The strong link between seed dispersal and seed predation is emphasized by the frequent distinction made between granivory occurring prior to dispersal from that occurring afterwards (Crawley, 2000; Hulme and Benkman, 2001). Granivores feeding on seeds and fruit prior to seed dispersal will probably have the greatest potential to structure the interaction between plants and seed-dispersers. These pre-dispersal seed predators influence seed dispersal through at least three mechanisms. First, by reducing crop sizes, predispersal seed predators decrease the probability of seeds being dispersed, irrespective of the dispersal mechanism (Janzen, 1970). Secondly, density-responsive seed-dispersers may visit fruit crops diminished by seed predators less frequently than they do larger crops (Sallabanks and Courtney, 1992). Thirdly, pre-dispersal granivores not only may cause the fruit they damage to be rejected by vertebrate seed-dispersers but may also lead to reduced feeding rates on undamaged fruits. This last case may be particularly true for avian seed-dispersers (e.g. Hubbard and McPherson, 1997; Garcia *et al.*, 1999), but possibly not mammals (e.g. Forget 1990; Steele *et al.*, 1993; Forget *et al.*, 1994). In contrast, the principal role of post-dispersal seed predators in the plant–disperser interaction is to act as a filter on the resulting seed shadow, modifying seed densities and distributions.

While the balance of evidence suggests that pre-dispersal granivores might influence the frugivore–plant interaction most strongly, such a conclusion depends on the relative intensity of pre- and post-dispersal seed predation. Surprisingly, although numerous studies have reported rates of pre- and post-dispersal seed predation, relatively few have simultaneously assessed both for the same plant species (Hulme and Benkman, 2001) (Table 17.1). Where such comparisons can be made, it is evident that post-dispersal predation is, on average, more severe than pre-dispersal seed predation (36% vs. 44% respectively, paired $t = 2.28$, d.f. = 27, $P < 0.05$). However, caution should be applied when assessing rates of pre-dispersal seed predation, because losses are easily under- or overestimated. For example, losses are most commonly underestimated when plants abort immature seeds and fruit that have been damaged by granivores (Andersen, 1988). However, losses may also be overestimated, when the presence of invertebrate eggs or larval exit holes is interpreted to result in seed death (e.g. Terborgh *et al.*, 1993; Forget *et al.*, 1994), since these 'parasitized' seeds are sometimes viable. Although germination rates of infested seeds are often less than those of uninfested seeds (Andersson, 1992;

Table 17.1. Percentage seed predation reported in field studies that have separately quantified pre- and post-dispersal seed removal.

		Pre-dispersal losses		Post-dispersal losses		
Species	Ecosystem	%	Agent	%	Agent	Author
Cirsium canescens	Temperate grassland	51.8	Tephritid flies	0–99.5	Rodents	Louda *et al.* (1990)
Cirsium vulgare	Temperate grassland	3–17	Moth larvae	21–66	Rodent	Klinkhammer *et al.* (1988)
Quercus robur	Deciduous forest	2–30	Weevils	100	Rodents	Crawley and Long (1995)
		25–80	Knopper galls			
Fagus crenata	Deciduous forest	36.9	Insects	12.3	Vertebrates	Homma *et al.* (1999)
Fagus sylvatica	Deciduous forest	3–17	Moth larvae	5–12	Mammals	Nilsson and Wästljung (1987)
Fraxinus excelsior	Deciduous forest	15–75	Moth larvae	25–75	Rodents	Gardner (1977)
Pinus sylvestris	Coniferous forest	80	Crossbills	67–96	Rodents	Castro *et al.* (1999)
Pistacia terebinthus	Mediterranean shrubland	4.6	Chalcidoid wasp	85	Rodents/ants	Traveset (1994)
		71.7	Birds			
Lobularia maritima	Mediterranean shrubland	80–99	*Messor* ants	40–95	*Messor* ants	Pico and Retana (2000)
Eucalyptus baxteri	Sclerophyllous woodland	66	Beetles, wasps, moths	90	Ants	Andersen (1989)
Leptospermum juniperinum	Sclerophyllous woodland	44	Beetles, wasps, moths	90	Ants	Andersen (1989)
Casuarina pusilla	Heathland	83	Beetles, wasps, moths	60	Ants	Andersen (1989)
Leptospermum myrsinoides	Heathland	64	Beetles, wasps, moths	90	Ants	Andersen (1989)
Acacia farnesiana	Deciduous tropical forest	0–37.8	Bruchid beetles	35.2–66	Rodents	Traveset (1990, 1991)
				6–26	Bruchids	
Astrocaryum mexicanum	Deciduous tropical forest	50	Squirrels	90	Mice	Sarukhan (1986)
Dryobalanops aromatica	Malaysian rain forest	20	Weevils	17	Vertebrates	Itoh *et al.* (1995)
Dryobalanops lanceolata	Malaysian rain forest	32.5	Weevils	9	Vertebrates	Itoh *et al.* (1995)
Tachigali versicolor	Panamanian rain forest	20	Bruchids	43	Rodents	Forget *et al.* (1999)
Cecropia shreberiana	Puerto Rican rain forest	6	Vertebrates, insects	9	Ants	Myster (1997)
Cecropia polyphlebia	Costa Rican rain forest	12	Vertebrates, insects	2	Ants	Myster (1997)
Gonzalagunia spicata	Puerto Rican rain forest	6	Vertebrates, insects	9	Ants	Myster (1997)
Gonzalagunia kallunkii	Costa Rican rain forest	6	Vertebrates, insects	5	Ants	Myster (1997)
Miconia racemosa	Puerto Rican rain forest	3	Vertebrates, insects	15	Ants	Myster (1997)
Ocotea leucoxylon	Puerto Rican rain forest	3	Vertebrates, insects	1	Ants	Myster (1997)
Palicourea riparia	Puerto Rican rain forest	1	Vertebrates, insects	9	Ants	Myster (1997)
Phytolacca rivinoides	Puerto Rican rain forest	4	Vertebrates, insects	1	Ants	Myster (1997)
Urera caracasana	Costa Rican rain forest	0		4	Ants	Myster (1997)
Witheringa cocroloboides	Costa Rican rain forest	8	Vertebrates, insects	3	Mammals	Myster (1997)

Mucunguzi, 1995), some studies have found similar (Steele *et al.*, 1993; Miller, 1994) or even improved germination (Hauser, 1994).

In general, pre-dispersal seed predation differs from post-dispersal seed predation in respect of the guilds of seed-feeding animals involved (Table 17.1) and the relative influence of different plant traits on rates of seed predation (Hulme and Benkman, 2001). It is therefore surprising that a significant positive correlation between rates of pre- and post-dispersal seed predation exists for the studies in Table 17.1 ($r = 0.857$, d.f. = 26, $P < 0.01$) (Fig. 17.2a). Clearly, certain characteristics of plants lead to similar susceptibilities to both pre- and post-dispersal granivores. This is unexpected, because most pre-dispersal granivores are specialists, feeding on only one or a few closely related plant species, whereas post-dispersal granivores tend to be polyphagous

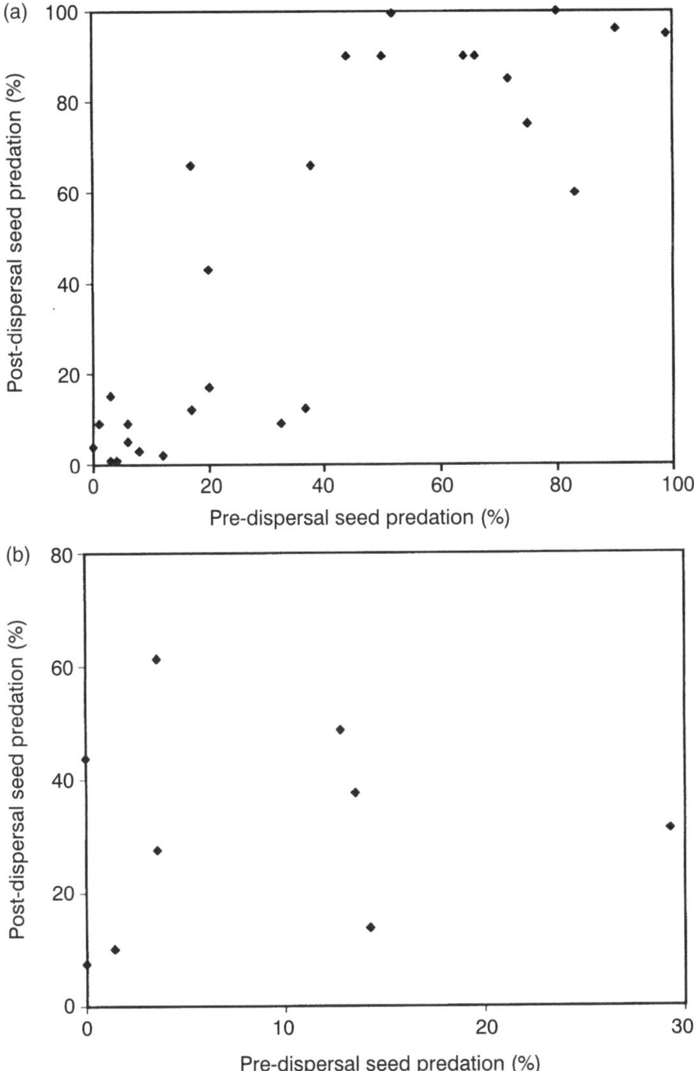

Fig. 17.2. Relationship between pre- and post-dispersal seed predation revealing: (a) a positive relationship among 23 species described in Table 17.1, and (b) no relationship among ten individuals of *Fraxinus excelsior* occurring in one woodland in north-eastern England (P.E. Hulme, unpublished data).

generalists. Possibly, plant species with few physical or chemical seed defences suffer considerable losses to both pre- and post-dispersal granivores, while seed defences appropriate against one guild of granivores (pre- or post-dispersal) also limit the impact of the other guild. Further studies that simultaneously examine both pre- and post-dispersal granivory are required to explore the generality of these preliminary findings.

Within a plant species, different individuals may experience markedly different rates of pre- and post-dispersal seed predation (Crawley and Long, 1995) (Fig. 17.2b). Selection of individual plants may be based on heritable traits (e.g. seed chemistry, crop size, phenology), but may also be determined by non-heritable factors (e.g. degree of exposure to elements, nutritional status, neighbouring vegetation structure). In contrast to the among-species comparisons in Table 17.1, within-species comparisons reveal no correlation between pre- and post-dispersal granivory (Fig. 17.2b). Most demographic studies have viewed variation in granivory among individual plants as noise that masks average trends. However, if variation is based on heritable traits, it would be of considerable significance to the evolution of plant–granivore interactions.

Distance, Density and Dispersal

Demographic consequences of the interactions between seed dispersal and granivory have been examined in detail by Janzen (1970). Janzen's model suggests that seed predators feeding where seed densities are highest (either through direct density dependence or by concentrating their foraging beneath fruiting individuals) might prevent competitive exclusion and promote tree species' coexistence. This seminal work has spawned considerable data (and debate) regarding the role of herbivores in the maintenance of species diversity in tropical forests (reviewed in Hammond and Brown, 1998). However, most attention has focused on predicted patterns (Wills *et al.*, 1997; Condit *et al.*, 2000; Harms *et al.*, 2000) rather than the underlying process that Janzen proposed. Thus, significant gaps exist in the current understanding of the underlying processes, especially with respect to the relative importance of pre- and post-dispersal seed predation and the impacts of vertebrate and invertebrate granivores.

Process-orientated studies have tended to focus on post-dispersal events (Hammond and Brown, 1998). Yet pre-dispersal predators may significantly reduce seed crop size (Table 17.1, Fig. 17.2a) which may have important consequences for post-dispersal seed fate. First, by shortening the tail of the dispersal curve (see previous section), pre-dispersal granivores may have additive or multiplicative effects through a reduction in the number of seeds that escape from distance-responsive seed predators. Secondly, by lowering the average density of the seed shadow, pre-dispersal granivores may have compensatory effects through an increase in the number of seeds that escape from density-dependent seed predators.

Within a single plant community, different plant species may often interact with distinct assemblage of granivores (Fig. 17.3). The balance of current evidence indicates that invertebrate post-dispersal seed predators tend to forage in a manner consistent with Janzen's model, whereas vertebrate granivores do not (Hammond and Brown, 1998). Nevertheless, most studies of post-dispersal seed fate fail to distinguish the impacts of different guilds of granivores (Hulme, 1998a). Since invertebrate and vertebrate granivores often differ in both the temporal and the spatial scales of their effects, as well as their preferences for plant species (Hulme, 1998a), separation of their impact is crucial for assessing the role seed predators play in plant community structure.

The weight of evidence suggests that, on average, seed predation rates of vertebrates are almost twice those of invertebrates (Hulme, 1998a). This difference suggests that any effects of invertebrate granivores are strongly modified by the greater impact of vertebrates. Therefore, if Janzen's model operates in species-rich tropical forests, these findings suggest that: (i) post-dispersal seed fate is strongly modified by pre-dispersal seed predators; (ii) the impact of invertebrate post-dispersal granivores is currently underestimated (e.g. Levey and Byrne, 1993); or

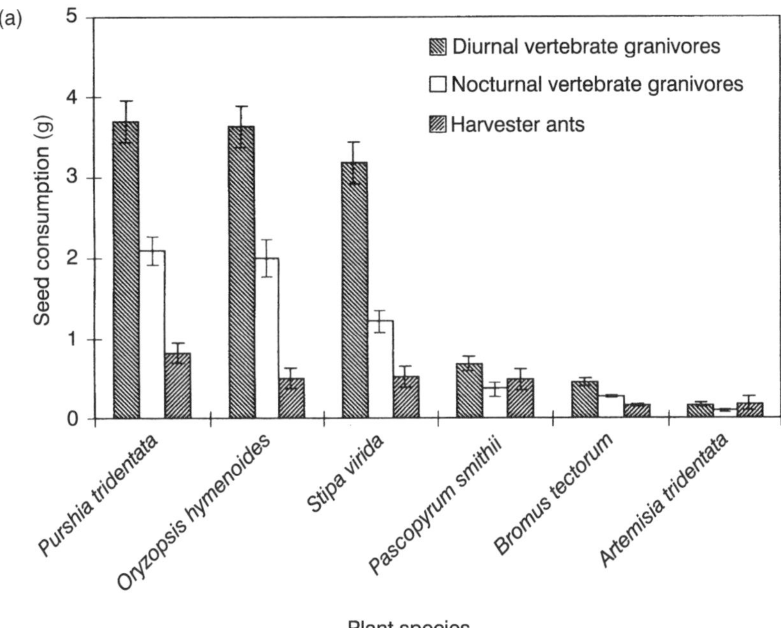

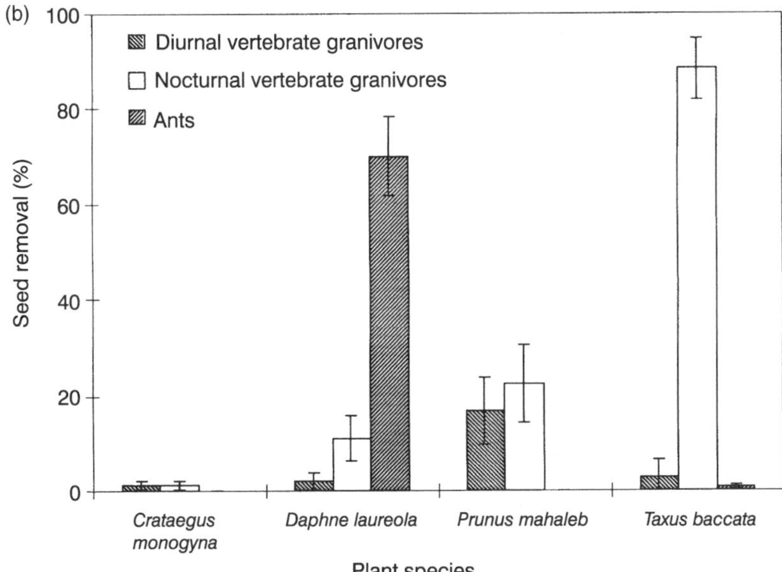

Fig. 17.3. Variation in impacts of different guilds of post-dispersal granivores feeding on plants co-occurring in the same community: (a) differences in absolute rates of seed removal but similar relative importance of different guilds of granivores for six species of desert plants (data from Kelrick *et al.*, 1986), and (b) differences in absolute rates of seed removal and relative importance of different guilds of granivores in four species of Mediterranean shrubs (data from Hulme, 1992, 1997). Error bars represent 1 SE.

(iii) seed predators are not as important as other mechanisms (e.g. fungal infection, seedling herbivory) in structuring these forests.

Process-orientated studies rarely examine how post-dispersal seed losses are influenced by microhabitat variation. The impact of distance-responsive and density-dependent seed predation is strongly mediated by the microhabitat characteristics of the seed shadow. Spatial heterogeneity in rates of post-dispersal seed predation is often considerable and rarely a simple function of the distance from the nearest fruiting adult (Hulme, 1998a; see also Schupp *et al.*, this volume). For example, while predation on the seeds of yew (*Taxus baccata*) is high beneath adults, dispersed seeds encounter significantly reduced predation only in open microhabitats or beneath the canopy of a different tree species, not under shrubs or in tall herbaceous vegetation (Fig. 17.4a). These patterns probably reflect some microhabitat preferences of post-dispersal granivores (Hulme, 1997). For example, rodents often prefer to forage in densely vegetated microhabitats, whereas thermophilic ants prefer open microhabitats (Fig. 17.4b). Such preferences lead to 'hot spots' of granivore activity, which are often independent of the density of the local seed rain or the distance from a fruiting adult.

Seeds dispersed away from their parent and which arrive beneath the canopy of a conspecific adult have little scope to escape either specialists or distance-responsive, generalist seed predators. Furthermore, if the conspecific adult is fruiting, opportunities to escape density-dependent seed predators are negated. Seeds dispersed beneath the canopy of another species may escape from specialist seed predators but not from generalists, particularly if generalist granivores are distance-responsive. In addition, if the plant is fruiting or is frequently used as a perching or roosting site, the local seed rain under it may be sufficient to attract both generalist and specialist density-dependent granivores.

Process-orientated studies that have assessed seed predation beneath and beyond canopies of fruiting trees tend to control, rather than account for, microhabitat variation. Seeds located experimentally away from a focal tree have often been placed distant from other trees in homogeneous, usually relatively open microhabitats. Such experimental designs may not adequately sample the seed shadow and may bias interpretation towards identifying significant distance effects. Future studies should assess the relative importance of both distance and seed density in relation to the characteristics of the seed shadow and especially variation among microhabitats.

Recruitment, Regeneration and Removal rates

High seed-predation rates do not necessarily translate into dramatic reductions in plant populations. For example, for the desert annual *Erodium circutarium*, consumption of over 95% of seeds by heteromyid rodents was predicted to reduce plant density by only 30% (Soholt, 1973). Average rates of pre- and post-dispersal seed predation are often much lower than 95% (Table 17.1). Thus, it may be possible that even moderately high seed losses have little impact on demography. Seed predation may be of limited importance in plant demography when: (i) plants regenerate primarily by vegetative means; (ii) seed losses to predators are buffered by the presence of a large persistent seed bank; (iii) seed predators are satiated by large seed crops; and/or (iv) regeneration is microsite- rather than seed-limited. I shall now explore each of these cases in relation to seed dispersal mechanism.

Three major, though not mutually exclusive, regenerative strategies have been identified for flowering plants (Hodgson *et al.*, 1994): (i) vegetative expansion through the formation of persistent rhizomes, stolons or suckers; (ii) regeneration by seeds from a persistent seed bank; and (iii) regeneration by seeds that do not form a persistent seed bank. For the vascular flora of the British Isles, the relative importance of these regenerative strategies is related to the seed-dispersal mechanism ($\chi^2 = 75.5$, d.f. = 15, $P < 0.01$) (Fig. 17.5a). In general, animal-dispersed species are relatively more dependent on regeneration by seed and less on vegetative regeneration than species with abiotic (wind and water) or unspecialized dispersal mechanisms. For animal-dispersed species, the

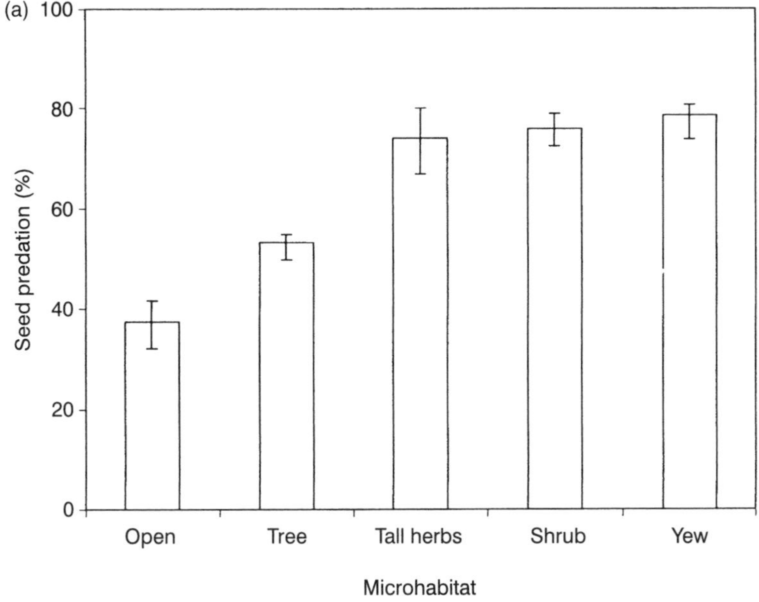

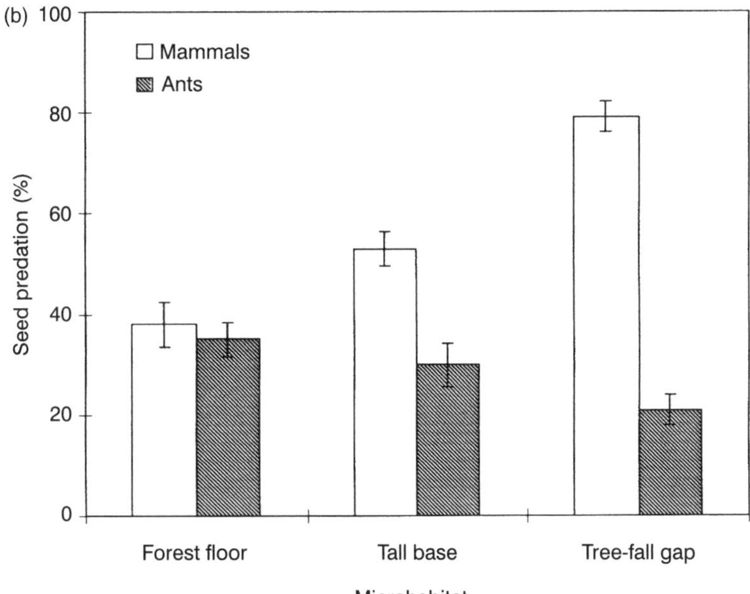

Fig. 17.4. Microhabitat variation in rates of post-dispersal seed removal, highlighting: (a) significant variation among five woodland microhabitats in the removal of *Taxus baccata* seeds with high rates of removal in several microhabitats distant from fruiting trees (data from Hulme, 1996), and (b) a significant interaction between microhabitat and granivore type in the removal of *Cicer arietinum* seeds in a Bornean rain forest (P.E. Hulme, unpublished data). Error bars represent 1 SE.

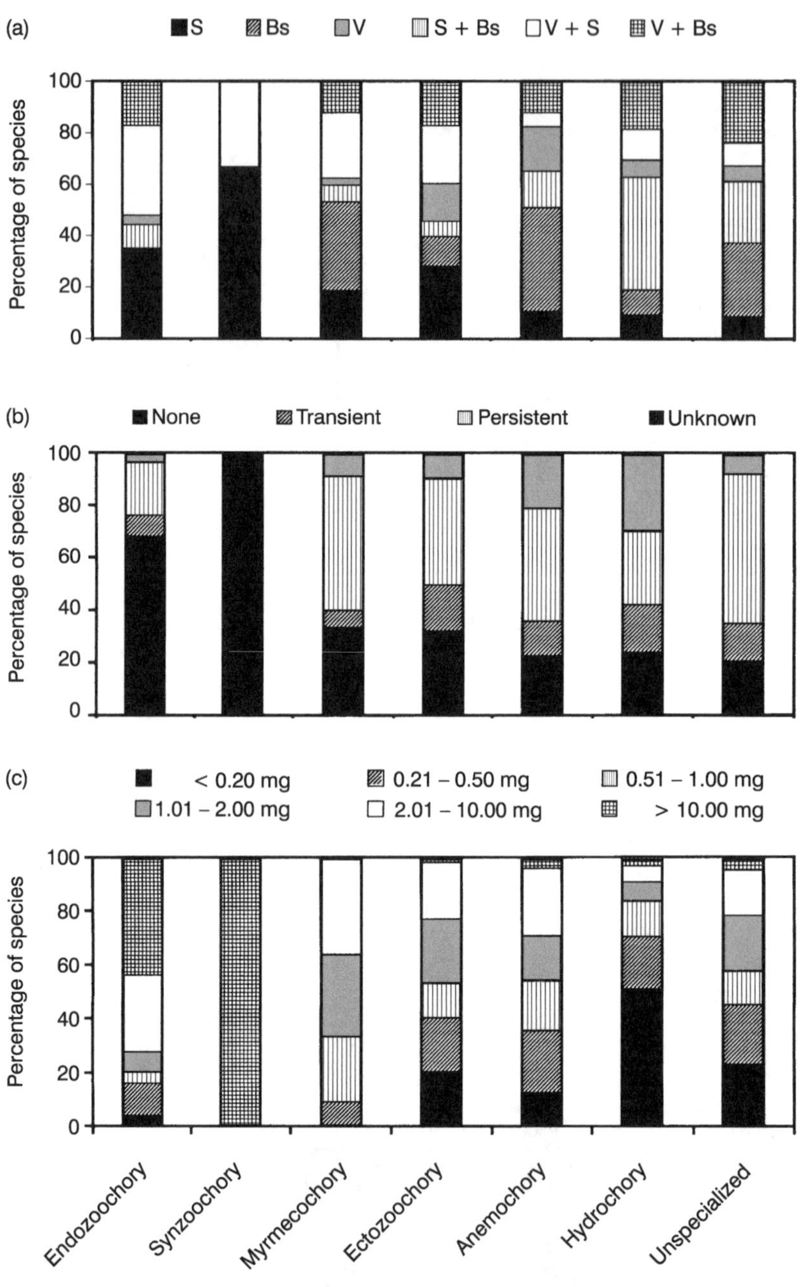

Fig. 17.5. Variation among different seed dispersal mechanisms in relation to regeneration attributes of north-west European plant species (data from Hodgson *et al.*, 1994). (a) By regeneration strategy; V, vegetative expansion through the formation of persistent rhizomes, stolons or suckers; Bs, regeneration by seeds that form a persistent bank of viable but dormant seeds in the soil; S, regeneration by seeds that do not form a persistent seed bank. (b) By seed-bank persistence: none, seeds remain viable in the soil < 1 year; transient, seeds remain viable in the soil for < 5 years; persistent, seeds remain viable in soil for > 5 years; or unknown, no data available. (c) By seed size. Seed-dispersal terminology:

continued

relative importance of regeneration by seed declines as follows: synzoochory (seed hoarding by vertebrates) > myrmecochory (dispersal by ants) > ectozoochory (dispersal on the exterior of animals, e.g. on fur or feathers) > endozoochory (ingestion of fruit by vertebrates and subsequent defecation or regurgitation of intact seeds).

The vascular flora of the British Isles has also been classified in relation to the size of the seed bank as follows (Hodgson *et al.*, 1994): none (seeds remain viable in the soil < 1 year), transient (seeds remain viable in the soil for < 5 years) or persistent (seeds remain viable in the soil for > 5 years). Again, the relative importance of a seed bank is related to the seed dispersal mechanism ($\chi^2 = 49.5$, d.f. = 9, $P < 0.01$) (Fig. 17.5b). In general, animal-dispersed species rarely have a persistent seed bank and are more likely to have no seed bank, compared with species with either abiotic or unspecialized dispersal mechanisms. For animal-dispersed species, the relative importance of a seed bank decreases as follows: myrmecochory > ectozoochory > endozoochory > synzoochory.

Seed production is usually less variable in plants dispersed by mutualistic frugivores than among those that are either synzoochorous or dispersed by abiotic means (Herrera *et al.*, 1998). This pattern is consistent with the assumption that species dispersed by frugivores should rarely produce excessively large fruit crops, since this might result in the frugivore population becoming satiated and less efficient in dispersal (Herrera *et al.*, 1998). In contrast, synzoochorous species may irregularly produce large seed crops. These mast crops are probably more successful at limiting the impacts of specialist pre-dispersal seed predators than they are at limiting the impacts of generalist post-dispersal seed predators (Gardner, 1977; De Steven, 1983; Nilsson and Wästljung, 1987; Crawley and Long, 1995). Vertebrates however, are often satiated when there is community-wide synchrony among plant species (Itoh *et al.*, 1995; Curran and Leighton, 2000) or when the plant community is dominated by one or a few species (Nilsson and Wästljung, 1987; Homma *et al.*, 1999). Densities of seedlings are often higher after mast years than after non-mast years (Gardner, 1977; Crawley and Long, 1995; Itoh *et al.*, 1995; Curran and Leighton, 2000), presumably because more seeds escape predation. This is not equivalent to stating that granivores have no effect during mast years, because the number of seedlings recruiting in the absence of predators is not known and could potentially be much greater. These findings for masting trees also suggest that their recruitment is seed-limited, since more seed production results in more seedlings.

Assessing whether plant recruitment is seed- or microsite-limited is fraught with difficulty, because microsite availability will vary both spatially and temporally. Generalizations from experimental sowing of additional seeds into natural plant communities are constrained by the limited spatial and temporal scales they address. Nevertheless, it appears that seed-limited recruitment is common in many plant communities, particularly in ruderal habitats, although within any community significant variation occurs among plant species (Turnbull *et al.*, 2000). As yet, seed-addition experiments have not been designed to examine the relative importance of microsite limitation to seeds dispersed in different ways (Turnbull *et al.*, 2000). The available data are rather inconclusive, with evidence for seed limitation in some vertebrate-dispersed species (e.g. *Sorbus aucuparia*, *Actaea spicata*, *Convallaria majalis*), but not others (e.g. *Viburnum opulus*, *Vaccinium* spp., *Frangula alnus*) (Eriksson and Ehrlen, 1992). It has been suggested that microsite limitation may be less common for relatively large-seeded species (Crawley, 1990). If this were true, then the general trend for larger seed size in animal-dispersed species ($\chi^2 = 80.5$, d.f. = 15, $P < 0.01$) (Fig. 17.5c; Leishman *et al.*, 1995) might indicate less

Fig. 17.5. (*continued*) endozoochory, ingestion of fruit by vertebrates and subsequent defecation or regurgitation of intact seeds; synzoochory, seed-hoarding by vertebrates; myrmecochory, dispersal by ants; ectozoochory, dispersal on the exterior of animals, e.g. on fur or feathers; anemochory, dispersal by wind; hydrochory, dispersal by water; or unspecialized.

microsite limitation on average than in species with either abiotic or unspecialized dispersal mechanisms.

Although the evidence presented above is only circumstantial, the consistent trends suggest that granivores may influence both the demography and evolution of animal-dispersed species more than they influence those processes in plant species with predominantly abiotic or unspecialized dispersal mechanisms. In general, animal-dispersed species tend to rely proportionally more on regeneration by seeds, often lack a persistent seed bank and are more limited in recruitment by seed abundance than by microsite abundance. These trends appear to be strongest for seeds actively dispersed by vertebrates. However, since the impact of granivores on synzoochorous species may be buffered by the irregular production of mast crops, this implies that endozoochorous species may be most prone to the deleterious impacts of seed predators. A further implication of these findings is that current knowledge of the impacts of granivores on plant demography may be somewhat biased, with an over-representation of studies on animal-dispersed species (Table 17.1) (Hulme, 1993, 1998a). Future research needs to address this imbalance and, more specifically, to focus on seed predation in plant species that differ in dispersal strategy.

Granivores and Seed Fate: Death or Dispersal?

Many seed-predation studies, especially those examining post-dispersal granivory, assume that removal of seeds by granivores ultimately results in the consumption of the seed. However, when seed predators cache seeds for later consumption and seed recovery is incomplete, the end result of 'granivory' may be seed dispersal. Because few studies follow the fate of seeds removed by vertebrate granivores, it is possible that the negative impact of granivores on plant demography may be overestimated. However, it is equally dangerous to assume that all granivores act as mutualists. If granivores act as mutualists, caching should facilitate dispersal away from the parent (enhancing both escape from natural enemies and opportunities for colonization) and/or dispersal into specific microsites.

The distance over which seeds might be successfully cached by granivores is often underestimated, due to the logistical difficulty of locating seeds transported more than several metres. Indeed, most studies that have attempted to estimate dispersal distances report relatively short mean dispersal distances: 4.1 m for *Gustavia superba* (Forget, 1992), 6.5 m for *Vouacapoua americana* (Forget, 1990), 7.8–9.5 m for *Purshia tridentata* (Vander Wall, 1994), 9.6 m for *Simplocarpus renifolius* (Wada and Uemura, 1994), 12.2–44.7 m for *Aesculus turbinata* (Hoshizaki *et al.*, 1999) and 15.3 m *for Quercus robur* (Jensen and Nielsen, 1986). These mean dispersal distances are similar to those identified for bird-dispersed seeds, but significantly less than mean distances frequently achieved via wind dispersal (Willson, 1993). These examples relate to seed dispersal by rodents; seed-dispersing corvids may transport seeds considerably further (Tomback and Linhart, 1990). Thus, at least for rodents, caching is unlikely to enhance opportunities for colonization and may not move seeds far enough from parent trees to escape distance-responsive seed predators. Furthermore, burial of several seeds within a cache will increase sibling competition among seedlings and may also attract density-dependent granivores.

Seed-dispersing granivores often select particular cache sites, presumably to increase their chance of seed recovery (Vander Wall, 1990). For example, Clark's nutcracker (*Nucifraga columbiana*) caches pine seeds on steep, south-facing slopes (Tomback and Linhart, 1990). South-facing slopes are often free of snow early in spring and thus facilitate cache recovery. Coincidentally these sites may also be particularly suitable for pine seed germination and seedling establishment. In contrast, the Eurasian nutcracker (*Nucifraga caryocatactes*) often buries pine seeds above the timberline, where pine is unable to establish (Tomback and Linhart, 1990). Rodents often use physical cues in the habitat to aid cache relocation and their caches are frequently made in the vicinity of fallen logs or at the base of trees (Vander Wall, 1990). While the selection of caches is certainly non-random, it is unclear whether these sites represent optimum microsites for plant establishment.

One of the key differences between caching and other forms of seed dispersal is the process of seed burial. For many large-seeded forest species, burial is a necessary requirement for germination, since otherwise seeds will often desiccate and die (e.g. Watt, 1919; Vander Wall, 1994). This benefit of burial declines with decreasing seed size and for small seeds, especially herbaceous species, burial at the depths normally found for caches can significantly reduce germination rates (Froud-Williams *et al.*, 1984). Nevertheless, burial also protects seeds from post-dispersal seed predation, particularly by invertebrates (Hulme, 1998b; Hulme and Borelli, 1999). Thus for large seeds the benefit of caching may simply be burial, independent of the distance from the parent plant.

Successful seed dispersal is more likely through scatter-hoarding than through larder-hoarding, because seeds are buried in many shallow caches and distributed among a variety of microhabitats. The large number of caches often results in incomplete seed recovery (Vander Wall, 1990). In contrast, larder hoards are often buried more deeply (frequently within animal burrows) and the single location makes recovery of seeds highly probable (Smith and Reichman, 1984). Even where recovery is incomplete, the depth of burial may prevent successful germination from larder hoards. Detailed studies have shown that in most cases survival and germination of seeds from scatter-hoarded caches is low. Germination rates reported for seeds cached by rodents include: 0.02% for *Oryzopsis hymenoides* (McAdoo *et al.*, 1983), 0.8% for *Pinus strobus* (Abbott and Quink, 1970), 0–2% for *Dipteryx panamensis* (Forget, 1993), 1.5–2.5% for *Q. robur* (Jensen and Nielsen, 1986), 0–4% for *Fagus sylvatica* (Jensen, 1982), 5–8.5% for *P. tridentata* (Vander Wall, 1994) and 0.75–10% for *G. superba* (Forget, 1992). The limited data on seedling emergence rates for seeds dispersed by other means show higher rates of seedling emergence. A comparable estimate of seedling emergence for *Phillyrea latifolia*, a bird-dispersed species, is 8% (Herrera *et al.*, 1994) and for two wind-dispersed *Acer* spp. 19–24% (Houle, 1995). It is clear that plants dispersed by granivores bear a high cost in terms of seeds consumed for a relatively low return in terms of seedling regeneration. However, even relatively low germination rates may translate into a substantial number of seedlings if seed abundance is high. Yet the few studies that have monitored the subsequent survival of seedlings record high seedling mortality (Forget, 1993; Vander Wall, 1994), reflecting that cache locations may not necessarily be suitable for establishment.

Few studies have shown unequivocally that seed dispersal by granivores is essential for plant regeneration. A clear test would be to examine whether plant regeneration is significantly reduced in the absence of seed-caching granivores. This is rarely feasible. However, it may be possible to predict when granivores most probably act as mutualists:

- Seeds will be scatter- rather than larder-hoarded. Marked taxonomic differences among granivores occur in the type of caching undertaken. Some seed-feeding corvids (e.g. jays and nutcrackers) generally scatter-hoard, whereas most granivorous mammals larder-hoard, with the important exceptions of Sciuridae (e.g. tree squirrels and chipmunks) and Dasyproctidae (e.g. agoutis) (Smith and Reichman, 1984).
- Synzoochory will probably be most important for plant species for which it is the primary (or only) seed-dispersal mechanism. Where plants utilize other seed-dispersal mechanisms (e.g. anemochory or endozoochory), it must be shown that synzoochory is either equally or more efficient at facilitating seedling establishment (e.g. Vander Wall, 1994).
- When food resources are scarce, vertebrate seed-caching appears more pronounced than when seeds are superabundant and consumption of cached seeds appears most intense (see Forget *et al.*, this volume). Plant species that maximize caching and minimize cache exploitation are likely to be those that periodically produce large seed crops, satiating granivores.

In summary, it remains unclear whether regeneration is higher in the presence of seed-eaters than in their absence. Nevertheless, caution must be applied when equating seed removal and seed predation; although most seeds removed are consumed, a small fraction may be dispersed to suitable microsites. Interpreting the consequences of seed removal will require knowledge of the granivores involved, their foraging behaviour and the reproductive traits of the plant. Thus rodent scatter-hoarding of large-seeded, mast-fruiting trees may be essential for regeneration of those trees. However, murid rodents that larder-hoard the small seeds of a bird-dispersed shrub may significantly reduce plant recruitment. While it is impossible to know the fate of all seeds removed by granivores, application of basic ecological knowledge can shed light on the probabilities of seed dispersal versus predation.

Conclusions

Integrating the study of granivores into research on plant–frugivore interactions is an essential step towards understanding the evolutionary and demographic role of seed-dispersal mutualisms. This integration is particularly relevant for animal-mediated seed dispersal, because the plants involved appear particularly prone to the impacts of granivores. Escape from seed predators is believed to be a major force in the evolution of seed dispersal (Howe and Westley, 1988) and yet the relative importance of pre- and post-dispersal seed predation for different dispersal mechanisms remains poorly understood. Granivores are influenced by spatial variation in seed density, while simultaneously influencing it themselves. Thus, seed shadows are far more complex than simply a negative exponential distribution from the parent. While distance from a fruiting parent remains an important determinant of seed distribution, considerable variation in seed density and predation will occur among microhabitats, irrespective of distance from the parent. Nevertheless, local seed rain is rarely taken into account when assessing the intensity of seed predation (but see Herrera *et al.*, 1994; Kollmann and Schill, 1996; Rey and Alcantara, 2000). Seed shadows are often perceived differently by various guilds of granivores, and their impacts on seed distribution can be quite distinct. However, few studies have attempted to quantify separately the impacts of different guilds of granivores. If various guilds of granivores use dissimilar microhabitats, have different seed preferences and/or are active in distinct seasons, failure to quantify their separate effects may mask general trends (Hulme, 1994, 1997). Furthermore, where certain granivores disperse seeds and others act as seed predators, separation of their individual impacts will be essential for an accurate assessment of seed dynamics. The new paradigms and approaches described above will facilitate improved understanding of the interaction between seed dispersal and predation and will provide a basis for accurately predicting the role of granivores in plant communities.

References

Abbott, H.G. and Quink, T.F. (1970) Ecology of eastern white pine seed caches made by small forest mammals. *Ecology* 51, 271–278.

Abrahamson, W.G. (1989) *Plant–Animal Interactions.* McGraw Hill, New York.

Andersen, A.N. (1988) Insect seed predators may cause far greater losses than they appear to. *Oikos* 52, 337–340.

Andersen, A.N. (1989) How important is seed predation to recruitment in stable populations of long-lived perennials? *Oecologia* 81, 310–315.

Andersson, C. (1992) The effect of weevil and fungal attacks on the germination of *Quercus robur* acorns. *Forest Ecology and Management* 50, 247–251.

Castro, J., Gomez, J.M., Garcia, D., Zamora, R. and Hodar, J.A. (1999) Seed predation and dispersal in relict Scots pine forests of southern Spain. *Plant Ecology* 145, 115–123.

Cipollini, M. and Levey, D.J. (1997) Why are some fruits toxic? Glycoalkaloids in *Solanum* and fruit choice by vertebrates. *Ecology* 78, 782–798.

Condit, R., Ashton, P.S., Baker, P., Bunyavejchewin, S., Gunatilleke, S., Gunatilleke, N., Hubbell, S.P., Foster, R.B., Itoh, A., LaFrankie, J.V., Lee, H.S., Losos, E., Manokaran, N., Sukumar, R. and Yamakura, T. (2000) Spatial patterns in the distribution of tropical tree species. *Science* 288, 1414–1418.

Crawley, M.J. (1990) The population dynamics of plants. *Philosophical Transactions of the Royal Society of London, Series B* 330, 125–140.

Crawley, M.J. (2000) Seed predators and plant population dynamics. In: Fenner, M. (ed.) *Seeds: the Ecology of Regeneration in Plant Communities*, 2nd edn. CAB International, Wallingford, UK, pp. 167–182.

Crawley, M.J. and Long, C.R. (1995) Alternate bearing, predator satiation and seedling recruitment in *Quercus robur* L. *Journal of Ecology* 83, 683–696.

Curran, L.M. and Leighton, M. (2000) Vertebrate responses to spatiotemporal variation in seed production of mast fruiting Dipterocarpaceae. *Ecological Monographs* 70, 101–128.

De Steven, D. (1983) Reproductive consequences of insect seed predation in *Hammamelis virginiana. Ecology* 64, 89–98.

Eriksson, O. and Ehrlen, J. (1992) Seed and microsite limitation in plant populations. *Oecologia* 91, 360–364.

Fenner, M. (ed.) (2000) *Seeds: the Ecology of Regeneration in Plant Communities*, 2nd edn. CAB International, Wallingford, UK.

Forget, P.M. (1990) Seed-dispersal of *Vouacapoua americana* (*Caesalpiniaceae*) by caviomorph rodents in French Guiana. *Journal of Tropical Ecology* 6, 459–468.

Forget, P.M. (1992) Seed removal and seed fate in *Gustavia superba* (*Lecythidaceae*). *Biotropica* 24, 408–414.

Forget, P.M. (1993) Postdispersal predation and scatterhoarding of *Dipteryx panamensis* (*Papilionaceae*) seeds by rodents in Panama. *Oecologia* 94, 255–261.

Forget, P.M., Munoz, E. and Leigh, E.G. (1994) Predation by rodents and bruchid beetles on seeds of *Scheelea* palms on Barro Colorado Island, Panama. *Biotropica* 26, 420–426.

Forget, P.M., Kitajima, K. and Foster, R.B. (1999) Pre- and post-dispersal seed predation in *Tachigali versicolor* (*Caesalpiniaceae*): effects of timing of fruiting and variation among trees. *Journal of Tropical Ecology* 15, 61–81.

Froud-Williams, R.J., Chancellor, R.J. and Drennan, D.S.H. (1984) The effects of seed burial and soil disturbance on emergence and survival of arable weeds in relation to minimal cultivation. *Journal of Applied Ecology* 21, 629–641.

Garcia, D., Zamora, R., Gomez, J.M. and Hodar, J.A. (1999) Bird rejection of unhealthy fruits reinforces the mutualism between juniper and its avian dispersers. *Oikos* 85, 536–544.

Gardner, G. (1977) The reproductive capacity of *Fraxinus excelsior* on the Derbyshire limestone. *Journal of Ecology* 65, 107–118.

Hammond, D.S. and Brown, V.K. (1998) Disturbance, phenology and life-history characteristics: factors influencing distance/density-dependent attack on tropical seeds and seedlings. In: Newberry, D.M., Prins, H.H.T. and Brown, N.D. (eds) *Dynamics of Tropical Communities.* Blackwell Science, Oxford, pp. 401–474.

Harms, K.E., Wright, S.J., Calderon, O., Hernandez, A. and Herre, E.A. (2000) Pervasive density-dependent recruitment enhances seedling diversity in a tropical forest. *Nature* 404, 493–495.

Hauser, T.P. (1994) Germination, predation and dispersal of *Acacia albida* seeds. *Oikos* 70, 421–426.

Herrera, C.M. and Pellmyr, O. (eds) (2001) *Plant–Animal Interactions.* Blackwell Science, Oxford, UK.

Herrera, C.M., Jordano, P., Lopez-Soria, L. and Amat, J.A. (1994) Recruitment of a mast-fruiting, bird-dispersed tree: bridging frugivore activity and seedling establishment. *Ecological Monographs* 64, 315–344.

Herrera, C.M., Jordano, P., Guitian, J. and Traveset, A. (1998) Annual variability in seed production by woody plants and the masting concept: reassessment of principles and relationship to pollination and seed dispersal. *American Naturalist* 152, 576–594.

Hodgson, J.G., Grime, J.P., Hunt, R. and Thompson, K. (1994) *The Electronic Comparative Plant Ecology.* Chapman & Hall, London, UK.

Homma, K., Akashi, N., Abe, T., Hasegawa, M., Harada, K., Hirabuki, Y., Irie, K., Kaji, M., Miguchi, H., Mizoguchi, N., Mizunaga, H., Nakashizuka, T., Natume, S., Niiyama, K., Ohkubo, T., Sawada, S., Sugita, H., Takatsuki, S. and Yamanaka, N. (1999) Geographical variation in the early regeneration process of Siebold's Beech (*Fagus crenata* Blume) in Japan. *Plant Ecology* 140, 129–138.

Hoshizaki, K., Wajirou, S. and Satohiko, S. (1999) Impacts of secondary seed dispersal and herbivory on seedling survival in *Aesculus turbinata. Journal of Vegetation Science* 8, 735–742.

Houle, G. (1995) Seed dispersal and seedling recruitment: the missing links. *Ecoscience* 2, 238–244.

Howe, H.F. and Westley, L.C. (1988) *Ecological Relationships of Plants and Animals.* Oxford University Press, New York.

Hubbard, J.A. and McPherson, G.R. (1997) Acorn selection by Mexican jays: a test of a tri-trophic symbiotic relationship hypothesis. *Oecologia* 110, 143–146.

Hulme, P.E. (1992) The ecology of a temperate plant in a Mediterranean environment: post-dispersal seed predation of *Daphne laureola*. In: Thanos, C.A. (ed.) *Plant–Animal Interactions in Mediterranean Type Ecosystems*. Athens University Press, Athens, Greece, pp. 281–286.

Hulme, P.E. (1993) Post-dispersal seed predation by small mammals. *Symposium of the Zoological Society of London* 65, 269–287.

Hulme, P.E. (1994) Rodent post-dispersal seed predation in grassland: magnitude and sources of variation. *Journal of Ecology* 82, 645–652.

Hulme, P.E. (1996) Natural regeneration of yew (*Taxus baccata* L): microsite, seed or herbivore limitation? *Journal of Ecology* 84, 853–861.

Hulme, P.E. (1997) Post-dispersal seed predation and the establishment of vertebrate dispersed plants in Mediterranean scrublands. *Oecologia* 111, 91–98.

Hulme, P.E. (1998a) Post-dispersal seed predation: consequences for plant demography and evolution. *Perspectives in Plant Ecology, Evolution and Systematics* 1, 32–46.

Hulme, P.E. (1998b) Post-dispersal seed predation and seed bank persistence. *Seed Science Research* 8, 513–519.

Hulme, P.E. and Benkman, C.W. (2001) Granivory. In: Herrera, C.M and Pellmyr, O. (eds) *Plant–Animal Interactions*. Blackwell Science, Oxford, UK.

Hulme, P.E. and Borelli, T. (1999) Variability in post-dispersal seed predation in deciduous woodland: relative importance of location, seed species, burial and density. *Plant Ecology* 145, 149–156.

Itoh, A., Yamakura, T., Ogino, K. and Lee, H.S. (1995) Survivorship and growth of seedlings of four dipterocarp species in a tropical rain-forest of Sarawak, East Malaysia. *Ecological Research* 10, 327–338.

Janzen, D.H. (1970) Herbivores and the number of tree species in tropical forests. *American Naturalist* 104, 501–528.

Jensen, T.S. (1982) Seed production and outbreaks of non-cyclic rodent populations in deciduous forests. *Oecologia* 54, 184–192.

Jensen, T.S. and Nielsen, O.F. (1986) Rodents as seed dispersers in a heath–oak wood succession. *Oecologia* 70, 214–221.

Jordano, P. (2000) Fruits and frugivory. In: Fenner, M. (ed.) *Seeds: the Ecology of Regeneration in Plant Communities*, 2nd edn. CAB International, Wallingford, pp. 125–165.

Karban, R. and Lowenberg, G. (1992) Feeding by seed bugs and weevils enhances germination of wild *Gossypium* species. *Oecologia* 92, 196–200.

Kelrick, M.I., MacMahon, J.A., Parmenter, R.R. and Sisson, D.V. (1986) Native seed preferences of shrub-steppe rodents, birds and ants: the relationships of seed attributes and seed use. *Oecologia* 68, 327–337.

Klinkhammer, P.G.L., de Jong, T.J. and van der Meijden, E. (1988) Production, dispersal and predation of seeds in the biennial *Cirsium vulgare*. *Journal of Ecology* 76, 403–414.

Kollmann, J. and Schill, H.-P. (1996) Spatial patterns of dispersal, seed predation and germination during colonisation of abandoned grassland by *Quercus petraea* and *Corylus avellana*. *Vegetatio* 125, 193–205.

Krefting, L.W. and Roe, E.I. (1949) The role of some birds and mammals in seed germination. *Ecological Monographs* 19, 270–286.

Leishman, M.R., Westoby, M. and Jurado, E. (1995) Correlates of seed size variation: a comparison among five temperate floras. *Journal of Ecology* 83, 517–529.

Levey, D.J. (1986) Methods of seed processing by birds and seed deposition patterns. In: Estrada, A. and Fleming, T.H. (eds) *Frugivores and Seed Dispersal*. W. Junk Publishers, Amsterdam, The Netherlands, pp. 147–159.

Levey, D.J. and Byrne, M.M. (1993) Complex ant–plant interactions: rain forest ants as secondary dispersers and post-dispersal seed predators. *Ecology* 74, 1802–1812.

Louda, S.M., Potvin, M.A. and Collinge, S.K. (1990) Predispersal seed predation, postdispersal seed predation and competition in the recruitment of seedlings of a native thistle in sandhills prairie. *American Midland Naturalist* 124, 105–113.

McAdoo, J.K., Evans, C.C., Roundy, B.A., Young, J.A. and Evans, R.A. (1983) Influence of heteromyid rodents on *Oryzopsis hymenoides* germination. *Journal of Range Management* 36, 61–64.

Miller, M.F. (1994) Large African herbivores, bruchid beetles and their interactions with *Acacia* seeds. *Oecologia* 97, 265–270.

Mucunguzi, P. (1995) Effects of bruchid beetles on germination and establishment of *Acacia* species. *African Journal of Ecology* 33, 64–70.

Myster, R.W. (1997) Seed predation, disease and germination on landslides in neotropical lower montane wet forest. *Journal of Vegetation Science* 8, 55–64.

Nilsson, S.G. and Wästljung, U. (1987) Seed predation and cross pollination in mast-seeding beech (*Fagus sylvatica*) patches. *Ecology* 68, 260–265.

Ollerton, J. and Lack, A. (1996) Partial predispersal seed predation in *Lotus corniculatus* L. (*Fabacaea*). *Seed Science Research* 6, 65–69.

Palacios, E., Rodriguez, A. and Defler, T.R. (1997) Diet of a group of *Callicebus torquatus lugens* (Humboldt, 1812) during the annual resource bottleneck in Amazonian Colombia. *International Journal of Primatology* 18, 503–522.

Pico, F.X. and Retana, J. (2000) Temporal variation in the female components of reproductive success over the extended flowering season of a Mediterranean perennial herb. *Oikos* 89, 485–492.

Rey, P.J. and Alcantara, J.M. (2000) Recruitment dynamics of a fleshy-fruited plant (*Olea europaea*): connecting patterns of seed dispersal to seedling establishment. *Journal of Ecology* 88, 622–633.

Sallabanks, R. and Courtney, S.P. (1992) Frugivory, seed predation, and insect–vertebrate interactions. *Annual Review of Ecology and Systematics* 37, 377–400.

Sarukhan, J. (1986) Studies on the demography of tropical trees. In: Tomlinson, P.B. and Zimmerman, M.H. (eds) *Tropical Trees as Living Systems.* Cambridge University Press, Cambridge, UK, pp. 163–184.

Smith, C.C. and Reichman, O.J. (1984) The evolution of food caching by birds and mammals. *Annual Review of Ecology and Systematics* 15, 329–351.

Snow, B. and Snow, D. (1988) *Birds and Berries.* T. & A.D. Poyser, Waterhouses, UK.

Soholt, L.F. (1973) Consumption of primary production by a population of kangaroo rats (*Dipodomys merriami*) in the Mojave desert. *Ecological Monographs* 43, 357–376.

Steele, M.A., Knowles, T., Bridle, K. and Simms, E.L. (1993) Tannins and partial consumption of acorns – implications for dispersal of oaks by seed predators. *American Midland Naturalist* 130, 229–238.

Terborgh, J., Losos, E., Riley, M.P. and Bolaños Riley, M. (1993) Predation by vertebrates and invertebrates on the seeds of five canopy tree species of an Amazonian forest. *Vegetatio* 107/108, 375–386.

Tomback, D.F. and Linhart, Y.B. (1990) The evolution of bird-dispersed pines. *Evolutionary Ecology* 4, 185–219.

Traveset, A. (1990) Postdispersal predation of *Acacia farnesiana* seeds by *Stator vachelliae* (Bruchidae) in Central America. *Oecologia* 84, 506–512.

Traveset, A. (1991) Pre-dispersal seed predation in Central American *Acacia farnesiana* – factors affecting the abundance of co-occurring bruchid beetles. *Oecologia* 87, 570–576.

Traveset, A. (1994) Cumulative effects on the reproductive output of *Pistacia terebinthus* (*Anacardiaceae*). *Oikos* 71, 152–162.

Turnbull, L.A., Crawley, M.J. and Rees, M. (2000) Are plant populations seed limited? A review of seed sowing experiments. *Oikos* 88, 225–238.

Vander Wall, S.B. (1990) *Food Hoarding in Animals.* University of Chicago Press, Chicago, Illinois.

Vander Wall, S.B. (1994) Seed fate pathways of antelope bitterbrush: dispersal by seed-caching yellow pine chipmunks. *Ecology* 75, 1911–1926.

Wada, N. and Uemura, S. (1994) Seed dispersal and predation by small rodents on the herbaceous understory plant *Symplocarpus renifolius. American Midland Naturalist* 132, 320–327.

Watt, A.S. (1919) On the causes of failure of natural regeneration in British oak-woods. *Journal of Ecology* 7, 173–203.

Wills, C., Condit, R., Foster, R.B. and Hubbell, S.P. (1997) Strong density- and diversity-related effects help to maintain tree species diversity in a neotropical forest. *Proceedings of the National Academy of Sciences USA* 94, 1252–1257.

Willson, M.F. (1993) Dispersal mode, seed shadows, and colonization patterns. *Vegetatio* 107/108, 261–280.

18 Plant–Animal Coevolution: Is it Thwarted by Spatial and Temporal Variation in Animal Foraging?

Colin A. Chapman[1,2] and Lauren J. Chapman[1,2]
[1]*Department of Zoology, University of Florida, Gainesville, FL 32611, USA;*
[2]*Wildlife Conservation Society, 185th Street and Southern Boulevard, Bronx, New York, NY 10460, USA*

Introduction

The last two decades have witnessed large changes in views on the evolution of seed-dispersal systems. Early theories generated straightforward, testable predictions based on several key assumptions (Snow, 1965; Howe and Estabrook, 1977; Howe, 1979). During the 1980s, there was a gradual accumulation of field studies that did not support these predictions or did so only in a very general way (Howe and Smallwood, 1982; Herrera, 1984, 1985; Howe, 1984). These developments coincided with the recognition that descriptions of tight coevolution, or at least mutual dependence of particular plants and dispersers, were anomalies that tended to involve either very large seeds and seed-dispersers (e.g. elephants, *Loxodonta africana*, and *Balanites wilsoniana* (Lieberman *et al.*, 1987; Chapman *et al.*, 1992); gorillas, *Gorilla gorilla*, and *Cola lizae* (Tutin *et al.*, 1991)) or island situations with depauperate disperser assemblages (e.g. *Lycopersicon esculentum* and Galapagos tortoises, *Testudo elephantopus* (Rick and Bowman, 1961; see also Temple, 1977)).

In 1985, Herrera provided a critical evaluation of early studies and their assumptions (see also Herrera, 1986). He concluded that coevolved plant–vertebrate seed-dispersal systems were, at best, very rare in nature. He suggested that numerous factors limit the potential for coevolution between plants and their animal dispersers. These factors include: inequality in the evolutionary lifespans of plant and animal taxa, difference in generation lengths of plants and their dispersers, extensive gene flow between plant populations, weak selective pressures on dispersers, ecological variables outside the control of the parent plant (e.g. the influence of other fruiting plants), unpredictability of germination sites, secondary dispersal and the lack of evolutionary plasticity (Wheelwright and Orians, 1982; Herrera, 1985, 1986, 1998; Fischer and Chapman, 1993; Chapman, 1995).

Foraging patterns of vertebrates can also constrain plant–animal coevolution. For example, Herrera (1985) emphasized that the identity of taxa dispersing a given plant species can change over relatively short distances. Because these different animal species will

 Seed Dispersal and Frugivory: Ecology, Evolution and Conservation (eds D.J. Levey, W.R. Silva and M. Galetti)

probably handle seeds in different ways and individual plant species will be responding evolutionarily to the integrated selective pressures of all dispersers, the direction and intensity of the overall selection pressure will probably be inconsistent and weak.

Weak selection pressure can also result from changes in the behaviour of a single disperser species; a particular species might be a reliable disperser at one time or at one location, but not at a different time or location. For example, Gautier-Hion *et al.* (1993) studied the foraging behaviour of *Cercopithecus pogonias* and *Cercopithecus wolfi* in Gabon and the Democratic Republic of Congo and found that they were mainly seed-dispersers in Gabon and mainly seed predators in the Congo. Despite such examples and frequent claims in the seed-dispersal literature of substantial temporal variation in plant–vertebrate interactions (Herrera, 1982, 1984, 1998; Howe, 1983, 1993; Schupp, 1990; Jordano, 1992; Herrera *et al.*, 1998), there are few studies documenting variation across several years or across different spatial scales (but see Herrera, 1998).

We examine the degree to which the diets of red colobus (*Procolobus badius*), a seed predator, and redtail monkey (*Cercopithecus ascanius*), a seed-disperser, vary over the following spatial and temporal scales: (i) groups of red colobus within Kibale National Park, Uganda, with overlapping home ranges; (ii) eight populations of red colobus and four populations of redtail monkeys, each separated by approximately 15 km within or near Kibale; (iii) distantly separated populations within three primate genera across Africa; and (iv) annual variation among 4 years for a single red colobus group. For each scale, general diet data (e.g. % of the diet composed of fruit) are presented to illustrate the degree of dietary variability, and specific examples are provided to demonstrate how a given plant–animal interaction can change.

Study Animals

Red colobus monkeys are large-bodied (8.2 kg), diurnal primates, found in social groups of between 25 and 40 monkeys (Struhsaker, 1975). Groups usually contain at least three adult males and many adult females; females are the dispersing sex. In all populations studied, young leaves are the most common food item. Fruits are also eaten on a seasonal basis. When red colobus eat fruits, the seeds are destroyed (no seeds have been found in 150 dung samples (T. Gillespie, Florida, 2000, personal communication)). During a single feeding bout, a large group of red colobus can dramatically reduce the number of fruits on a tree. Thus, they can be significant seed predators.

Redtail monkeys are small-bodied (3.6 kg) primates found in social groups that average 30–35 individuals and typically contain a single male (Struhsaker and Leland, 1979). Their diet is dominated by fruit and insects (Struhsaker and Leland, 1979). They can be significant seed-dispersers for some tree species, often processing fruits in their cheek pouches and spitting out seeds away from the parent tree (Chapman, 1995; Lambert, 1997).

Methods

We have studied red colobus and redtail monkeys in Kibale National Park (766 km^2; 0° 13′–0° 41′ N and 30° 19′–30° 32′ E) (Struhsaker, 1997; Chapman and Lambert, 1999) in western Uganda since 1994. Mean annual rainfall in the region (measured at Makerere University Biological Field Station) is 1778 mm (1990–1998). There is an elevational gradient from north to south, which reflects a north-to-south increase in temperature and decrease in rainfall.

Observations of diet were made over 4 complete years at one site and for 1 or 2 years at eight other sites, each separated by approximately 15 km within the same forest system. Behavioural observations of red colobus totalled 3355 h and of redtail monkeys 587 h. During each half-hour that the observer was with the group, five point samples were made of different individuals. If the animal was feeding, the species and the plant part (e.g. fruit, young leaf, leaf petiole) were recorded. The percentage of time spent feeding on a particular plant species or part was calculated as the number of scans spent eating that item, divided by the total number of scans in which animals were feeding. For detailed information on sampling methods, duration of sampling and locations, see Chapman *et al.* (1997, 2000). At each site, food availability was quantified with a series of 200 m by 10 m transects, monitored on a monthly basis to assess phenology (Chapman *et al.*, 1997, 1999).

Results

Spatial contrasts

Neighbouring groups

We quantified the diet of two groups of red colobus from May 1998 to June 1999. Group 1 (24 individuals) used an area of 26.4 ha, while Group 2 (48 individuals) used an area of 21.9 ha. Home-range overlap of these groups was 10.7 ha, which represented 41% of Group 1's home range and 49% of Group 2's home range. Group 2 spent 70% of its time in the area of overlap, whereas Group 1 spent 49% of its time in that area.

Despite this degree of overlap in home ranges, diets differed between the two groups with respect of plant parts consumed (Fig. 18.1) and species exploited (Table 18.1). For example, there was a small grove of *Prunus africana* in the area of home-range overlap. Group 1 was a significant seed predator of *P. africana*, eating its seeds for 31% of the time they ate seeds, compared with 1.6% for Group 2. This difference occurred despite the fact that Group 2 spent 70% of its time in this area, while Group 1 spent only 49% of its time there.

Interdemic contrasts

Both species exhibited high variation among populations separated by approximately 15 km in both the plant parts and species consumed. For red colobus, the largest difference was found in the amount of time spent eating young leaves (38.2% maximum difference); however, the amount of time spent preying on seeds also varied among populations from 1.9% to 17.2% (Table 18.2).

The plant species most important to red colobus differed among populations (Table 18.3). Much of this variation reflected differences in forest composition among sites (Table 18.4; Chapman *et al.*, 1997). Some foods

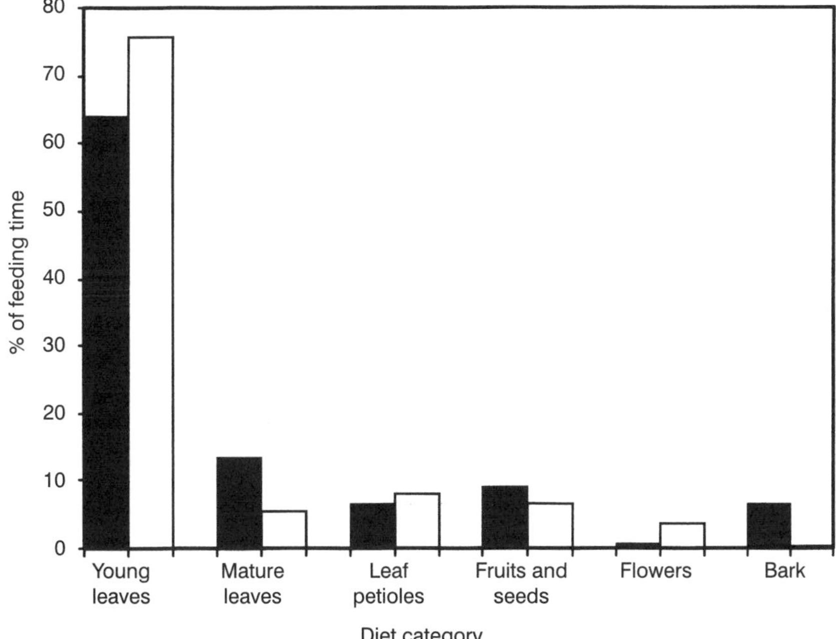

Fig. 18.1. The percentage of feeding time devoted to different plant parts by two neighbouring groups (solid bars vs. open bars) of red colobus monkeys (*Procolobus badius*) in Kibale National Park, Uganda.

Table 18.1. Density (trees ha^{-1}) and percentage of feeding time spent eating in 12 tree species used by two neighbouring groups of red colobus monkeys at the Kanyawara study area of Kibale National Park, Uganda.

		Group 2		Group 1	
Species	Family	Density %	Feeding time	Density %	Feeding time
Albizzia grandibracteata	*Leguminosae*	16	1.2	2	3.1
Bosqueia phoberos	*Moraceae*	45	7.8	34	4.1
Celtis africana	*Ulmaceae*	11	7.6	6	8.1
Celtis durandii	*Ulmaceae*	38	21.3	22	7.2
Ficus exasperata	*Moraceae*	8	2.1	4	0.5
Funtumia latifolia	*Apocynaceae*	35	8.2	25	8.9
Markhamia platycalyx	*Bignoniaceae*	32	9.0	16	3.2
Milletia dura	*Leguminosae*	9	1.3	1	0.6
Olea welwitchii	*Oleaceae*	4	0.9	2	2.2
Parinari excelsa	*Rosaceae*	2	6.7	2	8.1
Dombeya mukole	*Sterculiaceae*	2	7.5	1	2.3
Prunus africana	*Rosaceae*	1	4.1	3	17.2

Table 18.2. The percentage of scan samples in which red colobus were eating particular items at eight areas in or near Kibale National Park, Uganda. The values do not sum to 100% because of groups eating plant parts that are not listed below (e.g. pine needles).

Location	Young leaves	Mature leaves	Leaf petiole	Fruit/seeds	Flowers	Bark
Sebatoli	72.4	7.4	7.1	6.4	3.3	2.0
K-15	69.8	2.6	5.8	17.2	2.3	0.3
Mikana	87.0	2.0	4.2	3.0	2.2	0.0
K-30	57.6	9.9	14.2	6.7	2.0	4.1
Dura River	65.1	4.6	8.7	13.9	6.2	0.0
Mainaro	57.5	16.2	1.8	10.8	7.2	3.6
Nkuruba	67.3	18.4	2.8	1.9	2.3	6.4
Kahunge	48.8	21.0	0.0	3.1	22.7	2.7
Largest difference	38.2	19.0	14.2	15.3	20.7	6.4

important to particular populations were not present at other sites. The Kahunge group represents a striking example. These monkeys fed on *Acacia kirkii* 92% of the time, and this tree species was only found at this site. Similarly, *Cynometra alexandri* was eaten by the Mainaro population for 41% of that group's feeding time, and was only found at this site. In contrast, some of the observed differences could not be attributed to availability. For example, *Celtis africana* was eaten at six of the seven sites where it was found and was not eaten at Mainaro, despite the fact that it was common there. A dramatic example of interdemic variation in seed predation concerns red colobus feeding on the seeds of *Celtis durandii*. The K15 group ate *C. durandii* seeds for 15.4% of its foraging time. In contrast, the Mainaro group was never recording eating *C. durandii* seeds and yet the density of this tree was very similar at the two sites (K15 = 33.0 trees ha^{-1}, Mainaro = 33.8 trees ha^{-1}).

The time redtail monkeys spent feeding on different plant parts varied among sites (Fig. 18.2). For example, the animals at Kanyawara ate fruit for only 35.7% of their feeding time, while those at Mainaro ate fruit for 59.7% of their feeding time (Fig. 18.2). The amount of time redtail monkeys spent eating from particular plant species also varied among areas (Table 18.5). In some cases, the variation could be related to plant density, while in other cases there was no apparent relationship (Table 18.5). For example, *Mimusops bagshawei*

Table 18.3. The percentage of red colobus feeding time involving the top five most frequently eaten plant species (underlined) at each of eight sites in Kibale National Park, Uganda, and the corresponding use at the other sites. Species are listed in order of their overall frequency of use at all sites. Only four plant species were eaten at Kahunge, and two species tied for the fifth at Sebatoli.

Species		Sebatoli	K15	Mikana	K30	Dura	Mainaro	Nkuruba	Kahunge
Acacia hockii	*Leguminosae*	–	–	–	–	–	–	–	91.9
Celtis durandii	*Ulmaceae*	5.4	23.6	19.0	10.4	27.2	6.0	–	–
Celtis africana	*Ulmaceae*	4.3	12.2	13.7	9.9	1.5	–	19.1	–
Albizzia grandibracteata	*Leguminosae*	1	4.1	3.6	8.4	10.8	1.8	14.6	0.68
Prunus africanum	*Rosaceae*	5.9	1.7	3.2	13.0	–	2.4	16.3	–
Cynometra alexandri	*Leguminosae*	–	–	–	–	–	40.7	–	–
Funtumia latifolia	*Apocynaceae*	5.4	8.1	3.1	7.2	12.8	3.0	–	–
Aningeria altissima	*Sapotaceae*	8.7	8.7	0.3	0.9	14.9	–	–	–
Markhamia platycalyx	*Bignoniaceae*	3.1	10.2	6.1	9.2	1	–	0.9	–
Mimusops bagshawei	*Sapotaceae*	0.8	–	–	0.4	4.6	5.4	16.1	–
Strombosia scheffleri	*Olacaceae*	10.9	0.9	2.0	9.2	2.7	–	–	–
Dombeya mokole	*Sterculiaceae*	–	4.1	5.2	3.5	–	–	12.8	–
Olea welwitchii	*Oleaeceae*	5.1	1.5	13.5	3.9	–	–	0.2	–
Bosqueia phoberos	*Moraceae*	3.1	–	2.9	0.8	3.1	5.4	0.9	–
Newtonia buchananii	*Leguminosae*	11.2	–	0.4	–	–	–	–	–
Parinari excelsa	*Rosaceae*	–	–	5.3	0.1	–	–	–	–
Celtis zenkeri	*Ulmaceae*	–	–	–	–	–	5.4	–	–
Cola gigantea	*Sterculiaceae*	–	–	–	–	5.1	–	–	–
Sapium ellipticum	*Euphorbiaceae*	1.3	–	–	0.1	–	0.6	–	0.68
Bridelia micrantha	*Euphorbiaceae*	–	–	–	1.63	–	–	–	0.68

Table 18.4. The density (individuals ha^{-1}) of preferred red colobus food trees (top five most eaten species at any of the sites) found at seven sites in or near Kibale National Park, Uganda. The superscripts indicate the ranking of the five most commonly eaten species for sites where behavioural data were collected (if a superscript number is given twice, the species were tied). Densities of trees are not available for the Nkuruba and Mikana sites.

Species	Family	Sebatoli	K15	K30	Dura	Mainaro	Kahunge
Celtis durandii	*Ulmaceae*	2.5^{5}	33.0^{1}	47.1^{2}	63.8^{1}	33.8^{2}	–
Funtumia latifolia	*Apocynaceae*	25.0^{5}	27.0^{5}	33.8	43.8^{3}	2.5	–
Markhamia platycalyx	*Bignoniaceae*	38.8	43.0^{3}	50.0^{4}	8.8	1.3	–
Bosqueia phoberos	*Moraceae*	–	–	50.0	22.5	1.3^{4}	–
Cynometra alexandri	*Leguminosae*	–	–	–	–	63.8^{1}	–
Strombosia scheffleri	*Olacaceae*	36.3^{2}	1.0	12.5^{5}	2.5	–	–
Newtonia buchananii	*Leguminosae*	26.3^{1}	1.0	–	3.8	–	–
Aningeria altissima	*Sapotaceae*	23.8^{3}	2.0^{4}	1.7	2.5^{2}	–	–
Mimusops bagshawei	*Sapotaceae*	6.3	1.0	3.3	7.5	–^{4}	–
Acacia kirkii	*Leguminosae*	–	–	–	–	–	20.0^{1}
Celtis africana	*Ulmaceae*	–	7.0^{2}	4.2^{3}	–	1.3^{5}	–
Albizia grandibracteata	*Leguminosae*	–	–	1.3	1.3^{4}	–	10.0^{2}
Blighia sp.	*Sapindaceae*	7.5	2.0	0.8	1.3	–	–
Cola gigantea	*Sterculiaceae*	–	–	–	6.3^{5}	–	–
Prunus africana	*Rosaceae*	2.5^{4}	–	–^{1}	–	–	–
Sapium ellipticum	*Euphorbiaceae*	2.5	–	–	–	–	2.0^{2}
Total density		171.3	117.0	204.0	164.0	102.7	32.0
Cumulative dbh		9496	2759	5548	6708	4747	1765.0

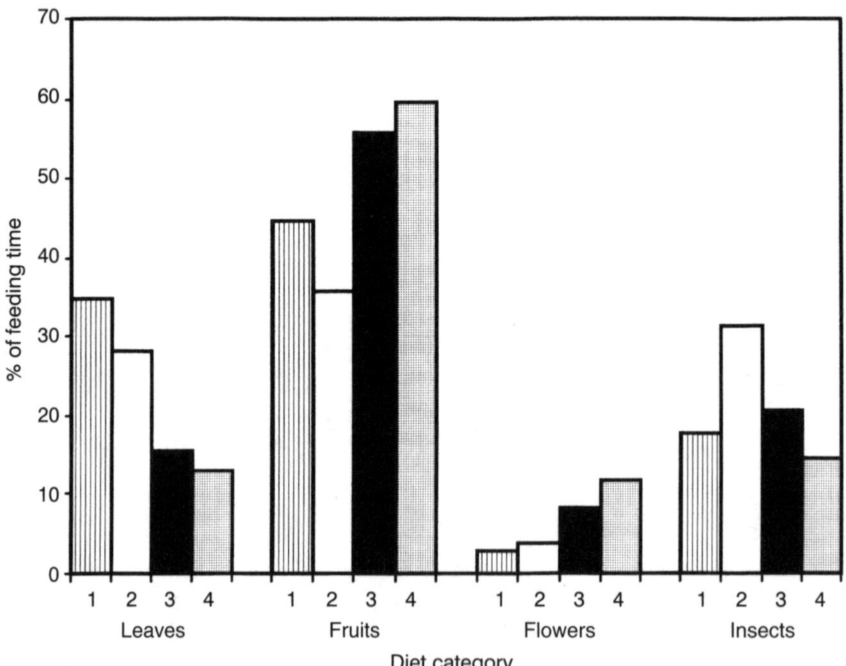

Fig. 18.2. The percentage of feeding time devoted to different plant parts by redtail monkeys (*Cercopithecus ascanius*) from four different populations (1, Sebatoli; 2, Kanyawara; 3, Dura; 4, Mainaro) in Kibale National Park, Uganda.

is a canopy-level or emergent tree with a drupe that averages 1.7 cm in length and contains an oval-shaped, 1.1 cm seed. Redtail monkeys are important dispersers of these seeds (Lambert, 1997). *Mimusops bagshawei* fruits were eaten at Kanyawara (1.8% of annual diet and 32% of the diet in the month when it fruited) and at Dura River (4.5% of total annual diet and 57% of the diet in the months it was available), but the redtail monkeys at Sebatoli were never observed to eat this fruit. At Dura River, *M. bagshawei* is relatively rare (< 1.3 trees ha^{-1}); it is more common at Kanyawara (3.3 trees ha^{-1}), but its availability is greatest at Sebatoli (6.3 trees ha^{-1}), where it was not eaten. Monthly monitoring of tree phenology indicates that the magnitude and duration of *M. bagshawei*'s fruiting was similar at all sites (Chapman *et al.*, 1999).

Distantly separated populations

Most study sites that provide detailed data on primate diets are widely separated. Thus, if one wants to compare widely separated sites, one first needs to find plant and animal species that occur over a wide area. While it is generally true that tropical trees do not have wide distributions, Africa is a bit of an exception in that many of the tree species range very widely and some are found in all major tropical-forest blocks (Richards, 1996). We take advantage of this and first contrast two sites where both diet data and plant lists are available. Subsequently, we contrast the published descriptions of diet of different populations or subspecies of red colobus (*P. badius*), different species of black-and-white colobus (*Colobus* spp.) and members of the 'Cephus' group of cercopithecine monkeys, which includes the redtail monkey. This 'Cephus' group is comprised of six closely related species that probably diverged from a common ancestor during recent isolation events associated with glaciation. Since species within these groups have recently diverged, it seems reasonable to expect that they might have similar dietary needs and consume similar foods.

First, to examine large-scale variation in primate seed dispersal and predation, we contrast the primate and tree communities from Kibale and Lopé, Gabon. The Lopé Reserve (5000 km^2; 0° 10′ S, 11° 35′ E) in central Gabon is similar to Kibale in several ways. Both areas receive similar levels of rainfall (Lopé = 1548 mm (Tutin *et al.*, 1997b); Kibale = 1778 mm (C.A. Chapman and L.J. Chapman, unpublished data)) and have similar seasonal cycles and temperature regimes. Lists of trees > 10 cm diameter at breast height (dbh) are available for both sites (Tutin *et al.*, 1991; Chapman *et al.*, 1999). While these lists are not totally comparable (e.g. the sampling areas differ, and Kibale's list does not include opportunistic collections), they do provide a general indication of similarity in the tree communities.

Thirteen per cent of the tree species found at Kibale (n = 109 tree species) also occurred at Lopé. The list of tree species at Lopé (n = 258) was greater than at Kibale; thus a smaller percentage of that flora was shared with Kibale (5.4%). Of the 14 plant species found at both sites, differences in use were documented. At Kibale, *C. ascanius* used *Spathodea campanulata* and *Symphonia globuliera,* while these plant species were not used by *Cercopithecus cephus* at Lopé. In contrast, *C. cephus* at Lopé ate *Irvingia gabonensis* and *Myrianthus arboreus,* while *C. ascanius* at Kibale ignored these species. There were also differences in the use of specific plant parts between closely related species at the two sites (Table 18.6). For example, *Colobus guereza* was rarely seen to prey on unripe seeds, while *Colobus satanus* did so regularly.

Kibale's primate biomass is eight times that of Lopé (total biomass Kibale = 2710 kg km^{-2}, Lopé = 319 kg km^{-2}; frugivore biomass Kibale = 634 kg km^{-2}, Lopé = 228 kg km^{-2}, folivore biomass Kibale = 2077 kg km^{-2}, Lopé = 91 kg km^{-2} (Table 18.6)). Such differences in the biomasses of dispersers and seed predators is likely to translate into differences in plant-taxa exploitation between sites.

Secondly, to examine large-scale variation in primate seed dispersal and predation in a more general way, we obtained diet data from 32 studies that used similar behavioural methods to collect feeding data (Table 18.7). Some populations of red colobus monkeys are primarily seed predators (the maximum % of time spent eating seeds was 54.4%), while others rarely eat seeds (5.6%) (Table 18.7). Different species of black-and-white colobus varied even

Table 18.5. The density of trees > 10 cm dbh and the percentage of time spent feeding by redtail monkeys on these trees at four sites in Kibale National Park, Uganda. Listed are those species that the redtail monkeys used for > 1% of their foraging effort at any site. The plant parts eaten for each tree species are listed in order of importance in their diet.

			Kanyawara (K30)		Sebatoli		Mainaro		Dura River	
Species	Family	Part	% Foraging	Density	% Foraging	Density	% Foraging	Density	% Foraging	Density
Celtis durandii	*Ulmaceae*	UF/RF/YL/LB	14.84	47.1	4.82	2.50	29.37	33.80	29.79	63.80
Chrysophyllum gorganusanum	*Sapotaceae*	RF/RL/UF/YL/SD	0.47	2.6	12.53	8.80	15.84	21.20	5.05	47.50
Celtis africana	*Ulmaceae*	YL/UF/RF/FL	7.66	4.2	4.34	0.00	1.65	1.30	0.00	0.00
Bosqueia phoberos	*Moraceae*	YL/RF/UF	3.89	50.0	1.93	0.00	0.00	0.00	5.92	22.50
Teclea nobilis	*Rutaceae*	RF/FL/YL/UF	7.07	17.1	0.00	0.00	1.32	0.00	2.26	0.00
Diospyros abyssinica	*Ebenaceae*	RF/UF/YL/FL	7.42	40.0	1.93	2.50	0.99	1.30	0.00	0.00
Prunus africana	*Rosaceae*	YL/RF/LB/BA	2.47	0.0	7.23	2.50	0.00	0.00	0.00	0.00
Warbugia stuhlmanni	*Canellaceae*	RF/UF/YL/FL	0.00	0.0	0.00	0.00	7.92	0.00	0.52	0.00
Croton sp.	*Euphorbiaceae*	RF/UF/FL/YL/BA	0.00	0.8	8.19	41.30	0.00	1.30	0.00	0.00
Uvariopsis congensis	*Annonaceae*	RF/UF	1.06	60.4	0.00	0.00	3.96	43.80	2.79	60.00
Maesa lanofolato	*Myrsinaceae*	RF	1.65	0.0	6.02	0.00	0.00	0.00	0.00	0.00
Albizia grandbracteata	*Leguminosae*	YL	0.94	1.3	4.58	0.00	0.00	0.00	1.05	1.30
Bequertiodendron oblanceolatum	*Sapotaceae*	YL/RF	0.00	0.0	0.00	0.00	0.00	0.00	6.27	57.50
Mimusops bagshawei	*Sapotaceae*	RF/UF/YL/FL	1.18	3.3	0.00	6.30	0.33	0.00	4.53	0.00
Ficus exasperata	*Moraceae*	RF/YL/UF	1.65	3.8	2.17	2.50	1.98	1.30	0.00	0.00
Cynometra alexandri	*Leguminosae*	FL/YL	0.00	0.0	0.00	0.00	4.29	63.80	0.00	0.00
Celtis mildbraedii	*Ulmaceae*	RF/YL/UF	0.00	0.0	0.00	0.00	4.29	32.50	0.00	0.00
Ficus natalensis	*Moraceae*	RF	0.00	0.4	0.00	0.00	0.00	0.00	3.83	0.00

Linociera johnsonii	*Oleaceae*	FL/YL	0.00	5.4	0.96	8.80	2.64	0.00	0.00	0.00
Markhamia platycalyx	*Bignoniaceae*	FL/LP/YL	0.59	50.0	1.45	38.80	0.00	1.30	1.05	8.80
Strychnos mitis	*Loganiaceae*	RF/TL/FL/UF	0.82	7.5	0.48	0.00	1.32	0.00	0.00	0.00
Olea welwitschii	*Oleaceae*	YL	2.00	3.3	0.00	0.00	0.00	0.00	0.35	1.30
Bridelia micrantha	*Euphorbiaceae*	RF/LB	0.82	0.0	1.45	0.00	0.00	0.00	0.00	0.00
Funtumia latifolia	*Apocynaceae*	RF/FL/YL/UF	0.12	33.8	0.72	2.50	0.00	2.50	1.39	43.80
Ficus sansibarica	*Moraceae*	RF	1.06	0.0	0.72	0.00	0.00	0.00	0.00	0.00
Monodora myristica	*Annonaceae*	FL	0.00	0.4	0.00	0.00	0.00	0.00	1.74	3.80
Fagara angolensis	*Rutaceae*	YL/FL	0.00	0.0	1.69	0.00	0.00	0.00	0.00	0.00
Pseudospondias microcarpa	*Anacardiaceae*	YL	0.12	1.7	0.48	0.00	0.00	0.00	1.05	3.80
Lovoa swynnertonni	*Meliaceae*	FL	0.00	0.8	0.00	3.80	0.00	0.00	1.57	3.80
Balanites wilsoniana	*Balanitaceae*	FL/RF/YL	0.00	1.7	1.20	0.00	0.33	0.00	0.00	1.30
Newtonia bucchanani	*Leguminosae*	YL	0.00	0.0	1.45	26.30	0.00	0.00	0.00	3.80
Chaetacme aristata	*Ulmaceae*	RF	0.35	17.1	0.00	0.00	0.00	0.00	1.05	3.80
Casearia sp.	*Flacourtiaceae*	RF/UF	0.00	1.3	0.00	0.00	0.00	0.00	1.39	0.00
Spathodea campanulata	*Bignoniaceae*	FL	0.00	0.8	0.00	0.00	0.00	0.00	1.22	0.00
Dombeya mukole	*Sterculiaceae*	YL/FL	1.18	9.2	0.00	0.00	0.00	0.00	0.00	1.30

dbh, diameter at breast height; RF, ripe fruit; UF, unripe fruit; YL, young leaves; FL, flower; LB, leaf bud; BA, bark; SD, seed.

more dramatically in the extent to which they were seed predators: populations are reported to eat seeds from 0% to 60.1% of their feeding time. The amount of time that different populations of monkeys within the 'Cephus' group ate fruit varied from 35.7% to 81.3%.

Temporal contrasts

Annual variation

Data on the diet of the same groups of red colobus were collected in 1994, 1995, 1996 and 1998. These data reveal considerable interannual variation in dietary components (Fig. 18.3). For example, in 1994 red colobus spent 55.8% of their feeding time eating young leaves and in 1998 they spent 75.8% of their feeding time doing so. Much of this interannual variation probably reflects interannual differences in food availability. Chapman *et al.* (1999) examined the phenology of 3793 trees from 104 species at two sites over 76 months and found marked variation among years in phenology for several species. However, some of the red colobus variation in diet is clearly not a function of availability (Table 18.8). For example, *C. durandii* fruits were available to red colobus every year, but

Table 18.6. Descriptions of the primate community found at Kibale, Uganda and Lopé, Gabon (annual rainfall = mm, biomass = kg km^{-2}, density = individuals km^{-2}).

	Density	Biomass	Leaves	Ripe fruit	Unripe fruit/seed	Insects
Kibale National Park, Uganda*						
Perodicticus potto	17.7	1.9				
Galagoides thomasi + *Galago matschiei*	79.5	12.6				
Lophocebus albigena	9.2	60	5	59	3	26
Papio anubis	–	–	–			
Cercopithecus ascanius	140	328	16	44	15	22
Cercopithecus mitis	41.8	133	21	45	13	20
Cercopithecus lhoesti	8	13				
Cercopithecus aethiops	rare	rare				
Procolobus badius	300	1760	75	6	16	3
Colobus guereza	58.1	317	76	13	2	0
Pan troglodytes	2.5	85	8	80	0	0 (12% THV)
Total density, ~ 656.8; total biomass, 2710; frugivore biomass, 633.5; folivore biomass, 2077.						
Lopé Reserve, Gabon†						
Cercopithecus nictitans	19.2	62.8				
Cercopithecus pogonias	4.6	10.1				
Cercopithecus cephus	5.1	10.2	11	49	5	35
Lophocebus albigena	8.1	33.7	30	36	4	28
Colobus satanas	10.8	90.7	4	60	26	–
Gorilla gorilla	0.6	45.3				
Pan troglodytes	0.6	22.5		–		
Mandrillus sphinx	3.8	43.9				
Total biomass, 318.6; frugivore biomass, 227.9; folivore biomass, 90.7.						

*Struhsaker (1975, 1978, 1980), Struhsaker and Leland (1979), Chapman and Wrangham (1993), Weisenseal *et al.* (1993), Chapman *et al.* (1995), Chapman (unpublished data).
†Primate density, mean of five neighbouring sites from, White (1994a, b), diet data for *L. albigena* from Ham (1994), diet data for *C. satanas* from Harrison (1986).

red colobus feeding time on this species varied from 1.3% in 1994 to 7.3% in 1995.

Discussion

The examination of the diet of the red colobus, a seed predator, and the redtail monkey, a seed-disperser, across different spatial and temporal scales demonstrates considerable variation among plant parts and species consumed. It seems likely that this variation will lead to spatial and temporal variation in selection pressures associated with the interaction between these monkeys and specific plant species. This variation may constrain coevolution of the participants (Herrera, 1988; Horvitz and Schemske, 1990; Jordano, 1993). This interpretation is open to debate, however. It is possible, although we regard it as unlikely, that successful recruitment at any particular site is very episodic and that these animals play a consistent role at these times. Regardless, the variation we describe will induce stochasticity in the number and species composition of recruiting seedlings.

Unfortunately, studies of a sufficiently long duration to document the temporal variability in frugivore behaviour are extremely rare (Herrera, 1998). Similarly, few studies document frugivore foraging and seed dispersal over a spatial scale where the same species of plants and animals are probably interacting, but where variation in frugivore foraging behaviour occurs (Chapman and Chapman, 1999). On the practical side, this shortcoming highlights the importance of long-term studies of frugivore–plant interactions across a range of spatial scales (Herrera, 1985, 1998; Jordano, 1993; Wilson and Whelan, 1993). The

Table 18.7. Percentage of feeding time devoted to different plant parts by red colobus, black-and-white colobus and redtail monkeys or 'Cephus' group from a variety of sites across Africa.

Red colobus* (*Procolobus badius*)	Young leaves	Mature leaves	Seeds	Flowers	Other
P.b. tholloni (1)	54.3	6.4	37.9	1.4	
P.b. badius (2)	31.7	20.2	31.2	16.1	
P.b. rufomitrata (3)	52.4	11.5	25.0	6.2	4.9
P.b. temminckii (4)	41.5	6.5	54.4	8.7	7.4
P.b. temminckii (5)	34.9	11.8	44.5	8.7	2.9
P.b. tephrosceles (6)	34.8	44.1	11.3	6.8	2.9
P.b. tephrosceles (7)	50.6	23.1	5.6	11.8	†
P.b. kirkii (8)	46.7	14.6‡	31.7	10.6	2.3
P.b. kirkii (8)	53.4	11.9§	31.2	5.4	1.3
P.b. tephrosceles (9)	46.8–87.1	2.0–21.0	1.9–17.2	2.0–22.7	
Black-and-white colobus*	**Young leaves**	**Mature leaves**	**Seeds**	**Flowers**	**Other**
C. angolensis (10)	21.2	6.4	66.7	5.9	
C. angolensis (11)	67.9‖		32.1		
C. angolensis (12)	24.9	38.9			
C. polykomos (13)	29.9	26.7	36.5	2.7	4.7
C. satanas (14)	23.0	19.0	58.0		
C. satanas (15)	23.0	3.0	64.2	5.3	4.4
C. guereza (16)	23.7	29.1	36.9	0.5	8.1
C. guereza (17)	29.7	28.0	46.6	2.9	14.5
C. guereza (18)	33.1	19.8	45.6	7.7	2.1
C. guereza (19)	36.9	24.8	37.6	8.9	2.6
C. guereza (20)	61.7	12.4	13.6	2.1	10.2
C. guereza (21)	80.1	5.8	9.8	0.1	4.2
C. guereza (22)	85.6	3.7	7.3	2.3	0.8

(Contd.)

Table 18.7. *Continued.*

Redtail monkeys (*Cercopithecus*) 'Cephus' group	Young leaves	Mature leaves	Fruit pulp	Flowers	Insects	Seeds
C. ascanius (23)	6.8	0.4	61.3	2.0	25.1	0.4
C. ascanius (24)	10.9	3.3	43.6	15.3	21.8	0.1
C. ascanius (25)	34.7	0.0	44.6	2.7	17.6	
C. ascanius (26)	27.8	0.4	35.7	3.7	31.2	
C. ascanius (27)	15.0	0.4	55.6	8.2	20.6	
C. ascanius (28)	12.2	0.7	59.7	11.6	14.5	
C. cephus (29)	6.1		81.3	–	12.6	
C. cephus (30)	11.4		67.0	5.7	9.1	6.8
C. cephus (31)	4.0		49.0	6.0	35.0	5.0

(1) Democratic Republic of Congo: Maisels *et al.* (1994); (2) Sierra Leone: Davies *et al.* (1999); (3) Kenya: Marsh (1981); (4) Senegal: Gatinot (1977); (5) Gambia: Davies (1994); (6) Tanzania: Clutton-Brock (1975, 1977); (7) Uganda: Struhsaker (1975); (8) Mturi 1993 two groups in the same area: Mturi (1993); (9) range of populations: this study, (10) Democratic Republic of Congo: Maisels *et al.* (1994); (11) Kenya: Moreno-Black and Maples (1977); (12) Rwanda: Fimbel *et al.* (unpublished data); (13) Sierra Leone: Dasilva (1992, 1994); (14) Cameroon: McKey *et al.* (1981), McKey and Waterman (1982); (15) Gabon: Harrison and Hladik (1986); (16) Kakamega, Kenya: Fashing (1999); (17) Ituri Forest, Democratic Republic of Congo: Bocian (1997); (18) Budongo, Uganda (logged area): Plumptre and Reynolds (unpublished data); (19) Budongo, Uganda (unlogged area): Plumptre and Reynolds (unpublished data); (20) Kibale, Uganda: Oates (1977), Struhsaker and Oates (1975); (21) Kibale, Uganda: this study (Group 1); (22) Kibale, Uganda: this study (Group 2); (23) Kakamega, Kenya: Cords (1986); (24) Kibale, Uganda (young leaves and leaf buds combined): Struhsaker (1978); (25) Kibale at Sebatoli, Uganda (young leaves, buds and petioles combined): this study; (26) Kibale at Kanyawara, Uganda (young leaves, buds and petioles combined): this study; (27) Kibale at Dura River, Uganda: this study; (28) Kibale at Mainaro, Uganda: this study; (29) Makokou, Gabon (all leaves assumed to be young): Gautier-Hion *et al.* (1980); (30) Lopé (continuous forest), Gabon (all leaves assumed to be young): Tutin *et al.* (1997a); (31) Lopé (forest fragment), Gabon (all leaves assumed to be young): Tutin (1999), Tutin *et al.* (1997b).

[*]For the colobine monkeys some studies listed fruit and seeds separately. Based on the fact that no seeds have been found in 270 *C. guereza* and *P. badius* fecal samples (T. Gillespie, Florida, 2000, personal communication), we assume that, when the colobines ingest fruit pulp, they are also ingesting the seeds and are acting as seed predators.

[†]10.4% leaves of unknown age.

[‡]Includes 7.3 on leaf stalks.

[§]Includes 5.6 on leaf stalks.

[||]Young and mature leaves.

investigations that have examined spatial and temporal variation in plant–animal interactions typically suggest that a single year's study at one site of how a particular frugivore disperses the seed of a specific plant may at best provide a snapshot of the interaction and at worst present a serious distortion or an erroneous picture (Herrera, 1998).

Studies such as this one and several that have preceded it (Herrera, 1985, 1998; Jordano, 1993) suggest that there is still much to be learned if we are to make advances in understanding the evolution of fruit morphology using ecological evidence, or in identifying important processes determining how seed dispersal contributes to the distribution of adult trees. These studies also stress the need to identify novel systems or approaches that can be used to identify selective pressures acting on fruit morphology and to determine how seed dispersal patterns influence the distribution of seedlings and, subsequently, adult trees. It is clear that 10+-year studies at a number of spatially separated sites will continue to be constrained by field logistics and time.

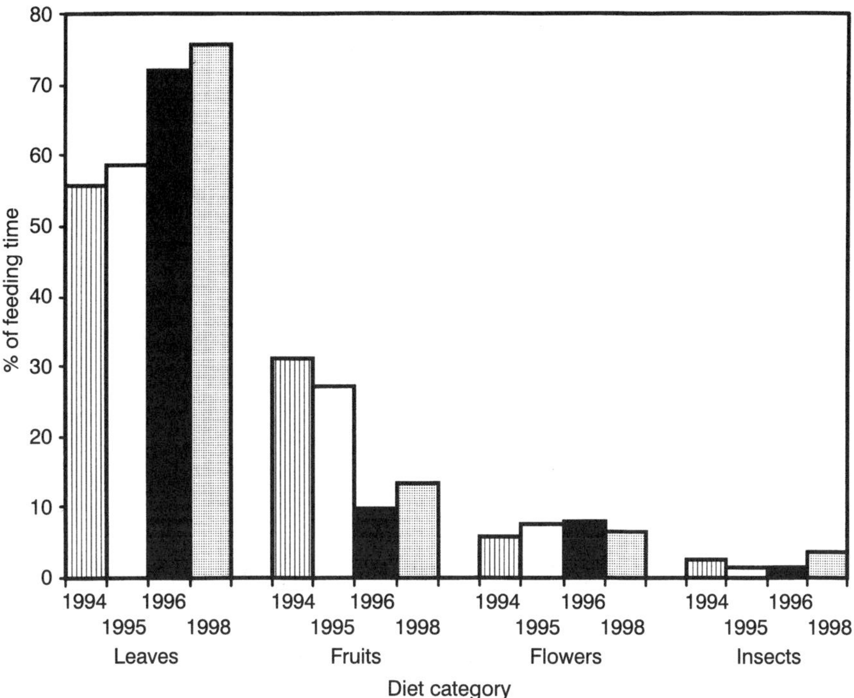

Fig. 18.3. The percentage of feeding time devoted to different plant parts by one group of red colobus monkeys (*Procolobus badius*) in Kibale National Park, Uganda, over 4 years.

Table 18.8. The percentage of time spent feeding from the five most important food species in the diet of the red colobus in K30 in each of the years of study (underlined) and the percentage of time eating these species in years when they were not in the top five.

Species	Family	1994	1995	1996	1998
Celtis durandii	*Ulmaceae*	11.8	16.3	17.2	21.3
Strombosia schlefferi	*Olacaceae*	9.5	8.0	10.6	10.2
Prunus africana	*Rosaceae*	16.4	11.6	4.5	4.1
Markhamia platycalyx	*Bignonaceae*	9.3	10.5	6.3	9.0
Celtis africana	*Ulmaceae*	12.2	8.1	6.1	7.5
Albizzia grandibracteata	*Leguminosae*	10.5	4.2	9.6	1.2
Dombeya mukole	*Sterculiaceae*	1.7	5.0	6.6	6.8
Bosqueia phoberos	*Moraceae*	–	1.0	2.9	7.8

Acknowledgements

Funding for this research was provided by the Wildlife Conservation Society, National Geographic Society, Lindbergh Foundation and National Science Foundation (grant numbers SBR-9617664, SBR-990899). Permission to conduct this research was given by the Office of the President, Uganda, the National Council for Science and Technology, the Uganda Wildlife Authority and the Ugandan Forest Department. Karyn Rode helped with the collection of literature data found in Table 18.6. We would like to thank Tom Gillespie, Ronda Green, Beth Kaplin and Doug Levey for helpful comments on this work.

References

Bocian, C.M. (1997) Niche separation of black-and-white colobus (*Colobus angolensis* and *C. guereza*) in the Ituri Forest. PhD thesis, City University of New York, New York, 202 pp.

Chapman, C.A. (1995) Primate seed dispersal: coevolution and conservation implications. *Evolutionary Anthropology* 4, 74–82.

Chapman, C.A. and Chapman, L.J. (1999) Implications of small-scale variation in ecological conditions for the diet and density of red colobus monkeys. *Primates* 40, 215–232.

Chapman, C.A. and Chapman, L.J. (2000) Constraints on group size in redtail monkeys and red colobus: testing the generality of the ecological constraints model. *International Journal of Primatology* 21, 565–585.

Chapman, C.A. and Lambert, J.E. (2000) Habitat alteration and the conservation of African primates: a case study of Kibale National Park, Uganda. *American Journal of Primatology* 50, 169–186.

Chapman, C.A. and Wrangham, R.W. (1993) Range use of the forest chimpanzees of Kibale: implications for the evolution of chimpanzee social organization. *American Journal of Primatology* 31, 263–273.

Chapman, C.A., Wrangham, R.W. and Chapman, L.J. (1995) Ecological constraints on group size: an analysis of spider monkey and chimpanzee subgroups. *Behavioral Ecology and Sociobiology* 36, 59–70.

Chapman, C.A., Chapman, L.J., Wrangham, R., Isabirye-Basuta, G. and Ben-David, K. (1997) Spatial and temporal variability in the structure of a tropical forest. *African Journal of Ecology* 35, 287–302.

Chapman, C.A., Wrangham, R.W., Chapman, L.J., Kennard, D.K. and Zanne, A.E. (1999) Fruit and flower phenology at two sites in Kibale National Park, Uganda. *Journal of Tropical Ecology* 15, 189–211.

Chapman, L.J., Chapman, C.A. and Wrangham, R.W. (1992) *Balanites wilsoniana*: elephant dependent dispersal? *Journal of Tropical Ecology* 8, 275–283.

Clutton-Brock, T.H. (1975) Feeding behavior of red colobus and black and white colobus in East Africa. *Folia Primatologica* 23, 165–207.

Clutton-Brock, T.H. (1977) Some aspects of intraspecific variation in feeding and ranging behavior in primates. In: Clutton-Brock, T.H. (ed.) *Primate Ecology: Studies of Feeding and Ranging Behavior in Lemurs, Monkeys, and Apes.* Academic Press, London, pp. 539–556.

Cords, M. (1986) Interspecific and intraspecific variation in the diet of two forest guenons, *Cercopithecus ascanius* and *C. mitis*. *Journal of Animal Ecology* 55, 811–827.

Dasilva, G.L. (1992) The western black-and-white colobus as a low-energy strategist: activity budgets, energy expenditure, and energy intake. *Journal Animal Ecology* 61, 79–91.

Dasilva, G.L. (1994) Diet of *Colobus polykomos* on Tiwai Island: selection of food in relation to its seasonal abundance and nutritional quality. *International Journal of Primatology* 15, 1–26.

Davies, A.G. (1994) Colobine populations. In: Davies, A.G. and Oates, J.F. (eds) *Colobine Monkeys: Their Ecology, Behavior and Evolution.* Cambridge University Press, Cambridge, pp. 285–310.

Davies, A.G., Oates, J.F. and Dasilva, G.A. (1999) Patterns of frugivory in three West African colobine monkeys. *International Journal Primatology* 20, 327–357.

Fashing, P.J. (1999) The behavioural ecology of an African colobine monkey: diet, range use, and patterns of intergroup aggression in eastern black-and-white colobus monkeys (*Colobus guereza*). PhD thesis, Columbia University, New York.

Fischer, K. and Chapman, C.A. (1993) Frugivores and fruit syndromes: differences in patterns at the genus and species levels. *Oikos* 66, 472–482.

Gatinot, B.L. (1977) Le régime alimentaire du colobe bai au Sénegal. *Mammalia* 41, 373–402.

Gautier-Hion, A., Emmons, L.H. and Dubost, G. (1980) A comparison of the diets of three major groups of primary consumers in Gabon (primates, squirrels, and ruminants). *Oecologia* 45, 182–189.

Gautier-Hion, A., Gautier, J.-P. and Maisels, F. (1993) Seed dispersal versus seed predation: an inter-site comparison of two related African monkeys. *Vegetatio* 107/108, 237–244.

Ham, R.M. (1994) Behaviour and ecology of grey-cheeked mangabeys (*Cercocebus albigena*) in the Lopé Reserve, Gabon. PhD thesis, Stirling University, Scotland.

Harrison, M.J.S. (1986) Feeding ecology of black colobus, *Colobus satanas*, in Gabon. In: Else, L. and Lee, P.C. (eds) *Primate Ecology and Conservation.* Cambridge University Press, Cambridge, pp. 31–37.

Harrison, M.J.S. and Hladik, C.M. (1986) Un primate granivore: le colobe noir dans la forêt du Gabon, potentialité d'évolution du comportement alimentaire. *Revue d'Ecologia* 41, 281–298.

Herrera, C.M. (1982) Seasonal variation in the quality of fruits and diffuse coevolution between plants and avian dispersers. *Ecology* 63, 773–785.

Herrera, C.M. (1984) Adaptations to frugivory of Mediterranean avian seed dispersers. *Ecology* 65, 609–617.

Herrera, C.M. (1985) Determinants of plant–animal coevolution: the case of mutualistic dispersal of seeds by vertebrates. *Oikos* 44, 132–141.

Herrera, C.M. (1986) Vertebrate-dispersed plants: why they don't behave they way they should. In: Estrada, A. and Fleming, T.H. (eds) *Frugivores and Seed Dispersal.* Dr W. Junk Publisher, Dordrecht, pp. 5–18.

Herrera, C.M. (1988) Variation in mutualisms: the spatio-temporal mosaic of an insect pollinator assemblage. *Biological Journal of the Linnean Society* 35, 95–125.

Herrera, C.M. (1998) Long-term dynamics of Mediterranean frugivorous birds and fleshy fruits: a 12-year study. *Ecological Monographs* 68, 511–538.

Herrera, C.M., Jordano, P., Guitian, J. and Traveset, A. (1998) Annual variability in seed production by woody plants and the mast concept: reassessment of principles and relationships to pollination and seed dispersal. *American Naturalist* 152, 576–594.

Horvitz, C.C. and Schemske, D.W. (1990) Spatiotemporal variation in insect mutualists of a neotropical herb. *Ecology* 71, 1085–1097.

Howe, H.F. (1979) Fear and frugivory. *American Naturalist* 114, 925–931.

Howe, H.F. (1983) Annual variation in a neotropical seed-dispersal system. In: Sutton, S.L., Whitmore, T.C. and Chadwick, C.A. (eds) *Tropical Rainforests: Ecology and Management.* Blackwell Scientific, Oxford, pp. 211–227.

Howe, H.F. (1984) Constraints on the evolution of mutualisms. *American Naturalist* 123, 764–777.

Howe, H.F. (1993) Aspects of variation in a neotropical seed dispersal system. *Vegetatio* 107/108, 149–162.

Howe, H.F. and Estabrook, G.F. (1977) On intraspecific competition for avian dispersers in tropical trees. *American Naturalist* 116, 817–832.

Howe, H.F. and Smallwood, J. (1982) Ecology of seed dispersal. *Annual Review of Ecology and Systematics* 12, 201–228.

Jordano, P. (1992) Fruits and frugivory. In: Fenner, M. (ed.) *Seeds: The Ecology of Regeneration in Plant Communities.* CAB International, Wallingford, UK, pp. 105–156.

Jordano, P. (1993) Geographical ecology and variation of plant–seed disperser interactions: southern Spanish junipers and frugivorous thrushes. *Vegetatio* 107, 85–104.

Lambert, J.E. (1997) Digestive strategies, fruit processing, and seed dispersal in the chimpanzees (*Pan troglodytes*) and redtail monkeys (*Cercopithecus ascanius*) of Kibale National Park, Uganda. PhD dissertation, University of Illinois at Urbana-Champaign.

Lieberman, D., Lieberman, M. and Martin, C. (1987) Notes on seeds in elephant dung from Bia National Park, Ghana. *Biotropica* 19, 365–369.

McKey, D.B. and Waterman, P.G. (1982) Ranging behavior of a group of black colobus (*Colobus satanas*) in the Douala-Edea Reserve, Cameroon. *Folia Primatologica* 39, 264–304.

McKey, D.B., Gartlan, J.S., Waterman, P.G. and Choo, C.M. (1981) Food selection by black colobus monkeys (*Colobus satanas*) in relation to plant chemistry. *Biological Journal of the Linnean Society* 16, 115–146.

Maisels, F., Gautier-Hion, A. and Gautier, J.-P. (1994) Diets of two sympatric colobines in Zaire: more evidence on seed-eating in forests on poor soils. *International Journal of Primatology* 15, 681–701.

Marsh, C.W. (1981) Ranging behavior and its relation to diet selection in Tana River red colobus (*Colobus badius rufomitratus*). *Journal of Zoology, London* 195, 473–492.

Moreno-Black, G.S. and Maples, W.R. (1977) Differential habitat utilization of four Cercopithecidae in a Kenyan forest. *Folia Primatologica* 27, 85–107.

Mturi, F.A. (1993) Ecology of the Zanzibar red colobus monkey, *Colobus badius kirkii* (Gray, 1968), in comparison with other red colobines. In: Lovett, J.C. and Wasser, S.K. (eds) *Biogeography and Ecology of the Rain Forest of Eastern Africa.* Cambridge University Press, Cambridge, pp. 243–266.

Oates, J.F. (1977) The guereza and its food. In: Clutton-Brock, T.H. (ed.) *Primate Ecology: Studies of Feeding and Ranging Behavior in Lemurs, Monkeys, and Apes.* Academic Press, London, pp. 275–321.

Richards, P.W. (1996) *The Tropical Rain Forest: An Ecological Study*, 2nd edn. Cambridge University Press, Cambridge.

Rick, C.M. and Bowman, R.I. (1961) Galapagos tomatoes and tortoises. *Evolution* 15, 407–417.

Schupp, E.W. (1990) Annual variation in seedfall, post-dispersal predation, and recruitment of a neotropical tree. *Ecology* 71, 504–515.

Snow, D.W. (1965) A possible selective factor in the evolution of fruiting seasons in tropical forests. *Oikos* 15, 274–281.

Struhsaker, T.T. (1975) *The Red Colobus Monkey.* University of Chicago Press, Chicago, 311 pp.

Struhsaker, T.T. (1978) Food habits of five monkey species in the Kibale Forest, Uganda. In: Chivers, D.J. and Herbert, J. (eds) *Recent Advances in Primatology*, Vol. 1. *Behavior.* Academic Press, New York, pp. 225–248.

Struhsaker, T.T. (1980) Comparison of the behavior and ecology of red colobus and red-tail monkeys in the Kibale Forest, Uganda. *African Journal of Ecology* 18, 33–51.

Struhsaker, T.T. (1997) *Ecology of an African Rain Forest: Logging in Kibale and the Conflict Between Conservation and Exploitation.* University Presses of Florida, Gainesville, Florida, 434 pp.

Struhsaker, T.T. and Leland, L. (1979) Socioecology of five sympatric monkey species in the Kibale Forest, Uganda. In: Rosenblatt, J., Hinde, R.A., Beer, C. and Busnel, M.C. (eds) *Advances in the Study of Behavior,* Vol. 9. Academic Press, New York, pp. 158–228.

Struhsaker, T.T. and Oates, J.F. (1975) Comparison of the behavior and ecology of red colobus and black-and-white colobus monkeys in Uganda: a summary. In: Tuttle, R.H. (ed.) *Socioecology and Psychology of Primates.* Mouton Publishers, The Hague, pp. 103–124.

Temple, S.A. (1977) Plant–animal mutualisms: coevolution with dodo leads to near extinction of plant. *Science* 197, 885–886.

Tutin, C.E.G. (1999) Fragmented living: behavioural ecology of primates in a forest fragment in the Lopé Reserve Gabon. *Primates* 40, 249–265.

Tutin, C.E.G., Williamson, E.A., Rogers, M.E. and Fernandez, M. (1991) A case study of a plant–animal interaction: *Cola lizae* and lowland gorillas in the Lope Reserve, Gabon. *Journal of Tropical Ecology* 7, 181–199.

Tutin, C.E.G., White, L.J.T., Williamson, E.A., Fernandez, M. and McPherson, G. (1994) List of plant species identified in the northern part of the Lope Reserve, Gabon. *Tropics* 3, 249–276.

Tutin, C.E.G., Ham, R.M., White, L.J.T. and Harrison, M.J.S. (1997a) The primate community of the Lope Reserve, Gabon: diets, responses to fruit scarcity, and effects on biomass. *American Journal of Primatology* 42, 1–24.

Tutin, C.E.G., White, L.J.T. and Mackanga-Missandzou, A. (1997b) The use of rainforest mammals of natural forest fragments in an equatorial African savanna. *Conservation Biology* 1190–1203.

Weisenseal, K., Chapman, C.A. and Chapman, L.J. (1993) Nocturnal primates of Kibale Forest: effects of selective logging on prosimian densities. *Primates* 34, 445–450.

Wheelwright, N.T. and Orians, G.H. (1982) Seed dispersal by animals: contrasts with pollen dispersal, problems of terminology, and constraints on coevolution. *American Naturalist* 119, 402–413.

White, L.J.T. (1994a) Biomass of rain forest mammals in the Lopé Reserve, Gabon. *Journal of Animal Ecology* 63, 499–512.

White, L.J.T. (1994b) The effects of commercial mechanized selective logging on a transect in lowland rainforest in the Lopé Reserve, Gabon. *Journal of Tropical Ecology* 10, 313–322.

Wilson, M.F. and Whelan, C.J. (1993) Variation of dispersal phenology in a bird-dispersed shrub, *Cornus drummondii. Ecological Monographs* 63, 151–172.

19 The Frugivorous Diet of the Maned Wolf, *Chrysocyon brachyurus*, in Brazil: Ecology and Conservation

José Carlos Motta-Junior and Karina Martins

Laboratório de Ecologia Trófica, Departamento de Ecologia, Instituto de Biociências da Universidade de São Paulo, 05508-900 São Paulo, São Paulo State, Brazil

Introduction

Studies of seed dispersal by frugivorous animals have focused on birds and bats (Jordano, 1992). The role of mammalian carnivores (order Carnivora) as seed-dispersers remains relatively unexplored, despite the fact that they often consume large quantities of fruit, have long gut passage times and move over large areas (Rogers and Applegate, 1983; Herrera, 1989; Willson, 1993; Traveset and Willson, 1997; Cypher, 1999). Willson (1993), for example, concluded that bears, procyonids, mustelids and canids are among the most important dispersal agents in North America. In South America, the situation is probably the same, but the literature remains fragmented and the existing studies are quite narrowly focused (Young, 1990; Bustamante *et al.*, 1992; Lombardi and Motta-Junior, 1993; Castro *et al.*, 1994; Courtenay, 1994; Dalponte and Lima, 1999).

We present a broad study of frugivory and seed dispersal by the maned wolf (*Chrysocyon brachyurus*), the largest canid (20–26 kg) in South America. It inhabits grasslands and savannah-like habitats in central South America, including all of Paraguay, north-eastern Argentina, north-western Uruguay, extreme south-eastern Peru and central regions of eastern Bolivia and southern Brazil (Langguth, 1975; Dietz, 1985; Mones and Olazarri, 1990; Nowak, 1999). Its diet is omnivorous and includes fruit (Dietz, 1984; Medel and Jaksic, 1988; Motta-Junior *et al.*, 1996).

We first analyse the wolf's diet in eight localities to document its degree of frugivory. We then test whether the maned wolf selects fruit species non-randomly and whether it defecates seeds of 16 species in viable condition. We focus attention on the relationship between the maned wolf and *Solanum lycocarpum* (*Solanaceae*), locally named *lobeira* ('wolf's fruit'). This species appears to be especially important to maned wolves (Dietz, 1984; Lombardi and Motta-Junior, 1993; Courtenay, 1994). Finally, we use satellite images of our study sites to estimate the proportion of wolf habitat (*cerrado*) in each. We combine these data with data on diet to determine the extent to which maned wolves depend on fruiting plants that occur only in *cerrado* vegetation.

Study Areas

We collected wolf faeces in eight localities, all in the *cerrado* region of south-eastern Brazil: the Ecological Stations of Itirapina (EITI) (2300 ha; 22° 13′ S, 47° 54′ W), Águas de Santa Bárbara (EASB) (2712 ha; 22° 48′ S, 49° 12′ W) and Jatai (EEJA) (5532 ha; 21° 35′ S, 47° 47′ W), the Experimental Station of Itapetininga (EITA) (6706 ha; 23° 41′ S, 47° 59′ W), the reserve of Universidade Federal de São Carlos (UFSC) (725 ha; 21° 58′ S, 47° 52′ W), a farm, Fazenda Fortaleza (FFOR) (10,000 ha; 21° 45′ S, 48° 02′ W), Parque Florestal Salto Ponte (PFSP) (1240 ha; 19° 10′ S, 48° 48′ W) and Parque Nacional Serra da Canastra (PNSC) (71,525 ha; 20° 15′ S, 46° 37′ W). The first six sites are located in São Paulo State (SP) and the last two are in Minas Gerais State (MG).

The vegetation of EITI and EASB is mostly grassland savannah or *campo cerrado*, while EITA is covered by pine plantations and remnants of natural habitats (*cerrado*, gallery forest) in similar proportions. EEJA is dominated by *cerradão*, a xeromorphic semideciduous forest (see Eiten, 1974, 1978), with some *campo sujo* (a more open grassland savannah). Although UFSC has *cerrado* and gallery forest, extensive areas within and outside its limits are covered by *Eucalyptus* spp. and sugarcane plantations. The FFOR farm is the most disturbed study site, because of a low proportion of natural habitats (mostly *cerrado* and dry forest) in relation to large *Eucalyptus* plantations and sugarcane fields outside its boundaries. PNSC is composed mostly of grassy fields and grassland savannahs; it was the largest and most pristine study site. PFSP is covered mostly by pine plantations, but almost all neighbouring landscapes are comprised of *cerrado* in the uplands and gallery forests and *Mauritia flexuosa* palm groves in the valleys on permanently marshy ground.

We included data on diet from other studies in central Brazil: Dietz (1984), Motta-Junior *et al.* (1996) and Jácomo (1999). These studies were selected because of similar methods and large sample sizes.

The general climate in these areas is between Koeppen's Aw and Cwa, tropical and warm, with distinct dry and wet seasons (Setzer, 1966; Eiten, 1978). Annual rainfall ranges from 750 to 2000 mm (Eiten, 1978; Nimer and Brandão, 1989), but, in most of the *cerrado* region, it is between 1100 and 1600 mm, with a single 5–6-month dry season (April/May–September/October). Rainfall in the driest month is only 10–30 mm. The annual average temperature is 20–26°C (Setzer, 1966; Eiten, 1974, 1978). Detailed descriptions of the study areas and the *cerrado* region can be found in Eiten (1974, 1978), Ratter (1980) and Dietz (1984).

Material and Methods

Diet and fruit selection

The diet of the maned wolf was evaluated by analysis of scats collected from December 1997 to February 2000. Trails and paths commonly used by wolves were regularly walked in search of scats, which could be attributed to the maned wolf by their typically large diameters (> 25 mm), shape, odour and association with wolf tracks and hairs (see Dietz, 1984; Motta-Junior *et al.*, 1996). Scats were fixed in a 9 : 1 solution of 70% alcohol and 10% formalin. To process, they were washed in water over a fine-mesh screen and oven-dried (50°C) for 2 days. Remains of seeds, teeth, bills, scales and exoskeletons were used to count and identify food items. The average number of seeds per fruit species was compared with the number of seeds in scats to estimate the number of fruits consumed and their biomass (see Castro *et al.*, 1994; Motta-Junior *et al.*, 1996). Whenever possible, fruit and prey in scats were identified to the species level, using reference collections. Some identifications were made by specialists in museums and herbaria, where voucher specimens were deposited. To describe the diet, we used the frequency of occurrence as a function of total occurrences (*sensu* Dietz, 1984) and estimated biomass consumption. Mean number of seeds per fruit species was obtained in the study sites in order to estimate the number of fruits consumed by counting seeds in faeces (Castro *et al.*, 1994; Motta-Junior *et al.*, 1996). The number of fruits consumed of each species was multiplied by the mean fruit mass to estimate the biomass consumption of fruits. Animal-prey biomass was estimated by counting the minimum number of individuals in faeces and then multiplying this number by the average mass of each species at the study sites (e.g. Emmons, 1987; Motta-Junior *et al.*, 1996).

To evaluate the availability of fruit species in each of five study areas, we randomly placed 42–50 10 m × 10 m quadrats along known paths of wolves. Within each quadrat, we recorded the presence or absence of all plant species with fruits consumed by wolves, as determined by faecal analysis. The size and number of quadrats in each area seemed adequate, as no new plant species were added to the list after sampling the 30–35th quadrats. The absolute observed frequencies of fruits in the diet were compared with absolute frequencies (expected) of fruiting-plant species, derived from quadrat data, through contingency tables (Jaksic, 1989; Zar, 1999). We assumed that plant occurrence reflected fruit occurrence, since almost all individual plants had mature fruits during 2–3 months.

Germination tests

To determine if wolves defecate seeds in viable condition and are therefore seed-dispersers, we conducted two types of germination trials. The first type tested whether seeds extracted from wolf faeces would germinate. The second tested whether germination is enhanced or inhibited by passage through the wolf's digestive system. For this trial, we used seeds taken directly out of fruits as a control. The germination of treatment and control seeds was compared through contingency tables (Zar, 1999). For both types of trials, seeds were taken from fresh wolf scats or fruits, rinsed with water and placed on moist filter-paper in covered Petri dishes. Germination tests were conducted in the laboratory, under light and temperature conditions that mimicked those that occur in the field. Germinated seeds were removed and the trials were run until no seed had germinated for 30 days. We tested ten species for the first type of trial and eight species for the second type.

Landscape analysis

To assess the association of plant species in the diet and vegetation cover, recent (1998/99) digital satellite images of the eight study areas were analysed. The phytophysiognomies in the study areas were categorized according to Eiten (1974, 1978). We estimated the proportion of *cerrado* (*sensu lato*) vegetation cover, excluding *cerradão* (a xeromorphic forest physiognomy). Based on our observations on the frequency of wolf tracks in different habitats and on accounts in the literature (e.g. Langguth, 1975; Dietz, 1984), the maned wolf's main habitat in Brazil is open savannah-like vegetation (i.e. *cerrado*). The program FRAGSTATS™ (McGarigal and Marks, 1995) was used to obtain the proportion of *cerrado* in the study sites. We used Spearman rank-order correlation, r_s (Siegel and Castellan, 1988), to determine the association between percentage of *cerrado* vegetation cover in the landscape and the frequency of *cerrado* items in the diet. If wolves opportunistically use *cerrado* resources, a significant positive correlation is expected between *cerrado* items in the diet and *cerrado* cover in the eight study areas. On the other hand, if wolves concentrate their foraging in *cerrado* patches, no correlation is expected.

Results and Discussion

Diet and fruit selection

All populations of maned wolves were highly omnivorous (Table 19.1). Combining data from all sites, fruits comprised 39.5% of total occurrences in scats, grains, foliage and stems comprised 13.0%, insects 3.7% and small vertebrates 43.8%. In total, 142 species of plants and animals were consumed (range = 34–59 across sites). Among animal prey, small mammals (rodents and opossums) were the most important by frequency of occurrence, but armadillos dominated the animal biomass (see Motta-Junior *et al.*, 1996; J.C. Motta-Junior, unpublished data). The occurrence of fruit species in diet ranged from 22.5% to 54.3% across sites, and fruit biomass ranged from 15.6% to 57.3%. Considering only the

Table 19.1. Diet of the maned wolf in ten localities in south-eastern and central Brazil. The figures are percentages of frequency of occurrence based on total occurrences of all items. A detailed list of fruit species is in the Appendix.

Food items	EEJA (SP)	EASB (SP)	EITI (SP)	EITA (SP)	UFSC (SP)	FFOR (SP)	PFSP (MG)	PNSC (MG)	PNSC (MG)*	FAL (DF)†	PNE (GO)‡
Fruits											
Anacardiaceae		0.5	0.6		0.5	1.9	0.2			0.7	1.5
Annonaceae	6.2	0.5	1.7	1.5	1.0	1.3	4.9	0.1	0.3	2.3	12.3
Bromeliaceae	1.4		4.9	2.8		1.0		1.1	2.1		0.9
Caricaceae						1.0					
Chrysobalanaceae							0.2	11.7			5.3
Cucurbitaceae	0.2		0.6				0.1	5.7	0.6	0.3	0.6
Ebenaceae							0.1				0.2
Guttiferae									0.3		
Hippocrateaceae			0.6							1.0	
Leguminosae							0.3				
Malpighiaceae	0.2					0.3		0.1			0.2
Melastomataceae							0.1	0.1			
Moraceae							0.1				
Myrtaceae	2.8	14.2	10.4	2.6	2.0	3.5	0.7	2.3	1.6	1.3	2.5
Palmae	0.6	3.6	2.9	8.4		1.0	0.2	2.3	1.2		10.0
Rubiaceae		3.6	1.2		0.5		1.0	1.0			0.1
Rutaceae	2.0		1.7	1.1	1.0	4.8					0.5
Sapindaceae											0.4
Solanaceae	31.3	18.3	11.3	24.7	23.5	7.7	27.3	11.1	32.6	25.7	18.0
Sapotaceae			1.7				0.2	2.7		3.6	1.7
Other non-identified fruits			0.3	1.8	2.0		0.9	5.6	1.0		0.1
Grains											
Helianthus, *Oryza*, *Phaseolus*, *Zea*	0.8					3.5			0.2		
Foliage, stems											
Sugarcane (*Saccharum*)	2.2	0.5				6.1					
Grasses (*Poaceae*, *Cyperaceae*)	12.9	19.3	12.7	11.9	13.0	17.0	17.6	5.8	11.1	11.8	3.2
Subtotal plants	60.6	60.5	50.6	54.8	43.5	49.1	53.9	49.6	51.0	46.7	57.5
Insecta	4.9	2.5	4.1	5.5	5.0	3.2	3.1	3.1	5.7	2.0	1.7
Pisces, Anura, Sauria	2.8	1.5	2.6	1.9	6.0	4.5	3.3	6.6	0.5	2.6	3.1
Aves	6.1	12.2	9.8	8.7	15.5	8.6	15.2	12.4	12.0	13.8	11.3
Mammalia	25.6	23.3	32.9	29.1	30.0	34.6	24.5	28.3	30.8	34.9	26.4
Subtotal animals	39.4	39.5	49.4	45.2	56.5	50.9	46.1	50.4	49.0	53.3	42.5
Number of total occurrences	650	197	347	1172	200	312	862	1113	2056	304	4540
Number of scats	191	77	96	397	52	80	248	310	740	105	1673
Number of exploited species	55	40	53	59	34	57	53	38	> 42	43	> 41

*Dietz (1984).
†DF, Distrito Federal (Brazil); FAL, Fazenda Agua Limpa: Motta-Junior *et al.* (1996).
‡GO, Golas State (Brazil); PNE, Parque Nacional das Emas: Jácomo (1999).

fruit component of the diet, the prevalence of *S. lycocarpum* was remarkable in most areas, ranging from 7.5% to 32.6% of total occurrences and 4.8% to 53.4% of total biomass. Fruits of *Annonaceae*, *Myrtaceae* and *Solanaceae* were exploited in all ten study sites (Table 19.1), and those weighing 100–1000 g made up the bulk of the fruits consumed in the eight study sites (Fig. 19.1).

The diet of the maned wolf was not only highly variable among sites but also variable within sites, by season. In almost all of the eight study areas, small mammals and *S. lycocarpum* fruits were consumed mostly in the dry season, while other fruits were mostly consumed in the rainy season. In contrast, Dietz (1984) and Motta-Junior *et al.* (1996) did not find seasonality in the consumption of *S. lycocarpum* fruits. They did, however, find that small mammals were consumed mostly in the dry season and other fruits in the rainy season.

We examined the seasonal pattern of fruit consumption in more detail at EEJA to gain insight into the relative importance of *S. lycocarpum* fruit and other fruits through the annual cycle (Fig. 19.2). We found a significant negative association between the frequency in the diet of *S. lycocarpum* fruit and of other fruits ($r_s = -0.733$; $P < 0.025$; $n = 9$). Thus, when wolves eat few *S. lycocarpum* fruits, they eat more other fruits, and vice versa. Likewise, consumption of *S. lycocarpum* fruit is negatively correlated with the fruiting level of other fruits ($r_s = -0.867$; $P < 0.003$; $n = 9$) and the consumption of other fruits is positively correlated with the fruiting level of other fruits ($r_s = 0.800$; $P < 0.01$; $n = 9$). These relationships suggest that wolves use *S. lycocarpum* fruit to maintain an approximately constant rate of fruit intake in the face of seasonal variation in community-wide fruit availability. This view is further supported by lack of correlation between fruiting levels of *S. lycocarpum* and its rate of consumption by the maned wolf ($r_s = 0.150$; $P > 0.50$; $n = 9$).

Despite the influence of season and of other fruits in determining consumption of *S. lycocarpum* fruit, wolves showed a strong and consistent preference for this fruit in all five sites in which we had sufficient data to perform a test (Table 19.2). Virtually all other fruit species appear to be either avoided or consumed in the same proportion that they occurred in the field (Table 19.2). These findings emphasize the important role that *S. lycocarpum* plays in the ecology of the maned wolf and support Courtenay's (1994) suggestion that *S. lycocarpum* is a potential indicator of a suitable maned wolf habitat. An important issue to be addressed in future studies is whether the occurrence of the maned wolf in Brazil is largely determined by the presence of *S. lycocarpum.*

Legitimacy of the maned wolf as a seed-disperser

With exception of *Allagoptera campestris*, whose seeds were heavily destroyed (87.5% of 522 seeds in faeces), all fruit species consumed by wolves had most seeds unharmed after passing through the wolf's gut (91.7–100% of seeds intact). Seed germination occurred in 14 of 16 species tested (Table 19.3). These results uphold the legitimacy of the maned wolf as a seed-disperser (*sensu* Herrera, 1989).

In a recent review on vertebrate effects on seed germination, Traveset (1998) listed only six carnivore species whose effect on seed germination had been studied. Of 28 total tests, the carnivores enhanced germination in eight. The remaining 20 tests yielded no significant difference between seeds extracted from scat versus those from pulp (control). These results are similar to ours. Of 26 tests on eight plant species (Table 19.3), gut passage enhanced the percentage of germination in seven and had no effect on 12. In contrast to the results of Traveset (1998), but in accordance with Cypher (1999), we found that gut treatment sometimes inhibited seed germination; in seven tests control seeds germinated better than seeds from faeces (Table 19.3).

Small seeds appeared in larger numbers per scat than large seeds (seed dry mass × number of seeds per scat: $r_s = -0.784$; $P < 0.001$; $n = 20$ species). A similar result was found for Mediterranean carnivores (Herrera, 1989) and this suggests that small-seeded species are clump-dispersed by the maned wolf, while some of the larger ones are probably scatter-dispersed (see Howe, 1989). The average

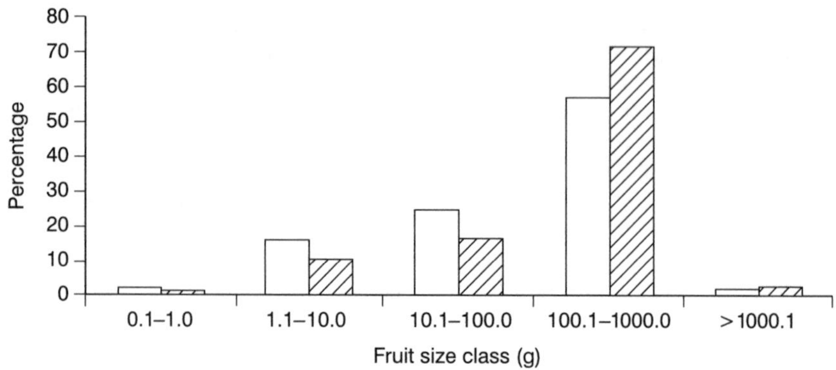

Fig. 19.1. Occurrences and estimated biomass consumption of fruits in the diet of the maned wolf as a function of fruit size class. Data gathered from the eight studied localities in south-eastern Brazil and from a locality in central Brazil (Motta-Junior *et al.*, 1996).

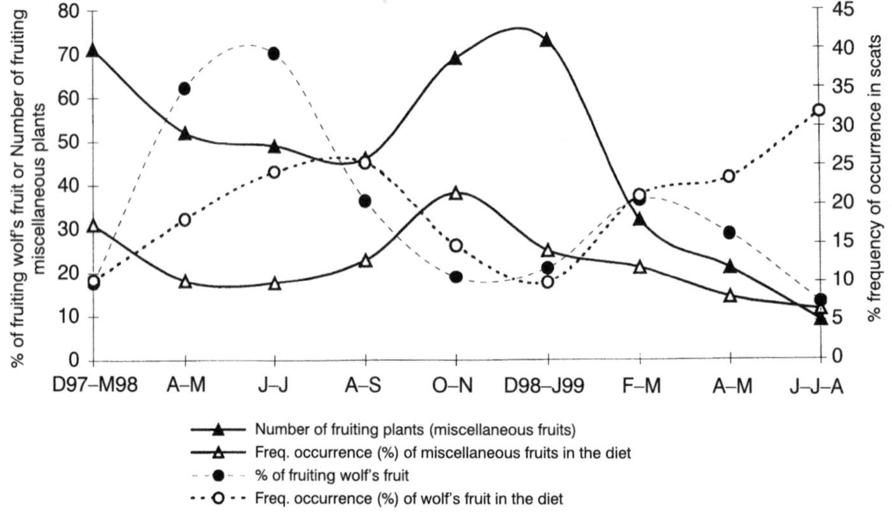

Fig. 19.2. Functional responses of the maned wolf in relation to the fruiting phenology of *S. lycocarpum* and other fruits from December 1997 to August 1999. Data collected at the Ecological Station of Jataí (SP), Brazil.

number of conspecific seeds per scat ranged from 5.0 to 3360.8, with a minimum of 1 (several species) and a maximum of 26,775 (*Solanum sisymbriifolium*).

Contrary to Herrera (1989), our results indicate that there is a relationship between the dry mass of individual seeds and the percentage of broken/cracked seeds in the faeces ($r_s = 0.453$; $P < 0.05$; $n = 20$ species). This means that the larger the seed, the higher the possibility of physical damage (see Appendix for seed mass data), although it is important to remember that in our study damage was always < 10% of seeds found in faeces (except for *A. campestris*).

Effectiveness as a seed-disperser

Although the maned wolf seems to have a short passage time of food through the gut – approximately 20–30 min (Barboza *et al.*,

Table 19.2. Analysis of fruit species selection (SEL) by the maned wolf in five localities, south-eastern Brazil. Goodness-of-fit tests were performed considering occurrences in scats as observed frequencies (OBS) and occurrences in quadrats, so providing expected frequencies at each site (EXP).

	PFSP (MG)			EEJA (SP)			EITA (SP)			EASB (SP)			EITI (SP)		
Fruit species	OBS	EXP	SEL	OBS	EXP	SEL	OBS	EXP	SEL	OBS	EXP	SEL	OBS	EXP	SEL
Alibertia sessilis	9	21.9	–							7	10.3	≅	4	12.2	–
Anacardium humile													1	11.4	–
Andira cf. *humilis*	3	24.7	–												
Annona cf. *cornifolia*	4	5.5	≅												
Annona crassiflora	8	11.0	≅										1	2.3	≅
Annona cf. *tomentosa*	7	21.9	–												
Annona spp.				6	3.8	≅	11	6.6	≅	1	9.6	–			
Bromelia cf. *balansae*				9	32.0	–							17	2.3	+
Brosimun gaudichaudii	1	11.0	–												
Byrsonima intermedia				1	50.9	–									
Campomanesia pubescens	6	5.5	≅	12	35.8	–	13	131.1	–	26	17.0	≅	21	18.3	≅
Diospyros hispida	1	38.4	–												
Duguetia furfuracea	23	65.8	–	34	71.6	–	6	32.8	–				4	3.8	≅
Mauritia flexuosa	2	27.4	–												
Melancium campestre	1	5.5	–	1	1.9	≅									
Miconia albicans	1	19.2	–												
Palmae sp. (*peaozinho*)										3	2.2	≅			
Parinari obtusifolia	2	13.7	–												
Peritassa campestris													2	5.3	–
Pouteria spp.													6	11.4	≅
Psidium cf. *cinereum*				6	5.7	≅	17	85.2	–	2	15.5	–	15	15.2	≅
Solanum crinitum				64	3.8	+									
Solanum lycocarpum	228	24.7	+	140	18.8	+	264	26.2	+	36	9.6	+	38	10.7	+
Solanum sisymbriifolium							23	26.2	≅						
Solanum spp.							3	6.6	≅						
Syagrus spp.				4	52.8	–	17	39.3	–	4	14.8	–	9	25.1	–
TOTAL	296	296		277	277		354	354		79	79		118	118	
χ^2; d.f.;	1839.2; 13;			1882.9; 9;			2038; 7;			106.1; 6;			194.7; 10;		
P	< 0.0001			< 0.0001			< 0.0001			< 0.0001			< 0.0001		

+, Preference; –, avoidance; ≅, no selection or rejection.

1994) – this is counterbalanced by its highly cursorial habits – it walks many kilometres during the night – and by a large home range (21–115 km^2) (Dietz, 1984; Carvalho and Vasconcellos, 1995).

By recording the location of tracks and faeces and by radiotelemetry data (Dietz, 1984), we have documented that the maned wolf prefers open *cerrado*. Therefore, only plant species that establish in open *cerrado* are likely to be effectively dispersed by wolves (see Bustamante *et al.*, 1992). Because most plant species we found in wolf scats came from *cerrado* vegetation (see Appendix), the wolf is probably an important disperser for many of these species. Further, its treatment of seeds may be necessary to remove compounds in the fruit pulp that inhibit germination, as is the case with *S. lycocarpum* (Flavia S. Pinto, personal communication).

Future studies on the maned wolf as a seed-disperser should assess microhabitat sites of faecal deposition versus plant establishment and seed predation (Bustamante *et al.*, 1992).

Table 19.3. Germination tests of seeds found in scats of the maned wolf collected in the field. Significant tests and associated probabilities are underlined. Treatments with highest percentage of germination are in bold.

			Chi-square	
Fruit species (locality)	Scat-derived % (*n*)	Fruit-derived % (*n*)	Yates's correction	*P*
Alibertia cf. *sessilis* (PNSC)	40.0 (25)			
Alibertia cf. *sessilis* (PNSC)	87.9 (33)			
Alibertia cf. *sessilis* (EASB)	97.0 (100)			
Allagoptera campestris (PNSC)	0.0 (8)			
Annona cf. *tomentosa* (PFSP)	0.0 (60)			
Bromelia cf. *balansae* (PNSC)	0.0 (60)			
Bromelia cf. *balansae* (PNSC)	9.5 (21)			
Bromelia balansae (EITI)	100.0 (20)			
Campomanesia sp. (PNSC)	77.0 (18)	97.0 (34)	3.06	> 0.05
Campomanesia pubescens (EEJA)	97.5 (120)	99.0 (120)	0.25	> 0.30
Campomanesia pubescens (EEJA)	98.0 (50)	100.0 (50)	0.00	> 0.90
Campomanesia pubescens (EASB)	75.0 (65)	**100.0** (65)	16.04	< 0.001
Campomanesia pubescens (EASB)	0.0 (50)	**98.0** (50)	92.20	< 0.0001
Duguetia furfuracea (EEJA)	17.5 (40)	**40.0** (40)	3.90	< 0.05
Duguetia furfuracea (EEJA)	14.0 (50)	22.0 (50)	0.61	> 0.30
Duguetia furfuracea (EEJA)	0.0 (20)	0.0 (20)		
Melancium campestre (PNSC)	21.9 (64)			
Melancium campestre (PFSP)	45.2 (147)	54.2 (147)	2.33	> 0.10
Myrtaceae sp. 1 (PNSC)	6.0 (50)			
Parinari obtusifolia (PNSC)	5.0 (20)			
Psidium sp. (PNSC)	92.7 (110)			
Psidium sp. (PNSC)	30.9 (136)			
Psidium guajava (FFOR)	97.0 (800)	95.0 (800)	3.66	> 0.05
Solanum sisymbriifolium (EITA)	**10.7** (300)	2.3 (300)	15.80	< 0.001
Solanum crinitum (EEJA)	30.0 (50)			
Solanum crinitum (EEJA)	42.0 (50)			
Solanum crinitum (EEJA)	**52.0** (100)	31.0 (100)	8.24	< 0.01
Solanum crinitum (EEJA)	**3.3** (720)	0.1 (720)	19.70	< 0.001
Solanum crinitum (EEJA)	17.7 (180)	16.0 (180)	0.08	> 0.70
Solanum lycocarpum (PFSP)	8.9 (180)	12.2 (180)	0.74	> 0.30
Solanum lycocarpum (UFSC)	**50.0** (40)	22.5 (40)	5.41	< 0.05
Solanum lycocarpum (UFSC)	**60.0** (30)	13.3 (30)	12.13	< 0.001
Solanum lycocarpum (UFSC)	30.0 (30)	**63.3** (30)	5.42	< 0.05
Solanum lycocarpum (UFSC)	11.7 (60)	**28.3** (60)	4.22	< 0.05
Solanum lycocarpum (UFSC)	43.3 (30)	36.7 (30)	0.07	> 0.70
Solanum lycocarpum (EEJA)	0.0 (180)	0.0 (180)	0.00	> 0.95
Solanum lycocarpum (EEJA)	26.0 (50)	**58.0** (50)	9.24	< 0.01
Solanum lycocarpum (EEJA)	28.0 (50)	**76.0** (50)	21.19	< 0.001
Solanum lycocarpum (EEJA)	64.0 (50)	82.0 (50)	3.25	> 0.05
Solanum lycocarpum (EEJA)	**68.0** (50)	44.0 (50)	4.91	< 0.05
Solanum lycocarpum (EEJA)	**86.0** (50)	58.0 (50)	8.38	< 0.01
Syagrus rommanzofiana (EITA)	24.6 (69)			

%, percentage of germination; *n*, number of seeds.

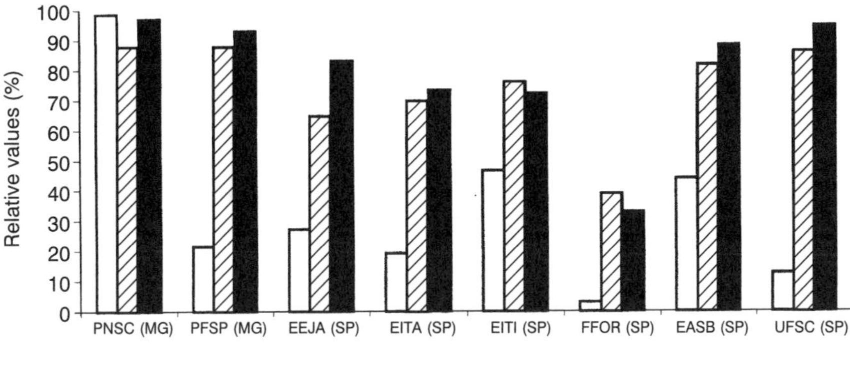

Fig. 19.3. Occurrence and estimated biomass of *cerrado* fruits in the diet of the maned wolf in relation to the *cerrado* vegetation land cover. Data are from eight areas in south-eastern Brazil.

Conservation Aspects

Although apparently expanding its range in some regions, such as eastern Minas Gerais State (Dietz, 1985), the maned wolf is vanishing in others, such as Rio Grande do Sul State in Brazil and Uruguay (Mones and Olazarri, 1990). The conservation status of the species remains 'threatened' in Brazil (Fonseca *et al.*, 1994) and 'near threatened' worldwide, according to the International Union for Conservation of Nature and Natural Resources (IUCN) (Nowak, 1999).

Despite high variation among our sites in per cent of *cerrado* habitat, maned wolves systematically consumed high proportions of fruits from *cerrado* habitats (Fig. 19.3), both in terms of frequency of occurrence in faeces ($r_s = 0.476$; $P > 0.10$; $n = 8$ localities) and estimated biomass ($r_s = 0.333$; $P > 0.10$; $n = 8$). This result is echoed by the wolf's preference for prey that inhabit *cerrado* habitats (Motta-Junior *et al.*, 1996). These findings suggested that wolves search for food resources more intensively in *cerrado* vegetation than expected by chance, confirming the importance of *cerrado* for maned wolf conservation (see Fonseca *et al.*, 1994). Yet Dias (1994) estimates that approximately 37% of the *cerrado* biome has been destroyed in Brazil and that only 6.6% is preserved in reserves.

Dietz (1984) and Courtenay (1994) document an expansion of the maned wolf into deforested areas that have become grassland. But their suggestion that these areas plus *S. lycocarpum* fruit are sufficient for the maintenance of wolves appears premature. Unquestionably, range expansions have occurred. We do not have any information on the viability of these populations, however. Because wolves are markedly omnivores, they require more than fruit; they also need animal matter. In poor-quality *cerrado*, they may prey on poultry, which brings them into conflict with humans.

Our diet study has relevance for maintenance techniques for captive maned wolves. In particular, captive wolves are often given a diet with a higher proportion of meat than consumed by their wild counterparts. The excessive intake of protein that results may exacerbate the development of cystinuria and gingivitis (Barboza *et al.*, 1994). Our study shows unequivocally that wild populations of wolves have very mixed diets, including high proportions of fruits. We suggest that meat intake of captive maned wolves be reduced (as also proposed by Barbosa *et al.*, 1994) and that the proportion of fruits in their diet should be increased to 30–50% of daily wet mass of food provided.

Acknowledgements

We would like to thank FAPESP (process no. 97/06090-7) and CNPq for financial support. Julio A. Lombardi, J.R. Stehmann and Rafaela C. Forzza identified some plants. Alexandre Percequillo, Márcio R.C. Martins and Luís F. Silveira identified some small mammals, reptiles and birds, respectively. Sergio T. Meirelles made helpful comments on statistical procedures. Jean Paul Metzger and Fernanda Othero provided the landscape analysis. The administrative personnel of the Instituto Florestal de São Paulo, A.W. Faber-Castell, RIPASA S/A, and Parque Ecológico de São Carlos, allowed us to work on their properties. We thank Douglas Levey, Mauro Galetti, Ramiro Bustamante and Vânia R. Pivello for invaluable suggestions for and criticism of the manuscript. We are grateful to Diego Queirolo, Sonia C.S. Belentani, Adriana A. Bueno, Edevaldo O. Aparecido and Magno C. Branco for their invaluable aid in data collection.

References

Barboza, P.S., Allen, M.E., Rodden, M. and Pojeta, K. (1994) Feed intake and digestion in the maned wolf (*Chrysocyon brachyurus*): consequences for dietary management. *Zoo Biology* 13, 375–381.

Bustamante, R.O., Simonetti, J.A. and Mella, J.E. (1992) Are foxes legitimate and efficient seed-dispersers? A field test. *Acta Oecologica* 13, 203–208.

Carvalho, C.T. and Vasconcellos, L.E.M. (1995) Disease, food and reproduction of the maned wolf – *Chrysocyon brachyurus* (Illiger) (Carnivora, Canidae) in southeast Brazil. *Revista Brasileira de Zoologia* 12(3), 627–640.

Castro, S.A., Silva, S.I., Meserve, P.L., Gutierrez, J.R., Contreras, L.C. and Jaksic, F.M. (1994) Frugivoría y dispersión de semillas de pimiento (*Schinus molle*) por el zorro culpeo (*Pseudalopex culpaeus*) en el Parque Nacional Fray Jorge (IV Región, Chile). *Revista Chilena de Historia Natural* 67, 169–176.

Courtenay, O. (1994) Conservation of the maned wolf: fruitful relations in a changing environment. *Canid News* 2, 41–43.

Cypher, B.L. (1999) Germination rates of tree seeds ingested by coyotes and racoons. *American Midland Naturalist* 142(1), 71–76.

Dalponte, J.C. and Lima, E.S. (1999) Disponibilidade de frutos e a dieta de *Lycalopex vetulus* (Carnivora – Canidae) em um cerrado de Mato Grosso, Brasil. *Revista Brasileira de Botânica* 22(2) (suppl.), 325–332.

Dias, B. (1994) Conservação da natureza no cerrado brasileiro. In: Pinto, M.N. (ed.) *Cerrado: caracterização, ocupação e perspectivas*, 2nd edn. Editora Universidade de Brasília, Brasilia, pp. 607–663.

Dietz, J.M. (1984) Ecology and social organization of the maned wolf (*Chrysocyon brachyurus*). *Smithsonian Contributions to Zoology* 392, 1–51.

Dietz, J.M. (1985) *Chrysocyon brachyurus. Mammalian Species* 234, 1–4.

Eiten, G. (1974) An outline of the vegetation of South America. In: *Symposia of the Congress of the International Primatological Society*, 5. Japan Science Press, Tokyo, pp. 529–545.

Eiten, G. (1978) A sketch of the vegetation of central Brazil. In: *Congresso Latino-Americano de Botânica*, 29. Brasília and Goiania, pp. 1–37.

Emmons, L.H. (1987) Comparative feeding ecology of felids in a neotropical rainforest. *Behavioural Ecology and Sociobiology* 20, 271–283.

Fonseca, G.A.B., Rylands, A.B., Costa, C.R.M., Machado, R.B. and Leite, Y.L.R. (eds) (1994) *Livro vermelho dos mamíferos brasileiros ameaçados de extinção*. Fundação Biodiversitas, Belo Horizonte.

Herrera, C.M. (1989) Frugivory and seed dispersal by carnivorous mammals, and associated fruit characteristics, in undisturbed Mediterranean habitats. *Oikos* 55, 250–262.

Howe, H.F. (1989) Scatter- and clump-dispersal and seedling demography: hypothesis and implications. *Oecologia* 79, 417–426.

Jácomo, A.T.A. (1999) Nicho alimentar do lobo-guará (*Chrysocyon brachyurus* Illiger, 1811) no Parque Nacional das Emas – GO. MSc thesis, Universidade Federal de Goiás, Goiania, Brazil.

Jaksic, F.M. (1989) Opportunism vs. selectivity among carnivorous predators that eat mammalian prey: a statistical test of hypotheses. *Oikos* 56, 427–430.

Jordano, P. (1992) Fruits and frugivory. In: Fenner, M. (ed.) *Seeds: the Ecology of Regeneration in Natural Plant Communities*. CAB International, Wallingford, UK, pp. 105–151.

Langguth, A. (1975) Ecology and evolution in the South American canids. In: Fox, M.W. (ed.) *The Wild Canids: Their Systematics, Behavioral Ecology and Evolution*. Van Nostrand Reinhold, New York, pp. 192–206.

Lombardi, J.A. and Motta-Junior, J.C. (1993) Seed dispersal of *Solanum lycocarpum* St. Hil. (*Solanaceae*) by the maned wolf, *Chrysocyon brachyurus* Illiger (Mammalia, Canidae). *Ciência and Cultura* 45, 126–127.

Lorenzi, H. (1998) *Árvores brasileiras: manual de identificação e cultivo de plantas arbóreas nativas do Brasil*, Vol. 2, 2nd edn. Editora Plantarum, Nova Odessa.

McGarigal, K. and Marks, B.J. (1995) *FRAGSTATS: Spatial Pattern Analysis Program for Quantifying Landscape Structure.* US Forest Service General Technical Report PNW 351, Portland, Oregon.

Medel, R.G. and Jaksic, F.M. (1988) Ecología de los cánidos sudamericanos: una revisión. *Revista Chilena de Historia Natural* 61, 67–79.

Mendonça, R.C., Felfili, J.M., Walter, B.M.I., Silva Junior, M.C.S., Rezende, A.V., Filgueiras, T.S. and Nogueira, P.E. (1998) Flora vascular do cerrado. In: Sano, S.M. and Almeida, S.P. (eds) *Cerrado: ambiente e flora.* EMBRAPA, Planaltina, pp. 289–556.

Mones, A. and Olazarri, J. (1990) Confirmación de la existencia de *Chrysocyon brachyurus* (Illiger) en el Uruguay (Mammalia: Carnivora: Canidae). *Comunicaciones Zoológicas del Museo de Historia Natural de Montevideo* 174(7), 1–6.

Motta-Junior, J.C., Talamoni, S.A., Lombardi, J.A. and Simokomaki, K. (1996) Diet of the maned wolf, *Chrysocyon brachyurus*, in central Brazil. *Journal of Zoology, London* 240, 277–284.

Nimer, E. and Brandão, A.M.P.M. (1989) *Balanço hídrico e clima da região dos cerrados.* Instituto Brasileiro de Geografia e Estatística, Rio de Janeiro.

Nowak, R.M. (1999) *Walker's Mammals of the World*, 6th edn. Johns Hopkins University Press, Baltimore, Maryland, 1921 pp.

Ratter, J.A. (1980) *Notes on the Vegetation of Fazenda Agua Limpa (Brasília, DF, Brazil).* Royal Botanical Garden, Edinburgh.

Rogers, L.L. and Applegate, R.D. (1983) Dispersal of fruit seeds by black bears. *Journal of Mammalogy* 64, 310–311.

Setzer, J. (1966) *Atlas climático e ecológico do estado de São Paulo.* Comissão Interestadual da Bacia do Paraná – Uruguai, São Paulo.

Siegel, S. and Castellan, N.J., Jr (1988) *Nonparametric Statistics for the Behavioral Sciences*, 2nd edn. McGraw-Hill, New York, 399 pp.

Traveset, A. (1998) Effect of seed passage through vertebrate frugivores' guts on germination: a review. *Perspectives in Plant Ecology, Evolution and Systematics* 1/2, 151–190.

Traveset, A. and Willson, M. (1997) Effects of birds and bears on seed germination in the temperate rainforest of southeast Alaska. *Oikos* 80, 89–95.

Willson, M. (1993) Mammals as seed-dispersal mutualists in North America. *Oikos* 67, 159–176.

Young, K.R. (1990) Dispersal of *Styrax ovatus* seeds by the spectacled bear (*Tremarctus ornatus*). *Vida Silvestre Neotropical* 2(2), 68–69.

Zar, J.H. (1999) *Biostatistical Analysis*, 4th edn. Prentice Hall, Upper Saddle River, New Jersey, 663 pp.

Appendix

Fruit species eaten by the maned wolf in south-eastern Brazil, with some of their characteristics. The scientific names and classifications by habitat are according to Mendonça *et al.* (1998), Lorenzi (1998) and Motta-Junior *et al.* (1996).

Family and species	Habitat	Life-form	Ripe fruit colour	Seeds per fruit	Fruit mass (g)
Anacardiaceae					
Anacardium humile	CER	Shrub	Orange–red	1	20.0
Mangifera indica	DIS	Tree	Yellow	1	250.0
Annonaceae					
Annona crassiflora	CFO	Tree	Green	119	1021.0
*Annona monticola**	CER	Shrub			
Annona tomentosa	CFO	Shrub	Green	111	325.0
Aff. *Annona cornifolia*	CER	Shrub	Orange	40.5	8.3
Annonaceae small–elliptic	UNK				
Annonaceae flat–elliptic	UNK				
Duguetia furfuracea	CER	Shrub	Red–Green	22.6	75.8
Bromeliaceae					
Bromelia balansae	CER	Herb	Yellow	24.5	18.33

Continued

Appendix *Continued.*

Family and species	Habitat	Life-form	Ripe fruit colour	Seeds per fruit	Fruit mass (g)
Caricaceae					
Carica papaya	DIS	Tree	Orange	710	500.0
Chrysobalanaceae					
Couepia cf. *grandiflora*	CFO	Shrub	Yellow	1	13.2
Parinari obtusifolia	CER	Shrub	Brown	1	10.2
Cucurbitaceae					
Melancium campestre	CER	Herb	Green	101.3	65.4
Ebenaceae					
Diospyros hispida	CER	Shrub–tree	Brown	7.5	64.9
Guttiferae (*Cluseaceae*)					
*Calophyllum brasiliense**	FOR	Tree			
Hippocrateaceae					
*Salacia crassifolia**	CER	Tree			
Leguminosae					
Andira cf. *humilis*	CER	Shrub–tree	Yellow–green	1	18.8
Malpighiaceae					
Byrsonima intermedia	CFO	Shrub	Yellow	1	0.4
Byrsonima sp.*	UNK				
Melastomataceae					
Miconia albicans	CFO	Shrub	Green	21.8	0.3
Miconia sp.	UNK	Shrub	Black		
Moraceae					
Brosimum gauchicaudi	CER	Shrub	Orange–red	1	5.8
Morus sp.*	DIS	Shrub	Black		
Myrtaceae					
Campomanesia pubescens	CER	Shrub	Yellow–green	3.1	2.0
*Eugenia uniflora**	FOR	Tree			
Myrciaria cf. *cauliflora*	FOR	Tree	Black	2.8	5.5
Psidium aff. *cinereum*	CER	Shrub	Red–green	10.3	25.0
Psidium guajava	FOR	Tree	Yellow	166.5	45.1
Palmae (*Arecaceae*)					
Astrocaryum sp.*	CER	Shrub			
Allagoptera campestris	CER	Shrub	Yellow–green	1	0.5
Mauritia flexuosa	FOR	Tree	Brown	1	51.1
Palmae sp. 1	UNK			1	
Palmae sp. 2	UNK			1	
Syagrus petraea	CER	Shrub	Yellow–green	1	4.1
Syagrus rommanzofiana	FOR	Tree	Orange–yellow	1	5.2
Rubiaceae					
Alibertia sessilis	CFO	Shrub–tree	Black	31.7	6.9
Palicourea sp.*	UNK	Shrub			
Rutaceae					
Citrus sp.	DIS	Tree	Orange	5.3	128.5
Sapindaceae					
Talisia sp.*	UNK	Shrub–tree			

Continued

Appendix *Continued.*

Family and species	Habitat	Life-form	Ripe fruit colour	Seeds per fruit	Fruit mass (g)
Sapotaceae					
Pouteria torta	CER	Shrub-tree	Yellow	1.3	15.0
Pouteria cf. *ramiflora*	FOR	Shrub-tree	Yellow–green	1.5	14.0
Solanaceae					
Solanum lycocarpum	CER	Shrub	Green	270.8	358.2
Solanum crinitum	DIS	Shrub	Green	1251	58.6
Solanum sisymbriifolium	DIS	Herb	Red	115	3.1
Solanum sp. 'jurubeba'	DIS	Herb	Orange–green	59.3	2.4
Solanaceae sp.	UNK				

*Fruit species reported in Dietz (1984), Motta-Junior *et al.* (1996) and Jácomo (1999).
CER, *cerrado* (*sensu lato*); CFO, *cerrado* and gallery forest; FOR, gallery forest; DIS, disturbed environments; UNK, unknown.

20 Frugivore-generated Seed Shadows: a Landscape View of Demographic and Genetic Effects

Pedro Jordano and José A. Godoy
Estación Biológica de Doñana, CSIC, Apdo. 1056, E-41080 Sevilla, Spain

Introduction

A seed shadow is the spatial pattern of seed distribution relative to parent trees and other conspecifics; it results from the process of seed dispersal and represents the starting template for plant regeneration. Janzen (1970) and Connell (1971) consider it the population recruitment surface. For animal-dispersed, endozoochorous species the seed shadow results primarily from movement patterns of frugivores. Presumably, frugivores can dramatically affect both the demography and genetic make-up of animal-dispersed plant species. These effects, however, have rarely been documented in an integrated way.

In this chapter we focus on how frugivores influence the number and spatial pattern of propagules that reach the soil, and their simultaneous influence on gene flow via seed dispersal. We advocate an integrated view of both demographic and genetic effects to understand the role of frugivores on plant recruitment (Alvarez-Buylla *et al.*, 1996). Given that multiple influences sequentially alter after this initial effect of frugivores (i.e. post-dispersal seed predation, germination, seedling mortality), we need to quantitatively assess the relative importance of dispersal by frugivores for plant population biology. Seed dispersal by frugivores is the link in the demographic transition between the ripe fruit crop on the trees and, after delivery, the subsequent stages of establishment of germinated seeds, seedlings, saplings and established adults, i.e. the whole recruitment cycle. Thus, seed dispersal may play a pivotal role in the demography of plant populations (Harper, 1977) by simultaneously influencing not only the numerical dynamics of recruitment from dispersal to establishment, but also the genetic make-up of the seed shadow.

The difficulty of tracking the origin of frugivore-dispersed seeds has precluded a robust analysis of vertebrate seed dispersal (Levey and Sargent, 2000). Indeed, the difficulties in measuring and analysing the dispersal of seeds in natural communities has been considered an unavoidable limitation of the field (Wheelwright and Orians, 1982). Recent developments in molecular biology (Carvalho, 1998), however, have resulted in a series of molecular tools based on DNA analysis that allow analysis of gene-flow patterns via seed dispersal (Ouborg *et al.*, 1999) and the statistical analysis of the resulting patterns of genetic structure (Schnabel *et al.*, 1998a; Luikart and England, 1999).

More specifically, for animal-dispersed species, gene flow via seeds can be estimated directly (see Ouborg *et al.*, 1999, for a review). Yet recent studies of gene flow in plants are primarily focused on pollen flow (Sork *et al.*, 1999); very few studies have used molecular markers to assess seed-dispersal patterns. Even fewer studies have linked genetic patterns in fleshy-fruited plant species with the behaviour of frugivores (Loiselle *et al.*, 1995a,b; Schnabel *et al.*, 1998b). Ideally, one should link detailed observations of bird foraging behaviour and movement to distinct types of landscape patches or microhabitats (Wenny and Levey, 1998; Jordano and Schupp, 2000), with monitoring of seed rain (i.e. using seed traps) (Kollmann and Goetze, 1997) and analysis of the genetic make-up of the seeds.

Frugivores thus have the potential to influence both plant demography and genetic structure. This influence, in turn, has an impact on two major arenas of seed dispersal ecology, namely seed-dispersal limitation and landscape patterns of gene flow via seeds.

Demographic effects: seed-dispersal limitation

Dissemination limitation is probably the major demographic effect that frugivore activity can have on plant populations. It occurs whenever seed delivery by frugivores is insufficient to saturate available microhabitats for establishment; in species that are dispersal-limited, increasing seed input would result in increased recruitment (Ehrlén and Eriksson, 2000; Jordano and Schupp, in prep.; see also Schupp *et al.*, this volume; Table 20.1). If we include delayed consequences of frugivore activity for plant recruitment (i.e. the quantity and quality components of disperser effectiveness) (Schupp, 1993; Jordano and Schupp, 2000), three major forms of limitation processes may operate through the dispersal stage of plant regeneration (Table 20.1; Jordano and Schupp, in prep.; see also Schupp *et al.*, this volume): seed source limitation, dissemination limitation and limitation of establishment. We now review these briefly.

A demographic limitation operating during the seed-dispersal and establishment stages simply represents a low realized recruitment relative to the maximum potential. Thus, a primary form of dissemination limitation arises whenever the seed crop is insufficient to reach all the available safe sites ('source limitation' *sensu* Clark *et al.* (1999a)). This type of limitation is clearly independent of frugivore activity; it simply results from low fruit production.

Table 20.1. Demographic effects of frugivores through limitation of seed dispersal. Frugivore activity can limit plant population recruitment by both direct influences on fruit-removal levels and seed-delivery patterns and indirect (delayed) effects via the influences on the survival prospects of dispersed seeds and established seedlings. A non-restrictive view of seed-dispersal limitation would have to include the multiple effects of frugivores at all the stages of plant recruitment (Jordano and Schupp, 2001; Schupp *et al.*, this volume).

Stage and limitation	Definition
1. Seed limitation	Seed production at population level is insufficient to saturate all available safe sites (source limitation, *sensu* Clark *et al.*, 1999a)
2. Dissemination limitation	
2.1. Quantitatively restricted seed dispersal	Independent of the quantity of seeds produced, disperser activity is insufficient to disperse all seeds away from parents
2.2. Distance-restricted seed dispersal	Independent of the quantity of seeds dispersed away from parents, most seed dispersal is short-distance
2.3. Spatially contagious seed dispersal	Independent of distance of seed dispersal, seeds are not spread evenly, but rather are deposited patchily (aggregated), with many seeds in some sites and few to none in most sites
3. Establishment limitation	Independent of number of seeds arriving in a site, biotic and abiotic factors limit establishment of new individuals (i.e. delayed consequences in respect of Stage 2)

A second aspect of seed-dispersal limitation, dissemination limitation, derives from the role of frugivores in seed dissemination (Table 20.1). It includes all processes associated with frugivore foraging that limit the number, distance and/or spatial distribution of seeds over the landscape. But, independently of the quantity of seeds successfully dispersed away from parent trees, dispersal can be limited by distance-restricted seed delivery (e.g. due to territory defence) (Snow and Snow, 1988) and/or spatially aggregated patterns of seed delivery (e.g. in the vicinity of parent trees or conspecifics, at lek sites, latrines, roosts, etc.) (Dinerstein and Wemmer, 1988; Izhaki *et al.*, 1991; Chapman and Chapman, 1995; Julliot, 1997; Wenny and Levey, 1998). Spatial aggregation is a characteristic feature of animal-generated seed shadows (Fig. 20.1; Harms *et al.*, 2000).

Finally, establishment limitation might occur as a consequence of frugivore activity if, independently of the number of seeds dispersed away from parents, frugivores fail to deliver seeds to the safest sites for germination and establishment (Table 20.1).

Genetic effects: identifying the source tree of dispersed seeds and characterizing the genetic make-up of the seed shadow

Frugivores influence the genetic make-up of the seed shadow, i.e. the particular combination of dispersed genotypes and their location relative to parent trees and other conspecifics. Thus, a sizeable fraction of gene flow in endozoochorous species can be attributed to frugivore activity but, as far as we know, no study has yet dissected the contributions of

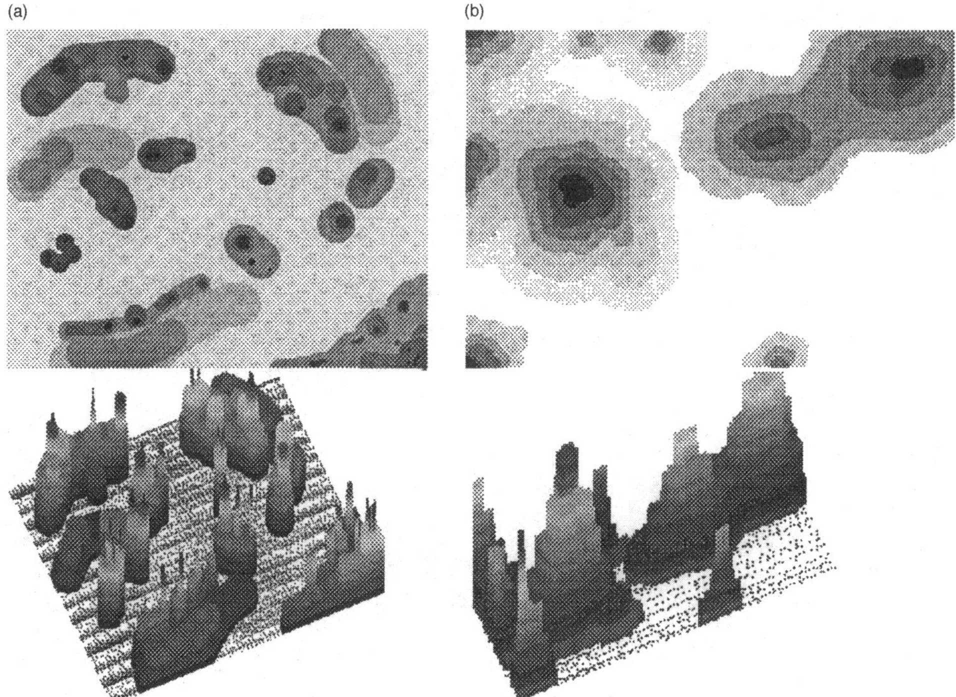

Fig. 20.1. Idealized representation of the landscape pattern of seed shadows in (a) animal-dispersed species (e.g. *P. mahaleb*) (Jordano and Schupp, 2000; J.L. García-Castaño and P. Jordano, unpublished), and (b) abiotically (wind)-dispersed species (e.g. *Acer mono*, *Betula grossa*) (Nakashizuka *et al.*, 1995). Darker shading or higher profile (in the lower panels) indicates greater seed density. Seed densities range from 1 to 140 seeds m^{-2} in (a) and from 1 to 60 seeds m^{-2} (*A. mono*) or 2 to > 5600 seeds m^{-2} (*B. grossa*) in (b).

pollination and seed dispersal to gene flow in animal-dispersed species. Gene flow via pollen is certainly extensive, especially in obligate outcrossed species, such as many woody endozoochorous species (Hamrick and Godt, 1997), but the scanty evidence available suggests that seed dispersal is also important (Dow and Ashley, 1996; Schnabel *et al.*, 1998b). By genotyping animal-dispersed seeds we can resolve several long-standing issues in seed-dispersal ecology, including the spatial relationships (e.g. distance, aggregation) between the maternal tree and its propagules, the full characterization of the seed shadow and the provenance of dispersed seeds. In addition, we can better understand gene-flow patterns in complex landscapes, where habitat heterogeneity may impose dramatic patterns of genetic structure (Cain *et al.*, 2000).

To identify the maternal tree of seeds dispersed by frugivores, we compare the multilocus genotype of the seed endocarp (which is diploid maternal tissue in angiosperm species with drupaceous fruits) (Roth, 1977) with the genotype of adult trees in the population, obtained from leaf tissue (Godoy and Jordano, 2001). Using this procedure, dispersal distances can be estimated for individual seeds defecated or regurgitated by frugivores, sampled in our regular monitoring of seed rain using seed traps scattered in the forest. This represents a major advance in seed dispersal, opening new avenues for research by combining ecological data with molecular tools (Godoy and Jordano, 2001).

Recent applications of hypervariable molecular markers, such as simple sequence repeats (SSR), or microsatellite loci, to seed-dispersal studies (Ouborg *et al.*, 1999) allow inference of a seed's or plant's maternal tree, providing a powerful tool in seed-dispersal ecology to assess dispersal distances directly. Using spatial autocorrelation techniques, a few studies have documented extreme clumping of progeny, either relative to the parents or to related individuals (Sork *et al.*, 1993; Loiselle *et al.*, 1995a; Schnabel *et al.*, 1998b; Streiff *et al.*, 1998; Ueno *et al.*, 2000). Such clumping is probably common in fleshy-fruited species with highly non-random and aggregated seed rain (Clark *et al.*, 1999a; Harms *et al.*, 2000). In most cases, however, the fine-scale patterns of genetic structuring cannot be linked to specific foraging modes or seed-deposition patterns by frugivores, and the role of dispersers is largely inferred.

Our goal in this chapter is to outline a protocol for an integrated analysis of the demographic and genetic effects of frugivores of plant populations. These two types of frugivore effects have rarely been considered together. We examine dispersal limitation and attempt to establish conceptual bridges with genetic effects to allow a more integrated understanding of frugivore effects in plant communities. We introduce molecular methods to unambiguously identify the maternal tree of seeds dispersed by frugivores and discuss the relevance of this approach for understanding the evolutionary effects of frugivores on plant populations. Finally, we discuss potential implications for future research.

Methods

Study site and species

This study was conducted in the Reserva de Navahondona-Guadahornillos (Parque Natural de las Sierras de Cazorla, Segura y las Villas, Jaén province, south-eastern Spain), with the main study site located at Nava de las Correhuelas (1615 m elevation). Deciduous vegetation covers deep soils. Small trees and shrubs are mixed with extensive patches of grassland, while adjacent rocky exposed slopes are dominated by open pine forest (*Pinus nigra*, subsp. *salzmannii*) with juniper. Extensive patches of open habitat (~66% of surface area), with either grassland, gravelly soil or rock outcrops, appear with shrubs and small trees, both isolated and clumped, giving way to open pine forests (Jordano, 1993; Jordano and Schupp, 2000, for detailed description). *Prunus mahaleb* is a small tree (2–10 m height) growing scattered at mid-elevations (1200–2000 m) in south-eastern Spanish mountains. It is relatively abundant at Nava de las Correhuelas, with an estimated population of ~180 reproductive individuals. *Prunus mahaleb* has insect-pollinated flowers; approximately equal proportions of solitary bees and flies act as pollinators (see Jordano, 1993, for details).

Frugivorous birds visiting *P. mahaleb* trees include legitimate seed-dispersers (warblers, *Sylvia* spp.; robin, *Erithacus rubecula*; thrushes, *Turdus* spp.; and redstarts, *Phoenicurus* spp.), which swallow fruits whole and defecate and/or regurgitate seeds, usually after leaving the tree. Some species, however, peck the fruit, tearing off the pulp and dropping the seed to the ground beneath the parent (e.g. tits, *Parus* spp., and chaffinch, *Fringilla coelebs*) (Jordano and Schupp, 2000, and references therein). Seed rain of *P. mahaleb* in the study area is generated by frugivorous birds and, marginally, by carnivorous mammals (J.L. García-Castaño and P. Jordano, 1999, personal observation). Seed rain is highly patchy, largely restricted to covered microhabitats beneath woody vegetation close to fruiting trees (Jordano and Schupp, 2000).

Frugivorous birds were watched during observation periods at focal trees in the study area. Basic data on feeding rates (i.e. the number of fruits taken per visit, etc.), the microhabitat type of the first perch used after leaving the feeding tree and its distance from the focal tree were recorded for each observation (for details, see Jordano and Schupp, 2000). Nine types of microhabitats were defined according to the presence of shrub cover, height of vegetation and whether or not a rocky substrate was present (Jordano and Schupp, 2000). For the present analyses we pooled some types, resulting into six distinct microhabitats: *P. mahaleb* fruiting trees, low shrubs (*Berberis vulgaris*, *Juniperus communis*), mid-height shrubs (*Crataegus monogyna*, *Lonicera arborea*, etc.), pine trees (including those with and without low shrub cover beneath them) and, as open substrates, deep soil and rocky substrates (including gravelly soil, rocks with soil and rock boulders). By combining the information on number of visits recorded, mean number of seeds dispersed per visit, proportion of exit flights to each microhabitat and seed-rain data, we were able to estimate the contribution of each main disperser species to the seed rain in each microhabitat (for details see Jordano and Schupp, 2000).

Methods: seed-addition experiments

Experimental design

We used a factorial design, with treatments for seed addition, exclusion of post-dispersal seed predators (rodents) and type of microhabitat. Seeds were sown at ten replicate locations in each of four microhabitat types (as defined in Jordano and Schupp, 2000). Two microhabitats had woody plant cover: beneath *P. mahaleb* trees and beneath pine trees with low shrubs (mostly *Juniperus* and *B. vulgaris*, < 1 m height); the other two, gravelly soil and deep soil with grassy cover, were open microhabitats with no shrub or tree cover.

The seed-addition treatment included two levels of seed-sowing density. The 'control' level had seed density adjusted to the median background seed density recorded for each microhabitat type in a concurrent sampling of seed-rain patterns (Jordano and Schupp, 2000). The 'added', or seed-addition, level had sowing density adjusted to the 95% percentile of the seed rain, roughly a threefold increase in the number of seeds sown in the control.

Finally, the predator exclusion treatment consisted of wire-mesh enclosures that were either closed to exclude predators or open to allow them access. Thus, each replicate plot in a given microhabitat type had four subplots for seed sowing, according to the combinations of the seed-addition ('added' and 'control' levels) and rodent-exclusion treatments ('open' and 'excluded' levels).

Methods: genotyping of individual dispersed seeds

Seed sampling

Seeds for genetic analyses were sampled with seed traps (aluminum trays protected with 0.8 cm wire mesh; see Jordano and Schupp, 2000) during the 1996 fruiting season. Sets of two seed traps were located beneath the canopy of each of five 'focal' *P. mahaleb* maternal trees selected as illustrative of the range of tree sizes and growing in sites and neighbourhood densities typical of the species in the study population. In addition, sets of two seed traps were located away from each focal tree in 2–3 replicate locations per focal tree, including two microhabitat types (as defined in Jordano and Schupp, 2000): beneath mid-height shrubs and beneath pine trees. Mid-height shrubs (e.g. *C. monogyna*, *Rosa canina* and *L. arborea*) were selected in the neighbourhood of the focal trees (within 10 m) or at more distant locations (> 10 m). Another set of two seed traps per focal tree was located beneath pine trees with low shrubs (two replicate locations) and beneath pine trees with open understorey (three replicate locations). The design included 20 replicate sampling locations totalling 38 seed traps, a stratified random sample of 481 sampling points used in a concurrent study of seed rain.

A total of 95 seeds sampled from the seed traps were analysed. Both defecated and regurgitated seeds were included in the sample, with 37 seeds from traps beneath *P. mahaleb* (sampled at random from the total seed sample captured in these traps) and 58 seeds from traps in other microhabitats. Seeds were kept dry and at room temperature until analysis.

The locations of seed traps and all the adult, reproductive trees in the population were mapped and recorded in a geographical information system (GIS) database. Leaf tissue from adult trees in the population (180 trees including ~100% of the potential maternal trees for 1996 progeny) was collected (see Jordano and Godoy, 2000), kept in liquid nitrogen within labelled duplicate cryotubes and stored at –80°C.

DNA extraction and amplification

DNA was extracted from 100–200 mg of fresh leaf tissue, using the rapid miniprep method of Cheung *et al.* (1993). Tissue was homogenized in 320 μl of extraction buffer (200 mM Tris-HCl pH 8.0, 70 mM ethylenediamine tetra-acetic acid (EDTA), 2 M NaCl, 20 mM sodium bisulphite) with an electric drill (560 W; full speed) with attached plastic disposable pestles. After homogenization, 80 μl of 5% sarcosyl was added and the sample was incubated at 65°C for 30 min and centrifuged at 16,000 *g* for 15 min to remove insoluble material. DNA was precipitated by the addition of 90 μl of 10 M ammonium acetate and 200 μl of isopropanol. The mixture was incubated at room temperature for 5 min and centrifuged for 15 min at 16,000 *g*. The resulting pellet was washed with 70% ethanol, dried and resuspended in 100 μl TE buffer.

This DNA, extracted from leaf tissue of all the potential maternal trees in the population, was used to construct a database of multilocus genotypes of adult trees. These genotypes were matched with those obtained from DNA extracted from the seed endocarp tissue, thereby allowing identification of the maternal tree for each seed (Godoy and Jordano, 2001). To extract DNA from the lignified seed endocarps, we used a similar protocol, with the following modifications: tissue was homogenized in 320 μl of extraction buffer and resuspended in 50 μl TLE (200 mM Tris-HCl pH 8.0, 70 mM EDTA). Additional details and conditions for amplification using the polymerase chain reaction (PCR) are given elsewhere (Godoy and Jordano, 2001).

We used a series of microsatellite primers designed for cultivated *Prunus* species (Abbot, 1998, personal communication; G. King, 1998, personal communication; Cipriani *et al.*, 1999; Downey and Iezzoni, 2000; Sosinski *et al.*, 2000). We tested a total of 43 primers, of which 16 showed polymorphic variation for *P. mahaleb*. We selected a subset of nine primers for use in this study (Godoy and Jordano, 2001).

Statistical analyses

Patterns of fine-scale genetic structure in the tree population were examined using the multilocus microsatellite genotypes of 180 trees based on nine polymorphic loci. A coancestry coefficient, f_{ij}, was estimated between all possible pairs of adult trees genotyped using program f_{ij} *Anal* from J. Nason (Sork *et al.*, 1998). Briefly, f_{ij} measures the correlation in the frequencies of homologous alleles, p_i and p_j, at a locus in pairs of mapped individuals i and j, revealing the degree of genetic similarity for pairs of adult trees growing at different distance intervals. Any spatial pattern of genetic structure in the population will show up in a plot of f_{ij} vs. distance (an autocorrelogram). The coancestry coefficient is well suited for examining spatial patterns of genetic variation; it assesses the autocorrelation structure of genetic affinity among coexisting individuals (Heywood, 1991; Loiselle *et al.*, 1995a). The significance of f_{ij} values was assessed with randomization tests (Slatkin and Arter, 1991). A plot of f_{ij} values as a function of increasing distance, for both the observed data and the 95% bootstrap estimate derived by randomization ($n = 5000$ resamplings) was examined for significant autocorrelation values at 5 m distance intervals. Assuming no adaptation to the conditions of local microsites, significant f_{ij} values are interpretable in terms of non-random gene flow via pollen and/or seeds, resulting in genetic structuring due to local processes of isolation by distance (Loiselle *et al.*, 1995a).

The relationships between individual dispersed seeds sampled in seed traps and the focal maternal trees were examined by comparing their multilocus microsatellite profiles. For each seed sampled in a seed trap, we examined the match between the multilocus genotype of the endocarp and the multilocus genotype of the focal tree (from leaf tissue) associated with the seed trap. Because the endocarp tissue in *Prunus* is diploid and maternally derived (Roth, 1977), such multilocus genotypes have to be fully matching alleles in all loci.

Matches between the maternal tree and seed-endocarp multilocus genotypes were found and their significance evaluated using the packages Kinship, version 1.3, and Relatedness, version 5.0.5 (Queller and Goodnight, 1989). The method tests the significance of a hypothesized mother–offspring pedigree relationship between a dispersed seed and an adult tree, based on the identity of the endocarp and leaf genotypes, given r_p and r_m, the probabilities that the individuals share an allele by direct descent from their father or mother, respectively. When comparing endocarp and a maternal tree (leaf) tissue, we used $r_p = 1.0$ and $r_m = 1.0$ (K.F. Goodnight, 2000, personal communication; see also Queller and Goodnight, 1989). The test uses the r values, the population allele frequencies and the multilocus genotypes to calculate the likelihood that the genotype combination could have been produced by the relationship specified ($r_p = 1.0$; $r_m = 1.0$). A randomization test is used to assess the significance of the ratio between this likelihood and the one based on a null hypothesis of no relationship. This comparison allowed identification or exclusion of the focal tree as the maternal tree for each seed and the assignment of the maternal tree from the population. The procedure provided significant ($P < 0.001$) and unambiguous assignments for $n = 78$ seeds (82.11% of the seeds sampled).

Results

Patterns of frugivore foraging and the seed shadow

Frugivorous birds feeding on *P. mahaleb* fruits forage non-randomly after leaving fruiting trees, resulting in an extremely patchy pattern of seed rain. Bird preference for patches covered with vegetation, either mid-height shrubs or low shrubs (Jordano and Schupp, 2000), was significant ($\chi^2 = 21.1$, d.f. = 1, $P < 0.0001$); open microhabitats and pines were avoided. In addition, flight distances to the first perch were very short, with 77.5% to perches within 30 m and most species, except *Turdus viscivorus* and *Turdus merula*, perching within 15 m of the focal tree. This resulted in seed densities under shrubs significantly exceeding those in open microhabitats, with the seed rain beneath pines being intermediate. Aggregation of seeds was particularly extreme in the neighbourhood of *P. mahaleb* trees, as a result of both high frequency of departure flights to other *P. mahaleb* trees after feeding and a trend for departure flights to end in perches < 15 m away from any *P. mahaleb* tree (> 92% exit flights of all species except the two *Turdus* species), irrespective of distance (Jordano and Schupp, 2000).

The resulting seed densities differed significantly among microhabitats ($F = 34.65$, d.f. = 8252, $P < 0.0001$), as expected from the highly non-random foraging by the main frugivores. Seed densities in open microhabitats (1.0 ± 0.4, 0.7 ± 0.6, 1.5 ± 0.6 and 5.8 ± 1.1 seeds m^{-2}, for deep soil, gravelly soil, soil with stones and rocks, respectively; means ± 1 SE) were much lower than those recorded in covered microhabitats (90.9 ± 11.4, 31.5 ± 6.4, 7.7 ± 2.5, 10.5 ± 2.8 and 36.7 ± 6.9 seeds m^{-2}, for *P. mahaleb*, mid-height shrubs, low shrubs, pines with low shrubs and pines, respectively) (Jordano and Schupp, 2000).

Table 20.2 summarizes the results of bird foraging observations and the estimates of specific contributions to the seed rain in different microhabitats. The seed rain to covered microhabitats was typically determined by more than three frugivore species, while seed rain to open microhabitats was determined by one or two species. No single species contributed more than 45% of the seed rain to covered microhabitats, and the seed rain to rocky substrates or beneath pines was determined mainly by *Phoenicurus ochruros* or *T. viscivorus*, respectively (Table 20.2). Because of differences in abundance and visitation rate, some species with low flight frequencies to a particular microhabitat have disproportionately large contributions to the seed rain in those patches. This pattern is illustrated by *T. viscivorus*, with a high contribution to the seed rain beneath *P. mahaleb* or to deep soil patches, despite a low frequency of flights to these microhabitat types. A similar trend occurs for *P. ochruros* to deep soil and rock substrates (Table 20.2).

Seed-addition experiments

The seed-addition treatment had a significant effect on the percentage of plots where at least one seedling emerged ($\chi^2 = 17.9$, $P < 0.0001$): 72.5% of the 'added' subplots recruited at least one seedling, while only 49.3% of the 'control' subplots did so. Among the subplots not recruiting any seedlings (39.1% of the total), 35.2% were seed-addition treatments and 64.8% were controls. The effect of the predator-exclusion treatment was similar; 70.0% of the 'excluded' subplots had at least one seedling emerging vs. 51.9% of the 'control' subplots. The response of seedling emergence to either seed addition or exclusion of post-dispersal seed predators was strongly dependent on the microhabitat type (open or with plant cover) (Fig. 20.2), indicated by a significant microhabitat × addition × exclusion interaction ($F = 8.86$, $P = 0.003$, d.f. = 1312). The effect of seed addition was particularly marked under enclosures beneath shrub cover (Fig. 20.2). When taking into account differences in seedling recruitment among sites, i.e. looking at the proportion of initial seeds that resulted in seedlings, the only significant result was for the exclusion treatment ($F = 26.76$, $P < 0.0001$, d.f. = 1312) and the site × exclusion interaction ($F = 3.83$, $P = 0.048$, d.f. = 1312).

Table 20.2. Exit flight frequencies to different microhabitat types and estimated relative contributions to the seed rain in these microhabitats by avian frugivores visiting *P. mahaleb* trees. For each bird species, figures indicate the percentage of flights to each type of microhabitat and, in parentheses, the estimated percentage of the seed rain in each microhabitat contributed by the species. See Jordano and Schupp (2000) for details.

Species	*Prunus*	Low shrubs	Tall shrubs	Pines*	Deep soil	Rock substrates†
Erithacus rubecula	6.1 (3.7)	43.1 (30.9)	49.3 (19.2)	1.5 (0.2)	0.0 (0.0)	0.0 (0.0)
Phoenicurus ochruros	20.8 (29.4)	4.9 (8.9)	16.8 (16.5)	13.9 (4.2)	1.9 (48.6)	41.7 (86.2)
Sylvia cantillans	28.6 (5.4)	14.3 (3.4)	42.8 (5.6)	14.3 (0.3)	0.0 (0.0)	0.0 (0.0)
Sylvia communis	15.4 (9.8)	34.6 (28.0)	34.6 (15.2)	15.4 (2.2)	0.0 (0.0)	0.0 (0.0)
Sitta europaea	55.6 (4.8)	0.0 (0.0)	11.1 (0.7)	33.3 (0.7)	0.0 (0.0)	0.0 (0.0)
Turdus merula	11.5 (13.5)	19.2 (28.8)	50.0 (40.7)	19.3 (4.6)	0.0 (0.0)	0.0 (0.0)
Turdus viscivorus	6.5 (33.4)	0.0 (0.0)	0.6 (2.1)	85.8 (87.8)	0.6 (51.4)	6.5 (13.8)

*Includes pines with and without low shrubs beneath.
†Includes gravelly soil, rocks with soil and rock substrates.

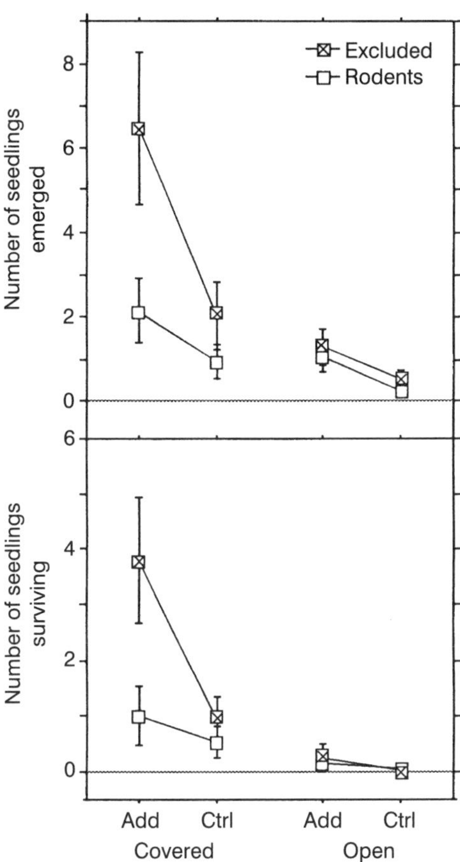

Fig. 20.2. Interaction charts for seed-addition experiments with *P. mahaleb* seeds sown in 'covered' microhabitats (combining sites beneath *P. mahaleb* trees and sites beneath pine trees with low juniper shrubs) and 'open' microhabitats (including sites with grassy cover on deep soil; and sites with gravelly and rocky soil). Shown are the mean number of seedlings emerged (± 1 SE) and mean number of seedlings surviving after the first summer (± 1 SE) in experimental blocks that combined the following treatments: 'control' seed density ('Ctrl') – the median seed density sampled with seed traps – and 'added' seed density ('Add') – the 95 percentile of the median density for each microhabitat type, i.e. approximately a threefold increase in the number of seeds sown. Each block included these treatments within exclusions of post-dispersal seed predators (rodents) ('Excluded') and in plots accessible to them ('Rodents').

DNA extraction and amplification from individual seeds

To check the accuracy of leaf and endocarp tissue comparison, endocarps and embryos of progeny obtained in diallel crosses (P. Jordano, unpublished data) were genotyped and compared to the genotype of their corresponding sire and dam trees. The multilocus genotypes of the seed endocarps were identical to those from leaves of the mother tree, as expected from the anatomical origin of the endocarp tissue, which, in *Prunus* drupes, derives from the carpelar wall (Roth, 1977) and is therefore diploid and maternally derived (Table 20.3; Godoy and Jordano, 2001). Therefore, the endocarp genotype of any *P. mahaleb* dispersed seed can be used to unambiguously identify its source tree. On the other hand, the genotypes of embryos were compatible with those of their sire and dam trees, i.e. for every locus one allele was contributed by the mother and the other by the father. Interestingly, by comparing the genotypes of the embryo and the endocarp of a dispersed seed, the haplotype of the male gamete can be easily and unambiguously inferred and the males with compatible genotypes in the population can be identified as putative fathers. Additional genotyping of the embryo can thus be used to identify the siring tree of dispersed trees with higher exclusion probabilities than possible if the mother were not identified. This approach would allow the concurrent analysis of seed dispersal and pollination and thus seed- and pollen-mediated gene flow (J.A. Godoy and P. Jordano, in preparation).

We were able to unambiguously assign the maternal trees for $n = 78$ seeds (82.1%); the maternal trees for the remaining 17 seeds may have been an unsampled adult tree from the local population or a tree from another population. As far as we know, our sampling of the adult trees in Nava de las Correhuelas was complete. Thus, we attribute this fraction of unassigned seeds (17.9%) to immigrants from other populations.

Local genetic structuring of adult trees

Significant genetic structure was evident in the spatial autocorrelation of the derived estimate of the coancestry coefficient among pairs of adult trees (Fig. 20.3). In particular, there was a significant peak in autocorrelation at the interval of 0–20 m. These results parallel those obtained previously with RAPD markers in the same population (Jordano, 2001); *Prunus* trees with close genetic distance grow close together. The autocorrelation coefficient between the genetic distance matrix derived from RAPD markers and a 'hypothesis' matrix

Table 20.3. Example of polymorphic variation in nine simple sequence repeat (SSR) or microsatellite loci of *P. mahaleb* leaves (maternal tree number 1927) and both endocarp and embryo tissues of two of its seeds. The size of alleles for each of the nine SSR loci is given for the leaf, endocarp and embryo tissue. Alleles are designed by the size (bp) of their products. In all cases the multilocus genotype for the endocarps matches the leaf genotype, as expected from the diploid maternal derivation of the endocarp tissue; however, the genotypes of the two embryos differ from either the maternal tissues of leaves and endocarps, with unmatching alleles shown in bold type. See Godoy and Jordano (2001) for details.

Locus	Leaves	Endocarp no. 1	Endocarp no. 2	Embryo no. 1	Embryo no. 2
UDP96-001	124/124	124/124	124/124	124/124	124/124
pchgms3	179/191	179/191	179/191	**191**/191	179/191
UDP96-018	246/246	246/246	246/246	246/246	246/246
UDP97-403	107/107	107/107	107/107	107/107	107/107
PS12A02	175/185	175/185	175/185	**185**/185	**175**/175
pchcms5	233/235	233/235	233/235	**235**/235	**235**/235
UDP98-406	98/102	98/102	98/102	**102**/102	98/102
UDP97-402	148/152	148/152	148/152	**148**/148	148/152
MS01A05	200/200	200/200	200/200	200/200	200/200

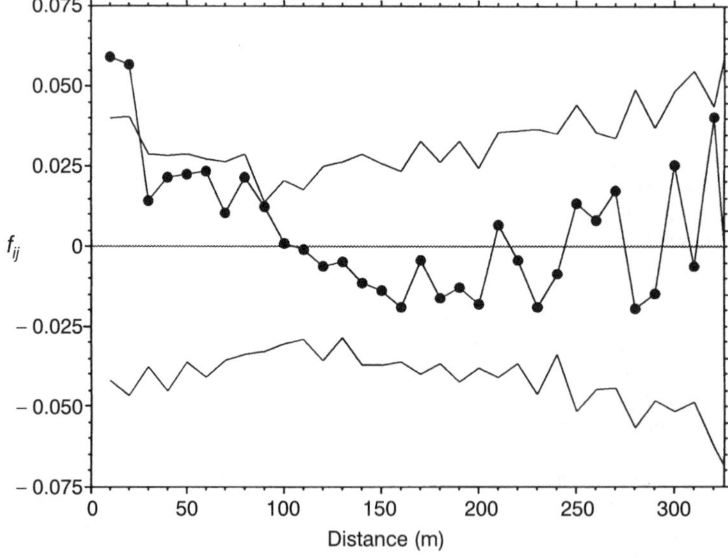

Fig. 20.3. Spatial autocorrelogram for the estimated coancestry values (f_{ij}) between pairs of adult *P. mahaleb* trees (n = 180) within 5 m distance intervals in the Nava de las Correhuelas population. Thin lines represent the bootstrap 95% confidence intervals about the hypothesis of no spatial genetic structure. High f_{ij} values outside the confidence intervals indicate significant positive autocorrelation in genetic similarity among individuals located up to 20 m apart; coancestry values were not significantly different from zero beyond this distance class.

about neighbourhood structure (i.e. trees located in the same patch having distance equal to zero, those growing in different patches having distance equal to one (see Jordano, 2001)) was significant ($r = 0.467$, $P = 0.044$, $n = 10{,}000$ randomizations; Mantel's test (Casgrain and Legendre, 1998)). Both microsatellite and RAPD data show a significant effect of isolation by distance in a pattern that parallels the strong structuring of the seed shadow generated by frugivores.

Maternity assignments for progeny of 'focal' trees

Focal trees differed in the proportion of their own seeds beneath their canopies (Fig. 20.4). For some trees most seeds deposited by frugivores beneath the canopy were their own. This contrasted with other trees in which seeds voided by frugivores were from several maternal trees. On average, 2.8 (range, 1–4) maternal trees contributed seeds beneath a given *P. mahaleb* tree (Table 20.4), a type of 'autodispersal' in which seeds are delivered away from the mother tree but beneath the canopy of conspecifics. When estimated over the entire sample, 70% of assigned seeds were located beneath their maternal plants (Table 20.4). Forty per cent of assigned seeds sampled beneath mid-height shrubs were from nearby focal trees. The proportion of seeds beneath mid-height shrubs and pine trees that were assignable to focal trees decreased dramatically at longer distances (> 10 m away from the focal trees) (Table 20.4). Sampling locations beneath mid-height shrubs had similar numbers of distinct maternal trees contributing seeds, whereas locations under pines had fewer contributing maternal trees at greater distances (Table 20.4). A relatively high number of maternal trees (up to five trees) contributed seeds to areas beneath shrub cover, irrespective of distance to source trees. The number of trees contributing seeds to a particular sampling location was positively correlated with the number of reproductive trees within 15 m radius of the sampling location ($r_s = 0.526$, $P = 0.002$, $n = 19$), an expected result given the markedly restricted flight patterns of birds after feeding in fruiting *P. mahaleb* trees.

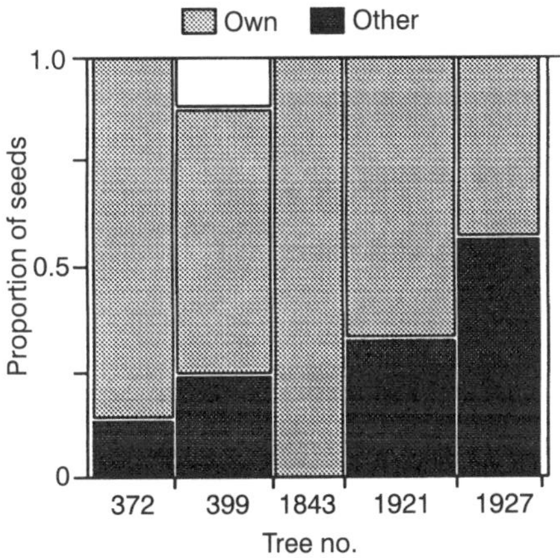

Fig. 20.4. Genetic composition of the dispersed seed shadow in seed traps beneath five 'focal' *P. mahaleb* trees. Seeds are divided into 'own' progeny (light grey) – seeds from the maternal focal tree – and those from 'other' trees (dark grey), mostly in the neighbourhood of the focal tree. The blank area represents unassigned seeds. Seeds were assigned to a particular maternal tree on the basis of the relatedness value estimated from the multilocus microsatellite genotype of the seed endocarp compared with the multilocus microsatellite genotypes of 180 *P. mahaleb* reproductive trees in the population.

Discussion

Seed dispersal by animals is a key process in plant population dynamics, one that subsumes both demographic and genetic effects. Our results show that frugivore activity can severely limit plant population recruitment and strongly influence local genetic structure. The main mechanism driving this process is frugivore foraging behaviour. In particular, the non-random pattern of frugivore movement in heterogeneous landscapes imposes markedly non-random patterns of seed rain among distinct microhabitat types. This, together with the fact that a given frugivore disperses many seed species, makes the seed rain generated by all frugivores highly heterogeneous and aggregated. Thus, most sites in the landscape receive few or no seeds despite copious fruit production and thorough fruit

Table 20.4. Summary of the genetic composition of the seeds sampled in seed traps at different replicate sampling locations for five types of microhabitats. Sampling sites away from *P. mahaleb* were located in relation to each of five 'focal' trees at close (< 10 m) or far (> 10 m) distance. Sites with *Pinus* were all > 10 m away from the focal tree.

Microhabitat type	No. of replicate locations (no. of seeds)*	'Own' progeny†	No. of trees contributing progeny‡	Distance to nearest *Prunus*§	No. of *Prunus* < 15 m‖
Beneath *Prunus*	5 (37)	26 (0.7)	2.8 (1–4)	0 m	2.8 (1–5)
Mid-shrubs, < 10 m	5 (31)	12 (0.4)	2.8 (1–5)	< 10 m	3.8 (3–5)
Mid-shrubs, > 10 m	4 (19)	1 (0.05)	3.0 (1–3)	> 10 m	6.8 (4–9)
Pinus + low shrubs	2 (4)	0 (0.0)	2.0 (1–2)	> 10 m	1.0 (1)
Pinus	3 (4)	0 (0.0)	0.7 (0–1)	> 10 m	0.3 (0–1)

* Number of replicate sampling locations of each microhabitat type and number of seeds genotyped. The number of replicate locations and type of microhabitat sampled for each tree varied depending on the characteristics of the site.
† Number (and proportion) of seeds assigned to the maternal focal tree.
‡ Mean number (range) of distinct maternal trees contributing progeny to the sampling location.
§ Distance category to focal tree.
‖ Mean number (range) of reproductive *P. mahaleb* trees within 15 m radius of seed-sampling location.

removal, a phenomenon that Jordano and Schupp (in prep.) define as dissemination limitation (Table 20.1).

Dissemination limitation is a rather general characteristic of frugivore-generated seed shadows (Wenny and Levey, 1998; Wenny, 2000; see also Schupp *et al.*, this volume). Most studies on this topic have focused on distance-restricted dispersal (e.g. Clark *et al.*, 1999a); consequences for plant population dynamics deserve further study. First, the fact that the diversity of frugivore species contributing seeds to a particular patch in the forest can vary depending on the type of microhabitat has potential consequences for population structure. Secondly, marked peaks and valleys in the landscape pattern of seed distribution, which depend on the particular landscape of the site, indicate that recruitment can be limited by seed dispersal. Potentially, seeds cannot reach microhabitats where establishment probability will be high. Thirdly, animal-created seed shadows can result in marked local genetic structuring of the population, which can then influence gene flow and recruitment.

Most recent analyses of endozoochorous seed dispersal focus on seed rain resulting from all dispersers. Disclosing the unique contributions of each disperser species to a seed shadow can only be completed by combining simultaneous analysis of frugivore activity, post-foraging movements and seed rain patterns (e.g. Wenny and Levey, 1998; Jordano and Schupp, 2000). When this is accomplished, it becomes evident that the seeds contributed to different portions of the seed shadow are delivered by different frugivore species. As a result, population recruitment can be attributable to the activity of only a limited set of species within a diverse frugivore assemblage. Thus these recent studies with Mediterranean and neotropical frugivorous birds have shown that, despite a high diversity of interactions, plant–frugivore mutualisms can be directed by a few key interactions, with disproportionate effects on seed recruitment.

Uneven seed delivery patterns invariably result in extremely heterogeneous seedling recruitment patterns. For both *P. mahaleb* and *Ph. latifolia* (Jordano and Herrara, 1995), most locations in the landscape had very small ratios of seedlings recruited to seeds dispersed. The main factor influencing this limited recruitment depended on microhabitat type. In general, dissemination limitation was higher in open microhabitats, where lack of seed arrival resulted in patches where seedling recruitment was impossible. Post-dispersal factors influencing germination and/or seedling survival can be more limiting in covered microhabitats, where seed delivery by frugivores is frequently high. For example, covered microhabitats for *P. mahaleb* typically show very high seed predation rates but high seedling survival rates (Schupp, 1995; Hulme, 1997). Whenever frugivorous birds are the primary dispersers, we might expect seed delivery to covered microhabitats and avoidance of more open patches, resulting in aggregated seed rain. In covered microhabitats, our seed-addition treatments disproportionately increased both seedling emergence and seedling survival only when seed predators were excluded, suggesting that post-dispersal seed predation, not failure of seeds to arrive, limits recruitment in these microhabitats. In open microhabitats, the addition of seeds resulted in more seedlings initially but not after the first summer drought. This suggests the final establishment of seedlings is strongly limited by adverse abiotic conditions, not by dissemination limitation. The significant three-way interaction of microhabitat type, seed-addition treatment and predator exclusion indicates that the influence of seed dispersal on initial seedling establishment in this system is strongly dependent on microhabitat type. Our central conclusion is that seed-dispersal limitation cannot be seen as a population-wide process unless one considers the summed contributions to recruitment of all patches.

Our genetic analyses demonstrate that frugivores also dramatically influence the spatial pattern of dispersed genotypes. Frugivores generate two types of shadows: seed shadows and genotype shadows. Our analysis of microsatellite loci variation reveals that genotype shadows can be as aggregated as seed shadows. First, the adult population showed a dramatic spatial pattern of genetic structure, with a strong peak in genetic similarity near (< 20 m) conspecifics, consistent with the pattern of seed rain estimated from observations of frugivorous birds (Jordano and Schupp, 2000). This short distance peak was also evident from RAPD analyses (Jordano, 2001) and confirms previous results with other animal-dispersed species (Loiselle *et al.*, 1995a; Aldrich *et al.*, 1998; Schnabel *et al.*, 1998b; Ueno *et al.*, 2000). Strongly distance-limited seed shadows, coupled with spatially aggregated seed delivery, probably result in highly structured genetic diversity within populations of animal-dispersed species. It is interesting that animal-dispersed species exhibit a very high proportion of genetic variation within populations, typically > 70%, as determined by allozyme data or analysis of molecular variance (AMOVA) (Hamrick *et al.*, 1993; Nybom and Bartish, 2000). Thus, despite the fact that individual populations of animal-dispersed species 'capture' most of the genetic diversity found at the regional level, our results show that patches of neighbouring adult trees can be very genetically homogeneous. This pattern is probably attributable to the effects of isolation by distance operating at the within-population level and, in the case of *P. mahaleb*, is associated with the short-distance movements made by frugivores near fruiting trees (Jordano and Schupp, 2000).

A technical advance of our research is the ability to link genotypes of the maternal trees with those of dispersed seeds (also see Godoy and Jordano, 2001). By focusing on 'focal' trees we corroborated a strongly contagious pattern of seed delivery; most seeds of a given plant can be found either beneath its canopy or in the immediate vicinity (in our system, under covered microhabitats within 10–15 m). Seeds from our focal trees did not appear in seed traps located > 15 m away or in microhabitats not frequented by frugivores. Previous analyses of seed shadows have likewise shown very limited dispersal distances and contagious seed delivery (Murray, 1988; Mack, 1995; Levey and Sargent, 2000) but may fail to detect long-distance dispersal events, which can now be tracked with genetic markers. Our analyses of genotypes of dispersed seeds in this population revealed disproportionately frequent short-distance dispersal events (< 5 m), combined with extremely infrequent events of long-distance dispersal (> 250 m) (Godoy and Jordano, 2001). We were able to assign maternity for 82.1% of the seeds sampled and estimate that most of the remaining 17.9% are attributable to long-distance dispersal from other *P. mahaleb* populations in the region.

We found enormous variation among focal trees in the genetic composition of the seeds delivered by frugivores beneath their canopies. Individual fruiting trees not only receive dispersed seeds from their own canopy, but also receive seeds dispersed from conspecific trees. We consider this to be a sort of 'autodispersal'; our data reveal that some trees can act as 'sinks' for seeds from other trees. For example, we found up to 60% of seeds from other trees beneath the canopy of particularly attractive trees, while other fruiting *P. mahaleb* had no seeds from neighbours or, typically, < 20%. These results, combined with our finding that up to five distinct maternal trees can contribute seeds to a single patch, suggest a complex pattern of overlapping seed shadows from different individual trees. Clearly, theoretical models attempting to estimate seed shadows by extrapolating from linear dispersal distances should take into account this intrinsic complexity of animal-generated seed shadows (Clark *et al.*, 1999a,b). This could be accomplished with relative ease for each sampling location by including estimates of the minimum number of maternal trees expected to contribute progeny, given the distribution of distances to adult trees and the type of microhabitat in the sampling location. Our preliminary data (Table 20.4; see also Godoy and Jordano, 2001) indicate that up to five or six trees can contribute progeny beneath covered microhabitats close to a fruiting tree, and up to three trees can contribute seeds to similar habitats at more distant locations from these microhabitats; finally, up to two trees can contribute seeds to less preferred microhabitats.

The use of molecular techniques to assess maternity of dispersed seeds offers new avenues for research in plant–frugivore mutualisms. Direct assessment of dispersal distances, analysis of the genetic diversity and make-up of the seed shadow at a microscale and estimation of both demographic and genetic effects of frugivores that differ in foraging modes are possible by using hypervariable markers, such as microsatellites. We see the most promising approach as the one that combines such data with careful observations of frugivore behaviour and demographic analyses of the stages in plant recruitment.

Conclusion and Perspectives

The activity of frugivores simultaneously influences the number and genetic make-up of seeds dispersed across the landscape. These two components of the seed shadow set the template for plant recruitment. Approaches to plant–frugivore interactions have moved towards integrating the net effects of frugivores on the entire sequence of recruitment stages following seed delivery (Schupp, 1993). Recent developments in molecular hypervariable markers allow unambiguous assessment of paternity and kinship relationships for seeds obtained in studies of seed rain, providing a powerful tool for directly assessing dispersal distances and spatial patterns of dispersal (Ouborg *et al.*, 1999; Godoy and Jordano, 2001). The key issue of whether frugivore activity limits dispersal in plant populations and constrains plant population dynamics can now be rigorously assessed. The natural recruitment cycle of many plant species can be seed limited, especially in situations of dramatic disturbance, such as severe habitat fragmentation. Local extinction of isolated populations in fragments can result not only from demographic bottlenecks originating from failure to disperse seeds, but also from severe genetic bottlenecks. If strongly aggregated seed dispersal of endozoochorous species is typical, then fragmentation will result in spatial isolation of close relatives, leading to severe reduction of genetic diversity. 'Traditional' approaches to the study of plant–frugivore interactions (Snow and Snow, 1988) are essentially blind to such scenarios and should be combined with both demographic and genetic studies to fully understand how frugivores affect plant communities.

Acknowledgements

The overview presented here of mechanisms causing recruitment limitation is based on ideas that have been jointly developed by P. Jordano and E.W. Schupp, and that will be published in full detail later.

We are indebted to Alicia Prieto and Rocío Requerey and especially Manolo Carrión, Juan Luis García-Castaño, Juan Miguel Arroyo, Myriam Márquez, Jesús Rodríguez and Joaquín Muñoz for generous help with field and laboratory work and making this study possible. Help, advice and discussions with Gene Schupp and Janis Boettinger were fundamental during many stages of this project. Birgit Ziegenhagen, Pere Arús and Rémy Petit and especially Graham King and Amy Iezzoni provided advice and useful discussion on DNA extraction protocols, *Prunus* microsatellites and related issues. John Nason allowed use of the statistical package for analysis of spatial patterns of relatedness. Gene Schupp, Mauro Galetti, Carol Horvitz, Luciana Griz and Bette Loiselle and, especially Doug Levey, made helpful suggestions and improvements to earlier versions of the manuscript. We thank Mauro Galetti, Wesley Silva and Doug Levey for making possible the attendance of P.J. at the symposium. The Agencia de Medio Ambiente, Junta de Andalucía, provided generous facilities, which made possible this study in the Sierra de Cazorla and authorized our work there. This work was supported by grants PB96-0857 and 1FD97-0743-C03-01 from the Comisión Interministerial de Ciencia y Tecnología, Ministerio de Educación y Ciencia and the European Commission and also by funds from the Consejería de Educación y Ciencia, Junta de Andalucía.

References

Aldrich, P.R., Hamrick, J.L., Chavarriaga, P. and Kochert, G. (1998) Microsatellite analysis of demographic genetic structure in fragmented populations of the tropical tree *Symphonia globulifera*. *Molecular Ecology* 7, 933–944.

Alvarez-Buylla, E.R., García Barrios, R., Lara Moreno, C. and Martínez Ramos, M. (1996) Demographic and genetic models in conservation biology: applications and perspectives for tropical rain forest tree species. *Annual Review of Ecology and Systematics* 27, 387–421.

Cain, M.L., Milligan, B.G. and Strand, A.E. (2000) Long-distance seed dispersal in plant populations. *American Journal of Botany* 87, 1217–1227.

Carvalho, G.R. (1998) Molecular ecology: origins and approach. *Advances in Molecular Ecology* 306, 1–23.

Casgrain, P. and Legendre, P. (1998) *The R Package for Multivariate and Spatial Analysis*, Version 4.0. University of Montreal, Montreal, Canada. http://alize.ere.umontreal.ca/~casgrain/R/

Chapman, C.A. and Chapman, L.J. (1995) Survival without dispersers: seedling recruitment under parents. *Conservation Biology* 9, 675–678.

Cheung, W.Y., Hubert, N. and Landry, B.S. (1993) A simple and rapid DNA microextraction method for plant, animal, and insect suitable for RAPD and PCR analyses. *PCR Methods and Applications* 3, 69–70.

Cipriani, G., Lot, G., Huang, W.G., Marrazzo, M.T., Peterlunger, E. and Testolin, R. (1999) AC/GT and AG/CT microsatellite repeats in peach [*Prunus persica* (L) Batsch]: isolation, characterisation and cross-species amplification in Prunus. *Theoretical and Applied Genetics* 99, 65–72.

Clark, J.S., Beckage, B., Camill, P., Cleveland, B., HilleRisLambers, J., Lichter, J., McLachlan, J., Mohan, J. and Wycoff, P. (1999a) Interpreting recruitment limitation in forests. *American Journal of Botany* 86, 1–16.

Clark, J.S., Silman, M., Kern, R., Macklin, E. and Hilleris-Lambers, J. (1999b) Seed dispersal near and far: patterns across temperate and tropical forests. *Ecology* 80, 1475–1494.

Connell, J.H. (1971) On the role of natural enemies in preventing competitive exclusion in some marine animals and in rain forest trees. In: Den Boer, P.J. and Gradwell, G. (ed.) *Dynamics of Populations*. Centre for Agricultural Publishing and Documentation (PUDOC), Wageningen, The Netherlands, pp. 298–312.

Dinerstein, E. and Wemmer, C.M. (1988) Fruits rhinoceros eat: dispersal of *Trewia nudiflora* (*Euphorbiaceae*) in lowland Nepal. *Ecology* 69, 1768–1774.

Dow, B.D. and Ashley, M.V. (1996) Microsatellite analysis of seed dispersal and parentage of saplings in bur oak, *Quercus macrocarpa*. *Molecular Ecology* 5, 615–627.

Downey, S.L. and Iezzoni, A.F. (2000) Polymorphic DNA markers in black cherry (*Prunus serotina*) are identified using sequences from sweet cherry, peach, and sour cherry. *American Society of Horticultural Science* 125, 76–80.

Ehrlén, J. and Eriksson, O. (2000) Dispersal limitation and patch occupancy in forest herbs. *Ecology* 81, 1667–1674.

Godoy, J.A. and Jordano, P. (2001) Seed dispersal by animals: exact identification of source trees with endocarp DNA microsatellites. *Molecular Ecology* 10, 2275–2283.

Hamrick, J.L. and Godt, M.J.W. (1997) Effects of life history traits on genetic diversity in plant species. In: Silvertown, J., Franco, M. and Harper, J.L. (eds) *Plant Life Histories. Ecology, Phylogeny and Evolution*. Cambridge University Press, Cambridge, UK, pp. 102–118.

Hamrick, J.L., Murawski, D.A. and Nason, J.D. (1993) The influence of seed dispersal mechanisms on the genetic structure of tropical tree populations. In: Fleming, T.H. and Estrada, A. (eds) *Frugivory and Seed Dispersal: Ecological and Evolutionary Aspects.* Kluwer Academic Publisher, Dordrecht, The Netherlands, pp. 281–297.

Harms, K.E., Wright, S.J., Calderón, O., Hernández, A. and Herre, E.A. (2000) Pervasive density-dependent recruitment enhances seedling diversity in a tropical forest. *Nature* 404, 493–495.

Harper, J.L. (1977) *Population Biology of Plants.* London, UK.

Heywood, J.S. (1991) Spatial analysis of genetic variation in plant populations. *Annual Review of Ecology and Systematics* 22, 335–355.

Hulme, P.E. (1997) Post-dispersal seed predation and the establishment of vertebrate dispersed plants in Mediterranean scrublands. *Oecologia* 111, 91–98.

Izhaki, I., Walton, P.B. and Safriel, U.N. (1991) Seed shadows generated by frugivorous birds in an eastern mediterranean scrub. *Journal of Ecology* 79, 575–590.

Janzen, D.H. (1970) Herbivores and the number of tree species in tropical forests. *American Naturalist* 104, 501–528.

Jordano, P. (1993) Pollination biology of *Prunus mahaleb* L.: deferred consequences of gender variation for fecundity and seed size. *Biological Journal of the Linnean Society* 50, 65–84.

Jordano, P. (2001) Conectando la ecología de la reproducción con el reclutamiento poblacional de plantas leñosas Mediterráneas. In: Zamora, R. and Pugnaire, F. (eds) *Aspectos ecológicos y funcionales de los ecosistemas mediterráneos.* Consejo Superior de Investigaciones Científicas, Madrid, pp. 183–211.

Jordano, P. and Godoy, J.A. (2000) RAPD variation and population genetic structure in *Prunus mahaleb* (*Rosaceae*), an animal-dispersed tree. *Molecular Ecology* 9, 1293–1305.

Jordano, P. and Schupp, E.W. (2000) Determinants of seed dispersal effectiveness: the quantity component in the *Prunus mahaleb*– frugivorous bird interaction. *Ecological Monographs* 70, 591–615.

Julliot, C. (1997) Impact of seed dispersal of red howler monkeys *Alouatta seniculus* on the seedling population in the understorey of tropical rain forest. *Journal of Ecology* 85, 431–440.

Kollmann, J. and Goetze, D. (1997) Notes on seed traps in terrestrial plant communities. *Flora* 192, 1–10.

Levey, D.J. and Sargent, S. (2000) A simple method for tracking vertebrate-dispersed seeds. *Ecology* 81, 267–274.

Loiselle, B., Sork, V.L., Nason, J. and Graham, C. (1995a) Spatial genetic structure of a tropical understory shrub, *Psychotria officinalis* (*Rubiaceae*). *American Journal of Botany* 82, 1420–1425.

Loiselle, B.A., Sork, V.L. and Graham, C. (1995b) Comparison of genetic variation in bird-dispersed shrubs of a tropical wet forest. *Biotropica* 27, 487–494.

Luikart, G. and England, P.R. (1999) Statistical analysis of microsatellite DNA data. *Trends in Ecology and Evolution* 14, 253–256.

Mack, A.L. (1995) Distance and non-randomness of seed dispersal by the dwarf cassowary *Casuarius bennetti. Ecography* 18, 286–295.

Murray, K.G. (1988) Avian seed dispersal of three neotropical gap-dependent plants. *Ecological Monographs* 58, 271–298.

Nakashizuka, T., Iida, S., Masaki, T., Shibata, M. and Tanaka, H. (1995) Evaluating increased fitness through dispersal: a comparative study on tree populations in a temperate forest, Japan. *Ecoscience* 2, 245–251.

Nybom, H. and Bartish, I.V. (2000) Effects of life history traits and sampling strategies on genetic diversity estimates obtained with RAPD markers in plants. *Perspectives in Plant Ecology, Evolution and Systematics* 3, 93–114.

Ouborg, N.J., Piquot, Y. and van Groenendael, J.M. (1999) Population genetics, molecular markers and the study of dispersal in plants. *Journal of Ecology* 87, 551–568.

Queller, D.C. and Goodnight, K.F. (1989) Estimating relatedness using genetic markers. *Evolution* 43, 258–275.

Roth, I. (1977) Fruits of angiosperms. In: Linsbauer, K. (ed.) *Encyclopedia of Plant Anatomy.* Gebrüder Borntraeger, Berlin, pp. 374–381.

Schnabel, A., Beerli, P., Estoup, A. and Hillis, D. (1998a) A guide to software packages for data analysis in molecular ecology. *Advances in Molecular Ecology* 306, 291–296.

Schnabel, A., Nason, J.D. and Hamrick, J.L. (1998b) Understanding the population genetic structure of *Gleditsia triacanthos* L.: seed dispersal and variation in female reproductive success. *Molecular Ecology* 7, 819–832.

Schupp, E.W. (1993) Quantity, quality, and the effectiveness of seed dispersal by animals. In: Fleming, T.H. and Estrada, A. (eds) *Frugivory and Seed Dispersal: Ecological and Evolutionary Aspects.* Kluwer Academic Publishers, Dordrecht, The Netherlands, pp. 15–29.

Schupp, E.W. (1995) Seed–seedling conflicts, habitat choice, and patterns of plant recruitment. *American Journal of Botany* 82, 399–409.

Slatkin, M. and Arter, H.E. (1991) Spatial autocorrelation methods in population genetics. *American Naturalist* 138, 499–517.

Snow, B.K. and Snow, D.W. (1988) *Birds and Berries.* T. & A.D. Poyser, Calton, UK.

Sork, V.L., Huang, S. and Wiener, E. (1993) Macrogeographic and fine-scale genetic structure in a North-American oak species *Quercus rubra. Annales des Sciences Forestières* 50(S), 128–136.

Sork, V.L., Campbell, D.R., Dyer, R., Fernández, J., Nason, J., Petit, R., Smouse, P. and Steinberg, E. (1998) *Gene Flow in Fragmented, Managed, and Continuous Populations.* Research paper No. 3. Available at: <http:www.nceas.ucsb.edu/papers/geneflow>. National Center for Ecological Analysis and Synthesis, Santa Barbara, California.

Sork, V.L., Nason, J., Campbell, D.R. and Fernández, J.F. (1999) Landscape approaches to historical and contemporary gene flow in plants. *Trends in Ecology and Evolution* 14, 219–224.

Sosinski, B., Gannavarapu, M., Hager, L.D., Beck, L.E., King, G.J., Ryder, C.D., Rajapakse, S., Baird, W.V., Ballard, R.E. and Abbott, A.G. (2000) Characterization of microsatellite markers in peach [*Prunus persica* (L.) Batsch]. *Theoretical and Applied Genetics* 101, 421–428.

Streiff, R., Labbe, T., Bacilieri, R., Steinkellner, H., Glossl, J. and Kremer, A. (1998) Within-population genetic structure in *Quercus robur* L. and *Quercus petraea* (Matt.) Liebl. assessed with isozymes and microsatellites. *Molecular Ecology* 7, 317–328.

Ueno, S., Tomaru, N., Yoshimaru, H., Manabe, T. and Yamamoto, S. (2000) Genetic structure of *Camellia japonica* L. in an old-growth evergreen forest, Tsushima, Japan. *Molecular Ecology* 9, 647–656.

Wenny, D.G. (2000) Seed dispersal, seed predation, and seedling recruitment of a neotropical montane tree. *Ecological Monographs* 70, 331–351.

Wenny, D.G. and Levey, D.J. (1998) Directed seed dispersal by bellbirds in a tropical cloud forest. *Proceedings of the National Academy of Sciences USA* 95, 6204–6207.

Wheelwright, N.T. and Orians, G.H. (1982) Seed dispersal by animals: contrasts with pollen dispersal, problems of terminology, and constraints on coevolution. *American Naturalist* 119, 402–413.

21 Contributions of Seed Dispersal and Demography to Recruitment Limitation in a Costa Rican Cloud Forest

K. Greg Murray[1] and J. Mauricio Garcia-C.[1,*]
[1]Department of Biology, Hope College, Holland, MI 49423, USA

Introduction

Explaining the coexistence of ecologically similar plant species in forest communities remains a key mystery of ecology. To solve it, many authors have focused on specializations of different species for particular edaphic conditions (e.g. Clark *et al.*, 1999) or microsites, especially those created by small-scale disturbance (e.g. Canham and Marks, 1985; Brokaw, 1987; Denslow, 1987, and references cited therein; Popma *et al.*, 1988; Webb, 1988; Nakashizuka, 1989; Peterson and Pickett, 1990; Davies *et al.*, 1998). These authors propose that coexistence is facilitated by variation in competitive asymmetries among species along environmental gradients, such that some species have an advantage under some conditions while others are favoured under other conditions. According to this viewpoint, species richness is largely dependent upon environmental heterogeneity and upon adaptations for particular combinations of physical conditions.

More recently, the role of recruitment limitation – the failure of all potentially colonizable sites to receive propagules of all competing species – has received increasing attention in the form of both theoretical treatments (Chesson and Warner, 1981, 1985; Shmida and Ellner, 1984; Pacala and Tilman, 1994; Clark and Ji, 1995; Hurtt and Pacala, 1995) and empirical studies (Streng *et al.*, 1989; Clark *et al.*, 1998; Hubbell *et al.*, 1999; Turnbull *et al.*, 2000, and references cited therein). The theoretical studies predict that a virtually unlimited number of similar species can be supported in a community if seed dispersal is limited to the extent that competitively inferior species sometimes end up in patches that fail by chance to receive seeds from competitive dominants (e.g. Hurtt and Pacala, 1995). Most empirical studies have shown that most sites do indeed fail to receive propagules of all species. The recruitment-limitation hypothesis thus enjoys considerable support.

One weakness of some previous studies is that they examine only one of the above hypotheses, but present evidence consistent with both. Rarely have both biological and stochastic processes been examined in the same study (but see Dalling *et al.*, 1998a). For example, finding that all species fail to reach all sites certainly suggests dispersal limitation, but with

*Present address: Apartado 1179-2100, Guadalupe, San Jose, Costa Rica.

large numbers of species this finding becomes virtually inevitable. The failure of seeds of all species to reach all sites does not preclude the possibility that competitive hierarchies determine success in sites reached by more than one nor that these hierarchies differ at sites with different characteristics. A better test of the recruitment-limitation hypothesis would focus on a particular regeneration guild of plants, i.e. a group within which competition for space is likely to be most intense.

In this chapter we examine the spatial occurrence of seeds of pioneer species, a well-defined guild in cloud forest at Monteverde, Costa Rica. As a group, pioneers are shade-intolerant and have life-history traits characteristic of 'r-strategists' (*sensu* MacArthur and Wilson, 1967): rapid growth, early maturation, high fecundity and well-developed seed dispersal mechanisms (Swaine and Whitmore, 1988). Physiological mechanisms that restrict germination to environmental conditions present in gaps, and those that allow long-term survival in a dormant state are also thought to characterize many pioneer species (Vázquez-Yanes, 1980; Hubbell and Foster, 1987; Murray, 1988); indeed, tropical forest soils contain abundant dormant seeds of pioneer species (e.g. Garwood, 1989, and references cited therein).

To determine whether recruitment limitation might operate among pioneer plants at Monteverde, we first examine levels of spatial heterogeneity in the soil seed bank. We then compare levels of heterogeneity in the seed bank with those in the recent seed rain to determine whether the patchiness of the seed bank is due simply to that produced by seed-dispersers or whether seed predators and pathogens modify spatial patterns among surviving seeds once they reach the soil. We also present experimental data from predator- and pathogen-exclusion experiments to tease apart the relative importance of different sources of mortality in modifying spatial patterns in the seed bank. Finally, because accumulation of seeds in the soil over long periods of time is thought to be a key component of many pioneer species' recruitment strategies, and because accumulation can alter spatial heterogeneity of the seed bank, we also estimate seed longevity of Monteverde pioneer species and compare it with that of pioneers from elsewhere.

Study Site and Methods

Study site

Data were collected in the Monteverde Cloud Forest Reserve, hereafter referred to as MCFR, between approximately 1500 and 1800 m in the Cordillera de Tilarán of north-western Costa Rica (10° 18′ N, 84° 48′ W). The Monteverde area lies on a gently sloping plateau, straddling the continental divide. Local climate and vegetation are strongly influenced by the prevailing north-east trade winds. Blowing mist, especially in forests near the divide, prevents serious water stress even during the dry season (January–early May). Forest types in the area range from lower montane rain forest (Holdridge life zone classification system (Holdridge, 1967)) near the continental divide, through lower montane wet forest to lower montane moist forest and premontane wet forest along the lower edge of the plateau. Clark *et al.* (2000) and Haber (2000) provide extensive treatments of the physical environment and vegetation of the area, respectively.

Sampling for the study was concentrated along five transects, each 500 m long, through pristine lower montane wet forest and lower montane rain forest.

Seed rain measurements

To determine the density of seed input at the leaf litter surface, we placed seed traps at 50 m intervals along each of the five transects (55 traps total) in March 1993 and collected seeds at monthly intervals until July 1996 (42 months total). Seed traps consisted of 50 cm × 50 cm polyvinyl chloride (PVC) tubing (*c.* 20 mm diameter) frames with fibreglass window-screen attached at all edges and hanging down approximately 30 cm at the frame centre. The depression thus formed prevented large seeds from bouncing out of the traps. Over the fibreglass screen on each trap we placed a polyester fabric liner of sufficiently tight weave to catch even the smallest seeds dispersed by animals at Monteverde (< 1 mm). Finally, all traps were elevated approximately 50 cm above the soil surface on legs of PVC tubing to deter seed predation by rodents, and the legs were treated with Tanglefoot® or automotive grease to deter arthropods from crawling into the traps.

Fabric liners were collected monthly from all traps, transported to the laboratory in separate plastic bags and dried (if needed) in a 40°C oven to facilitate separation of seeds from debris. All debris was sorted carefully under a dissecting microscope and all seeds recovered were preserved in 70% ethanol. Identifications were made using a reference collection for the area.

Standing crops of seeds in forest soils

Soil cores, each 25 cm × 25 cm × 10 cm deep, were collected (without overlying leaf litter) each March and July from 1993 to 1996 at six equally spaced sites along each of the five transects. Cores were collected within 1 m of the seed traps described above. Samples were spread approximately 3 cm deep in plastic trays in a greenhouse and watered regularly to stimulate germination. Seedlings were collected from each sample for a total of 6 months, as soon as they were large enough to be identified by comparison with a reference collection. The soil in each tray was turned every 2 months to stimulate germination of any remaining seeds.

Seed removal from soil

To understand the roles played by seed predators and pathogens in the demography of pioneer plants, we followed the fates of marked seeds under field conditions over a 3-year period. We concentrated on a subset of the most common pioneers at Monteverde: *Cecropia polyphlebia, Urera elata, Guettarda poasana, Phytolacca rivinoides, Bocconia frutescens* and *Witheringia meiantha.* These species comprise a diverse group in every way – in growth form, in seed size and in taxonomic affinity (see Table 21.2). They are also important components of the seed bank; together, they comprised over half of the total seed bank in each of the six sampling periods (K.G. Murray and J.M. Garcia-C., unpublished data).

We obtained seeds from ripe fruits of each species in and around the MCFR from March to June 1993. Seeds were removed from fruit pulp by hand, washed, air-dried and stored in a drying oven at approximately 25°C until needed. Before placement in the field, seeds were lightly misted with fluorescent orange spray paint (Rustoleum™) and allowed to air-dry for 1 day. The fluorescent paint rendered seeds very easy to relocate with ultraviolet light, even when mixed with soil. Pilot experiments with marked and unmarked seeds revealed no differences in either nocturnal or diurnal removal rates, at least over 24 h periods (K.G. Murray, unpublished data). Square grids for removal experiments were located at three sites along each of the five transects (50, 250 and 450 m, regardless of site characteristics). Seed stations on each of the grids were located 50 cm apart on a 2.0 m × 2.0 m grid (25 stations per grid), and each station was marked at the centre with a PVC tubing stake approx. 2.2 cm diameter × 25 cm long. At each station, we buried ten *U. elata*, ten *C. polyphlebia*, ten *W. meiantha*, eight *B. frutescens*, ten *P. rivinoides* and four *G. poasana* seeds within 3 cm of the PVC stake on 15 July 1993. All seeds were buried approximately 1 cm deep.

We recovered two soil samples from each of the grids after 1, 2, 4, 8, 12, 24 and 36 months. Samples were collected from randomly chosen stations on each grid with a soil extractor (15 cm diameter × 10 cm deep) centred on the marking PVC tube and were transported to the laboratory in plastic bags. Samples were then dried in an oven at approximately 40°C until all clumps of soil could be broken apart and easily sorted under ultraviolet light. Marked seeds were identified to species and counted under a dissecting microscope. Seed fragments were noted but not included in counts of surviving seeds. We did not assess seeds from these experiments for viability – merely for remaining intact while exposed to macroscopic predators. Viability of seeds protected from such predators was assessed in pathogen-exposure and inherent viability experiments, described below.

Pathogen-exposure experiments

We assessed effects of soil microorganisms on seeds protected from macroscopic predators by placing seeds (collected at the same time as those used in experiments described above) in mesh bags of mosquito netting, layered within fibreglass window-screen at the litter–soil interface. Bags were placed within sheet metal exclosures (25 cm × 25 cm) covered with hardware cloth (1 cm mesh). Approximately 10 cm of each exclosure extended below ground and approximately 20 cm extended above ground. Seeds were thus protected from rodents and large arthropods, but exposed to fungi and bacteria, as well as to very small invertebrates, such as nematodes. Exclosures were placed at two sites along each of the transects. In each of ten mesh bags per site we placed three *B. frutescens* seeds and ten seeds each of *U. elata, C. polyphlebia, W. meiantha, P. rivinoides* and *G. poasana.* One bag was recovered from each site after 8, 18, 24 and 36 months and the seeds placed on top of filter-paper in 10 cm Petri dishes under fluorescent grow lights at approximately 30°C. The number of seeds germinating was scored every other day for 5 weeks. This period of time was sufficient to detect most germination; few seeds of these species germinated after 4 weeks. The initial viability of seeds used in the experiment was assessed in the same way, using 50 seeds of each species.

'Inherent viability' experiments

We assessed the survival rates of seeds protected from both macroscopic predators and soil microorganisms by placing seeds (collected at the same time as seeds used in other experiments) in 1.5 ml centrifuge tubes without caps or soil and then plugging the tube opening with loosely wadded nylon mosquito netting. To protect seeds from macroscopic predators and soil microbes while maintaining exposure to ambient temperature, humidity and gas concentrations, we buried the tubes upside down approximately 10 cm deep at one site on each of two transects. This treatment may not have provided perfect protection, but we are confident that it at least impeded predators and pathogens. Fifty *U. elata,* 50 *C. polyphlebia,* 50 *P. rivinoides,* 50 *W. meiantha,* 30 *B. frutescens* or 15 *G. poasana* seeds were placed into each tube, one tube per species. We recovered one tube of seeds of each species from each site after 6, 12, 24 and 36 months and assessed viability as in the pathogen-exposure experiments (see above). Because seeds used in these experiments were from the same batch as that used in pathogen-exposure experiments, we used the same estimates of initial viability.

Data analyses

All univariate statistical analyses reported here were performed with SYSTAT 8.0 (SPSS Inc., 233 South Wacker Drive, Chicago, IL 60606-6307, USA). Ordinations of species composition (detrended correspondence analyses) were performed with DECORANA (Microcomputer Power, 111 Clover Lane, Ithaca, NY 14850, USA).

Results

Monteverde soils contain high but variable densities of dormant seeds, as do most tropical soils. Overall densities (i.e. all species, including those with seasonal and transient seed banks (*sensu* Garwood, 1989) as well as those with persistent ones) ranged from 3000 to over 6000 m^{-2}; the typical 25 cm × 25 cm sample contained 20–25 species (Table 21.1).

Spatial heterogeneity of Monteverde pioneer seed rain and seed banks

Community-level patterns

To evaluate spatial heterogeneity in species composition of pioneers, we performed ordinations (detrended correspondence analyses (Gauch, 1982)) on the seed-bank data from the March 1995 collection and the previous year's seed rain at the same 30 sites. At least 35–45 pioneer species occurred in our seed-trap and/or soil samples, but because we could not distinguish between the seedlings of some of them, the analyses below include data from just the 23 species that we could recognize unambiguously as both seeds and seedlings (Table 21.2). Because it remains unclear which approach is more appropriate in the context of recruitment limitation, we performed the analysis twice: once using the actual numbers of individuals of each species occurring in each sample and again using presence/absence data only. We found that the species composition of both the seed bank and seed rain are indeed spatially heterogeneous, as we expected (Fig. 21.1). Furthermore, we found that the degree of spatial patchiness is similar in the seed rain and seed bank; the points for the 30 seed-bank samples in Fig. 21.1 are perhaps slightly more tightly clustered than are those for the seed rain at the same sites, but not markedly so. The seed bank is thus only slightly (if at all) less patchy than the seed rain,

Table 21.1. Summary statistics on standing crops of seeds in Monteverde soils. Mean species richness based on 30 625 $cm^2 \times 10$ cm deep samples per collection date; density adjusted to m^{-2} basis. Both include species with transient as well as persistent seed banks, i.e. they are not restricted to just the selected pioneer species in Table 21.2.

Collection date	Seeds m^{-2}	Species richness
July 1993	3049 ± 686.7	24.9 ± 1.39
March 1994	3544 ± 371.2	25.6 ± 1.11
July 1994	3861 ± 681.0	21.1 ± 1.18
March 1995	5506 ± 631.0	24.6 ± 0.88
July 1995	4819 ± 713.4	23.5 ± 0.99
March 1996	6496 ± 1297.1	25.3 ± 1.02

Table 21.2. Pioneer species included in subsequent analyses. To be included, a species had to be recognizable from both the seed and young seedling. The six-letter codes are those used in Figs 21.2, 21.3 and 21.4. Seed sizes are given as longest dimension × shortest dimension.

Family	Species	Code	Approximate seed size (mm)
Actinidiaceae	*Saurauia* spp.	Sauspp	1.1 × 1.1
Asteraceae	*Clibadium* spp.	Clispp	2.0 × 1.0
	Neomirandea angularis	Neoang	2.8 × 0.5
Caprifoliaceae	*Viburnum venustum*	Vibven	4.7 × 3.0
Cecropiaceae	*Cecropia polyphlebia*	Cecpol	3.0 × 0.8
Euphorbiaceae	*Sapium* spp.	Sapspp	4.5 × 2.5
	Tetrorchidium costaricensis	Tetcos	6.0 × 3.0
Heliconiaceae	*Heliconia tortuosa*	Heltor	9.0 × 4.5
Malvaceae	*Hampea appendiculata*	Hamapp	9.5 × 6.0
Myrsinaceae	*Myrsine coriacea*	Myrcor	3.5 × 2.6
Papaveraceae	*Bocconia frutescens*	Bocfru	4.8 × 2.8
Phytolaccaceae	*Phytolacca rivinoides*	Phyriv	2.5 × 1.0
Piperaceae	*Pothomorphe umbellatum*	Potumb	0.8 × 0.5
Poaceae	*Lasiacis nigra*	Lasnig	3.1 × 1.8
Rubiaceae	*Gonzalagunia rosea*	Gonros	1.5 × 1.3
	Guettarda poasana	Guepoa	7.0 × 5.0
Solanaceae	*Lycianthes multiflora*	Lycmul	2.0 × 0.4
	Solanum aphyodendron	Solaph	2.0 × 0.4
	Solanum chrysotrichum	Solchr	2.0 × 0.4
	Solanum siparunoides	Solsip	4.5 × 1.0
Tiliaceae	*Heliocarpus americanus*	Helame	1.2 × 1.0
Ulmaceae	*Trema micrantha*	Tremic	1.9 × 1.7
Urticaceae	*Urera elata*	Ureela	0.9 × 0.9

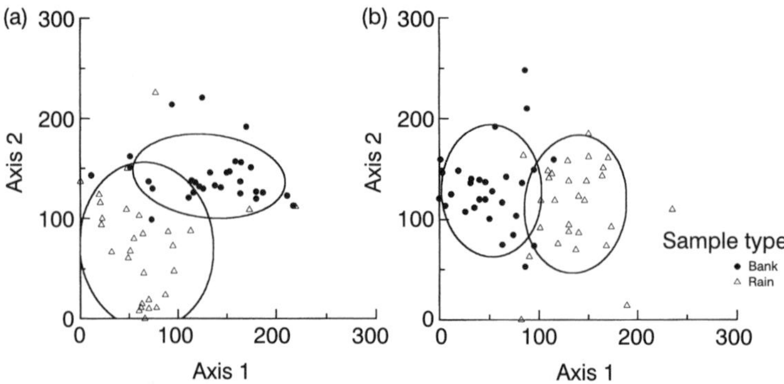

Fig. 21.1. Detrended correspondence analyses for March 1995 seed-bank and previous year's (April 1994–March 1995) seed-rain samples at 30 sites. (a) Based upon all numerical data on 23 selected pioneer species. (b) Based upon presence/absence data only for the 23 pioneer species. Locations of seed-rain and seed-bank samples in ordination space are determined by their species composition, so that symbols closer to one another reflect greater similarity. Ellipses are bivariate confidence ellipses, containing 70% of the data points. Whether represented by presence/absence or all numerical data, species composition in the seed bank is distinct from that in the previous year's seed rain.

despite the fact that seeds of some species may accumulate in the soil for many years.

Interestingly, we found that the species composition of the seed rain and seed bank differed substantially, regardless of whether we used presence/absence or all numerical data – the distributions of points for the seed-rain and seed-bank samples barely overlapped in ordination space (Fig. 21.1). Clearly, then, the species composition of the seed bank is not simply a function of the species composition in the seed rain; post-dispersal factors (e.g. seed demography) also play a role.

Within-species patterns

Because spatial heterogeneity at the community level probably reflects different dispersal and demographic patterns among species, we also examined spatial heterogeneity within species, using two different estimators. The first is simply the mean frequency of occurrence of each pioneer species in the seed bank, defined as the number of sites out of 30 with at least one germinating seed, averaged over the six sampling dates. For the seed rain, the frequency of occurrence was defined as the number of traps receiving at least one seed of the species over the 3-year period from August 1993 to July 1996. For a second estimator of spatial heterogeneity in the seed bank, we computed the variance : mean ratio for each species among the 30 sites and then averaged this ratio over the six sampling dates. For the seed rain, we computed the variance : mean ratio for total seeds of each species deposited in the 30 traps over the 3-year study period. Random occurrence among the 30 sites would follow a Poisson distribution and give a variance : mean ratio of 1.0. Values less than 1.0 indicate a regular dispersion pattern and values greater than 1.0 indicate clumping.

We present both presence/absence and variance : mean ratio data because we do not know which is the more appropriate estimator of spatial heterogeneity for pioneer seedlings competing for space in recent tree-fall gaps. On the one hand, if the surviving seedlings in recently created gaps are simply random subsets of seeds present in the seed bank (the 'lottery' model (Chesson and Warner, 1981, 1985)), presenting all numerical data and representing heterogeneity as a variance : mean ratio is most appropriate. However, if strict, transitive competitive hierarchies exist among pioneer species, then even a single seedling of a dominant species may succeed in excluding large numbers of seedlings of other species. In this case, presence/absence data may be a more appropriate estimator (Turnbull *et al.*, 2000).

An additional caution is in order here as well. Spatial dispersion patterns of the seed rain and seed bank are likely to be scale-dependent, just as they are for most organisms.

Sampled with seed traps and soil cores, as in this study, heterogeneity is likely to decrease with increasing surface area sampled. And, because our seed traps sampled four times the surface area sampled by our soil cores, our estimates of spatial heterogeneity are probably biased towards higher values in the seed bank. This bias is minimized by using presence/absence data to estimate patchiness, but is not eliminated entirely.

Regardless of which metric we use to represent it, spatial patterns varied widely among the 23 pioneer species. Some species, e.g. *U. elata* and *C. polyphlebia*, were present in the soil at all sites in every sampling period, and their frequency of occurrence seems largely due to their ubiquity in the seed rain: virtually all seed traps received seeds of both species during the study (Fig. 21.2b). Other species, e.g. *Clibadium* sp., were very infrequent in the seed bank, despite the fact that many sites received seed input. Still others, e.g. *B. frutescens*, *G. poasana* and *P. rivinoides*, were far more frequent in the seed bank than in the seed rain. The median frequency of the 23 species in both the seed rain and seed bank was 23%, or seven of 30 sites.

Using variance : mean ratio as a heterogeneity measure, virtually all of the pioneers show moderately to highly clumped patterns in both the seed bank and seed rain (Fig. 21.2a). Note that the *y* axis of Fig. 21.2a is on a log scale to allow one to visualize all the species on the same graph. For half of the pioneer species, the seed rain was substantially more patchy than the seed bank, often by more than an order of magnitude. For the other half, the seed bank was somewhat more patchy than the seed rain. The great spatial heterogeneity that characterizes the seed banks of most species is thus due in part to a patchy seed rain, but other factors (e.g. variation in post-dispersal seed predation and pathogen attack) are also at work, increasing patchiness in some species while decreasing it in others.

Seed longevity and its consequences for recruitment limitation

The longevity of dormant pioneer seeds in the soil can affect not only total seed densities but also the spatial heterogeneity of the seed bank, if longevity varies among species or among sites. We estimated the longevity of pioneer seeds in the soil by comparing the mean density of viable seeds in the soil with the mean annual seed rain at the same sites. Among the 23 focal pioneer species, we find at least two distinct patterns. *Cecropia polyphlebia* and *U. elata*, the two most common species in the seed bank at Monteverde, have about the same density of seeds in the top 10 cm of soil as in the annual seed rain (Fig. 21.3). This finding implies that these species' seed banks turn over rapidly and that colonization of gaps is from a short-term seed bank or even from seeds arriving after a gap is formed. The second pattern is more common and is best illustrated by species like *P. rivinoides* and *G. poasana*, in which the soil seed bank contains many years' worth of seed rain. In the most extreme cases (*B. frutescens*, *Heliconia tortuosa* and *Trema micrantha*), seeds were common in the soil but never occurred in over 3 years of seed rain at the same 30 sites. The opposite extreme – the annual seed rain greatly exceeding the seed bank – occurs only rarely among pioneers at Monteverde. Only the herb *Pothomorphe umbellatum* and the common tree *Hampea appendiculata* seem to lack the dormant-seed strategy completely.

To determine whether long-term accumulation in the soil tends to reduce the spatial heterogeneity of the seed bank, we computed a seed-bank: seed-rain heterogeneity index for each species by dividing the variance : mean ratios in Fig. 21.2a for the seed bank by those for the seed rain. Values greater than 1.0 indicate greater spatial heterogeneity in the seed bank than in the seed rain, while values less than 1.0 indicate greater heterogeneity in the seed rain. We plotted the index for each species against its density in the seed bank relative to annual seed rain (from Fig. 21.3). Species with higher bank : rain ratios are those that accumulate over long periods of time, with a value of 1.0 indicating an average seed-bank density equal to 1 year's accumulation of seed rain.

Changes in the spatial distributions of seeds in the seed bank and seed rain were very different for species that accumulate in the soil compared with those that turn over more

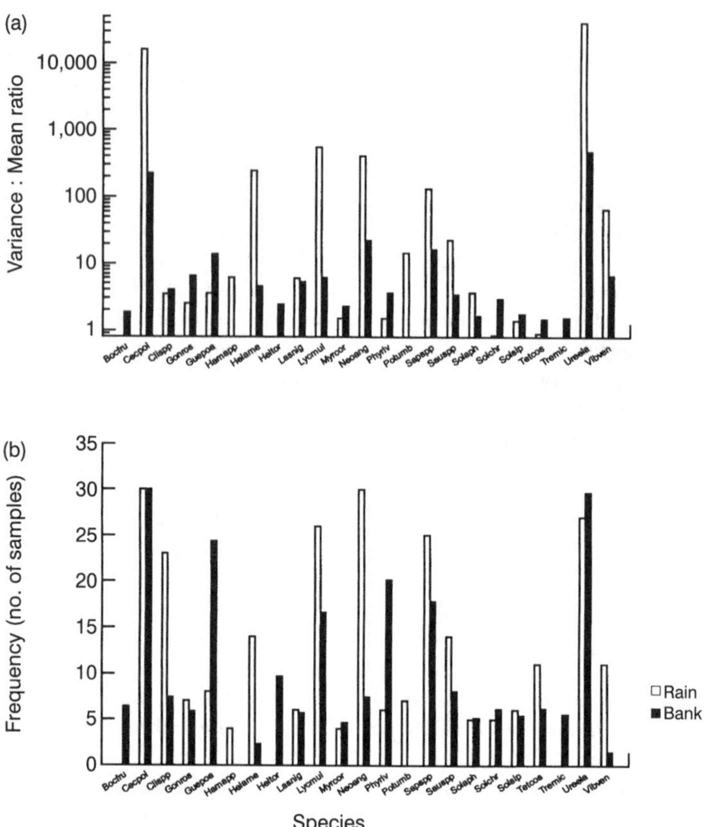

Fig. 21.2. Spatial heterogeneity of Monteverde pioneer seed banks and seed rain. For codes, see Table 21.2. (a) Average variance : mean ratios of selected pioneer species' seeds in the soil and 40-month cumulative seed rain. For the seed bank, values represent the average variance : mean ratio of seeds among 30 sites over the six sampling dates from July 1993 to March 1996. Seed-rain values represent the variance : mean ratio of seeds deposited in seed traps at the same 30 sites over the period from April 1993 to July 1996. Note the log scale for the *y* axis. (b) Average frequency of selected pioneer species' seeds in the soil and seed rain. For the seed bank, values for each species represent the mean number of sites (out of 30 sampled) receiving at least one seed over the six sampling dates. Seed-rain values are the number of sites receiving at least one seed of that species over the period April 1993–July 1996. All 23 species were patchily distributed in both the seed rain and the seed bank. Heterogeneity was higher in the seed rain for about half of the species, but higher in the seed bank for the other half.

rapidly (Fig. 21.4). Species that accumulate in the soil for long periods have high bank : rain patchiness indices, indicating that their distributions are far more heterogeneous in the seed bank than in the seed rain. In contrast, species with short lifespans in the soil display the opposite pattern: their distributions in the soil are far more uniform than in the recent seed rain. Thus, the establishment of long-term seed banks appears to increase seed-bank patchiness to a significant degree, while short residence times in the soil are accompanied by decreasing patchiness.

The role of seed predators and pathogens

Mortality rates attributable to different causes varied widely among the six focal species. Detailed results will be presented elsewhere, but data from three of the species demonstrate most of the range of variation. When

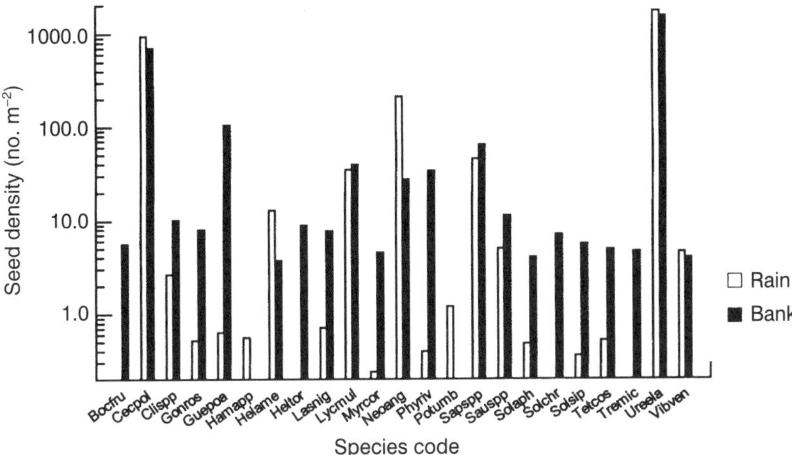

Fig. 21.3. Mean annual seed-rain vs. seed-bank density for selected pioneer species. For codes, see Table 21.2. Note that the *y* axis is on a log scale. While some species lack a seed bank entirely and others maintain seed banks similar in size to their annual seed rain, most species accumulate seed banks equal to at least several times their annual seed rain.

U. elata seeds were protected from predators and pathogens, no substantial decrease in viability of seeds was apparent over at least 3 years (Fig. 21.5). Seeds exposed to predators are removed rapidly, however, and those exposed to pathogens lose viability rapidly, with the result that *Urera* maintains a soil seed bank that is only about as large as the annual seed rain (Fig. 21.3). In *C. polyphlebia*, animals removed seeds moderately rapidly, but we see little evidence for pathogen attack because the seeds lose viability very rapidly, even when protected from both threats (Fig. 21.5). As a result, *Cecropia* also maintains a soil seed bank of about the same size as the annual seed rain (Fig. 21.3), but for a different reason. Seeds of *P. rivinoides* have high survival in the soil, whether protected from predators and pathogens or not (Fig. 21.5). The result is that *Phytolacca* maintains a soil seed bank approximately two orders of magnitude larger than the annual seed rain (Fig. 21.3).

Discussion

Monteverde soils contain dense, species-rich assemblages of dormant seeds. Densities averaged 3000–6500 seeds m^{-2} in the top 10 cm of soil (Table 21.1), which exceeds all but a few estimates reported for tropical forest soils in studies reviewed by Garwood (1989). A few estimates for tropical soils are higher (e.g. Butler and Chazdon, 1998; Dupuy and Chazdon, 1998), but these are for second-growth, rather than primary, forest sites. Within-sample species richness was also high (typically 24–25 species per 25 cm × 25 cm × 10 cm deep sample), such that all gaps can be assumed to be colonized by a dense and diverse array of pioneer plant species (Table 21.1). And, if most colonization of tree-fall gaps is from buried dormant seeds rather than those that arrive after gap formation (e.g. Putz and Appanah, 1987; Young *et al.*, 1987; Lawton and Putz, 1988; Murray, 1988), competition (both intra- and interspecific) for space in recent tree-fall gaps is probably intense.

Spatial heterogeneity of Monteverde seed banks and seed rain

Despite the high density and species richness of Monteverde seed banks, high spatial heterogeneity in species composition was the rule. Of the 23 pioneer species that we could reliably recognize at both the seed and seedling stage, the median frequency of occurrence was just six of 30 soil samples (20%) collected

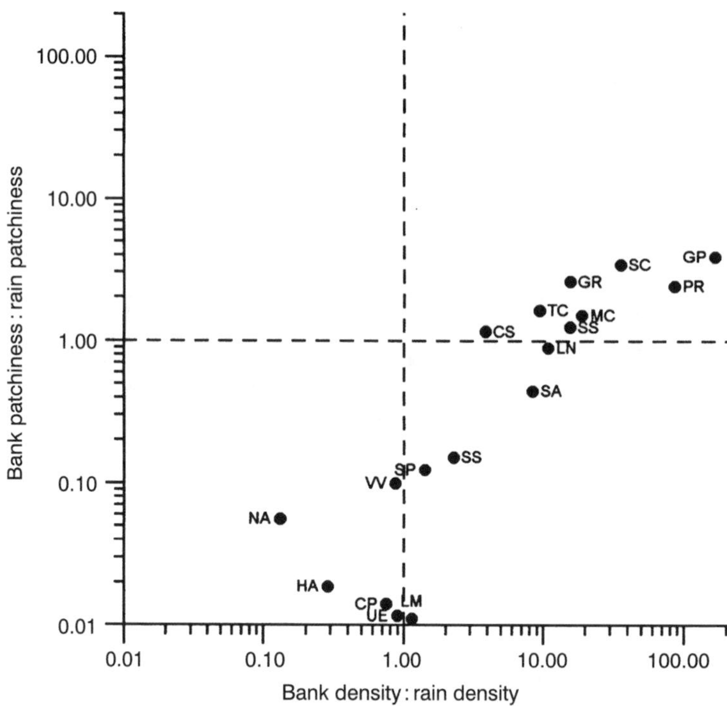

Fig. 21.4. Relative spatial heterogeneity of selected pioneer species' seed banks and seed rain vs. seed-bank : seed-rain ratio. Patchiness of seed bank and seed rain consists of variance : mean ratios. Points are labelled by first letters of genus and species (see Table 22.2), except for *Clibadium* spp. (CS), *Sapium* spp. (SP) and *Saurauia* spp. (SS). *Trema micrantha*, *Heliconia tortuosa* and *Bocconia frutescens* are not included because they never occurred in the seed rain at the 30 sites sampled, and *Hampea appendiculata* and *Pothomorphe umbellatum* are not included because they did not occur in the seed bank at those sites. Note that both axes are on log scales. Species that accumulate in the soil over long periods tend to become more patchily distributed than in the seed rain, while those with small seed banks become less patchy.

on each of the six sampling dates spread over 3 years (Fig. 21.2b). Only two species (*C. polyphlebia* and *U. elata*) were predictably present in virtually all soil samples. Heterogeneity as estimated by the variance : mean ratio was also high for nearly all pioneer species; the median ratio for numbers of seeds among the 30 sites sampled on each sampling date was approximately 3.0 and some species had ratios as high as 300 (Fig. 21.2a). As a result of this high degree of spatial heterogeneity, nearly all pioneer species at Monteverde can be considered recruitment-limited at the seed-bank stage: that is, most species are unlikely to be present in the soil (and hence unlikely to colonize a new tree-fall gap) at any particular location. *Cecropia polyphlebia* and *U. elata* are the only two competitors that a newly germinated seedling will predictably encounter in a recent tree-fall gap. High spatial patchiness of particular species has also been found by Butler and Chazdon (1998), Dalling and Denslow (1998) and Dalling *et al.* (1998b). Working with a group of 24 pioneer species on the 50 ha forest plot on Barro Colorado Island in Panama, Dalling *et al.* (1998a) also found evidence of patchy spatial distributions of young seedlings among gaps.

Comparison of heterogeneity measures in the seed bank with those for the seed rain over the same period clearly shows that much of the patchiness of the seed bank is produced at the time of seed deposition (Fig. 21.2). Median frequency of occurrence of the 23 selected pioneer species in the seed rain over the entire 3-year study was just seven of 30 sites (23%),

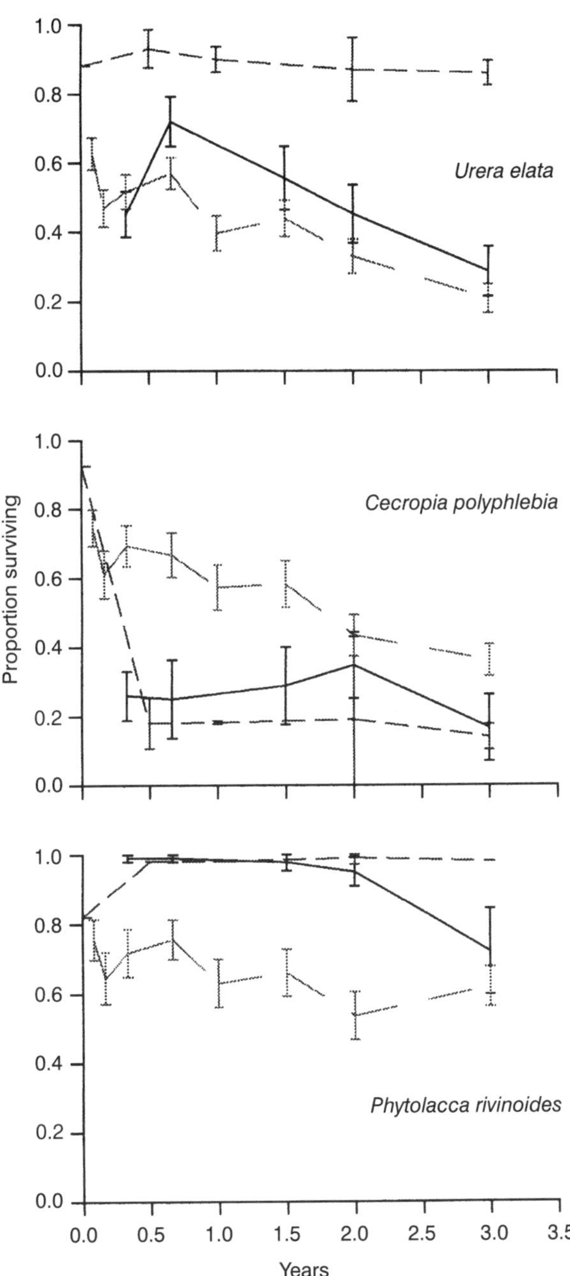

Fig. 21.5. Survivorship of representative pioneer species' seeds in the soil at Monteverde. Solid lines represent seeds exposed to both pathogens and all seed predators. Long-dashed lines represent seeds protected from macroscopic predators (see methods) but exposed to soil pathogens. Short-dashed lines represent seeds protected from both predators and pathogens. The mortality of seeds in this treatment thus probably represents limits to viability inherent in the species. Error bars represent 1 SEM, and $n = 15$ for all means. Even among these three species, a wide range of vulnerability to seed predators and pathogens is apparent.

and the median variance : mean ratio was 3.7. Clearly, the spatial patterns produced by seed-dispersers are important in determining membership in the assemblage of colonists in new gaps. Studies at other sites have found similar levels of patchiness in the seed rain (e.g. Brokaw, 1986; Martínez-Ramos and Alvarez-Buylla, 1986). On Barro Colorado Island in Panama, for example, Hubbell *et al.* (1999) showed that most species (including shade-tolerant and pioneer species) failed to reach 90% of 200 seed traps over a 10-year period.

The role of seed predators and pathogens

The patchy seed distributions produced by dispersers were modified by differences among species in post-dispersal survival. Regardless of whether we use variance : mean ratio or frequency of occurrence as a measure of spatial heterogeneity, the comparison of seed-bank with seed-rain heterogeneity in Fig. 21.2 shows that, for roughly half of the 23 selected pioneer species, the seed bank is more patchy than the seed rain (e.g. *G. poasana, P. rivinoides*). For the other half, the opposite is true (e.g. *C. polyphlebia, U. elata, Lycianthes multiflora*). One result of the differences in seed survival among species is the marked difference in species composition between the seed bank and the previous year's seed rain, shown in Fig. 21.1. Clearly, then, post-dispersal processes are important determinants of species composition in the seed bank, and thus probably influence species composition in newly created treefall gaps. We are aware of no other studies that compare such patterns to assess the importance of post-dispersal survival for species composition, but we suspect that these patterns will be found to be general.

Both macroscopic predators and pathogens, as well as properties inherent in the plant species themselves, generate the differences in survivorship that affect seed longevity. Among the three species for which data are presented here, important differences in sources of mortality are evident: *U. elata* appears to be attacked by both predators and pathogens, *C. polyphlebia* suffers mortality through rodents or invertebrates but not significantly through pathogens and *P. rivinoides* is resistant to both (Fig. 21.5). *Cecropia polyphlebia* loses viability rapidly, even in the absence of both external sources of mortality, suggesting that it lacks the physiological capacity to remain dormant for long periods. In general, the degree to which seeds accumulate in the soil (Fig. 21.4) is consistent with the patterns of survivorship shown in Fig. 21.5. *Urera* and *Cecropia* suffered heavy mortality in experiments and maintain seed banks equal to approximately 1 year of seed rain. In contrast, *Phytolacca* suffered little mortality in experiments and maintains a seed bank equivalent to about 85 years of seed rain.

Few other seed-bank studies provide comparable data on sources of mortality for other pioneer species. As noted by Alvarez-Buylla and Martínez-Ramos (1990), most direct studies of seed longevity rely on methodologies that protect seeds from macroscopic predators, soil microbes or both. In contrast, our methods allowed us to distinguish between the different sources of mortality and have the added advantage of being conducted in virtually undisturbed soil. Detailed studies of the demography of *Cecropia obtusifolia* in Mexico by Alvarez-Buylla and her colleagues also distinguished among sources of mortality and found that both removal by predators and inherent loss of viability were rapid. Turnover times for seeds in the soil were just 1.02–1.07 years (Alvarez-Buylla and Martínez-Ramos, 1990); the large seed bank of this species at their site is maintained only by copious seed production from numerous individuals. Our data suggest the same for *C. polyphlebia.* Dalling *et al.* (1998b) found rapid loss of viability of *Cecropia insignis* and *Miconia argentea* seeds on Barro Colorado Island in Panama, and used fungicide treatments to show that soil fungi were major sources of mortality for both species. In addition, *M. argentea* suffered considerable mortality from litter-dwelling ants. Interestingly, ants appear to be less important as seed predators at Monteverde than at most lowland sites, probably due to the relatively high elevation. The abundance of litter-dwelling ants is substantially lower in the portion of the MCFR where this study was conducted than at lower elevations nearby on both the Pacific and Caribbean slopes (D. McKenna and K.G. Murray, unpublished data).

Seed longevity and its consequences for recruitment limitation

Pioneer plants at Monteverde exhibit greater diversity in reproductive ecology than previously appreciated, particularly with respect to the establishment and longevity of a dormant seed bank. Theoretical studies demonstrate the importance of seed dormancy for species with spatially or temporally unpredictable establishment opportunities (e.g. Venable, 1989, and studies cited therein), and the majority of tropical-forest pioneer species examined appear to have well-developed mechanisms for 'enforced' (*sensu* Harper, 1977) dormancy. On the other hand, Murray (1988) suggested that periods of dormancy longer than 2 years would confer little additional fitness advantage over shorter-term dormancy, due to an increase in generation time. Garwood's review of seed banks in tropical forests summarizes reports of seed-bank : seed-rain ratios (Garwood, 1989, Table 5). Ratios for pioneer species ranged from 15 to 25 in Venezuelan forest (Uhl and Clark, 1983) to 0 to 25 in Thailand (Cheke *et al.*, 1979). At Monteverde, most species (e.g. *P. rivinoides* and *G. poasana*) maintain seed banks many times the density (some to more than 100×) of the mean annual seed rain, suggesting long-term accumulation (Fig. 21.3). In the most extreme cases (*B. frutescens*, *H. tortuosa* and *T. micrantha*), seeds were common in the soil but were never found in the seed rain, despite more than 3 years of sampling at the same 30 sites. A smaller number of species (e.g. *C. polyphlebia* and *U. elata*) maintain seed banks of about the same density as the annual seed rain (Fig. 21.3). A few species even lacked a dormant seed bank altogether (e.g. *H. appendiculata*), despite the fact that they are extremely common and successful pioneers at Monteverde. Overall, the ratio of seed-bank : annual seed-rain density varied by more than three orders of magnitude among the 23 focal pioneer species, with a median of 8.4.

Our data thus support neither the early contention that virtually all pioneer regeneration is from seed banks accumulated over long periods of time nor the more recent hypothesis that regeneration of most species is from recent seed rain and a short-term seed bank only (Alvarez-Buylla and Martínez-Ramos, 1990; Alvarez-Buylla and García-Barrios, 1991). Rather, pioneer species fall along a continuum between both possible extremes: from virtually no seed bank to a very long-term one. If our data from Monteverde are representative of a broad spectrum of neotropical pioneers, we propose that accumulations over 5 or more years are the rule rather than the exception.

Does the accumulation of seeds in the soil enhance or erode the spatial heterogeneity of the seed bank? If seed predators and pathogens act in a density-dependent manner, their actions might be expected to render the seed bank more homogeneous over time. On the other hand, if they act in a density-independent manner, they may enhance patchiness by entirely eliminating the less common species from many areas. Our data suggest that both patterns may occur at Monteverde, but among different groups of pioneer species. Spatial distributions of species with short lifespans in the seed bank (e.g. *C. polyphlebia*, *U. elata* and *Heliocarpus americanus*) become far less patchy over time, suggesting strong density-dependent mortality (Fig. 21.4). In contrast, species that accumulate in the soil over long periods of time (e.g. *G. poasana*, *P. rivinoides* and *Gonzalagunia rosea*) become much more patchily distributed, as though mortality is density-independent or even negatively density-dependent. It is important to note, however, that accumulation in the soil could lead to greater heterogeneity even without the actions of predators and pathogens. If the seed rain at different sites is highly autocorrelated among years, seed-bank variance : mean ratios will by necessity increase as seeds accumulate in the soil. Currently, we cannot determine how much of the increased patchiness among seeds that accumulate over long periods is due to such temporal autocorrelation and how much is due to the actions of seed predators and pathogens. Dalling *et al.* (1997) found strong evidence of density-dependent seed mortality in the two most common pioneers on Barro Colorado Island, Panama, but seed 'footprints' of one of them (*T. micrantha*) persisted in the soil seed bank for years after the parent tree had died.

Implications for conservation

The great differences in species composition between the seed bank at Monteverde and the previous year's seed rain (Fig. 21.1), as well as the differences in longevity of pioneer species' seeds (Fig. 21.3), suggest an important influence of seed predators and pathogens. Indeed, our detailed experiments with marked seeds provide abundant evidence that post-dispersal predators and pathogens exert profound effects on the survival of pioneer seeds (Fig. 21.5), and our species-by-species comparisons of spatial heterogeneity in the seed bank and seed rain (Fig. 21.2) show that these effects have different consequences for different species. Removal of seed predators or enhancement of their populations via forest fragmentation could have profound effects on pioneer-plant seed-bank densities and heterogeneity.

Other investigators have shown increased densities of granivorous rodents in fragmented landscapes (e.g. Malcolm, 1988; Gascon *et al.*, 1999; Bayne and Hobson, 2000; Umapathy and Kumar, 2000), but we are aware of no studies that examine the consequences of such population changes for seed demography. Given that an increasing proportion of the world's biological diversity will come to reside in such fragments, studies that integrate seed dispersal and demography with the population dynamics of seed predators would greatly aid in the development of management strategies for fragmented landscapes.

Acknowledgements

We are especially grateful to Rodrigo Solano and Benito Guindon for assistance with all phases of data collection for this study; their patience and attention to detail were truly phenomenal. Dr Willam Haber and Willow Zuchowski spent many hours teaching us to identify the plants; without them, this study (and many others at Monteverde) simply could not have been done. Many others helped with parts of the field or laboratory work as well, among them Dr Kathy Winnett-Murray and students from Hope College too numerous to name here. The directors and staff of the Monteverde Cloud Forest Preserve, especially Bob Carlson, were especially helpful in facilitating our work at Monteverde. Many helpful suggestions on an earlier version of the manuscript were made by Jim Dalling, Peter Feinsinger, Doug Levey and Kathy Winnett-Murray. The study was funded by a Hope College Faculty Development Grant and National Science Foundation (NSF) grants BSR-9006734 and DEB-9208229 to K.G. Murray and by NSF-REU grant DBI-9322220 to Dr Christopher Barney.

References

Alvarez-Buylla, E.R. and García-Barrios, R. (1991) Seed and forest dynamics: a theoretical framework and an example from the Neotropics. *American Naturalist* 137, 133–154.

Alvarez-Buylla, E.R. and Martínez-Ramos, M. (1990) Seed bank versus seed rain in the regeneration of a tropical pioneer tree. *Oecologia* 84, 314–325.

Bayne, E. and Hobson, K. (2000) Relative use of contiguous and fragmented boreal forest by red squirrels (*Tamiasciurus hudsonicus*). *Canadian Journal of Zoology* 78, 359–365.

Brokaw, N.V.L. (1986) Seed dispersal, gap colonization, and the case of *Cecropia insignis*. In: Estrada, A. and Fleming, T.H. (eds) *Frugivores and Seed Dispersal.* W. Junk, Dordrecht, The Netherlands, pp. 322–331.

Brokaw, N.V.L. (1987) Gap-phase regeneration of three pioneer tree species in a tropical forest. *Journal of Ecology* 75, 9–19.

Butler, B.J. and Chazdon, R.L. (1998) Species richness, spatial variation and abundance of the soil seed bank of a secondary tropical rain forest. *Biotropica* 30, 214–222.

Canham, C.D. and Marks, P.L. (1985) The response of woody plants to disturbance patterns of establishment and growth. In: Pickett, S.T.A. and White, P.S. (eds) *The Ecology of Natural Disturbance and Patch Dynamics.* Academic Press, Orlando, Florida, pp. 197–217.

Cheke, A.S., Nanakorn, W. and Yankoses, C. (1979) Dormancy and dispersal of seeds of secondary forest species under the canopy of a primary tropical rainforest in northern Thailand. *Biotropica* 11, 88–95.

Chesson, P.L. and Warner, R.R. (1981) Environmental variability promotes coexistence in lottery competitive systems. *American Naturalist* 117, 923–943.

Chesson, P.L. and Warner, R.R. (1985) Coexistence mediated by recruitment fluctuations: a field guide to the storage effect. *American Naturalist* 125, 769–787.

Clark, D.B., Palmer, M.W. and Clark, D.A. (1999) Edaphic factors and the landscape-scale distributions of tropical rain forest trees. *Ecology* 80, 2662–2675.

Clark, J.S. and Ji, Y. (1995) Fecundity and dispersal in plant populations: implications for structure and diversity. *American Naturalist* 146, 72–111.

Clark, J.S., Macklin., E. and Wood, L. (1998) Stages and spatial scales of recruitment limitation in southern Appalachian forests. *Ecological Monographs* 68, 213–235.

Clark, K.L., Lawton, R.O. and Butler, P.R. (2000) The physical environment. In: Nadkarni, N.M. and Wheelwright, N.T. (eds) *Monteverde: Ecology and Conservation of a Tropical Cloud Forest.* Oxford University Press, New York, pp. 3–38.

Dalling, J.W. and Denslow, J.S. (1998) Soil seed bank composition along a forest chronosequence in seasonally moist tropical forest, Panama. *Journal of Vegetation Science* 9, 669–678.

Dalling, J.W., Swaine, M.D. and Garwood, N.C. (1997) Soil seed bank community dynamics in seasonally moist lowland tropical forest, Panama. *Journal of Tropical Ecology* 13, 659–680.

Dalling, J.W., Hubbell, S.P. and Silvera, K. (1998a) Seed dispersal, seedling establishment and gap partitioning among tropical pioneer trees. *Journal of Ecology* 86, 674–689.

Dalling, J.W., Swaine, M.D. and Garwood, N.C. (1998b) Dispersal patterns and seed bank dynamics of pioneer trees in moist tropical forest. *Ecology* 79, 564–578.

Davies, S.J., Palmiotto, P.A., Ashton, P.S., Lee, H.S. and LaFrankie, J.V. (1998) Comparative ecology of 11 sympatric species of *Macaranga* in Borneo: tree distribution in relation to horizontal and vertical resource heterogeneity. *Journal of Ecology* 86, 662–673.

Denslow, J.S. (1987) Tropical rain forest gaps and tree species diversity. *Annual Review of Ecology and Systematics* 18, 431–452.

Dupuy, J.M. and Chazdon, R.L. (1998) Long-term effects of forest regrowth and selective logging on the seed bank of tropical forests in NE Costa Rica. *Biotropica* 30, 223–237.

Garwood, N.C. (1989) Tropical soil seed banks: a review. In: Leck, M.A., Simpson, R.L. and Parker, V.T. (eds) *Ecology of Soil Seed Banks.* Academic Press, New York, pp. 149–209.

Gascon, C., Lovejoy, T.E., Bierregaard, R.O. Jr, Malcolm, J.R., Stouffer, P.C., Vasconcelos, H.L., Laurance, W.F., Zimmerman, B., Tocher, M. and Borges, S. (1999) Matrix habitat and species richness in tropical forest remnants. *Biological Conservation* 91, 223–229.

Gauch, H.G., Jr (1982) *Multivariate Analysis in Community Ecology.* Cambridge University Press, Cambridge, UK.

Haber, W.A. (2000) Plants and vegetation. In: Nadkarni, N.M. and Wheelwright, N.T. (eds) *Monteverde: Ecology and Conservation of a Tropical Cloud Forest.* Oxford University Press, New York, pp. 39–94.

Harper, J.L. (1977) *Population Biology of Plants.* Academic Press, New York.

Holdridge, L.R. (1967) *Life Zone Ecology.* Tropical Science Center, San Jose, Costa Rica.

Hubbell, S.P. and Foster, R.B. (1987) Canopy gaps and the dynamics of a neotropical forest. In: Crawley, M.J. (ed.) *Plant Ecology.* Blackwell Scientific Publishing, Oxford, UK, pp. 77–96.

Hubbell, S.P., Foster, R.B., O'Brien, S.T., Harms, K.E., Condit, R., Wechsler, B., Wright, S.J. and Loo de Lao, S. (1999) Light-gap disturbances, recruitment limitation, and tree diversity in a neotropical forest. *Science* 283, 554–557.

Hurtt, G.C. and Pacala, S.W. (1995) The consequences of recruitment limitation: reconciling chance, history and competitive differences between plants. *Journal of Theoretical Biology* 176, 1–12.

Lawton, R.O. and Putz, F.E. (1988) Natural disturbance and gap-phase regeneration in a wind-exposed tropical cloud forest. *Ecology* 69, 764–777.

MacArthur, R.H. and Wilson, E.O. (1967) *The Theory of Island Biogeography.* Princeton University Press, Princeton, New Jersey.

Malcolm, J.R. (1988) Small mammal abundances in isolated and non-isolated primary forest reserves near Manaus, Brazil. *Acta Amazonica* 18, 67–83.

Martínez-Ramos, M. and Alvarez-Buylla, E. (1986) In: Estrada, A. and Fleming, T.H. (eds) *Frugivores and Seed Dispersal.* W. Junk, Dordrecht, The Netherlands, pp. 333–346.

Murray, K.G. (1986) Consequences of seed dispersal for gap-dependent plants: relationships between seed shadows, germination requirements, and forest dynamic processes. In: Estrada, A. and Fleming, T.H. (eds) *Frugivores and Seed Dispersal.* W. Junk, Dordrecht, The Netherlands, pp. 187–198.

Murray, K.G. (1988) Avian seed dispersal of three neotropical gap-dependent plants. *Ecological Monographs* 58, 271–298.

Nakashizuka, T. (1989) Role of uprooting in composition and dynamics of an old-growth forest in Japan. *Ecology* 70, 1273–1278.

Pacala, S.W. and Tilman, D. (1994) Limiting similarity in mechanistic and spatial models of plant competition in heterogeneous environments. *American Naturalist* 143, 222–257.

Peterson, C.J. and Pickett, S.T.A. (1990) Microsite and elevational influences on early forest regeneration after catastrophic windthrow. *Journal of Vegetation Science* 1, 657–662.

Popma, J., Bongers, F., Martínez Ramos, M. and Veneklaas, E. (1988) Pioneer species distribution in treefall gaps in neotropical rainforest: a gap definition and its consequences. *Journal of Tropical Ecology* 4, 77–88.

Putz, F.E. and Appanah, S. (1987) Buried seeds, newly dispersed seeds, and the dynamics of a lowland forest in Malaysia. *Biotropica* 19, 326–333.

Shmida, A., and Ellner, S. (1984) Coexistence of plant species with similar niches. *Vegetatio* 58, 29–55.

Streng, D.R., Glitzenstein, J.S. and Harcombe, P.A. (1989) Woody seedling dynamics in an East Texas floodplain forest. *Ecological Monographs* 59, 177–204.

Swaine, M.D. and Whitmore, T.C. (1988) On the definition of ecological species groups in tropical rain forests. *Vegetatio* 75, 81–86.

Turnbull, L.A., Crawley, M.J. and Rees, M. (2000) Are plant populations seed-limited? *Oikos* 88, 225–238.

Uhl, C. and Clark, K. (1983) Seed ecology of selected Amazon Basin successional species. *Botanical Gazette* 144, 419–425.

Umapathy, G. and Kumar, A. (2000) The occurrence of arboreal mammals in the rain forest fragments in the Anamalai Hills, south India. *Biological Conservation* 92, 311–319.

Vázquez-Yanes, C.R. (1980) Notas sobre la autoecologia de los árboles pioneros de rápido crecimiento de la selva tropical lluviosa. *Tropical Ecology* 21, 103–112.

Venable, D.L. (1989) Modelling the evolutionary ecology of seed banks. In: Leck, M.A., Simpson, R.L. and Parker, V.T. (eds) *Ecology of Soil Seed Banks.* Academic Press, New York, pp. 67–87.

Webb, S.L. (1988) Windstorm damage and microsite colonization in two Minnesota forests. *Canadian Journal of Forest Research* 18, 1186–1195.

Young, K.R., Ewel, J.J. and Brown, B.J. (1987) Seed dynamics during forest succession in Costa Rica. *Vegetatio* 71, 157–174.

22 A Meta-analysis of the Effect of Gut Treatment on Seed Germination

Anna Traveset[1] and Miguel Verdú[2]

[1]*Institut Mediterrani d'Estudis Avançats (CSIC-UIB), C/ Miquel Marqués 21, 07190-Esporles, Mallorca, Balearic Islands, Spain;* [2]*Centro de Investigaciones sobre Desertificación (CSIC-UV), Camí de La Marjal s/n, Apdo. Oficial, 46470-Albal, Valencia, Spain*

Introduction

The dispersal of seeds by vertebrate frugivores is a process that usually implies the consumption of fruit pulp and the regurgitation or defecation of viable seeds (Ridley, 1930). An important advantage of seed ingestion by frugivores is a presumed increase in germination percentage (germinability) and rate (speed) (Krefting and Roe, 1949; van der Pijl, 1982). Recent analyses show, however, that such enhancement is far from universal and that several variables intrinsic to the plant or to the frugivore can influence the response of seeds to gut treatment (Traveset, 1998).

Three mechanisms determine how frugivores can directly affect seed germination: (i) possible mechanical and/or chemical scarification of the seed-coat, which may depend upon gut retention time and on the type of food ingested with seeds (e.g. Agami and Waisel, 1988; Barnea *et al.*, 1990, 1991; Izhaki and Safriel, 1990; Yagihashi *et al.*, 1998); (ii) separation of seeds from pulp because germination is reduced or precluded if seeds remain associated with pulp (e.g. Rick and Bowman, 1961; Temple, 1977; Izhaki and Safriel, 1990; Barnea *et al.*, 1991; Engel, 2000; Traveset *et al.*, 2001); and (iii) the effect that results from faecal material surrounding the seeds, which may influence germination and/or future seedling growth. For example, seedlings emerging from ingested seeds tend to be more vigorous than those emerging from uningested ones because remaining faecal material fertilizes the seedlings, especially in the case of large-mammalian faeces, which often take a long time to decompose (e.g. Dinerstein and Wemmer, 1988; Quinn *et al.*, 1994; Ocumpaugh *et al.*, 1996; Paulsen, 1998; Traveset *et al.*, 2001; T.R. Paulsen, unpublished).

The importance of the first mechanism – the modification of seed-coat traits (e.g. permeability of the coat to water and gases) after gut treatment, which changes the capacity of germination and/or the speed at which seeds germinate – is the focus of this study. A review of this effect has recently been published (Traveset, 1998). In this review, Traveset analysed the results of studies on nearly 200 plant species in 68 families, using a slight variation of the 'vote-counting' method (Light and Pillemer, 1971; Hedges and Olkin, 1980). This method is conservative and has the advantage of being simple, but it has the disadvantage of low statistical power (Cooper, 1998). Another shortcoming of the method is that all study cases are given the same weight, and thus the magnitude of the effect is the same for all studies. This equal weighting can generate biases when the number of seeds and replicates varies tremendously among studies. Thus, in this chapter we explored the effect of gut treatment on germination using a statistically powerful approach, meta-analysis, which combines results from independent studies and which is not sensitive to sample-size effects.

Meta-analysis requires the extraction of a common metric of effect size from each study included in the analysis. Choosing the best metric is of crucial importance for obtaining correct ecological inferences (Osenberg *et al.*, 2000). Estimates of effects from individual studies are combined into a pooled estimate of the overall strength of the effect, which is then used to assess significance (e.g. Xiong and Nilsson, 1999; Gurevitch *et al.*, 2000; Osenberg *et al.*, 2000).

Our analysis includes studies reviewed in Traveset (1998), plus studies that were missed in that review or were published more recently. We tested the following hypotheses:

1. Seed passage through a frugivore's gut increases both the germinability of seeds and the rate of germination.

2. Different taxonomic groups have different effects on seed germination. We expect such differences to result from differences in:
(a) seed retention times in digestive tracts, and/or
(b) chemical composition of the food ingested with the seeds.

3. Seeds from fleshy-fruited and dry-fruited species differ in their response to gut passage. This hypothesis assumes that pulp texture generally affects seed retention time in the gut (Levey, 1986) and, ultimately, germination patterns.

4. Large seeds will be affected differently from small seeds by gut passage, given that seed size affects retention time in the gut (e.g. Levey, 1986; Levey and Grajal, 1991; Gardener *et al.*, 1993), which presumably determines the degree of seed-coat scarification.

5. The effect on germination will vary depending upon plant life-form. Different life-forms have different frequencies of seed dormancy (Baskin and Baskin, 1998), which might reflect differences in traits such as coat thickness or sculpture.

6. Differences among habitats (cultivated areas, grasslands, shrublands and woodlands) may differ in the magnitude of the frugivores' effect on germination. As with the previous hypothesis, seed dormancy frequencies might vary depending upon the ecological conditions of the habitat where the species usually live (Baskin and Baskin, 1998, Ch. 12).

7. The effect of gut passage on germination differs between temperate and tropical plant species. This hypothesis is based on the prediction by Izhaki and Safriel (1990) that germination enhancement is adaptive in unpredictable or less constant environments.

8. Because the studies used to test the above hypotheses were conducted under a wide range of conditions (laboratory, greenhouse, field), we also tested whether the experimental conditions affected the probability of finding a significant effect.

The meta-analysis we performed, as with any type of statistical analysis, was limited to the data sets available. We have information on seed responses to gut passage (in particular, germinability data) from more than 100 frugivore species belonging to approximately 50 families, but these data are somewhat biased towards studies performed with either birds, mostly from the temperate zones, or non-flying mammals. Other frugivore groups are underrepresented (e.g. bats, reptiles and fishes). We do not include fish studies in our analyses because of the especially small number of studies available. Likewise, most data are from fleshy-fruited species; little is known about the responses of seeds in dry fruits to the passage through frugivores' digestive tracts.

Methods

We used results of 351 experiments, from 83 studies covering a total of 213 plant species, all of which considered seeds removed from pulp as 'controls' and used a minimum of 24 seeds. 'Treated' (ingested) seeds were typically seeds that had been defecated, except for several cases in which data were given for regurgitated seeds only. We excluded studies in which it was not clear whether ingested and control seeds came from the same population and those that presented no sample sizes. Most studies provided only the final percentage germination and not the germination rate. When graphs of cumulative germination were presented, the rate was obtained directly from the curves. The entire database is available from the authors upon request.

We first calculated an effect size for each experiment. Because the end-point of seed germination experiments is binary (germinated versus non-germinated), we considered the log-transformed odds ratio (lnOR) to be the most appropriate metric of effect size (Egger *et al.*, 1997). Data were organized in a 2 × 2 contingency table; columns represented treatments and rows represented possible outcomes (Table 22.1). The odds of an event (i.e. seed germination) are the probability of the event occurring divided by the probability that the event does not occur. This odds ratio estimates the probability of germination after gut passage, relative to its probability in the control group. It was calculated following the Mantel–Haenszel procedure (Rosenberg *et al.*, 2000).

We next combined individual effect sizes, weighted by the reciprocal of their sampling variances, into a cumulative effect size representing the overall magnitude of the effect of gut treatment on seed germination, using the Mantel–Haenszel procedure (Rosenberg *et al.*, 2000). This estimate of effect size and its variance allows us to calculate confidence intervals (CIs) and the total heterogeneity, Q_T, of effect sizes (Rosenberg *et al.*, 2000). The effect of gut treatment on seed germination percentage is considered significant if zero is not included in the CI of the cumulative effect size. The heterogeneity among the n individual effect sizes was tested with the statistic Q_T against a χ^2_{n-1} distribution. A significant Q_T means that the variance among effect sizes is greater than expected from sampling error, and indicates that there may be underlying structure to the data and, therefore, that other explanatory variables should be taken into account.

If Q_T is statistically significant, the next step is to incorporate the data structure in the meta-analysis by including mediating variables (predictors). To test the hypotheses formulated in the Introduction, our predictors were: (i) the taxonomic group of the frugivore (birds, non-flying mammals, bats and reptiles); (ii) fruit type (dry and fleshy); (iii) seed size (small (< 5 mm), medium (5–10 mm) and large (> 10 mm)); (iv) plant life-form (herb, shrub and tree); (v) habitat (cultivated, grassland, shrubland and woodland); (vi) zone (temperate and tropical); and (vii) experimental condition (laboratory, field and greenhouse). A cumulative effect size with its variance and 95% CI was calculated for each group within a given predictor. Also, the heterogeneity within the group was tested with the statistic Q_W, against a χ^2_{n-1} distribution. A significant Q_W means that further heterogeneity remains unexplained within that group. To test differences in effect sizes among groups, the total heterogeneity Q_T was partitioned into between-group heterogeneity (Q_B) and within-group heterogeneity (Q_W). A significant Q_B means there are differences among groups in their cumulative effect sizes, also noted by non-overlapping 95% CIs (Rosenberg *et al.*, 2000). Fixed-effects analyses were used for all statistical tests.

A single study often contained several experiments. Therefore, experiments within a single study may be interdependent, leading to pseudo-replication. To control for this bias, we conducted an additional meta-analysis at the

Table 22.1. Contingency table representing the arrangement of data on seed germination for the calculation of odds ratios.

	Ingested	Control	Total
Germinated	A	B	A + B
Non-germinated	C	D	C + D
Total	$n_t = A + C$	n_c = B + D	N = A + B + C + D

A, B, C and D represent the number of seeds observed in each cell.

study level by pooling all experiments within a single study (see Xiong and Nilsson, 1999, for a similar procedure). Thus, each study contributed only one degree of freedom to the analysis, except in the few cases where experiments were reported for more than one frugivore group (i.e. bats and birds) and the experiments were pooled for each frugivore group. Predictors were assigned to each study only when more than 80% of the experiments had the same predictor (for example, when nine out of ten experiments from a given study were performed with small-seeded species).

Because some unpublished studies were included in the meta-analysis, a graphical method was performed to detect possible publication bias (e.g. if studies showing a significant effect of gut treatment on seed germination have greater probability of being published than those showing non-significant effects). We used a weighted histogram, which plots the effect size of the experiments against their weighted frequency (weights = 1/variance). If publication bias exists, the histogram will be depressed around the region of no-effect size.

Only three published studies (Barnea *et al.*, 1991; Iudica and Bonaccorso, 1997; Mas and Traveset, 1999) and unpublished studies by A. Traveset *et al.* had replicates to obtain means and standard deviations of the speed of germination, measured as time elapsed until reaching 50% germination (T_{50}). These studies contained a total of 30 experiments with 16 plant species. The effect-size metric for the study of the effect of gut treatment on germination rate was Hedges' *d*. This metric is an estimate of the standardized mean difference, unbiased by small sample sizes (Hedges and Olkin, 1985). We rejected an alternative metric, the response ratio, because effect sizes calculated with it were not normally distributed (Hedges *et al.*, 1999). Because of the small number of studies reporting data appropriate for a meta-analysis of germination rate, it was not possible to run an additional meta-analysis at the study level to control for pseudoreplication. For this reason, results of this meta-analysis must be interpreted with caution. We performed another meta-analysis on germination rate with a non-parametric estimate of the variance (i.e. variance calculated with sample sizes instead of standard deviations) (see Rosenberg *et al.*, 2000). In this analysis we could include 27 more experiments (and thus had a total of 57 observations). Effect sizes weighted by the non-parametric variance were bootstrapped 5000 times to estimate the 95% CI. All meta-analyses were run on METAWIN 2.0 (Rosenberg *et al.*, 2000).

Colinearity among variables was investigated by constructing a weighted least-squares multiple-regression model, following modifications proposed by Hedges (1994) for meta-analysis. Effect size, weighted by the inverse of its variance, was included as the dependent variable and predictors (transformed into dummy variables) were the independent variables. This model was run (SPSS 9.0) in a stepwise form to test the significance of each predictor, while controlling for effects of all other predictors. To determine significance, the difference between the sum of squares of the regression with (Q_R) and without (Q_r), the predictor ($Q_R - Q_r = Q_{\text{partial}}$) was tested against a chi-square distribution.

Results

Effect on final percentage germination

The overall effect of gut treatment on seed germination percentage was significantly positive (mean effect size lnOR = 0.29; 95% CI: (0.27–0.31)). The magnitude of this effect is considered 'moderate' in the social sciences (Cohen, 1988). A similar value was found when meta-analysis was carried out at the study level (lnOR = 0.25; (0.23–0.27)), which avoids pseudo-replication effects. The heterogeneity test of Q_T was highly significant ($P < 0.00001$), implying that other variables (predictors) account for some of the variation among studies.

When taxon is included as a predictor, the heterogeneity between groups, Q_B, is strongly significant ($P < 0.00001$), indicating that the effect of gut treatment on seed germination percentage differs among taxa. As shown in Fig. 22.1, all taxa except reptiles (whose 95% CI contains zero) have significant and positive effects on seed germination percentage. Taxa with overlapping 95% CIs do not differ

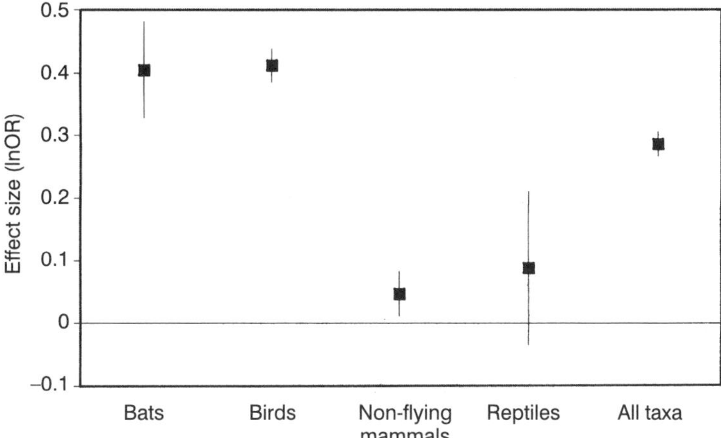

Fig. 22.1. Effect size of seed passage through the digestive system of bats (n = 19 experiments), birds (n = 180), non-flying mammals (n = 113), reptiles (n = 39) and all taxa (n = 351) on percentage germination. Effect size is calculated as the logarithm of the odds ratio (lnOR). Error bars indicate 95% confidence interval. An effect size is statistically significant if its error bar does not intersect the zero line. Two taxa are significantly different if their confidence intervals do not overlap.

statistically in effect size. Therefore, taxa may be grouped into two groups: (i) bats and birds, which have a similar effect size; and (ii) non-flying mammals and reptiles, which have little or no effect on seed germination percentage, respectively. The same groups were found when meta-analysis was carried out at the study level, although the effect of non-flying mammals shifted from being significant to being non-significant.

When considering fruit type, significant differences ($P < 0.00001$) were found between fleshy and dry fruits. While seeds from fleshy fruits ingested by frugivores germinated in higher percentages than control seeds, germination of seeds from dry fruits was negatively affected by gut treatment (Fig. 22.2a). The same results were found when meta-analysis was performed at the study level.

Seeds of all sizes were positively affected by gut treatment, although to different degrees ($P < 0.00001$). The effect on germination of large seeds was three to four times higher than the effect on small and medium seeds (Fig. 22.2b). Effect size for medium and small sizes did not differ. At the study level, the highest effect size was found for large seeds, but effect sizes for medium-sized seeds became non-significant and significantly lower than those for small seeds.

Frugivores had a more than four times stronger effect on seed germinability for trees than for shrubs or herbs ($P < 0.00001$) (Fig. 22.2c). When performing the meta-analysis at the study level, the strongest effect was still for trees and the effect for shrubs became non-significant.

Seeds from plants living in different habitats also responded differently to gut treatment ($P < 0.00001$) (Fig. 22.2d). The effect was positive for all habitats, but the magnitude of the effect ranged from strong to weak in the following order: cultivated > grassland > woodland > shrubland. The same order was found in the meta-analysis at the study level, although the effect for shrublands became significantly negative.

The effect of gut treatment on percentage germination was twice as large in the tropics as in the temperate zones ($P < 0.00001$) (Fig. 22.3a). This result was the same when the meta-analysis was performed at the study level. This finding, however, appears to be generated by the biased proportion of trees represented in tropical studies (80%). When we analysed the data comparing only seeds from trees, we found that temperate species were significantly more affected by gut treatment than were tropical species (lnOR = 0.75 vs. lnOR = 0.51, with positive and non-overlapping CIs) (Fig. 22.3b).

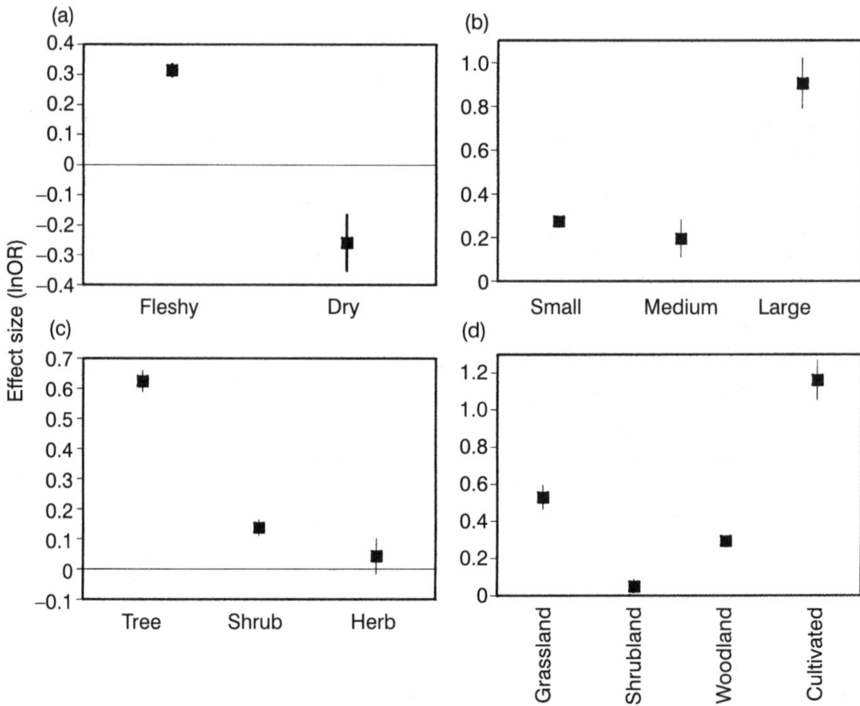

Fig. 22.2. Effect size of seed passage through frugivores' guts on percentage germination of seeds (a) from fleshy (n = 319 experiments) and dry (n = 32) fruits; (b) small (n = 253), medium (n = 43) and large (n = 37) sizes of seeds; (c) from trees (n = 116), shrubs (n = 154) and herbs (n = 66); and (d) from plants inhabiting grasslands (n = 25), shrublands (n = 79), woodlands (n = 214) and cultivated areas (n = 25). Data are mean effect sizes (logarithm of the odds ratio (lnOR)) and 95% confidence intervals.

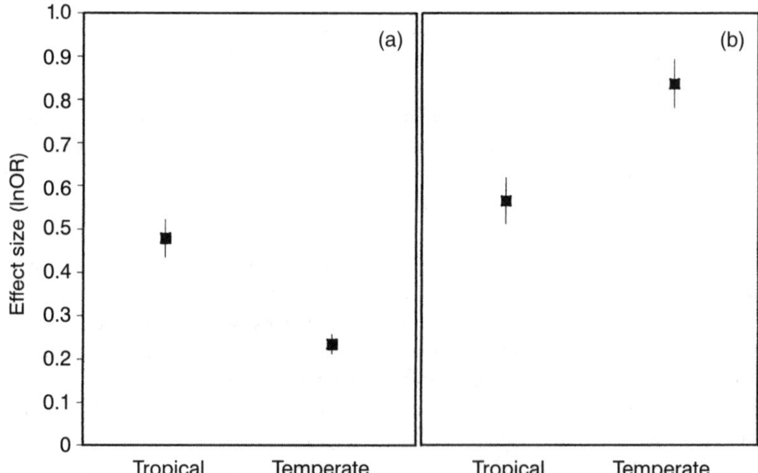

Fig. 22.3. Effect size of seed passage through frugivores' guts on percentage germination of seeds from (a) plants living in tropical (n = 82 experiments) and temperate (n = 269) zones, and (b) only trees tested in the two zones. Data are mean effect sizes (logarithm of the odds ratio (lnOR)) and 95% confidence intervals.

The experimental conditions under which seed germination was tested also influenced the results significantly ($P < 0.00001$) (Fig. 22.4). Experiments performed in the laboratory (usually in growth chambers) or in the field showed similar effect sizes, but greenhouse experiments seemed not to detect gut-treatment effects. The meta-analysis performed at the study level gave the same results, although field studies showed a slightly greater effect size than laboratory studies.

In all cases, the heterogeneity within a group Q_W (i.e. bats, small seeds, tropical zones, etc.) was statistically significant, suggesting that, even though our predictors explained a portion of the total variance, another fraction remains unexplained within groups. The regression model including all predictors was highly significant ($\chi^2 = 1339$; d.f. = 14; $P << 0.001$), although it only explained 13.5% of the variance. The error term was also highly significant ($\chi^2 = 5907$; d.f. = 275; $P < 0.001$) indicating that more predictors should be added to the model. All predictors remained highly significant after statistically controlling for the rest of the predictors, which indicates that the portion of the variance explained by each predictor is independent of that explained by the rest of the predictors (Table 22.2). Publication bias did not affect our analysis. The weighted histogram did not show smaller frequencies in the zone of no effect (Fig. 22.5).

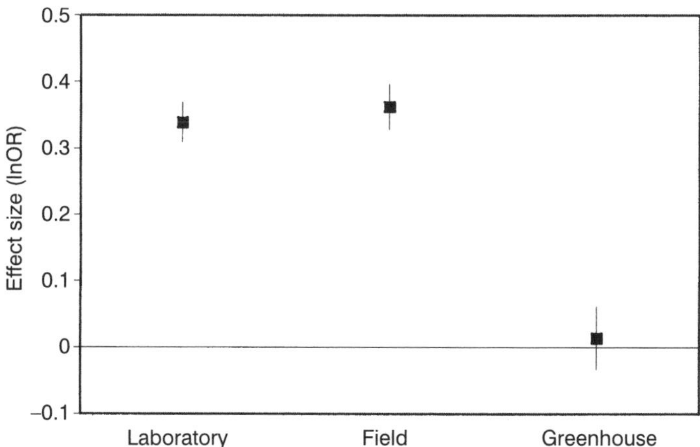

Fig. 22.4. Differences in the effect size of frugivores' guts on seed germination percentage among the three common germination conditions in which the experiments were run: greenhouse (n = 57), laboratory (n = 179) and field (n = 115). Data are mean effect sizes (logarithm of the odds ratio (lnOR)) and 95% confidence intervals.

Table 22.2. Results of the weighted regression model of effect size vs. predictors. The significance of each predictor after statistical control of the remaining predictors is tested. $Q_{partial}$ is calculated as the difference between the sum of squares of the regression with (Q_R) and without (Q_r) the predictor ($Q_R - Q_r = Q_{partial}$) and tested against a χ^2 distribution with the degrees of freedom shown in parentheses.

Predictor	Q_R (d.f.)	Q_r (d.f.)	$Q_{partial}$ (d.f.)	P
Frugivore taxa	1339.2 (14)	1230.5 (11)	100.8 (3)	10^{-23}
Fruit type	1339.2 (14)	1250.6 (13)	50.9 (1)	10^{-21}
Seed size	1339.2 (14)	1318.5 (12)	25.6 (2)	10^{-5}
Life-form	1339.2 (14)	1263.5 (12)	140.1 (2)	10^{-17}
Habitat	1339.2 (14)	953.6 (11)	243.1 (3)	10^{-83}
Zone	1339.2 (14)	1327.4 (13)	2.3 (1)	10^{-4}
Experimental condition	1339.2 (14)	1287.6 (12)	41.5 (2)	10^{-12}

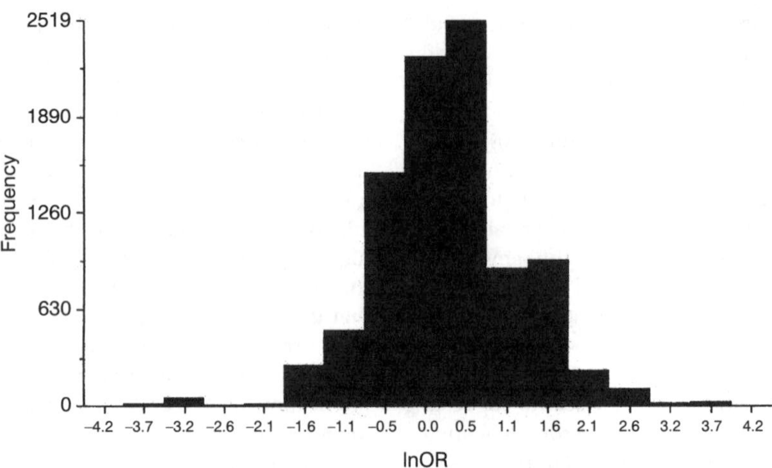

Fig. 22.5. Weighted histogram of effect size of the experiments against their weighted frequency (weights = 1/variance) to detect publication bias. lnOR, logarithm of the odds ratio.

Effect on rate (velocity) of germination

The speed at which seeds germinate was also significantly affected by gut treatment (Hedges' $d = -0.3843$; (−0.5716 to −0.1970)). The negative effect size means that the time needed for gut-treated seeds to germinate was lower than that for control seeds. Thus, frugivores significantly accelerated seed germination. The heterogeneity test, Q_T, was highly significant ($P < 0.00001$) and therefore other explanatory variables (predictors) should be investigated. This is not possible, however, because of the small number of replicated experiments reporting germination curves. Conclusions and limitations were the same when the sample size was increased from 30 to 57 experiments by performing a meta-analysis weighted by a non-parametric estimate of the variance (Hedges' $d = -6.73$; (−20.8 to −0.25) bias-corrected bootstrap 95% CI).

The difference between control and ingested seeds on germination rate (T_{50}) averaged only a few days. It was similar among the taxonomic groups (1.8 ± 14.7 days for birds, $n = 63$; 1.5 ± 14.8 days for non-flying mammals, $n = 55$; and −1.4 ± 1.0 days for bats, $n = 4$; means ± SD). Only one observation existed for reptiles; it showed a difference of 1 day. Differences in time to first germination (or dormancy length) between ingested and control seeds also averaged a few days (1.5 ± 16.8 days, $n = 66$); the range, however, was quite wide, from a delay (for ingested seeds) of 110 days to an acceleration of 37 days.

Discussion

What affects seed germination?

Our analysis confirms that seed passage through the digestive tract of vertebrate frugivores, regardless of taxa, influences germination. Ingested seeds germinate in greater numbers and take less time to germinate than uningested seeds. As expected, the magnitude of the effect differed significantly among frugivore groups. In contrast to Traveset's (1998) results, however, birds and bats had a significantly greater positive effect on germination than either non-flying mammals or reptiles. We attribute this result to the shorter gut-passage times, usually associated with body size (Karasov, 1990), in the former group compared with the latter. Moreover, birds and bats void the indigestible mass (seeds) as quickly as possible to reduce the high energetic costs of ballast during flight, a selective pressure that is non-existent in non-volant vertebrates (Levey and Grajal, 1991). Seed retention time in the gut presumably determines the abrasive effect on the seed-coat, and we have some evidence showing that germination success decreases as retention time increases (Murray *et al.*, 1994). Besides having different seed retention times, the two groups of frugivores may differ significantly in the chemical composition of ingested food (with variable acidity, water content, secondary compounds, etc.), which can also affect the degree of seed-coat scarification and, thus, germination (Murray *et al.*, 1994; Witmer, 1996; Cipollini and Levey, 1997; Wahaj *et al.*, 1998). The food's chemical composition can, in turn, affect gut passage time (Witmer, 1996; Wahaj *et al.*, 1998; Levey and Martínez del Río, 1999). A change in diet from insects to fruits, for instance, appears to decrease seed retention time in American robins, *Turdus migratorius* (Levey and Karasov, 1994).

We found a striking difference between the effect of gut treatment on seeds of fleshy and dry fruits; the effect was positive for the former but negative for the latter. This finding suggests that frugivores might act as a selective agent on the type (texture, chemical composition, etc.) of pericarp within which seeds are embedded. Endozoochory is in fact evolutionarily associated with the production of fleshy pulp. The negative effect found on seeds from dry fruits indicates that their ingestion by frugivores is not always beneficial to the plant. It would be interesting to know how seeds from fleshy and dry fruits differ in traits relevant to germination, such as coat structure and thickness, and in retention time in vertebrate guts.

The size of seeds proved to be important in determining the magnitude of the germination effect. Small seeds are often retained for longer periods in an animal's digestive tract than are large seeds (e.g. Garber, 1986; Levey and Grajal, 1991; Gardener *et al.*, 1993; Izhaki *et al.*, 1995) and thus they may be more likely to be excessively abraded. Seed-coat thickness, possibly associated with seed size, might also account for this finding.

Traveset (1998) concluded that seeds of trees were more frequently affected by gut passage than were seeds of other life-forms. Although this result was confirmed by our meta-analysis, the factors that produce it are unknown. It may be that some seed traits differ among life-forms, making them differently 'sensitive' to mechanical and/or chemical abrasion. Interestingly, a seed trait that varies in the same direction as the effect of life-form is dormancy. Trees in the temperate zone have a greater frequency of non-dormancy (51%) than other life-forms (42% for shrubs and 37% for herbs) (Baskin and Baskin, 1998).

When comparing habitat types, cultivated species appeared to be much more affected than species in the other habitats, regardless of life-form. The explanation for this result may be related to the fact that seed dormancy usually decreases in cultivated species (Baskin and Baskin, 1998). In fact, human selection may have caused a reduction in the thickness of seed-coat, which would probably mean a loss of dormancy (Baskin and Baskin, 1998).

If the effect of gut treatment on germination is in fact associated with seed dormancy, we might also expect that species in the tropics, with a lower dormancy frequency (Baskin and Baskin, 1998), would be more strongly affected than species in the temperate zone. However, the pattern we observed is the opposite, at least for trees. We have no explanation for this result.

Differences in germination percentage between ingested and uningested seeds may depend on the type of experimental conditions. The probability of detecting a difference appeared to be greater when seeds were planted in the natural habitat (or simply outdoors) and when simply placed in Petri dishes than when planted in a greenhouse. In the few studies that performed the same test under different conditions, either a greater effect in laboratory compared with field experiments (Bustamante *et al.*, 1992, 1993; Figueiredo and Perin, 1995; Yagihashi *et al.*, 1998) or a similar result between the two conditions was found (Figueiredo and Perin, 1995; Figueiredo and Longatti, 1997). Traveset *et al.* (2001) performed simultaneous tests in the three conditions and found similar results in the laboratory and greenhouse, detecting differences between ingested and control seeds only in the field. Other studies on plant responses to particular effects have also reported significantly different results depending on environmental conditions (e.g. Curtis and Wang, 1998).

Adaptive significance of germination enhancement

The enhancement of germination of seeds passed through a frugivore's gut can be adaptive only if it translates into an increase in plant fitness. High germinability is presumably necessary to maximize reproductive success and seedling recruitment (Harper, 1977). An acceleration of germination, in contrast, may not always be beneficial, and can even be detrimental if other factors affecting seedling recruitment (pathogen attack, herbivory, water limitation, light conditions, etc.) favour dormancy (e.g. Janzen 1981; Jones and Sharitz, 1989). The potential advantages of fast germination probably vary among species, depending on the type of seed dormancy they have and on the ecological conditions prevailing in the habitat. Non-dormant species might be expected to benefit from fast germination more than dormant species, because such a response might reduce mortality, due to factors such as seed predation (e.g. Schupp, 1993), sibling competition (e.g. Hyatt and Evans, 1998) or shade intolerance (Jones *et al.*, 1997). For dormant species, the risk of mortality may be spread over time to maximize plant fitness, by being differently affected by each vertebrate frugivore that consumes their fruits (Izhaki and Safriel, 1990; Barnea *et al.*, 1991).

Seed passage through the gut of a vertebrate can probably break only seed-coat dormancy (so-called functional dormancy) and not physiological (internal or embryological) dormancy, as differences in germination rate between ingested and uningested seeds are usually only a few days and more rarely a few weeks. Such a relatively short period is unlikely to result in important differences in seedling survival, as recently confirmed with *Myrtus communis* by Traveset *et al.* (2001).

Avenues of Future Research

Given that seed germination responses can vary significantly due to a variety of factors, both extrinsic and intrinsic to the plant, future tests on the effect of gut treatment on germination need to be tightly controlled. In particular, control seeds should be collected from the same populations where fresh faeces are gathered and at the same time. When performing experiments with captive frugivores, we recommend that fruits be mixed and randomly sampled before being fed to the frugivores; this will ensure that seed traits of the two groups (control and treatment) differ minimally. Due to the differences we documented among different environmental conditions, we also recommend the tests be done in the species' natural habitat. If that is not possible, it would be prudent to perform tests in more than one condition. To avoid planting seeds that appear intact but are not viable, seed viability should be tested prior to germination trials.

The speed of germination, and not simply the final percentage germination, should be more closely examined, as differences may exist in the former that could pass undetected. We also recommend that germination be monitored for a period long enough to allow most seeds to germinate.

Our data set shows that we especially need much more information on the effect of gut treatment on: (i) seeds from dry fruits; and (ii) seeds from life-forms other than trees in the tropics. We also need more studies that examine the effect of bats and reptiles on seed germination.

It is also of crucial importance to determine the mechanism/s by which seed germination behaviour is modified after seeds are ingested by an animal. Future research should investigate the changes that seed-coat traits (thickness, porosity, etc.) undergo after seeds pass through a frugivore's gut. It would also be interesting to know how often seed dormancy is related to seed-coat thickness (Baskin and Baskin, 1998) and how this affects the probability of germination enhancement after gut treatment.

Finally, to elucidate the consequences of seed ingestion by frugivores for plant recruitment and, ultimately, the consequences of frugivory for plant reproduction, we need more studies that focus on the transition stage from seed to seedling. Moreover, the importance of seed ingestion by frugivores on germination performance and plant fitness can only be determined by evaluating the effect of all other factors, biotic and abiotic, that influence germination and establishment success.

Acknowledgements

We are grateful to J. Gurevitch, J. Sánchez-Meca and M.S. Rosenberg for their help with the statistical procedures, to Javier Rodríguez for computer assistance and to J. Sánchez-Meca, Carlos Martínez del Río, Patrick Jansen, Thomas Engel, Patricio García-Fayos and especially Doug Levey for comments on the manuscript. We acknowledge the great editorial work done by the latter. We also deeply thank all people (too many to be listed here) who have shared information and unpublished data to fill cells in our database. Our study was framed within projects PB97-1174 and 1FD97-0551.

References

Agami, M. and Waisel, Y. (1988) The role of fish in distribution and germination of seeds of the submerged macrophytes *Najas marina* L. and *Ruppia maritima* L. *Oecologia* 76, 83–88.

Barnea, A., Yom-Tov, Y. and Friedman, J. (1990) Differential germination of two closely related species of *Solanum* in response to bird ingestion. *Oikos* 57, 222–228.

Barnea, A., Yom-Tov, Y. and Friedman, J. (1991) Does ingestion by birds affect seed germination? *Functional Ecology* 5, 394–402.

Baskin, C.C. and Baskin, J.M. (1998) *Seeds: Ecology, Biogeography, and Evolution of Dormancy and Germination*, 1st edn. Academic Press, San Diego, 666 pp.

Bustamante, R.O., Simonetti, J.A. and Mella, J.E. (1992) Are foxes legitimate and efficient seed-dispersers? A field test. *Acta Oecologica* 13, 203–208.

Bustamante, R.O., Grez, A., Simonetti, J.A., Vásquez, R.A. and Walkowiak, A.M. (1993) Antagonistic effects of frugivores on seeds of *Cryptocarya alba* (Mol.) Looser (*Lauraceae*): consequences on seedling recruitment. *Acta Oecologica* 14, 739–745.

Cipollini, M.L. and Levey, D.J. (1997) Secondary metabolites of fleshy vertebrate-dispersed fruits: adaptive hypotheses and implications for seed dispersal. *American Naturalist* 150, 346–372.

Cohen, J. (1988) *Statistical Power Analysis for the behavioral Sciences*, 2nd edn. Hillsdale, Erlbaum.

Cooper, H. (1998) *Synthesizing Research: A Guide for Literature Reviews*, 3rd edn. Thousand Oaks, Sage, 201 pp.

Curtis, P.S. and Wang, X. (1998) A meta-analysis of elevated CO_2 effects on woody plant mass, form, and physiology. *Oecologia* 113, 299–313.

Dinerstein, E. and Wemmer, C.M. (1988) Fruits *Rhinoceros* eat: dispersal of *Trewia nudiflora* (*Euphorbiaceae*) in lowland Nepal. *Ecology* 69, 1768–1774.

Egger, M., Smith, G.D. and Phillips, A.N. (1997) Meta-analysis. *Principles and Procedures. British Medical Journal* 315, 1533–1537.

Engel, T.R. (2000) Seed dispersal and forest regeneration in a tropical lowland biocoenosis (Shimba Hills, Kenya). PhD thesis, University of Bayreuth, Germany, 344 pp.

Figueiredo, R.A. and Longatti, C.A. (1997) Ecological aspects of the dispersal of a *Melastomataceae* by marmosets and howler monkeys (Primates: Platyrrhini) in a semideciduous forest of Southeastern Brazil. *Revue d'Ecologia (Terre Vie)* 52, 3–8.

Figueiredo, R.A. and Perin, E. (1995) Germination ecology of *Ficus luschnathiana* drupelets after bird and bat ingestion. *Acta Oecologica* 16, 71–75.

Garber, P.A. (1986) The ecology of seed dispersal in two species of Callitrichid primates (*Saguinus mystax* and *Saguinus fuscicollis*). *American Journal of Primatology* 10, 155–170.

Gardener, C.J., McIvor, J.G. and Janzen, A. (1993) Passage of legume and grass seeds through the digestive tract of cattle and their survival in faeces. *Journal of Applied Ecology* 30, 63–74.

Gurevitch, J., Morrison, J.A. and Hedges, L.V. (2000) The interaction between competition and predation: a meta-analysis of field experiments. *American Naturalist* 155, 435–453.

Harper, J.L. (1977) *Population Biology of Plant.* Academic Press, London, 892 pp.

Hedges, L.V. (1994) Fixed effect models. In: Cooper, H.M. and Hedges, L.V. (eds) *The Handbook of Research Synthesis.* Russell Sage Foundation, New York, pp. 285–289.

Hedges, L.V. and Olkin, I. (1980) Vote-counting methods in research synthesis. *Psychological Bulletin* 88, 359–369.

Hedges L.V. and Olkin, I. (1985) *Statistical Methods for Meta-analysis.* Academic Press, New York, 369 pp.

Hedges, L.V., Gurevitch, J. and Curtis, P.S. (1999) The meta-analysis of response ratios in experimental ecology. *Ecology* 80, 1150–1156.

Hyatt, L.A. and Evans, A.S. (1998) Is decreased germination fraction associated with risk of sibling competition? *Oikos* 83, 29–35.

Iudica, C.A. and Bonaccorso, F.J. (1997) Feeding of the bat, *Sturnira lilium*, on fruits of *Solanum riparium* influences dispersal of the pioneer tree in forests of northwestern Argentina. *Studies on Neotropical Fauna and Environment* 32, 4–6.

Izhaki, I. and Safriel, U.N. (1990) The effect of some mediterranean scrubland frugivores upon germination patterns. *Journal of Ecology* 78, 56–65.

Izhaki, I., Korine, C. and Arad, Z. (1995) The effect of bat (*Rousettus aegyptiacus*) dispersal on seed germination in eastern Mediterranean habitats. *Oecologia* 101, 335–342.

Janzen, D.H. (1981) *Enterolobium cyclocarpum* seed passage rate and survival in horses, Costa Rican Pleistocene seed dispersal agents. *Ecology* 62, 593–601.

Jones, K.N., Allen, B.C. and Sharitz, R.R. (1997) Analysis of pollinator foraging: test for non-random behaviour. *Functional Ecology* 11, 255–259.

Jones, R.H. and Sharitz, R.R. (1989) Potential advantages and disadvantages of germinating early for trees in floodplain forests. *Oecologia* 81, 443–449.

Karasov, W.H. (1990) Digestion in birds: chemical and physiological determinants and ecological implications. *Studies in Avian Biology* 13, 391–415.

Krefting, L.W. and Roe, E. (1949) The role of some birds and mammals in seed germination. *Ecological Monographs* 19, 284–286.

Levey, D.J. (1986) Methods of seed processing by birds and seed deposition patterns. In: Estrada, A. and Fleming, T.H. (eds) *Frugivores and Seed Dispersal.* Junk, Dordrecht, pp. 147–158.

Levey, D.J. and Grajal, A. (1991) Evolutionary implications of fruit-processing limitations in cedar waxwings. *American Naturalist* 138, 171–189.

Levey, D.J. and Karasov, W.H. (1994) Gut passage of insects by European starlings and comparison with other species. *The Auk* 111, 478–481.

Levey, D.J. and Martínez del Río, C. (1999) Test, rejection, and reformulation of a chemical reactor-based model of gut function in a fruit-eating bird. *Physiological and Biochemical Zoology* 72, 369–383.

Light, R.J. and Pillemer, D.B. (1971) Accumulating evidence: procedures for resolving contradictions among different research studies. *Harvard Educational Review* 41, 429–471.

Mas, R. and Traveset, A. (1999) Efectes de l'ingestió per ocells sobre la germinació de dues espècies de *Solanum. Butlletí de la Societat d'Història Natural de les Balears* 42, 69–77.

Murray, K.G., Russell, S., Picone, C.M., Winnett-Murray, K., Sherwood, W. and Kuhlmann, M.L. (1994) Fruit laxatives and seed passage rates in frugivores: consequences for plant reproductive success. *Ecology* 75, 989–994.

Ocumpaugh, W.R., Archer, S. and Stuth, J.W. (1996) Switchgrass recruitment from broadcast seed vs. seed fed to cattle. *Journal of Range Management* 49, 368–371.

Osenberg, C.W., Sarnelle, O. and Cooper, S.D. (2000) Effect size in ecological experiments: the application of biological models in meta-analysis. *American Naturalist* 150, 798–812.

Paulsen, T.R. (1998) *Turdus* spp. and *Sorbus ocuparia* seeds: effect of ingestion on seed mass, germination and growth. In: Adams, N.J. and Slotow, R.H. (eds) *Ostrich. Proceedings of the 22 International Ornithological Congress.* Durban, p. 301.

Quinn, J.A., Mowrey, D.P., Emanuele, S.M. and Whalley, R.D.B. (1994) The 'foliage is the fruit' hypothesis: *Buchloe dactyloides* (*Poaceae*) and the shortgrass prairie of North America. *American Journal of Botany* 81, 1545–1554.

Rick, C.M. and Bowman, R.I. (1961) Galápagos tomatoes and tortoises. *Evolution* 15, 407–417.

Ridley, H.N. (1930) *The Dispersal of Plants Throughout the World.* Reeve, Ashford, UK, 744 pp.

Rosenberg, M.S., Adams, D.C. and Gurevitch, J. (2000) *Metawin: Statistical Software for Meta-Analysis,* Version 2.0. Sinauer Associates, Sunderland, Massachusetts, 128 pp.

Schupp, E.W. (1993) Quantity, quality and the effectiveness of seed dispersal by animals. *Vegetatio* 107/108, 13–29.

Temple, S.A. (1977) Plant–animal mutualism: coevolution with dodo leads to near extinction of plant. *Science* 197, 885–886.

Traveset, A. (1998) Effect of seed passage through vertebrate frugivores' guts on germination: a review. *Perspectives in Plant Ecology, Evolution and Systematics* 1/2, 151–190.

Traveset, A., Bermejo, T. and Willson, M.F. (2001) Effect of manure composition on seedling emergence and growth of two common shrub species of southeast Alaska. *Plant Ecology* (in press).

Traveset, A., Riera, N. and Mas, R. (2001) Ecology of the fruit-colour polymorphism in *Myrtus communis* and differential effect of mammals and birds on seed germination and seedling growth. *Journal of Ecology* (in press).

van der Pijl, L. (1982) *Principles of Dispersal in Higher Plants.* Springer-Verlag, Berlin, 199 pp.

Wahaj, S.A., Levey, D.J., Sanders, A.K. and Cipollini, M.L. (1998) Control of gut retention time by secondary metabolites in ripe *Solanum* fruits. *Ecology* 79, 2309–2319.

Witmer, M.C. (1996) Do some bird-dispersed fruits contain natural laxatives? A comment. *Ecology* 77, 1947–1948.

Xiong, S. and Nilsson, C. (1999) The effects of plant litter on vegetation: a meta-analysis. *Journal of Ecology* 87, 984–994.

Yagihashi, T., Hayashida, M. and Miyamoto, T. (1998) Effects of bird ingestion on seed germination of *Sorbus commixta. Oecologia* 114, 209–212.

23 Effectiveness of Seed Dispersal by *Cercopithecus* Monkeys: Implications for Seed Input into Degraded Areas

Beth A. Kaplin[1] and Joanna E. Lambert[2]

[1]*Department of Environmental Biology, Antioch New England Graduate School, 40 Avon Street, Keene, NH 03431, USA;* [2]*Department of Anthropology, University of Oregon, Eugene, OR 97403-1218, USA*

Introduction

Primates comprise a large proportion of the frugivore biomass in tropical forests (Eisenberg and Thorington, 1973; Terborgh, 1983) and are known to defecate or spit large numbers of seeds. Thus, they are likely to play a role in seed dispersal and tropical-forest regeneration (Chapman, 1995). However, there have been relatively few studies of seed dispersal by one of the most widely distributed and frugivorous genera of primates in Africa, *Cercopithecus* (Gautier-Hion, 1984; Kingdon, 1989; Gautier-Hion *et al.*, 1993; Wrangham *et al.*, 1994; Fairgreves, 1995; Plumptre *et al.*, 1997; Kaplin and Moermond, 1998; Lambert, 1999).

Commonly named guenons, *Cercopithecus* monkeys were believed to be mainly seed predators (Happel, 1988; Rowell and Mitchell, 1991), until recent studies documented that they defecate seeds in viable condition (Fairgreves, 1995; Kaplin and Moermond, 1998; Lambert, 1999). They are a particularly interesting group in which to study and contrast the process of frugivory and seed dispersal with that by other primates. In particular, guenons possess cheek pouches, a characteristic feature of the Cercopithecinae (the Old World subfamily to which they belong) (Fleagle, 1999), and they typically spit seeds. In contrast, apes and New World monkeys lack cheek pouches and tend to defecate seeds that are generally larger than those dispersed by guenons (Rowell and Mitchell, 1991; Wrangham *et al.*, 1994; Lambert, 1999). Guenons consume more foliage and have longer gut retention times than their New World frugivorous counterparts (Lambert, 1998). They may be either fully arboreal or semiterrestrial; this locomotor diversity may influence seed dispersal across the landscape (Fimble, 1994; Kaplin, 2001). Because of their wide geographical distribution, relatively large body size, high levels of frugivory and exploitation of both arboreal and terrestrial habitats, *Cercopithecus* monkeys have the potential to be effective seed-dispersers and may be particularly important in the regeneration of degraded forest.

We have three objectives in this chapter. First, we review the effectiveness of guenons as

 Seed Dispersal and Frugivory: Ecology, Evolution and Conservation (eds D.J. Levey, W.R. Silva and M. Galetti)

seed-dispersers, using Schupp's (1993) quantity and quality components of seed dispersal. The quantitative components we assess include fruit consumption, number and reliability of visits to fruiting trees and number of seeds dispersed per visit. The qualitative components we assess include the quality of seed treatment and deposition by guenons. We then assess their role in regeneration of degraded and human-disturbed forests. Finally, we suggest avenues for future research. As case studies, we focus on the blue monkey (*Cercopithecus mitis doggetti*) and mountain monkey (*Cercopithecus lhoesti*), studied in the Nyungwe Forest, Rwanda, and the redtail monkey (*Cercopithecus ascanius*) and blue monkey (*C. mitis stuhlmanni*), studied in Kibale Forest, Uganda.

Quantitative Traits of Seed-dispersal Effectiveness

Fruit consumption and reliability of visits

Quantity of dispersal is a function of the number of visits made by a disperser and the number of seeds dispersed per visit (Schupp, 1993). Here we consider the importance of fruit in the diet and reliability of visitation, both of which influence number of visits.

Fleshy fruit is a component of the diet of nearly all *Cercopithecus* species studied thus far (Rudran, 1978; Struhsaker, 1978; Gautier-Hion, 1980, 1983; Cords, 1986; Lawes, 1991; Fairgreves, 1995; Kaplin and Moermond, 1998). Guenon species differ in the amount of fruit eaten and in the choice of individual fruit species. Mountain monkeys in the Nyungwe Forest, Rwanda, for example, consume an overall mean of 24.5% fruit (± 9.8; $n = 9$ months) in their diet, while blue monkeys consume nearly twice as much fruit (47.4% ± 15.8; $n = 10$ months) (Kaplin and Moermond, 1998). In the Kibale Forest, Uganda, redtails consume an overall mean of 42% fruit in their diet (± 10.9; $n = 12$ months), while the blue monkeys consume a mean of 28.5% fruit (± 12.2; $n = 10$ months) (Lambert, 1997; J.E. Lambert, unpublished). A single fruit species may comprise a large proportion of the diet when it is locally abundant. For example, blue monkeys will spend up to 50% of their feeding time on the fruit of *Beilschmiedia troupinii* (*Lauraceae*) (Kaplin *et al.*, 1998) and redtails can spend up to 33% of their feeding time on the fruit of *Diospyros abyssinica* (*Ebenaceae*) (J.E. Lambert, unpublished). However, when fruit is temporarily scarce, guenons will switch to a diet of seeds or leaves (e.g. Gautier-Hion, 1988; Kaplin and Moermond, 1998; Wrangham *et al.*, 1998). This dietary flexibility is highlighted by population-level differences in diet composition; where fleshy fruit species are rare in a region due to low tree densities, guenon diets are mostly comprised of seeds and young leaves (Maisels and Gautier-Hion, 1994).

Disperser reliability includes a temporal component, the pattern of daily plant visitation, and a spatial component, in which a reliable disperser will dependably visit all individuals of all populations throughout the range of the plant species (Schupp, 1993). Guenons appear to demonstrate consistency in fruit selection over time in some cases, which may translate into visitor reliability. Based on dung sampled in the Nyungwe Forest, for instance, both blue and mountain monkeys dispersed the seeds of *Ficus oreodryadum* (*Moraceae*) and *Balthasarea schliebenii* (*Theaceae*) in dung throughout the study period. However, there are few quantitative data on the reliability of visits to fruiting trees. Guenons are most likely to be reliable visitors to species with relatively large fleshy fruits or asynchronous fruiters (e.g. figs). Their fruit choice is dependent on the array of fleshy fruits available at any time (Kaplin and Moermond, 1998). Chapman and Chapman (this volume) found not only intraspecific variation in the diet composition of guenon populations through time, but also interspecific variation in fruit consumption: even though certain fruit species were available to several different guenon groups, some groups consumed the fruits while others did not.

Patterns of visitation to fruiting plants can be assessed by considering the assemblage of dispersers that visit trees. Guenons may be the primary dispersers of a specific subset of tropical-forest tree species. Data from several studies of frugivores in the Nyungwe Forest (Kristensen, 1993; Sun and Moermond, 1997; Sun *et al.*, 1997; M. Masozera, unpublished data) suggest that two species of mature forest trees (*B. troupinii* and *Strombosia scheffleri*) with relatively large seeds are not visited by turacos, understorey birds or chimpanzees, but are visited regularly by guenons, which disperse their seeds (Table 23.1). Lacking from this analysis are data for visitation patterns of the black-and-white casqued hornbill (*Ceratogymna subcylindricus*), which can undoubtedly handle large fruits, and frugivorous bats.

Similar findings come from studies conducted in Malawi and Uganda. Fruiting-tree censuses and miscellaneous observations over 16 months at the main study site and an additional 18 months of opportunistic frugivory observations across all forested areas in Malawi showed that, of 134 fruit-producing trees observed, fruits of five species (*Aningeria adolfi-friedericii, Chrysophyllum gorungosanum, Ocotea usambarensis, Parinari excelsa* and *Ensete ventricosum*) were not eaten by any birds, but were eaten by guenons (Dowsett-Lemaire, 1988). Results from transects, primate-group scan sampling and dung analysis of two species of guenons, a colobus monkey, chimpanzees, hornbills and turacos in the Budongo Forest, Uganda, suggest that guenons were the only frugivores to consume the ripe flesh of six fruit-producing tree species (*Alaphia landolphioides, Alstonia boonei, Chrysophyllum perpulchrum, Cola gigantea, Guarea cedrata* and *Ricinodendron heudelotii*) (Plumptre *et al.*, 1997). More detailed studies are needed for estimates of visit reliability.

Seed handling and the probability of dispersal

We now discuss a second quantitative component of dispersal effectiveness: number of seeds dispersed per visit. The number of seeds dispersed is a product of the number of seeds handled and the probability that a handled seed is dispersed, both of which are affected by handling method (Schupp, 1993). Guenons handle seeds in a variety of ways; they can spit or drop seeds after removing the pulp, swallow seeds whole and defecate them or place fruits in their cheek pouches and remove them later to consume the pulp and drop or swallow the seeds (Kaplin and Moermond, 1998; Lambert, 1999).

Cheek pouches are an important component of seed handling in guenons (Corlett and Lucas, 1990; Rowell and Mitchell, 1991; Lambert, 1999; B.A. Kaplin, personal observation). Fairgreves (1995) documented 14 tree species whose seeds were spat from blue-monkey cheek pouches after the pulp had been removed (seed size ranged from 3 to 30 mm). In the Kibale Forest, seed spitting was the most common pattern of seed handling year round, and seeds that were spat had usually been stored in cheek pouches (Lambert, 1999). *Cercopithecus* monkeys tend to spit seeds > 10 mm under canopies of fruiting trees; thus, species with large seeds may have a low probability of dispersal by guenons. However, fruits of these species may be placed into cheek pouches and thereby dispersed 50 m or more from the parent tree (Rowell and Mitchell, 1991; Lambert, 1999). Lambert (1999) found that, while a majority of seeds in cheek pouches were spat under the parent tree (83.3%), some seeds (16.7%) were spat as far as 100 m from the parent tree.

In addition to dropping or spitting seeds, guenons defecate both large and small seeds intact. In the Nyungwe Forest, all mountain-monkey dung collected ($n = 58$) and 94% of blue-monkey dung ($n = 50$) contained intact seeds. Blue monkeys and mountain monkeys dispersed a mean of 2.3 and 6.4 seeds > 2 mm, respectively, per dung sample, with a maximum of 105 seeds > 2 mm in a single dung. Fairgreves (1995) found a mean of 6.12 seeds > 2 mm per blue-monkey dung sample ($n = 147$, SD = 12.76, range = 0–92) in the Budongo Forest, Uganda, and a maximum of 92 seeds > 2 mm of a single species in a single dung sample. Eighty-five per cent of these dung piles contained intact seeds. Fifty-five per cent of the Kibale redtail dung samples ($n = 135$) contained seeds, but most of these seeds (84%) were small-sized *Ficus* spp. (~2 mm). Blue monkeys in Kibale had seeds > 5 mm in only 16% (range = 5–16, mean = 9.3) of their dung samples ($n = 63$). Sixteen per cent of the dung samples with seeds contained only destroyed seeds, while 83.7% contained only intact seeds.

Differences in seed-spitting and defecation rates among sites may be due to feeding-behaviour differences among species of

Table 23.1. Observations from selected members of the frugivore community in the Nyungwe Forest, Rwanda. Dispersal distances are in metres, gut passage rates in hours for primates, minutes for birds.

		Tree species and mean seed size (mm)							
Frugivore	Disperser attributes	*Strombosia scheffleri* 18.30	*Beilschmeidia troupinii* 14.65	*Chrysophyllum rwandense* 13.25	*Syzygium parvifolium* 13.00	*Ilex mitis* 5.00	*Ocotea michelsonii* 3.50	*Harungana montana* 2.00	*Balthasarea schliebenii* 1.50
Cercopithecus species*	Seed handling	Defecate	Defecate	Defecate and spit	Defecate	Defecate	Defecate	Defecate	Defecate
	Dispersal distance	Unknown	Unknown	Unknown	Unknown	Unknown	Unknown	Unknown	Unknown
	Gut passage rate†	28 ± 4.75	28 ± 4.75	28 ± 4.75	28 ± 4.75	28 ± 4.75	28 ± 4.75	28 ± 4.75	28 ± 4.75
Chimpanzees‡	Seed handling	Not eaten	Not eaten	Defecate and spit	Defecate	Not eaten	Not eaten	Not eaten	Not eaten
	Dispersal distance	n/a	n/a	Unknown	Unknown	n/a	n/a	n/a	n/a
	Gut passage rate§	31 ± 12	31 ± 12	31 ± 12	31 ± 12	31 ± 12	31 ± 12	31 ± 12	31 ± 12
Turacos‖	Seed handling	Not eaten	Defecate rarely	Not eaten	Dropped	Defecate	Defecate	Defecate rarely	Defecate
	Dispersal distance	n/a	Unknown	n/a	Unknown	198 ± 63.4	Unknown	Unknown	153 ± 49
	Gut passage rate	n/a	Unknown	n/a	n/a	108 ± 15.3	Unknown	Unknown	55 ± 9.3
Understorey birds¶	Defecate seed	Not eaten	Not eaten	Not eaten	Not eaten	Defecate	Not eaten	Defecate	Not eaten
	Dispersal distance	n/a	n/a	n/a	n/a	20	n/a	19	n/a
	Gut passage rate	n/a	n/a	n/a	n/a	12 ± 0.8	n/a	12 ± 0.8	n/a

**Cercopithecus mitis doggetti* and *Cercopithecus lhoesti*; from Kaplin and Moermond, 1998; Kaplin *et al.*, 1998.
†From Lambert, 1998.
‡*Pan troglodytes*; M. Masozera, unpublished data.
§From Lambert, J.E. (2001) Digestive retention times in forest guenons with reference to chimpanzees. *International Journal of Primatology* 22(4), 521–548.
‖*Corythaeola cristata*, *Musophaga johnstoni* and *Tauraco schuettii*; from Sun *et al.*, 1997.
¶Bulbuls, family Pycnonotidae; from Kristensen, 1993.
n/a, not applicable.

Cercopithecus (Gautier-Hion *et al.*, 1993), different phenological patterns or different plant species at the respective sites. Primates may switch diets during periods of food shortage; these changes can influence rates of seed dispersal and seed predation. For example, the level of seed predation by primates in Gabon at a given time depends on community-wide availability of succulent fruits at that time (Tutin *et al.*, 1996). Gautier-Hion *et al.* (1993) identified seasonal shifts in guenon diets, as well as differences between the diets of closely related *Cercopithecus* species. In particular, *Cercopithecus pogonius* in Gabon dispersed the seeds of succulent fruits while *Cercopithecus wolfi* in Zaïre were mainly seed predators. The number of species and density of fruiting trees were lower at the Zaïre site, and *C. wolfi* shifted from being a seed predator to leaf eater to aril eater, depending on the season. Similarly, in the Nyungwe Forest, guenons alternated between acting predominantly as seed predators, seed defecators and seed droppers (or spitters) during the course of the study, depending on phenology patterns and resource availability (Fig. 23.1; Kaplin and Moermond, 1998).

Guenons also handle certain seed species in several different ways, each method having a different effect on the probability of dispersal. For example, three of the 39 species consumed by redtails were handled in two different ways; sometimes seeds were spat, sometimes

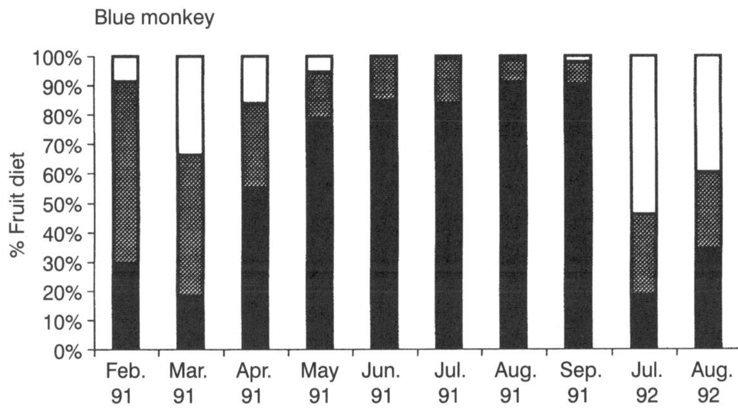

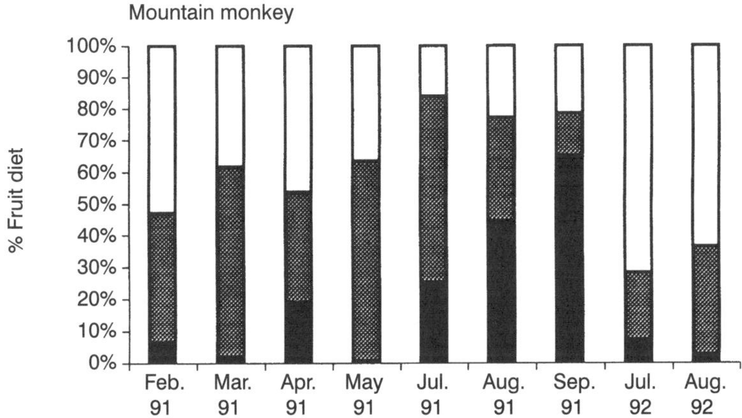

Fig. 23.1. Temporal shifts in seed-handling behaviours in the blue and mountain monkeys of Nyungwe Forest, Rwanda. Bars show percentage of fruit in diet with seeds handled in one of three ways: black bars = seed dropped; grey bars = seed defecated; clear bars = seed predated. July and August 1992 were periods of low fruit availability.

swallowed. Even when the redtails were found to be primarily spitters of a particular seed species, they occasionally swallowed those seeds. For example, redtails swallowed seeds of *Aphania senegalensis* (1.6%; 2/129 fruit-eating events (FEEs) observed), *Celtis durandii* (12.6%; 35/278 FEEs) and *Pseudospondias microcarpa* (1.5%; 3/204), although spitting seeds predominated for these fruit species. Of the 17 seed species defecated by blue monkeys in Nyungwe Forest, two were also chewed and digested and nine were also spat or dropped from parent canopies or carried in cheek pouches. Of the seed species defecated by mountain monkeys, two were chewed and destroyed and six were spat or dropped from the parent canopy. Although there is variability and even overlap in how seeds are handled relative to the probability of dispersal, there is also some consistency in handling methods for some fruit species. For example, in both the Kibale and Nyungwe Forests, seeds of all fig species (*Ficus* spp.) were swallowed whole and were the most common species found intact in the dung of all four guenon species.

Additional quantitative traits influencing the probability of seed dispersal

In addition to seed-handling methods, length of stay within a fruiting tree, number of fruits eaten and gut passage rates will affect the number of seeds dispersed per visit. Swallowed seeds may be regurgitated or defecated, and gut passage time for defecation varies considerably (Schupp, 1993). Of importance in evaluating the effectiveness of guenon seed dispersal is the frequency of visits and the number of seeds dispersed per visit. The average fruit-feeding rate of redtails in the Kibale Forest was 1.57 ± 0.47 fruit min^{-1}, and they spent approximately 2 h day^{-1} consuming fruit. Based on these data and the density of redtail monkeys (no. individuals km^{-2}), we calculate that the redtail monkey population in Kibale removes approximately 24,492 large seeds km^{-2} day^{-1}. In comparison, we calculate that chimpanzees remove approximately 1398 large seeds km^{-2} day^{-1}. Because redtails move out of the parent crown with seeds in their cheek pouches in 8.7% of the total fruit-feeding events (FEEs), an estimated 2130 seeds km^{-2} day^{-1} are moved away from the parent crown by redtails. In comparison, chimpanzees swallow 77% of the seeds they put in their mouths and remove an estimated 1076 seeds from the parent crown km^{-2} day^{-1}.

Quantity of seed dispersal: a summary

Dispersers are typically deficient in at least one quantitative trait (Schupp, 1993); *Cercopithecus* monkeys are no exception. In evaluating effectiveness we are interested in the major determinant of quantitative importance of a disperser to a plant. We do not yet have this kind of information for guenons. Based on avian frugivore studies, Schupp (1993) suggested that the number of visits by a disperser serves as a better predictor of total quantity of seed dispersed than does the number of seeds dispersed per visit. Guenons probably make fewer daily visits to a fruiting tree than most avian frugivores; however, groups of up to 40 individuals will often stay in fruiting-tree crowns for ≥ 30 min, almost continuously (B.A. Kaplin and J.E. Lambert, unpublished data) consuming fruit, at an average rate of 1.57 fruit min^{-1} (J.E. Lambert, unpublished data). These features of guenon feeding behaviour lead us to predict that the number of seeds dispersed per visit is more highly correlated with total number of seeds dispersed than is the number of visits.

Quality of Seed Dispersal

Seed treatment

Quality of seed dispersal is a function of the treatment a seed receives in the mouth and gut and the quality of seed deposition (i.e. the probability that a deposited seed survives and produces a new adult) (Schupp, 1993). Questions relevant to the quality of treatment include the following. Do the animals disperse intact seeds or do they destroy them? What is the probability that a seed handled in a certain manner will produce a new reproductive adult? Most seeds, regardless of guenon handling methods (spat, dropped or defecated), are intact and viable following deposition (Chapman, 1995; Lambert, 1997; Kaplin and Moermond, 1998). Contrary to the suggestion that guenons destroy most of the seeds they ingest (Rowell and Mitchell, 1991), we have demonstrated that they have the capability of dispersing a relatively large number of seeds away from parent crowns via defecation, placement in cheek pouches and spitting. Furthermore, the method of seed handling relative to the probability of dispersal is variable through time, depending on the array of resources available; guenons may spit most seeds in one month and defecate most seeds in another (Kaplin and Moermond, 1998).

Although guenons spit seeds under parent-tree crowns, their use of cheek pouches and the virtually unstudied role of secondary dispersal suggest that spitting should not imply a poor probability of the seed producing an adult plant. Spat seeds are viable (Kaplin and Moermond, 1998; Lambert, 2001) and, for some plant species, removal of pulp by guenons may be beneficial even if the seeds are not carried away from the parent tree. For example, in the Kibale Forest redtail monkeys spat out cleaned seeds of *Strychnos mitis* fruit (*Loganiaceae*) (seed dimensions: 1 cm × 1 cm × 0.5 cm) in a majority (88%) of the *S. mitis* fruit-eating observations (FEEs). In 83% of the *S. mitis* FEEs (*S. mitis* observed in 542 of 2930 FEEs), redtail monkeys spat out seeds within 10 m of the removal site and in 56% of the FEEs they moved < 1 m before spitting seeds. Most (83%) of *S. mitis* seeds spat out by redtail monkeys germinated, while only 12% of unprocessed seeds survived to germination (d.f. = 1; $\chi^2 = 91.2$; $P < 0.001$). Of the processed seeds that germinated, 60% survived to the seedling stage, whereas only 5% of the unprocessed seeds survived to seedling establishment (d.f. = 1; $\chi^2 = 605.39$; $P < 0.001$). Unprocessed seeds were also more likely to be attacked by seed predators (d.f. = 1; $\chi^2 = 11.77$; $P < 0.001$) and fungus (d.f. = 1; $\chi^2 = 156.88$; $P < 0.001$) (Lambert, 2001).

Quality of deposition

The quality of seed deposition is a function of the seed shadows produced by dispersers and the survival probabilities associated with these shadows (Schupp, 1993). Attributes of guenon behaviour that determine where a seed will fall include habitat selection and movement patterns. Although the locations of 'suitable' sites are unpredictable, some sites are predictably associated with higher probabilities of survival than others (Schupp *et al.*, 1989). While many seeds are deposited by primates in sites with low survival rates (e.g. Chapman, 1989; Andresen, 1999), several studies have found that primates occasionally deposit seeds in 'suitable' sites. For example, Rogers *et al.* (1998) suggest that seeds in gorilla dung deposited in gorilla nests may experience increased seedling survival over seeds deposited in dung along gorilla trails or that germinate under parent trees. Julliot (1997) found that an aggregated pattern of seedling distribution correlated with the clumped pattern of seed dispersal produced by howler monkeys at sleeping sites.

Guenons have the potential to disperse seeds to a variety of habitats within their home range. They have relatively long daily travel distances (Nyungwe mountain monkey mean = 2092 ± 333.8 m, Nyungwe blue monkey mean = 1307 ± 335.6 m; Kibale redtail mean = 1178 ± 399.8 m), coupled with long gut-retention times, which together determine seed-deposition patterns. In addition, seeds placed in cheek pouches can be carried 30–100 m from the parent tree before being spat (Rowell and Mitchell, 1991; Lambert, 1999; Voysey *et al.*, 1999). Guenons typically enter several habitat types in a day (Kaplin, 2001). In particular, mountain monkeys commonly use revegetating landslides, forest selectively logged at least 30 years earlier and road and trail edges (Kaplin and Moermond, 2000).

To determine whether guenons deposit seeds in specific habitat types, we recorded the microhabitat into which seeds were dropped, spat or defecated by Kibale redtails. 'Microhabitat' in this study was a description of the vegetation structure within a 2 m radius around a deposited seed and served as a broad measure of the forest structure and light regime of the dispersal microhabitat. Each microhabitat was assigned a vegetation structure type (VST) score of 1–5: 1 = closed upper canopy with a dense, herbaceous understorey; 2 = open upper canopy with an understorey consisting mostly of grasses; 3 = open upper canopy with a dense, herbaceous understorey containing no grass species; 4 = closed upper canopy with an open understorey; and 5 = open upper canopy and an open understorey (e.g. large forest gap, trail or road).

Redtail monkeys spat or defecated seeds primarily into VSTs 1, 2 and 3, which we lumped into a 'primate-microhabitat' category. Redtails never spat seeds into VST categories 4 or 5 (open understorey areas). Of a total of 1143 seed-spitting events, redtails spat seeds into VST 1 in 52% of the total FEEs, into VST 2 in 23% and into VST 3 in 25%.

Seeds ($n = 8600$) of seven large-seeded species known to be dispersed by redtails were set into each of the VSTs and monitored for rates of: (i) damage by rodent gnawing or insect boring; (ii) attack by fungus; (iii) germination; (iv) root establishment; (v) establishment followed by mortality; and (vi) removal. These seed-fate trials indicated that some seed species experience greatest damage when in 'non-primate microhabitats' (i.e. VST 4, VST 5 or areas with an open understorey). For example, *A. senegalensis, C. durandii, P. microcarpa, Linociera johnsonii, Monodora microcarpa* and *Uvariopsis congensis* all experienced less seed damage and seed removal by rodents when in a primate microhabitat (VSTs 1, 2 and 3) than when in non-primate VSTs (Lambert, 1997). *Uvariopsis congensis* seeds were also more likely to germinate in primate microhabitat than elsewhere. These results demonstrate that many seeds are viable after being spat by redtails and, moreover, seeds spat into 'primate microhabitats' have a greater rate of survival.

Cercopithecus monkeys move through disturbed forest, where they also deposit large seeds in widely distributed dung piles. These seeds may survive and germinate, potentially contributing to regeneration in degraded forest. Dung piles of equivalent sizes collected from Nyungwe Forest blue or mountain monkeys were examined for seeds > 2 mm and then placed in three different disturbed forest habitats: secondary forest, landslides and degraded hilltops. Secondary forest was dominated by dense stands of pioneer trees, landslides were large areas on slope sides dominated by grasses and other herbs but without trees, and degraded hilltops had been selectively logged such that canopy height was low and vegetation was dominated by understorey shrubs. Table 23.2 shows the four main large seed species present in the dung ($n = 186$ seeds total) and the percentages that germinated across all the disturbed habitats. Seeds persisted in the dung piles for a mean of 180.6 ± 45.4 days and 4.6% of these seeds germinated. Seedlings were present for a mean of 11.5 months (± 0.98; range = 9–13 months, the last date checked).

Table 23.2. Fates of seeds > 2 mm in *Cercopithecus* dung piles placed in secondary and disturbed forest.*

Species	% occurrence in dung†	Total no. seeds‡	Mean no. seeds/dung	% germinated§	No. days to germination‖	Mean no. months seedling persisted
Canthium sp. B	27.0	46	5.8 ± 3.1	23.9	120.8	11.4 ± 3.38
Embellia sp.	23.3	59	11.8 ± 16	6.8	224	13.5 ± 1
Galiniera coffeoides	16.7	57	8.1 ± 5.2	22.8	171.5	9.34 ± 1.94
Rapanea melanophloeios	10	24	8.0 ± 3.0	2.5	206	12 ± 2

*Ten dung piles were placed in each of three different disturbed habitats (see text). Not all seed species were present in all dung piles. Results summed across habitats.
†The percentage of dung piles containing seeds of indicated species.
‡Total no. seeds present in all 30 dung piles.
§Based on total no. seeds of given species.
‖Determined by presence of first germinating seed.

Seed spitting is a prominent feature of seed handling by guenons and the importance of seeds spat or dropped by monkeys to forest regeneration is only just beginning to receive attention. Spat seeds are generally large and are often spat under the parent canopy. Chapman and Chapman (1996) found a positive relationship between the proportion of the fruit crop removed from parent crowns and both seed disappearance and seedling survival under the parent crown; those species that had a lower percentage of the fruit crop removed (e.g. more seeds dropped under parent crowns) had lower rates of seed disappearance and higher rates of seedling survival under those crowns. This suggests that some species have seeds that can both disperse away from the parents and recruit under or near the parent crowns. Forget (1992) suggested that the benefit of seed dispersal for a large-seeded tree is probably greatest when its seeds are dispersed near the parent, in a habitat where food availability satiates predators and seedling survival is promoted. This hypothesis is supported by the observation that many tropical tree species have aggregated distributions (Ashton, 1969; Hubbell, 1979).

The length of time that a seed is maintained in a frugivore's gut will also influence seed shadows; longer retention times may result in greater dispersal distances. Mean retention times in guenons are longer than in many other frugivorous primates of similar body size, such as *Ateles* or *Cebus*, and more similar to those of folivorous New World primates (Lambert, 1998). Mean gut-retention time from experiments with three captive *Cercopithecus* species (*C. mitis stuhlmanni*, *C. neglectus* and *C. ascanius schmidti*) was 28.8 ± 4.75 h (Lambert, 1997). Similarly, Maisels (1993) found a mean retention time in guenons of 26.7 ± 3.7 h (*C. ascanius* and *C. erythrotis*) and 26.9 ± 6.7 h for *C. pogonius*.

There have been few studies of seed fate following processing by Old World primates (although see Lambert, this volume). In the New World, most seeds dispersed by primates are killed by seed predators and/or buried by dung beetles (Chapman, 1989; Estrada and Coates-Estrada, 1991; Andresen, 1999). In a study of gorilla seed dispersal in Gabon, Voysey *et al.* (1999) found that, while rates of predation on seeds in gorilla dung piles depended on seed species and year, gorilla nests served as 'safe' sites for seed germination and vigorous seedling growth. These findings highlight the importance of assessing the effects of post-dispersal factors on seed fate.

Implications for Conservation: Seed Input into Degraded Forest

Deforestation and fragmentation are causing widespread changes in the forests of sub-Saharan Africa. Vertebrate seed-dispersers are believed to play an important role in the regeneration of degraded forest habitats; they are probably the main dispersers of large-seeded primary-forest trees and can transport these seeds into degraded habitats (Bawa and Hadley, 1990). In particular, guenons regularly use abandoned farm clearings and young, regenerating forest (Fimble, 1994), secondary forest (Thomas, 1991) and pine and cypress plantations adjacent to forest (Maganga and Wright, 1991). Guenons also enter agricultural fields to raid crops (Naughton-Treves *et al.*, 1998; B.A. Kaplin and J.E. Lambert, personal observation). We now evaluate the role guenons may play in regeneration of degraded forest, a qualitative aspect of dispersal effectiveness. We examined the use of disturbed forest sites by guenons and the distribution of saplings of large-seeded or mature forest species known to be eaten and dispersed by guenons in these sites.

We documented blue- and mountain-monkey use of forest habitat by determining the number of times groups entered each of the different habitat types within their home ranges during complete day follows (Kaplin *et al.*, 1998; Kaplin, 2001). The more terrestrial mountain monkeys included 21 ha of landslide in their home range, which was used by the group in 17.4% of the habitat-type entries recorded, while the blue monkeys were never observed to enter landslide habitat during the study. Blue- and mountain-monkey groups selected habitat types non-randomly during the study (blue monkey: $\chi^2 = 546.55$, $P < 0.001$, d.f. = 12; mountain monkey: $\chi^2 = 13018.64$, $P < 0.001$, d.f. = 12) (Kaplin, 2001). Mountain monkeys entered all habitat types available in their home range less than expected, except for secondary forest, which they entered most frequently. Blue monkeys used wet valleys and steep slopes less than expected; mature forest was used most frequently.

We identified eight different habitat types (wet valleys, secondary ridge tops, landslides, open areas, secondary degraded hilltops, upper slopes, secondary forest and closed canopy or mature forest) and conducted stratified random sampling of trees (> 10 cm diameter at breast height (dbh)) and saplings within the home ranges of the guenon groups (Kaplin, 2001). Only two of the six species present as saplings in these habitats did not show a positive correlation between sapling density and adult tree density within habitat types (*Ilex mitis*, *Aquifoliaceae*, Pearson correlation coefficient $r = -0.14$, d.f. = 4, $P > 0.10$; and *S. scheffleri*, *Olacaceae*, Pearson correlation coefficient $r = 0.37$, d.f. = 4, $P > 0.10$) (Fig. 23.2). The presence of all other sapling species was positively correlated with adult conspecific presence (Fig. 23.2). Figure 23.3 shows the distribution of saplings of *I. mitis* and *S. scheffleri* in each of the eight forest habitats, including degraded habitats where no or very few adult conspecifics exist. *Ilex mitis* is dispersed by guenons, turacos and small, understorey birds, while *S. scheffleri* is apparently not eaten by any of the frugivorous birds in the Nyungwe Forest study, but is eaten by guenons. Furthermore, all the species of saplings we found in degraded habitats have seeds dispersed by guenons; certain frugivorous birds and bats may also disperse them.

The dispersal of large seeds is of particular interest in forest restoration because large birds and mammals are the primary dispersers of these seeds (Bawa and Hadley, 1990). Small-sized seeds are probably dispersed by a suite of birds and bats that are unlikely to disperse the large seeds characteristic of primary-forest tree species (Foster and Janson, 1985; Moermond *et al.*, 1993). Chapman and Onderdonk (1998) found a decrease in the recruitment of large-seeded species in forest fragments in Kibale Forest and suggested it was due to decreased primate populations in the fragments.

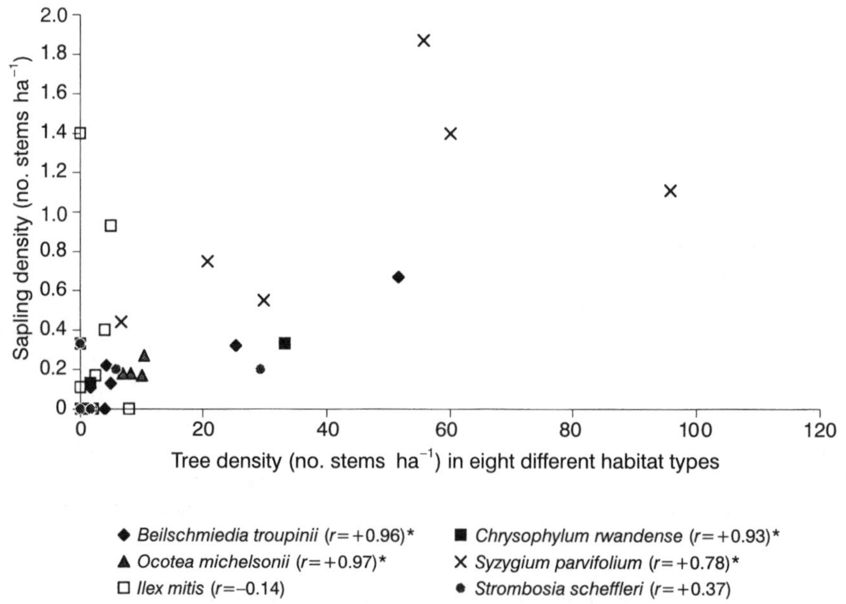

Fig. 23.2. Correlation between sapling and adult conspecific densities of tree species with fruits known to be eaten by guenons across eight different habitat types in the Nyungwe Forest, Rwanda. See text for description of habitat types. Pearson correlation coefficient values marked with * are significant ($P < 0.05$).

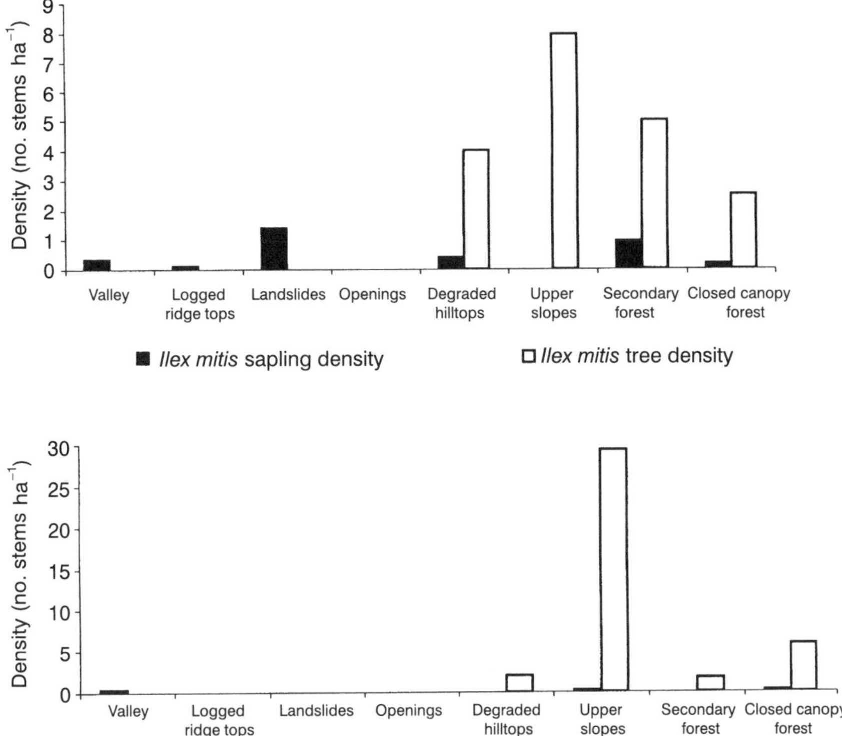

Fig. 23.3. The presence of saplings and adults of two tree species with fruits eaten by guenons in the Nyungwe Forest, Rwanda. Top = *Ilex mitis* (*Aquifoliaceae*) and bottom = *Strombosia scheffleri* (*Olacaceae*).

Large seeds arrive in disturbed or early-successional habitats by several routes. They may be present in the soil seed bank prior to disturbance or exist as seedlings that are released following disturbance. Large seeds may also be directly dispersed into degraded habitats. Uhl (1987), comparing soil seed banks, stump sprouting and germination of seeds arriving through seed dispersal in abandoned farms in Amazonia, found that seed dispersal is a critical first step in the establishment of primary-forest species on these disturbed sites. His experiments demonstrated that primary-forest tree seedlings will survive and grow in abandoned farms as long as shade is available. Although considered shade-tolerant, primary-forest seedling species can coexist with pioneer (shade-intolerant) seedling species. Guevara *et al.* (1986) found mature-forest (shade-tolerant) species mixed with pioneer (shade-intolerant) species in early-successional sites. Simultaneous colonization of such species is probably common in forests recovering from disturbances (Whitmore, 1989). Uhl (1987) also found that the probability that a seed of a primary-forest species would escape predation and germinate increased as seed size increased and, once seeds germinated, they stood a high chance of survival.

There are obvious conservation implications from these types of studies. In a recent analysis of forest succession in abandoned agricultural areas surrounding Kibale Forest, Duncan and Chapman (1999) found that half of the *Musa* plants they examined were visited by primates capable of dispersing seeds of forest species rarely dispersed by bats and birds. Restoration and regeneration in degraded landscapes is likely to be partially limited by lack of seed rain by dispersers of large seeds.

Based on their high rates of frugivory, their handling of large seeds, their long mean retention times and their use of degraded areas, *Cercopithecus* monkeys are likely to play a role in the establishment of large-seeded species in degraded habitats. This should be a profitable and important area for further research.

Epilogue and Avenues for Future Research

Empirical studies of primate seed dispersal have mostly focused on the quantity of seeds dispersed, although more recent efforts have dealt with qualitative aspects of dispersal effectiveness (e.g. Julliot, 1997; Rogers *et al.*, 1998; Lambert, 1999; Voysey *et al.*, 1999). *Cercopithecus* monkeys have relatively long gut-retention times, have long daily travel distances and may pass through a variety of habitat types during the course of a day; certain *Cercopithecus* species will reliably enter openings or degraded forest during the course of a day. As the proportion of tropical forest becoming degraded increases, we should be more concerned with the interactions of degraded forest and seed-dispersers; *Cercopithecus* monkeys are an ideal taxon for these kinds of studies.

There are myriad anatomical, ecological and physiological variables that influence the effectiveness of seed dispersal. Effectiveness must be evaluated through a combination of quantitative and qualitative attributes and should include plant demography (Schupp, 1993). Although we have documented many attributes of *Cercopithecus* seed-dispersal behaviour that determine effectiveness, there are many aspects that remain little studied. In particular, future studies should focus on the impact of secondary dispersal agents and the fate of dispersed seeds (especially in degraded versus intact forest), broad patterns of community-wide frugivory and the relative effectiveness of *Cercopithecus* monkeys, the role of fruit morphology and chemical defences in dispersal patterns, the ecomorphology of feeding, including the impact of oral and digestive adaptations (e.g. cheek pouches), qualitative aspects of seed dispersal, such as deposition patterns (e.g. comparisons of clumped and low-density seed dispersal, presence of safe sites, microsite analyses), and the demographics of large-seeded, 'primate-dispersed' tree species.

Acknowledgements

We are grateful to Michel Masozera for sharing his unpublished data with us. We thank Doug Levey, Gene Schupp and Colin Chapman, for helpful comments on earlier drafts of this manuscript, and the organizers of the Third Frugivory and Seed Dispersal Symposium for bringing the meeting and this manuscript to fruition, especially Doug Levey, Mauro Galetti and Wesley Silva.

References

Andresen, E. (1999) Seed dispersal by monkeys and the fate of dispersed seeds in a Peruvian rain forest. *Biotropica* 31, 145–158.

Ashton, P.S. (1969) Speciation among tropical forest trees: some deductions in the light of recent evidence. *Biological Journal of the Linnean Society* 1, 55–96.

Bawa, K.S. and Hadley, M. (1990) *Reproductive Ecology of Tropical Forest Plants*, Vol. 7. Man and Biosphere Series, Parthenon Publishing Group, Park Ridge, New Jersey, 415 pp.

Chapman, C.A. (1989) Primate seed dispersal: the fate of dispersed seeds. *Biotropica* 21, 148–154.

Chapman, C.A. (1995) Primate seed dispersal: coevolution and conservation implications. *Evolutionary Anthropology* 4, 74–82.

Chapman, C.A. and Chapman, L.J. (1996) Frugivory and the fate of dispersed and non-dispersed seeds of six African tree species. *Journal of Tropical Ecology* 12, 491–504.

Chapman, C.A. and Onderdonk, D.A. (1998) Forests without primates: primate/plant codependency. *American Journal of Primatology* 45, 127–141.

Cords, M. (1986) Interspecific and intraspecific variation in diet of two forest guenons, *Cercopithecus ascanius* and *C. mitis*. *Journal of Animal Ecology* 55, 811–827.

Corlett, R.T. and Lucas, P.W. (1990) Alternative seed-handling strategies in primates: seed-spitting by long-tailed macaques (*Macaca fascicularis*). *Oecologia* 82, 166–171.

Dowsett-Lemaire, F. (1988) Fruit choice and seed dissemination by birds and mammals in the evergreen forests of upland Malawi. *Revue Ecologie (Terre et Vie)* 43, 251–285.

Duncan, R.S. and Chapman, C.A. (1999) Seed dispersal and potential forest succession in abandoned agriculture in tropical Africa. *Ecological Applications* 9, 998–1008.

Eisenberg, J.F. and Thorington, R.W. (1973) A preliminary analysis of a neotropical mammal fauna. *Biotropica* 5, 150–161.

Estrada, A. and Coates-Estrada, R. (1991) Howler monkeys (*Alouatta palliata*), dung beetles (Scarabaeidae) and seed dispersal: ecological interactions in the tropical rain forest of Los Tuxtlas, Mexico. *Journal of Tropical Ecology* 7, 459–474.

Fairgreves, C. (1995) The comparative ecology of blue monkeys (*Cercopithecus mitis stuhlmanni*) in logged and unlogged Forest, Budongo Forest Reserve, Uganda. PhD dissertation, Scottish Primate Research Group, University of Edinburgh, Edinburgh, Scotland.

Fimble, C. (1994) Ecological correlates of species success in modified habitats may be disturbance- and site-specific: the primates of Tiwai Island. *Conservation Biology* 8, 106–113.

Fleagle, J.G. (1999) *Primate Adaptation and Evolution*, 2nd edn. Academic Press, New York.

Forget, P.-M. (1992) Seed removal and seed fate in *Gustavia superba* (*Lecythidaceae*). *Biotropica* 24, 408–414.

Foster, S.A. and Janson, C.H. (1985) The relationship between seed size and establishment conditions in tropical woody plants. *Ecology* 66, 773–780.

Gautier-Hion, A. (1980) Seasonal variations of diets related to species and sex in a community of *Cercopithecus* monkeys. *Journal of Animal Ecology* 49, 237–269.

Gautier-Hion, A. (1983) Leaf consumption by monkeys in western and eastern Africa: a comparison. *African Journal of Ecology* 21, 107–113.

Gautier-Hion, A. (1984) La dissemination des graines par les cercopithécides forestiers africains. *Revue Ecologie (Terre et Vie)* 39, 159–165.

Gautier-Hion, A. (1988) The diet and dietary habits of forest guenons. In: Gautier-Hion, A., Bourlière, F. and Gautier, J.-P. (eds) *A Primate Radiation: Evolutionary Biology of the African Guenons*. Cambridge University Press, Cambridge, pp. 257–283.

Gautier-Hion, A., Gautier, J.-P. and Maisels, F. (1993) Seed dispersal versus seed predation: an inter-site comparison of two related African monkeys. *Vegetatio* 107/108, 237–244.

Guevara, S., Purata, S.E. and Van der Maarel, E. (1986) The role of remnant trees in tropical forest secondary succession. *Vegetatio* 66, 77–84.

Happel, R. (1988) Seed-eating by west African cercopithecines, with reference to the possible evolution of bilophodont molars. *American Journal of Physical Anthropology* 75, 303–317.

Hubbell, S.P. (1979) Tree dispersion, abundance, and diversity in a tropical dry forest. *Science* 203, 1299–1309.

Julliot, C. (1997) Impact of seed dispersal by red howler monkeys *Alouatta seniculus* on the seedling population in the understory of tropical rain forest. *Journal of Ecology* 85, 431–440.

Kaplin, B.A. (2001) Ranging behaviour of two species of forest guenons (*Cercopithecus lhoesti* and *C. mitis doggetti*) in the Nyungwe Forest Reserve, Rwanda. *International Journal of Primatology* 22(4), 521–548.

Kaplin, B.A. and Moermond, T.C. (1998) Variation in seed handling by two species of forest monkeys in Rwanda. *American Journal of Primatology* 45, 83–101.

Kaplin, B.A. and Moermond, T.C. (2000) Foraging ecology of the mountain monkey (*Cercopithecus l'hoesti*): implications for its evolutionary history and use of disturbed forest. *American Journal of Primatology* 50, 227–246.

Kaplin, B.A., Munyaligoga, V. and Moermond, T.C. (1998) The influence of temporal changes in fruit availability on diet composition and seed handling in blue monkeys (*Cercopithecus mitis doggetti*). *Biotropica* 30, 56–71.

Kingdon, J. (1989) *Island Africa*. Princeton University Press, Princeton, New Jersey, 287 pp.

Kristensen, K.A. (1993) Human disturbance and seed dispersal by understory avian frugivores: key elements in the maintenance of tropical forest diversity. Masters thesis, Institute of Environmental Studies, University of Wisconsin–Madison, Madison, Wisconsin.

Lambert, J.E. (1997) Digestive strategies, fruit processing, and seed dispersal in the chimpanzees (*Pan troglodytes*) and redtail monkeys (*Cercopithecus ascanius*) of Kibale National Park, Uganda. PhD dissertation, University of Illinois, Urbana–Champaign, Illinois.

Lambert, J.E. (1998) Primate digestion: interactions among anatomy, physiology, and feeding ecology. *Evolutionary Anthropology* 7, 8–20.

Lambert, J.E. (1999) Seed handling in chimpanzees (*Pan troglodytes*) and redtail monkeys (*Cercopithecus ascanius*): implications for understanding hominoid and Cercopithecine fruit-processing strategies and seed dispersal. *American Journal of Physical Anthropology* 109, 365–386.

Lambert, J.E. (2001) Red-tailed guenons (*Cercopithecus ascanius*) and *Strychnos mitis*: evidence for plant benefits beyond seed dispersal. *International Journal of Primatology* 22(2), 189–201.

Lawes, M.J. (1991) Diet of samango monkeys (*Cercopithecus mitis erythrarchus*) in the Cape Vidal dune forest, South Africa. *Journal of Zoology* 224, 149–173.

Maganga, S.L.S. and Wright, R.G. (1991) Bark-stripping by blue monkeys in a Tanzanian forest plantation. *Tropical Pest Management* 37, 169–174.

Maisels, F. (1993) Gut passage rate in guenons and mangabeys: another indicator of a flexible feeding niche? *Folia Primatologia* 61, 35–37.

Maisels, F. and Gautier-Hion, A. (1994) Why are *Caesalpinioideae* so important for monkeys in hydromorphic rainforests of the Zaire Basin? In: *Advances in Legume Systematics*, Vol. 5: *The Nitrogen Factor.* Royal Botanical Gardens, Kew, UK, pp.189–204.

Moermond,T.C., ka Kajondo, K., Sun C., Kristensen, K., Munyaligoga, V., Kaplin, B.A., Graham, C. and Mvukiyumwami, J. (1993) Avian frugivory in a montane rain forest: tree visitation patterns. In: Wilson, R.T. (ed.) *Proceedings VIII Pan-African Ornithological Conference.* Royal Museum of Central Africa, Tervuren, Belgium, pp. 421–439.

Naughton-Treves, L., Treves, A., Chapman, C.A. and Wrangham, R. (1998) Temporal patterns of crop-raiding by primates: linking food availability in croplands and adjacent forest. *Journal of Applied Ecology* 35, 596–606.

Plumptre, A.J., Reynolds, V. and Bakuneeta, C. (1997) *The Effects of Selective Logging in Monodominant Tropical Forests on Biodiversity.* Final Report Overseas Development Administration Forestry Research Programme No. R6057, Institute of Biological Anthropology Report, Oxford, 129 pp.

Rogers, M.E., Voysey, B.C., McDonald, K.E., Parnell, R.J. and Tutin, C.E.G. (1998) Lowland gorillas and seed dispersal: the importance of nest sites. *American Journal of Primatology* 45, 45–68.

Rowell, T.E. and Mitchell, B.J. (1991) Comparison of seed dispersal by guenons in Kenya and capuchins in Panama. *Journal of Tropical Ecology* 7, 269–274.

Rudran, R. (1978) *Socioecology of Blue Monkeys (*Cercopithecus mitis stuhlmanni*) of the Kibali Forest, Uganda.* Smithsonian Institution Press, Washington, DC, 88 pp.

Schupp, E.W. (1993) Quantity, quality and the effectiveness of seed dispersal by animals. *Vegetatio* 107/108, 15–29.

Schupp, E.W., Howe, H.F., Augspurger, C.K. and Levey, D.J. (1989) Arrival and survival in tropical treefall gaps. *Ecology* 70, 562–564.

Struhsaker, T.T. (1978) Food habits of five monkey species in the Kibale Forest, Uganda. In: Chivers, D.J. and Herbert, J. (eds) *Recent Advances in Primatology.* Academic Press, New York, pp. 225–245.

Sun, S. and Moermond, T.C. (1997) Foraging ecology of three sympatric turacos in a montane forest in Rwanda. *The Auk* 114, 396–404.

Sun, S., Ives, A.R., Kraeuter, H.J. and Moermond, T.C. (1997) Effectiveness of three turacos as seed dispersers in a tropical montane forest. *Oecologia* 112, 94–103.

Terborgh, J. (1983) *Five New World Primates.* Princeton University Press, Princeton, New Jersey, 260 pp.

Thomas, S.C. (1991) Population densities and patterns of habitat use among anthropoid primates in the Ituri Forest, Zaire. *Biotropica* 23, 68–83.

Tutin, C.E.G., Parnell, R.J. and White, F. (1996) Protecting seeds from primates: examples from *Diospyros* spp. in the Lope Reserve, Gabon. *Journal of Tropical Ecology* 12, 371–384.

Uhl, C. (1987) Factors controlling succession following slash-and-burn agriculture in Amazonia. *Journal of Ecology* 73, 377–407.

Voysey, B.C., McDonald, K.E., Rogers, M.E., Tutin, C.E.G. and Parnell, R.J. (1999) Gorilla and seed dispersal in the Lope Reserve, Gabon. I: Gorilla acquisition by trees. *Journal of Tropical Ecology* 15, 23–38.

Whitmore, T.C. (1989) Canopy gaps and the two major groups of forest trees. *Ecology* 70, 536–538.

Wrangham, R.W., Chapman, C.A. and Chapman, L.J. (1994) Seed dispersal by forest chimpanzees in Uganda. *Journal of Tropical Ecology* 10, 355–368.

Wrangham, R.W., Conklin-Brittain, N.L. and Hunt, K.D. (1998) Dietary response of chimpanzees and cercopithecines to seasonal variation in fruit abundance. I. Antifeedants. *International Journal of Primatology* 19, 949–970.

24 Exploring the Link between Animal Frugivory and Plant Strategies: the Case of Primate Fruit Processing and Post-dispersal Seed Fate

Joanna E. Lambert
Department of Anthropology, University of Oregon, Eugene, OR 97403-1218, USA

Introduction

Frugivores have evolved a diversity of fruit-processing strategies in which seeds may be dropped, masticated and destroyed, regurgitated, spat or swallowed and then defecated intact (Moermond and Denslow, 1985; Levey, 1986; Corlett and Lucas, 1990; Rowell and Mitchell, 1991; Kaplin and Moermond, 1998; Lambert, 1999). Because these fruit-processing behaviours influence seed deposition patterns, vertebrate fruit processing can determine the degree to which post-dispersal factors, such as seed predation and pathogen attack, will influence seed and seedling survivorship and, ultimately, plant reproductive strategies.

Given the enormous variability in animal foraging and feeding strategies, models linking animal feeding and seed dispersal to seed fate and plant demography are difficult to generate (especially across taxa) and are therefore few. One notable exception is Howe's (1989) 'scatter- and clump-dispersal' hypothesis. In this model, Howe hypothesizes that selection pressures from vertebrates that habitually scatter seeds ('scatter dispersal') result in different seed and seedling traits from those of vertebrates that habitually leave many seeds in large faecal clumps ('clump dispersal'). Howe predicts that 'scatter-dispersed' plant species are unlikely to evolve tolerance against density-dependent factors (i.e. predators or pathogens) because they will recruit to the seedling stage as isolated individuals, while 'clump-dispersed' plant species are likely to be under selective pressure to evolve chemical and/or mechanical defences against competitors, seed predators and pathogens.

Ultimately, an animal's impact as a disperser should be measured by the relative survival of dispersed seeds versus non-dispersed seeds. Thus, to build and test models linking animal attributes to plant adaptations (e.g. scatter- vs. clump-dispersal hypothesis), we need data on both fruit-processing behaviours of animals and the post-dispersal fate of seeds. Unfortunately, such a 'marriage of animal foraging with plant demography is rarely consummated' (Howe, 1989, p. 417).

Data linking foraging behaviour with plant demography are lacking for primates, despite the fact that primates have been more thoroughly studied than any other order of tropical mammals (Peres and Janson, 2000). Moreover, most primate species are at least partially frugivorous, and many are highly so. Yet, with a few exceptions, fruit-processing behaviour and the post-dispersal fate of seeds remain unquantified for primate species.

My goals are thus twofold. First, I shall provide a review of the fruit-processing behaviour of primates. This discussion is based on published, but largely anecdotal, reports of seed-handling behaviour. I augment these reports with data from my own detailed case-study on fruit processing by two African primates – one cercopithecine monkey, *Cercopithecus ascanius*, and one ape, *Pan troglodytes*. Secondly, I present data from experiments designed to evaluate the influence of cercopithecine and chimpanzee seed-handling behaviours and seed deposition on seed fate. In particular, I examine how fruit processing and seed deposition affect post-dispersal fungal attack, predation and germination. Because one of the primate species in the case-study is a scatter-disperser and the other a clump-disperser, these seed-fate experiments allow evaluation of the scatter- vs. clump-dispersal hypothesis (Howe, 1989).

Fruit-processing Behaviour in Primates

Fruit processing refers to the behavioural methods by which a primate removes nutritive or energy-rich pulp or aril from a seed (Lambert, 1997). Differences occur both within and among primate species, depending upon fruit and seed size, ripeness, pericarp morphology, fruit availability and the anatomy and behaviour of the frugivore (McKey, 1978; Corlett and Lucas, 1990; Kinzey and Norconk, 1990; Davies, 1991; Rowell and Mitchell, 1991; Gautier-Hion *et al.*, 1993; Leighton, 1993; Maisels *et al.*, 1994; Ungar, 1995; Tutin *et al.*, 1996; Kaplin and Moermond, 1998; Lambert, 1999). Primates do not regurgitate seeds in the manner described in birds and duikers (Levey, 1986; Feer, 1995). Thus, among primates there are three broad categories of oral processing: swallowing, spitting and masticating seeds (Corlett and Lucas, 1990). No primate is strictly a seed predator, seed swallower or seed spitter; indeed, most employ a variety of fruit-processing behaviours (e.g. Corlett and Lucas, 1990; Rowell and Mitchell, 1991; Kaplin and Moermond, 1998; Lambert and Garber, 1998).

Primates as taxonomically and geographically diverse as ceboids, colobines and cercopithecines masticate and thereby destroy seeds. In lowland Brazil, for example, brown capuchins (*Cebus apella*) kill seeds of several species (Peres, 1991). In Borneo, leaf monkeys (*Presbytis rubicunda*) chisel fruit pulp off seeds, drop the pulp and then break the seed-coat and swallow the endosperm (Davies, 1991). Gautier-Hion *et al.* (1993) demonstrate that, in the Democratic Republic of Congo (formally Zaïre), Wolf's guenons (*Cercopithecus wolfi*) are seed predators when fleshy fruit is unavailable.

Some primates have anatomical adaptations for seed predation. Members of the subfamily Pitheciinae, for example, have a derived anterior dentition, including large, laterally splayed canines and anteriorly inclined upper and lower incisors. The inclined incisors allow for nipping and cropping of plant material, while the splayed canines are used for piercing and opening tough pericarps (Kinzey and Norconk, 1990; Kinzey, 1992). Colobines, too, are often seed predators, with dental and digestive specializations for masticating seeds and detoxifying secondary metabolites (Kay and Davies, 1994; Oates, 1994; Lambert, 1998a).

Many primates swallow and defecate intact seeds. For example, Wrangham *et al.* (1994) document that chimpanzees in the Kibale National Park, Uganda, swallow and disperse seeds of 59 species. These seeds range in size from < 1 mm (*Ficus* spp.) to 2.7 cm (*Cordia millenii*); 53% of these species are large seeded (i.e. > 0.5 cm). Gibbons (*Hylobates mulleri*) (McConkey, 2000), bonobos (*Pan paniscus*) (Idani, 1986), orang-utans (*Pongo pygmaeus*) (Galdikas, 1982) and gorillas (*Gorilla gorilla gorilla*) (Voysey *et al.*, 1999) are likewise reported to swallow and defecate large seeds. Indeed, Tutin *et al.* (1991) report that gorillas in Gabon swallow and defecate intact seeds of at least 65 fruit species, 66% of which are large-seeded.

However, seed swallowing is not restricted to large-bodied primates, such as apes. Garber (1986) reports that Peruvian tamarins (*Saguinus fuscicollis* and *Saguinus mystax*) swallow and pass intact seeds of many tree and liana species, including seeds as large as 2 cm. Spider monkeys (*Ateles geoffroyi*) also swallow seeds intact (Muskin and Fischgrund, 1981). Van Roosmalen (1980) states that: 'In the spider monkey, frugivory is characterized most strikingly by the fact that it mostly swallows the seeds of fruit rather than discards them.' Several cercopithecines, including guenons (*Cercopithecus pogonias, C. wolfi, Cercopithecus nictitans, Cercopithecus cephus, Cercopithecus neglectus*), and baboons (*Papio anubis*) are known to swallow the seeds of fruit, particularly small-seeded species (Lieberman *et al*., 1979; Gautier-Hion, 1980, 1984; Gautier-Hion *et al*., 1993).

Finally, primates may spit seeds. Long-tailed macaques (*Macaca fasicularis*) fill their cheek pouches with fruit, subsequently manipulating seeds back into the mouth one by one; pulp is scraped off and the cleaned seed spat out (Corlett and Lucas, 1990). Seed spitting in this fashion has been reported in several guenon species (*C. cephus, C. nictitans, C. pogonias, Cercopithecus mitis, Cercopithecus l'hoesti* and *C. ascanius*) (Gautier-Hion, 1980; Rowell and Mitchell, 1991; Kaplin and Moermond, 1998; Lambert, 1999, 2001).

How a primate processes fruit will directly influence both the distance the seed is dispersed from the parent tree and the number of seeds per deposition. For example, assuming that fruit pulp is removed soon after a fruit is harvested, seeds that are spat and dropped are more likely to be deposited in close proximity to the parent tree than seeds that are ingested (Rowell and Mitchell, 1991; Lambert, 1999). Likewise, swallowed seeds are most likely to be dispersed some distance away from parent trees (Garber, 1986; Stevenson, 2000). When fruits are processed one by one, spat or dropped seeds are likely to be deposited on the forest floor singly in a pattern of scatter dispersal either below the tree's crown or within 10–30 m of the parent tree (Rowell and Mitchell, 1991; Lambert, 1997). Swallowed seeds, however, can be deposited in a faecal clump of many or few seeds, depending on the animal's digestive kinetics and body mass, faecal size, the intensity of the feeding bout during which the seeds were swallowed and the height of tree from which seeds are defecated (Howe, 1989; Corlett and Lucas, 1990; Andresen, 1999; Lambert, 1999).

A case-study of primate fruit processing

As illustrated by the preceding review, primates handle fruits and seeds in diverse ways. Published accounts, however, are largely descriptive. Thus, we have little detailed understanding of how fruit-processing behaviours affect the seed-dispersal effectiveness (*sensu* Schupp, 1993) of a given primate species. To address this lacuna, I quantified the fruit-processing behaviours of chimpanzees (*P. troglodytes*) and redtail monkeys (*C. ascanius*) in Kibale National Park, Uganda. Redtail monkeys have an annual diet comprising up to 40% fruit and chimpanzees up to 74% (Wrangham *et al*., 1998; J.E. Lambert, unpublished). In Kibale, the fruit portions of these species' diets can overlap substantially (Lambert, 1997, 1999). Between 1993 and 1994, in the Kanyawara study area of Kibale (for site details, see Struhsaker, 1997; Chapman and Lambert, 2000), focal redtails and chimpanzees were observed for the duration of a single fruit-eating event (FEE). A FEE began when the focal animal picked a fruit (either manually or orally) and ended when that fruit was ingested and the seed either dropped, spat, swallowed or destroyed.

Details of the study are published elsewhere (Lambert and Garber, 1998; Lambert, 1999, 2001; see also Kaplin and Lambert, this volume). Here, I provide a summary of the results to illustrate the broad differences between sympatric redtail monkeys and chimpanzees. Overall, I found that chimpanzees are seed swallowers and thus may be categorized as 'clump-dispersers' in Howe's (1989) terminology. Seventy-seven per cent of 1046 chimpanzee FEEs resulted in a seed being swallowed. Chimpanzees swallowed seeds more frequently than did redtail monkeys ($\chi^2 = 691$; $P < 0.01$; d.f. = 1). Even when chimpanzees spat seeds rather than swallowing them (23% of the total FEEs), they typically spat them out in wads of fibre and masses of seeds.

Redtail monkeys, on the other hand, were seed spitters and thus 'scatter-dispersers'. Redtails spat seeds in 61% of the 2955 FEEs, and did so more often than chimpanzees ($\chi^2 = 642$; d.f. = 1; $P < 0.01$). In 71% of FEEs in which redtails spat seeds, fruit was placed into the cheek pouch and the seed was not spat until after the monkey had moved within the tree (79.8%) or to another tree (20.2%). Redtails swallowed seeds in 29.5% of FEEs, although almost all swallowed seeds were very small in size (< 3 mm); they destroyed seeds in 9.5% of total FEEs.

Most seeds swallowed by chimpanzees were defecated at least a kilometre away from the parent tree (an approximation based on daily range behaviour) (Lambert, 1997). All chimpanzee dung samples ($n = 81$) contained seeds. Ninety-three per cent of these samples contained large seeds (> 0.5 cm). The number of conspecific large seeds found in a single chimpanzee dung sample ranged from one to 149, with an average of 37; 11% of chimpanzee dung samples contained at least two species of large seeds. Fifty-five per cent of redtail dung samples ($n = 135$) contained seeds, but most (84%) were *Ficus* spp. (1–2 mm in size). Redtails typically deposited large seeds in low-density clusters, and only 0.08% of their dung samples had seeds ≥ 0.5 cm. The number of large seeds found in a single redtail monkey defecation ranged from one to six, with an average of 3.5. No redtail monkey dung contained more than one species of large seed.

To summarize, redtail monkeys typically scatter-disperse large seeds – either as single, spat seeds or in very low-density dung clusters. These results are similar to findings on other cercopithecine monkeys (e.g. *M. fasicularis* (Lucas and Corlett, 1998); *C. mitis* and *C. l'hoestis* (Kaplin and Moermond, 1998)). Indeed, Corlett and Lucas (1990) have categorized primates as being either 'seed swallowers', 'seed spitters' or 'seed destroyers' and only members of the subfamily Cercopithecinae were categorized as 'seed spitters'. Seed spitting is particularly common when cercopithecines exploit large-seeded species. Virtually all seeds swallowed by redtail monkeys in Kibale were small-seeded; < 1% of dung samples contained larger seeds. Conversely, 90% of seeds swallowed by redtails were *Ficus* spp. (1–2 mm in size), mirroring the pattern found in dung samples; 84% of seeds in redtail dung were *Ficus* spp.

Chimpanzees, on the other hand, were seed swallowers and clump-dispersers. In contrast to redtails, seed swallowing in chimpanzees was not limited to small seeds (e.g. *Monodora myristica* seeds were regularly recovered in dung samples, even though these seeds have a length > 2 cm). As mentioned previously, the swallowing of large seeds is common among apes.

Relating Primate Fruit-processing Behaviour to Seed Fate

How do these patterns of primate fruit processing relate to seed fate? And is there any evidence that clump dispersal is more likely to result in recruitment in some plant species and scatter dispersal more likely to result in recruitment in others (Howe, 1989)? To evaluate these questions, I placed 8095 seeds on the forest floor in treatments that simulated the patterns in which redtail monkeys and chimpanzees were observed to deposit seeds. I made no attempt to evaluate the influence of distance removed from parent tree on seed fate but, following Howe (1989), focused on the number of seeds deposited. All data were collected at the Kanyawara study site of the Kibale National Park, Uganda, June 1993–August 1994.

Methods

I used seeds of seven species: *Uvariopsis congensis*, *Pseudospondias microcarpa*, *Cordia abyssinica*, *Celtis durandii*, *Linociera johnsonii*, *M. myristica* and *Aphania senegalensis* (see Table 24.1 for tree characteristics). These species were chosen because they occur commonly in the study area, are large-seeded and had large fruit crops during the study period and because their fruits are readily consumed by both chimpanzees and redtails.

Four treatments simulated the range of typical dispersal seed densities created by chimpanzees and redtails. At 5 m intervals along transects, seeds were placed in densities of one, five or 30 seeds. The one-seed treatment simulated seed spitting by redtail monkeys, the five-seed treatment simulated defecation by

Table 24.1. Characteristics of seven study species of tree: diameter at breast height (dbh) (cm), crown diameter (m), fruit width and length (cm), seed width and length (cm) and number of seeds per fruit. For all measurements, $n = 60$, except where noted; mean followed by range (in parentheses).

Fruit species	DBH ($n = 6$) (cm)	Crown diameter (m)	Fruit length (cm)	Fruit width (cm)	Seed length (cm)	Seed width (cm)	Seeds per fruit
Uvariopsis congensis	22.9 (19.3–26)	5.79	2.4 (2.27–3.52)	1.9 (1.7–2.1)	1.31 (1.1–1.8)	0.83 (0.73–0.98)	5.9 (3–8)
Celtis durandii	34.7 (17.5–45.5)	8.25	0.52 (0.39–0.53)	0.48 (0.41–0.51)	0.45 (0.39–0.51)	0.39 (0.31–0.41)	1
Cordia abyssinica	47.6 (34.5–62)	12.15	1.3 (1–1.6)	0.9 (0.7–1.1)	0.9 (0.7–1.1)	0.7 (0.6–0.9)	1
Monodora myristica	88.5 (69–138.8)	13.59	18.5 (13.1–24.2)	14.1 (11–16.7)	2 (1.7–2.2)	1.4 (1.1–1.5)	290; $n = 10$ (160–341)
Aphania senegalensis	34.8 (16–53)	8.01	1.6 (1.1–1.9)	1.1 (1–1.4)	1.2 (0.9–1.6)	0.8 (0.6–1)	1
Linociera johnsonii	28.7 (16–39.8)	6.53	2.1 (1.7–2.5)	1.5 (1.3–1.6)	1.8 (1.4–2.1)	1.1 (1–1.4)	1
Pseudospondias microcarpa	104.7 (89–148.5)	12.8	2.4 (2–2.7)	1.6 (1.4–1.9)	1.7 (1.1–2.2)	0.9 (0.8–1.1)	1

redtails and the 30-seed treatment simulated defecation by chimpanzees. Ripe fruit were collected directly from trees and pulp was manually removed from the seeds. For all species except *A. senegalensis*, each treatment was replicated 30 times, so that for the one-seed treatment there was a total of 30 seeds per species, for the five-seed treatment there was a total of 150 seeds per species and for the 30-seed treatment there was a total of 900 seeds per species. Fewer *A. senegalensis* fruit were available; thus, for this species, the one-seed treatment was replicated 15 times, the five-seed treatment 16 times and the 30-seed treatment 18 times. To test the influence of primate dung on seed fate, I also included a treatment of five seeds in dung, again replicating the treatment 30 times per species (except in *A. senegalensis*, which I replicated 16 times). For the dung treatment, seeds were collected from chimpanzee and redtail dung and placed into fresh chimpanzee dung. All treatments were set up along transects in the K-30 forestry compartment of the Kanyawara study area of Kibale National Park – a compartment with no history of recent logging disturbance (Struhsaker, 1997).

To minimize human disturbance in the forest understorey, transects were placed 2 m from and parallel to trails. Six 600 m transects and one 325 m transect (*A. senegalensis*) were established (one species per transect). Steep hills were avoided in order to prevent seeds from being washed away. Each seed station was randomly assigned one seed treatment. At each station seeds were placed in a small (*c.* 15 cm × 15 cm), cleared area. Stations were marked with a piece of flagging tape approximately 2 m away. Rodents may have learned to associate the flagging tape with food and/or may have used the transects as foraging trap lines. However, this is unlikely, given the distance between seed stations and the fact that rodents rely more on olfaction than on vision (Hulme, 1993).

Stations were checked 1 week after being established, and then twice monthly for up to 13 months. For each station, I recorded the number of seeds that: (i) were damaged or removed by rodents; (ii) were attacked by fungus; (iii) had germinated; (iv) had established roots; and (v) had established and then died. Note that these categories are not mutually exclusive. For example, some seeds were recorded as attacked by fungus and then several weeks later recorded as damaged. Although unlikely, some damaged seeds may have survived despite their damage (Norconk *et al.*, 1998). Fungi generally belonged to one of two morphological types. One fungus type was pubescent, and consisted of a sheath of furry hairs, grey-white in colour; the other was a grey–yellow mould, which appeared as spots on the seed-coat. Germinating seeds had cotyledon leaves and a radicle. Established seedlings were those seeds that had roots spreading into the topsoil; roots always preceded the appearance of non-cotyledon leaves. Most established seedlings died. Two types of post-establishment seedling death were noted: (i) herbivory, which was apparent from nibbled leaves; and (ii) loss of turgor pressure and eventual desiccation.

In the case of removed seeds, I searched the area around the station (a circle, *c.* 10 m diameter) for missing seeds. If they were not found, I assumed rodents had killed them (De Steven and Putz, 1984). The role of duikers as seed-dispersers in Kibale is unknown and there is no evidence of myrmecochory. Rodents in North America and the neotropics disperse seeds by caching and failing to retrieve them (e.g. see Forget *et al.*, this volume; Vander Wall, this volume; Jansen *et al.*, this volume). However, despite considerable study (e.g. Basuta, 1979; Kasenene, 1980, 1984; Basuta and Kasenene, 1987; Lwanga, 1994; Struhsaker, 1997), the intensity of caching as observed in the neotropics has not been documented in the Kibale rodent community. These studies do indicate, however, that the most abundant species (*Hybomys univittatus, Hylomyscus stella, Mus minutoides, Praomys jacksoni*) are major seed predators.

Results: does primate seed treatment influence seed fate?

Across species

Regardless of seed species and treatment, most seeds were either damaged or removed. Overall, 27% of the seeds were damaged and 61% were removed. Very few seeds germinated (179/8095 = 2%) and, of these, almost half (86/179 = 48%) died within 13 months. Regardless of species, 37% of seeds were attacked by fungus.

Species patterns

Patterns of post-dispersal fate differed among species (Table 24.2). For example, the percentage of seeds damaged by rodents ranged from < 1% (*P. microcarpa*) to 59% (*C. durandii*). There was a smaller range in species' variation in seed removal; the percentage of removed seeds ranged from 35% (*M. myristica*) to 100% (*P. microcarpa*). The percentage of seeds attacked by fungus ranged from 0% (*A. senegalensis, C. durandii* and *L. johnsonii*) to 95% (*M. myristica*). The percentage of seeds that germinated ranged from 0% (*C. abyssinica* and *P. microcarpa*) to 7% (*M. myristica*).

Influence of seed density

Data were aggregated and chi-square analyses were used to test whether the percentage of seeds damaged, removed, attacked by fungus and/or germinated differed among the treatments. Except for *P. microcarpa*, all species had seeds damaged *in situ*, although overall a greater percentage of seeds were damaged in higher-density stations (Table 24.3). *Celtis durandii* ($\chi^2 = 7.78$; d.f. = 2; $P = 0.02$), *M. myristica* ($\chi^2 = 8.7$; d.f. = 2; $P = 0.01$) and *U. congensis* ($\chi^2 = 27.06$; d.f. = 2; $P < 0.01$) had a greater percentage of seeds damaged in seed treatments of five and 30, and *L. johnsonii* had a greater percentage of seeds damaged in depots of 30 ($\chi^2 = 142$; d.f. = 2; $P < 0.01$). All species had many seeds removed by rodents, but a greater percentage of seeds were removed in lower-density seed stations (Table 24.4). For example, *C. durandii* ($\chi^2 = 20.38$; d.f. = 2; $P < 0.01$), *M. myristica* ($\chi^2 = 31.88$; d.f. = 2; $P < 0.01$) and *U. congensis* ($\chi^2 = 7.6$; d.f. = 2; $P = 0.02$) had a greater percentage of seeds removed when in stations with one seed, and *L. johnsonii* ($\chi^2 = 125.8$; d.f. = 2; $P = 0.01$) had a greater percentage of seeds removed in seed stations of one or five seeds. *Pseudospondias microcarpa* was the most heavily hit by rodent predators and had all of its seeds either completely consumed *in situ* or removed. In short, there was a general pattern for a greater percentage of seeds to be removed when in lower-density seed treatments and damaged when in higher-density stations.

As noted earlier, several species (*A. senegalensis, C. durandii* and *L. johnsonii*) experienced no fungal attack (Table 24.5). Of those species that did experience fungal attack, there was an overall pattern for seeds to be attacked by fungus in higher-density seed stations. In *U. congensis* ($\chi^2 = 24.7$; d.f. = 2; $P < 0.01$) and *M. myristica* ($\chi^2 = 253$; d.f. = 2; $P \leq 0.01$), there was significantly more fungal attack in seed stations with 30 seeds, and *C. abyssinica* ($\chi^2 = 9.35$; d.f. = 2; $P < 0.01$) had more seeds attacked by fungus in stations with five or 30 seeds.

Table 24.2. Number of seeds damaged, removed, attacked by fungus and germinated. All seeds within species pooled, regardless of treatment. The first number indicates the number of seeds affected by the variable (i.e. damage, removal, fungus, germination) in each sample; the second number represents sample size. The number in parentheses represents the proportion of seeds affected. Note that categories are not mutually exclusive; some seeds were recorded in more than one category.

Species	Total seeds damaged	Total seeds removed	Total seeds with fungus	Total seeds germinated
Aphania senegalensis	80/715 (0.11)	674/715 (0.94)	0/715 (0.00)	3/715 (0.004)
Celtis durandii	734/1230 (0.59)	483/1230 (0.39)	0/1230 (0.00)	1/1230 (0.0008)
Cordia abyssinica	77/1230 (0.06)	483/1230 (0.39)	1084/1230 (0.88)	1/1230 (0.0)
Linociera johnsonii	636/1230 (0.52)	499/1230 (0.41)	0/1230 (0.0)	65/1230 (0.05)
Monodora myristica	151/1230 (0.12)	433/1230 (0.35)	1172/1230 (0.95)	91/1230 (0.07)
Pseudospondias microcarpa	11/1230 (0.009)	1230/1230 (1.0)	11/1230 (0.009)	0/1230 (0.0)
Uvariopsis congensis	527/1230 (0.43)	659/1230 (0.53)	759/1230 (0.62)	13/1230 (0.01)

Table 24.3. Percentage of seeds damaged under three seed densities. Chi-square (χ^2) analyses were used to test whether the percentage of seeds damaged differed among the three treatments.

Species	One-seed station	Five-seed station	30-seed station	χ^2 *P* value
Aphania senegalensis	0.07	0.09	0.12	$\chi^2 = 0.827$ $P > 0.6$
Celtis durandii	0.3	0.55	0.65	$\chi^2 = 7.786$ $P = 0.02$
Cordia abyssinica	0.03	0.04	0.08	$\chi^2 = 3.13$ $P > 0.2$
Linociera johnsonii	0.17	0.17	0.65	$\chi^2 = 142.19$ $P < 0.01$
Monodora myristica	0.03	0.11	0.12	$\chi^2 = 8.719$ $P = 0.01$
Pseudospondias microcarpa	0.0	0.0	0.01	χ^2 = n/a
Uvariopsis congensis	0.2	0.63	0.44	$\chi^2 = 27.06$ $P < 0.01$

n/a, chi-square test not appropriate because one or more cells were 0.

Table 24.4. Percentage of seeds removed under three seed densities. Chi-square (χ^2) analyses were used to test whether the percentage of seeds damaged differed among the three treatments.

Species	One-seed station	Five-seed station	30-seed station	χ^2 *P* value
Aphania senegalensis	0.93	1.0	0.94	$\chi^2 = 1.496$ $P > 0.4$
Celtis durandii	0.6	0.51	0.33	$\chi^2 = 20.38$ $P < 0.01$
Cordia abyssinica	0.84	0.81	0.87	$\chi^2 = 3.325$ $P = 0.18$
Linociera johnsonii	0.83	0.7	0.28	$\chi^2 = 125.84$ $P < 0.01$
Monodora myristica	0.87	0.45	0.66	$\chi^2 = 31.88$ $P < 0.01$
Pseudospondias microcarpa	1.0	1.0	1.0	χ^2 = n/a
Uvariopsis congensis	0.73	0.55	0.5	$\chi^2 = 7.636$ $P = 0.02$

n/a, chi-square test not appropriate because one or more cells were 0.

Very few seeds germinated. There appeared to be little influence of seed density on percentage of seeds germinating, although only two species had sample sizes sufficient to test (Table 24.6). Neither *L. johnsonii* ($\chi^2 = 1.7$; d.f. = 2; $P > 0.5$) nor *M. myristica* ($\chi^2 = 4.19$; d.f. = 2; $P > 0.1$) showed a significant association between seed density and likelihood of germination. By the end of study, all seedlings had perished.

To assess the influence of primate dung on seed removal and damage, I examined the percentage of damaged and removed seeds per station when in dung versus when not in dung. Density was held constant at five seeds per station (Table 24.7). For five of the seven species, primate dung did not affect probability of seed damage or removal. The patterns for the remaining species suggest that primate dung may serve as a foraging cue for rodents. When in dung, the seeds of *U. congensis* ($\chi^2 = 8.49$; d.f. = 1; $P \leq 0.04$) and the seeds of *C. abyssinica* ($\chi^2 = 13.8$; d.f. = 1; $P \leq 0.05$) were more likely to be removed. However, *U. congensis* seeds suffered more *in situ* rodent damage ($\chi^2 = 3.4$; d.f. = 1; $P \leq 0.01$) when not in dung.

Table 24.5. Percentage of seeds attacked by fungus under three seed densities. Chi-square (χ^2) analyses were used to test whether the percentage of seeds damaged differed among the three treatments.

Species	One-seed station	Five-seed station	30-seed station	χ^2 P value
Aphania senegalensis	0.0	0.0	0.0	χ^2 = n/a
Celtis durandii	0.0	0.0	0.0	χ^2 = n/a
Cordia abyssinica	0.73	0.91	0.9	$\chi^2 = 9.357$ $P < 0.01$
Linociera johnsonii	0.0	0.0	0.0	χ^2 = n/a
Monodora myristica	0.53	0.82	1.0	$\chi^2 = 253.4$ $P \leq 0.01$
Pseudospondias microcarpa	0.0	0.01	0.01	χ^2 = n/a
Uvariopsis congensis	0.27	0.53	0.68	$\chi^2 = 24.729$ $P < 0.01$

n/a, chi-square test not appropriate because one or more cells were 0.

Table 24.6. Percentage of seeds that germinated under three seed densities. Chi-square (χ^2) analyses were used to test whether the percentage of seeds damaged differed among the three treatments.

Species	One-seed station	Five-seed station	30-seed station	χ^2 P value
Aphania senegalensis	0.0	0.01	0.002	χ^2 = n/a
Celtis durandii	0.0	0.0	0.01	χ^2 = n/a
Cordia abyssinica	0.0	0.0	0.0	χ^2 = n/a
Linociera johnsonii	0.03	0.07	0.05	$\chi^2 = 1.171$ $P > 0.5$
Monodora myristica	0.03	0.03	0.08	$\chi^2 = 4.189$ $P > 0.1$
Pseudospondias microcarpa	0.0	0.0	0.0	χ^2 = n/a
Uvariopsis congensis	0.0	0.006	0.008	χ^2 = n/a

n/a, chi-square test not appropriate because one or more cells were 0.

Table 24.7. Number of seeds damaged and removed in dung and no dung treatments (each treatment included five seeds). Note that damage and remove categories are not mutually exclusive; some seeds were recorded in more than one category (e.g. damaged and then later removed). Chi-square analyses were used to test whether the percentage of seeds damaged or removed differed between the two treatments.

Species	Damage (in dung)	Damage (not in dung)	Remove (in dung)	Remove (not in dung)
Aphania senegalensis	10/80 (0.13)	7/80 (0.09)	70/80 (0.88)	80/80 (1.0)
Celtis durandii	69/150 (0.46)	83/150 (0.55)	77/150 (0.51)	77/150 (0.51)
Cordia abyssinica	2/150 (0.01)	6/150 (0.04)	140/150 (0.93)*	122/150 (0.81)
Linociera johnsonii	22/150 (0.15)	25/150 (0.17)	115/150 (0.77)	105/150 (0.7)
Monodora myristica	14/150 (0.09)	16/150 (0.11)	66/150 (0.44)	67/150 (0.45)
Pseudospondias microcarpa	0/150 (0.0)	0/150 (0.0)	130/150 (0.87)	150/150 (1.0)
Uvariopsis congensis	35/150 (0.23)	94/150 (0.63)*	104/150 (0.69)*	83/150 (0.55)

*$P \leq 0.05$.

Discussion

The study of seed dispersal has the potential to serve as 'a unifying theme in population ecology' (Schupp and Fuentes, 1995, p. 267). Yet most seed-dispersal research fails to adequately evaluate patterns of deposition and seed fate. Because fruit processing directly influences seed deposition, animal behavioural ecologists have the opportunity to contribute directly to this 'unifying theme'. However, primate biologists have (historically) focused on questions that lie outside the realm of plant–animal interactions and/or community ecology. Thus, models addressing primate (and other vertebrate) seed-dispersal effectiveness often make unwarranted generalizations about fruit-processing behaviour. Quantifying patterns of seed processing and using the results to design experiments that explore their consequences for seed fate provide a new approach.

Did scatter vs. clump dispersal (Howe, 1989) differentially influence seed fate in our study species? For three of seven species, seed densities did not appear to have an effect on seed fate – *A. senegalensis* had virtually no germinating seeds and no fungal attack, and seeds in all treatments were equally vulnerable to predation. Likewise, seed treatment did not differentially influence the fate of *C. abyssinica* seeds; there was virtually no germination, and seeds were highly vulnerable to fungal attack and predation, independent of treatment. *P. microcarpa*, too, was highly desirable to rodents, regardless of seed treatment.

For another three of the seven species, seed treatment did affect seed fate, though the effects offer little support for Howe's (1989) model. *Monodora myristica* was particularly vulnerable to fungal attack in larger clumps of seeds. This vulnerability might suggest it is adapted to scatter dispersal. However, single seeds of this species were also more likely to be killed by rodents than clumps of seeds. Thus, pathogens and predation in this species offset any selective advantage of the number of seeds in a seed-dispersal event – with more pathogen attack in clumps of seeds and more predation on single seeds.

There were also seed-treatment effects on the fate of *L. johnsonii* and *C. durandii*, although, again, the effects did not clearly correspond to those predicted from the scatter- vs. clump-dispersal model. Both species had more seeds damaged when in higher densities. But they also had more seeds removed when in less dense seed stations. Overall, high predation swamped any effect(s) of seed deposition patterns. Although I did not explicitly test distance effects in this analysis, I found that seeds succumbed to severe post-dispersal predation regardless of whether they were under parent trees or in transects far from mature conspecifics (Lambert, 1997). Indeed, data indicate that rodent seed predation in Kibale during 1993/94 was so intense that seeds never remained in clumps long enough to produce saplings. These data support predictions from models that evaluate rodent caching versus predation based on seed size, season and phenology (P.-M. Forget, Paris, 2000, personal communication). In summary, the lack of significant effects on seed fate, regardless of seed density, suggests that the difference between scatter and clump dispersal is largely irrelevant.

However, the seed-fate results of one species, *U. congensis*, suggest some capacity to recruit in clumps. It had a significantly higher percentage of seeds taken when seeds were 'scatter-dispersed' (i.e. in experimental seed depots of one or five). This species also had a significantly higher percentage of germinating seeds when in clumps of 30, probably because seeds survived longer. *Uvariopsis congensis* is a highly favoured fruit species in Kibale and is readily consumed by all five frugivorous primate species there. Redtail monkeys and other cercopithecines spit out its seeds within approximately 3 m of the parent tree (J.E. Lambert, unpublished data). Chimpanzees defecate its seeds in dense clumps. In both deposition patterns, seeds must recruit near conspecifics – either siblings or parents. Interestingly, *U. congensis* is one of the most common trees in the study area and is found in dense groves. For example, *U. congensis* adults have significantly more conspecifics of all life stages within 50 m than eight other species in Kibale (Lambert, 1997). Similarly, of 25 species, Chapman and Chapman (1995) found *U. congensis* to have the greatest number of conspecifics under the crown.

Thus, in *U. congensis* there is an apparent correspondence between seed arrival (below parents or near conspecifics), seed deposition (clumps) and adult recruitment (dense groves). Interpreting this correspondence is challenging. Ecological data linking behaviour and extant distribution need not necessarily reflect long-term coevolutionary relationships (Herrera, 1985; Howe, 1989; Lambert and Garber, 1998). Moreover, a correspondence between deposition and adult recruitment may be less indicative of the importance of dispersal than in systems where there is no such correspondence (Schupp and Fuentes, 1995). None the less, regardless of whether animal fruit processing selects for seed adaptations, and regardless of whether seed deposition and adult recruitment are causally linked, data presented here do suggest that *U. congensis* seed survivorship is greater in the presence of conspecifics. Similarly, in comparisons of seed fate under versus away from parent trees, both Chapman and Chapman (1996) and I (J.E. Lambert, unpublished data) found higher *U. congensis* seed predation when seeds were dispersed away from parents. Because rodent predation accounts for most seed mortality, it would be interesting to test if rodents are readily satiated in the presence of abundant *U. congensis* seeds.

In this review and case-study I demonstrate high variation in seed-handling behaviour among primates. Primates also exhibit enormous variability in ranging behaviour, habitat utilization, dietary flexibility, positional behaviour and digestive strategies. For example, I have argued elsewhere that the seed-spitting behaviour of cercopithecines can be explained by a complex interplay between anatomical and physiological traits (Lambert, 1999). Such complexity suggests that the seed-dispersal effectiveness of a primate species in one forest may be very different from the effectiveness provided by another species in the same forest, or even by the same species in another habitat or in a different season. This variation appears to be particularly high in the palaeotropics, where we find seed spitting by cercopithecines ('cheek-pouched' monkeys). In the neotropics, most seed-dispersing primates are seed swallowers, which suggests that clump dispersal is common in Central and South America. In Africa and Asia, cercopithecines often co-occur with apes (seed swallowers) and colobines (seed predators). Because apes and cercopithecines often eat the same fruit species (Lambert, 1997; Wrangham *et al.*, 1998), a given plant species may be subject to at least two (swallowing, spitting) and probably three (swallowing, spitting and seed predation) seed-handling patterns. High variation in seed handling has an important evolutionary impact: it precludes plants from adapting to differences between scatter- and clump-dispersal strategies. Indeed, my results suggest that post-dispersal effects (i.e. rodent predation and fungal attack) may have more impact on selection than seed deposition patterns.

Conservation Implications

In most tropical forests, primates comprise the largest percentage of arboreal, mammalian biomass (Terborgh, 1983) and move large numbers of seeds (Wrangham *et al.*, 1994; Chapman and Chapman, 1995; Julliot, 1997; Overdorff and Strait, 1998; Andresen, 1999; Lambert, 1999; McConkey, 2000; Stevenson, 2000). The role of primates as seed-dispersers is probably particularly important for large-seeded or hard- husked fruit species, which may be inaccessible to smaller, arboreal taxa (Andresen, 2000; see also Kaplin and Lambert, this volume). The conservation of primates is thus key to maintaining effective seed dispersal of some species (Andresen, 2000; Entwistle *et al.*, 2000).

While it is difficult to quantify the direct and indirect ecological impacts of primate extinction, a glimpse of what may happen is visible. For example, Chapman and Chapman (1995) estimated the potential loss in plant biodiversity that would result if all the large-bodied seed-dispersers (i.e. primates) were removed from the Kanyawara study area of Kibale. They assumed this would result in all fruit dropping below parent trees. On the basis of presence or absence of seedlings and saplings under adults, they concluded that 60% of the 25 tree species they studied would ultimately be lost if large seed-dispersing animals – such as primates – were removed. Along similar lines, Chapman and Onderdonk (1998) evaluated intact forest (with complete primate communities) and fragments around Kibale in which there were no primate seed-dispersers. They found fewer seedlings, fewer species and a higher percentage of small-seeded species in the forest fragments. A reduction in plant species richness not only has long-term ecological implications, but may also have more immediate, practical considerations. More than a third of all forest tree species in Kibale have seeds dispersed by primates and 42% of primate-dispersed species have some direct utility to local people, including food, medicine and fodder (Lambert, 1998b). These results demonstrate the complex links among plant species, dispersers and the human populations that rely on forest, forest edge and forest fragments. The conservation of primate species is an important goal in itself. In working to ensure their protection, we gain indirect, concomitant benefits by maintaining seed dispersal and the regeneration of economically important trees.

Future Research

No individual can paint the entire picture of the seed-dispersal process. Before we can make the link between animal frugivory and plant demography, we must have information for all phases of a plant's life history (Chambers and MacMahon, 1994; Schupp, 1995; Garber and Lambert, 1998). Indeed, while I have attempted here to experimentally evaluate the link or 'consummate the marriage' between animal frugivory, seed deposition and seed fate, the best site for seed survival may not necessarily be the best site for seedling establishment, which may not, in turn, be the ideal location for sapling recruitment (Schupp, 1995; Schupp and Fuentes, 1996). Detailed observations of fruit consumption must be collected alongside data on seed fate, as well as parallel data on additional life stages of plant populations. Such studies will allow a link or 'interface' (*sensu* Herrera *et al.*, 1994) between animals and plant demography. For example, one rather glaring lacuna in the information presented here is the fate of removed seeds. Given the extremely high percentage of seed removal in this study and others, more attention needs to be focused on secondary (and tertiary and quaternary) dispersal and seed predation in Kibale and, more generally, in palaeotropical forests. From the perspective of someone who has spent many hours focusing on the animal end of fruit–frugivore interactions, I suggest that such an enterprise be done collaboratively, with animal ecologists and plant population biologists working together.

Acknowledgements

I thank the Office of the President (Uganda), the Uganda National Council for Science and Technology, the Uganda Forestry Department, Gilbert Isabirye Basuta and John Kasenene for granting permission to work in Uganda and in Kibale Forest. I am grateful to Doug Levey, Mauro Galetti and Wesley Silva for inviting me to participate in the Frugivores and Seed Dispersal Symposium in Brazil. Colin Chapman has provided intellectual and logistical support since my initial foray into the world of primate–plant interactions. Doug Levey, Colin Chapman, Beth Kaplin, Stephen Wooten, Britta Torgrimson and Pierre-Michel Forget provided critical feedback in this research and Agaba Erimosi was essential in the data collection. This research was supported by a Makerere University Grant for Biological Research, Sigma Xi, the American Society of Primatologists, University of Illinois, and the University of Oregon.

References

Andresen, E. (1999) Seed dispersal by monkeys and the fate of dispersed seeds in Peruvian rainforest. *Biotropica* 31, 145–158.

Andresen, E. (2000) Ecological roles of mammals: the case of seed dispersal. In: Entwistle, A. and Dunstone, N. (eds) *Priorities for the Conservation of Mammalian Diversity.* Cambridge University Press, Cambridge, UK, pp. 12–26.

Basuta, G.M.I. (1979) The ecology and biology of small rodents in the Kibale Forest, Uganda. MSc thesis, Makerere University, Kampala, Uganda.

Basuta, G.M.I. and Kasenene, J.M. (1987) Small rodent populations in selectively felled and mature tracts of Kibale Forest, Uganda. *Biotropica* 19, 260–266.

Chambers, J.C. and MacMahon, J.A. (1994) A day in the life of a seed: movements and fates of seeds and their implications for natural and managed systems. *Annual Review of Ecology and Systematics* 25, 263–292.

Chapman, C.A. and Chapman, L.J. (1995) Survival without dispersers: seedling recruitment under parents. *Conservation Biology* 9, 675–678.

Chapman, C.A. and Chapman, L.J. (1996) Frugivory and the fate of dispersed and non-dispersed seeds of six African tree species. *Journal of Tropical Ecology* 12, 1–14.

Chapman, C.A. and Lambert, J.E. (2000) Habitat alteration and the conservation of African primates: a case study of the Kibale National Park, Uganda. *American Journal of Primatology* 50, 169–185.

Chapman, C.A. and Onderdonk, D.A. (1998) Forests without primates: primate/plant codependency. *American Journal of Primatology* 45, 127–141.

Corlett, R.T. and Lucas, P.W. (1990) Alternative seed-handling strategies in primates: seed-spitting by long-tailed macaques (*Macaca fasicularis*). *Oecologia* 82, 166–171.

Davies, G. (1991) Seed-eating by red leaf monkeys (*Presbytis rubicunda*) in dipterocarp forest of northern Borneo. *International Journal of Primatology* 12, 119–144.

De Steven, D. and Putz, F.E. (1984) Impact of mammals on early recruitment of a tropical canopy tree, *Dipteryx panamensis*, in Panama. *Oikos* 4, 207–216.

Entwistle, A.C., Mickleburgh, S. and Dunstone, N. (2000) Mammal conservation: current contexts and opportunities. In: Entwistle, A. and Dunstone, N. (eds) *Priorities for the Conservation of Mammalian Diversity.* Cambridge University Press, Cambridge, UK, pp. 2–7.

Feer, F. (1995) Seed dispersal in African forest ruminants. *Journal of Tropical Ecology* 11, 683–689.

Galdikas, B.M. (1982) Orang utans as seed dispersers at Tanjung Puting, Central Kalimantan: implications for conservation. In: Boer, L.E.M. (ed.) *The Orang Utan: its Biology and Conservation.* W. Junk Publishers, The Hague, pp. 285–298.

Garber, P.A. (1986) The ecology of seed dispersal in two species of callitrichid primates (*Saguinus mystax* and *Saguinus fuscicollis*). *American Journal of Primatology* 10, 155–170.

Garber, P.A. and Lambert, J.E. (1998) Primates as seed dispersers: ecological processes and directions for future research. *American Journal of Primatology* 45, 3–8.

Gautier-Hion, A. (1980) Seasonal variation of diet related to species and sex in a community of *Cercopithecus* monkeys. *Journal of Animal Ecology* 49, 237–269.

Gautier-Hion, A. (1984) La dissemination des graines par les cercopithecides forestiers Africains. *Revue d'Ecologie (La Terre et la Vie)* 39, 159–165.

Gautier-Hion, A., Gautier, J.P. and Maisels, F. (1993) Seed dispersal versus seed predation: an inter-site comparison of two related African monkeys. *Vegetatio* 107/108, 237–244.

Herrera, C.M. (1985) Determinants of plant–animal coevolution: the case of mutualistic dispersal of seeds by vertebrates. *Oikos* 44, 132–141.

Herrera, C.M., Jordano, P., Lopez-Soria, L. and Amat, J.A. (1994) Recruitment of a mast-fruiting, bird-dispersed tree: bridging frugivore activity and seed establishment. *Ecological Monographs* 64, 315–344.

Howe, H.F. (1989) Scatter- and clump-dispersal and seedling demography: hypotheses and implications. *Oecologia* 79, 417–426.

Hulme, P.E. (1993) Post-dispersal seed predation by small mammals. *Symposium Proceedings of the Zoological Society, London* 65, 269–287.

Idani, G. (1986) Seed dispersal by pygmy chimpanzees (*Pan paniscus*): a preliminary report. *Primates* 27, 411–447.

Julliot, C. (1997) Seed dispersal by red howling monkeys (*Alouatta seniculus*) in the tropical rain forest of French Guiana. *International Journal of Primatology* 17, 239–258.

Kaplin, B.A. and Moermond, T.C. (1998) Variation in seed handling by two species of forest monkeys in Rwanda. *American Journal of Primatology* 45, 83–102.

Kasenene, J.M. (1980) Plant regeneration and rodent populations in selectively felled and unfelled areas of the Kibale Forest, Uganda. MSc thesis, Makerere University, Kampala, Uganda.

Kasenene, J.M. (1984) The influence of selective logging on rodent populations and the regeneration of selected tree species in the Kibale Forest, Uganda. *Tropical Ecology* 5, 179–195.

Kay, R.N.B. and Davies, A.G. (1994) Digestive physiology. In: Davies, A.G. and Oates, J.F. (eds) *Colobine Monkeys: their Ecology, Behaviour and Evolution.* Cambridge University Press, Cambridge, UK.

Kinzey, W.G. (1992) Dietary and dental adaptations in the Pitheciineae. *American Journal of Physical Anthropology* 88, 499–514.

Kinzey, W.G. and Norconk, M.A. (1990) Hardness as a basis of fruit choice in two sympatric primates. *American Journal of Physical Anthropology* 81, 5–15.

Lambert, J.E. (1997) Digestive strategies, fruit processing, and seed dispersal in the chimpanzees (*Pan troglodytes*) and redtail monkeys (*Cercopithecus ascanius*) of Kibale National Park, Uganda. PhD thesis, University of Illinois, Urbana-Champaign.

Lambert, J.E. (1998a) Primate digestion: interactions among anatomy, physiology, and feeding ecology. *Evolutionary Anthropology* 7, 8–20.

Lambert, J.E. (1998b) Primate frugivory in Kibale National Park, Uganda, and its implications for human use of forest resources. *African Journal of Ecology* 36, 234–240.

Lambert, J.E. (1999) Seed handling in chimpanzees (*Pan troglodytes*) and redtail monkeys (*Cercopithecus ascanius*): implications for understanding hominoid and cercopithecine fruit processing strategies and seed dispersal. *American Journal of Physical Anthropology* 109, 365–386.

Lambert, J.E. (2001) Red-tailed guenons (*Cercopithecus ascanius*) and *Strychnos mitis*: evidence for plant benefits beyond seed dispersal. *International Journal of Primatology* 22, 189–201.

Lambert, J.E. and Garber, P.A. (1998) Evolutionary and ecological implications of primate seed dispersal. *American Journal of Primatology* 45, 9–28.

Leighton, M. (1993) Modeling dietary selectivity by Bornean orangutans: evidence for integration of multiple criteria in fruit selection. *International Journal of Primatology* 14, 257–314.

Levey, D.J. (1986) Methods of seed processing by birds and seed deposition patterns. In: Estrada, A. and Fleming, T.H. (eds) *Frugivores and Seed Dispersal.* Dr W Junk Publishers Dordrecht, The Netherlands, pp. 147–158.

Lieberman, D., Hall, J.B., Swaine, M.D. and Lieberman, M. (1979) Seed dispersal by baboons in the Shai Hills, Ghana. *Ecology* 60, 65–75.

Lucas, P.W. and Corlett, R.T. (1998) Seed dispersal by long-tailed macaques. *American Journal of Primatology* 45, 29–44.

Lwanga, J.S. (1994) The Role of Seed and Seeding Predators and Browsers on the Regeneration of Two Forest Canopy Species (*Minusops bagshawei* and *Strombosia scheffleri*) in Kibale Forest Reserve, Uganda. PhD thesis, University of Florida, Gainesville, Florida.

McConkey, K.R. (2000) Primary seed shadow generated by gibbons in the rain forests of Barito Ulu, Central Borneo. *American Journal of Primatology* 52, 13–30.

McKey, D.B. (1978) Soils, vegetation, and seed-eating by black colobus monkeys. In: Montgomery, G.G. (ed.) *The Ecology of Arboreal Folivores.* Smithsonian Institution Press, Washington, DC, pp. 423–437.

Maisels, F., Gautier-Hion, A. and Gautier, J.-P. (1994) Diets of two sympatric colobines in Zaire: more evidence on seed-eating in forests on poor soils. *International Journal of Primatology* 15, 681–701.

Moermond, T.C. and Denslow, J.S. (1985) Neotropical frugivores: patterns of behaviour, morphology, and nutrition with consequences for fruit selection. In: Buckley, P.A., Foster, M.S., Morton, E.S., Ridgely, R.S. and Buckley, F.G. (eds) *Neotropical Ornithology.* AOU Monographs, Washington, DC, pp. 865–897.

Muskin, A. and Fischgrund, A.J. (1981) Seed dispersal of *Stemmadenia* (*Apocynaceae*) and sexually dimorphic feeding strategies by *Ateles* in Fikal, Guatemala. *Biotropica* 13, 78–80.

Norconk, M.A., Grafton, B.W. and Conklin-Brittain, N.L. (1998) Seed dispersal by neotropical seed predators. *American Journal of Primatology* 45, 103–126.

Oates, J.F. (1994) The natural history of Africa colobines. In: Davies, A.G. and Oates, J.F. (eds) *Colobine Monkeys: Their Ecology, Behaviour and Evolution.* Cambridge University Press, Cambridge, UK, pp. 75–128.

Overdorff, D.J. and Strait, S.G. (1998) Seed handling by three prosimian primates in southeastern Madagascar: implications for seed dispersal. *American Journal of Primatology* 45, 69–82.

Peres, C.A. (1991) Seed predation of *Cariniana micrantha* (*Lecythidaceae*) by brown capuchin monkeys in central Amazonia. *Biotropica* 23, 262–270.

Peres, C.A. and Janson, C.H. (2000) Species coexistence, distribution, and environmental determinants of neotropical primate richness: a community-level zoogeographic analysis. In: Fleagle, J.G., Janson, C. and Reed, K.E. (eds) *Primate Communities.* Cambridge University Press, Cambridge, pp. 55–73.

Rowell, T.E. and Mitchell, B.J. (1991) Comparison of seed dispersal by guenons in Kenya and capuchins in Panama. *Journal of Tropical Ecology* 7, 269–274.

Schupp, E.W. (1993) Quantity, quality, and the effectiveness of seed dispersal by animals. In: Fleming, T.H. and Estrada, A. (eds) *Frugivory and Seed Dispersal: Ecological and Evolutionary Aspects.* Kluwer Academic Publishers, Boston, Massachusetts, pp. 15–30.

Schupp, E.W. (1995) Seed–seedling conflicts, habitat choice, and patterns of plant recruitment. *American Journal of Botany* 82, 399–409.

Schupp, E.W. and Fuentes, M. (1995) Spatial patterns of seed dispersal and the unification of plant population ecology. *Ecoscience* 2, 267–275.

Stevenson, P. (2000) Seed dispersal by woolly monkeys (*Lagothrix lagothricha*) at Tinigua National Park, Columbia: dispersal distance, germination rates, and dispersal quality. *American Journal of Primatology* 50, 275–289.

Struhsaker, T.T. (1997) *Ecology of an African Rain Forest: Logging in Kibale and the Conflict between Conservation and Exploitation.* University Press of Florida, Gainesville, Florida.

Terborgh, J. (1983) *Five New World Primates.* Princeton University Press, Princeton, New Jersey.

Tutin, C.E.G., Williamson, E.A., Rogers, M.E. and Fernandez, M. (1991) A case study of a plant–animal relationship: *Cola lizae* and lowland gorillas in the Lope Reserve, Gabon. *Journal of Tropical Biology* 7, 181–199.

Tutin, C.E.G., Parnell, R.J. and White, F. (1996) Protecting seeds from primates: examples from *Diospyros* spp. in the Lope Reserve, Gabon. *Journal Tropical Ecology* 12, 371–384.

Ungar, P.S. (1995) Fruit preferences of four sympatric primate species at Ketambe, Northern Sumatra, Indonesia. *International Journal of Primatology* 16, 221–245.

van Roosmalen, M.G.M. (1980) Habitat preferences, diet, feeding strategy and social organization of the black spider monkey (*Ateles paniscus paniscus*) in Surinam. PhD thesis, Rijksuniversiteit voor Natuurbeheer, Leersum.

Voysey, B.C., McDonald, K.E., Rogers, M.E., Tutin, C.E.G. and Parnell, R.J. (1999) Gorillas and seed dispersal in the Lope Reserve, Gabon. I: Gorilla acquisition by trees. *Journal of Tropical Ecology* 15, 23–38.

Wrangham, R.W., Chapman, C.A. and Chapman, L.J. (1994) Seed dispersal by forest chimpanzees in Uganda. *Journal of Tropical Ecology* 10, 355–368.

Wrangham, R.W., Conklin-Brittain, N.L. and Hunt, K.D. (1998) Dietary response of chimpanzees and cercopithecines to seasonal variation in fruit abundance. I. Antifeedants. *International Journal of Primatology* 19, 949–970.

25 Extinct Pigeons and Declining Bat Populations: Are Large Seeds Still being Dispersed in the Tropical Pacific?

Kim R. McConkey[1] and Donald R. Drake[2]

[1]*School of Biological Sciences, Victoria University of Wellington, PO Box 600, Wellington, New Zealand;* [2]*Botany Department, University of Hawaii, 3190 Maile Way, Honolulu, HI 96822, USA*

The islands of Polynesia have a long history of faunal decline and extinction. Ever since humans first reached the region some 3000 years ago, the native forests have provided them with animals for food and land for cultivation. More than 2000 species of birds, many of them much larger than extant species, have become extinct in the Pacific Islands (Steadman, 1995a). On many islands only fragmented forests persist to support the remaining species. As a result, the guilds of birds and bats available for seed dispersal of forest plants are impoverished and some plants may no longer be adequately dispersed. In this chapter, we explain the ecological history of one archipelago, the islands of Tonga in western Polynesia, and then examine whether the largest remaining frugivore, the insular flying fox (*Pteropus tonganus*), is adequately dispersing large-seeded species.

Human History and Avian Extinctions

Humans first arrived in the Pacific Islands approximately 30,000 years ago (Fig. 25.1). The Solomon Islands were the first to be inhabited, and from here people of the Lapita culture spread to West Polynesia and Micronesia around 3500 years ago, reaching all of Oceania by 1000 years ago (Dye and Steadman, 1990). Although they brought with them taro, yams, breadfruit and chickens (Kirch, 1994; Kirch *et al.*, 1995), their survival depended on their ability to gather food from the forests and reefs of their new home (Dye and Steadman, 1990). Having evolved in the absence of humans and other mammalian predators, the island birds were easy prey for the hunters and most species were hunted to extinction (Steadman, 1993, 1995a,b). Equally devastating was the impact of introduced mammals (rats, dogs and pigs), which either competed with or ate the native birds. Forest reduction also contributed to bird declines (Steadman, 1995a; Freifeld, 1999). Our only evidence of many island bird species is prehistoric bones from archaeological and palaeontological sites, but these are only a subset of the biota that must have formerly existed (Steadman, 1995a).

Situated in western Polynesia, the islands of Tonga (Fig. 25.1) share much of their fauna with Fiji to the west and Samoa to the north, but

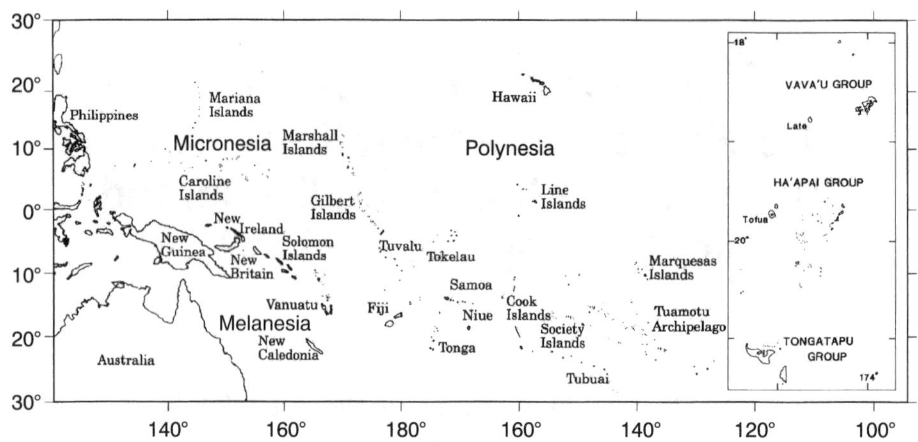

Fig. 25.1. Map of the Pacific Islands, indicating the regions of Polynesia, Micronesia and Melanesia. The islands of Tonga are shown in more detail.

lack the diversity of these larger neighbours (Dye and Steadman, 1990). Before humans reached Tonga about 3000 years ago, there was very little turnover in the species composition of land birds. After colonization, however, hunting and habitat loss resulted in the extinction of at least six pigeon species, three of which were larger than any extant birds (Steadman, 1997; Table 25.1); three other frugivorous bird species disappeared as well (Steadman, 1993). These extinctions mean that today's frugivore guild is only 40% as diverse as it was 3000 years ago. The smallest and largest bird species were disproportionately affected (Steadman, 1993). Only one pigeon species (*Ducula pacifica*) remains in Tonga, sharing the forest with three doves and several non-specialized feeders (Table 25.1). Flying foxes are the only other surviving frugivores. One species, *Pteropus samoensis*, has recently become extinct in Tonga (Koopman and Steadman, 1995), but another species, *Pteropus tonganus*, is still relatively common (Grant, 1998).

This devastation is typical of the Pacific region, where at least nine species of pigeons and doves have become extinct and numerous island populations have become locally extinct (Steadman, 1997). Whereas a typical East Polynesian island supported at least five or six species of columbids in three or four genera before the arrival of humans, these islands now support only zero to three species in zero to three genera. Similarly, a typical West Polynesian island supported six or seven species in four or five genera and now has only one to six species in one to five genera (Steadman, 1997).

The Impact on Seed Dispersal

The loss of vertebrates in Tonga over the past 3000 years has almost certainly had an impact on natural forest processes, such as seed dispersal. Seed-dispersal patterns have probably also been disrupted by reductions in native habitat and the foraging of introduced seed predators. The possible effects of each of these factors are discussed below.

Species reduction

Plants on isolated Pacific Islands are extremely vulnerable to extinction because so few pollinators and seed-dispersers survive on these islands today (Rainey *et al.*, 1995). Large-seeded species may be especially vulnerable, since a disproportionate number of frugivorous bird extinctions in the Pacific are of species larger than those surviving today (Steadman, 1993). For example, the remains of the extinct *Ducula* sp. nova from Tonga suggest that this pigeon was capable of swallowing seeds up to 62.5 mm in diameter, and even the slightly smaller *Ducula david* could probably have swallowed seeds up to 50 mm in diameter (Steadman, 1989; H. Meehan and K.R. McConkey, unpublished data). Extant species of *Ducula*

Table 25.1. Extant and extinct fruit-eating vertebrates of Tonga. Total body length (bill tip to tail tip) is given for birds and body mass for mammals. (From Watling, 1982; Steadman, 1989, 1993, 1995a, 1997, 1998; King, 1990; Koopman and Steadman, 1995; Banack, 1998; Steadman and Freifeld, 1998.)

Extant species	Common name	Body size	Extinct species	Common name	Body size
Birds					
Gallicolumba stairii	Shy-ground dove	26 cm	*Caloenas canacorum*	Giant 'Nicobar' pigeon	nm
Ducula pacifica[I]	Pacific pigeon	40 cm	*Ducula david*	David's pigeon	
Ptilinopus perousii	Many-coloured fruit dove	23 cm	*Ducula latrans*[E]	Peale's pigeon	4.1 cm[T]
Ptilinopus porphyraceus	Purple-capped fruit dove	23 cm	*Ducula* sp. nova	Immense pigeon	
Vini australis	Blue-crowned lorikeet	19 cm	*Didunculus* sp.	Tongan tooth-billed pigeon	40 cm
Prosopeia tabuensis[I]	Red-shining parrot	45 cm	*Columba vitiensis*[E]	White-throated pigeon	
Pycnonotus cafer[I]	Red-vented bulbul	20 cm	*Turdus poliocephalus*[E]	Island thrush	5.7 cm[T]
Aplonis tabuensis	Polynesian starling	19 cm	*Eclectus* sp. nova	'Eclectus' parrot	nm
Lalage maculosa	Polynesian triller	15 cm	**Phigys solitarius*[E]	Collared lory	40 cm
Foulehaio caruncalata	Wattled honeyeater	19 cm			22 cm
					nm
Mammals					20 cm
Pteropus tonganus	Tongan fruit-bat	428 g	*Pteropus samoensis*[E]	Samoan fruit-bat	
Rattus exulans[I]	Polynesian rat	92 g			
Rattus norvegicus[I]	Norway rat	215 g			
Rattus rattus[I]	Ship rat	140 g			379 g

* Steadman (1993) refers to this species as *Vini solitarius*.
I, introduced; E, locally extinct; T, length of tibiotarsus (length from distal end of fibular crest to distal knob) – for comparison, the tibiotarsus of *D. pacifica* is 3.1 cm; nm, no measurements available.

are considered to be effective seed-dispersers in the Old World (Leighton and Leighton, 1983) and it is reasonable to assume that the two extinct species swallowed seeds intact and defecated or regurgitated them at sites far from the source tree.

A relative of Samoa's tooth-billed pigeon (*Didunculus strigirostris*) also became extinct in Tonga following human colonization. Tonga's species (*Didunculus* sp. nova) is estimated to be 18–39% larger than the Samoan species (Steadman, 1997). *Didunculus strigirostris* has a parrot-like bill and feeds on seeds and fruit on the ground, but its role as a potential disperser is not known (Ramsay, 1864; for a review, see Steadman and Freifeld, 1999).

The Pacific pigeon (*D. pacifica*) is the only pigeon surviving in Tonga. It colonized Tonga at approximately the same time as humans (Steadman, 1997), perhaps in response to environmental modification or the extinction of local *Ducula* species. It is well adapted to living among forest patches and flies frequently from patch to patch or from island to island (Steadman, 1993), but still requires native habitat to survive (Franklin and Steadman, 1991; Steadman, 1998; Freifeld, 1999). Since the extinction of the large native *Ducula* species, *D. pacifica* is probably the only bird that can provide successful long-distance dispersal of large-seeded species. However, with a gape width of only 9.4 mm (Powlesland *et al.*, 2000) to 14.5 mm (H. Meehan and K.R. McConkey, unpublished data), the largest fruits it can swallow are those of *Myristica hypargyraea* (for a review, see Steadman and Freifeld, 1999), which, at 27 mm in diameter, are two-fifths the size of fruits that were supposedly dispersed by the extinct pigeons (62.5 mm) (H.J. Meehan, K.R. McConkey and D.R. Drake, unpublished data; Fig. 25.2). *Ducula* species are able to swallow very large seeds because they can stretch the base of their jaws to accommodate fruits that are much larger than their physical gape (Leighton, 1982). *Ducula pacifica* regurgitates seeds of this size, rather than passing them through the gut (H. Meehan, unpublished data), so seeds are unlikely to be dispersed very far. The largest seeds that have been found in *D. pacifica* faeces are those of *Terminalia litoralis* (26 mm × 14 mm) and *Guettarda speciosa* (20 mm × 19 mm) (H. Meehan and K.R. McConkey, unpublished data).

Flying foxes (Pteropodidae) are the only other large, highly mobile frugivores surviving in Tonga. In Samoa, flying foxes are

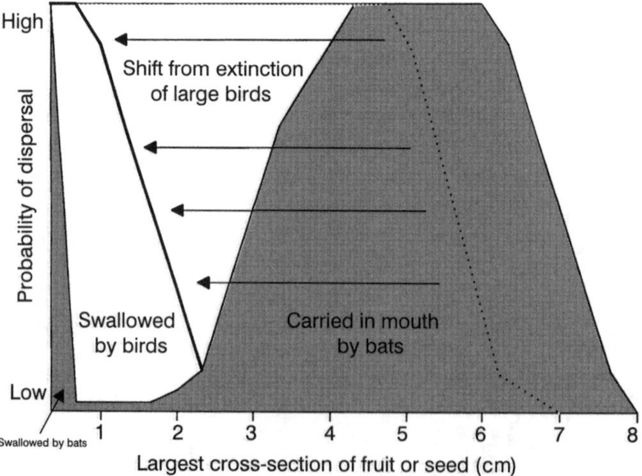

Fig. 25.2. A graphical model of fruit or seed size versus the probability of dispersal away from the parent tree for extinct and extant frugivorous birds and flying foxes (adapted from Rainey *et al.*, 1995). The grey area indicates the fruit sizes dispersed by bats, which is compared with the sizes predicted to have been dispersed by large extinct pigeons (dotted line) and those dispersed by extant birds (solid line). The shift from the dotted line to the solid line indicates the fruit sizes that will no longer be dispersed by birds, following the extinction of the largest avian frugivores.

generalists, feeding on 60% of the forest species and 79% of the canopy species. Thus, they are potential seed-dispersers for a wide range of plants (Banack, 1998). Although they fly regularly between islands, flying foxes are unable to swallow seeds larger than the guava (*Psidium guajava*) (3.7 mm × 3.2 mm) due to a very narrow oesophagus (Richards, 1990; Fig. 25.2). They do, however, regularly carry large fruit considerable distances (e.g. 40 mm diameter fruits have been observed being carried for more than 1.25 km in Samoa), though inter-island dispersal has not yet been recorded (Rainey *et al.*, 1995). With this pattern of seed dispersal, mid-sized fruits (approximately 5–20 mm in diameter) probably have a very low dispersal rate away from the parent tree. They are not frequently carried and cannot be swallowed (Rainey *et al.*, 1995). Consequently, if they are eaten, they are more likely to be consumed in the fruiting tree than are larger fruits.

Given such an impoverished frugivore guild, how is seed dispersal likely to have been affected in Tonga? Large seeds (> 30 mm in diameter) have only one vertebrate vector – the bat. Tonga's flora is composed of about 150 species of upper- and mid-canopy trees and lianas, 104 of which (69.3%) possess characteristics suggesting vertebrate dispersal (D.R. Drake, unpublished data). A survey of fruit sizes (from Smith, 1979, 1981, 1985, 1988, 1991, supplemented by unpublished data of H. Meehan *et al.*) reveals that there are at least 13 (12.5% of the 104) plant species whose fruit are too large to be swallowed regularly by *D. pacifica* (excluding large, but soft, small-seeded species, which can be consumed in pieces (Table 25.2)). Although *D. pacifica* can probably swallow fruit in the lower size range of all species noted as being attractive to pigeons, it is unlikely that these items feature regularly in their diet, given that most individual fruit will be too large for the pigeons' gape to accommodate. For at least seven of the 13 species with large fruit, the bat was probably always the only frugivore eating the fruit, while another three plant species have fruit that exhibit general features usually more attractive to birds than to bats (Table 25.2). Although fruit from the latter group are occasionally eaten by bats, they may lack a regular seed-disperser today. It is important to note, however, that bats

Table 25.2. Large fruit from the Tongan flora. The most likely disperser of these fruit in prehuman times is indicated. *Ducula pacifica* can probably swallow fruit in the lower size range of all species suggested to be attractive to pigeons (P) (fruit up to 30 mm in diameter).

Species	Length (mm)	Width (mm)	Colour	Habit	Main disperser*
Burckella richii †B	49	44	Green	Tree	B + W
Calophyllum neo-ebudicum B	17–37	13–35	Purple–black	Tree	P > B
Diospyros major	60	30	Yellow–pink	Tree	P = B
Diospyros samoensis B	20–40	20–35	Pink	Tree	P > B
Inocarpus fagifer †B	39–94	22–80	Green	Tree	B
Melodinus vitiensis	60	100	Green–brown	Liana	B
Neisosperma oppositifolium †B	62–81	45–66	Yellow	Tree	B + W
Pandanus tectorius †B	48–52	35–42	Yellow	Tree	B + W
Planchonella garberi B	20–40	22–40	Purple	Tree	P > B
Planchonella grayana †B	26–46	30–42	Green	Tree	B
Salacia pachycarpa	–	45	Green	Liana	B
Syzigium quandrangulatum	–	25–35	–	Tree	P = B
Terminalia catappa †B	47–71	32–44	Red–pink	Tree	P = B + W

* This list is generated by direct observation of the fruit types selected by bats and extant pigeons, or what is suggested by the characters of the fruit (van der Pijl, 1982; Marshall, 1983; Levey *et al.*, 1994). Noted is whether the fruit are clearly attractive to bats only (B), more likely to attract pigeons than bats (P > B) or if it is not obvious (P = B); B + W indicates the fruit is both bat- and water-dispersed; P = B + W indicates that the fruit is dispersed in all three ways.

† Measured by H. Meehan *et al.* (unpublished data). All other measurements from Smith (1979, 1981, 1985, 1988, 1991).

B Observed being eaten by bats (Brooke *et al.*, 2000; K.R. McConkey, personal observation).

(*P. samoensis* and *P. tonganus*) from Samoa eat fruit with more varied physical characteristics than do *Pteropus* species in South-East Asian forests, which have more plant species (Banack, 1998; M. Gumal, personal communication). Banack (1998) described the Samoan flying foxes as 'sequential specialists', because they specialize on only a few species at any one time, but eat a wide range of species over a longer period. Such sequential foraging is thought to be due to the lower variety of food resources available at any one time on small islands (Banack, 1998), making it difficult to predict what would be the main disperser of specific plant species on oceanic islands. A further three Tongan tree species could not be distinguished as bat- or bird-dispersed, but are probably eaten by one or both frugivores. This is a very preliminary survey, however, and considerably more data are required to confirm suspected relationships.

Habitat reduction

Habitat alteration is a serious problem in Tonga, as the islands are small and easily accessible to people. Crane (1979) estimated that the islands of Tonga retained only 10% of their original forest cover approximately 20 years ago; deforestation has continued since then (Steadman and Freifeld, 1998; Franklin *et al.*, 1999). Two national parks exist in Tonga, but one of these is confined to a small patch of forest close to one of the main town centres and is highly disturbed (Franklin *et al.*, 1999). The other is more extensive and contains 450 ha of relatively undisturbed forest (Drake *et al.*, 1996). What forest remains outside the parks is mostly confined to steep cliffs or slopes; most flat land has been converted to agriculture (D.R. Drake, personal observation). As forest becomes more fragmented, the remaining patches become increasingly isolated, and seed and pollen dispersal (i.e. gene flow) between patches may become more critical for long-term species survival (Aizen and Feinsinger, 1994). Yet the loss of mobile dispersers of large seeds means that long-distance dispersal may be increasingly less likely to occur.

Introduction of seed predators

Before humans settled in Polynesia, the only seed predators were various parrot (Steadman, 1993), crab and insect species. Thus, the Pacific Island flora evolved in the absence of granivorous mammals. It now has to contend with three non-native rat species (*Rattus exulans, Rattus norvegicus* and *Rattus rattus*). All three native predator taxa are known to destroy considerable numbers of seeds elsewhere (Janzen, 1971; Louda and Zedler, 1985; O'Dowd and Lake, 1990; Galetti and Rodrigues, 1992), but rats appear to be particularly destructive where they occur (Becker and Wong, 1985; Howe *et al.*, 1985; Chapman, 1989; Sánchez-Cordero and Martínez-Gallardo, 1998).

Seed predation by rodents has not been well documented in tropical Polynesia. In New Zealand, rat predation does not appear to be extremely high (compared with other areas), despite rats not being a part of the original fauna (Moles and Drake, 1999). Nevertheless, by targeting seeds of selected species, rats have significantly altered forest composition in New Zealand (Daniel, 1973; Campbell and Atkinson, 1999).

In Tonga, we have found evidence of rats feeding on seeds of 20 species and the hermit crab (*Coenobita* spp.) feeding on seven species in the same time period (5 months) (K.R. McConkey and N. Parsons, unpublished data). While rats fed on both naked seeds and seeds in whole fruit, crabs were only observed taking seeds from whole fruit. If the presence of rats increases the intensity of seed predation, then the role of Tonga's remaining seed-dispersers may be especially vital to plant reproduction, since predation is often most intense among poorly dispersed seeds near the parent tree (Howe *et al.*, 1985; Howe, 1993; Fragoso, 1997).

Flying Foxes as 'Keystone' Seed-dispersers in the Pacific

Following severe impoverishment in frugivore guilds, the flying fox has been considered a 'keystone' seed-disperser in the Pacific Islands (Bräutigam and Elmqvist, 1990; Cox *et al.*, 1991; Fujita and Tuttle, 1991; Mickelburgh *et al.*, 1992; Rainey *et al.*, 1995; Banack, 1998).

Pteropus tonganus, for example, is the only remaining species of a guild of at least four frugivores that fed on large fruit 3000 years ago in Tonga.

There are 35 flying-fox species (Pteropodidae) in the Pacific (Flannery, 1995). Practically all have suffered severe declines, particularly *Pteropus mariannus*, which remains only as a remnant population of a few hundred individuals in Guam (Wiles and Payne, 1986; Mickelburgh *et al.*, 1992). The main cause of its decline is hunting, but deforestation has also played a role. In Samoa, cultivated land provides the bat with mangoes, papayas and other fruits, but flying foxes prefer primary-forest fruit (Banack, 1998).

Reductions in bat densities are predicted to cause reduced dispersal of large seeds away from parent trees (Cox *et al.*, 1991; Fujita and Tuttle, 1991; Rainey *et al.*, 1995). Due to the foraging behaviour of flying foxes, however, this reduction in seed dispersal may be more severe than initially thought. In particular, some species defend established areas within feeding trees (Richards, 1990; M.Y. Gumal, personal communication). At low bat density, bats forage indefinitely within a single tree and discard seeds directly beneath it. At high density, defence of feeding territories by early-arriving bats forces other bats to remove fruits to other trees for processing, which presumably results in more effective seed dispersal (Richards, 1990; Rainey *et al.*, 1995). If this behaviour is widespread, we would expect a non-linear relationship between bat density and the number of seeds being dispersed beyond the parent crown. Richard's and Rainey *et al.*'s observations suggest that there is a threshold of bat density below which no dispersal occurs, since all feeding bats can be accommodated within the fruiting tree (Fig. 25.3). Detection of this threshold is important to ensure that declining flying fox populations are maintained at a level that will enable continued dispersal of large seeds.

Are Large Seeds Still being Dispersed? Evidence from Tonga

In spite of its diminishing area of rain forest, Tonga supports a healthy population of *P. tonganus* (Grant, 1998), one of the most widespread species in the south Pacific (Flannery, 1995). Grant's (1998) partial survey of the Vava'u Islands, Tonga, in 1995, generated mean density estimates of at least 42 bats km^{-2}, with density varying among islands depending on the presence or absence of colonial roosts. We took advantage of this geographical variation in bat densities in the Vava'u Islands to investigate the effect of high and low bat density on seed dispersal of large-seeded rain-forest trees.

The Vava'u group consists of *c.* 58 raised limestone islands lying on a single submarine platform (Nunn, 1994; Fig. 25.4). The largest, 'Uta Vava'u (96 km^2), has cliffs up to 213 m

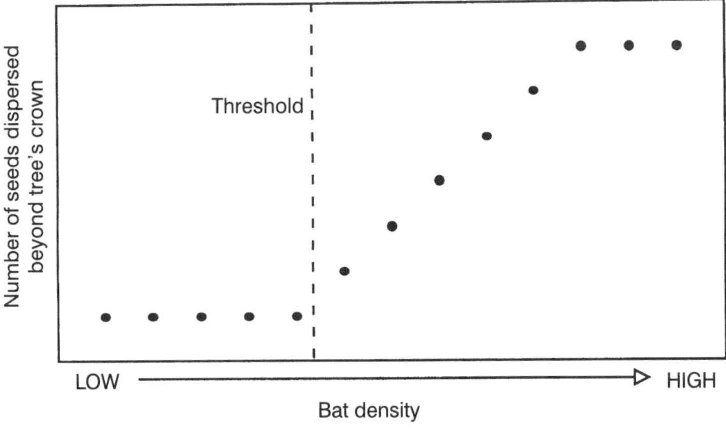

Fig. 25.3. The predicted relationship between bat density and the number of seeds expected to be dispersed beyond the crown.

high that drop steeply to the sea along most of its northern coast. South of ‘Uta Vava‘u lie smaller islands, ranging up to 9.3 km^2 and 100 m elevation. Coastal substrates consist of limestone or sand. Inland soils are derived from andesitic tephra (Orbell *et al.*, 1985).

Mean annual rainfall in Vava‘u is 2256 mm, half of which falls between January and April (Thompson, 1986). Cyclones occur between November and April; northern Tonga is affected on average by one cyclone per year. Diurnal temperature variation is greater than seasonal variation, with mean daily maximum and minimum temperatures of 30.2°C and 23.5°C in February and 26.6°C and 20.0°C in August, respectively (Thompson, 1986).

In 1993 the 14 inhabited islands in the Vava‘u group supported a combined human population of about 16,000 (Christopher, 1994). The uninhabited islands vary in degree of anthropogenic disturbance; many are cultivated or sustain populations of goats. Typical canopy height in forest patches is 10–20 m.

We selected seven sites for the study (Fig. 25.4). One was situated east of Utula‘aina Point on the northern coast of the main island, ‘Uta Vava‘u, in one of the best remaining forest patches. This site was visited three times, with 3–4 months between visits. The other sites were smaller islands that retain significant forest cover. Each of these islands was visited once. Only 11% of the main island has forest cover, while the other six islands range from 24% (Vaka‘eitu Island) to 86% (‘Euakafa Island) forest cover (Franklin *et al.*, 1999). A visit to an island entailed measuring seed dispersal by bats from selected trees over a 5-day period and estimating food (flower and fruit) and bat

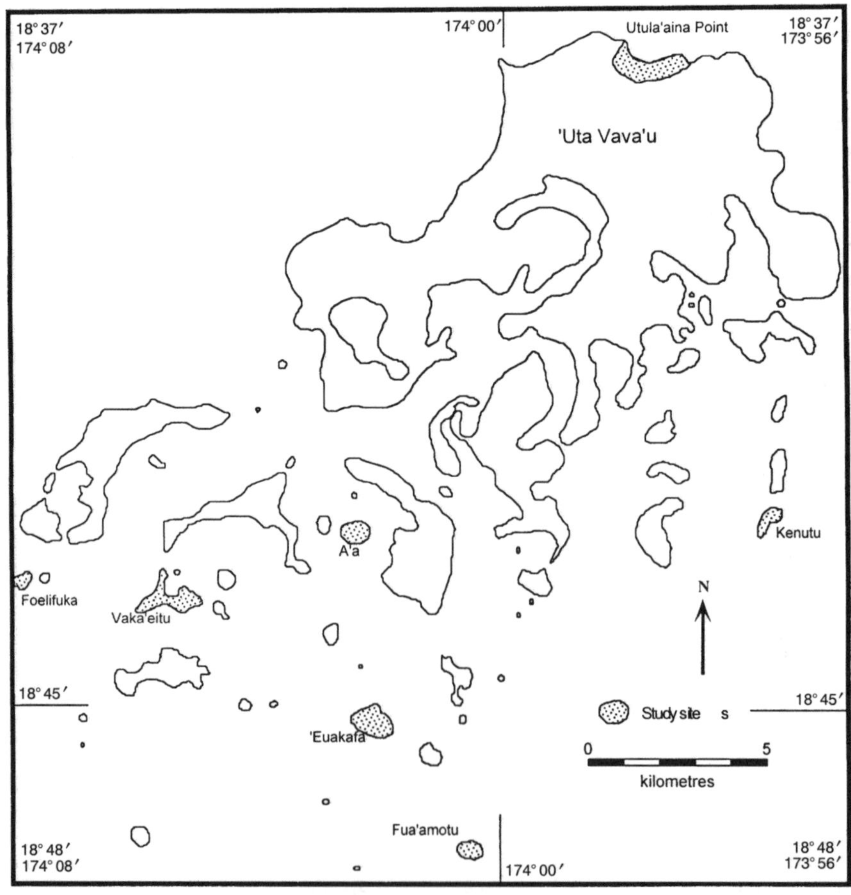

Fig. 25.4. Map of the Vava‘u Islands, Tonga, showing the study sites.

abundance. Because food and bat abundance varied temporally as well as spatially, the three visits to the site on the main island are treated as separate sites in the analyses.

Measuring bat abundance

Bat abundance at each study site was estimated along a transect, 600–700 m long, cleared of obstructing vegetation and marked every 20 m. Transects were 40 m wide, which was the width necessary to ensure easy detection of all bats, regardless of habitat structure (K.R. McConkey, personal observation). Observations began 1.5 h before dark and finished when it became too dark to easily see bats. Flying foxes on islands with a low faunal diversity tend to be more diurnal (Rainey *et al.*, 1995), and thus were fully active during this period. Transects were walked in both directions and the distance of bats from the transect estimated within 10 m. The number of fruit and flower sources available to the bats was counted from 100 canopy-height trees, selected using the point-centred quarter method (Mueller-Dombois and Ellenberg, 1974). Only food sources known to be selected by flying foxes were used in the estimate. An index of bat abundance (based on the number of bats per hour seen along the transects divided by the number of food trees) was used to compare study sites.

Measuring seed rain of study trees

We studied the seed rain of 53 trees of 13 species having medium- to large-sized fruits (13–71 mm in diameter) (Table 25.3). At each site, the forest floor was scanned for bat ejecta (wads of fruit fibre spat by bats after swallowing the juice), and seeds dropped by bats – easily identified by the bats' distinctive teeth marks in the remaining pulp around the seeds (Banack, 1998). For each tree, 2 m wide transects, radiating 40 m at right angles to each other, were cleared. If fruiting conspecifics were closer than 50 m apart, the transect extended to halfway between the fruiting trees. Transects were checked every day, for 3–5 consecutive days. All seeds were removed and their distance from the nearest edge of the tree crown was measured. For the analyses we only used data from seeds that had been dispersed by bats; we did not include fruit that fell from the tree naturally or that had been jostled loose by bats. The crown spread of the trees was measured along the four transects. Mean canopy diameter was 7.57 m (range 2.5–20 m).

Data analysis

We categorized seeds into three zones of dispersal: under the canopy (under), < 5 m beyond the canopy (edge) and between 5 and 40 m from the canopy (distant). The edge zone was used to distinguish fruit that had possibly bounced away from the source tree.

We estimated dispersal effectiveness by calculating seed density in each zone and by calculating the total number of seeds dispersed to each zone. Seed density in each zone was the number of bat-dispersed seeds found in a zone divided by the area searched in that zone. Total seed number was calculated by multiplying the density within a zone by the total area occupied by that zone (i.e. 'under' includes the total area under the canopy, 'edge' is the total circular area between 0 and 5 m from the canopy edge and 'distant' is the total area between 5 and 40 m from the canopy edge).

To determine if patterns of seed deposition in the three zones differed as a function of bat density (effect variable), we correlated the mean seed density, or total number of seeds dispersed to each zone at each site, with bat abundance at that site. The non-parametric Spearman rank correlation was used (Statsoft, 1994). For the analysis with total number of seeds, we used proportional data (number of seeds found in each zone relative to the total number of seeds found in all three zones) to standardize differences in fruit production between trees.

Results: are large seeds still being dispersed in Tonga?

Indices for bat abundance ranged from 0.08 to 0.93 (Table 25.4). Highest seed densities and

Table 25.3. Fruit characteristics of the plant species studied. All measurements are done on fresh fruit and seeds.

Species	Family	No. study trees	Colour	Pulp type	Fresh fruit mass (g)	Fresh fruit size (mm)	Pulp weight (g)	Mean seed no.	Fresh seed weight (g)	Fresh seed size (mm)
Burckella richii	*Sapotaceae*	1	Green	Firm–moist	44	49 × 44 × 42	31	1	13	35 × 30 × 29
Chionanthus vitiensis	*Oleaceae*	1	Yellow	Soft–moist	–	30 × 17 × 14	–	1	–	–
Diospyros elliptica	*Ebenaceae*	4	Orange–red	Soft–dry	2	20 × 13 × 13	1	2.4	0.8	13 × 7 × 6
Faradaya amicorum	*Verbenaceae*	1	Red	Soft–moist	15	46 × 23 × 24	7	1	8	35 × 30 × 29
Guettarda speciosa	*Rubiaceae*	3	White	Firm–moist	9	23 × 27 × 26	3	1	6	20 × 22 × 22
Inocarpus fagifer	*Papilionaceae*	3	Green	Fibrous–dry	138	80 × 71 × 53	55	1	95	65 × 71 × 42
*Mangifera indica**	*Anacardiaceae*	2	Yellow–green	Soft–moist	42	49 × 40 × 37	36	1	6	35 × 18 × 13
Pandanus tectorius	*Pandanaceae*	5	Yellow	Fibrous–moist	35	51 × 38 × 28	3	9.4	–	–
Planchonella grayana	*Sapotaceae*	1	Green	Firm–moist	29	34 × 37 × 37	25	3.7	2	19 × 10 × 6
Pleiogynium timoriense	*Anacardiaceae*	23	Red	Soft–moist	5	18 × 21 × 20	2	5–12	33†	14 × 17 × 17†
Syzygium clusiifolium	*Myrtaceae*	4	Red–pink	Soft–dry	10	34 × 26 × 26	8	1	2	19 × 13 × 14
Terminalia catappa	*Combretaceae*	3	Red–pink	Soft–moist	30	57 × 38 × 31	15	1	15	54 × 33 × 25
Terminalia litoralis	*Combretaceae*	2	Red	Soft–moist	4	24 × 17 × 16	2.2	1	1.8	2.3 × 1.4 × 1.3

* This species is not native to Tonga; immature fruit were measured, since this was the condition of the fruit taken by bats.
† Includes entire multiseeded diaspore.

seed numbers were consistently found under the canopy, with decreasing densities in the next two zones (Fig. 25.5 and Table 25.4). There was no significant relationship between bat abundance and seed number or density in the 'under' and 'edge' zones (Table 25.5). However, seeds had a significantly higher chance of being dispersed to the distant zone when bat abundance was high. This was true when both absolute seed density and total seed number were considered. However, occasional seed dispersal to the 'distant' zone happened at all bat densities (Fig. 25.5 and Table 25.4), indicating no threshold in bat abundance that must be reached to achieve effective seed dispersal.

Conclusions

The probability of seed dispersal beyond the parent tree is positively correlated to bat abundance. Although our results reveal no threshold bat density for seeds to be dispersed > 5 m from the parent tree, such dispersal is sporadic and infrequent at low bat densities. It may occur only for particularly favoured fruiting trees in which bats would be expected to congregate, creating an artificially high bat density at a very local scale. At higher bat densities (indices > 0.4) seed dispersal > 5 m beyond the crown of a fruiting tree occurs with more predictable regularity and a threshold may be detectable with larger sample sizes.

Maintaining bat populations at densities sufficient for frequent seed dispersal may be crucial to maintain natural patterns of forest dynamics and seed-mediated gene flow of large-seeded species in the tropical Pacific. Thus, while it is unlikely that tree species would become extinct without dense bat populations, populations of large-seeded species may become very localized. There could be a shift in forest composition to smaller-seeded plants, as is found in areas of Uganda where primate populations have been reduced (Chapman and Onderdonk, 1998).

Perspectives for Future Research

An important area of future research is to determine the extent to which forest structure would change following reductions in bat populations. Certain areas of the Pacific, such as Guam, have suffered very severe bat losses in recent decades. Rainey *et al.* (1995) compared the percentage of bat-dispersed seeds found in transects in Guam and Western Samoa. Seed dispersal by bats in Guam was practically nil, while in Western Samoa, where bat numbers are relatively high, 37% of seeds found in the transects were bat-dispersed. The potential effect of this difference on forest structure could

Table 25.4. Summary data for dispersal of seeds in the three zones. The index of bat abundance is based on the number of bats per hour seen along 600–700 m transects divided by the number of food trees found from 100 canopy trees checked for fruits and flowers.

		Bat abundance			Mean seed density (seeds m^{-2})			Mean seed number (%)		
Site	Number of study trees	Bats h^{-1}	Number of food trees	Index	Under	Edge	Distant	Under	Edge	Distant
Utula'aina Point 1	4	14	15	0.93	0.04	0.01	0.002	23	19	58
Foelifuka Island	5	15.6	19	0.82	0.32	0.03	0.002	29	39	32
'Euakafa Island	8	10	15	0.67	0.48	0.03	0.002	50	36	13
Utula'aina Point 3	9	5	8	0.63	0.26	0.03	0.001	31	62	7
Vaka'eitu Island	5	1	2	0.50	0.41	0.04	0.001	20	51	28
Utula'aina Point 2	7	7.2	15	0.48	0.95	0.12	0.001	20	46	33
Kenutu Island	4	8	24	0.33	0.46	0.01	0	79	21	0
Fua'amotu Island	4	5	19	0.26	0.32	0.02	0.0005	75	19	6
A'a Island	7	1	13	0.08	0.51	0.06	0.0006	29	68	3

be investigated by determining the role that dispersal (or lack of dispersal) plays in recruitment limitation of forest species (Clark *et al.*, 1999; Nathan and Muller-Landau, 2000).

Pollination on oceanic islands in the Pacific also depends on flying foxes (Cox *et al.*, 1991; Rainey *et al.*, 1995). Therefore, reductions in bat populations may be detrimental to pollination rates (i.e. seed set and pollen-mediated gene flow). Fitness consequences of pollinator loss should be evaluated.

Many questions remain about the seed-dispersal role of extinct frugivores in Tonga.

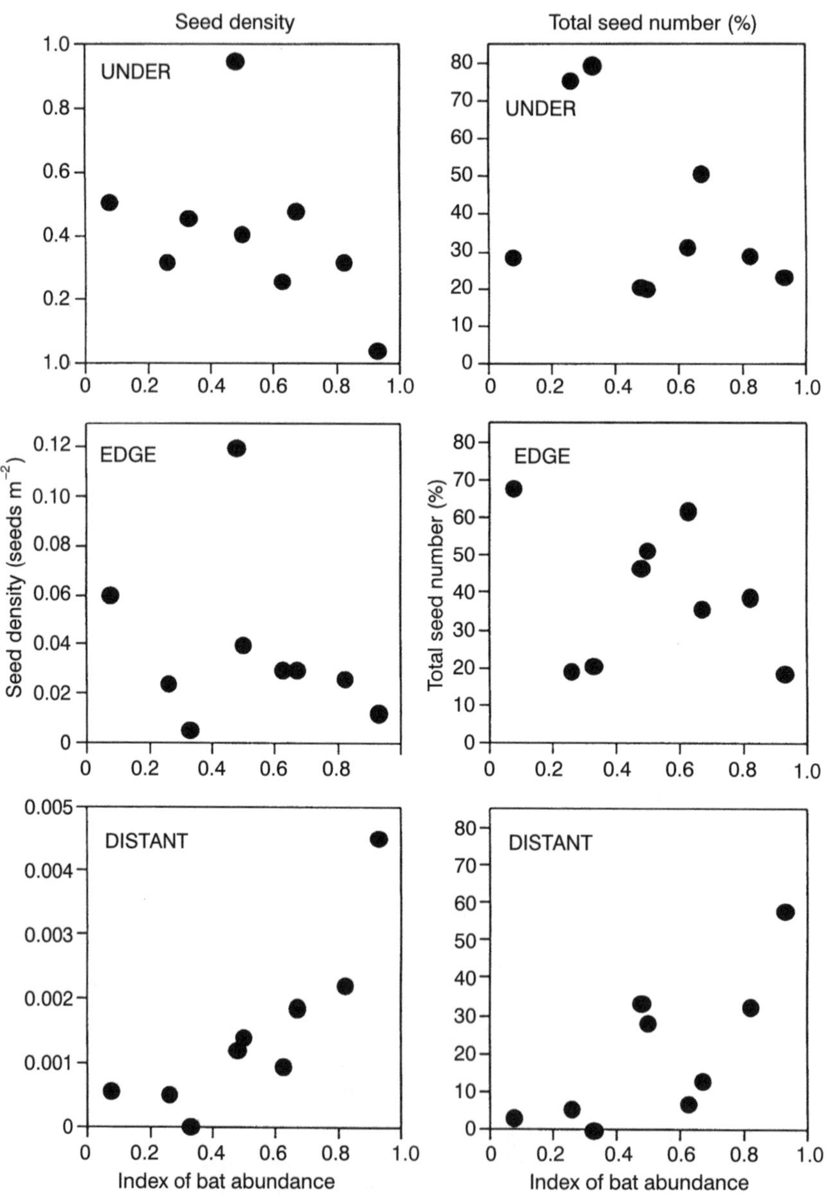

Fig. 25.5. The relationship between bat abundance and the pattern of seed deposition in the three distance zones from the parent trees. Absolute seed density and total seed number (%) are shown separately. Note that the scales on seed-density graphs are not consistent.

Table 25.5. Results of Spearman rank correlations between bat abundance and frequency of dispersal in the three zones.

	Spearman rank	*P* value
Seed density		
Under	–0.55	0.12
Edge	–0.26	0.50
Distant	0.88	0.002
Total seed number		
Under	–0.22	0.57
Edge	–0.32	0.41
Distant	0.73	0.02

What did they disperse? Are there fruit species that have no dispersers in the Tongan forests today? We are currently investigating this topic by matching the flora with past and present fauna. Our focus will be the three fruit species that are too large for *D. pacifica* to swallow, but which exhibit characteristics that are usually more attractive to birds than bats. The number of bird extinctions in Tonga is not unique to the Pacific region, however, and the effect of these extinctions on seed-dispersal processes in more floristically rich islands remains to be investigated.

Acknowledgements

We thank Nola Parsons, Nic Gorman, Defini Tau'alupe and Hayley Meehan for their help in collecting data. Logistical support was provided by Pat and Keith McKee, Aleiteisi Tangi, Tavake Tonga and Marlene Moa. B. Loiselle, J. Lord, W.R. Sykes, J. Franklin, B.D. Bell, R. Fitzjohn, K. McAlpine and M. Gumal provided constructive criticism on earlier drafts. R. Wear helped with the identification of the hermit crabs. We would like to thank the Tongan government for permission to carry out the research. Financial support was provided by the Wildlife Conservation Society, Percy Sladen Memorial Trust, Victoria University of Wellington and Polynesian Airlines.

This is contribution number 9 to the Island Biology Research Programme at Victoria University of Wellington.

References

Aizen, M.A. and Feinsinger, P. (1994) Forest fragmentation, pollination, and plant reproduction in a chaco dry forest, Argentina. *Ecology* 75(2), 330–351.

Banack, S. (1998) Diet selection and resource use by flying foxes (genus *Pteropus*). *Ecology* 79(6), 1949–1967.

Becker, P. and Wong, M. (1985) Seed dispersal, seed predation, and juvenile mortality of *Aglaia* sp. (*Meliaceae*) in lowland dipterocarp rain forest. *Biotropica* 17(3), 230–237.

Bräutigam, A. and Elmqvist, T. (1990) Conserving Pacific Island flying foxes. *Oryx* 24, 81–89.

Brooke, A.P., Solek, C. and Tualaulelei, A. (2000) Roosting behavior of colonial and solitary flying foxes in American Samoa (Chiroptera: Pteropodidae). *Biotropica* 32(2), 338–350.

Campbell, D.J. and Atkinson, I.A.E. (1999) Effects of kiore (*Rattus exulans* Peale) on recruitment of indigenous coastal trees on northern offshore islands of New Zealand. *Journal of the Royal Society of New Zealand* 29(4), 265–290.

Chapman, C.A. (1989) Primate seed dispersal: the fate of dispersed seeds. *Biotropica* 21(2), 148–154.

Chapman, C.A. and Onderdonk, D.A. (1998) Forests without primates: primate/plant codependency. *American Journal of Primatology* 45(1), 127–141.

Christopher, B. (1994) *The Kingdom of Tonga: a Geography Resource for Teachers.* Friendly Islands Book Shop, Nuku'alofa, Tonga.

Clark, J.S., Beckage, B., Camill, P., Cleveland, J., Lambers, J., Lichter, J., McLachlan, J., Mohan, J. and Wyckoff, P. (1999) Interpreting recruitment limitation in forests. *American Journal of Botany* 86, 1–16.

Cox, P.A., Elmqvist, T., Pierson, E.D. and Rainey, W.E. (1991) Flying foxes as strong interactors in South Pacific Island ecosystems: a conservation hypothesis. *Conservation Biology* 5(4), 448–454.

Crane, E.A. (1979) *The Geography of Tonga.* Wendy Crane Publisher, Nuku'alofa, Tonga.

Daniel, M.J. (1973) Seasonal diet of the ship rat (*Rattus r. rattus*) in lowland forest in New Zealand. *Proceedings of the New Zealand Ecological Society* 20, 21–30.

Drake, D.R., Whistler, W.A., Motley, T.J. and Imada, C.T. (1996) Rain forest vegetation of 'Eua Island, Kingdom of Tonga. *New Zealand Journal of Botany* 34, 65–77.

Dye, T. and Steadman, D.W. (1990) Polynesian ancestors and their animal world. *American Scientist* 78, 207–215.

Flannery, T. (1995) *Mammals of the South-West Pacific and Moluccan Islands.* Reed Books, Chatswood, Australia.

Fragoso, J.M.V. (1997) Tapir-generated seed shadows: scale dependent patchiness in the Amazon rainforest. *Journal of Ecology* 85, 519–529.

Franklin, J. and Steadman, D. (1991) The potential for conservation of Polynesian birds through habitat mapping and species translocation. *Conservation Biology* 5(4), 506–521.

Franklin, J., Drake, D.R., Bolick, L.A., Smith, D.S. and Motley, T.J. (1999) Rain forest composition and patterns of secondary succession in the Vava'u Island Group, Kingdom of Tonga. *Journal of Vegetation Science* 10, 61–64.

Freifeld, H.B. (1999) Habitat relationships of forest birds on Tutila Island, American Samoa. *Journal of Biogeography* 26, 1191–1213.

Fujita, M.S. and Tuttle, M.D. (1991) Flying foxes (Chiroptera: Pteropodidae): threatened animals of key ecological and economic importance. *Conservation Biology* 5(4), 455–463.

Galetti, M. and Rodrigues, M. (1992) Comparative seed predation on pods by parrots in Brazil. *Biotropica* 24, 222–224.

Grant, G.S. (1998) Population status of *Pteropus tonganus* in Tonga. *Atoll Research Bulletin* 454, 1–13.

Howe, H.F. (1993) Aspects of variation in a neotropical seed dispersal system. *Vegetatio* 107/108, 149–162.

Howe, H.F., Schupp, E.W. and Westley, L.C. (1985) Early consequences of seed dispersal for a neotropical tree (*Virola surianemsis*). *Ecology* 66(3), 781–791.

Janzen, D.H. (1971) Seed predation by animals. *Annual Review of Ecology and Systematics* 2, 465–491.

King, C.M. (1990) *The Handbook of New Zealand Mammals.* Oxford University Press, Auckland, New Zealand.

Kirch, P.V. (1994) *The Wet and the Dry: Irrigation and Agricultural Intensification in Polynesia.* University of Chicago Press, Chicago, Illinois.

Kirch, P.V., Steadman, D.W., Butler, V.L., Hather, J. and Weisler, M.I. (1995) Prehistory and human ecology in eastern Polynesia: excavations at Tangatatau Rockshelter, Mangaia, Cook Islands. *Archaeology in Oceania* 30, 47–65.

Koopman, K.F. and Steadman, D.W. (1995) Extinction and Biogeography of Bats on 'Eua, Kingdom of Tonga. *American Museum Noviates, No. 3125.* American Museum of Natural History, New York.

Leighton, M. (1982) Fruit resources and patterns of feeding grouping among sympatric hornbills (Bucerotidae). PhD thesis, University of California, Davis.

Leighton, M. and Leighton, D.R. (1983) Vertebrate responses to fruiting seasonality within a Bornean rainforest. In: Sutton, S.L., Whitmore, T.C. and Chadwick, A.C. (eds) *Tropical Rainforest: Ecology and Management.* Blackwell Scientific Publications, Oxford, UK, pp. 181–196.

Levey, D.J., Moermond, T.C. and Denslow, J.S. (1994) Frugivory: an overview. In: Hespenheide, H.A. and Hartshorn, G. (eds) *La Selva: Ecology and Natural History of a Neotropical Rainforest.* Chicago University Press, Chicago, Illinois, pp. 282–294.

Louda, S.M. and Zedler, P.H. (1985) Predation in insular plant dynamics: an experimental assessment of postdispersal fruit and seed survival, Enewetak Atoll, Marshall Islands. *American Journal of Botany* 72(3), 438–445.

Marshall, A.G. (1983) Bats, flowers and fruit: evolutionary relationships in the Old World. *Biological Journal of the Linnean Society* 20, 115–135.

Mickelburgh, S.P., Hutson, A.M. and Racey, P.A. (1992) *Old World Fruit Bats: an Action Plan for their Conservation.* IUCN, Gland, Switzerland.

Moles, A.T. and Drake, D.R. (1999) Post-dispersal seed predation on eleven large-seeded species from the New Zealand flora: a preliminary study in secondary forest. *New Zealand Journal of Botany* 37, 679–685.

Mueller-Dombois, D. and Ellenberg, H. (1974) *Aims and Methods of Vegetation Ecology.* John Wiley & Sons, Toronto, Canada.

Nathan, R. and Muller-Landau, H.C. (2000) Spatial patterns of seed dispersal, their determinants and consequences for recruitment. *Trends in Ecology and Evolution* 15(7), 278–285.

Nunn, P.D. (1994) *Oceanic Islands.* Blackwell Scientific, Oxford, UK.

O'Dowd, D.J. and Lake, P.S. (1990) Red crabs in rain forest, Christmas Island: differential herbivory of seedlings. *Oikos* 58, 289–292.

Orbell, G.E., Rijkse, W.C., Laffan, M.D. and Blakemore, L.C. (1985) Soils of part Vava'u Group, Kingdom of Tonga. *New Zealand Soil Survey Report* 66, 1–48.

Powlesland, R.G., Hay, J.R. and Powlesland, M.H. (2000) Bird fauna of Niue Island in 1994–95. *Notornis* 47, 39–53.

Rainey, W.E., Pierson, E.D., Elmqvist, T. and Cox, P.A. (1995) The role of flying foxes (Pteropodidae) in oceanic island ecosystems of the Pacific. *Symposium of the Zoological Society of London* 67, 79–96.

Ramsay, E.P. (1864) On the *Didunculus strigirostris,* or tooth-billed pigeon from Upolo. *Ibis* 6, 98–100.

Richards, G.C. (1990) The spectacled flying-fox, *Pteropus conspicillatus* (Chiroptera: Pteropodidae), in north Queensland. 2. Diet, seed dispersal and feeding ecology. *Australian Mammalogy* 13, 25–31.

Sánchez-Cordero, V. and Martínez-Gallardo, R. (1998) Postdispersal fruit and seed removal by forest dwelling rodents in a lowland rainforest in Mexico. *Journal of Tropical Ecology* 14, 139–151.

Smith, A.C. (1979, 1981, 1985, 1988, 1991) *Flora Vitiensis Nova: A New Flora of Fiji*, Vols. 1–5. Pacific Tropical Botanical Gardens, Lawai, Hawaii.

Statsoft. (1994) *STATISTICA for the MacIntosh*, Release 4.1, Vol. 1. Statsoft, Tulsa, Oklahoma.

Steadman, D.W. (1989) New species and records of birds (Aves: Megapodiidae, Columbidae) from an archaeological site on Lifuka, Tonga. *Proceedings of the Biological Society of Washington* 102(3), 537–552.

Steadman, D.W. (1993) Biogeography of Tongan birds before and after human impact. *Proceedings of the National Academy of Sciences USA* 90, 818–822.

Steadman, D.W. (1995a) Prehistoric extinctions of Pacific Island birds: biodiversity meets zooarchaeology. *Science* 267, 1123–1131.

Steadman, D.W. (1995b) Extinction of birds on tropical Pacific Islands. In: Steadman, D.W. and Mead, J.M. (eds) *Late Quaternary Environments and Deep History: A Tribute to Paul S. Martin*. The Mammoth Site of Hot Springs, South Dakota, Scientific Papers, Hot Springs, South Dakota, pp. 33–49.

Steadman, D.W. (1997) The historic biogeography and community ecology of Polynesian pigeons and doves. *Journal of Biogeography* 24, 737–753.

Steadman, D.W. (1998) Status of land birds on selected islands in the Ha'apai group, Kingdom of Tonga. *Pacific Science* 52(1), 14–34.

Steadman, D.W. and Freifeld, H.B. (1998) Distribution, relative abundance, and habitat relationships of landbirds in the Vava'u group, Kingdom of Tonga. *The Condor* 100, 609–628.

Steadman, D.W. and Freifeld, H.B. (1999) The food habits of Polynesian pigeons and doves: a systematic and biogeographic review. *Ecotropica* 5, 13–33.

Thompson, C.S. (1986) *The Climate and Weather of Tonga*. Miscellaneous Publication 188, New Zealand Meteorological Service, Wellington, New Zealand.

van der Pijl, L. (1982) *Principles of Dispersal in Higher Plants*. Springer-Verlag, New York.

Watling, D. (1982) *Birds of Fiji, Tonga and Samoa*. Millwood Press, Wellington, New Zealand.

Wiles, G.J. and Payne, N.H. (1986) The trade in fruit-bats *Pteropus* spp. on Guam and other Pacific Islands. *Biological Conservation* 38, 143–161.

26 Potential Consequences of Extinction of Frugivorous Birds for Shrubs of a Tropical Wet Forest

Bette A. Loiselle and John G. Blake

Department of Biology and International Center for Tropical Ecology, University of Missouri – St Louis, 8001 Natural Bridge Road, St Louis, MO 63121-4499, USA

Introduction

Aldo Leopold wrote that habitat conservation requires that we save 'every cog and wheel' (Leopold, 1949). The basis for his statement is that species interact in ways both positive and negative, direct and indirect. When interactions are strong, loss of one species may affect abundance of other species in the community (e.g. Paine, 1966; Terborgh, 1986). Diffuse interactions, although much less strong, may also affect species' abundances and distributions. Seed dispersal by animals is an example of a diffuse interaction, because many animals typically disperse the seeds of a given plant species (e.g. McDiarmid *et al.*, 1977; Lambert, 1989). It is also an important mutualism, especially in tropical wet forests (Gentry, 1982). In this chapter we explore the potential consequences to plants of losing one or more seed-dispersers from the community in a system characterized by diffuse interactions between fleshy-fruited plants and fruit-eating birds.

In tropical wet forests of Costa Rica, a relatively small set of avian frugivores remove and disperse most seeds of some understorey shrubs (Loiselle and Blake, 1990, 1999). Given the large-scale disturbances that have occurred in the region, population declines and the local extinction of birds are highly probable (Levey and Stiles, 1994). The local extinction of any one of these species might significantly alter recruitment patterns of these plants and, over the long term, influence plant community structure. Here, we investigate how extinction of four avian frugivores might affect four shrub species whose fruits are commonly eaten by these birds. We evaluate the potential impacts of frugivore loss by comparing seed shadows produced by the existing frugivore community with seed shadows that we predict would result if one or more bird species were extinct. We then modify predicted seed shadows by imposing four different seed-survival models. If the current frugivore community creates a significantly different spatial distribution of dispersed seeds from that created by a community that lacks one or more frugivores, then extinction of those frugivores may result in qualitative and quantitative changes to plant recruitment patterns. Thus, this study provides a glimpse of how plant populations might be altered following extinction of one or more species of frugivorous birds.

We start by providing a brief review of factors that influence how much plant populations may be affected by the loss of a seed-disperser. Frugivores directly affect the number of viable seeds removed from a parent plant and the spatial dispersion of those seeds in the environment (Schupp, 1993). Whether local extinction of a seed-disperser affects plant populations probably depends on four factors: (i) number of disperser species; (ii) ecological redundancy of dispersal roles; (iii) density compensation following extinction; and (iv) disperser substitution or shift (Fig. 26.1). We examine each of these factors separately.

When the set of potential seed-dispersers is small, loss of a single species is more likely to reduce the fitness of the plant(s) it disperses (Howe, 1984). The number of disperser species depends on many factors, including density of a plant species, fruit-crop size, fruit and seed size and species richness of the frugivore community (e.g. Snow, 1971; Martin, 1985; Wheelwright, 1985; Howe and Westley, 1988; Pizo, 1997; Hamann and Curio, 1999). In the same community, seeds from large fruits are typically dispersed by fewer species than are seeds from small fruits (Snow, 1971; Wheelwright, 1985). In addition, plants in old-growth forests probably have fewer seed-dispersers, on average, than do plants in second-growth forests, even after controlling for effects of seed size (Hamann and Curio, 1999). Moreover, plants on islands or in fragments of old-growth forest probably have relatively depauperate disperser communities and, thus, probably have fewer disperser species, on average, than their counterparts on the mainland or in continuous forests (Cox *et al.*, 1991; Pizo, 1997).

The extent to which dispersers are ecologically redundant will affect how much their extinction may alter plant populations (Fig. 26.1). The ecological roles of seed-dispersers can be divided into quantitative and qualitative components (Schupp, 1993). If species are essentially performing the same ecological function in terms of both quantity and quality of dispersal, then they are 'redundant' and loss of a single disperser species will probably not have a noticeable impact on plant populations (although the quantity of seeds

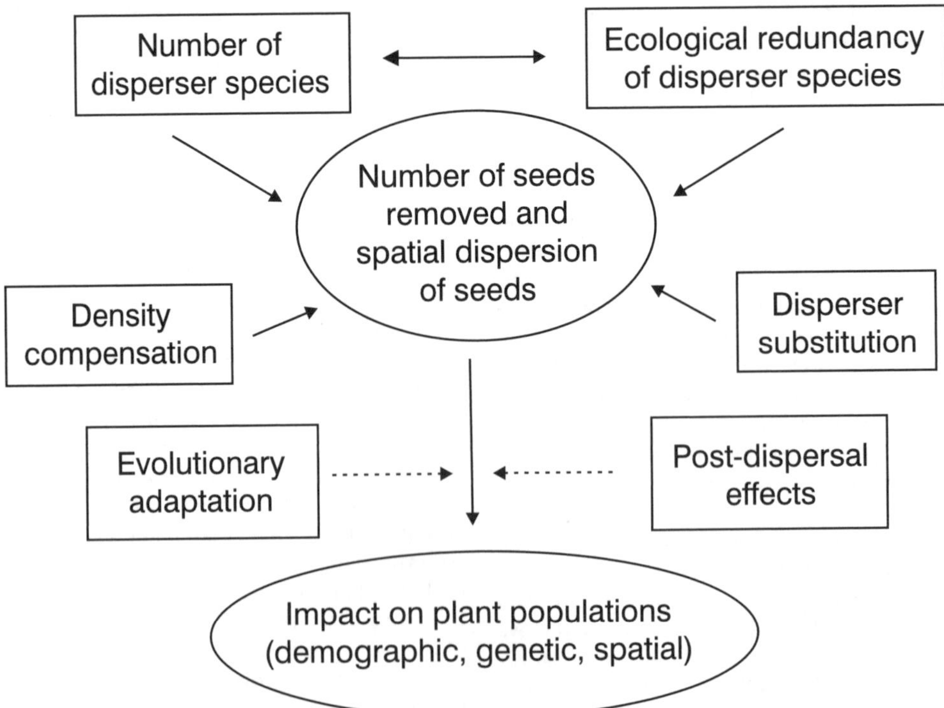

Fig. 26.1. Factors affecting the degree of impact which local extinction of seed-dispersers may have on plant populations.

removed may decline). The few studies that have determined individual roles of disperser species have shown that dispersers have different quantitative and qualitative effects on dispersal (e.g. Howe, 1980; Howe and Vande Kerckhove, 1980; Bond and Slingsby, 1984; Hoppes, 1988; Murray, 1988; Alcántara *et al.*, 2000). The extent of ecological redundancy among species is probably related to the number of disperser species in the system; systems with few disperser species probably have low overlap in the ecological roles of those species.

In cases where some overlap occurs in the ecological roles of disperser species, the impact on plant populations may depend on the degree to which remaining species increase their densities following extinction of other ecologically similar species. There is some evidence for density compensation on islands and in fragments of tropical forests following loss of species (see Wiens, 1989, for review; Renjifo, 1999). Although the number of bird species recorded in fragmented forests declined in Andean forests, the number of individuals per given sample effort did not (Renjifo, 1999). Whether density compensation counteracts loss of a seed-disperser depends on the behaviours of species that increase in abundance in response to that loss. In particular, if loss of a dispersal agent is compensated for by increases in abundance of species within the same guild (e.g. loss of one large, arboreal frugivore compensated for by increases in another), there will probably be fewer changes in seed-dispersal patterns than if the species that increased in abundance belonged to a different guild.

Following extinction of a seed-disperser, remaining or newly introduced species may shift their behaviour and, in essence, act as ecological substitutes in seed dispersal. Disperser substitution or shift is distinguished from ecological redundancy in that the behaviour of the extant member changes following extinction of the other mutualist. An example of the former is where exotic species have substituted as pollinators or seed-dispersers following extinctions or declines of native mutualists (Cox, 1983; Bond and Slingsby, 1984; Cox *et al.*, 1991; Lord, 1991; Aizen and Feinsinger, 1994; Traveset, 1999; see also Janzen and Martin, 1982; Renner, 1998).

Few studies have measured the impact of species loss on plant fitness, but reduced fitness seems apparent in some cases (Bond and Slingsby, 1984; Cox *et al.*, 1991; but see Aizen and Feinsinger, 1994). Of course, dispersers only leave the template of seeds in the environment. Post-dispersal events and stochasticity shape those initial seed shadows and thereby influence recruitment patterns of plants (Schupp, 1993). In only the rarest of circumstances would loss of a seed-disperser be uncoupled with other changes in the system that might directly or indirectly affect plant populations (e.g. abundance of seed predators (Pizo, 1997)). Consequently, new selective forces are also likely to operate and, ultimately, impacts on plant populations will result from changes in seed dispersal and post-dispersal events and from responses (or lack thereof) of plants to these new selective forces (see Keeler-Wolf, 1988; Cox *et al.*, 1991).

We now explore the consequences of frugivore extinction for plant populations by considering the number of disperser species and the degree to which they are ecologically redundant. As the extinctions are only hypothetical, we are unable to evaluate whether disperser substitution/shift or density compensation would occur and counteract the loss of the dispersal services of the extinct species.

Methods

We used data on fruit consumption and habitat use of four avian frugivores (*Pipra mentalis*, *Corapipo leucorrhoa*, Pipridae; *Hylocichla mustelina*, Turdidae; and *Chlorothraupis carmioli*, Thraupidae) to define their contributions to dispersal of seeds from four shrubs or treelet species in the *Melastomataceae* (*Clidemia densiflora* (Standl.) Gleason, *Henriettea tuberculosa* (Donn. Sm.) L.O. Wms., *Miconia simplex* Tr. and *Ossaea macrophylla* (Benth.) Cogn.) (all species hereafter referred to by genus). Data were collected on a 10 ha study plot during late wet and dry seasons (December–April) from 1988 to 1993 at La Selva Biological Station in north-east Costa Rica (10° 26′ N, 83° 59′ W).

To evaluate the impact of the local extinction of one or more bird species on plant populations, we needed to combine data on quantity of seeds removed by each bird species and on where those seeds are deposited (i.e. which habitat). In addition, we needed to evaluate where seedling recruitment is most likely. Because we did not have data on seed and seedling survival, we quantified the relative distribution of adults by habitat. Estimates on the quantity of seeds dispersed came from faecal samples of birds, whereas information on where seeds were deposited was based on captures in mist nets (60 nets, located 40 m apart on a 10 net × 6 net grid). We tallied number of seeds collected from captured birds at each net site separately for each plant and bird species. Distribution of fruiting adults was determined by sampling two 12.5 m × 2 m transects located parallel to and 1 m away from each side of each mist net. Mist-net locations were classified by habitat: (i) upland flat and ridge sites; (ii) upper, middle and lower slopes; and (iii) bottom-land or swale sites. Results presented here assume that capture at a net site reflects activity of birds and seed-deposition patterns (see Levey, 1988a). More complete description of methods, assumptions and study site are in McDade *et al.* (1994) and Loiselle and Blake (1990, 1991, 1993, 1999).

To determine the potential impact that local extinction of one or more of these four frugivores could have on the distribution of dispersed seeds, we used GEODISTN.BAS, a QuickBasic 4.50 program that tests for differences between two spatial distributions (Syrjala, 1996). This non-parametric test evaluates the null hypothesis that two populations have the same spatial distributions. We used the Cramér–von Mises test statistic, because it is relatively insensitive to a small number of extreme observations (Syrjala, 1996); 1000 permutations were run to determine the test statistic. Specifically, we compared the spatial distribution of seeds derived from the present frugivore community (i.e. distribution pattern of seeds among mist-net sites, determined from faecal samples of captured birds) with the spatial distribution of seeds that would result if seeds deposited by one or more selected frugivore species were eliminated from the data matrix. A significant test statistic indicates that the seed-deposition pattern produced by a given subset of frugivores (i.e. with one or more of the four frugivores eliminated) differed from that of the entire frugivore community. Because seeds are not expected to survive equally well at all sites or in all habitats, we also applied three different survival models to the seed shadow (see next paragraph for description of models) to evaluate the robustness of our results. Following application of a seed-survival model, we then re-examined the relationship between the spatial distribution of seeds left by an intact frugivore community with one in which one or more frugivores were removed. Finally, to help interpret the results, we compared the spatial distribution of seeds from a given plant species left by the entire frugivore community with the actual distribution of fruiting adults for each of the seed-survival models. A significantly positive overlap is indicative of which survival model best matches reality.

Five conditions were tested for each of the four plant species and for each of the seed-survival models: *Chlorothraupis* extinct; *Corapipo* extinct; *Hylocichla* extinct; *Chlorothraupis, Corapipo* and *Hylocichla* extinct; and *Pipra* extinct. Among these combinations, the least likely to occur is the one in which *Pipra* is extinct and, for this reason, we kept its removal as a separate case. The four seed-survival models were as follows: (i) seeds had equal probability of survival (1%) at all net sites (SEED RAIN, equivalent to species-specific seed rain); (ii) survival of seeds was weighted by the relative number of fruiting adults occurring in three major habitats as determined from transect data (i.e. seeds in habitats with more adults were given greater survival probabilities under the assumption that such habitats were more favourable for recruitment and that the favourability was positively and linearly related to the relative frequency of adults; ranges of seed survival were 0–2.35% for *Miconia,* 0.18–2.88% for *Ossaea,* 0.72–1.22% for *Henriettea* and 0.69–1.33% for *Clidemia*) (HABITAT); (iii) survival of seeds was increased fivefold (5%) at net sites where adults occurred and was 1% otherwise (POSITIVE SITE); and (iv) survival of seeds was decreased fivefold (0.2%) at net sites where adults occurred and was 1% otherwise (NEGATIVE SITE). We used these models of survival because we do not have information on recruitment requirements of these plants. The SEED RAIN model is essentially the null model and reflects equal probability of recruitment across the environment. The HABITAT model assumes that plant recruitment responds to conditions at the macro scale, such that seeds in habitats where the majority of adults occur are likely to recruit more seedlings than those in habitats where few adults occur. The POSITIVE and NEGATIVE SITE models are exactly contrary to model recruitment on a local scale. In the POSITIVE model, local site conditions are presumed to be 'favourable' because of the presence of an adult, whereas in the NEGATIVE model the presence of the adult is modelled to have a negative impact on recruitment probabilities. The latter case may reflect recruitment patterns generated by distance-dependent mortality. Empirical studies in tropical forests have found seed and seedling survival rates to be within the range modelled here (0.2–5.0%) or higher (e.g. Augspurger, 1984; Kitajima and Augspurger, 1989; E. Wiener, Peru, 1999, personal communication).

Results

Based on seeds recovered from faecal samples of all captured birds, we estimate that our four focal species removed 98.5% of seeds from *Miconia*, 95.5% from *Clidemia*, 87.8% from *Ossaea* and 73.5% from *Henriettea* (Loiselle and Blake, 1999). Because the four shrubs produce fruit in the understorey, mist nets are likely to provide a reliable estimate of the bird species that feed on their fruits. In particular, it is unlikely that bird species that typically forage above net level would remove many seeds from these plants. Further, we have frequently observed these bird species eating fruit from these plant species (unpublished data). Consequently, a small part of the entire frugivore assemblage at La Selva has a potentially large impact on recruitment of these plant species.

The hypothetical seed shadows created by each bird species is illustrated in Plate 2a (see Frontispiece) for *Clidemia*. Although all species apparently deliver a relatively large quantity of seeds to the north-western section of the plot, considerable differences exist in the spatial pattern of *Clidemia* seeds in other sections. For example, *Pipra* appears to be primarily responsible for seeds deposited in the northeast corner of the plot. An illustration of how composite seed shadows would be altered following bird extinctions is provided in Plate 2b (see Frontispiece) for the HABITAT survival model. The extinction of *Chlorothraupis* would cause elimination of some high-density seed areas and contraction of others (Colour Plate 2b).

If seeds have an equal chance of survival (SEED RAIN model), then loss of *Chlorothraupis* would have an impact on *Henriettea* (Table 26.1). Extinction of *Pipra* alone, as well as the combined loss of *Chlorothraupis*, *Corapipo* and *Hylocichla* would significantly alter distributions of *Clidemia*, *Henriettea* and *Miconia* (Table 26.1). Although loss of the latter three frugivores appears, at first glance, to have the same consequence as the loss of *Pipra*, their effects are indeed different. The spatial distribution of seeds left by all frugivores except *Chlorothraupis*, *Corapipo* and *Hylocichla* are significantly different from distributions left by communities with all species except *Pipra* (Table 26.2). In contrast, the spatial distribution of *Ossaea* seeds appears to be insensitive to the composition of frugivore species present (Table 26.1). *Hylocichla* and, to a lesser extent, *Corapipo* appear to be unimportant seed-dispersers; their extinction (alone) did not affect any plant species under any survival condition (Tables 26.1 and 26.3–26.5), although *Corapipo* may affect *Henriettea*. Yet, when loss of these two species was combined with loss of *Chlorothraupis*, significant effects on seed distributions often occurred, suggesting that dispersal impacts of these species are additive.

The impact of extinctions under the HABITAT model largely matched those of the SEED RAIN model (Table 26.3). With the HABITAT model, however, only the extinction of *Pipra* altered the distribution of *Miconia*. Similarly, under the POSITIVE SITE and NEGATIVE SITE models, *Pipra* emerged as a significant seed-disperser for all plant species except *Ossaea* (Tables 26.4 and 26.5). In contrast, loss of no other single frugivore had an effect on distribution patterns of seeds, although loss of three frugivores did affect seed distributions of *Clidemia* in the POSITIVE SITE model

(Table 26.4) and those of *Clidemia*, *Henriettea* and *Miconia* in the NEGATIVE SITE model (Table 26.5).

To test which survival model best reflected the observed distribution of adult fruiting plants, we compared spatial distribution of seeds left by all frugivores following application of the survival models to that of the adult fruiting plants. Among the four survival models, the HABITAT model provided the best match to adult fruiting plants, except for *Clidemia* (Table 26.6). The NEGATIVE SITE model provided the closest match ($P = 0.84$) for *Clidemia*. The POSITIVE SITE and HABITAT models were virtually identical ($P = 0.81$ in both) for *Ossaea*. If we assume that current dispersal processes are similar to those of the recent past, these results suggest that the HABITAT model most closely reflects survival probabilities of seeds for three plant species and that the NEGATIVE SITE model most closely matches seed and seedling survival of the remaining species. Thus, impacts of local

Table 26.1. *P* values of Cramer–von Mises tests comparing spatial distribution of seeds from frugivore communities after applying the model assuming equal probability of survival (SEED RAIN) for four understorey plants in the *Melastomataceae*. Comparisons are of entire frugivore community vs. frugivore community with one or more bird species removed.

	Bird species removed				
Plant species	CHCA	COLE	HYMU	ALL3	PIME
Clidemia	0.22	0.54	0.96	0.002	0.004
Henriettea	0.04	0.09	0.58	0.04	0.02
Miconia	0.47	0.14	0.36	0.03	0.001
Ossaea	0.53	0.78	0.58	0.79	0.81

CHCA, COLE, HYMU and PIME, *Chlorothraupis*, *Corapipo*, *Hylocichla* and *Pipra* extinct, respectively; ALL3, *Chlorothraupis*, *Corapipo* and *Hylocichla* extinct.

Table 26.2. *P* values of Cramer–von Mises tests of differences between seed shadows with PIME extinct and ALL3 extinct (see Table 26.1 for codes) for each of the four plant-survival models and plant species.

	Plant species			
Survival model	*Clidemia densiflora*	*Henriettea tuberculosa*	*Miconia simplex*	*Ossaea macrophylla*
SEED RAIN	0.001	0.02	0.006	0.88
HABITAT	0.001	0.02	0.03	0.39
POSITIVE SITE	0.002	0.06	0.009	0.82
NEGATIVE SITE	0.004	0.02	0.02	0.87

Table 26.3. *P* values of Cramer–von Mises tests of differences between seed shadows left by frugivore communities after applying the HABITAT model of survival for four understorey plants in the *Melastomataceae*. Comparisons are of entire frugivore community vs. frugivore community with one or more bird species removed. See Table 26.1 for codes.

	Bird species removed				
Plant species	CHCA	COLE	HYMU	ALL3	PIME
Clidemia	0.06	0.79	0.98	0.001	0.005
Henriettea	0.05	0.07	0.51	0.03	0.01
Miconia	0.99	0.22	0.53	0.13	0.005
Ossaea	0.60	0.35	0.65	0.36	0.36

extinction of frugivores on the four focal plant species are best represented by these two models.

Discussion

Our results suggest that loss of frugivores, especially *Pipra*, would alter seed rain and eventually alter distributions of three of the four melastome species included in this study. Our estimated effect is largely qualitative, because the models do not specifically test the quantitative component of seed dispersal (i.e. how many seeds go undispersed when a frugivore becomes extinct). Yet we chose the four bird species in this study precisely because of their quantitative importance to one or more of the plant species, as indicated by the high proportion of *Melastomataceae* seeds these birds defecated upon capture in mist nets. Thus, the impacts outlined here may be conservative. Our results assume that the ecological role of an extinct seed-disperser is not compensated for by another species. Moreover, we assume that an extant species would not change its behaviour (i.e. disperser shift/substitution) to replace the ecological

Table 26.4. *P* values of Cramer–von Mises tests of differences between seed shadows left by frugivore communities after applying POSITIVE SITE model for four understorey plants in the *Melastomataceae*. Comparisons are of entire frugivore community vs. frugivore community with one or more bird species removed. See Table 26.1 for codes.

	Bird species removed				
Plant species	CHCA	COLE	HYMU	ALL3	PIME
Clidemia	0.30	0.92	0.99	0.005	0.02
Henriettea	0.08	0.08	0.70	0.06	0.04
Miconia	0.49	0.22	0.51	0.06	0.003
Ossaea	0.56	0.70	0.62	0.76	0.72

Table 26.5. *P* values of Cramer–von Mises tests of differences between seed shadows left by frugivore communities after applying NEGATIVE SITE model for four understorey plants in the *Melastomataceae*. Comparisons are of entire frugivore community vs. frugivore community with one or more bird species removed. See Table 26.1 for codes.

	Bird species removed				
Plant species	CHCA	COLE	HYMU	ALL3	PIME
Clidemia	0.16	0.13	0.82	0.001	0.009
Henriettea	0.06	0.39	0.45	0.04	0.02
Miconia	0.37	0.19	0.26	0.04	0.004
Ossaea	0.42	0.77	0.58	0.81	0.81

Table 26.6. *P* values of Cramer–von Mises tests of differences between seed shadows with all frugivores extant against observed distributions of adult fruiting plants.

	Plant species			
Survival model	*Clidemia densiflora*	*Henriettea tuberculosa*	*Miconia simplex*	*Ossaea macrophylla*
SEED RAIN	0.04	0.81	0.04	0.35
HABITAT	0.04	0.85	0.16	0.81
POSITIVE SITE	0.002	0.14	0.14	0.81
NEGATIVE SITE	0.84	0.57	0.07	0.28

role of the extinct species. We have little evidence that our focal species compete for fruits; individuals of different species often co-occur and even feed together in the same plant without evidence of aggressive interactions (personal observations). Competition, if present, might lead to altered patterns of fruit consumption and habitat use following extinction of a competitor.

The threat of local extinction or severe population declines is evident for three of these four bird species. In 1960, La Selva was surrounded by tropical wet forest (see Plate 9 in Slud, 1960), but since then the landscape has been greatly altered. La Selva (approximately 1500 ha) now sits at the northern end of a peninsula of forest, surrounded on three sides by plantations, pastures or other human-modified habitats; it abuts Braulio Carrillo National Park (approximately 44,000 ha) to the south. Associated with these habitat changes, several bird species have declined at La Selva; some are probably locally extinct (Levey and Stiles, 1994). *Chlorothraupis* has declined in abundance at La Selva in the past decade and only a few family flocks may remain (Levey and Stiles, 1994; Bette A. Loiselle and John G. Blake personal observations). *Chlorothraupis* is still common at higher elevations (e.g. 500 m) in Braulio Carrillo National Park. *Corapipo* is an altitudinal migrant that breeds in foothill regions of Braulio Carrillo (300–700 m) and descends to lowland forest for several months each year (Stiles, 1988; Loiselle and Blake, 1991, 1992; Levey and Stiles, 1994). Long-term persistence at La Selva may depend on the forested connection to Braulio Carrillo. Continued deforestation around La Selva and lower elevations of Braulio Carrillo may, in the long term, affect populations of this species. *Hylocichla* is a latitudinal migrant from eastern deciduous forests of North America, where declines in breeding populations have been reported (Peterjohn *et al.*, 1995; Robinson *et al.*, 1995). Among the frugivores and plant species examined here, *Hylocichla* appears to be the least important member of the frugivore community. The single most important frugivore, *Pipra*, does not appear to be in any danger of local extinction at La Selva. Indeed, it is one of the most abundant birds in forest understorey (Levey, 1988a,b; Blake and Loiselle, 1991; Graham, 1996).

Fruit–frugivore mutualisms are common in tropical wet forests. The ecology of these mutualisms is complex, because many species and processes interact to affect seed removal, seed viability and seedling establishment. For some plants, the suite of animal dispersers is large, suggesting that the impact of any single disperser may be small. Yet dispersers are not equivalent in their ecological roles; a few species may have disproportionate impacts on plant fitness (Howe, 1980, 1993; Murray, 1988; Schupp, 1993). Extinction of such dispersers could affect seed removal, seed-deposition patterns and, consequently, plant recruitment (Howe, 1984). At La Selva, loss of *Chlorothraupis* is probable within the next few decades and loss of *Corapipo* and *Hylocichla* are possible. In most cases, loss of one of these species would not significantly alter the distribution of seeds but their combined loss would probably alter seed shadows significantly for two or three of the four species of melastomes discussed here.

We caution that results of our extinction models are not necessarily predictive. Indeed, we shall never be certain of extinction consequences until after they occur. None the less, our results clearly indicate that, even in diffuse plant–animal mutualisms, loss of a seed-disperser species can have potentially large impacts on patterns of seed deposition. Over time, these changes may have other effects on community structure and function, providing further justification for the preservation of 'every cog and wheel' (see Leopold, 1949).

Acknowledgements

We thank the Organization for Tropical Studies, especially Deborah and David Clark, for continued assistance and encouragement with our studies at La Selva Biological Station. Support for this research has come from National Geographic Society and Douroucouli Foundation awards to J.G.B. and National Science Foundation (DEB-9110698) and University of Missouri research awards to B.A.L.

References

Aizen, M.A. and Feinsinger, P. (1994) Habitat fragmentation, native insect pollinators, and feral honey bees in Argentine 'Chaco Serrano'. *Ecological Applications* 4, 278–392.

Alcántara, J.M., Rey, P.J., Valera, F. and Sáanchez-Lafuente, A.M. (2000) Factors shaping seedfall pattern of a bird-dispersed plant. *Ecology* 81, 1937–1950.

Augspurger, C.K. (1984) Seedling survival of tropical tree species: interactions of dispersal distance, light gaps, and pathogens. *Ecology* 65, 1705–1712.

Blake, J.G. and Loiselle, B.A. (1991) Variation in resource abundance affects capture rates of birds in three lowland habitats in Costa Rica. *Auk* 108, 114–127.

Bond, W.J. and Slingsby, P. (1984) Collapse of an ant–plant mutualism: the Argentine ant (*Iridomyrmex humilis*) and myrmecochorous Proteaceae. *Ecology* 65, 1031–1037.

Cox, P.A. (1983) Extinction of the Hawaiian avifauna resulted in a change of pollinators for the Ieie, *Freycinetia arborea*. *Oikos* 41, 195–199.

Cox, P.A., Elmqvist, T., Pierson, E.D. and Rainey, W.E. (1991) Flying foxes as strong interactors in South Pacific island ecosystems: a conservation hypothesis. *Conservation Biology* 5, 448–454.

Gentry, A.H. (1982) Patterns of neotropical plant species diversity. *Evolutionary Biology* 15, 1–84.

Graham, D.L. (1996) Interactions of understory plants and frugivorous birds in a lowland Costa Rican rainforest. PhD dissertation, University of Miami, Coral Gables, Florida, USA.

Hamann, A. and Curio, E. (1999) Interactions among frugivores and fleshy fruit trees in a Philippine submontane forest. *Conservation Biology* 13, 766–773.

Hoppes, W.G. (1988) Seedfall pattern of several species of bird-dispersed plants in an Illinois woodland. *Ecology* 69, 320–329.

Howe, H.F. (1980) Monkey dispersal and waste of a neotropical fruit. *Ecology* 61, 944–959.

Howe, H.F. (1984) Implications of seed dispersal by animals for tropical reserve management. *Biological Conservation* 30, 261–281.

Howe, H.F. (1993) Specialized and generalized dispersal systems: where does 'the paradigm' stand? *Vegetatio* 107/108, 3–13.

Howe, H.F. and Vande Kerckhove, G.A. (1980) Nutmeg dispersal by tropical birds. *Science* 210, 925–927.

Howe, H.F. and Westley, L.C. (1988) *Ecological Relationships of Plants and Animals.* Oxford University Press, New York, 273 pp.

Janzen, D.H. and Martin, P.S. (1982) Neotropical anachronisms: the fruits the gomphotheres ate. *Science* 215, 19–27.

Keeler-Wolf, T. (1988) Fruit and consumer differences in three species of trees shared by Trinidad and Tobago. *Biotropica* 20, 38–48.

Kitajima, K. and Augspurger, C.K. (1989) Seed and seedling ecology of a monocarpic tropical tree, *Tachigalia versicolor. Ecology* 73, 2129–2144.

Lambert, F. (1989) Fig-eating by birds in a Malaysian lowland rain forest. *Journal of Tropical Ecology* 5, 401–412.

Leopold, A. (1949) *A Sand County Almanac and Sketches Here and There.* Oxford University Press, New York, 256 pp.

Levey, D.J. (1988a) Tropical wet forest treefall gaps and distributions of understory birds and plants. *Ecology* 69, 1076–1089.

Levey, D.J. (1988b) Spatial and temporal variation in Costa Rican fruit and fruit-eating bird abundance. *Ecological Monographs* 58, 251–269.

Levey, D.J. and Stiles, F.G. (1994) Birds: ecology, behavior, and taxonomic affinities. In: McDade, L.A., Bawa, K.S., Hespenheide, H.A. and Hartshorn, G.S. (eds) *La Selva: Ecology and Natural History of a Neotropical Rain Forest.* University of Chicago Press, Chicago, Illinois, pp. 217–228.

Loiselle, B.A. and Blake, J.G. (1990) Diets of understory fruit-eating birds in Costa Rica. *Studies in Avian Biology* 13, 91–103.

Loiselle, B.A. and Blake, J.G. (1991) Temporal variation in birds and fruits along an elevational gradient in Costa Rica. *Ecology* 72, 180–193.

Loiselle, B.A. and Blake, J.G. (1992) Population variation in a tropical bird community: implications for conservation. *BioScience* 42, 838–845.

Loiselle, B.A. and Blake, J.G. (1993) Spatial dynamics of understory avian frugivores and fruiting plants in lowland wet tropical forest. *Vegetatio* 107/108, 177–189.

Loiselle, B.A. and Blake, J.G. (1999) Dispersal of melastome seeds by fruit-eating birds of tropical forest understory. *Ecology* 80, 330–336.

Lord, J.M. (1991) Pollination and seed dispersal in *Freycinetia baueriana*, a dioecious liana that has lost its bat pollinator. *New Zealand Journal of Botany* 29, 83–86.

McDade, L.A., Bawa, K.S., Hespenheide, H.A. and Hartshorn, G.S. (eds) (1994) *La Selva: Ecology and Natural History of a Neotropical Rain Forest.* University of Chicago Press, Chicago, Illinois, 486 pp.

McDiarmid, R.W., Ricklefs, R.E. and Foster, M.S. (1977) Dispersal of *Stemmadenia donnell-smithii* (*Apocynaceae*) by birds. *Biotropica* 9, 9–25.

Martin, T.E. (1985) Resource selection by tropical frugivorous birds: integrating multiple interactions. *Oecologia* 66, 563–573.

Murray, K.G. (1988) Avian seed dispersal of three neotropical gap-dependent plants. *Ecological Monographs* 58, 271–298.

Paine, R.T. (1966) Food web complexity and species diversity. *American Naturalist* 100, 65–75.

Peterjohn, B.G., Sauer, J.R. and Robbins, C.S. (1995) Population trends from the North American Breeding Bird Survey. In: Martin, T.E. and Finch, D.M. (eds) *Ecology and Management of Neotropical Migratory Birds.* Oxford University Press, New York, pp. 3–39.

Pizo, M.A. (1997) Seed dispersal and predation in two populations of *Cabralea cajerana* (*Meliaceae*) in the Atlantic forest of southeastern Brazil. *Journal of Tropical Ecology* 13, 559–578.

Renjifo, L.M. (1999) Effects of the landscape matrix on composition and conservation of bird communities. Unpublished PhD dissertation, University of Missouri – St Louis.

Renner, S.S. (1998) Effects of habitat fragmentation on plant pollinator interactions in the tropics. In: Newberry, D.M., Prins, H.H.T. and Brown, N. (eds) *Dynamics of Tropical Communities.* Blackwell Science, Oxford, UK, pp. 339–360.

Robinson, S.K., Rothstein, S.I., Brittingham, M.C., Petit, L.J. and Grzybowski, J.A. (1995) Ecology and behavior of cowbirds and their impact on host populations. In: Martin, T.E. and Finch, D.M. (eds) *Ecology and Management of Neotropical Migratory Birds.* Oxford University Press, Oxford, UK, pp. 428–460.

Schupp, E.W. (1993) Quantity, quality and the effectiveness of seed dispersal by animals. *Vegetatio* 107/108, 15–29.

Slud, P. (1960) The birds of Finca 'La Selva', a tropical wet forest locality. *Bulletin of the American Museum of Natural History* 121, 49–148.

Snow, D.W. (1971) Evolutionary aspects of fruit-eating in birds. *Ibis* 113, 194–202.

Stiles, F.G. (1988) Altitudinal movements of birds on the Caribbean slope of Costa Rica: implications for conservation. In: Almeda, F. and Pringle, C.M. (eds) *Tropical Rainforests: Diversity and Conservation.* California Academy of Sciences and Pacific Division, American Association for Advancement of Science, San Francisco, California, pp. 243–258.

Syrjala, S.E. (1996) A statistical test for a difference between the spatial distributions of two populations. *Ecology* 77, 75–80.

Terborgh, J. (1986) Keystone plant resources in the tropical forest. In: Soulé, M.E. (ed.) *Conservation Biology: the Science of Scarcity and Diversity.* Sinauer Associates, Sunderland, Massachusetts, pp. 330–344.

Traveset, A. (1999) La importancia de los mutualismos para conservación de la biodiversidad en ecosistemas insulares. *Revista Chilena de Historia Natural* 72, 527–538.

Wheelwright, N.T. (1985) Fruit size, gape width, and the diets of fruit-eating birds. *Ecology* 66, 808–818.

Wiens, J.A. (1989) *The Ecology of Bird Communities,* Vol. 2, *Processes and Variations.* Cambridge University Press, Cambridge, UK, 316 pp.

27 Primate Frugivory in Two Species-rich Neotropical Forests: Implications for the Demography of Large-seeded Plants in Overhunted Areas

Carlos A. Peres[1] and Marc van Roosmalen[2]

[1]*School of Environmental Sciences, University of East Anglia, Norwich NR4 7TJ, UK;* [2]*Departamento de Botânica, Instituto Nacional de Pesquisa da Amazônia, Caixa Postal 478, Manaus, Amazonas 69011-970, Brazil*

Introduction

Successful seedling recruitment in higher plants involves a sequence of nested regeneration stages, including the production and pollination of flowers, fruit development, dispersal and germination of viable seeds, seedling emergence, and growth to reproductive maturity. A disruption in any of these processes or a reduction in the transition probability between consecutive stages could result in a seedling recruitment bottleneck. However, the transition from viable seeds to seedlings is perhaps the most crucial to the population dynamics of most plant species (Harper, 1977), a maxim that has been confirmed yet again in a long-term study of seedling recruitment of 53 species of Panamanian trees (Harms *et al.*, 2000).

Dispersers can limit seedling recruitment if the probability of a non-dispersed seed becoming an established plant is considerably lower than that of dispersed seeds. This need not mean that the final spatial pattern of recruited seedlings will be congruent with that of dispersed seeds because the survival requirements of seeds and seedlings are often different (e.g. Rey and Ancántara, 2000). Nevertheless, seeds not removed by frugivores almost always fall immediately underneath the parent tree and have a near-zero survival probability, even in the short term (Augspurger, 1984; Chapman and Chapman, 1996). In the last 15 years, several community-wide studies have provided evidence of strong compensatory effects of density-dependent mortality for both low-diversity (e.g. Penfold and Lamb, 1999) and high-diversity (e.g. Terborgh *et al.*, this volume) tropical forest systems, confirming the importance of vertebrate seed-dispersers.

Dispersal of seeds away from parent plants is therefore often essential for the successful recruitment in a large number of tropical forest plants (e.g. Howe and Smallwood, 1982) and frequency-dependent effects produced by pathogens and seed predators appear to be widespread (e.g. Wills and Condit, 1999; see also Terborgh *et al.*, this volume). Ingestion of whole seeds by vertebrates is often the predominant mechanism of seed dispersal, particularly in closed-canopy tropical evergreen forests

 Seed Dispersal and Frugivory: Ecology, Evolution and Conservation (eds D.J. Levey, W.R. Silva and M. Galetti)

(van Roosmalen, 1985a; Charles-Dominique, 1993). Which vertebrates ingest which seeds depends on a suite of fruit morphological, nutritional and phenological traits (Howe and Smallwood, 1982; van Schaik *et al.*, 1993).

Tropical forest plants vary considerably in their seed size (Foster and Janson, 1985) and degree of consumer specificity (Peres, 2000a), and yet few studies have examined the structure of fruit–frugivore matrices in relation to seed and fruit morphology (but see Gautier-Hion *et al.*, 1985; Pratt and Stiles, 1985; Hamann and Curio, 1999). Fleshy-fruited plants usually produce large crops that are dispersed by a coterie of vertebrate species (Jordano, 1992). However, greater specificity should be expected between frugivores and zoochorous plants with increasingly larger seeds, simply because fewer frugivore species will be able to handle and/or ingest the seeds. This narrower spectrum of gut-dispersal services for large-seeded plants should be further aggravated where large-bodied species have been selectively extirpated by subsistence hunting (Peres, 2000b).

In this study we compare the nature of interactions between primates and plant species in two species-rich neotropical forests – Urucu, Amazonas, Brazil, and Raleighvallen-Voltzberg Reserve, central Suriname. We focus primarily on primates because they represent a large proportion of the frugivore biomass at these sites and in other non-hunted forests (Peres, 1999a), they rely heavily on fruits (Peres, 1994a; Mittermeier and van Roosmalen, 1981) and they were studied by virtually identical methods. In particular, we build on detailed information on the feeding ecology of the two large-bodied ateline primates occurring at these sites (*Lagothrix* and *Ateles*), which in many respects are functionally equivalent, particularly where they are not sympatric. We then briefly review the impact of subsistence hunting on Amazonian forest primates, and particularly the atelines, which are known to provide high-quality seed-dispersal services for large-seeded plants (Chapman, 1989; Julliot, 1997; Stevenson, 2000). We derive empirical estimates of seed-dispersal probabilities for plant species of varying seed sizes and make predictions on how the quality of dispersal services will be reduced in areas where vertebrate faunas have been selectively truncated in their body-size distribution by subsistence hunters. Finally, we identify a functionally cohesive guild of neotropical forest plant species that are strong candidates for succumbing to seedling recruitment bottlenecks in overhunted forests because they rely heavily on dispersal services provided by frugivores that are typically extirpated by subsistence hunters.

Study Sites and Methods

Urucu forest, Brazil

The Urucu forest (4° 50′ S, 65° 16′ W) is representative of most undisturbed *terra firme* forests in remote interfluvial regions of Amazonia rarely visited by hunters (Peres, 1990). Approximately 93% of the 900 ha study area consisted of high *terra firme* forest of low structural heterogeneity. *Sapotaceae* are the most species-rich tree family, with 29–30 species of trees ≥ 10 cm diameter at breast height (dbh) ha^{-1} (n = 2241 trees ≥ 10 cm dbh in three 1 ha plots) (Amaral, 1996), followed by the *Chrysobalanaceae, Lecythidaceae* and *Caesalpiniaceae.* A 10 m × 1000 m plot in the study area contained 332 species of trees ≥ 10 cm dbh (Amaral, 1996), near the maximum recorded for any 1 ha plot of Amazonian forests sampled to date. A more detailed description of the climate, habitat types, forest structure and composition and plant phenology at this site can be found elsewhere (Peres, 1991, 1993a,b, 1994a,b, 1999a; Amaral, 1996).

Frugivorous birds and mammals were studied by three independent observers during *c.* 25 days per month between February 1988 and October 1989, following an initial survey in April–May 1987. Several sampling techniques were used, including line-transect censuses (Peres, 1993b, 1999a) and systematic observations of five primate species: saddleback and moustached tamarins (Peres, 1991, 1993a), buffy sakis (Peres, 1993c), brown capuchins (Peres, 1994a) and woolly monkeys (Peres, 1994b, 1996). Frugivory data presented here were largely biased to these species, although feeding records were obtained whenever possible from all 12 sympatric primate species (Table 27.1).

Table 27.1. Size aggregation and adult body mass of primate species occurring (•) in each study area.

English name	Latin name	Urucu, Brazil	Voltzberg, Suriname	Body mass (kg)
Small-bodied species (< 1.5 kg)				
Pygmy marmoset	*Cebuella pygmaea*	•	–	0.15
Saddleback tamarin	*Saguinus fuscicollis*	•*	–	0.39
Moustached tamarin	*Saguinus mystax*	•*	–	0.51
Golden-handed tamarin	*Saguinus midas midas*	–	•*	0.53
Common squirrel monkey	*Saimiri sciureus*	–	•	0.85
Squirrel monkey	*Saimiri ustus*	•†	–	0.94
Red titi monkey	*Callicebus cupreus*	•	–	1.05
Collared titi monkey	*Callicebus torquatus*	•	–	1.20
Red-necked night monkey	*Aotus nigriceps*	•	–	1.00
Medium-bodied species (1.5–4.0 kg)				
Buffy saki monkey	*Pithecia albicans*	•*	–	2.10
White-faced saki monkey	*Pithecia pithecia*	–	•*	1.80
Bearded saki monkey	*Chiropotes satanas*	–	•*	2.70
Brown capuchin	*Cebus apella*	•*	•	2.95
White-fronted capuchin	*Cebus albifrons*	•	–	2.70
Wedge-capped capuchin	*Cebus olivaceus*	–	•	2.70
Large-bodied species (> 4 kg)				
Red howler monkey	*Alouatta seniculus*	•	•	6.80
Grey woolly monkey	*Lagothrix lagotricha cana*	•*	–	8.71
Black spider monkey	*Ateles paniscus*	–	•*	8.50
Black-faced spider monkey	*Ateles chamek*	•‡	–	8.90

*Those species that were most intensively studied.
†Although the Urucu forest lies within the presumed geographical range of *Saimiri boliviensis*, this species was phenotypically undistinguishable from *S. ustus.*
‡This species was restricted to the *igapó* forest along the Urucu River some 4 km from the study area (Peres, 1993b).

Raleighvallen-Voltzberg Reserve, Suriname

High upland (*terra firme*) forest was the predominant forest type, followed by low *terra firme*, liana, mountain savannah and *Pina* palm (*Euterpe oleracea*) swamp forest. The flora of high *terra firme* forest was very species-rich, including 331 of a total of 486 tree and liana species producing edible fruits recorded for Suriname (van Roosmalen, 1985a). During 3 years (1976–1978), data on eight primate species were collected in the 56,000 ha protected Raleighvallen-Voltzberg Reserve (hereafter, Voltzberg), currently part of the much larger Central Suriname Nature Reserve. Special emphasis was placed on the ecology of black spider monkeys and on how all primate species (Table 27.1) partitioned habitat and food resources.

In its first year, the study focused on all large-bodied frugivorous vertebrates, particularly primates. Two observers simultaneously walked transects of 8 km for 10 days (0600–1500 h) per month, and these surveys continued throughout the second and third years. In addition, several surveys were conducted throughout the 3-year study in other parts of Suriname, when complementary fruit consumption and seed-handling data were recorded. Further details on the plant and primate assemblage at Voltzberg can be found in Mittermeier and van Roosmalen (1981) and van Roosmalen (1985a,b).

Fruit and seed collections

At each study site, all fruits and seeds found on the ground or consumed by frugivores or seed predators were collected, measured, weighed and stored in Whril-Pak© bags containing a preservative (FAA; formaldehyde, alcohol and acetic acid). Collecting was done systematically by two or three spatially independent observers over a 21-month period at Urucu and by one to three observers over a 3-year period at Voltzberg. Whole fruits or fruit fragments found on the forest floor were collected over a 900 ha trail grid (120 m × 120 m) at Urucu and a 300 ha trail grid (100 m × 100 m) at Voltzberg. In all cases, three seed dimensions were measured (length, width and depth), using fresh samples of at least ten seeds per species. Seed dimensions presented here are mean values. The longest seed dimension is assumed to be the most important measure of seed size across all species, which is justifiable considering that mean seed length explained 76% of the total between-species variation in mean seed width at Urucu ($r = 0.87$, $P < 0.001$, $n = 592$ species) and 57% of that at Voltzberg ($r = 0.75$, $P < 0.001$, $n = 332$ species). Additional data collected on all fruits include wet mass, husk hardness, seed-dispersal type, number and wet mass of seeds and fruiting period.

Data on fruit and seed morphology and fruit–frugivore interactions were obtained from feeding observations of all obligate and facultative frugivores during community-wide vertebrate surveys and systematic follows of habituated primate groups of several diurnal species. During these follows, the seed content of a large number of dung samples was inspected and deposition sites were marked for subsequent monitoring of seed germination success. Observational effort at each forest site was biased to one or two small-bodied (*Saguinus* spp.), one mid-sized (e.g. *Cebus apella, Chiropotes satanas*) and one large-bodied species (*Lagothrix lagotricha* and *Ateles paniscus*). Finally, full-day vigils of individual fruiting trees and lianas were carried out on several occasions throughout the year, augmenting data on fruit–frugivore interactions. All preserved fruits and seed vouchers collected at Urucu and Voltzberg are deposited at the Botany Department, Instituto Nacional de Pesquisas da Amazônia, Manaus, and the University of Utrecht herbarium, respectively.

During all fruit-feeding bouts, we recorded the method of seed treatment and the seed-dispersal outcome (i.e. seed ingested whole, dropped intact underneath parent crown, dropped away from parent or visibly damaged). Seeds of a given species were considered to be successfully ingested when they were found intact in fresh dung samples inspected during primate-group follows or when ingestion of whole seeds was unmistakably recorded during fruit-feeding bouts. Modal seed-treatment outcomes of species interactions in the two fruit–primate matrices thus reflect whether seeds of a given species were typially ingested intact by a given primate species.

Primate responses to hunting pressure

A large-scale standardized series of repeated line-transect censuses were carried out by one of us (C.A.P.) over the last 13 years at 30 Amazonian forest sites. These sites have been subjected to varying degrees of hunting pressure (none, light, moderate and heavy) but otherwise remained essentially undisturbed. Survey methodology followed Peres (1999b). A compilation of primate population densities at 26 additional forest sites under varying degrees of hunting pressure (Peres, 1999c) was also used. Here we briefly review the effects of subsistence hunting on large-bodied species; more detailed analyses are provided elsewhere (Peres, 1999c, 2000b,c).

Results

Fruit and seed size

Seed dispersal in the two sites was predominantly animal-mediated. Over four-fifths of all woody plant species sampled at Urucu (81.0%, $n = 689$) were zoochorous; this proportion was even higher at Voltzberg (86.7%, $n = 332$). Mean length of 592 morphospecies of mature fruits handled by frugivores and/or seed predators at Urucu ranged between 4 and 198 mm (mean ± SD = 32.2 ± 23.9 mm), whereas those of 332 species collected at Voltzberg ranged between 5 and 580 mm (mean ± SD = 80.2 ± 65.1 mm). Mean seed length across all edible fruit morphospecies at Urucu ranged from 1 to 97 mm (mean = 15.9 ± 10.7 mm, $n = 592$) and from 1 to 65 mm at Voltzberg (mean = 15.7 ± 11.8 mm, $n = 332$ species) (Fig. 27.1). In particular, 94 (15.9%) tree and liana species at Urucu and 65 (19.6%) at Voltzberg produced large seeds (here defined as those ≥ 25 mm in mean length). Large-seeded species thus represented nearly one-fifth of the woody flora (including tree, liana, epiphyte and hemiepiphyte species) in these species-rich neotropical forests, and over half of those species were classified as gut-dispersed (61.7% and 55.4% at Urucu and Voltzberg, respectively). Moreover, large seeds were more likely to be produced by large canopy trees and high-climbing lianas than by understorey plants (see Foster and Janson, 1985). For example, there was a weak positive correlation between mean dbh of fertile (fruiting) trees and mean seed length across all fruit species consumed by primates at Urucu ($r = 0.372$, $n = 195$, $P < 0.01$), where tree dbh explained 65% and 54% of the variation in tree height and crown volume (m^3), respectively, in a random set of 996 trees (Peres, 1991).

Patterns of primate frugivory

Plant species relying primarily on gut dispersal by primates in the two forests typically produced the following fruit types, all of which are less commonly consumed by other frugivores: hard- or thick-husked, large-seeded berries, indehiscent and thick-husked capsules, thick-husked and fleshy drupes, baccate and fleshy syncarps, indehiscent fleshy pods, dehiscent pods with edible gum and fleshy pseudofruits. On the other hand, seeds associated with other fruit types – including figs, arillate capsules, dehiscent berries, fleshy fruiting spikes, fleshy perianths, dry pods, dry follicles and dry samaras – were consumed either by a small group of non-primate species or by a broader spectrum of frugivore species, and thus relied to a lesser extent on primate dispersal. Although primates could be loosely assigned to different fruit syndromes, primate body size was a poor predictor of the size of fruits they handled. There was only a weak relationship between primate body mass and mean length ($r^2 = 0.047$, $P < 0.001$, $n = 656$ species interactions) and mean width ($r^2 = 0.031$, $P = 0.002$, $n = 645$ species interactions) of fruits consumed at Urucu, and these relationships were not significant at Voltzberg (fruit length, $r^2 = 0.001$, $P = 0.491$; fruit width, $r^2 = 0.11$, $P = 0.061$; $n = 425$ species interactions).

Seed size and redundancy of dispersal services

As expected, larger-bodied primates were capable of ingesting a broader range of seed sizes than their small and mid-sized counterparts. Primate body mass explained a weak but significant proportion of the variation in mean length of all seed species ingested at both Urucu ($r^2 = 0.108$, $P < 0.001$, $n = 589$ species interactions) and Voltzberg ($r^2 = 0.090$, $P < 0.001$, $n = 425$ interactions) (Fig. 27.2). It is noteworthy that both tamarin species at Urucu were occasionally capable of ingesting seeds as large as 21 mm in length, about twice the size of those typically ingested by their congeners at Voltzberg. This difference can be partly attributed to differences between the two plant communities and the more intensive sampling effort on tamarins at the Urucu (Peres, 1993a). At the other end of the primate size spectrum, adults of *Ateles* and *Lagothrix* were occasionally capable of ingesting seeds as large as 45–50 mm, which is remarkable considering that members of both of these genera rarely exceed 9 kg in wild populations (Peres, 1994c).

Considering all plant species producing fruits handled by primates, we examined the relationship between mean seed size and the number of primate species known to ingest or defecate undamaged seeds at each site. While seeds ingested intact by five or more primate species were invariably small (mean length < 15 mm), large-seeded plants (≥ 25 mm) were ingested by only one or two primate species (Fig. 27.3). In general, large-bodied primates were capable of ingesting both small and large seeds, whereas small primates typically ingested primarily small seeds. Mid-sized species, such as capuchins (*C. apella* and *Cebus albifrons* at Urucu; *C. apella* and *Cebus olivaceus* at Voltzberg) were frequently capable of ingesting medium-sized seeds. Large seeds were almost invariably ingested only by howlers and woolly monkeys at Urucu and howlers and spider monkeys at Voltzberg. At Urucu, woolly monkeys were the dispersal vector of all 30 large-seeded species (≥ 25 mm) known to be ingested by primates, three of which were also dispersed by howlers. At Voltzberg, spider monkeys dispersed 21 of 33 large-seed species ingested by primates, 12 of which, however, were also dispersed by howlers. Slightly smaller-seeded species (20–25 mm) were often also dispersed by either species of capuchin monkeys occurring at each site and, in rare cases, by primates as small as tamarins. We have no evidence that seeds exceeding 45–50 mm could be ingested by any of the primate species. Rather, they tended to be scatter-hoarded on the forest floor by large rodents (e.g. van Roosmalen, 1985a; Peres and Baider, 1997),

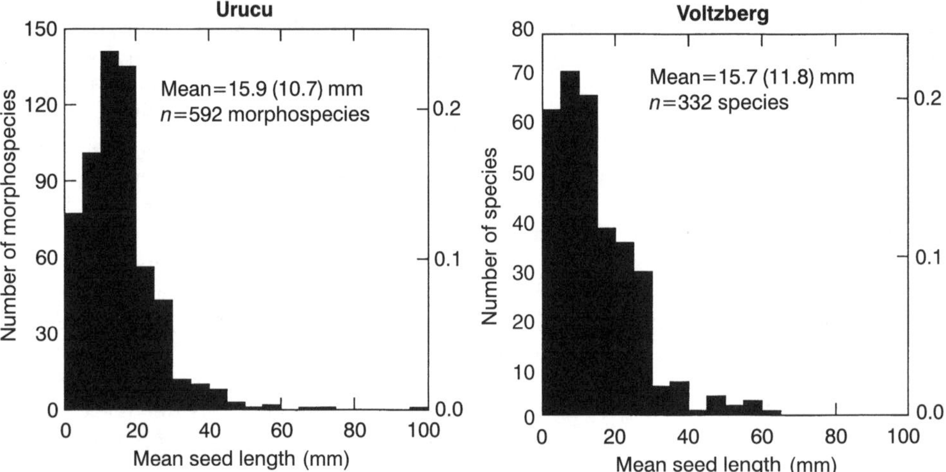

Fig. 27.1. Frequency distribution of mean seed length for all tree, liana and epiphyte species producing fruits that were known to be handled by vertebrate frugivores and seed predators at Urucu, Brazil, and Voltzberg, Suriname. Values in parentheses indicate standard errors.

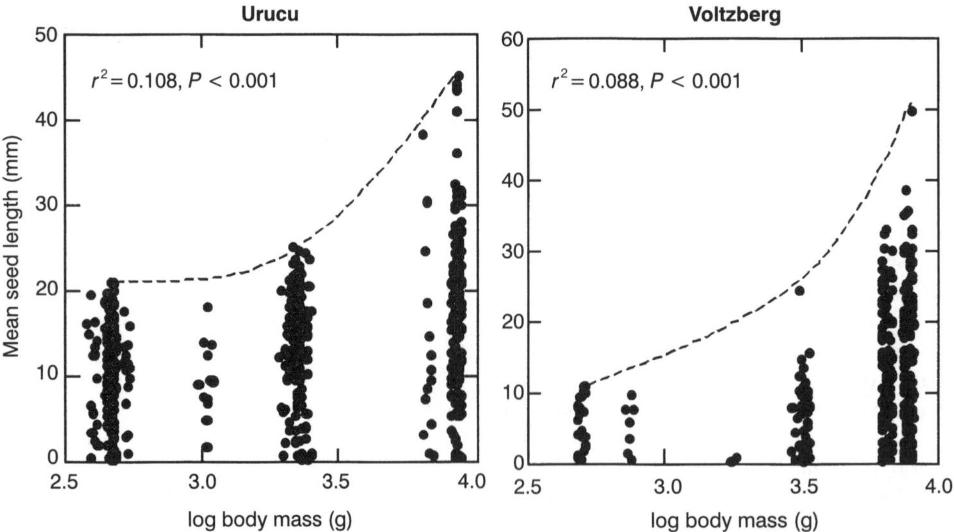

Fig. 27.2. Mean adult body mass of primate species in the two study sites and the mean length of seeds that they were typically capable of ingesting for all fruit species consumed.

but these species probably generated a more restricted seed shadow than did ateline primates (Stevenson, 2000). Large-seeded gut-dispersed plants were thus associated with a substantially lower redundancy of disperser species and were heavily reliant on the largest primates occurring at each site.

Effects of subsistence hunting on ateline primates

Densities of the three Amazonian ateline genera (*Alouatta*, *Lagothrix* and *Ateles*) at non-hunted sites are consistently higher than those at persistently hunted sites. In *terra firme* forests of Amazonia and the Guianan shields, the aggregate biomass of ateline primates at non-hunted sites (152 ± 17 kg km^{-2}, n = 9) declines to 125 ± 21 kg km^{-2} (n = 10) in lightly hunted sites, and is significantly lower in moderately (54 ± 14 kg km^{-2}, n = 7) and heavily hunted sites (19 ± 6 kg km^{-2}, n = 10) (Peres, 1999c). As a consequence, the contribution of these large-bodied species to the total primate biomass at any one site declines sharply from an average of 60% in non-hunted sites to only 15% in heavily hunted sites. This results in profound shifts in the size structure of primate assemblages across varying degrees of hunting pressure. For instance, the body mass of primates occurring at different sites is reduced from a mean of 2550 g (range = 1790–3420 g, n = 9) at non-hunted sites to only 1020 g (range = 700–1500 g, n = 10) at persistently hunted sites.

The impact of hunting is particularly severe on the two largest ateline genera because they are more attractive to hunters, have a lower reproductive rate, live in larger groups and cannot adjust to persistent hunting pressure by becoming more behaviourally inconspicuous, as do howlers (C.A. Peres, unpublished data). For several reasons, however, howlers appear to be less effective dispersers of very large seeds than are either woolly or spider monkeys. For example, because of the spatial structure of howler monkey groups, the seed shadows they generate are considerably more clumped at small spatial scales (Chapman, 1989; Julliot, 1997), which can result in greater post-dispersal seed predation (Andresen, 1999) and lower seedling survival.

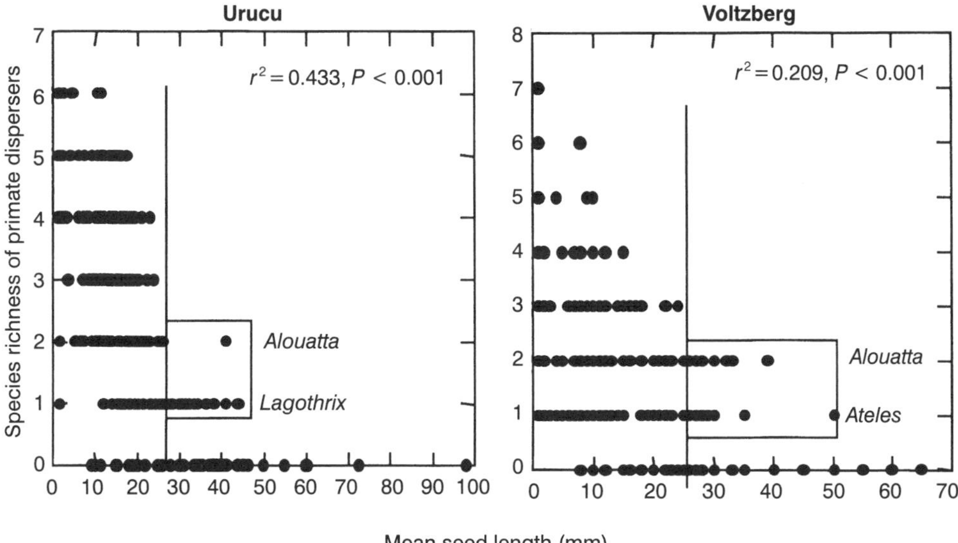

Fig. 27.3. Mean seed size of different plant species and the species richness of primate dispersers (defined as the number of primate species known to normally ingest seeds – see text). Small-seeded species are associated with a far greater redundancy of dispersal services at the level of species. Rectangles indicate those species that were exclusively dispersed by the ateline primates occurring at each forest site.

Seed size and dispersal probabilities in overhunted forests

A logistical regression on the fruit vs. primate species matrix at each study area reveals the thresholds beyond which large seeds are no longer effectively dispersed by the primate species expected to be selectively extirpated or severely reduced in numbers in overhunted forests. For gut-dispersed seeds, successful dispersal events (dependent variable) were defined as those in which seeds of a given species were typically ingested and eliminated intact by a given primate species. Dispersal events were defined as unsuccessful if seeds within a given pairwise interaction (fruit × primate species) were typically dropped or spat out underneath parent crowns or at short distances away from them, or otherwise had their embryos damaged or destroyed by the consumer.

Both seed size and consumer body size had significant effects on the dispersal fate of seeds from mature fruits handled by primates at the two sites (Table 27.2). For instance, these parameters could predict the dispersal probabilities of large seeds (≥ 25 mm) with reasonably good success; between 74 and 91% of the observed interaction outcomes were predicted accurately by the models. For these large-seeded plants, the probability of effective dispersal at Urucu is expected to be reduced to only 20% in persistently overhunted (but otherwise similar) forests in which both mid-sized and large-bodied primates (e.g. capuchins and the atelines) have been extirpated or severely reduced in numbers (Fig. 27.4). A dispersal probability > 50% at this site would require the persistence of populations of all primate species > 1.6 kg (95% confidence interval (CI) = 1.3–1.9 kg). This reduction in dispersal probability was even more severe at Voltzberg, where large-seeded plants would have only a 5% chance of being dispersed under the same defaunation scenario. A dispersal probability > 50% at this site would require the population persistence of all primates > 5.1 kg (95% CI = 4.48–5.77 kg), which includes the largest species, wich are most sensitive to hunting. To some degree, these between-site differences may be due to the fact that rare ingestion events of large seeds were given slightly greater weight in the fruit–frugivore matrix at Urucu than those at Voltzberg, which always reflected the modal seed-treatment pattern. The most realistic probabilities of effective primate dispersal of large seeds in defaunated neotropical forests are therefore closer to our predicted values for Voltzberg than those for the Urucu forest.

Discussion

Our results for two species-rich neotropical forests are consistent with the widely held view that seed size is a key life-history trait, strongly related to plant dispersal ability (e.g. Westoby *et al.*, 1992). In both sites, small-seeded species far outnumbered those with large seeds, but a substantial proportion of each woody flora was represented by a guild of large-seeded zoochorous plants almost exclusively dispersed by large-bodied arboreal vertebrates. While small seeds could be ingested by the entire range of consumer species, the size spectrum of potential dispersers became gradually narrower for increasingly larger seeds; the largest seeds were exclusively ingested by large frugivores.

Table 27.2. Logistic regression analysis showing how primate body mass and mean seed length affect the probability of ingestion of seeds from mature fruits that were consumed by any of the primate species occurring at the Urucu forest and the Voltzberg Reserve.

	Parameter estimates (*P* value)					
Forest site	Body mass*	Seed length*	Rho-sq.†	*P*	*n*‡	Percentage correct
Urucu	6.414 (0.000)	−22.14 (0.000)	0.570	0.000	646	91.2
Voltzberg	4.455 (0.000)	−3.818 (0.000)	0.369	0.000	703	73.7

*All independent variables were $\log_{10}$-transformed.
†McFadden's rho-squared value.
‡Number of modal species interactions accepted into the model.

The 'triangular envelope' relationship between seed size and primate species richness (Fig. 27.3) is also likely to hold for less observable frugivorous taxa. Increasingly larger-seeded plants are thus expected to interact with fewer species providing effective seed-dispersal services.

The specificity between large seeds and frugivores should be further intensified in areas where large-bodied frugivores have been selectively driven to local extinction by hunting, forest fragmentation or both. For example, in the Indo-Malayan region, large-seeded fruits are consumed by progressively fewer dispersers, and the largest depend on a few mammal and bird species that are highly vulnerable to hunting and habitat fragmentation (Corlett, 1998). The same could be argued for other Amazonian forests (Peres, 2000b,c) and species-rich frugivore assemblages elsewhere in the humid tropics (e.g. Gautier-Hion *et al.*, 1985).

Primates and large-seeded plants

Throughout the Guianas, 74% of the woody flora in all habitat types require animals for adequate seed dispersal (van Roosmalen, 1985a). However, this proportion increases to 87–90% in the upland high *terra firme* and mountain savannah forests of the Voltzberg study area, where approximately 40% of all woody plant species are dispersed by primates. Similar figures also apply to the Urucu woody flora, which exhibits a functionally comparable but richer species composition (Peres, 1991; Amaral, 1996). In this study, 340 (49%) of all 689 fruit morphospecies collected at Urucu were known to be ingested and dispersed by primates.

Woolly monkeys at Urucu and spider monkeys at Voltzberg consumed at least 193 (Peres, 1994b) and 171 species of fruits (van Roosmalen, 1985b), respectively, and they dispersed the vast majority of these. Mature seeds that were largely or exclusively dispersed by large-bodied primates (*Alouatta* and *Lagothrix* at Urucu; *Alouatta* and *Ateles* at Voltzberg) were associated with indehiscent, thick- or hard-husked fruits, containing one or a few large seeds, which tended to have a soft coat, a chemically protected endosperm and a tegument or seed testa that strongly adhered to a sweet pulp. These morphological traits appear to comprise a general 'large-primate fruit syndrome'; they are consumed almost exclusively by a small guild of dispersers that are capable of swallowing large seeds from indehiscent husks.

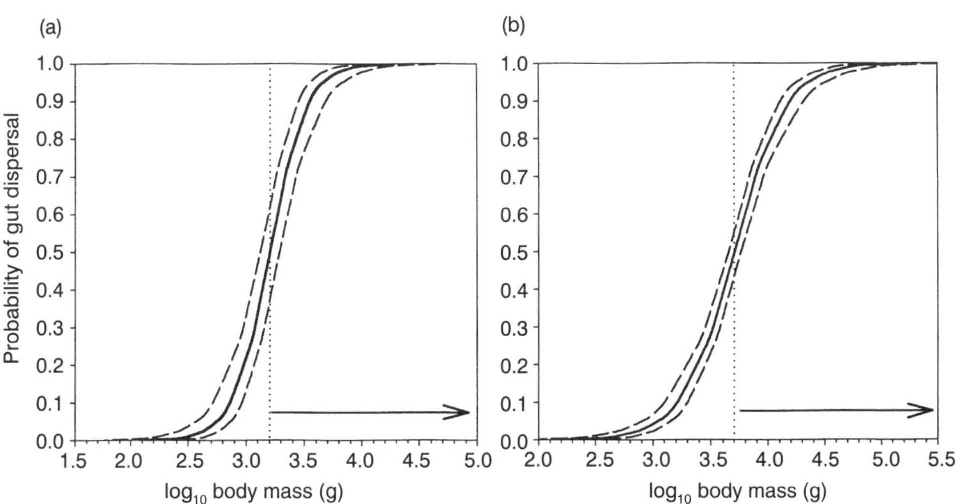

Fig. 27.4. Mean (± 95% CI) probability of gut dispersal for seeds of 25 mm in length as a function of primate body mass at (a) Urucu and (b) Voltzberg, predicted using logistic regression models based on the observed outcomes of seed processing in the entire primate by fleshy-fruit matrix at each forest site. Dotted lines indicate the body-mass extinction threshold (arrows) beyond which effective dispersal probabilities would fall below 50%.

Moreover, these large-seeded plants are often associated with tree and liana species that are restricted to mature (or late-successional) forests and capable of recruiting viable saplings in the shaded understorey, as opposed to small-seeded pioneer species, which are often associated with tree-fall gaps and young second growth (Denslow, 1980; Foster and Janson, 1985). Such specialized dispersal services could, therefore, be critical to the long-term persistence of the full integrity of plant communities throughout the entire forest-succession process.

Ateles and *Lagothrix* appear to be the highest-quality dispersers for this group of seeds, because they can ingest large seeds undamaged (this study), carry a huge volume of seeds in their guts (Peres, 1994b), have long retention times (Milton, 1984; Stevenson, 2000), cover large areas daily, range in spatially uncohesive social groups (van Roosmalen, 1985b; Peres, 1996) and scatter seeds over large areas of leaf litter during single deposition events (C.A. Peres and M. van Roosmalen, unpublished data). Capuchins (*Cebus* spp.) are endowed with excellent manual dexterity but are unable to ingest seeds as large as those swallowed by the large atelines, whereas howlers have good seed-swallowing abilities but have poorer manipulative skills than do capuchins and woolly and spider monkeys.

Although ingestion of large seeds was primarily restricted to large-bodied primates, seeds as long as 24 mm were occasionally observed in faecal samples of saddleback and moustached tamarins at Urucu. Such remarkable seed-swallowing capabilities of tamarins in relation to their small body size is supported by observations of congeners ingesting seeds larger than 20 mm in other Amazonian forests (Garber and Kitron, 1997; Knogge, 1999), confirming the important role of *Saguinus* as seed-dispersers. This is particularly the case with indehiscent fruits produced by many understorey tree and scandent species, which are almost exclusively consumed by tamarins, even in non-hunted, species-rich primate assemblages (Terborgh, 1983; Peres, 1993a).

Although large seed size has probably coevolved with particular groups of animal dispersers (Jordano, 1995), there are important morphological distinctions between large-seeded species that are primarily dispersed by large primates and those primarily scatter-hoarded by large rodents (e.g. Forget, 1990; Peres and Baider, 1997; Wright *et al.*, 2000). Fruits of rodent-dispersed trees tend to be sclerocarpic, and many species invest in physical rather than chemical protection of the large endosperm. But these species are unlikely to succumb to persistent recruitment failures in continuous forests, because large dasyproctid rodents (*Myoprocta* and *Dasyprocta*) are fairly resilient to hunting pressure and are rarely extirpated by hunters (Peres, 2000c). Indeed, both agoutis and acouchies can persist even in areas that have been overharvested by hunters assisted by dogs for many years (C.A. Peres, unpublished data). Moreover, lower dasyproctid densities could result in lower predation rates of post-dispersed seeds (Wright *et al.*, 2000). Therefore, large-seeded plants dispersed by rodents could probably maintain a viable recruitment in overhunted areas, whereas this is unlikely for species exclusively dispersed by large primates.

The fate of large-seeded plants in overhunted forests

Several studies indicate that hunting of neotropical primates is widespread and often intense (Freese, 1978; Peres, 1990; Ráez-Luna, 1995; Fa and Peres, 2001). In demographic terms, this harvest is almost invariably unsustainable and can reach into the core of even the largest and least accessible nature reserves (C.A. Peres and I. Lake, unpublished). Understanding higher-order interactions between large-bodied frugivore removal and long-term forest dynamics will become increasingly important in managed tropical forests, where the residual primate fauna will consist predominantly of small-bodied species (Peres and Dolman, 2000). Even if local extirpation does not occur, large primates may no longer provide effective seed-dispersal services in both quantitative and qualitative terms once they have been reduced to meagre numbers or driven to ecological extinction (Redford and Feinsinger, 2001).

Given a sufficiently long period, several large-seeded indehiscent fruit species that are heavily reliant on primate seed dispersal could experience chronic dispersal limitation and reduced regeneration where large primates have been severely overhunted (see Howe, 1984; Chapman and Onderdonk, 1998). We now provide a preliminary check-list of the most likely genera to experience dispersal limitation in Amazonian and Guianan forests (Table 27.3). This list includes only those species that apparently cannot normally recruit underneath parent plants, and thus appear to be dependent on animal vectors. Both the predominantly long nearest-neighbour distances between conspecific adults (see Terborgh *et al.*, this volume) and the strong role of density dependence in seedling recruitment in mature forests (Harms *et al.*, 2000) suggest that recruitment under parents is rarely possible, supporting the importance of seed-dispersers. To date, however, only a few studies have examined the effects of frugivore defaunation on tropical plant communities, and an adequate test of demographic bottlenecks has yet to address any species depending largely or entirely on primate dispersal (Table 27.3).

At the community level, seedling density in overhunted forests can be greater, indistinguishable or lower than that in non-hunted forests (Panama: Wright *et al.*, 2000; Mexico: Dirzo and Miranda, 1991; and Uganda: Chapman and Onderdonk, 1998). Patterns of seedling recruitment may well depend on the species being hunted, the redundancy of seed vectors for different plant species and the regeneration strategy of a given plant species. These conflicting results for entire seedling assemblages are likely to remain inconclusive until such effects are teased apart. However, the generally greater reliance of large-seeded species on biotic processes for successful recruitment (Jordano, 1995) suggests that they are more vulnerable to disequilibrium of faunal communities. Moreover, several studies suggest that plant population dynamics can be profoundly modified in the absence of key dispersers. In Thailand, seedling recruitment rates of two vertebrate-dispersed species in abandoned agricultural clearings are severely limited by low rates of seed dispersal (Hardwick *et al.*, 1997). In the island of La Réunion, large-seeded (> 20 mm), fleshy-fruit species have very low colonization rates (Thébaud and Strasberg, 1997) and are missing from recent lava flows, perhaps because of the large number of large-frugivore extinctions in the last three centuries (Cox *et al.*, 1991). Chapman and Onderdonk (1998) found that seedling recruitment for large-seeded species was dramatically reduced by primate population declines in Kibale, Uganda. In the Brazilian Atlantic Forest, some 39% of the tree species may eventually face severe recruitment bottlenecks resulting from limited seed-dispersal services for large-seeded plants following large frugivore extinctions (Silva and Tabarelli, 2000). Similar dispersal-limitation scenarios may occur in other regions where large frugivore assemblages in small forest fragments form nested subsets of the species still retained in large fragments (e.g. Kattan and Alvarez-Lopez, 1996).

It remains to be seen, however, to what degree large-seeded plants can adjust to the loss of key dispersers. Wright *et al.* (2000) examined how hunting alters seed dispersal, seed predation and seedling recruitment for two rodent-dispersed Panamanian palms, and found that, in non-hunted areas, most seeds were eaten by rodents after being dispersed away from parents. In contrast, few seeds had been dispersed where mammal abundance had been reduced by hunters, but those that had often escaped rodent predation, resulting in seedling densities three to five times higher than those in non-hunted sites.

In summary, we have shown that large-bodied primates in species-rich neotropical forests can provide effective dispersal services for a wide range of endozoochorous, large-seeded plants that produce thick-husked, indehiscent fruits, which are rarely or never handled by other frugivores. Primate body size was directly related to the size spectrum of seeds ingested; small seeds were dispersed by all primate species, whereas large seeds were dispersed only by large species. Because large-bodied frugivores, particularly ateline primates, are often driven to local extinction in persistently overhunted forests, loss of specialized dispersal services could substantially reduce recruitment of an important guild of gut-dispersed large-seeded plants. Further work is now required to evaluate the demographic consequences of this form of dispersal limitation.

Table 27.3. Large-seeded trees and woody lianas that could eventually face low seedling recruitment rates in persistently overhunted forests of Amazonia and the Guianan shield.

Family	Genera
Anacardiaceae	*Spondias*, *Anacardium*
Annonaceae	*Annona*, *Rollinia*, some *Duguetia*, *Ephedranthus* and *Fusaea*
Araceae	*Heteropsis*, *Philodendron*, *Monstera*
Bombacaceae	*Catostemma*, *Matisia*
Hippocrateaceae	*Cheiloclinium*, *Peritassa*, *Tontelea*, *Salacia*
Chrysobalanaceae	*Couepia*, *Parinari*, several *Licania*
Clusiaceae	*Calophyllum*, *Moronobea*, *Platonia*, *Rheedia*, *Symphonia*, *Tovomita*
Convolvulaceae	*Lysiostyles*, *Dicranostyles*, *Maripa*
Cucurbitaceae	*Cayaponia*
Ebenaceae	*Diospyros*
Euphorbiaceae	*Omphalea*
Flacourtiaceae	*Casearia*
Gnetaceae	*Gnetum*
Humiriaceae	*Sacoglottis*, *Schistostemon*, *Vantanea*
Icacinaceae	*Discophora*, *Emmotum*, *Leretia*, *Poraqueiba*
Lauraceae	*Licaria*, *Nectandra*, *Rhodostemonodaphne*, *Ocotea*, a few large-seeded *Aniba*
Lecythidaceae	*Gustavia*
Leguminosae Caesalpiniaceae	*Dimorphandra*, *Hymenaea*, *Mora*, *Peltogyne*, some *Swartzia*, some *Cassia*, *Senna* and *Dialium*
Leguminosae, Mimosaceae	*Inga*, *Enterolobium*, *Stryphnodendron*, *Zygia*, some indehiscent *Abarema*
Loganiaceae	*Strychnos*
Melastomataceae	Some large-seeded *Mouriri*
Meliaceae	Several *Guarea*
Menispermaceae	*Caryomene*, *Orthomene*, some *Abuta*, all *Anomospermum*
Moraceae	*Clarisia*, *Helicostylis*, *Maquira*, *Naucleopsis*, *Perebea*, *Trymatococcus*, some *Brosimum*, some *Coussapoa*
Myristicaceae	*Osteophloem*, some *Iryanthera*
Myrtaceae	Some *Eugenia*
Olacaceae	*Minquartia*, *Ptychopetalum*
Palmae	*Socratea*, *Syagrus*
Passifloraceae	*Dilkea*, some *Passiflora* with hard, thick husk
Polygalaceae	*Moutabea*
Quiinaceae	*Lacunaria*, *Quiina*
Rubiaceae	*Duroia*
Sapindaceae	*Talisia*, some *Paullinia*
Sapotaceae	*Chrysophyllum*, *Manilkara*, *Pouteria*, *Pradosia*, *Ecclinusa*, some *Micropholis*
Simaroubaceae	*Simaba*, *Simarouba*
Sterculiaceae	*Guazuma*, all *Theobroma*
Tiliaceae	*Apeiba*
Ulmaceae	*Ampelocera*
Violaceae	*Leonia*

Acknowledgements

This Urucu study was funded by the Wildlife Conservation Society, the Josephine Bay and Michael Paul Foundation, and the Center for Applied Biodiversity Sciences (CABS) of Conservation International. The Voltzberg study was funded by The Netherlands Foundation for Tropical Research. We thank Doug Levey and an anonymous reviewer for constructive comments on the manuscript.

References

Amaral, I.L. (1996) Diversidade floristica em floresta de *terra firme* na região do Rio Urucu – AM. MSc thesis, Instituto Nacional de Pesquisas da Amazônia, Manaus, Brazil.

Andresen, E. (1999) Seed dispersal by monkeys and the fate of dispersed seeds in a Peruvian rain forest. *Biotropica* 31, 145–158.

Augspurger, C.K. (1984) Seedling survival among tropical tree species: interactions of dispersal distance, light-gaps, and pathogens. *Ecology* 65, 1705–1712.

Chapman, C.A. (1989) Primate seed dispersal: the fate of dispersed seeds. *Biotropica* 21, 148–154.

Chapman, C.A. and Chapman, L.J. (1995) Survival without dispersers – seedling recruitment under parents. *Conservation Biology* 9, 675–678.

Chapman, C.A. and Chapman, L.J. (1996) Frugivory and the fate of dispersed and non-dispersed seeds in six African tree species. *Journal of Tropical Ecology* 12, 491–504.

Chapman, C.A. and Onderdonk, D.A. (1998) Forests without primates: primate/plant codependency. *American Journal of Primatology* 45, 127–141.

Charles-Dominique, P. (1993) Speciation and coevolution: an interpretation of frugivory phenomena. *Vegetatio* 107/108, 75–85.

Corlett, R.T. (1998) Frugivory and seed dispersal by vertebrates in the Oriental (Indomalayan) Region. *Biological Reviews* 73, 413–448.

Cox, P.A., Elmqvist, T., Pierson, E.D. and Rainey, W.E. (1991) Flying foxes as strong interactors in South Pacific Island ecosystems: a conservation hypothesis. *Conservation Biology* 5, 448–454.

Denslow, J.S. (1980) Gap partitioning among tropical rainforest trees. *Biotropica* 12 (suppl.), 47–55.

Dirzo, R. and Miranda, A. (1991) Altered patterns of herbivory and diversity in the forest understory: a case study of the possible consequences of contemporary defaunation. In: Price, P.W., Lewinsohn, T.M., Fernandes, G.W. and Benson, W.W. (eds) *Plant–Animal Interactions: Evolutionary Ecology in Tropical and Temperate Regions.* John Wiley & Sons, New York, pp. 273–287.

Fa, J.E. and Peres, C.A. (2001) Game vertebrate extraction in African and neotropical forests: an intercontinental comparison. In: Reynolds, J., Mace, G., Robinson, J.G. and Redford, K. (eds) *Conservation of Exploited Species.* Cambridge University Press, Cambridge.

Forget, P.M. (1990) Seed-dispersal of *Voucapoua americana* (*Caesalpinaceae*) by caviomorph rodents in French Guiana. *Journal of Tropical Ecology* 6, 459–468.

Foster, S.A. and Janson, C.H. (1985) The relationship between seed size and establishment conditions in tropical woody plants. *Ecology* 66, 773–780.

Freese, C.H., Heltne, P.G., Castro, N. and Whitesides, G. (1982) Patterns and determinants of monkey densities in Peru and Bolivia with notes on distributions. *International Journal of Primatology* 3, 53–90.

Garber, P.A. and Kitron, U. (1997) Seed swallowing in tamarins: evidence of a curative function or enhanced foraging efficiency? *American Journal of Primatology* 18, 523–538.

Gautier-Hion, A., Duplantier, J.-M., Quris, R., Feer, F., Sourd, C., Decoux, J.-P., Dubost, G., Emmons, L., Erard, L., Hecketsweiler, P., Moungazi, A., Roussilhon, C. and Thiollay, J.-M. (1985) Fruit characters as a basis of fruit choice and seed dispersal in a tropical forest vertebrate community. *Oecologia* 65, 324–337.

Hamann, A. and Curio, E. (1999) Interactions among frugivores and fleshy fruit trees in a Philippine submontane rainforest. *Conservation Biology* 13, 766–773.

Hardwick, K., Healey, J., Elliott, S., Garwood, N. and Anusarnsunthorn, V. (1997) Understanding and assisting natural regeneration processes in degraded seasonal evergreen forests in northern Thailand. *Forest Ecology and Management* 99, 203–214.

Harms, K.E., Wright, S.J., Calderon, O., Hernandez, A. and Herre, E.A. (2000) Pervasive density-dependent recruitment enhances seedling diversity in a tropical forest. *Nature* 404, 493–495.

Harper, J.L. (1977) *The Population Biology of Plants.* Academic Press, New York.

Howe, H.F. (1984) Implications of seed dispersal by animals for tropical reserve management. *Biological Conservation* 30, 264–281.

Howe, H.F. and Smallwood, J. (1982) Ecology of seed dispersal. *Annual Review of Ecology and Systematics* 13, 201–218.

Jordano, P. (1992) Fruits and frugivory. In: Fenner, M. (ed.) *Seeds: the Ecology of Regeneration in Natural Plant Communities.* CAB International, Wallingford, UK, pp. 105–156.

Jordano, P. (1995) Angiosperm fleshy fruits and seed dispersers: a comparative-analysis of adaptation and constraints in plant–animal interactions. *American Naturalist* 145, 163–191.

Julliot, C. (1997) Impact of seed dispersal by red howler monkeys *Alouatta seniculus* on the seedling population in the understorey of tropical rain forest. *Journal of Ecology* 85, 431–440.

Kattan, G. and Alvarez-Lopez, H. (1996) Preservation and management of biodiversity in fragmented landscapes in the Colombian Andes. In: Schellas, J. and Greenberg, R. (eds) *Forest*

Patches in Tropical Landscapes. Island Press, London, pp. 3–18.

Knogge, C. (1999) *Tier-Pflze-Interaktionen im Amazonasregenwald. Samenausbreitung durch die sympatrischen Tamarine* Saguinus mystax *und* Saguinus fuscicollis *(Callitrichidae, Primates).* Schüling, Münster.

Milton, K. (1984) The role of food processing factors in primate food choice. In: Rodman, P.S. and Cant, J.G.H. (eds) *Adaptations for Foraging in Nonhuman Primates.* Columbia University Press, New York, pp. 249–279.

Mittermeier, R.A. and van Roosmalen, M.G.M. (1981) Preliminary observations on habitat utilization and diet in eight Suriname monkeys. *Folia Primatologica* 36, 1–39.

Muchaal, P.I. and Ngandjui, G. (1999) Impact of village hunting on wildlife populations in the Western Dja Reserve, Cameroon. *Conservation Biology* 13, 385–396.

Penfold, G.C. and Lamb, D. (1999) Species coexistence in an Australian subtropical rain forest: evidence from compensatory mortality. *Journal of Ecology* 87, 316–329.

Peres, C.A. (1990) Effects of hunting on western Amazonian primate communities. *Biological Conservation* 54, 47–59.

Peres, C.A. (1991) Ecology of mixed-species groups of tamarins in Amazonian *terra firme* forests. PhD thesis, University of Cambridge, Cambridge, UK.

Peres, C.A. (1993a) Diet and feeding ecology of saddle-back and moustached tamarins in an Amazonian *terra firme* forest. *Journal of Zoology* 230, 567–592.

Peres, C.A. (1993b) Structure and spatial organization of an Amazonian *terra firme* forest primate community. *Journal of Tropical Ecology* 9, 259–276.

Peres, C.A. (1993c) Notes on the ecology of buffy saki monkeys (*Pithecia albicans,* Gray 1860): a canopy seed-predator. *American Journal of Primatology* 31, 129–140.

Peres, C.A. (1994a) Primate responses to phenological changes in an Amazonian *terra firme* forest. *Biotropica* 26, 98–112.

Peres, C.A. (1994b) Diet and feeding ecology of gray woolly monkeys (*Lagothrix lagotricha cana*) in central Amazonia: comparisons with other atelines. *International Journal of Primatology* 15, 333–372.

Peres, C.A. (1994c) Which are the largest New World monkeys? *Journal of Human Evolution* 26, 245–249.

Peres, C.A. (1996) Use of space, foraging group size, and spatial group structure in gray woolly monkeys (*Lagothrix lagotricha cana*) at Urucu, Brazil: a review of the Atelinae. In: Norconk, M.A., Rosenberger, A.L. and Garber, P.A. (eds) *Adaptive Radiations in Neotropical Primates.* Plenum, New York, pp. 467–488.

Peres, C.A. (1999a) Nonvolant mammal community structure in different Amazonian forest types. In: Eisenberg, J.F. and Redford, K.H. (eds) *Mammals of the Neotropics: the Central Neotropics.* University of Chicago Press, Chicago, pp. 564–581.

Peres, C.A. (1999b) General guidelines for standardizing line-transect surveys of tropical forest primates. *Neotropical Primates* 7, 11–16.

Peres, C.A. (1999c) Effects of hunting and habitat quality on Amazonian primate communities. In: Fleagle, J.G., Janson, C. and Reed, K.E. (eds) *Primate Communities.* Cambridge University Press, Cambridge, pp. 268–283.

Peres, C.A. (2000a) Identifying keystone plant resources in tropical forests: the case of *Parkia* pod gums. *Journal of Tropical Ecology* 16, 287–317.

Peres, C.A. (2000b) Effects of subsistence hunting on vertebrate community structure in Amazonian forests. *Conservation Biology* 14, 240–253.

Peres, C.A. (2000c) Evaluating the impact and sustainability of subsistence hunting at multiple Amazonian forest sites. In: Robinson, J.G. and Bennett, E.L. (eds) *Hunting for Sustainability in Tropical Forests.* Columbia University Press, New York, pp. 83–115.

Peres, C.A. and Baider, C. (1997) Seed dispersal, spatial distribution, and size structure of Brazil-nut trees (*Bertholletia excelsa, Lecythidaceae*) at an unharvested stand of eastern Amazonia. *Journal of Tropical Ecology* 13, 595–616.

Peres, C.A. and Dolman, P. (2000) Density compensation in neotropical primate communities: evidence from 56 hunted and non-hunted Amazonian forests of varying productivity. *Oecologia* 122, 175–189.

Pratt, T.K. and Stiles, E.W. (1985) The influence of fruit size and structure on composition of frugivore assemblages in New Guinea. *Biotropica* 17, 314–321.

Ráez-Luna, E.F. (1995) Hunting large primates and conservation of the Neotropical rain forests. *Oryx* 29, 43–48.

Redford, K. and Feinsinger, P. (2001) The half-empty forest. In: Reynolds, J., Mace, G., Robinson, J.G. and Redford, K. (eds) *Conservation of Exploited Species.* Cambridge University Press, Cambridge.

Rey, P.J. and Alcántara, J.M. (2000) Recruitment dynamics of a fleshy-fruited plant (*Olea europaea*): connecting patterns of seed dispersal to seedling establishment. *Journal of Ecology* 88, 622–633.

Silva, J.M.C. and Tabarelli, M. (2000) Tree species impoverishment and the future flora of the Atlantic forest of northeast Brazil. *Nature* 404, 72–74.

Stevenson, P.R. (2000) Seed dispersal by woolly monkeys (*Lagothrix lagothricha*) at Tinigua National Park, Colombia: dispersal distance, germination rates, and dispersal quantity. *American Journal of Primatology* 50, 275–289.

Terborgh, J. (1983) *Five New World Primates: a Study in Comparative Ecology*. Princeton University Press, Princeton.

Thébaud, C. and Strasberg (1997) Plant dispersal in fragmented landscapes: a field study of woody colonization in rainforest remnants of the Mascarene Archipelago. In: Laurance, W.F. and Bierregard, R.O., Jr (eds) *Tropical Forest Remnants: Ecology, Management, and Conservation of Fragmented Communities*. University of Chicago Press, Chicago, Illinois, pp. 321–332.

van Roosmalen, M.G.M. (1985a) *Fruits of the Guianan Flora*. Veenman, Wageningen, The Netherlands.

van Roosmalen, M.G.M. (1985b) Habitat preferences, diet, feeding strategy and social organization of the black spider monkey (*Ateles paniscus paniscus* Linnaeus 1758), in Suriname: a socioecological field study. *Acta Amazônica* 15, 1–238.

van Schaik, C.P., Terborgh, J.W. and Wright, S.J. (1993) The phenology of tropical forests: adaptive significance and consequences for primary consumers. *Annual Review of Ecology and Systematics* 24, 353–377.

Westoby, M., Jurado, E. and Leishman, M. (1992) Comparative evolutionary ecology of seed size. *Trends in Ecology and Evolution* 7, 368–372.

Wills, C. and Condit, R. (1999) Similar non-random processes maintain diversity in two tropical rainforests. *Proceedings of the Royal Society of London Series B – Biological Sciences* 266, 1445–1452.

Wright, S.J., Zeballos, H., Dominguez, I., Gallardo, M.M., Moreno, M.C. and Ibáñez, R. (2000) Poachers alter mammal abundance, seed dispersal and seed predation in a neotropical forest. *Conservation Biology* 14, 227–239.

28 Patterns of Fruit–Frugivore Interactions in Two Atlantic Forest Bird Communities of South-eastern Brazil: Implications for Conservation

Wesley R. Silva,[1] Paulo De Marco Jr,[2] Érica Hasui[3] and Verônica S.M. Gomes[3]

[1]*Laboratório de Interações Vertebrados-Plantas, Depto Zoologia, UNICAMP, 13083-970 Campinas, São Paulo State, Brazil;* [2]*Laboratório de Ecologia Quantitativa, Universidade Federal de Viçosa, 36570-000 Viçosa, Minas Gerais State, Brazil;* [3]*Graduate Program in Ecology, IB UNICAMP, 13083-970 Campinas, São Paulo State, Brazil*

Introduction

Plant–frugivore interactions are key components of complex forest communities. For several reasons, the modification or loss of such interactions may have profound implications for conservation, especially in the tropics. First, some species of fruiting plant may be critical for maintaining frugivore populations during periods when other species are not producing fruit (Howe, 1977, 1981; Terborgh, 1986a,b; Peres, 2000). Secondly, some plant species are likely to experience changes in dispersal and recruitment if populations of frugivores are reduced (Pizo, 1997; Asquith *et al.*, 1999; Wright *et al.*, 2000). In fact, because frugivores appear to be especially sensitive to habitat fragmentation (Loiselle and Blake, 1992; Kattan *et al.*, 1994), long-term maintenance of fruiting plants in habitat fragments may be especially difficult. Thirdly, seed dispersal by frugivores can contribute to the restoration and management of degraded habitats. For example, fruit-eating birds are readily attracted to remnant trees and even artificial perches in degraded areas, thereby facilitating regeneration by increasing seed rain (McDonnell and Stiles, 1983; Guevara and Laborde, 1993; McClanahan and Wolfe, 1993; Silva *et al.*, 1996; Holl, 1998; Duncan and Chapman, 1999).

Despite the importance of fruit–frugivore interactions on a community level, there are few conservation-orientated studies that include more than several species of fruiting plants and frugivores (Silva and Tabarelli, 2000; Galetti, 2001). Thus, most studies overlook a myriad of interactions (e.g. by inconspicuous species) that contribute to the recruitment dynamics of plants and to the food-web structure of frugivores. Here we present a community-level study of fruit–frugivore interactions in the Atlantic Forest of Brazil. This biome, formerly widespread along the Brazilian coast, represents one of the most threatened in the world and is reduced to a series of fragments (Fonseca, 1985; SOS Mata Atlântica and INPE, 1992; Ranta *et al.*, 1998), jeopardizing endemic species of fauna and flora (Terborgh, 1992; Brooks and Balmford, 1996).

We focus on frugivorous birds because they are important mutualists to fruiting plants, are diurnal, have conspicuous foraging behaviour and can be easily captured. We quantified general patterns of frugivory in two nearby communities, one suffering greater impact by humans than the other. In each, we recorded observations of fruit consumption by birds. A record of a fruit species eaten by a bird species constitutes an 'interaction' between those species. We tallied and analysed the number of unique interactions in each community. Our goal was to assign a conservation value to each species of plant and bird, under the assumption that the more interactions a species is part of, the more important it is in the community-wide maintenance of fruit–frugivore interactions.

Study Site and Methods

Study site

Data were collected at Intervales State Park, a 50,000 ha preserve in south-eastern São Paulo State. Intervales is part of a reserve network that totals 120,000 ha of continuous Atlantic Forest. It is characterized by a mosaic of pristine forest and second-growth vegetation along an altitudinal gradient of 60–1200 m. Our study sites are 7 km apart and at the same altitudinal range (800–1000 m). They are connected by forest vegetation but differ in the degree of disturbance. The first site (Sede) houses the park's administration facilities and is covered by second-growth vegetation and open areas, surrounding some tracts of mature forest, mostly dominated by bamboo thickets, with many roads and trails crossing the different habitats. The second site (Barra Grande) is mainly composed of mature forest with less dense bamboo thickets. The occurrence of a few trails does not seem to have a severe impact on this area. The sites exhibit similar seasonality: December to February are the hottest and wettest months, June to August are the coldest and driest. The frequency of frost varies among years; some frosts are extremely severe. Previous studies in Intervales have revealed that its avian species composition is largely intact, especially with respect to frugivores (Galetti, 1996; Aleixo and Galetti, 1997; Vielliard and Silva, 2001).

Feeding data

We visited Intervales monthly from January 1999 to March 2000, spending 3–5 days censusing each site. A 5–8 km transect along trails crossing different habitats was walked slowly by a trained person for approximately 8 non-consecutive hours daily. During the censuses, we recorded bird-feeding behaviour at understorey and canopy fruiting trees, as sight permitted. In addition to these transects, we conducted observations on focal trees. To avoid sampling biases towards more conspicuous plant species, we limited observations on focal trees to a maximum of 3 non-consecutive hours. We also captured birds in mist nets (2700 and 4500 net-hours for Sede and Barra Grande, respectively) to collect seeds from faecal samples. Seeds were identified using a reference collection. These three procedures have been widely used to identify interactions between species of frugivorous birds and plants, although they are each subject to sampling biases (see Wheelwright *et al.*, 1984).

The relative importance of a seed-disperser species to a plant is a function of several qualitative and quantitative variables (*sensu* Schupp, 1993). We considered an 'interaction' as our sample unit. A feeding record of one bird species on one plant species generates an interaction. Interactions are binary; either a given bird species eats a given fruit species or it does not. Although we made no distinction among frugivores that are likely to vary in their efficiency as seed-dispersers (Moermond and Denslow, 1985; Levey, 1987), we eliminated all feeding records involving probable seed predators (e.g. parrots and parakeets) (Janzen, 1981; Galetti and Rodrigues, 1992), as well as those in which birds consistently took very small portions of fruit pulp without ingesting seeds.

All interactions were arranged in a bird species by plant species matrix, which was used to calculate the number of interactions and an 'importance index' for both plant and bird species. The importance index, I_j, developed by Murray (2000), weights the contribution of each bird species relative to that of the alternative dispersers of each of its food plants, and is given by the equation

$$I_j = \sum_{i=1}^{s} \left(\frac{C_{i,j} / T_i}{S} \right)$$

where T_i is the total number of bird species feeding on plant species i and S is the total number of plant species included in the sample. $C_{i,j} = 1$ if bird species j consumes fruits of plant species i or 0 if it does not. Note that this index allows one to estimate the community-level contribution of a species as a function not only of the number of interactions it performs, but also of the number of other taxa interacting with the same species. Thus, for example, if two bird species in a community of frugivores take part in exactly the same number of interactions, their potential importance as seed-dispersers will differ according to how many other frugivores feed on the same set of plant species. Values of I_j can range from 0.0 for a bird that consumes no fruits whatsoever to 1.0 for a bird that is the sole consumer of all plant species in its community.

The same equation was used to estimate the relative importance of the different plant species in the birds' diet (I_j), in which case T_i is the total number of plant species fed on by bird species i, S is the total number of bird species included in the sample, and $C_{i,j} = 1$ if plant species j is included in the diet of bird species i or 0 if it is not. Plant species with the highest values of I are those that produce fruit consumed by many bird species that eat few other species of fruits.

Fruiting plant and frugivorous bird species compositions in both study sites were compared using Sorensen's similarity coefficient (Krebs, 1989). This index ranges from 0 (no species are found in both communities) to 1.0 (all species are found in both communities). Differences in the average number of interactions per bird and plant family were tested by Mann–Whitney tests and differences in the frequency distribution of interactions between families of frugivores by a chi-square non-parametric test.

Bird nomenclature follows Sick (1993) and plant taxonomy follows Cronquist (1981).

Results

The frugivores

The frugivorous birds recorded in this study comprise the typical suite of frugivores for Intervales (Vielliard and Silva, 2001). A total of 68 species of frugivores accounted for 397 interactions unevenly distributed between sites (Table 28.1). Similarity in species composition between sites was relatively high (Sorensen index = 0.70) and the average number of interactions per bird species was also very similar (Sede = 4.1 ± 4.2, $n = 233$; Barra Grande = 3.8 ± 4.0, $n = 178$; Mann–Whitney test, $P > 0.05$).

The number of Emberizidae species was quite similar between the sites (18 in Sede and 19 in Barra Grande), although the species composition was slightly different. Tyrant flycatchers (Tyrannidae) had significantly more species in Sede than in Barra Grande (17 vs. 6, respectively; $\chi^2 = 6.9$, d.f. = 1, $P = 0.01$).

Five of 13 families (Emberizidae, Muscicapidae, Pipridae, Tyrannidae and Cotingidae) accounted for 83% of total interactions ($n = 397$). Excluding those families and subfamilies with records in only one site, the difference in the distribution of frequencies of interactions among bird families between sites was significant ($\chi^2 = 27.2$, d.f. = 9, $P = 0.01$). Tyrannidae and Emberizidae had more interactions than expected by chance in the more disturbed site (Sede), whereas Cotingidae and Ramphastidae had more interactions in the less disturbed site (Barra Grande). Less than 5% of total interactions were by species rarely seen eating fruit (Table 28.1), although some species contributing to these interactions were fairly common. This reflects an opportunistic pattern of fruit exploitation, at least for our sites.

The fruiting plants

A total of 103 plant species took part in 397 interactions with birds in both sites (Table 28.2). Similarity of species composition

Table 28.1. Number of species in families of frugivorous birds and total number of interactions observed between bird and fruiting-plant species in each bird family, performed in Barra Grande and Sede, Intervales State Park, south-east Brazil.

	No. of species				No. of interactions			
Bird families	Unique to Sede	Unique to Barra Grande	Common to both sites	Total	Unique to Sede	Unique to Barra Grande	Common to both sites	Total
Columbidae	–	–	1	1	1	1	1	3
Cotingidae	2	2	2	6	16	25	–	41
Cracidae	–	–	2	2	10	9	1	20
Emberizidae (Cardinalinae)	–	–	1	1	6	4	–	10
Emberizidae (Coerebinae)	–	1	–	1	–	1	–	1
Emberizidae (Emberizinae)	–	1	–	1	–	1	–	1
Emberizidae (Icterinae)	–	1	1	2	1	2	–	3
Emberizidae (Thraupinae)	3	1	13	17	86	45	7	138
Mimidae	1	–	–	1	2	–	–	2
Muscicapidae	–	–	4	4	22	24	2	48
Phasianidae	–	1	–	1	–	1	–	1
Picidae	–	1	–	1	–	1	–	1
Pipridae	1	1	2	4	21	21	3	45
Ramphastidae	1	–	2	3	4	9	–	13
Trogonidae	–	1	2	3	8	8	–	16
Tyrannidae	12	1	5	18	36	7	–	43
Vireonidae	–	–	2	2	6	5	–	11
Total	20	11	37	68	219	164	14	397

was low (Sorensen index = 0.28). Despite the difference in plant species composition and total number of interactions between the two communities, the average number of interactions per plant species was very similar between sites (Sede = 3.5 ± 3.3, $n = 233$; Barra Grande = 3.4 ± 3.6, $n = 178$; Mann–Whitney, $P > 0.05$).

Proceeding the same way as with birds and excluding those families with records in only one site, the difference in the distribution of interactions among families between sites was marginally significant ($\chi^2 = 25.7$, d.f. = 15, $P = 0.05$), but only *Euphorbiaceae* and *Rosaceae* accounted for more interactions than expected by chance in Sede.

Species of *Melastomataceae* accounted for most records of fruit consumption in both sites (19.7% in Sede, $n = 233$; 23.5% in Barra Grande, $n = 178$). Melastomes in Sede were more abundant along forest edges and in second growth, whereas in Barra Grande they were mainly trees and shrubs of the canopy and understorey. Among other plant families, pioneer and second-growth species formed the bulk of interactions in Sede (*Anacardiaceae, Flacourticacea, Euphorbiaceae, Rosaceae, Myrsinaceae, Solanaceae* and *Ulmaceae*), whereas canopy and shade-tolerant tree species contributed the most interactions in Barra Grande (*Lauraceae, Myrtaceae, Sapindaceae, Symplocaceae, Loranthaceae*). Five families (*Anacardinaceae, Aquifoliaceae, Flacourtiaceae, Piperaceae* and *Ulmaceae*) interacted with birds only in Sede, where they accounted for 19.7% of the local interactions ($n = 233$).

Table 28.2. Number of species in families of bird-dispersed plants and total number of interactions observed between bird and fruiting-plant species in each plant family in Sede and Barra Grande, Intervales State Park, south-east Brazil.

	No. of species				No. of interactions			
Plant families	Unique to Sede	Unique to Barra Grande	Common to both sites	Total	Unique to Sede	Unique to Barra Grande	Common to both sites	Total
Anacardiaceae	2	–	–	2	15	–	–	15
Annonaceae	–	1	–	1	–	1	–	1
Apocynaceae	1	–	–	1	1	–	–	1
Aquifoliaceae	1	–	–	1	8	–	–	8
Araliaceae	–	1	–	1	–	4	–	4
Arecaceae	1	1	–	2	4	2	–	6
Boraginaceae	1	–	–	1	3	–	–	3
Burseraceae	–	1	–	1	–	4	–	4
Cactaceae	–	1	–	1	–	3	–	3
Cecropiaceae	1	1	–	2	1	4	–	5
Celastraceae	–	1	–	1	–	3	–	3
Commelinaceae	1	–	–	1	1	–	–	1
Euphorbiaceae	2	–	1	3	20	7	–	27
Flacourtiaceae	2	–	–	2	5	–	–	5
Lauraceae	4	7	–	11	12	16	–	28
Liliaceae	1	–	–	1	1	–	–	1
Loranthaceae	1	1	3	5	8	14	1	23
Melastomataceae	8	4	6	18	40	36	6	82
Moraceae	1	–	1	2	10	8	–	18
Myrsinaceae	–	1	1	2	12	7	4	23
Myrtaceae	5	6	2	13	13	17	1	31
Nyctaginaceae	–	–	1	1	2	2	–	4
Onagraceae	–	–	1	1	2	1	1	4
Phytolaccaceae	1	–	–	1	3	–	–	3
Piperaceae	3	–	–	3	6	–	–	6
Rosaceae	1	–	1	2	8	–	1	9
Rubiaceae	4	3	–	7	8	7	–	15
Rutaceae	1	–	–	1	1	–	–	1
Sapindaceae	1	2	–	3	11	17	–	28
Solanaceae	5	2	–	7	9	6	–	15
Symplocaceae	1	2	–	3	1	5	–	6
Ulmaceae	1	–	–	1	12	–	–	12
Zingiberaceae	1	–	–	1	2	–	–	2
Total	51	35	17	103	219	164	14	397

Patterns of interactions

Twenty per cent of all recorded bird and plant species in both communities were responsible for the majority of the interactions.

The rank distribution of interactions among bird species was right-skewed; the vast majority of bird species took part in a few interactions, whereas a few bird species took part in many (Fig. 28.1). The hooded berry-eater, *Carpornis cucullatus*, a canopy and mid-level species, had the highest number of interactions for both sites combined ($n = 34$), followed closely by an understorey species, the blue manakin, *Chiroxiphia caudata* ($n = 32$). The ruby-crowned tanager, *Tachyphonus coronatus* ($n = 28$), and the rufous-bellied thrush, *Turdus rufiventris* ($n = 24$), were also commonly seen eating fruit, most often along forest edges or in second growth.

The same pattern of rank distribution was observed for plant species (Fig. 28.1), except

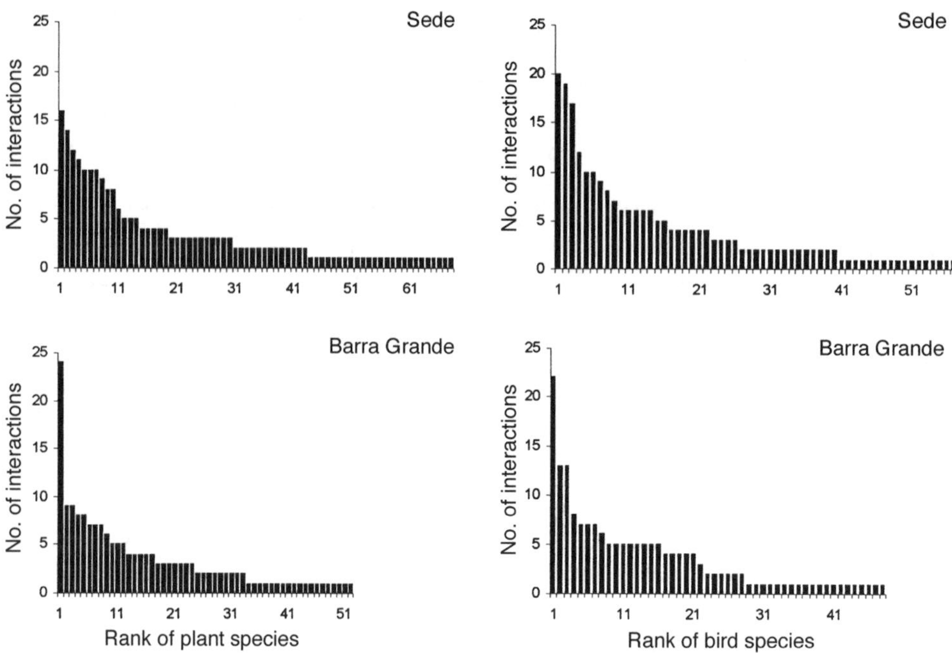

Fig. 28.1. Rank-ordered distribution of interactions for plant species (left panel) and bird species (right panel) in the Sede and Barra Grande regions of Intervales State Park.

that one melastome species, *Miconia pusilliflora*, in Barra Grande had more than twice the number of interactions ($n = 24$) than any of its co-occurring species. In Sede, *Myrsine coriacea* (*Myrsinaceae*) and *Schinus terebenthifolius* (*Anacardiaceae*) ranked first and second in interactions with frugivores ($n = 16$ and 14, respectively).

Importance indices

The 15 bird species with highest indices at Sede and Barra Grande are shown in Fig. 28.2. They performed 63.4% of all interactions in the matrix and comprised three different groups: (i) canopy or mid-level species that feed on a wide variety of fruit types, but mainly large drupes, berries and arillate seeds (e.g. *C. cucullatus, Penelope obscura, Pipile jacutinga, Turdus albicollis, Platycichla flavipes, Selenidera maculirostris, Trogon* spp., *Euphonia pectoralis*); (ii) understorey species that feed mainly on small drupes and berries (e.g. *C. caudata, Ilicura militaris, Schiffornis virescens, Trichothraupis melanops*); and (iii) forest-edge and second-growth species that feed on small drupes, berries and arillate seeds (e.g. *T. rufiventris, T. coronatus, Thraupis* spp., *Stephanophorus diadematus, Cissopis leveriana, Saltator similis*). Among these 15 species, seven are common to both sites. The remaining eight species in Sede are mostly edge species, whereas in Barra Grande they are mostly understorey or mid-level species (Fig. 28.2). Although abundant and speciose, tanagers (Thraupinae) had their importance diluted by more strictly frugivorous taxa, such as cotingids, manakins, thrushes, guans and toucans. In general, bird species with more strictly frugivorous diets had higher values in the less disturbed site, as shown remarkably by the hooded berry-eater (*C. cucullatus*) (Fig. 28.2).

The 15 most important plant species are shown in Fig. 28.3. They accounted for 60.4% of all interactions in the matrix and, like the bird species, can be classified into three groups: (i) canopy or mid-level species, usually shade-tolerant, with a wide variety of fruit types (e.g. *M. pusilliflora, Matayba guianensis, Cupania vernalis, Euterpe edulis, Ocotea puberula, Persea pyrifolia, Myrcia grandiflora*); (ii) forest-edge

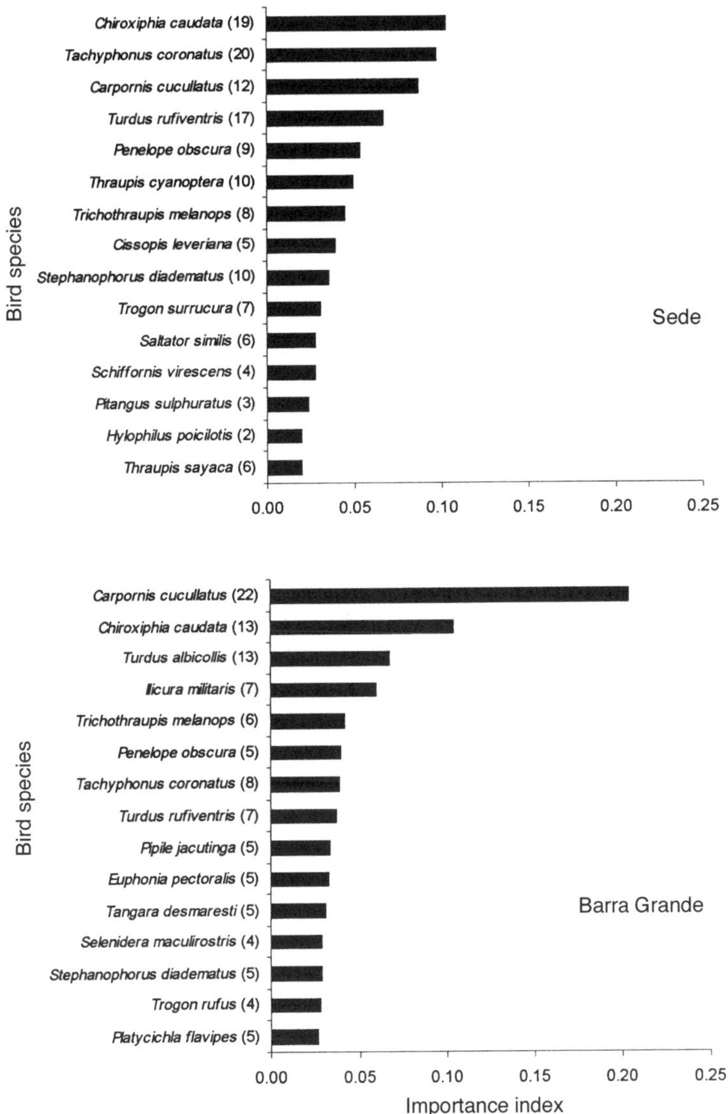

Fig. 28.2. Importance values for the 15 frugivorous birds with highest importance indices in the Sede and Barra Grande regions of Intervales State Park. The number of interactions of each species is given in parentheses.

and second-growth species, with small to medium drupes, berries and arillate seeds (e.g. *Trema micrantha, M. coriacea, S. terebinthifolius, Sapium glandulatum, Alchornea triplinervia, Tetrorchidium rubrivenium, Leandra dasytricha, Ossaea amygdaloides, Ficus luschnatiana, Allophyllus edulis, Ilex brevicuspis*); and (iii) hemi-parasitic mistletoes with small tough-skinned, mucilaginous fruits (e.g. *Struthantus* spp.). The higher values were assigned to the plant species of the first group in the less disturbed site (Barra Grande).

Discussion

The frugivore profile

The average number of interactions that we recorded between frugivorous birds and

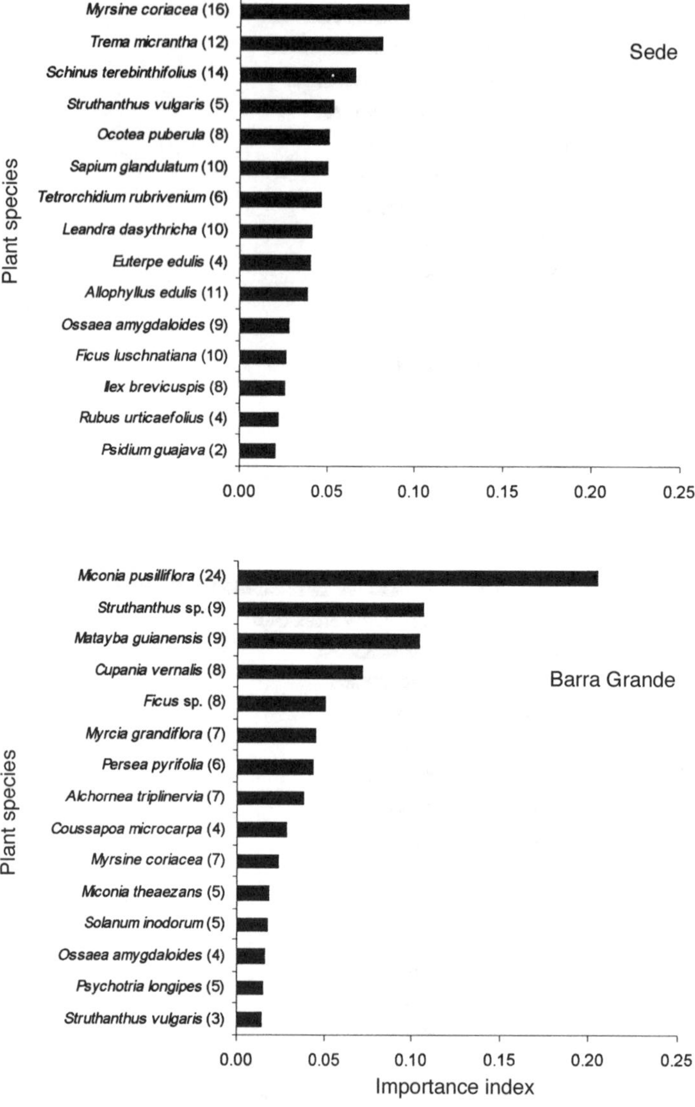

Fig. 28.3. Importance values for the 15 ornithochorous plants with highest importance indices in the Sede and Barra Grande regions of Intervales State Park. The number of interactions of each species is given in parentheses.

fruiting plants was similar to other studies conducted with equivalent methods (Galetti and Pizo, 1996), but lower than reported in a study of a more species-rich community (Wheelwright *et al.*, 1984). However, the average number of interactions and number of species in a community may vary independently from each other (Jordano, 1987). Although our dietary records for each bird species are not exhaustive, we assume that the interactions we report are an unbiased sample and thus that the patterns we found provide an accurate picture of how interactions are organized in both communities.

Fruit-eating bird communities may change in species composition and dynamics according to variation in spatial and temporal components of the habitat (Levey, 1988; Loiselle and Blake, 1991). Despite the difference in plant composition and degree of disturbance between Sede and Barra Grande, their bird communities were quite similar, indicating that frugivorous birds in the Atlantic Forest are flexible in habitat choice, at least within the range of habitats we studied. We detected no absence of large-bodied frugivorous species in the more disturbed site (Sede), as one might expect (Galetti and Aleixo, 1998). The major difference between sites was due to the dominance of edge-habitat tyrant-flycatchers in the Sede site (Table 28.1), a group that often relies heavily on fruit (Morton, 1977; Fitzpatrick, 1980; Sick, 1993). Together with tanagers, saltators and icterines, flycatchers play an important role in the seed-dispersal systems of several plant species in disturbed areas in the Atlantic Forest of Brazil (de Souza *et al.*, 1992; Rodrigues, 1995) and elsewhere in the neotropics (Snow and Snow, 1971).

The higher number of interactions involving cotingas and toucans in the more preserved site (Barra Grande) is difficult to explain, as these taxa were commonly seen in both sites. Some species of cotingas and toucans forage in open areas adjacent to forest (Estrada *et al.*, 1993; Guevara and Laborde, 1993). At Intervales, however, these species seem to prefer fruiting plants associated with less disturbed habitats. The number of interactions performed by a cotingid, the hooded berry-eater (*C. cucullatus*), was especially high. Although some cotingid species are known to migrate altitudinally (Sick, 1993), *Carpornis* is probably a resident species, because we recorded it in both sites during the entire study period. Its diverse diet is probably a result of its permanent residency and its willingness to eat fruits of both primary and secondary forest (W.R. Silva, personal observation). This species is found in the lowlands of Intervales only in severe winters (Galetti, 1996).

Thrushes and manakins were common frugivores in both sites. The rufous-bellied thrush (*T. rufiventris*) is an important seed-disperser in forest-edge habitats in Sede, whereas the white-necked thrush (*T. albicollis*) plays the same role in the forest understorey. The blue manakin (*C. caudata*), found mainly in the understorey, is one of the most active frugivores at both sites (W.R. Silva, personal observation).

Trogons usually include animal matter in their otherwise frugivorous diet (Remsen *et al.*, 1993) and sometimes seem to ignore nearby fruits while searching for animal prey (W.R. Silva, personal observation). Their number of interactions was equally distributed in the two study sites. Like the hooded berry-eater, the three trogon species at Intervales foraged in a variety of habitats and strata, taking several fruit types. Vireos, although primarily insectivorous in Brazil (Sick, 1993), can also exploit fruit sources opportunistically (Greenberg *et al.*, 1995; Pizo, 1997).

The number of interactions recorded for guans was certainly much lower than reported in a study that focused on guans in a nearby site (Galetti *et al.*, 1997). Like other frugivores recorded in this study, guans foraged for fruits in mature as well as in second-growth vegetation. This group, represented at Intervales by the dusky-legged guan (*P. obscura*) and the jacutinga (*P. jacutinga*), is a key focus for conservation efforts, because they are often sensitive to hunting pressure and habitat degradation (Galetti *et al.*, 1997; reviewed in Strahl *et al.*, 1997). We found guans equally common at our two study sites.

Overall, the weak-to-moderate level of habitat degradation at one of our sites appears to have had little impact on the avifauna, although abundance of some species may be affected (Galetti and Aleixo, 1998). Given that most remaining Atlantic Forest in south-eastern Brazil consists of a mosaic of different successional types similar to those that occur at Intervales (Fonseca, 1985; Tabarelli and Mantovani, 1997), it is likely that the coteries of avian frugivores will be similar, with a pervasive influence of facultative frugivores, such as tyrant-flycatchers, tanagers, thrushes, trogons and manakins.

The fruiting plants

The richness of ornithochorous plant species in both Sede and Barra Grande is difficult to compare with that of other sites because we made no attempt to survey the entire plant community, leaving many ornithochorous species unreported because no interaction was assigned to them. This is a common bias when working in a very speciose community and when depending on fortuitous observations of organisms that are often difficult to spot.

The plant families with the most interactions are commonly reported as important to birds in other neotropical sites (Leck, 1972; Snow, 1981; Wheelwright *et al.*, 1984; Blake and Loiselle, 1992). The difference in the species composition of the two plant communities used by birds in our study probably reflects the effect of human disturbance in the Sede site. However, despite the prevalence of second-growth species in Sede, those species were exploited by essentially the same bird species as in Barra Grande. This suggests that frugivorous birds in the Atlantic Forest readily forage across plant successional stages and can switch their food preferences according to local fruit availability.

The importance of fruiting *Melastomataceae* species to birds is well documented throughout the neotropics (Willis, 1966; Stiles and Rosselli, 1993; Galetti and Stotz, 1996). At our site, melastomes accounted for approximately 20% of all interactions and were eaten by almost all frugivorous species. The role of other plant families varied between sites. Although some families were consumed significantly more often in one site, a few accounted for the larger number of interactions in Sede. When families of edge and second-growth habitats, such as *Piperaceae, Rosaceae, Anacardiaceae, Flacourtiaceae* and *Ulmaceae*, are removed from the sample, the number of interactions becomes roughly equivalent between the two sites (Table 28.2), suggesting that the differences in the patterns of frugivory among Atlantic Forest sites are shaped by habitat structure.

The meaning of asymmetry in the interaction pattern

We found asymmetrically shaped distributions of interactions, in which a few species of birds or plants were responsible for the vast majority of interactions. Such a distribution seems typical of other tropical frugivore communities (Pérez, 1976; Wheelwright *et al.*, 1984; Blake and Loiselle, 1992; Galetti and Pizo, 1996; Murray, 2000). Asymmetry in interactions among species is expected for mutualistic systems and is a characteristic feature of multispecies assemblages (Herrera, 1984; Jordano, 1987). No matter what metric of interaction is used (*sensu* Herrera and Jordano, 1981), some plant or bird species will be disproportionately important to their counterparts in the community. Thus, a plant species can be of great significance to a bird species, which in turn can be unimportant as a seed-disperser to the plant because its contribution is 'diluted' by that of other frugivores exploiting the same species. Likewise, the long tail of the distribution suggests that a plant species can be relatively unimportant to birds but may none the less be highly dependent on a given species for dispersal.

The asymmetrical distribution persists even when applying the 'importance index' to a small fraction of the plant/bird community. This trend not only reinforces the occurrence of 'diffuse coevolution' among sets of fruits and frugivores (Herrera, 1985; Jordano, 1987), but also has implications in the conservation of such communities, because it suggests alternative routes for system response in cases where one interacting partner becomes extinct (Jordano, 1987).

Conservation Perspective

Frugivory in the neotropics is often portrayed as an interaction between spectacular birds and conspicuous fruits. This is a misconception, based on narrowly focused case-studies (e.g. D.W. Snow, 1962; B.K. Snow, 1970; Skutch, 1980; Wheelwright, 1983). On a community level, the role of dull-coloured and inconspicuous fruits and frugivores is not fully appreciated (Westcott and Graham, 2000). The importance rankings of the birds and plants that we report make it clear that many fruit–frugivore mutualisms in neotropical forests involve relatively inconspicuous species, as also noted by Murray (2000) in Costa Rica. For birds, these species are usually small, facultative frugivores that are widespread in second-growth habitats. For plants, many are widespread shrubs of successional habitats, bearing small berries and drupes with small and often numerous seeds, or species of mistletoes using particular sets of small frugivores. Perhaps none of them would fit into a category used to justify conservation efforts (Noss, 1990).

The importance values we assigned to species indirectly reflect the lack of redundancy in the interacting systems to which they belong (Walker, 1992), because higher values are a function of the interactions not shared with other species. Therefore, from a functional standpoint, species with high values are important for maintaining the resilience of their communities. This resilience may be particularly important during times of general fruit scarcity (e.g. the dry season at our site), when bird species with different degrees of fruit dependence rely on the fruits of several melastomes and other plant species that grow along edge and successional vegetation, a type of habitat that represents an important source of resources in tropical areas (Foster, 1980; Levey, 1988, 1990).

Our approach does not ignore the value of keystone resources but rather offers an alternative way to detect the 'mobile links' (*sensu* Gilbert, 1980) within a fruit–frugivore community and to evaluate their importance from a conservation perspective.

Acknowledgements

This study was supported by Foundation MB and BIOTA/FAPESP – the Biodiversity Virtual Institute Program (www.biota.org.br). We are grateful to A.M. Rosa, W. Zaca and B.A. Oliveira for their valuable help in fieldwork, and to K.G. Murray, D.J. Levey, M. Galetti and M.A. Pizo for insightful comments on an earlier draft of this manuscript. We are also in debt to M.L. Kawazaki, J. Tamashiro, R.A. Oliveira and R. Goldemberg for the difficult task of plant identification. The Fundação Florestal do Estado de São Paulo provided the logistical support at Intervales State Park.

References

Aleixo, A. and Galetti, M. (1997) The conservation of the avifauna in a lowland Atlantic forest in south-east Brazil. *Bird Conservation International* 7, 235–261.

Asquith, N.M., Terborgh, J., Arnold, A.E. and Riveros, C.M. (1999) The fruits the agouti ate: *Hymenaea courbaril* seed fate when its disperser is absent. *Journal of Tropical Ecology* 15, 229–235.

Blake, J.G. and Loiselle, B.A. (1992) Fruits in the diets of neotropical migrant birds in Costa Rica. *Biotropica* 24, 200–210.

Brooks, T. and Balmford, A. (1996) Atlantic forest extinctions. *Nature* 380, 115.

Cronquist, A. (1981) *An Integrated System of Classification of Flowering Plants.* Columbia University Press, New York, 1262 pp.

de Souza, F.L., Roma, J.C. and Guix, J.C. (1992) Consumption of *Didymopanax pachycarpum* unripe fruits by birds in southeastern Brazil. *Miscellania Zoologica* 16, 246–248.

Duncan, R.S. and Chapman, C.A. (1999) Seed dispersal and potential forest succession in abandoned agriculture in tropical Africa. *Ecological Applications* 9, 998–1008.

Estrada, A., Coates-Estrada, R., Meritt, D., Jr, Montiel, S. and Curiel, D. (1993) Patterns of frugivore species richness and abundance in forest islands and agricultural habitats at Los Tuxtlas, Mexico. In: Fleming, T.H. and Estrada, A. (eds) *Frugivores and Seed Dispersal: Ecological and Evolutionary Aspects.* Kluwer Academic Publishers, Dordrecht, pp. 245–257.

Fitzpatrick, J.W. (1980) Foraging behavior of neotropical tyrant flycatchers. *Condor* 82, 43–57.

Fonseca, G.A.B. (1985) The vanishing Brazilian Atlantic forest. *Biological Conservation* 34, 17–34.

Foster, R.B. (1980) Heterogeneity and disturbance in tropical vegetation. In: Soulé, M.E. and Wilcox, B.A. (eds) *Conservation Biology, an Evolutionary–Ecological Perspective.* Sinauer, Sunderland, pp. 75–92.

Galetti, M. (1996) Fruits and frugivores in a Brazilian Atlantic forest. PhD thesis, University of Cambridge, Cambridge.

Galetti, M. (2001) The future of the Atlantic Forest. *Conservation Biology* 14, 4.

Galetti, M. and Aleixo, A. (1998) Effects of palm heart harvesting on avian frugivores in the Atlantic rain forest of Brazil. *Journal of Applied Ecology* 35, 286–293.

Galetti, M. and Pizo, M. (1996) Fruit eating by birds in forest fragment in southeastern Brazil. *Ararajuba* 4, 71–79.

Galetti, M. and Rodrigues, M. (1992) Comparative seed predation on pods by parrots in Brazil. *Biotropica* 24, 222–224.

Galetti, M. and Stotz, D. (1996) *Miconia hypoleuca* (*Melastomataceae*) como espécie-chave para aves frugívoras no sudeste do Brasil. *Revista Brasileira de Biologia* 56, 435–439.

Galetti, M., Martuscelli, P., Olmos, F. and Aleixo, A. (1997) Ecology and conservation of the jacutinga *Pipile jacutinga* in the Atlantic forest of Brazil. *Conservation Biology* 82, 31–39.

Gilbert, L.E. (1980) Food web organization and the conservation of neotropical diversity. In: Soulé, M.E. and Wilcox, B.A. (eds) *Conservation Biology: an Evolutionary–Ecological Perspective.* Sinauer, Sunderland, Massachusetts, pp. 11–33.

Greenberg, R., Foster, M.S. and Marquez-Vadelamar, L. (1995) The role of the White-eyed Vireo in the dispersal of *Bursera* fruit on the Yucatan Peninsula. *Journal of Tropical Ecology* 11, 619–639.

Guevara, S. and Laborde, J. (1993) Monitoring seed dispersal at isolated standing trees in tropical pastures: consequences for local species availability. In: Fleming, T.H. and Estrada, A. (eds) *Frugivores and Seed Dispersal: Ecological and Evolutionary Aspects.* Kluwer Academic Publishers, Dordrecht, pp. 319–338.

Herrera, C.M. (1984) A study of avian frugivores, bird-dispersed plants, and their interaction in Mediterranean scrublands. *Ecological Monographs* 54, 1–23.

Herrera, C.M. (1985) Determinants of plant–animal coevolution: the case of mutualistic dispersal of seeds by vertebrates. *Oikos* 44, 132–141.

Herrera, C.M. and Jordano, P. (1981) *Prunus mahaleb* and birds: the high efficiency seed dispersal system of a temperate fruiting tree. *Ecological Monographs* 51, 203–221.

Holl, K.D. (1998) Do bird perching structures elevate seed rain and seedling establishment in abandoned tropical pasture? *Restoration Ecology* 6, 253–261.

Howe, H.F. (1977) Bird activity and seed dispersal of a tropical wet forest tree. *Ecology* 58, 539–550.

Howe, H.F. (1981) Dispersal of a neotropical nutmeg (*Virola sebifera*) by birds. *The Auk* 98, 88–98.

Janzen, D.H. (1981) *Ficus ovalis* seed predation by an Orange-chinned Parakeet (*Brotogeris jugularis*) in Costa Rica. *The Auk* 98, 841–844.

Jordano, P. (1987) Patterns of mutualistic interactions in pollination and seed dispersal: connectance, dependence asymmetries, and coevolution. *American Naturalist* 129, 657–677.

Kattan, G.H., Alvarez-Lopez, H. and Giraldo, M. (1994) Forest fragmentation and bird extinctions: San Antonio 80 years later. *Conservation Biology* 8, 138–146.

Krebs, C.J. (1989) *Ecological Methodology.* Harper Collins Publishers, New York, 654 pp.

Leck, C.F. (1972) Seasonal changes in feeding pressures of fruit- and nectar-eating birds in Panama. *Condor* 74, 54–60.

Levey, D.J. (1987) Seed size and fruit-handling techniques of avian frugivores. *American Naturalist* 129, 471–485.

Levey, D.J. (1988) Spatial and temporal variation in Costa Rican fruit and fruit-eating bird abundance. *Ecological Monographs* 58, 251–269.

Levey, D.J. (1990) Habitat-dependent fruiting behaviour of an understory tree, *Miconia centrodesma*, and tropical treefall gaps as keystone habitats for frugivores in Costa Rica. *Journal of Tropical Ecology* 6, 409–420.

Loiselle, B.A. and Blake, J.G. (1991) Temporal variation in birds and fruits along an elevational gradient in Costa Rica. *Ecology* 72, 180–193.

Loiselle, B.A. and Blake, J.G. (1992) Population variation in a tropical bird community. *BioScience* 42, 838–845.

McClanahan, T.R. and Wolfe, R.W. (1993) Accelerating forest succession in a fragmented landscape: the role of bird and perches. *Conservation Biology* 7, 279–288.

McDonnell, M.J. and Stiles, E.W. (1983) The structural complexity of old field vegetation and the recruitment of bird dispersed plant species. *Oecologia* 56, 109–116.

Moermond, T.C. and Denslow, J.S. (1985) Neotropical avian frugivores: patterns of behavior, morphology, and nutrition, with consequences for fruit selection. In: Buckley, P.A., Foster, M.S., Morton, E.S., Ridgely, R.S. and Buckley, F.G. (eds) *Neotropical Ornithology.* Ornithological Monographs No. 36, American Ornithologist Union, Washington, DC, pp. 865–897.

Morton, E.S. (1977) Intratropical migration in the Yellow-green Vireo and Piratic Flycatcher. *The Auk* 94, 97–106.

Murray, K.G. (2000) The importance of different bird species as seed dispersers. In: Nadkarni, N.M. and Wheelwright, N.T. (eds) *Monteverde: Ecology and Conservation of a Tropical Cloud Forest.* Oxford University Press, New York, pp. 294–295.

Noss, R.F. (1990) Indicators for monitoring biodiversity: a hierarchical approach. *Conservation Biology* 44, 335–364.

Peres, C.A. (2000) Identifying keystone plant resources in tropical forests: the case of gums from *Parkia* pods. *Journal of Tropical Ecology* 16, 287–317.

Pérez, L.T. (1976) Diseminacion de semillas por aves en 'Los Tuxtlas', Ver. In: Gómez-Pompa, A., Vásquez-Yanes, C., Rodrigues, S.A. and Cervera, A.B. (eds) *Regeneracion de selvas.* Compañia Editorial Continental, Mexico, pp. 447–470.

Pizo, M.A. (1997) Seed dispersal and predation in two populations of *Cabralea canjerana* (*Meliaceae*) in the Atlantic Forest of southeastern Brazil. *Journal of Tropical Ecology* 13, 559–578.

Ranta, P., Blom, T., Niemela, J., Joensuu, E. and Siitonen, M. (1998) The fragmented Atlantic rain forest of Brazil: size, shape and distribution of forest fragments. *Biodiversity and Conservation* 7, 385–403.

Remsen, J.V., Jr, Hyde, M.A. and Chapman, A. (1993) The diets of neotropical trogons, motmots, barbets and toucans. *Condor* 95, 178–192.

Rodrigues, M. (1995) Spatial distribution and food utilization among tanagers in southeastern Brazil (Passeriformes: Emberizidae). *Ararajuba* 3, 27–32.

Schupp, E.W. (1993) Quantity, quality and the effectiveness of seed dispersal by animals. In: Fleming, T.H. and Estrada, A. (eds) *Frugivores and Seed Dispersal: Ecological and Evolutionary Aspects.* Kluwer Academic Publishers, Dordrecht, pp. 15–29.

Sick, H. (1993) *Birds in Brazil, a Natural History.* Princeton University Press, Princeton, New Jersey, 703 pp.

Silva, J.M.C. and Tabarelli, M. (2000) Tree species impoverishment and the future of the Atlantic Forest of northeast Brazil. *Nature* 404, 72–74.

Silva, J.M.C., Uhl, C. and Murray, G. (1996) Plant succession, landscape management, and the ecology of frugivorous birds in abandoned Amazonian pastures. *Conservation Biology* 10, 491–503.

Skutch, A.F. (1980) Arils as food of tropical American birds. *Condor* 82, 31–42.

Snow, B.K. (1970) A field study of the Bearded Bellbird in Trinidad. *Ibis* 114, 139–162.

Snow, B.K. and Snow, D.W. (1971) The feeding ecology of tanagers and honeycreepers in Trinidad. *The Auk* 88, 291–322.

Snow, D.W. (1962) A field study of the Golden-headed Manakin, *Pipra erythrocephala*, in Trinidad, W.I. *Zoologica (New York)* 47, 183–198.

Snow, D.W. (1981) Tropical frugivorous birds and their food plants: a world survey. *Biotropica* 13, 1–14.

SOS Mata Atlântica and INPE (1992) *Atlas da evolução dos remanescentes florestais e ecossistemas associados do domínio da Mata Atlântica no período de 1985–1990.* Fundação SOS Mata Atlântica, São Paulo.

Stiles, F.G. and Rosselli, L. (1993) Consumption of fruits of the *Melastomataceae* by birds: how diffuse is coevolution? In: Fleming, T.H. and Estrada, A. (eds) *Frugivores and Seed Dispersal: Ecological and Evolutionary Aspects.* Kluwer Academic Publishers, Dordrecht, pp. 57–73.

Strahl, S.D., Beaujon, S., Brooks, D.M., Begazo, A.J., Sedaghatkish, G. and Olmos, F. (1997) *The Cracidae: their Biology and Conservation.* Hancock House, Surrey, 506 pp.

Tabarelli, M. and Mantovani, W. (1997) Colonização de clareiras naturais na floresta atlântica do sudeste do Brasil. *Revista Brasileira de Botânica* 20, 57–66.

Terborgh, J. (1986a) Keystone plant resources in the tropical forests. In: Soulé, M.E. (ed.) *Conservation Biology: the Science of Scarcity and Diversity.* Sinauer, Sunderland, Massachusetts, pp. 330–344.

Terborgh, J. (1986b) Community aspects of frugivory in tropical forests. In: Estrada, A. and Fleming, T.H. (eds) *Frugivores and Seed Dispersal.* Dr W. Junk, Dordrecht, pp. 371–384.

Terborgh, J. (1992) Maintenance of diversity in tropical forests. *Biotropica* 24, 283–292.

Vielliard, J. and Silva, W.R. (2001) Avifauna. In: Secretaria de Estado do Meio Ambiente, Fundação para a Conservação e a Produção Florestal do Estado de São Paulo (eds) *Intervales.* Fundação Florestal, São Paulo, pp. 123–145.

Walker, B.H. (1992) Biodiversity and ecological redundancy. *Conservation Biology* 6, 18–23.

Westcott, D.A. and Graham, D.L. (2000) Patterns of movement and seed dispersal of a tropical frugivore. *Oecologia* 122, 249–257.

Wheelwright, N.T. (1983) Fruits and the ecology of the resplendent quetzals. *The Auk* 100, 286–301.

Wheelwright, N.T., Haber, W.A., Murray, K.G. and Guindon, C. (1984) Tropical fruit-eating birds and their food plants: a survey of a Costa Rican lower montane forest. *Biotropica* 16, 173–191.

Willis, E.O. (1966) Competitive exclusion and birds at fruiting trees in western Colombia. *The Auk* 83, 479–480.

Wright, S.J., Zeballos, H., Domínguez, I., Gallardo, M.M., Moreno, M.C. and Ibáñez, R. (2000) Poachers alter mammal abundance, seed dispersal, and seed predation in a Neotropical forest. *Conservation Biology* 14, 227–239.

29 Limitations of Animal Seed Dispersal for Enhancing Forest Succession on Degraded Lands

R. Scot Duncan[1] and Colin. A. Chapman[1,2]

[1]Department of Zoology, PO Box 118525, University of Florida, Gainesville, FL 32611, USA; [2]Wildlife Conservation Society, 185th Street and Southern Boulevard, Bronx, NY 10460, USA

Introduction

Each year, approximately 127,300 km^2 of tropical forests are cleared, 55,000 km^2 are logged and as much as 30,000 km^2 are burned (Food and Agriculture Organization, 1999). Thus, the amount of tropical forest either deforested, logged or burned annually is approximately 212,000 km^2. Against these statistics is the hope that, with enough biological knowledge and political and economic change, forests can be restored on degraded lands. These restored forests can provide such ecosystem services as soil and water conservation, CO_2 sequestration and extractive use (Brown and Lugo, 1990, 1994). Because degradation is often severe, intervention is probably needed to restore forests in a time frame useful to current conservation goals (Parrotta, 1993).

There are two major challenges to 'regreening' the tropics. First, little is known about the process of forest succession on degraded lands. Most of what is known is limited to descriptions of vegetative change after land abandonment; studies of the underlying processes are few (Chazdon, 1994). Secondly, widespread forest restoration on degraded lands may be cost-prohibitive. Though there may be future subsidies for restoration via international agreements, such as the Kyoto Protocol to the United Nations Framework Convention on Climate Change, restoration strategies using natural processes (e.g. animal-mediated seed dispersal) will be the most widely applicable and affordable.

For the purposes of this review, we define degraded land as formerly forested area where human activities have reduced tree cover more than 90%. Forest recovery on these lands requires abiotic resources (e.g. soil nutrients), the absence of recurrent disturbances (e.g. fire) and adequate floristic resources (e.g. soil seed bank, seedlings, adult trees) (Uhl *et al.*, 1981; Brown and Lugo, 1994). Generally, as the severity of land degradation increases, the available floristic resources decrease (Uhl *et al.*, 1982, 1988; Nepstad *et al.*, 1991; Brown and Lugo, 1994) and post-disturbance seed dispersal becomes increasingly important for forest regrowth (Da Silva *et al.*, 1996; Nepstad *et al.*, 1996). In many degraded systems, lack of seed dispersal to degraded areas is considered one of the leading obstacles to forest regrowth (Nepstad *et al.*,

 Seed Dispersal and Frugivory: Ecology, Evolution and Conservation (eds D.J. Levey, W.R. Silva and M. Galetti)

1991; Holl, 1999). Given that most tree species in the moist tropics produce animal-dispersed seeds (Howe and Smallwood, 1982; Chapman and Chapman, 1999), frugivores may be especially important in restoration efforts.

We review mechanisms that limit animal-mediated seed dispersal during forest succession on degraded tropical lands. We illustrate these limitations with examples from the system we know best, Kibale National Park in Uganda, and relate our findings to work done throughout the moist tropics. We then evaluate several proposed strategies to overcome seed-dispersal and recruitment limitations on degraded lands, and conclude with questions we think are important to address in the near future. While our focus is on the moist tropics, the principles we discuss are relevant to forest succession in other tropical and temperate regions.

Limitations to the Role of Seed Dispersal in Forest Succession

Challenges to tree recruitment on degraded lands

To understand the role that animal seed dispersal plays in forest succession on degraded lands, it is important to consider the harsh conditions facing trees establishing from seeds in this environment. Degraded tropical areas are typically hot and dry, and soil quality is often poor (Adedeji, 1984; Aide and Cavelier, 1994; Brown and Lugo, 1994). These conditions can make it difficult for tree seeds to germinate and seedlings to establish, even for many early-successional tree species ('pioneers'). However, fast-growing grasses and vines often establish in such environments and can dominate or 'arrest' succession (Borhidi, 1988; Uhl *et al.*, 1988). These species often limit tree seedling establishment and growth through competition (Nepstad *et al.*, 1991) and by creating a flammable habitat in which periodic fires kill the few tree seedlings and saplings that have established (Uhl and Buschbacher, 1985; Uhl and Kauffman, 1990; Kuusipalo *et al.*, 1995).

The numbers of seeds dispersed into degraded areas

One of the major constraints on regeneration of forest on degraded lands is that few animal-dispersed seeds are brought into disturbed areas by frugivores even when forest is nearby (Aide and Cavelier, 1994; Holl, 1998, 1999). This constraint is illustrated by three studies at Kibale. In the first, we examined seed rain by bats and birds below isolated trees within 150 m of forest edge (Duncan and Chapman, 1999). Of 1593 tree seeds collected, only 0.93% were of forest species and, of 11 tree species collected, only three were forest species. In a second, ongoing, study, we are examining seeds dispersed by birds in a successional forest at Kibale. Our preliminary results show that, of the 11,685 seeds we collected from mist-netted birds, only 11.2% were from plant species not fruiting on the site. Of these, 10.4% were of a single early-successional, light-demanding tree (*Maesa lanceolata*), whose seeds were probably unable to establish in the understorey. In a third study, Zanne and Chapman (2001) found that recruitment densities of animal-dispersed species declined with increasing distance from the forest edge in softwood plantations within Kibale. In addition, a plantation 15 km away from the forest had much lower recruitment densities of animal-dispersed stems than plantations within the park (see also Parrotta *et al.*, 1997; Wunderle, 1997).

There are several possible reasons why few forest seeds are dispersed into degraded areas. Typically, there is little incentive for forest-dwelling frugivores to visit degraded areas because fruit abundance there is usually low (Da Silva *et al.*, 1996). Forest frugivores may also be more at risk of predation in open degraded areas than in forest (Janzen *et al.*, 1976; Wegner and Merriam, 1979; Janzen, 1990). Similarly, disturbance-dwelling frugivorous species may avoid the unfamiliar habitat of the forest. These factors probably explain why many researchers have found a steep decline in seed dispersal with increasing distance from the forest edge (Charles-Dominique, 1986; Gorchov *et al.*, 1993; Aide and Cavelier, 1994; Holl, 1998, 1999).

The low numbers of seeds dispersed into degraded areas may be insufficient to initiate widespread forest succession. Our exploration of seed fate at Kibale suggests that hundreds of seeds may be needed for one seedling to establish. In another ongoing study, we are examining the fates of tree seeds dispersed into a 5-year-old successional forest on a logged cypress (*Cupressus lusitanica*) plantation. We monitored eight seed species (25 stations per species, five to 12 seeds per station) for 10 months. Because seed predation was very high, we caged some seed stations to exclude vertebrate granivores and allow seedling germination. We removed cages several weeks after germination to allow vertebrate herbivory. At open stations, only 4.8% of seeds germinated and 98.7% of seeds and germinated seedlings disappeared or were eaten (Table 29.1). In contrast, survival of seedlings germinating below cages was high (44%) at 10 months after dispersal, suggesting that in some systems seed predation may be a greater barrier to establishment than seedling mortality. Duncan and Duncan (2000) found that, of 1600 seeds (of eight species) set out in seed stations in grass-dominated abandoned agricultural areas at Kibale, only 20% of seeds survived 6 months. While 95% of seedlings (of six species) planted in the grassland survived 6–8 months, their growth was extremely slow, averaging only 1.8 cm year^{-1}. Chapman and Chapman (1999) sowed seeds of four early-successional species (one to four seeds per square metre) into grass-dominated abandoned cropland. After 4 years, densities of these species ranged from 0 to 0.027 seedlings m^{-2}, and were no greater than densities of these species in control plots. These low densities may have been due to the aggressiveness of the dominant grass species (*Pennisetum purpureum*) or to rodent seed predation. These experiments suggest that seeds dispersed into successional habitats face very low probabilities of recruitment.

Together, these results suggest that across different stages of succession few seeds are dispersed into degraded areas, and those seeds that do arrive face high seed predation, moderate seedling mortality and slow seedling growth. Others have also found that seed

Table 29.1. Results of seed predation and seedling survival experiments (mean ± SD). Open stations allowed access to seeds by seed predators. Caged stations protected seeds from predators to allow germinations; cages were removed soon after germination. Seed survival averages include seeds and seedlings that survived to 10 months. For the smaller-seeded *T. orientalis* and *C. africana*, seed number was not monitored. No *M. myristica* or *T. orientalis* seeds germinated from caged stations.

		Open stations			
Species	Initial seed no. per station	Mean no. seeds surviving at 10 months	Mean no. total germinations per station	No. seedlings surviving to 10 months	Caged stations: time to 50% mortality (months)
Bridelia micrantha (*Euphorbiaceae*)	5	0.16 ± 0.37	0.96 ± 1.10	0.16 ± 0.37	2.9
Celtis africana (*Ulmaceae*)	5	–	0.54 ± 1.30	0.14 ± 0.47	6.7
Diospyros abyssinica (*Ebenaceae*)	5	0.08 ± 0.4	0.04 ± 0.20	0	4.5
Mimusops bagshawei (*Sapotaceae*)	10	0	0	0	> 13.0
Monodora myristica (*Annonaceae*)	10	0.26 ± 1.25	0	0	–
Prunus africana (*Rosaceae*)	5	0	0.33 ± 1.13	0	4.0
Trema orientalis (*Ulmaceae*)	12	–	0	0	–
Uvariopsis congensis (*Annonaceae*)	10	0.22 ± 1.04	0.09 ± 0.42	0.09 ± 0.42	> 11.0

predation is high in degraded tropical lands and that most seeds are depredated within the first few weeks after dispersal (Nepstad *et al.*, 1991, 1996; Whelan *et al.*, 1991; Osunkoya, 1994). Seedling survival and growth rates are often low in other degraded systems, and it has been suggested that water stress and low levels of soil nutrients are to blame (Nepstad *et al.*, 1991, 1996; Gerhardt, 1993; Aide and Cavelier, 1994).

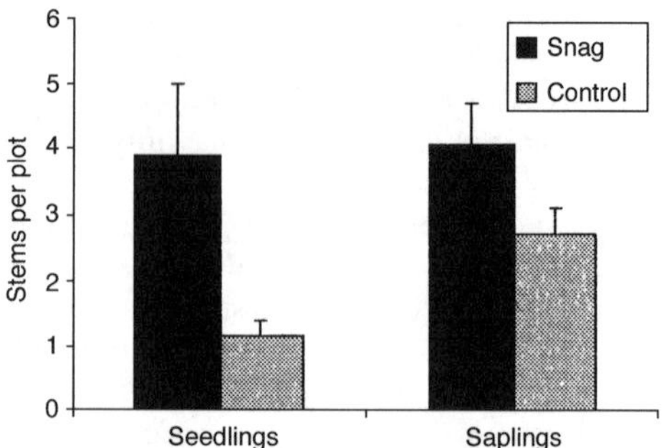

Fig. 29.1. Mean (+ SE) seedling and sapling numbers below snags (n = 20) and in adjacent open areas (control) in an early successional habitat at Kibale. All shrubs and trees of animal-dispersed species were counted in 2 m^2 plots. Snag plots had more seedlings and saplings than control plots (P = 0.014 and 0.016, respectively).

Location of seeds dispersed into degraded areas

A second constraint on animal seed dispersal into degraded areas is that there tends to be little dispersal into open areas where tree recruitment is most needed. Most animal-mediated seed dispersal tends to be concentrated below emergent structures (e.g. remnant or pioneer trees). Duncan and Chapman (1999) found that seed rain below isolated trees in an agricultural area was 90-fold greater than in adjacent grassland. Such patterns seem to affect seedling recruitment. In an ongoing study, we are examining seedling recruitment of animal-dispersed species below isolated snags and into adjacent areas (5 m away) in a 2-year-old succession. Densities of animal-dispersed seedling (height $\leq$ 0.25 m) and sapling (height $\geq$ 1.0 m) species were greater below snags than in adjacent plots (paired T-test, P = 0.018) (Fig. 29.1).

Greater seed rain below trees in degraded areas has been found in several tropical systems, and is most probably a result of seed dispersal by bats and birds (Guevara *et al.*, 1986; Uhl, 1987; Willson and Crome, 1989; Guevara and Laborde, 1993; Vieira *et al.*, 1994; Da Silva *et al.*, 1996; Nepstad *et al.*, 1996; Harvey, 2000). Others have also shown that seedling recruitment is greater in the shade below trees, probably due to lower temperatures, higher humidity and reduced competition with grasses (Guevara *et al.*, 1986; Vieira *et al.*, 1994; Toh *et al.*, 1999).

While seed rain to trees in degraded areas is frequently lauded as important for initiating forest succession on degraded lands and use of artificial perches has been proposed as a management tool, there may be limitations to the effectiveness of increasing seed rain. Trees in heavily degraded areas are often at low densities; thus their site-wide effect on tree seedling establishment will be small. In addition, areas with low tree densities are more vulnerable to fires, which kill young trees (Kuusipalo *et al.*, 1995; Toh *et al.*, 1999). Finally, where recruitment is successful below trees, the spread of vegetation into adjacent open areas may be slow, and decades or centuries may be needed for canopy closure (Toh *et al.*, 1999).

Seeds are sometimes dispersed to locations much less favourable to seedling establishment than below trees. This became evident in our survey of recruitment from seed dispersal by baboons (*Papio anubis*) into young successions on logged conifer plantations at Kibale. Baboons occasionally visit disturbed areas, where they feed, socialize and rest. During their visits they can disperse seeds of large-seeded mature forest species that are rarely or never dispersed by smaller-bodied, more abundant frugivores (e.g. birds). Thus, we expected baboon dispersal to contribute an important component to the species composition of recruited seedlings. We conducted surveys in these young successional areas to quantify the numbers and locations of baboon defecations and the seeds they contained. Defecation sites were subsequently monitored to quantify seedling establishment. We found that 62% of 52 defecations were in locations where conditions for seedling establishment and survival seemed unfavourable (Fig. 29.2). To date (~24 survey months), surveys have failed to find any tree seedlings recruiting from either favourable or unfavourable locations. Thus, most of the seeds dispersed by baboons into these degraded areas are placed into unfavourable microhabitats, where subsequent recruitment is unlikely.

The composition of seeds dispersed into degraded areas

Ideally, the composition of seeds dispersed into degraded areas would have high proportions of tree seeds, including early-, mid- and late-successional species. The early-successional species would quickly establish and, as they matured, provide an understorey environment conducive for survival and growth of mid- and late-successional species. However, the species composition of seeds dispersed into degraded areas often contains few tree seeds, and these are predominantly early-successional species. In our study of seed dispersal to trees in degraded agricultural lands adjacent to forest (Duncan and Chapman, 1999), only 14% of seeds were tree species; the majority were hemi-epiphytic figs, unable to germinate from the ground (Fig. 29.3). Furthermore, 99.9% of seeds recovered were early-successional species already present and fruiting on the site. A high proportion of early-successional species may be expected, since these species often produce larger numbers of seeds compared with mid- and late–successional species. However, the recruitment of these latter species is very important for advancing forest succession past the pioneer stage. From our ongoing study of seeds dispersed by birds in a 5-year-old successional forest, 21% of recovered seeds were tree seeds, and 47% of these were one early-successional species that could no longer germinate in the shaded understorey of the site. Most seeds were early-successional shrub species fruiting in the site (Fig. 29.3). Finally, in our study of baboon seed dispersal into logged plantations, 16% of seeds were tree species (Fig. 29.3). Ninety-seven per cent of these were mature forest species, but none have established in 24 months. Most seeds were mature forest herbs (99%), which also have low establishment rates.

In some regions, a majority of seeds dispersed into degraded areas can be tree species, but most of these are early-successional species, such as *Cecropia* spp. (Nepstad *et al.*, 1991; Guevara and Laborde, 1993; Da Silva *et al.*, 1996). For example, Holl and Lulow (1997) reported that tree seeds comprised 46% of seeds collected below artificial perches in abandoned pasture in Costa Rica, but only 29% of species and roughly 29% of seeds were species only found in forest.

Where there is poor establishment or limited dispersal of tree seeds, shrub seed dispersal to degraded areas may be important in initiating forest succession (Nepstad *et al.*, 1991). Vieira *et al.* (1994) found that shrubs

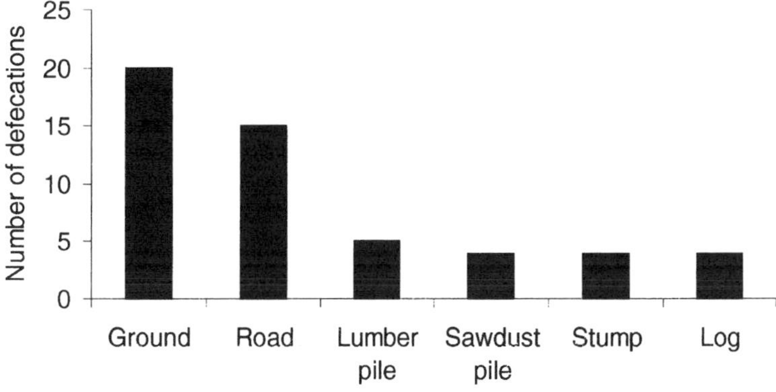

Fig. 29.2. Locations where baboon defecations were found in young successional habitat at Kibale. Only the ground location seems potentially favourable for seedling establishment and survival.

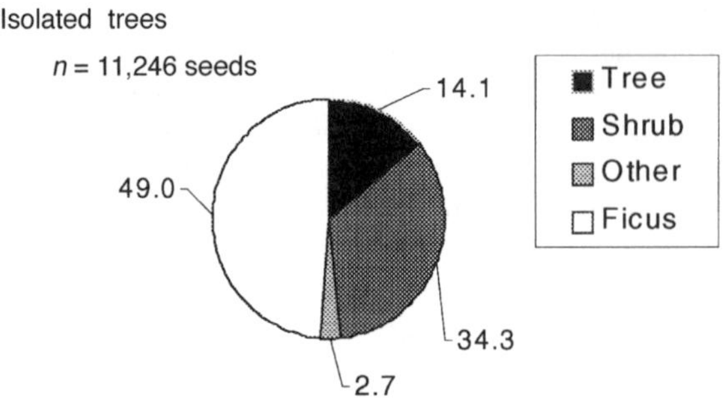

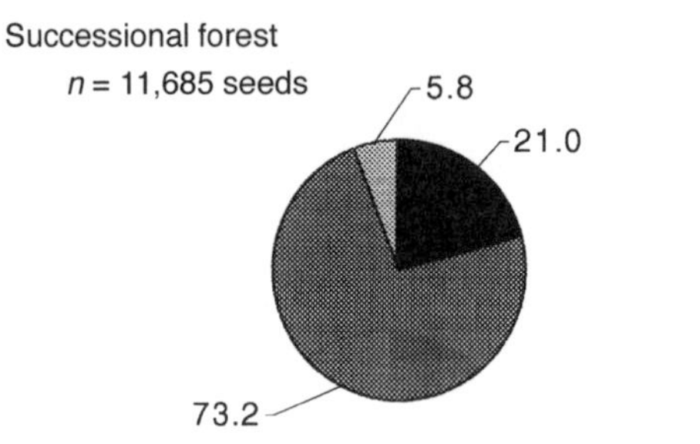

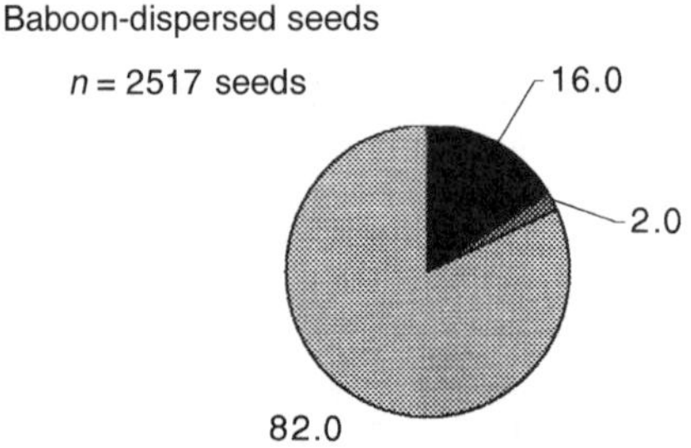

Fig. 29.3. Categories of seeds dispersed to isolated trees in agriculture near forest, by birds into a 5-year-old successional forest and by baboons (*Papio anubis*) into recently logged (< 2 years) conifer plantations.

recruiting on degraded pasture in Brazil attracted seed-dispersing birds and provided a microclimate conducive to the establishment of the tree seeds they dispersed. Zahawi and Augspurger (1999) found greater tree establishment in abandoned pastures with guava shrubs than in pastures without shrub cover. Establishment of shrubs as cover prior to the establishment of tree seedlings may be a successful successional pathway towards secondary forest in many degraded areas where tree recruitment is initially poor.

The timing of seed dispersal into degraded areas

Dispersal of seeds to degraded areas may or may not result in stem recruitment. Early-successional species must recruit soon after disturbance, when sunlight is abundant and before aggressive grasses or vines take over (Uhl and Clark, 1983). Mid- and late-successional species may only recruit after development of a shady, humid understorey. Their establishment may be necessary to prevent reinvasion of aggressive vines, shrubs and grasses after the short-lived pioneers begin to senesce.

We examined timing of seed deposition by birds into a chronosequence of successions (1, 2 and 4–6 years old) on logged *Pinus* spp. and *C. lusitanica* plantations. We quantified the seed banks present before logging, estimated seed deposition rates and surveyed seedling densities (height ≤ 0.25 m) of animal-dispersed species after logging. We found two seedling recruitment pulses, one at 1 year and one at 4–6 years after logging (Fig. 29.4). The first pulse seemed to be mostly due to recruitment from the soil seed bank, not seeds dispersed after logging. However, the second pulse seemed mainly dependent on post-disturbance seed dispersal. During the second year, seed dispersal into the succession was high, but seedling recruitment was very low. These results suggest there may be windows of time when seedling recruitment in successions may be more or less likely.

More evidence for the importance of timing for seed dispersal comes from our study of dispersal and recruitment into a 5-year-old succession on a logged cypress plantation. The early-successional tree *Trema orientalis* comprised 47% of 2455 tree seeds we recovered from captured birds. However, no *T. orientalis* seeds from caged stations protected from rodents (25 stations, 250 seeds) have germinated in this site after 21 months. We presume this early-successional species is unable to germinate in the shaded understorey of this successional forest. In addition, the poor establishment of mid- and late-successional tree seeds dispersed by baboons into recently logged plantations may be due to the harsh abiotic conditions facing seeds and seedlings in these cleared areas. These results suggest that the probability for recruitment of species varies through time and may often be a result of establishment limitations, not necessarily seed-dispersal limitations (for a temperate system, see Kollmann, 1995).

Strategies to Overcome Seed Dispersal and Recruitment Limitations

The four limitations described above provide a framework for understanding why seed dispersal often fails to promote forest succession of degraded lands. Consequently, land managers have considered ways of overcoming barriers to forest regrowth. For any given location, designing the best strategy will depend on the local ecological, economic and political conditions, and how they interact with management goals.

Sites with sufficient resources for forest cover to quickly establish may not need much intervention during the first few years. In more heavily degraded sites, the first priority may be slowing soil erosion and/or reducing the probability of fire. It can take many decades for topsoil to develop, and most tree species are probably incapable of establishing without it. Fire-breaks may be needed to prevent fires from quickly destroying years of restoration and perpetuating the dominance of grasses. In

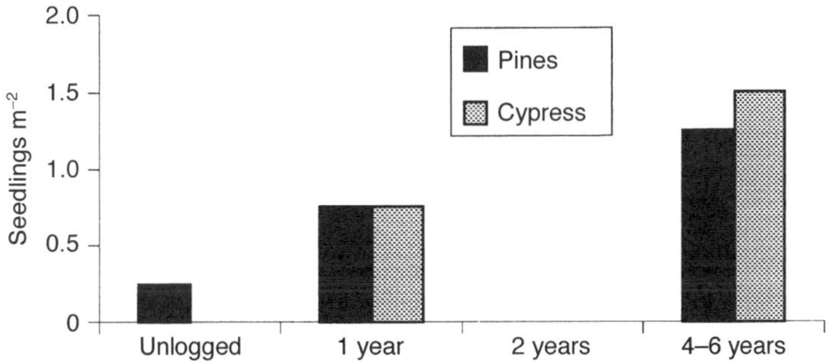

Fig. 29.4. Changes in median seedling density of animal-dispersed trees and shrubs during the first years of forest succession following harvest of exotic pine (*Pinus caribeae* and *Pinus patula*) and cypress (*Cupressus lusitanica*) plantations at Kibale.

1994, fire destroyed native trees planted by the Forest Absorbing Carbon Dioxide Emission (FACE) foundation in 500 ha of degraded grassland at Kibale.

When the area is secured from the threat of fire or other recurrent disturbances, the next steps towards reafforestation will depend on financial resources and the time frame within which forest cover is needed. Now, we consider the advantages and disadvantages of several management strategies.

Constructing perches or planting trees to increase seed rain

Animal-dispersed seed rain in degraded areas tends to be concentrated below the few trees present, and the microclimate below them can be better for seedling establishment than in adjacent open areas. Thus, some researchers have suggested that planted trees or artificial perches will initiate forest succession in degraded areas (McClanahan and Wolfe, 1987, 1993; Aide and Cavelier, 1994; Holl, 1998; Toh *et al.*, 1999). Trees planted in clumps may be more attractive to seed-dispersers than single trees (Toh *et al.*, 1999). Furthermore, planted trees or perches could be placed in the landscape to increase seed-disperser movement between isolated forest patches.

While vegetation can establish below and spread away from trees or perches in degraded tropical systems, it may take many years before canopy closure over a large area is attained (Toh *et al.*, 1999). Artificial perches would be needed in high densities to enhance recruitment over a large area. This will probably be expensive and may dilute seed rain among perches. Also, artificial perches may do little for recruitment if their shade is inadequate to reduce the density of herbaceous vegetation and the threat of fire around them. Holl (1998) investigated recruitment below artificial perches of two types in abandoned pasture and found no difference in recruitment below these perches relative to recruitment in nearby open areas. If trees are planted to become perches, several years may be needed before they attract seed rain and are no longer vulnerable to fires (Chapman and Chapman, 1999). Despite these drawbacks, the use of artificial or natural perches may be effective in small degraded areas or when financial resources are limited and complete canopy cover is not needed for many years.

Seed sowing

Sowing seeds is another means of increasing seed density in degraded areas (Sun *et al.*, 1995; Pinard *et al.*, 1996; Hau, 1997; Chapman and Chapman, 1999). Seeds of many species can be easily collected, prepared and dispersed over a large area. Managers can select which species to sow and where to sow them.

Chapman and Chapman (1999) had limited success with distributing seeds of early- and mid-successional species into recently abandoned cropland in Kibale. Others have also found limited or no success with seed sowing, or found that it was less effective than planting seedlings (Ray and Brown, 1995; Pinard *et al.*, 1996; Parrotta and Knowles, 1999). The timing of seed sowing may be important to consider, especially for early-successional species that may have a short window of time during which they can establish. Site preparation, especially removal of grasses or vines, may be important for successful seed sowing in many systems (Sun *et al.*, 1995; Sun and Dickinson, 1996; Sarmiento, 1997). Because rodent seed predation in degraded areas can be very high (Osunkoya, 1994; Nepstad *et al.*, 1996), distributing well-protected or large seeds less vulnerable to small rodents may be successful (Uhl, 1987; Hau, 1997). In summary, seed sowing may work in some systems with some species, but trials will be needed to identify the species and site-preparation strategies that will be successful.

Tree plantings

Planting native trees in degraded areas is a more direct way of facilitating succession (Tucker and Murphy, 1997). Managers can choose the species to plant and where to plant them. But the greatest advantage of this strategy is that it avoids the stages of seed dormancy, germination and establishment, when mortality is typically highest.

Unfortunately, trial plantings will be needed, because little is known about propagation of native species in most regions. Restoration ecologists could learn from agroforestry trials that identify native species with commercial value for propagation in agricultural areas (Davidson *et al.*, 1998; Stanley and Montagnini, 1999; Eibl *et al.*, 2000). Once species are identified, the construction and maintenance of propagation facilities may be costly. In many systems, continued interventions may be needed to ensure the survival of planted seedlings. For example, every 6 months the FACE project at Kibale must cut grasses around planted seedlings and saplings to reduce competition and fire-risk. Fire-breaks must also be maintained. The combined costs of native-species reafforestation via planting efforts may be great. Parrotta and Knowles (1999) report that reafforestation by planting on degraded mines in Brazil costs US\$250,000 km^{-2}.

In light of obstacles to reafforestation with native species, many have suggested using plantations of exotic fast-growing fuel wood or timber species to catalyse succession on degraded lands (Parrotta *et al.*, 1997). Many such species (e.g. *Acacia* spp., *Pinus* spp., *Eucalyptus* spp.) grow well on degraded lands, compete successfully with aggressive herbaceous growth and reduce the threat of fire within a few years of planting (Lugo, 1997). Furthermore, the protocols for successful planting of such species are well developed. But the main reason why plantations have attracted so much attention from restoration ecologists is that they can attract frugivores that disperse seeds into an understorey whose microclimate is suitable for the establishment of these species (Wunderle, 1997). Mature plantations within Kibale and throughout the world often harbour a dense and species-rich population of native stems, especially if remnant forest is nearby as a seed source (Chapman and Chapman, 1996; Fimbel and Fimbel, 1996). Finally, the harvest and sale of these timber species could help pay for the restoration effort (Chapman and Chapman, 1996).

Unfortunately, there are numerous drawbacks to using exotic timber plantations to facilitate forest succession. Some timber species may invade adjacent undisturbed habitats (Binggeli, 1989), although many frequently planted timber species do not seem to be invasive (e.g. pine, eucalyptus). The species recruiting from animal seed dispersal into plantations tend to be small-seeded early-successional species (Wunderle, 1997). If mid- and late-successional species fail to recruit in sufficient densities, then, when the timber species are gone (via harvesting or natural death), remaining stands of early-successional species may eventually be replaced by arrested successions (Wunderle, 1997). Thus, plantations may only be successful when they are in close proximity to primary forest (Parrotta *et al.*, 1997). Another limitation arises because timber species are often planted in high densities. As they are harvested or die and fall, native species in the understorey may be crushed.

The goals of managing plantations for profit versus enhancing native tree establishment can conflict. Native stems in plantation understories could slow the growth of timber species and reduce their market value (Lamb, 1998). Thus, some managers may keep plantations clear of native stems. Even if native stems are allowed to grow, they may be severely damaged during harvest, as in Kibale, where most native stems are cut to roll logs off-site. Less destructive harvesting techniques (e.g. directional felling) typically increase harvesting costs. After harvest, managers may want to replant these areas with timber species, thus preventing forest succession to more mature phases. Finally, if reafforestation with timber plantations becomes widespread and demand for plantation timber does not grow accordingly, the surplus of timber on the global market could further reduce the financial benefits of this strategy (Leslie, 1999).

Despite these potential pitfalls, reafforestation via timber production remains one of the most widely discussed methods of restoring biological productivity on degraded lands and may be one of the most practical strategies yet proposed. Lamb (1998) and Wunderle (1997) discuss strategies for designing plantations in ways that would help overcome some of the biological obstacles. Ashton *et al.* (1997, 1998) have studied ways of integrating commercial harvest of plantation species and plantings of native stems for forest restoration. However, there are still many unaddressed questions about the practicality of plantations in restoring forests, among which seed-dispersal-orientated questions figure prominently.

Directions for Future Research

Many questions need to be addressed regarding the role of seed dispersal in forest restoration on degraded tropical lands. First, many of the points raised in this chapter have only been investigated by a few studies that inadequately represent the diversity of habitats and geographical areas where reafforestation is needed. More work is needed to test the generality of the existing patterns. Here we outline additional research directions that may be important for realizing the potential of animal-mediated seed dispersal in ecological restoration.

To attract seed-dispersing animals from the forest into degraded areas, some have suggested planting fast-growing, fleshy-fruit-producing species (Green, 1993; Whittaker and Jones, 1994; Sarmiento, 1997; Tucker and Murphy, 1997; Duncan and Chapman, 1999; but see Toh *et al.*, 1999; Zahawi and Augspurger, 1999). These species could be established as plantations, in small clusters or as corridors to draw animals from the forest (Toh *et al.*, 1999). For example, abandoned *Musa* plantations on degraded land in Kibale have facilitated the establishment (almost to canopy closure) of native tree saplings dispersed from the forest in only 6 years. More research is needed to investigate these and other possible ways to increase use of degraded lands by forest frugivores.

However, the risks of drawing animals away from the forest need investigation. Such a strategy could disrupt processes within the forest if frugivores spend much time in degraded areas. For example, seeds of invasive exotic species may be brought back into the forest from degraded areas (Binggeli, 1989). Forest animals in degraded areas may be exposed to increased predation and hunting (Janzen *et al.*, 1976; Wegner and Merriam, 1979; Janzen, 1990). Animals may raid nearby agricultural areas, thus souring farmers' attitudes towards preserves and restoration (Naughton-Treves, 1998; Naughton-Treves *et al.*, 1998). Finally, frequent use of degraded pastures by forest species may increase potential for disease transmission between native and domestic animal species.

To avoid these risks, managers could rely on seed dispersal by frugivores that live in degraded habitats (Da Silva *et al.*, 1996; Holl, 1998). Managers could plant selected native trees in and near degraded areas to provide these frugivores with seeds to disperse and perches to use. Ecologists and agroforesters could collaborate to find native tree species useful to people and good for establishing in degraded areas.

These suggestions and much of the work done in restoration ecology in the tropics have focused on getting a first cohort of trees established on degraded lands. However, we also need to better understand the role of animal-mediated seed dispersal and plant recruitment in older successional forests (Finegan, 1996; Guariguata *et al.*, 1997; Holl and Kappelle, 1999). Where early-successional tree species are able to establish, will their dominance be followed by that of mid- and late-successional species dispersed to the site? If succession is failing, then to successfully intervene we shall need to know if the problem is a lack of dispersers, poor dispersal quality or poor establishment of dispersed seeds. Finding cost-effective ways of helping succession overcome these obstacles would be of great use to managers. We also need to know whether/how manipulations of secondary forests can increase forest regrowth. For example, Chapman *et al.* (2001) removed grasses and shrubs in 5-year-old successional forest on a former conifer plantation and found that reduced competition did not benefit the remaining trees (but see Guariguata, 1999).

Another area needing attention is the impact of woody-stemmed, invasive exotic species in secondary forest succession (Brown *et al.*, 1998; Russell *et al.*, 1998; Stadler *et al.*, 2000). Because many invasive exotics are early-successional species, they may become an important component of secondary forests. Examples from Africa include *Cecropia obtusifolia* (Richards, 1996), *Maesopsis eminii* (Binggeli, 1989) and *Cedrela odorata* (Zanne and Chapman, 2001). It is unknown how much of a threat these species are to secondary forest development or whether they can be used as part of a successful restoration strategy (Kuusipalo *et al*., 1995; Otsamo, 2000).

Finally, where secondary forests will be managed to benefit populations of rare plants and animals, we need to know how to enhance these habitats to benefit such species. This will require knowledge about the habitat requirements of these species and an understanding of how management of successional forest may affect them.

Acknowledgements

Funding for this research was provided by the Wildlife Conservation Society, National Geographic Society, the Explorer's Club, the Lindbergh Foundation, the University of Florida, a National Science Foundation (NSF) Graduate Fellowship and NSF grant numbers SBR-9617664 and SBR-990899. Permission to conduct this research was given by the Office of the President, Uganda, the National Council for Science and Technology, the Uganda Wildlife Authority and the Ugandan Forest Department. We would like to thank Virginia Duncan, Doug Levey, Helene Muller-Landau and Greg Murray for helpful comments on this work.

References

Adedeji, F.O. (1984) Nutrient cycles and successional changes following shifting cultivation practice in moist semi-deciduous forests in Nigeria. *Forest Ecology and Management* 9, 87–99.

Aide, T.M. and Cavelier, J. (1994) Barriers to lowland tropical forest restoration in the Sierra Nevada de Santa Marta, Colombia. *Restoration Ecology* 2, 219–229.

Ashton, P.M.S., Gamage, S., Gunatilleke, I. and Gunatilleke, C.V.S. (1997) Restoration of a Sri Lankan rainforest: using Caribbean pine *Pinus caribaea* as a nurse for establishing late-successional tree species. *Journal of Applied Ecology* 34, 915–925.

Ashton, P.M.S., Gamage, S., Gunatilleke, I. and Gunatilleke, C.V.S. (1998) Using Caribbean pine to establish a mixed plantation: testing effects of pine canopy removal on plantings of rain forest tree species. *Forest Ecology and Management* 106, 211–222.

Binggeli, P. (1989) The ecology of *Maesopsis* invasion and dynamics of the evergreen forest of the East Usambaras, and their implications for forest conservation and forestry practices. In: Hamilton, A.C. and Bensted-Smith, R. (eds) *Forest Conservation in the East Usambara Mountains Tanzania.* IUCN Tropical Forest Programme, Tanzania, pp. 269–300.

Borhidi, A. (1988) Vegetation dynamics of the savannization process on Cuba. *Vegetatio* 77, 177–183.

Brown, J.R., Scanlan, J.C. and McIvor, J.G. (1998) Competition by herbs as a limiting factor in shrub invasion in grassland: a test with different growth forms. *Journal of Vegetation Science* 9, 829–836.

Brown, S. and Lugo, A.E. (1990) Tropical secondary forests. *Journal of Tropical Ecology* 6, 1–32.

Brown, S. and Lugo, A.E. (1994) Rehabilitation of tropical lands: a key to sustaining development. *Restoration Ecology* 2, 97–111.

Chapman, C.A. and Chapman, L.J. (1996) Exotic tree plantations and the regeneration of natural forests in Kibale National Park, Uganda. *Biological Conservation* 76, 253–257.

Chapman, C.A. and Chapman, L.J. (1999) Forest restoration in abandoned agricultural land: a case study from East Africa. *Conservation Biology* 13, 1301–1311.

Chapman, C.A., Chapman, L.J., Zanne, A.E. and Burgess, M. (2001) Does weeding promote regeneration of an indigenous tree community in felled pine plantations in Uganda? *Restoration Ecology* (in press).

Charles-Dominique, P. (1986) Inter-relations between frugivorous vertebrates and pioneer plants: *Cecropia*, birds and bats in French Guyana. *Vegetatio* 66, 119–135.

Chazdon, R.L. (1994) The primary importance of secondary forests in the tropics. *Tropinet* 5, 1.

Da Silva, J.M.C., Uhl, C. and Murray, G. (1996) Plant succession, landscape management, and the ecology of frugivorous birds in abandoned Amazonian pastures. *Conservation Biology* 10, 491–503.

Davidson, R., Gagnon, D., Mauffette, Y. and Hernandez, H. (1998) Early survival, growth and foliar nutrients in native Ecuadorian trees planted on degraded volcanic soil. *Forest Ecology and Management* 105, 1–19.

Duncan, R.S. and Chapman, C.A. (1999) Seed dispersal and potential forest succession in abandoned agriculture in tropical Africa. *Ecological Applications* 9, 998–1008.

Duncan, R.S. and Duncan, V.E. (2000) Forest succession and distance from forest edge in an Afro-tropical grassland. *Biotropica* 32, 33–41.

Eibl, B., Fernandez, R.A., Kozarik, J.C., Lupi, A., Montagnini, F. and Nozzi, D. (2000)

Agroforestry systems with *Ilex paraguariensis* (American holly or yerba mate) and native timber trees on small farms in Misiones, Argentina. *Agroforestry Systems* 48, 1–8.

Fimbel, R.A. and Fimbel, C.C. (1996) The role of exotic conifer plantations in rehabilitating degraded forest lands: a case study from the Kibale Forest in Uganda. *Forest Ecology and Management* 81, 215–226.

Finegan, B. (1996) Pattern and process in neotropical secondary rain forest: the first 100 years of succession. *Trends in Ecology and Evolution* 11, 119–124.

Food and Agriculture Organization (1999) *State of the World's Forests.* Food and Agriculture Organization of the United Nations, Rome, 154 pp.

Gerhardt, K. (1993) Tree seedling development in tropical dry abandoned pasture and secondary forest in Costa Rica. *Journal of Vegetation Science* 4, 95–102.

Gorchov, D.L., Cornejo, F., Ascorra, C. and Jaramillo, M. (1993) The role of seed dispersal in the natural regeneration of rain forest after strip-cutting in the Peruvian Amazon. *Vegetatio* 107/108, 339–349.

Green, R. (1993) Avian seed dispersal in and near subtropical rainforests. *Wildlife Research* 20, 535–557.

Guariguata, M.R. (1999) Early response of selected tree species to liberation thinning in a young secondary forest in northeastern Costa Rica. *Forest Ecology and Management* 124, 255–261.

Guariguata, M.R., Chazdon, R.L., Denslow, J.S., Dupuy, J.M. and Anderson, L. (1997) Structure and floristics of secondary and old-growth forest stands in lowland Costa Rica. *Plant Ecology* 132, 107–120.

Guevara, S. and Laborde, J. (1993) Monitoring seed dispersal at isolated standing trees in tropical pastures: consequences for local species availability. *Vegetatio* 107/108, 319–338.

Guevara, S., Purata, S.E. and Vandermaarel, E. (1986) The role of remnant forest trees in tropical secondary succession. *Vegetatio* 66, 77–84.

Harvey, C.A. (2000) Windbreaks enhance seed dispersal into agricultural landscapes in Monteverde, Costa Rica. *Ecological Applications* 10, 155–173.

Hau, C.H. (1997) Tree seed predation on degraded hillsides in Hong Kong. *Forest Ecology and Management* 99, 215–221.

Holl, K.D. (1998) Do bird perching structures elevate seed rain and seedling establishment in abandoned tropical pasture? *Restoration Ecology* 6, 253–261.

Holl, K.D. (1999) Factors limiting tropical rain forest regeneration in abandoned pasture: seed rain, seed germination, microclimate, and soil. *Biotropica* 31, 229–242.

Holl, K.D. and Kappelle, M. (1999) Tropical forest recovery and restoration. *Trends in Ecology and Evolution* 14, 378–379.

Holl, K.D. and Lulow, M.E. (1997) Effects of species, habitat, and distance from edge on post-dispersal seed predation in a tropical rainforest. *Biotropica* 29, 459–468.

Howe, H.F. and Smallwood, J. (1982) Ecology of seed dispersal. *Annual Review of Ecology and Systematics* 13, 201–228.

Janzen, D.H. (1990) An abandoned field is not a tree fall gap. *Vida Silvestre Neotropical* 2, 64–67.

Janzen, D.H., Miller, G.A., Hackforth-Jones, J., Pond, C.M., Hooper, K. and Janos, D.P. (1976) Two Costa Rican bat-generated seed shadows of *Andira inermis* (*Leguminosae*). *Ecology* 57, 1068–1075.

Kollmann, J. (1995) Regeneration window for fleshy-fruited plants during scrub development on abandoned grassland. *Ecoscience* 2, 213–222.

Kuusipalo, J., Adjers, G., Jafarsidik, Y., Otasmo, A., Tuomela, K. and Vuokko, R. (1995) Restoration of natural vegetation in degraded *Imperata cylindrica* grassland: understorey development in forest plantations. *Journal of Vegetation Science* 6, 205–210.

Lamb, D. (1998) Large-scale ecological restoration of degraded tropical forest lands: the potential role of timber plantations. *Restoration Ecology* 6, 271–279.

Leslie, A.J. (1999) For whom the bell tolls. *Tropical Forest Update* 9, 13–15.

Lugo, A.E. (1997) The apparent paradox of reestablishing species richness on degraded lands with tree monocultures. *Forest Ecology and Management* 99, 9–19.

McClanahan, T.R. and Wolfe, R.W. (1987) Dispersal of ornithochorous seeds from forest edges in central Florida. *Vegetatio* 71, 107–112.

McClanahan, T.R. and Wolfe, R.W. (1993) Accelerating forest succession in a fragmented landscape: the role of birds and perches. *Conservation Biology* 7, 279–288.

Naughton-Treves, L. (1998) Predicting patterns of crop damage by wildlife around Kibale National Park, Uganda. *Conservation Biology* 12, 156–168.

Naughton-Treves, L., Treves, A., Chapman, C. and Wrangham, R. (1998) Temporal patterns of crop-raiding by primates: linking food availability in croplands and adjacent forest. *Journal of Applied Ecology* 35, 596–606.

Nepstad, D.C., Uhl, C. and Serrao, E.A.S. (1991) Recuperation of a degraded Amazonian landscape: forest recovery and agricultural restoration. *Ambio* 20, 248–255.

Nepstad, D.C., Uhl, C., Pereira, C.A. and Da Silva, J.M.C. (1996) A comparative study of tree establishment in abandoned pasture and mature forest of eastern Amazonia. *Oikos* 76, 25–39.

Osunkoya, O.O. (1994) Postdispersal survivorship of north Queensland rainforest seeds and fruits: effects of forest, habitat, and species. *Australian Journal of Ecology* 19, 52–64.

Otsamo, R. (2000) Secondary forest regeneration under fast-growing forest plantations on degraded *Imperata cylindrica* grasslands. *New Forests* 19, 69–93.

Parrotta, J.A. (1993) Secondary forest regeneration on degraded tropical lands: the role of plantations as 'foster ecosystems'. In: Lieth, H. and Lohmann, M. (eds) *Restoration of Tropical Forest Ecosystems*, proceedings of the symposium held on 7–10 October 1991. Kluwer Academic Publishers, Dordrecht, The Netherlands, pp. 63–73.

Parrotta, J.A. and Knowles, O.H. (1999) Restoration of tropical moist forests on bauxite-mined lands in the Brazilian Amazon. *Restoration Ecology* 7, 103–116.

Parrotta, J.A., Turnbull, J.W. and Jones, N. (1997) Introduction – catalyzing native forest regeneration on degraded tropical lands. *Forest Ecology and Management* 99, 1–7.

Pinard, M., Howlett, B. and Davidson, D. (1996) Site conditions limit pioneer tree recruitment after logging of dipterocarp forests in Sabah, Malaysia. *Biotropica* 28, 2–12.

Ray, G.J. and Brown, B.J. (1995) Restoring Caribbean dry forests – evaluation of tree propagation techniques. *Restoration Ecology* 3, 86–94.

Richards, P.W. (1996) *The Tropical Rain Forest: an Ecological Study*. Cambridge University Press, New York, 575 pp.

Russell, A.E., Raich, J.W. and Vitousek, P.M. (1998) The ecology of the climbing fern *Dicranopteris linearis* on windward Mauna Loa, Hawaii. *Journal of Ecology* 86, 765–779.

Sarmiento, F.O. (1997) Landscape regeneration by seeds and successional pathways to restore fragile Tropandean slopelands. *Mountain Research and Development* 17, 239–252.

Stadler, J., Trefflich, A., Klotz, S. and Brandl, R. (2000) Exotic plant species invade diversity hot spots: the alien flora of northwestern Kenya. *Ecography* 23, 169–176.

Stanley, W.G. and Montagnini, F. (1999) Biomass and nutrient accumulation in pure and mixed plantations of indigenous tree species grown on poor soils in the humid tropics of Costa Rica. *Forest Ecology and Management* 113, 91–103.

Sun, D. and Dickinson, G.R. (1996) The competition effect of *Brachiaria decumbens* on the early growth of direct-seeded trees of *Alphitonia petriei* in tropical north Australia. *Biotropica* 28, 272–276.

Sun, D., Dickinson, G.R. and Bragg, A.L. (1995) Direct seeding of *Alphitonia petriei* (*Rhamnaceae*) for gully revegetation in tropical northern Australia. *Forest Ecology and Management* 73, 249–257.

Toh, I., Gillespie, M. and Lamb, D. (1999) The role of isolated trees in facilitating tree seedling recruitment at a degraded sub-tropical rainforest site. *Restoration Ecology* 7, 288–297.

Tucker, N.I.J. and Murphy, T.M. (1997) The effects of ecological rehabilitation on vegetation recruitment: some observations from the wet tropics of North Queensland. *Forest Ecology and Management* 99, 133–152.

Uhl, C. (1987) Factors controlling succession following slash-and-burn agriculture in Amazonia. *Journal of Ecology* 75, 377–407.

Uhl, C. and Buschbacher, R. (1985) A disturbing synergism between cattle ranch burning practices and selective tree harvesting in the eastern Amazon. *Biotropica* 17, 265–268.

Uhl, C. and Clark, K. (1983) Seed ecology of selected Amazon basin successional species. *Botanical Gazette* 144, 419–425.

Uhl, C. and Kauffman, J.B. (1990) Deforestation, fire susceptibility, and potential tree responses to fire in the Eastern Amazon. *Ecology* 71, 437–449.

Uhl, C., Clark, K., Clark, H. and Murphy, P. (1981) Early plant succession after cutting and burning in the upper Rio Negro region of the Amazon Basin. *Journal of Ecology* 69, 631–649.

Uhl, C., Jordon, C., Clark, K., Clark, H. and Herrera, R. (1982) Ecosystem recovery in Amazon caatinga forest after cutting, cutting and burning, and bulldozer clearing treatments. *Oikos* 38, 313–320.

Uhl, C., Buschbacher, R. and Serrao, E.A.S. (1988) Abandoned pastures in eastern Amazonia I. Patterns of plant succession. *Journal of Ecology* 76, 663–681.

Vieira, I.C.G., Uhl, C. and Nepstad, D. (1994) The role of the shrub *Cordia multispicata* Cham. as a 'succession facilitator' in an abandoned pasture, Paragominas, Amazonia. *Vegetatio* 115, 91–99.

Wegner, J.F. and Merriam, G. (1979) Movements by birds and small mammals between a wood and adjoining farmland habitats. *Journal of Applied Ecology* 16, 349–357.

Whelan, C.J., Willson, M.F., Tuma, C.A. and Souza-Pinta, I. (1991) Spatial and temporal patterns of post dispersal seed predation. *Canadian Journal of Botany* 69, 428–436.

Whittaker, R.J. and Jones, S.H. (1994) The role of frugivorous bats and birds in the rebuilding of a tropical forest ecosystem, Krakatau, Indonesia. *Journal of Biogeography* 21, 245–258.

Willson, M.F. and Crome, F.H.J. (1989) Patterns of seed rain at the edge of a tropical Queensland rain forest. *Journal of Tropical Ecology* 5, 301–308.

Wunderle, J.M. (1997) The role of animal seed dispersal in accelerating native forest regeneration on degraded tropical lands. *Forest Ecology and Management* 99, 223–235.

Zahawi, R.A. and Augspurger, C.K. (1999) Early plant succession in abandoned pastures in Ecuador. *Biotropica* 31, 540–552.

Zanne, A.E. and Chapman, C.A. (2001) Expediting forest regeneration in tropical grasslands: distance and isolation from seed sources in plantations. *Ecological Applications* (in press).

30 Frugivory and Seed Dispersal in Degraded Tropical East Asian Landscapes

Richard T. Corlett
Department of Ecology and Biodiversity, University of Hong Kong, Pokfulam Road, Hong Kong, China

Introduction

Nowhere has the recent human impact on tropical forests been greater than in tropical East Asia (Fig. 30.1), where the area that has been deforested – mostly in the past century – is now greater than the area of forest remaining (FAO, 1997). Moreover, this region continues to have the highest rates of forest loss in the tropics, much of the remaining forest has already been heavily exploited, and most of the deforested area has suffered further degradation, hindering forest regeneration. Ecological research in the region has been concentrated in the best remaining areas of forest; ecological processes in the deforested areas have only recently attracted attention (Corlett, 2000). In much of the region, however, native biodiversity will survive, if at all, in a mosaic of primary and secondary forest patches, plantations, shrublands, grassland, agriculture and human habitations (Whitmore, 1999).

A distinctive characteristic of tropical and subtropical East Asia is that, without people, it would be almost entirely covered in forest (Kira, 1995). Apart from tiny areas above the climatic tree line and on several active volcanoes, the only permanent open habitats would have been on coastal cliffs, beaches and seasonally flooded river plains. An important consequence of this is that there is no major regional source of open-country plants and animals. For example, the open-country biota of equatorial Singapore, which was entirely forested two centuries ago, consists largely of exotics and native coastal species (Corlett, 1992a), while the majority of the native terrestrial biota – in all groups of organisms for which we have data – is confined to the small remaining area of forest (Corlett, 2000). Contrast this with tropical South Asia, Australia, Africa and some parts of the neotropics, where the biota of deforested regions is drawn from a rich pool of species in natural (or, at least, ancient) savannahs and woodlands. We might therefore predict that the consequences of deforestation for native biological diversity, on both local and regional scales, would be more severe in tropical East Asia than elsewhere.

Landscape degradation can be defined as a decrease in biomass, biodiversity, structural complexity and soil fertility as a result of human impact (Corlett, 2000). Landscape recovery is an increase in the same parameters when human impact is reduced. Seed dispersal is a key process in plant-species persistence in degraded landscapes because poorly dispersed or undispersed species are trapped in the shrinking area of primary forest and are thus vulnerable to extinction. Seed dispersal is also

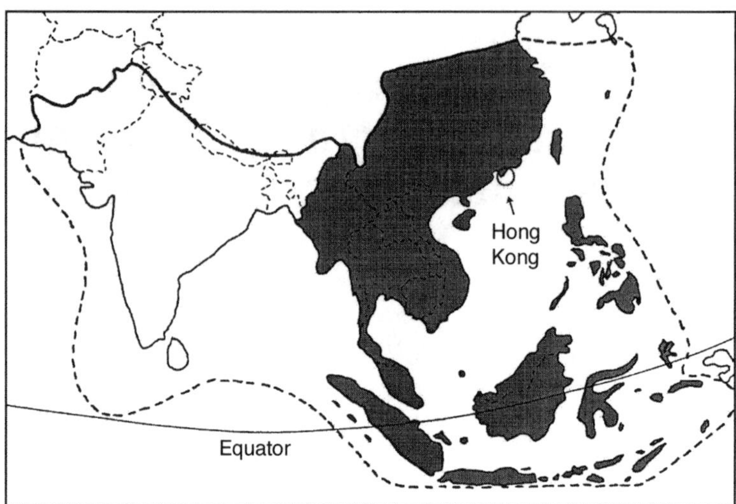

Fig. 30.1. Map of the Oriental Region showing (shaded) the region referred to as tropical East Asia in this chapter.

a key process in forest recovery on degraded sites, because the seed bank and other sources of regeneration (e.g. stumps and roots) have typically been eliminated by prolonged cultivation, cutting or fire (Nepstad *et al.*, 1990).

Most woody species in tropical East Asia are dispersed by animals (Corlett, 1998), so understanding plant–frugivore relationships in degraded tropical landscapes is essential if we are to conserve existing forests and accelerate the process of reafforestation. The major aims of this chapter are therefore: (i) to summarize our current understanding of seed-dispersal relationships in the region's forests; (ii) to review what is known about seed dispersal in deforested and degraded landscapes; (iii) to discuss the implications of this information for landscape restoration; and (iv) to suggest priorities for future research. The introduction and spread of exotic species is potentially a major threat to native biodiversity, particularly in human-dominated landscapes, so both the role of native birds in dispersing exotic plants and the effects of the regional trade in wild animals are also briefly reviewed. Except for a concluding discussion, in which I compare deforested regions in East Asia and elsewhere in the tropics, my focus is on tropical East Asia.

The relative neglect of degraded landscapes by ecologists in tropical East Asia means that most detailed examples have come from two small, densely populated areas, Singapore (650 km^2) and Hong Kong (1100 km^2), where the absence of extensive primary forests has focused research on the habitats and species that have survived. These two areas also provide an interesting historical comparison (Corlett, 2000). Singapore was deforested during the 19th century and still has some fragments of primary forest, albeit with an impoverished fauna. Hong Kong, in contrast, was largely deforested by the 17th century and has lost almost all its forest-dependent fauna.

Seed Dispersal in Forested Landscapes

Corlett (1998) offers a comprehensive review of frugivory and seed dispersal in the Oriental Region. Although this review covered a larger area than that considered here, most of the records came from tropical East Asia and most of these from forests. There remain several major gaps in our knowledge of which animals in the region are important seed-dispersal agents. The most conspicuous of these gaps are: the role of scatter-hoarding by rodents (but see Yasuda *et al.*, 2000), the role of large terrestrial herbivores and the importance of the ubiquitous and abundant, but inconspicuous, babblers (Sylviidae – Timaliini) in the forest understorey. With these exceptions, the major seed-dispersing taxa in the region are clearly apparent and will now be summarized (Table 30.1).

Seed-dispersing birds in tropical East Asia range in size from 5 g flowerpeckers (Dicaeidae) to 2–3 kg hornbills (Bucerotidae). In general, non-passerine frugivores are larger, have bigger gapes and have a higher proportion of fruits in their diets than passerines, although there are exceptions in both groups (Corlett, 1998; Heindl and Curio, 1999). Among the non-passerines, the most important dispersers in forests (based largely on the quantity of fruits they remove) are the hornbills, fruit pigeons (i.e. *Ducula* and *Ptilinopus* (Columbidae)) and barbets (Megalaimidae).

Table 30.1. Estimated relative importance of the vertebrate families responsible for most seed dispersal in intact forest and degraded landscapes of tropical East Asia (from +, minor or local importance, to ++++, very important everywhere). Blanks indicate that seed-dispersing members of the family are usually absent. Based on Corlett (1998), with additional unpublished data for degraded sites.

Family	Intact forest	Degraded land	Common name*
Mammals			
Pteropodidae	++++	+++	Fruit-bats
Cercopithecidae	+++	++	Macaques
Hylobatidae	++++		Gibbons
Hominidae	+++		Orang-utan
Canidae	++	+	Canids
Ursidae	++		Bears
Mustelidae	+	+	Mustelids
Viverridae	++++	+++	Civets
Herpestidae	+	+	Mongooses
Elephantidae	++		Elephants
Tapiridae	++		Tapirs
Rhinocerotidae	++		Rhinoceroses
Suidae	+	+	Pigs
Tragulidae	+		Mouse-deer
Cervidae	+	+	Deer
Bovidae	+		Cattle
Sciuridae	+		Squirrels
Muridae	++	++	Rats
Birds			
Phasianidae	++		Pheasants
Picidae	+		Woodpeckers
Megalaimidae	++++	+	Barbets
Bucerotidae	++++		Hornbills
Cuculidae		+	Cuckoos
Columbidae†	++++		Fruit pigeons†
Eurylaimidae	++		Broadbills
Irenidae	++		Fairy bluebirds, leafbirds
Corvidae	++	++	Corvids
Muscicapidae	++	++	Thrushes, robins
Sturnidae	+	+++	Starlings, mynas
Pycnonotidae	+++	++++	Bulbuls
Zosteropidae	+	+++	White-eyes
Sylviidae	++	++	Babblers, laughing thrushes
Nectariniidae	++	++	Flowerpeckers

*Common names of seed-dispersing members of the family.
†Pigeons, except *Ducula* and *Ptilinopus*, are largely seed predators.

Among the passerines, the most important groups are the bulbuls (Pycnonotidae), *Calyptomena* broadbills (Eurylaimidae) and, possibly, the babblers, with smaller contributions from many other groups (Table 30.1).

In contrast to the diversity of fruit acquisition and processing techniques shown by neotropical frugivorous birds (Levey *et al.*, 1994), the great majority of oriental birds take fruits from perches and swallow them whole. Thus fruit consumption and seed dispersal by these birds are limited principally by their maximum gape widths (Corlett, 1998; Heindl and Curio, 1999). All the above-mentioned birds can potentially eat the smallest fruits in the region, although harvesting them may not be economical for the largest species. Fruits with an equatorial diameter of more than 8–10 mm are too big for the smallest frugivorous birds, such as white-eyes and flowerpeckers. Progressively larger fruits are dispersed by progressively fewer bird species, unless, like many figs, they are soft enough for small pieces to be pecked out and contain small seeds distributed throughout the flesh. Fruits > 30 mm diameter are probably unavailable to all birds except for the larger hornbills, fruit pigeons and barbets. Note, however, that the most specialized fruit–frugivore relationship in the region is that between the tiny flowerpeckers and small-fruited mistletoes (Reid, 1991).

The major mammalian seed-dispersal agents in tropical East Asia are the Old World fruit-bats (Pteropodidae), which range in size from 15 to 1500 g, primates (macaques, gibbons and the orang-utan) and civets (Viverridae), with additional contributions of uncertain magnitude from large terrestrial herbivores (elephants, tapirs, rhinoceroses and, possibly, deer and pigs), a variety of carnivores and, probably, terrestrial rodents (Table 30.1). Because all mammalian frugivores have teeth, they are less likely to be gape-limited than their avian counterparts; fruits too big for most birds are still available to most mammals. Seed size, however, often has a strong influence on seed fate during fruit consumption. In contrast to the variety of fruit-processing techniques shown by New World fruit-bats (Fleming, 1986), all Old World fruit-bats that have been studied swallow only the smallest seeds and eject larger ones in a fibrous wad below feeding roosts (Corlett, 1998). For primates, the contrast is reversed, with fruit-processing techniques more diverse in tropical Asia than in the neotropics. Gibbons and orang-utans swallow most seeds intact, while macaques spit out all but the smallest ones, although larger seeds are often transported some distance in the cheek pouches before being spat (Lucas and Corlett, 1998). In general, the larger the seed, the more likely it is to be dropped, spat out or, in the case of birds, regurgitated, and the fewer the species that regularly defecate it.

In tropical East Asia, protected fruits – those with a thick, inedible, indehiscent rind – are concentrated in a few genera (e.g. *Aglaia*, *Garcinia* and *Nephelium*) and are consumed largely by primates (Corlett, 1998). Many other fruit characteristics, including phenology, presentation, colour and pulp chemistry, are likely to influence fruit choice and hence seed dispersal, but evidence for this is largely anecdotal in the region. Observations at many sites have shown that frugivores are not interchangeable. In particular, some large fruits seem to be taken only by large birds, primates or large fruit-bats, while others may depend on large terrestrial herbivores (Corlett, 1998).

Dispersal Agents in Degraded Landscapes

It is impossible to rank seed-dispersal agents by their general vulnerability to human impacts because of the great variety of such impacts. Hunting in otherwise intact forest selectively reduces or eliminates populations of large vertebrates with low intrinsic rates of population increase (Bennett and Robinson, 2000). Primates and carnivores are particularly vulnerable, but even ungulates, which are relatively fast-breeding, can be hunted to very low population densities. Large fruit-bats are also widely hunted and many of the largest species are exceptionally vulnerable, because they roost communally (Mickleburgh *et al.*, 1992). Large canopy birds, including hornbills and fruit pigeons, are difficult to hunt by traditional means, but are vulnerable where guns are available (Bennett *et al.*, 2000; O'Brian and Kinnaird, 2000). The impact of forest fragmentation, in the absence of hunting, is more complex and less selective, but large frugivores are again particularly sensitive, probably because of their large area requirements (Corlett, 2000).

Deforestation, fragmentation and hunting typically occur together. The result in the most degraded landscapes is a vertebrate fauna dominated by vagile habitat generalists, which can move between widely separated habitat patches, do not have narrow dietary or nesting requirements and are not hunted (Corlett, 2000; Table 30.1). Large fruit-bats, gibbons, orang-utans, elephants, tapirs, rhinoceroses, large deer, cattle, pheasants, hornbills, fruit pigeons, broadbills, most barbets and a wide range of smaller forest specialists survive only in the largest and most remote forest fragments. There are now large parts of the region with none of these taxa. Few species from the forest fauna have adapted to the deforested landscape, and the open-country fauna is dominated by species from the relatively small area of naturally open coastal and riverine habitats, which have spread through the region following deforestation (Hails, 1987; Corlett, 1992a; Wells, 1999). In some cases, this spread has been directly assisted by humans. Hong Kong's impoverished and generally small-bodied community of seed-dispersers (Table 30.2) is typical of the most highly degraded areas.

Bulbuls (Pycnonotidae) are abundant and ubiquitous in non-forest habitats. They are almost certainly the most important seed-dispersal agents in deforested tropical and subtropical East Asia (Corlett, 1998). The bulbul species involved vary over the region, but one or two are usually numerically dominant in any one area and several are very widespread. These common species are insectivore–frugivores, which typically take fruits from a perch and swallow them whole, defecating viable seeds. For those species for which information is available, the maximum gape is 13–15 mm, although some species will also peck pulp from large fruits (Corlett, 1996, 1998). In Hong Kong, 86% of the 255 fruit species studied had a mean diameter of < 13 mm – within the gape limits of light-vented and red-whiskered bulbuls (*Pycnonotus sinensis* and *Pycnonotus jocosus*), the commonest avian seed-dispersers (Corlett, 1996; Fig. 30.2).

Several other passerine families provide species that disperse seeds in some, but not all, degraded landscapes in tropical East Asia (Tables 30.1 and 30.2). White-eyes (Zosteropidae) can swallow fruits up to 10 mm. Some corvids (Corvidae), starlings (Sturnidae), thrushes (Muscicapidae–Turdinae) and laughing thrushes (*Garrulax*, Sylviidae) can swallow whole fruits with diameters in the 15–20 mm range, well above the size available to bulbuls (Corlett, 1998; R.T. Corlett, unpublished data). These species may therefore extend the range of fruit sizes, and thus plant species, that are dispersed in a particular area. A few species of frugivorous non-passerines also occur in some degraded landscapes, including a frugivorous cuckoo, the Asian koel (*Eudynamys scolopacea*), which has a maximum gape of > 20 mm (Corlett and Ko, 1995). However, the pigeons and doves of deforested areas, including the highly frugivorous green pigeons (*Treron* spp.), all belong to taxa considered to be largely seed predators, although at least some seeds survive gut passage intact in most species that have been studied and their role in seed dispersal may have been underestimated (Lambert, 1989; Corlett, 1998).

Although large fruit-bats (Pteropodidae) have been eliminated from much of the region, small (30–80 g) species, particularly in the genus *Cynopterus*, thrive in human-dominated landscapes (Mickleburgh *et al.*, 1992). The role of fruit-bats in degraded Asian landscapes may be limited, however, because the majority of pioneer trees and shrubs, including the important genus *Macaranga*, are bird-dispersed. One exception is *Adinandra dumosa* (*Theaceae*), which dominates early forest succession on highly degraded sites in Singapore and has small, green fruits, with tiny seeds that are defecated in flight by the bat *Cynopterus brachyotis* (Phua and Corlett, 1989).

The dispersal role of rodents in degraded landscapes is unclear. Small squirrels persist as long as there is woody vegetation and rats occur in all non-forest habitats. Apart from the recent evidence for scatter-hoarding mentioned above, intact small seeds are frequently found in rat droppings, and it is therefore likely that the importance of rats as seed-dispersal agents in habitats with no other non-flying mammals has been underestimated

Table 30.2. Seed-dispersal agents in Hong Kong, excluding rare species or those that eat very little fruit. The most important families and genera are in bold type and numbers of species are in parentheses. Information is from faecal analysis and direct observations, both published (Corlett, 1996) and unpublished.

Family	Genus	Common name	Body mass* (g)
Mammals			
Pteropodidae	***Cynopterus*** (1)	Fruit-bat	40–50
	Rousettus (1)	Fruit-bat	60–110
Cercopithecidae	***Macaca*** (1†)	Macaque	3000–6000
Viverridae	***Paguma*** (1)	Civet	3600–5000
	Viverricula (1)	Civet	2000–4000
Muridae	*Bandicota* (1)	Rat	500–700
	Niviventer (1)	Rat	70
	Rattus (1)	Rat	120
Birds			
Megalaimidae	*Megalaima* (1)	Barbet	200
Cuculidae	*Eudynamys* (1)	Koel	240
Corvidae	*Dendrocitta* (1)	Treepie	120
	Oriolus (1)	Oriole	90
	Pica (1)	Magpie	240
	Urocissa (1)	Magpie	170
Muscicapidae			
Turdinae	*Monticola* (1)	Thrush	60
	Myophonus (1)	Thrush	160
	Turdus (4)	Thrushes	50–70
	Zoothera (2)	Thrushes	70–150
Saxicolini	***Luscinia*** (2)	Robins	15–25
	Tarsiger (1)	Robin	15
Sturnidae	*Acridotheres* (1)	Myna	120
	Sturnus (4)	Starlings	40–160
Pycnonotidae	*Hemixos* (1)	Bulbul	50
	Hypsipetes (1)	Bulbul	40
	Pycnonotus (3)	Bulbuls	30–35
Zosteropidae	***Zosterops*** (1)	White-eye	10
Sylviidae			
Garrulacinae	***Garrulax*** (4)	Laughing thrushes	50–130
Timaliini	***Leiothrix*** (2)	Babblers	20–30
	Pomatorhinus (1)	Babbler	30
Nectariniidae	*Dicaeum* (3)	Flowerpeckers	5–7

*Bird masses from M.R. Leven, personal communication; others from various unpublished sources.
†Plus smaller numbers of several other species and their hybrids in mixed groups.

(Corlett, 1998). In Hong Kong, rats are the dominant grassland vertebrates and may be important in the initial establishment of small-seeded woody species (Corlett and Hau, 2000). Conversely, rats are also the major predators of larger seeds in these grasslands (Hau, 1997).

Civets are the most frugivorous family of Old World carnivores, with fruit a significant component in the diet of most species that have been studied (Corlett, 1998). Civets have wide gapes, swallow fruits more or less whole, travel long distances and defecate seeds intact, often selecting open sites to do so. When they are not hunted, several species (e.g. *Paguma larvata, Paradoxurus hermaphroditus, Viverricula indica*) appear to thrive in degraded landscapes, where they are probably very important dispersal agents for large fruits with large seeds that none of the previously mentioned taxa

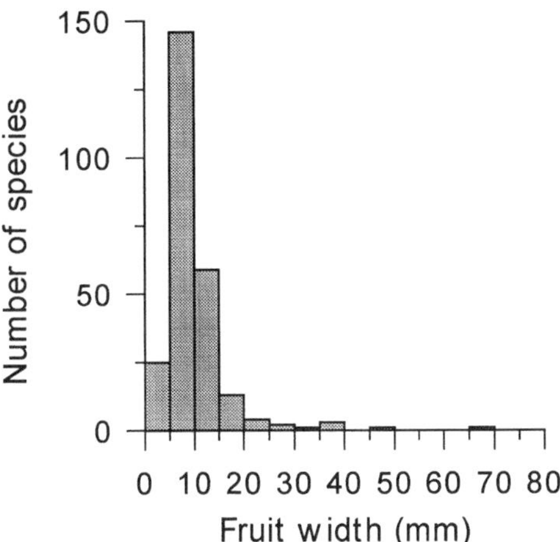

Fig. 30.2. Distribution of mean fruit diameters for 255 native woody-plant species in Hong Kong (redrawn from Corlett, 1996).

disperse (e.g. *Gnetum luofuense* in Hong Kong). However, civets are often highly favoured by hunters and are easily eliminated or reduced to very low densities by heavy trapping (Schreiber *et al.*, 1989).

Although macaques are also often a preferred prey of hunters and have been eliminated as pests from some areas, people from a variety of cultures believe that hunting such human-like animals is wrong. Where such beliefs are dominant, the more tolerant macaque species (particularly the long-tailed macaque, *Macaca fascicularis*, and rhesus macaque, *Macaca mulatta*) can thrive. Indeed, they reach their highest densities in degraded landscapes with at least some woody vegetation, and are often the largest wild vertebrates in such areas (Lucas and Corlett, 1998). Macaques are probably less effective seed-dispersal agents than other large vertebrates, because they spit most seeds. None the less, they may become the only dispersal agents for plants with large fruits and seeds in areas where other large vertebrates have been eliminated. A wide variety of other mammalian taxa may also disperse some seeds in some degraded landscapes. These include small carnivores (Canidae, Mustelidae, Herpestidae), pigs and small deer (Corlett, 1998).

Consequences of Disperser Loss during Landscape Degradation

As summarized above, the most vulnerable disperser species are the large bats, primates, large terrestrial herbivores and large non-passerine birds, which, in intact forest communities, are responsible for the dispersal of most plants with large fruits and seeds. In contrast, the overwhelmingly dominant dispersers in overhunted forests and in deforested landscapes are small passerines, mostly with gape limits < 20 mm. Plants with small fruits or with large soft fruits with many small seeds thus have abundant dispersal agents in any habitat in the region, while plants with large seeds generally lack dispersers.

The consequences of this loss of large-gaped dispersal agents are most obvious in the 'defaunated' but otherwise intact forests, which are becoming increasingly common in the remoter parts of the region. In such areas, one can find piles of large, uneaten fruits under trees that were themselves presumably dispersed by the now eliminated disperser fauna (Ng, 1983; Corlett and Turner, 1997). In deforested landscapes, in contrast, it is difficult to separate the impact of disperser loss from the numerous other drastic changes in the biological and physical environment that influence post-dispersal survival. In particular, large seeds are generally associated with shade tolerance, whereas small seeds are typical of woody pioneers. Thus the absence of plants with large seeds and fruits in secondary shrubland and forest cannot, without appropriate experiments, be attributed solely to their failure to get there.

The consequences of depending on habitat-generalist insectivore–frugivores for seed dispersal are not all negative, however. Such species visit and disperse seeds between a wide range of habitat types, including the edges and upper canopy of primary-forest fragments, exotic plantations with no fleshy fruits and isolated trees and shrubs in grassland. In Singapore, flocks of the Asian glossy starling (*Aplonis panayensis*) and the introduced white-vented myna (*Acridotheres cinereus*) feed in small-fruited canopy trees in the largest remaining primary forest fragment, as well as in a variety of open habitats, including parks and gardens (Corlett and Lucas, 1989). In contrast, the few surviving specialist frugivores in Singapore (e.g. red-crowned barbet, *Megalaima rafflesii*, and Asian fairy bluebird, *Irena puella*) are confined to the forest (Lim, 1997). *Cynopterus* fruit-bats, although more dependent on fruits than the open-country passerines, show a similar disregard for habitat boundaries, feeding both inside primary forest and on isolated trees in open habitats (Tan *et al.*, 1998). The more tolerant civet species also hunt rodents and deposit seeds in habitats, such as grasslands, where no fruits are available. Seeds of forest plants are thus dispersed by non-specialists to habitats that forest-dependent dispersal agents do not enter.

On a finer scale, seed deposition by birds at open sites is limited by perch availability, since most seeds are defecated or regurgitated from perches. Perch preferences, in terms of height, diameter, branching and other factors, are likely to differ between bird species (Slocum and Horvitz, 2000), but there are no regional data available on this. The tiny seeds that are swallowed by fruit-bats are more uniformly distributed, since they are defecated in flight, while the larger ones dropped under feeding roosts are more patchily distributed, since the same roosts are used repeatedly (Phua and Corlett, 1989; Corlett, 1998).

All local tree and shrub floras in the tropics contain species that are effectively dispersed in the most degraded landscapes, so it is unlikely that the initial establishment of woody plant cover is ever prevented by the characteristics of the available disperser fauna. The main impact of disperser loss will probably be on the diversity of this plant cover. This impact is likely to increase in the later stages of succession, when larger-seeded, shade-tolerant species would be expected to invade the site if dispersal agents were available (Hamann and Curio, 1999). Evidence that the species composition of late-secondary forests is limited by the local extinction of dispersal agents comes from Singapore, which has lost approximately half of its forest bird and mammal fauna, including most large and large-gaped species (Castelletta *et al.*, 2000; Corlett, 2000). Even after a century of regrowth, Singapore's secondary forests are formed almost exclusively of small-fruited species and are much less diverse floristically than contiguous fragments of primary forest (Corlett, 1991; Turner *et al.*, 1997). Large-fruited animal-dispersed families, such as the *Anacardiaceae*, *Burseraceae*, *Meliaceae* and *Sapotaceae*, are either very rare or represented only by exceptionally small-seeded species (e.g. *Campnospermum auriculatum*, *Anacardiaceae*; *Santiria apiculata*, *Burseraceae*). An important family of lowland rain-forest trees, the *Myristicaceae*, which depend for seed dispersal on large frugivorous birds that are locally extinct, is under-represented not only in secondary forest, but also in recruitment in the largest remaining primary-forest fragment (Ercelawn *et al.*, 1998).

Hong Kong also has many woody species that are apparently un- or under-dispersed in the modern landscape (Zhuang and Corlett, 1996). These include both large, single-seeded fruits (particularly in the family *Lauraceae*), which were probably dispersed by locally extinct large-gaped avian frugivores, and 35 species of *Fagaceae*, for which the original dispersal agents are unknown. The success of several *Fagaceae* species when planted as seedlings into degraded grasslands provides experimental support for the hypothesis that their current distribution in Hong Kong is limited by lack of seed dispersal (B.C.H. Hau, Hong Kong, 2000, personal communication).

Dispersal of Exotic Plant Species

Only a small minority of the exotic plant species now naturalized in tropical East Asia have fleshy fruits and are dispersed by vertebrates (Corlett, 1992b). The most widespread and abundant species is the tropical American shrub *Lantana camara*, whose fruits are a preferred food of open-country insectivore–frugivores throughout the region (R.T. Corlett, personal observations). There are also several widely naturalized, bird-dispersed species of *Solanum* and *Passiflora*, again from tropical America. The proportion of vertebrate-dispersed species is higher in woody plants, particularly among the most invasive species. Several tropical American pioneers are locally established in the region, with the bat-dispersed trees, *Cecropia peltata* and *Piper aduncum*, both potentially capable of displacing native pioneers (Rogers and Hartemink, 2000), and the relatively shade-tolerant, bird-dispersed shrub, *Clidemia hirta*, the only widespread exotic invader of undisturbed primary forest (Ickes and Williamson, 2000).

The Impact of the Trade in Live Animals

Most bird communities in the region have been influenced in some way by the huge regional trade in live birds. This trade is dominated by insectivore–frugivores (bulbuls, white-eyes, starlings, mynas and laughing thrushes), so an impact on seed dispersal is possible. Although a few species dominate, hundreds of others are traded in small numbers. Even common species, such as the red-whiskered bulbul (*P. jocosus*) in parts of Thailand, can be driven to local extinction by trapping (Kanjanavanit, 2000). The converse is that huge numbers of captive birds are released outside their natural ranges, largely for religious reasons (Severinghaus, 1999). Many feral populations of common cage birds (e.g. the red-whiskered bulbul) have become established in the region (Long, 1981) and in some places, such as in Hong Kong and Singapore, they are among the dominant open-country frugivores (Hails, 1987; Carey *et al.*, 2001).

The impact of these losses and 'gains' of bird species on seed dispersal in degraded landscapes is probably small, because of the lack of specialization in the dispersal relationships between small-fruited plants and small, fruit-eating birds (Corlett, 1996, 1998). However, the introduction and establishment of large-gaped frugivores may have a significant impact by restoring dispersal of fruits too large for the extant avifauna. In Hong Kong, a recent and rapidly spreading introduction is the greater necklaced laughing thrush (*Garrulax pectoralis*), which was probably a member of the primeval forest bird fauna (Carey *et al.*, 2001). It is now the largest-gaped avian frugivore in many areas of shrubland and secondary forest, capable of swallowing fruits that are too big for any other common bird species (R.T. Corlett, personal observations). An extreme example of the same phenomenon was observed in Singapore, where a feral pair of oriental pied hornbills (*Anthracoceros albirostris*) are the only birds I have ever seen consuming the large arillate seeds of fruiting *Myristicaceae*. Increasing wealth is fuelling a taste for more exotic cage-birds and there is now a risk that non-Asian species will become established in the region, just as Asian bulbuls and white-eyes are already established in several places outside Asia (Long, 1981).

In most cases, the effects of the trade in live mammals are less obvious than for birds, but primates are still being captured, illegally, in many parts of the region, and the introduction of both macaques and civets to several islands east of Wallace's line (Flannery, 1995), where there are no native equivalents, must have had a large impact on seed dispersal. Recent reports of an established population of feral long-tailed macaques (*M. fascicularis*) on the Indonesian half of the island of New Guinea, near the capital, Jayapura, are particularly worrisome (Anon., 2000). These macaques are generalist frugivores with catholic tastes and destructive feeding habits (Lucas and Corlett, 1998). They are likely to have an impact on native dispersal agents both directly, by plundering nests (Safford, 1997), and indirectly, by competing for fruit. Such transfers of ecologically important taxa across major biogeographical barriers are the experimental ecologist's dream, but the conservationist's nightmare.

Seed Dispersal and Landscape Restoration

General prescriptions for human-assisted forest restoration in the human-occupied tropics are impossible, since social and economic considerations are at least as important as ecological ones, and the humid tropics encompass the full range of social and economic conditions found on earth. It is possible, however, to suggest a range of potential strategies, based on what is known about seed dispersal in degraded landscapes.

Although much attention has focused on tree-planting as a means of restoring degraded tropical landscapes, the areas in need of restoration are increasing much faster than the current rate of such reafforestation. Forest succession, with or without human assistance, will continue to be the main mechanism of forest restoration in the tropics. Even in the most degraded grasslands, simply controlling fires is usually sufficient to initiate succession, at least in areas with relatively high rainfall (e.g. for Hong Kong, see Zhuang, 1997). Management actions that increase the number and diversity of seeds dispersed into a site are likely to increase the diversity of regenerating species, if not necessarily the rate of succession. Reduction of hunting pressure on large frugivores is an obvious first step, and the reintroduction of locally extinct seed-dispersal agents is worth considering, given the success of the many unplanned bird introductions in the region. At the same time, it is essential that remnant trees and surviving forest fragments, however small, are protected as future seed sources (Turner and Corlett, 1996).

The diverse native understorey that develops in most plantations in the region (e.g. Zhuang and Yau, 1999) suggests that decades of experience with establishing exotic monocultures could be utilized in landscape restoration. There is increasing evidence that such plantations can accelerate natural forest succession by ameliorating harsh soil and microclimate conditions, suppressing grasses and attracting seed-dispersers (Lugo, 1997; Parrotta *et al.*, 1997). In Hong Kong, plantation monocultures of the Australian exotic, *Lophostemon confertus*, support a much lower density of insectivore–frugivores than adjacent stands of native secondary forest of similar age, but seed dispersal has been sufficient for native trees and shrubs to invade and diversify these plantations over several decades (Kwok and Corlett, 2000).

Depending entirely on natural seed dispersal to bring tree species to a site will, however, result in a secondary forest dominated by a well-dispersed subset of the forest flora (Corlett, 1991). Unassisted succession is thus better at restoring biomass than biodiversity, particularly in areas that are highly degraded or distant from forest seed sources. Direct seeding (i.e. scattering or planting seeds) of poorly dispersed tree species is potentially the cheapest way of supplementing natural seed dispersal, but there is very little evidence for its effectiveness in the tropics. In Hong Kong, it has proved successful only with a native conifer, *Pinus massoniana* (Corlett, 1999). Further research aimed at identifying and, if possible, overcoming barriers to successful direct seeding should be a priority.

Planting seedlings of mixed, native species is the quickest way to restore forest cover, but planting large areas can be prohibitively expensive. In such situations, the most cost-effective method of accelerating and diversifying secondary succession in grassland is probably to plant poorly dispersed tree species as small clumps scattered over the site (Goosem and Tucker, 1995) or as linear wind-breaks (Harvey, 2000). Such plantings are expected to increase the diversity of the resulting secondary forest both directly, by adding species that would not arrive on their own accord, and indirectly, by acting as foci for biotic seed dispersal and seedling establishment. A major problem with this approach in many areas, however, is likely to be fire control, since small clumps or rows of trees are less easily protected than large blocks and will take longer to suppress fire-promoting grasses.

Research Needs for Tropical East Asia

Despite many publications on the biology of small pteropodid fruit-bats, no study in the Oriental Region has focused on their role in dispersing seeds into open habitats. Neotropical fruit-bats (which belong to an entirely independent radiation) are known to be major dispersers of this type because many woody pioneers in the neotropics are bat-dispersed, fruit-bats in general seem more willing to cross open areas then frugivorous birds, and defecation of small seeds in flight ensures a more even spread of dispersed seeds (e.g. Medellin and Gaona, 1999; Harvey, 2000). Bats may be less important in the early stages of succession in tropical East Asia, because most woody pioneers are dispersed by birds, but this assumption needs more thorough investigation. The role of Asian fruit-bats in dispersing larger-fruited primary-forest species between forest fragments and into pioneer woody vegetation may be more important but, again, good data are lacking.

More generally, no study in the region has looked at patterns of overall seed deposition across the full range of habitats in degraded landscapes. A comparison between the local seed rain and the woody flora would help answer a key question: to what extent does seed dispersal limit the rate of succession and the composition of the resultant woody secondary vegetation in degraded landscapes? Grasslands and exotic plantations offer the possibility of controlling for local vegetation structure and fruit supply, while investigating the effect of differences in disperser fauna and the availability of seed sources. Indeed, it should also be possible to use such simplified plant communities to make comparisons between different regions of the tropics.

Information on spatial patterns of seed rain needs to be supplemented by studies of movements by individual frugivores. There is evidence that landscape structure on a variety of scales can influence patterns of frugivore, and thus seed, movement in degraded landscapes (Metzger, 2000). Short-term radio-tracking is now practical for all mammals and for birds > 10 g (e.g. Double and Cockburn, 2000). There are preliminary data available on the movements of civets (Rabinowitz, 1991) and of several bird species within forests (Lambert, 1989). Similar information for vertebrates in degraded landscapes would be particularly valuable in understanding likely dispersal distances. Conversion of movement patterns into seed-dispersal patterns also requires information on the rates at which seeds move through the gut, which is currently available for very few Asian frugivores. Recent reports that Old World fruit-bats regularly retain a proportion of viable small seeds in their guts for > 12 h (Shilton *et al.*, 1999) demonstrate the need for great care in designing studies of gut passage rates.

Finally, seed dispersal is only one of the potential barriers – or filters – to secondary succession in deforested sites (e.g. Nepstad *et al.*, 1990). Seeds that are deposited in open sites often face intense seed predation (e.g. Hau, 1997), as well as poor conditions for germination and subsequent growth. Knowing which plant species are dispersed where and by which animals will have little predictive value unless we also understand the major factors controlling post-dispersal seed fates (Rey and Alcantara, 2000). The typically high mortalities suffered by seeds and, often, seedlings in degraded sites mean that large sample sizes are needed to obtain meaningful results. These problems appear to be tractable, at least for studies of single species (e.g. Jordano and Herrera, 1995; Rey and Alcantara, 2000). Early indications are that the biology of post-dispersal seed fates will be as diverse and interesting as the better-known biology of seed dispersal.

Degraded Landscapes Elsewhere in the Tropics

Some features of seed dispersal in degraded tropical landscapes – such as the relative success of small-seeded, small-fruited plants, and the very low seed deposition rates in treeless grasslands – appear to be universal, whereas others are likely to depend on the unique characteristics of the local disperser fauna. Differences between fruit- and seed-processing techniques of Old and New World birds, bats and primates have already been mentioned, but even the presence or absence of single species may be important. For example, in tropical Australia, cassowaries provide a unique dispersal service for very large fruits and seeds, so their survival and behaviour in degraded landscapes is likely to have a crucial influence on seed-dispersal patterns (Stocker and Irvine, 1983).

Differences in the aims and methodologies of seed-dispersal studies and in the physical environments of study sites make direct comparisons between different parts of the tropics difficult. In general, biological differences appear to be greatest between the Old and New World tropics, with Australia and New Guinea apparently distinct from both. Most groups of important dispersers in degraded landscapes are shared between Africa and tropical Asia, including the bulbuls, civets and pteropodid fruit-bats. Differences within the Old World tropics seem most often to reflect differences in climate. Tropical South Asia (the Indian subcontinent and adjacent areas) has a very similar biota to tropical East Asia, but its drier climate means that closed forests have always been less extensive. An apparent consequence of this difference is that forest vertebrates in South Asia, including seed-dispersal agents, seem to be more tolerant of landscape degradation and more willing to enter open areas. For example, the Indian grey hornbill (*Ocyceros birostris*) is the only Asian hornbill that thrives in non-forest habitats (Kemp, 1995). Tropical Africa, with its vast extent of non-forest vegetation, has a far richer open-country vertebrate fauna than any Asian site, and some of its large forest vertebrates also seem relatively likely to enter open habitats (Tutin *et al.*, 1997; Chapman and Chapman, 1999). This would be expected to result in more efficient dispersal of large-fruited, large-seeded plant species into and within degraded landscapes, although currently available evidence does not support this (Duncan and Chapman, 1999).

In the neotropics, in contrast, most seed dispersal in degraded landscapes is carried out by entirely different groups of birds and mammals, although the Corvidae, Muscicapidae and Sturnidae (including Mimidae) are pantropical. Phyllostomid fruit-bats and small emberizid passerines are major seed-dispersal agents in open neotropical habitats; both tend to extract and drop large seeds before swallowing the pulp (Levey *et al.*, 1994; Da Silva *et al.*, 1996). It may therefore be no coincidence that the major woody pioneers in the neotropics typically have much smaller seeds than those in Africa (Chapman *et al.*, 1999) and Asia (R.T. Corlett, personal observations). Bats may be most important in the earliest stages of woody succession in neotropical pastures, with birds becoming more significant as trees become established and provide perches (Da Silva *et al.*, 1996). The larger fruits and seeds of late-successional trees may be less effectively dispersed in degraded neotropical landscapes, because the non-flying forest animals seem generally less willing to enter open habitats than they are in Africa and Asia. For example, in Costa Rica, white-faced capuchins (*Cebus capucinus*) and howler monkeys (*Aloutta palliata*) fed only in fruit trees that were close enough to the forest edge to be reached by leaping (Slocum and Horvitz, 2000), while many African and several Asian primates will cross open ground to reach isolated fruit trees (Chapman and Onderdonk, 1998; Corlett, 1998). This contrast between the willingness of Old and New World forest mammals to enter non-forest habitats is also supported by the extensive literature on severe crop damage by forest primates, ungulates and elephants in Africa and Asia (e.g. Naughton-Treves, 1998) and the absence of any equivalent reports from the neotropics.

Despite these, apparently major, interregional differences in dispersal biology, there is no evidence for a significant effect on forest succession in degraded landscapes. Such effects could easily be overlooked, particularly if they only appear in the later stages of succession, when dispersal limitation may be more likely. Instead, the strongest patterns that emerge from the literature relate to the degree of site degradation and to climate. Throughout the tropics, the rate at which woody vegetation develops on cleared sites is inversely related to the duration and intensity of previous disturbance (e.g. Nepstad *et al.*, 1990; Corlett, 1991). When the comparison is restricted to highly degraded sites, high annual rainfall (> 2000 mm) seems to favour relatively rapid restoration of forest cover (Corlett, 1991; Zhuang, 1997), while in drier areas the barriers to tree establishment are much higher and grasslands can persist for decades in climates that could otherwise support closed forest (Da Silva *et al.*, 1996; Chapman and Chapman, 1999). Comparisons between sites with carefully matched climates and site histories are needed to reveal any effects of differences in the disperser fauna.

Acknowledgements

This chapter has benefited from comments from far too many people to list here. However, I am particularly grateful to David Dudgeon, Scot Duncan, Kwok Hon Kai, Billy Hau, Michael Leven and Douglas Levey.

References

Anon. (2000) Monkey business in Irian Jaya. *South Pacific Currents* 7, 2.

Bennett, E.L. and Robinson, J.G. (2000) Hunting for sustainability: the start of a synthesis. In: Robinson, J.G. and Bennett, E.L. (eds) *Hunting for Sustainability in Tropical Forests.* Columbia University Press, New York, pp. 499–520.

Bennett, E.L., Nyaoi, A.J. and Sompud, J. (2000) Saving Borneo's bacon: the sustainability of hunting in Sarawak and Sabah. In: Robinson, J.G. and Bennett, E.L. (eds) *Hunting for Sustainability in Tropical Forests.* Columbia University Press, New York, pp. 305–324.

Carey, G.J., Chalmers, M.L., Diskin, D.A., Kennerley, P.R., Leader, P.J., Leven, M.R., Lewthwaite, R.W., Melville, D.S., Turnbull, M. and Young, L. (2001) *The Avifauna of Hong Kong.* Hong Kong Bird Watching Society, Hong Kong.

Castelletta, M., Sodhi, N.S. and Subaraj, R. (2000) Heavy extinctions of forest avifauna in Singapore: lessons for biodiversity conservation in Southeast Asia. *Conservation Biology* 14, 1870–1880.

Chapman, C.A. and Chapman, L.J. (1999) Forest restoration in abandoned agricultural land: a case study from East Africa. *Conservation Biology* 13, 1301–1311.

Chapman, C.A. and Onderdonk, D.A. (1998) Forest without primates; primate/plant codependency. *American Journal of Primatology* 45, 127–141.

Chapman, C.A., Chapman, L.J., Kaufman, L. and Zanne, A.E. (1999) Potential causes of arrested succession in Kibale National Park: growth and mortality of seedlings. *African Journal of Ecology* 37, 81–92.

Corlett, R.T. (1991) Plant succession on degraded land in Singapore. *Journal of Tropical Forest Science* 4, 151–161.

Corlett, R.T. (1992a) The ecological transformation of Singapore, 1819–1990. *Journal of Biogeography* 19, 411–420.

Corlett, R.T. (1992b) The naturalized flora of Hong Kong: a comparison with Singapore. *Journal of Biogeography* 19, 421–430.

Corlett, R.T. (1996) Characteristics of vertebrate-dispersed fruits in Hong Kong. *Journal of Tropical Ecology* 12, 819–833.

Corlett, R.T. (1998) Frugivory and seed dispersal by vertebrates in the Oriental (Indomalayan) Region. *Biological Reviews* 73, 413–448.

Corlett, R.T. (1999) Environmental forestry in Hong Kong: 1871–1997. *Forest Ecology and Management* 116, 93–105.

Corlett, R.T. (2000) Environmental heterogeneity and species survival in degraded tropical landscapes. In: Hutchings, M.J., John, E.A. and Stewart, A. (eds) *The Ecological Consequences of Environmental Heterogeneity.* Blackwell Science, Oxford, pp. 333–355.

Corlett, R.T. and Hau, B.C.H. (2000) Seed dispersal and forest restoration. In: Elliott, S., Kerby, J., Blakesley, D., Hardwick, K., Woods, K. and Anusarnsunthorn, V. (eds) *Forest Restoration for Wildlife Conservation.* International Tropical Timber Organization and Forest Restoration Research, Chiang Mai University, Thailand, pp. 317–325.

Corlett, R.T. and Ko, W.P. (1995) Frugivory by koels in Hong Kong. *Memoirs of the Hong Kong Natural History Society* 20, 221–222.

Corlett, R.T. and Lucas, P.W. (1989) Consumption of *Campnospermum auriculatum* (*Anacardiaceae*) fruit by vertebrates in Singapore. *Malayan Nature Journal* 42, 273–276.

Corlett, R.T. and Turner, I.M. (1997) Long-term survival in tropical forest remnants in Singapore and Hong Kong. In: Laurance, W.F. and Bierregaard, R.O. (eds) *Tropical Forest Remnants: Ecology, Management and Conservation of Fragmented Communities.* University of Chicago Press, Chicago, Illinois, pp. 333–345.

Da Silva, J.M.C., Uhl, C. and Murray, G. (1996) Plant succession, landscape management, and the ecology of frugivorous birds in abandoned Amazonian pasture. *Conservation Biology* 10, 491–503.

Double, M. and Cockburn, A. (2000) Pre-dawn infidelity: females control extra-pair mating in superb fairy-wrens. *Proceedings of the Royal Society of London, Series B, Biological Sciences* 267, 465–470.

Duncan, R.S. and Chapman, C.A. (1999) Seed dispersal and potential forest succession in abandoned agriculture in tropical Africa. *Ecological Applications* 9, 998–1008.

Ercelawn, A.C., LaFrankie, J.V., Lum, S.K.Y. and Lee, S.K. (1998) Short-term recruitment of trees in a forest fragment in Singapore. *Tropics* 8, 105–115.

FAO (1997) *State of the World's Forests.* FAO, Rome, 200 pp.

Flannery, T.F. (1995) *Mammals of the South-West Pacific and Moluccan Islands.* Comstock, Ithaca, 464 pp.

Fleming, T.H. (1986) Opportunism versus specialization: the evolution of feeding strategies in frugivorous bats. In: Estrada, A. and Fleming, T.H. (eds) *Frugivores and Seed Dispersal.* Dr W. Junk Publishers, Dordrecht, The Netherlands, pp. 105–118.

Goosem, S. and Tucker, N.I.J. (1995) *Repairing the Rainforest: Theory and Practice of Rainforest Re-establishment in North Queensland's Wet Tropics.* Wet Tropics Management Authority, Queensland, Australia, 72 pp.

Hails, C. (1987) *Birds of Singapore.* Times Editions, Singapore.

Hamann, A. and Curio, E. (1999) Interactions among frugivores and fleshy fruit trees in a Philippine submontane forest. *Conservation Biology* 13, 766–773.

Harvey, C.A. (2000) Windbreaks enhance seed dispersal into agricultural landscapes in Monteverde, Costa Rica. *Ecological Applications* 10, 155–173.

Hau, C.H. (1997) Tree seed predation on degraded hillsides in Hong Kong. *Forest Ecology and Management* 99, 215–221.

Heindl, M. and Curio, E. (1999) Observations of frugivorous birds at fruit-bearing plants in the North Negros Forest Reserve, Philippines. *Ecotropica* 5, 167–181.

Ickes, K. and Williamson, G.B. (2000) Edge effects and ecological processes: are they on the same scale? *Trends in Ecology and Evolution* 15, 373.

Jordano, P. and Herrera, C.M. (1995) Shuffling the offspring: uncoupling and spatial discordance of multiple stages in vertebrate seed dispersal. *EcoScience* 2, 230–237.

Kanjanavanit, R. (2000) The decline of red-whiskered bulbuls in Thailand. *Oriental Bird Club Bulletin* 31, 66–67.

Kemp, A. (1995) *The Hornbills.* Oxford University Press, Oxford, 302 pp.

Kira, T. (1995) Forest ecosystems of East and Southeast Asia in a global perspective. In: Box, E.O. (ed.) *Vegetation Science in Forestry.* Kluwer Academic Publishers, Dordrecht, pp. 1–21.

Kwok, H.K. and Corlett, R.T. (2000) The bird communities of a natural secondary forest and a *Lophostemon confertus* plantation in Hong Kong, South China. *Forest Ecology and Management* 130, 227–234.

Lambert, F.R. (1989) Pigeons as seed predators and dispersers of figs in a Malaysian lowland forest. *Ibis* 131, 521–527.

Levey, D.J., Moermond, T.C. and Denslow, J.S. (1994) Frugivory: an overview. In: McDade, L.A., Bawa, K.S., Hespenheide, H.A. and Hartshorn, G.S. (eds) *La Selva: Ecology and Natural History of a Neotropical Rain Forest.* University of Chicago Press, Chicago, pp. 287–294.

Lim, K.S. (1997) Bird biodiversity in the Nature Reserves of Singapore. *Gardens Bulletin Singapore* 49, 225–244.

Long, J.L. (1981) *Introduced Birds of the World.* Reed, Sydney, 518 pp.

Lucas, P.W. and Corlett, R.T. (1998) Seed dispersal by long-tailed macaques. *American Journal of Primatology* 45, 29–44.

Lugo, A.E. (1997) The apparent paradox of reestablishing species richness on degraded lands with tree monocultures. *Forest Ecology and Management* 99, 9–19.

Medellin, R.A. and Gaona, O. (1999) Seed dispersal by bats and birds in forest and disturbed habitats of Chiapas, Mexico. *Biotropica* 31, 478–485.

Metzger, J.P. (2000) Tree functional group richness and landscape structure in a Brazilian tropical fragmented landscape. *Ecological Applications* 10, 1147–1161.

Mickleburgh, S.P., Hutson, A.M. and Racey, P.A. (1992) *Old World Fruit Bats: an Action Plan for their Conservation.* IUCN, Gland, 222 pp.

Naughton-Treves, L. (1998) Predicting patterns of crop damage by wildlife around Kibale National park, Uganda. *Conservation Biology* 12, 156–158.

Nepstad, D., Uhl, C. and Serrão, E.A. (1990) Surmounting barriers to forest regeneration in abandoned, highly degraded pastures: a case study from Paragominas, Pará, Brazil. In: Anderson, A.B. (ed.) *Alternatives to Deforestation: Steps Towards Sustainable Use of the Amazon Rain Forest.* Columbia University Press, New York, pp. 215–229.

Ng, F.S.P. (1983) Ecological principles of tropical lowland rain forest conservation. In: Sutton, S.L., Whitmore, T.C. and Chadwick, A.C. (eds) *Tropical Rain Forest: Ecology and Management.* Blackwell Scientific, Oxford, pp. 359–375.

O'Brian, T.G. and Kinnaird, M.F. (2000) Differential vulnerability of large birds and mammals to hunting in North Sulawesi, Indonesia, and the outlook for the future. In: Robinson, J.G. and Bennett, E.L. (eds) *Hunting for Sustainability in Tropical Forests.* Columbia University Press, New York, pp. 199–213.

Parrotta, J.A., Turnbull, J.W. and Jones, N. (1997) Catalyzing native forest regeneration on degraded tropical lands. *Forest Ecology and Management* 99, 1–7.

Phua, P.B. and Corlett, R.T. (1989) Seed dispersal by the lesser short-nosed fruit bat (*Cynopterus brachyotis,* Pteropodidae, Megachiroptera). *Malayan Nature Journal* 42, 251–256.

Rabinowitz, A.R. (1991) Behaviour and movements of sympatric civet species in Huai Kha Khaeng Wildlife Sanctuary, Thailand. *Journal of Zoology* 223, 281–298.

Reid, N. (1991) Coevolution of mistletoes and frugivorous birds. *Australian Journal of Ecology* 16, 457–469.

Rey, P.J. and Alcantara, J.M. (2000) Recruitment dynamics of a fleshy-fruited plant (*Olea europaea*): connecting patterns of seed dispersal to seedling establishment. *Journal of Ecology* 88, 622–633.

Rogers, H.M. and Hartemink, A.E. (2000) Soil seed bank and growth rates of an invasive species, *Piper aduncum,* in the lowlands of Papua New Guinea. *Journal of Tropical Ecology* 16, 243–251.

Safford, R.J. (1997) Nesting success of the Mauritius Fody *Foudia rubra* in relation to its use of exotic trees as nest sites. *Ibis* 139, 555–559.

Schreiber, A., Wirth, R., Riffel, M. and Van Rompaey, H. (1989) *Weasels, Civets, Mongooses and their Relatives: an Action Plan for the Conservation of Mustelids and Viverrids.* IUCN, Gland, 98 pp.

Severinghaus, L.L. (1999) Prayer animal release in Taiwan. *Biological Conservation* 89, 301–304.

Shilton, L.A., Altringham, J.D., Compton, S.G. and Whittaker, R.J. (1999) Old World fruit bats can be long distance seed dispersers through extended retention of viable seeds in the gut. *Proceedings of the Royal Society of London, Series B, Biological Sciences* 266, 219–223.

Slocum, M.G. and Horvitz, C.C. (2000) Seed arrival under different genera of trees in a neotropical pasture. *Plant Ecology* 149, 51–62.

Stocker, G.C. and Irvine, A.K. (1983) Seed dispersal by cassowaries (*Casuarius casuarius*) in North Queensland's rainforests. *Biotropica* 15, 170–176.

Tan, K.H., Zubaid, A. and Kunz, T.H. (1998) Food habits of *Cynopterus brachyotis* (Muller) (Chiroptera: Pteropodidae) in Peninsular Malaysia. *Journal of Tropical Ecology* 14, 299–307.

Turner, I.M. and Corlett, R.T. (1996) The conservation value of small, isolated fragments of lowland tropical rain forest. *Trends in Ecology and Evolution* 11, 330–333.

Turner, I.M., Wong, Y.K., Chew, P.T. and Ibrahim, A.B. (1997) The species richness of primary and old secondary tropical forest in Singapore. *Biodiversity and Conservation* 6, 537–543.

Tutin, C.E.G., White, L.J.T. and Mackanga-Missandzou, A. (1997) The use by rain forest mammals of natural forest fragments in an equatorial African savanna. *Conservation Biology* 11, 1190–1203.

Wells, D.R. (1999) *The Birds of the Thai–Malay Peninsula*, Vol. 1, *Non-passerines.* Academic Press, London.

Whitmore, T.C. (1999) Arguments on the forest frontier. *Biodiversity and Conservation* 8, 865–868.

Yasuda, M., Miura, S. and Hussein, N.A. (2000) Evidence for food hoarding behaviour in terrestrial rodents in Pasoh Forest Reserve, a Malaysian lowland rain forest. *Journal of Tropical Forest Science* 12, 164–173.

Zhuang, X. (1997) Rehabilitation and development of forest on degraded hills of Hong Kong. *Forest Ecology and Management* 99, 197–201.

Zhuang, X. and Corlett, R.T. (1996) The conservation status of Hong Kong's tree flora. *Chinese Biodiversity* 4 (suppl.), 36–43.

Zhuang, X. and Yau, M.L. (1999) The role of plantations in restoring degraded lands in Hong Kong. In: Wong, M.H., Wong, J.W.C. and Baker, A.J.M. (eds) *Remediation and Management of Degraded Lands.* Lewis Publishers, Boca Raton, Florida, pp. 201–208.

31 Behavioural and Ecological Considerations for Managing Bird Damage to Cultivated Fruit

Michael L. Avery
Wildlife Services, USDA-APHIS, National Wildlife Research Center, 2820 East University Avenue, Gainesville, FL 32641, USA

Introduction

Many bird species eat fruits and, likewise, many plant species are dependent on birds for the dispersal of seeds. Through cultivation and selective breeding, attributes of wild fruit have been changed to make fruit more palatable to humans. For example, cultivated species bear fruits that are often thinner-skinned, are more succulent, have fewer seeds and are easier to pick than non-cultivated species. These same changes, however, have also increased the attractiveness of fruit to avian consumers. Ecological relationships that have developed across evolutionary time between wild plants and frugivores become emphasized by the introduction of cultivated fruits that have been carefully bred, unknowingly and unintentionally, with bird-friendly traits.

Understanding depredations to fruit crops and developing effective means to reduce the impacts of depredating birds require an appreciation of the evolutionary and ecological bases for the birds' feeding behaviour. Unfortunately, research on avian depredation problems has seldom incorporated behavioural ecology. Rather, emphasis is often on development of methods that will mitigate a specific depredation problem in the short term, not on strategies that will effect durable, long-lasting solutions. The latter requires not only knowledge of immediate, local circumstances and management constraints (monetary, legal, societal), but also understanding behavioural and physiological adaptations of frugivorous birds. The array of fruit–frugivore interactions, particularly aspects such as optimal diet, flock dynamics and nutritional ecology, creates opportunities for wildlife managers and behavioural ecologists to collaborate in applying basic knowledge to important management issues.

On a national or regional scale, the economic impact of bird damage can be substantial. For example, a recent survey by the US Department of Agriculture produced an estimate of $41 million lost annually to wildlife damage in apples, grapes and blueberries (USDA, 1999). Most of the loss was attributable to birds. In addition, growers reported spending nearly $10 million annually to prevent wildlife damage, so the total economic impact currently exceeds $50 million annually for just three crops.

Whereas a loss of $41 million to birds is not trivial, it represents just 1% of the total annual apple, blueberry and grape production in the USA (USDA, 1999). If the losses were distributed evenly across all producers, there would be no bird problem. This is not the case, however. Bird damage is highly skewed, with most producers incurring little or no loss and a few producers having heavy losses (Hothem *et al.*, 1988; Johnson *et al.*, 1989). The percentage of the crop damaged by birds might be less than 5% overall, but this means little to a producer with losses of 20–25%. For extreme cases of bird damage, the most appropriate response might be to exclude birds from the crop with netting. This is also one of the costlier methods. Nevertheless, for certain commodities, including early-ripening blueberries and wine grapes, netting can be cost-effective (Fuller-Perrine and Tobin, 1993).

Damage by birds to cultivated fruit occurs worldwide. I shall not attempt a comprehensive review of all fruits affected and bird species involved, nor shall I attempt to describe the myriad of visual and aural bird deterrents that have been tested and are being marketed for control of bird damage to fruit crops (Avery *et al.*, 1988; Tobin *et al.*, 1988; Tipton *et al.*, 1989). Rather, I shall first discuss the use of non-lethal approaches to bird-damage management based on concepts of optimal feeding behaviour. This will be illustrated with the specific case of blueberry damage by cedar waxwings, *Bombycilla cedrorum*, in northern Florida, USA. Then, I shall consider population reduction as a possible component of integrated bird-management strategies and propose potentially useful areas for future research.

Feeding Behaviour and Ecology

Successful management of bird damage to cultivated fruits can be viewed within a conceptual framework largely derived from optimal foraging theory (Pyke *et al.*, 1977). Inherent in this framework is the idea of costs and benefits. To make a bird give up its preferred source of food, the fruit crop, the relative costs to the bird from feeding on the crop must increase to the point that alternative food sources become more profitable. The availability of alternative food sources is crucial. If the relative values of the alternative food and crop are similar, then the bird should readily abandon the crop for the alternative. If, however, the crop is substantially more valuable to birds than the alternative food, discouraging birds from feeding on the crop will be more difficult. For a variety of bird species, cultivated fruits provide nutritious, easily obtained food. With such great benefits there must be commensurately high potential costs to discourage birds from feeding on cultivated fruits.

Chemical Repellents and Crop Protection

Application of a chemical repellent to the crop is one non-lethal means of raising costs to depredating birds. There are two broad categories of avian repellents, primary and secondary, based upon their modes of action.

Primary repellents

Primary repellents are painful or irritating upon contact. The bird responds reflexively without having to learn an avoidance response. Many primary repellent compounds have relevance in interactions between birds and their natural prey (Clark, 1998). In the USA, one primary repellent compound, methyl anthranilate (MA), is the active ingredient in various formulated products marketed under the trade names of Bird Shield® and ReJeX-iT® (Avery *et al.*, 1996). These products are registered as bird repellents for use on cherries, blueberries and grapes.

MA is a naturally occurring compound used extensively in the food industry to give a grape or fruity flavour to sweets, chewing gum, soft drinks and other food items. Even though MA is safe and palatable to humans, birds do not like it. The repellence and mode of action of MA have been demonstrated experimentally through behavioural trials with nerve-cut and control birds (Mason *et al.*, 1989). Irritation and pain from MA are detected via the trigeminal nerve; all avian species tested so far perceive MA as an irritant, not as a taste repellent *per se*. The strong grapelike odour of MA is not aversive to birds (Clark, 1996). Birds must contact the MA-treated food in their mouths to experience the irritant effects. Rejection of

MA-treated food is contingent upon the options available to the bird. With no alternative food or with a relatively unattractive alternative food available, birds will continue to eat MA-treated food. If, however, MA-treated food is offered with untreated food of the same type, rejection of treated food occurs at much lower treatment levels (Avery *et al.*, 1995a). Because the irritation caused by MA may not be a very strong aversive stimulus, birds tend to return and resample the treated food. Thus, losses can accumulate even after the repellent is applied.

Secondary repellents

Secondary repellents are not immediately aversive but produce illness or discomfort after ingestion. Successful use of these compounds depends on the bird acquiring a learned avoidance response (Rogers, 1978). The bird must associate an adverse post-ingestional consequence with the appearance, smell or taste of the food, thereby learning to avoid it. For a bird, the consequences of ingesting a secondary repellent are potentially more dire than those of contacting a primary repellent. For this reason, an avoidance response produced by a secondary repellent is probably more robust than that produced by a primary repellent (Alcock, 1970; Rogers, 1974). A potential disadvantage to secondary repellents is that they are toxic and, for some compounds, there is not a great difference between a repellent dose and a lethal dose. The avoidance response is affected by various factors, such as the bird's prior experience with the food item, the strength of the post-ingestional discomfort and the availability of alternative food (Alcock, 1970).

Methiocarb (3,5-dimethyl-4-(methylthio) phenyl methylcarbamate) is an effective secondary repellent that has been used successfully in a variety of agricultural applications. As with other carbamates, its mode of action is via inhibition of acetylcholinesterase at synapses in the nervous system. Unlike many cholinesterase-inhibiting compounds, however, the effects of methiocarb are rapidly reversible, so disruption of the nervous system is only transitory. Applied properly, methiocarb is safe with regard to target and non-target species (Dolbeer *et al.*, 1994). Free-feeding birds acquire a repellent dose and stop feeding long before a lethal dose is ingested. The chemical has been tested extensively in many agricultural applications, including newly seeded and sprouted crops, ripening grain crops and soft fruits (Bailey and Smith, 1979; Conover, 1982; Porter, 1982). It was commercially sold as Mesurol® and formerly registered in the USA as a bird repellent on cherries, grapes and blueberries. In the USA, however, there is no current registration because of human health and safety concerns related to the cholinesterase-inhibiting action of the compound.

Another secondary avian repellent with potential utility in cultivated fruit is 9,10-anthraquinone. Birds that ingest food treated with the compound subsequently vomit and experience gastrointestinal discomfort (Avery *et al.*, 1997). Affected birds are not incapacitated, however, and there is no known effect on the nervous system. It is interesting that 9,10-anthraquinone has structural similarities to emodin, a powerful antifeedant found in fruits of *Rhamnus cathartica* (Sherburne, 1972; Avery *et al.*, 1997). In the USA, a formulated product called Flight Control® contains 50% anthraquinone and is currently registered for use as turf treatment to deter geese and other grazing birds (Blackwell *et al.*, 1999).

For fruit crops, test results with anthraquinone look promising. To examine frugivore responses to the repellent under controlled conditions, we conducted a feeding trial to expose cedar waxwings to technical-grade anthraquinone. We mist-netted 28 cedar waxwings in a blueberry field near Gainesville, Florida. Birds were caged individually and randomly assigned to four test groups of seven birds each. We quantified their consumption of a banana-mash diet (Denslow *et al.*, 1987) during 4 pretreatment days, and then assigned each group to receive one of four dietary concentrations of anthraquinone: 0, 500, 1000 or 10,000 p.p.m. in the banana-mash diet. As during pretreatment, birds were offered one cup containing the test diet for 3 h on four consecutive mornings. We videotaped one bird in the 10,000 p.p.m. group on the final pretreatment day and on each treatment day.

Consumption data were analysed in a repeated-measures analysis of covariance, with the birds' pretreatment consumption as the covariate. Over the 4-day treatment period, consumption varied ($F_{3, 24} = 162.21$; $P < 0.001$) among treatment groups (Fig. 31.1). Mean consumption by the 10,000 p.p.m. group ($\bar{x} = 4.11$ g, SE = 1.11) and by the 1000 p.p.m. group ($\bar{x} = 10.51$ g, SE = 1.62) was reduced ($P < 0.05$) relative to the 0 p.p.m. group ($\bar{x} = 19.41$ g, SE = 1.54). There was no interaction between day and test group ($F_{9, 71} = 1.82$; $P = 0.079$), as the birds responded very quickly to the adulterated diet. On the final pretreatment day, the videotaped bird averaged 5.5 bites (SE = 0.6) from the food cup during 15 feeding bouts and averaged 437 s (SE = 49) between bouts. When the anthraquinone treatment was added to the diet, the number of bites averaged 2.4 (SE = 0.6) during 12 feeding bouts. The mean interval between bouts remained the same (439 s), but there was considerably more variation (SE = 129 s). The range of inter-bout intervals during pretreatment was 134–697 s, compared with 71–1497 s during the initial treatment day. The greater variation in intervals between feeding bouts on the treatment day reflects uncertainty by the bird as it unexpectedly experienced post-ingestional discomfort after feeding where it had previously encountered only palatable food. On subsequent treatment days, the number of feeding bouts seen on videotape was 0, 2 and 0, respectively.

Limited field trials of the anthraquinone product, Flight Control®, in table and wine grapes in New Zealand and in cherries in the north-western USA have also produced encouraging results. Federal registration for these and other food-crop uses awaits further regulatory approval.

Increasing Costs to the Bird through Selective Crop Breeding

Reducing the quality of the crop as a food source for birds is potentially accomplished by altering attributes of the fruit through selective breeding. The objective of selective breeding is to increase the effort the bird has to expend to feed on the crop. Costs can be increased in different ways.

Food handling

Manipulation of the food item is an important commitment of time and effort (Pyke *et al.*, 1977). Intuitively, as the potential value of a food item increases, in terms of caloric value or nutrient content, so should the amount of time the bird is willing to spend manipulating and consuming it.

In northern Florida, the recent introduction of early-ripening varieties of blueberries, *Vaccinium* spp., has created an abundant food source in March, April and May, which overlaps the period of cedar waxwing occurrence in Florida (Nelms *et al.*, 1990). In addition, the

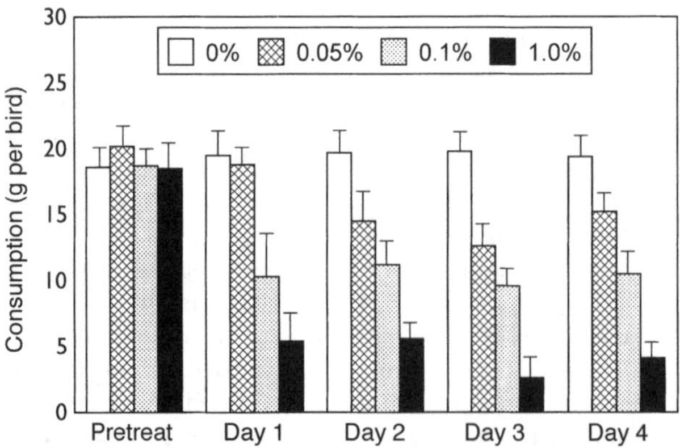

Fig. 31.1. Mean daily consumption by cedar waxwings ($n = 7$ per treatment) of banana mash treated with technical anthraquinone. Birds received one cup for 3 h on four consecutive mornings. Pretreatment values represent mean consumption of untreated banana mash during four daily 3 h trials.

availability of naturally occurring berries is particularly low in March in northern Florida (Skeate, 1987). The result is that blueberries (or other cultivated fruits) can represent an important food source for waxwings prior to their northward migration. We examined whether berry size and maturation date affect cedar waxwing damage to blueberries.

At the Horticultural Unit of the University of Florida (Gainesville, Florida, USA), we selected several cultivars with varying ripening dates and berry sizes. Following standard procedures (Nelms *et al.*, 1990), we evaluated berry loss from test bushes and assigned each blueberry cultivar to one of five damage categories: 0–20%, 21–40%, 41–60%, 61–80% and > 80% of fruits removed. We then determined the mean ripening date and berry size for each of the damage categories (Fig. 31.2). Results showed that varieties that produce small berries and that ripen early incur the greatest losses. The high level of loss among the earliest varieties is not surprising (Tobin *et al.*, 1991). For migrant and wintering birds at this time of year, there are few wild sources of fruit in northern Florida. Damage becomes less intense as wild fruits ripen in subsequent weeks.

The apparent berry-size selectivity demonstrated by birds in the field could be an artefact of early varieties being small-berried. To test directly whether cedar waxwings prefer small berries, we conducted a series of feeding trials with captive birds in which each bird was offered two berries that differed in size (Avery *et al.*, 1993). We recorded the fruit that was taken first and the time that the bird took to handle and swallow or drop the berry. We found that cedar waxwings do indeed prefer smaller-sized berries. The birds are almost perfect in their handling of the small berries; they drop very few and the time to swallow them is very short. In contrast, as berry size increases, the risk of dropping the fruit increases, as does the time it takes to swallow the fruit. The net result is that cedar waxwings do best, in terms of rate of energy gain, with smaller blueberries, even though larger fruits contain greater caloric rewards.

Breeding for larger fruit size might contribute to reduced berry loss, particularly if depredating birds have alternate food sources that are more efficiently handled and eaten. Alternatively, if waxwings persist in attempts to eat the larger fruit, they might actually damage more fruit by repeatedly plucking and dropping the big berries as they unsuccessfully attempt to consume the fruit.

A similar situation exists in Spanish olive orchards, where cultivated olives are twice as large as native olives. The larger size makes swallowing the fruit difficult or impossible for smaller frugivores, so bird species such as the blackcap, *Sylvia atricapilla*, opt to peck the fruit instead (Rey and Gutiérrez, 1996). If switching from swallowing to pecking is a widespread response by frugivorous species, then increased numbers of pecked fruit would probably negate any advantage of selectively breeding for larger fruit size.

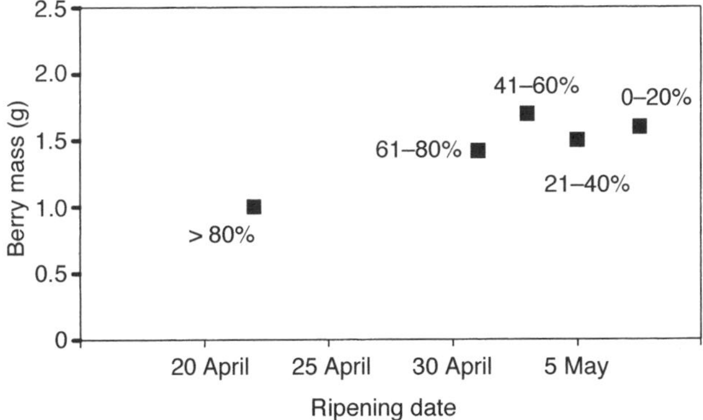

Fig. 31.2. Blueberry mass and ripening date relative to five categories of bird damage (0–20, 21–40, 41–60, 61–80, > 81% crop removed) in north-central Florida.

Digestive constraints

After ingestion, a food item still has to be digested and assimilated for the bird to benefit. Modification of the food item so that it is rendered more difficult to digest will reduce its attractiveness to depredating birds. Some frugivorous bird species, including major crop-depredating species, such as the American robin, *Turdus migratorius*, and the European starling, *Sturnus vulgaris*, possess a physiological constraint that makes it impossible for them to digest sucrose, a common constituent of many fruits (Martínez del Rio, 1990). These bird species lack the intestinal enzyme sucrase, which hydrolyses the 12-carbon sucrose molecule, which cannot be assimilated, into the six-carbon sugars glucose and fructose, which are assimilable. Means of exploiting this digestive constraint so that cultivated fruits will be less susceptible to bird damage include using sucrose as a spray on ripening fruit (Socci *et al.*, 1997) and manipulating the sugar composition of ripening fruit to produce elevated, bird-resistant levels of sucrose (Darnell *et al.*, 1994). Laboratory feeding trials have confirmed the potential usefulness of the latter approach to bird-damage reduction (Brugger *et al.*, 1993), but practical application remains to be tested. Furthermore, some frugivorous species when confronted with sucrose-rich fruit might consume more rather than less fruit. For example, cedar waxwings are able to digest sucrose, but relatively inefficiently, due to rapid gut passage rate (Martínez del Rio *et al.*, 1989). Consequently, to obtain the same energetic benefit, a cedar waxwing must consume more high-sucrose fruit than fruit that contains only glucose and fructose (Avery *et al.*, 1995b).

Alternative Sources of Food

The failure to appreciate the need for alternative feeding sites or food sources is a major impediment to initiating effective, ecologically based avian pest-management systems. Birds have to eat, and, as long as basic physiological needs are met, they will follow the path of least resistance. Application of virtually any method to protect a valuable crop from bird depredation will be more effective if alternative food is available. Novel though it might sound, the provision of such alternative food should be seriously considered and should be factored in as a cost of production by growers faced with persistent bird problems. For example, planting small-berried blueberry cultivars as alternative food sources for depredating cedar waxwings might fit well within an integrated bird-damage management plan. There is currently little interest on the part of blueberry producers in implementing this approach, however, and maintenance of the smaller-berry alternative bushes is a cost to producers that is not easy to bear. Establishment of feeding sites specifically for pest birds is probably not intuitively pleasing to most producers, and the effectiveness of this management approach needs to be experimentally tested.

Population Dynamics – Lethal Control of Problem Birds

Reducing the number of birds in the depredating population is seemingly a logical way to reduce crop damage. In the USA, lethal control has been facilitated by exempting some crop-depredating species, such as the red-winged blackbird, *Agelaius phoeniceus*, from protection under federal laws. Non-indigenous bird species in the USA, such as the European starling, are likewise not protected by federal laws. Thus, farmers can use lethal measures on some bird species as long as their actions are in accordance with local statutes and regulations. Most problem species, however, such as the American robin and cedar waxwing, are federally protected, so lethal control measures are only available under special permits, which are often difficult to obtain.

One of the major objections to lethal control is that it might not be effective in reducing damage. There is merit to this objection, as there are very few studies that clearly demonstrate an economic benefit to lethal control of depredating birds. Elliott (1964) reported that, during 1963, over 110,000 starlings were trapped and removed in eastern Washington, and that this effort 'practically eliminated' damage to the cherry crop in the Yakima Valley. During a 4-month period, Larsen and Mott

(1970) reported trapping over 3500 house finches, *Carpodacus mexicanus*, from a 0.4 ha blueberry planting near Portland, Oregon. There was no quantitative assessment of crop loss, but the grower felt that damage was 'considerably less' than in previous years. Palmer (1970) reported that bird damage at a California fig orchard dropped from 11% in 1967, when no control was applied, to 2.4% and 1.4% in 1968 and 1969, respectively, following the imposition of a trapping and poisoning programme. During the 2-year lethal-control effort, an estimated 53,000 house finches were removed. In Israel, mist-nets were used for 10 days in a 10 ha vineyard to remove about 2700 house sparrows, *Passer domesticus* (Plesser *et al.*, 1983). As a result, bird damage, which totalled $4500 in the previous year, was eliminated.

In Belgium more unconventional means of lethal control have been employed. Between 1972 and 1978, Ministry of Agriculture personnel used dynamite to destroy 22 starling roosts, killing an estimated 750,000 starlings (Tahon, 1980). The short-term impact of the programme provided some protection for the second half of the cherry season, although no crop-loss data are provided. In the long term, there was no measurable effect on the starling population from year to year, and the ultimate cost-effectiveness of the roost destruction programme was undetermined.

In the Rio Grande Valley of south Texas, great-tailed grackles, *Quiscalus mexicanus*, cause millions of dollars in damage to citrus crops by pecking holes in the skin of the fruit (Johnson *et al.*, 1989). Several non-lethal methods have been tried to reduce such damage, but none has proved practical or cost-effective (Tipton *et al.*, 1989). As a result, attention has shifted to lethal control. In particular, recent evaluations of improved trapping methods and baiting using the toxicant DRC-1339 (3-chloro-4-methylbenzamine hydrochloride) have proved promising for reducing local grackle populations in late summer, when damage problems are greatest (Glahn *et al.*, 2000). The baiting strategy involves putting the toxicant in water melon on elevated bait platforms, thereby providing the grackles with an irresistible food and water source during a very dry time of year. Field trials of the trapping and baiting methods not only showed their effectiveness for removing grackles, but also demonstrated that both control methods pose little danger to non-target species (Glahn *et al.*, 2000). Nevertheless, conclusive data on levels of damage reduction remain elusive.

Efficacy of lethal control

It is evident that large numbers of crop-depredating birds can be killed relatively quickly through judicious use of traps, poison and explosives, and, in fact, most lethal control programmes have focused more on documenting the numbers of dead birds than on quantifying the effects on crop damage. Although it seems reasonable that local, short-term crop protection can be achieved through reduction in depredating bird populations, quantification of the relationship between the number of birds killed and the associated reduction in crop damage is lacking. The prevailing attitude seems to be: 'A dead bird does not eat fruit.' A corollary is that the best damage-control strategy is to kill as many birds as possible.

It is hard to argue against the tenet that dead birds do not eat fruit, but it should be possible to devise a more scientifically based approach for lethal management. A lethal control programme ought to start with a clearly defined objective regarding the number of birds that are to be killed. I am aware of no instance in which an a priori analysis of the crop-damage situation has been conducted and a goal established for damage reduction through the removal of a specified target number of birds. In principle, at least, it should not be difficult to determine the amount of damage that can be accepted by a grower in a particular vineyard or orchard. Then, by applying appropriate techniques, the population could be reduced to the specified target level corresponding to the amount of expected damage.

I pose a simple hypothetical example to illustrate this point. Assume that a blueberry producer harbours 5000 house finches on a 25 ha farm. Further assume production of 2000 kg blueberries ha^{-1} and that one house finch can consume 1 kg of blueberries per growing season. Thus, if unchecked, the 5000-bird flock will consume 5000 kg, or 10%

of the expected blueberry production. The grower cannot accept this level of loss but is willing to accept a 2% loss, which corresponds to 1000 kg of blueberries. Under these conditions, the house-finch population should be reduced to 1000 birds, which means that 4000 birds have to be removed. A lethal control programme would then be devised to accomplish this objective and the progress of the programme monitored throughout the control period to evaluate its effectiveness in achieving the target mortality level.

Conservation and Management Implications

Non-native species

Introduced species play an important role in fruit-crop depredations. In the USA, the European starling is the major avian pest to crops of apples, blueberries and grapes (Avery *et al.*, 1994; USDA, 1999). A concerted, coordinated effort to reduce starling populations nationwide would not only provide relief from crop damage but would probably benefit native cavity-nesting birds, which must compete with the starling for limited nest sites (Weitzel, 1988). Another non-indigenous species, the monk parakeet, *Myiopsitta monachus*, is not a widespread problem in crops at this time, but damage by it to tropical fruit in south Florida is locally serious (Tillman *et al.*, 2001). Initiation of a population-reduction programme for monk parakeets before major depredation problems develop would be prudent.

Lethal control of native species is often difficult to justify, because such species possess beneficial attributes as part of the natural avifauna. Nevertheless, lethal control of native birds should be considered when sufficient information exists that economic losses are occurring and when reasonable target levels of mortality can be specified and achieved without jeopardizing non-target species.

Scale of management

The scale of the management effort is an important but neglected aspect of bird damage control. Depredation problems are at the field or orchard level – the scale at which we normally attempt to solve problems. The birds that are causing problems, however, can cover much more territory in a day. Because of their mobility, it might be most appropriate to design management strategies at the landscape level, taking into account movements and habitat use of the depredating species, as well as the temporal and spatial distribution of requisite resources. Much damage to fruit crops is done by large post-breeding flocks dominated by juvenile birds. If a broader temporal perspective to damage management is adopted, then perhaps measures could be initiated earlier in the year to limit reproduction by the target species so that fewer offspring are produced and the size of depredating flocks is reduced.

Improved methods

Tools at the field level will still be needed even if a landscape approach to management is adopted. To protect non-target species, nonlethal methods are preferred. Safer, more cost-efficient chemical repellents would help ease the depredation pressure experienced by growers and reduce the demand for lethal control. Repellents will not be the sole answer to bird depredations in fruit crops (Crabb, 1979), but they do represent an important component of an integrated programme.

Another non-lethal crop-protection method, the development of fruit cultivars with bird-resistant traits, has received little attention to date. One intriguing possibility is to develop fruit varieties that possess bird-resistant chemical defence compounds that are gradually deactivated as the fruit becomes ripe and ready to harvest. This is apparently the defence strategy that has developed in *R. cathartica* (Sherburne, 1972), and there is a precedent for it in crop breeding. In response to bird depredation, varieties of sorghum were developed that contained bird-resistant levels of tannins during early stages of grain development but which ripened into nutritional, palatable grain (Bullard and York, 1996). Successful application of this model to cultivated fruit would be a major breakthrough.

The usefulness of naturally occurring defensive compounds is largely unexplored. An example that merits further exploration centres on the damage done to pear buds by bullfinches, *Pyrrhula pyrrhula* (Greig-Smith *et al.*, 1983). These birds display preferences for certain pear cultivars over others, depending on the chemical constituents within the flower-buds (Greig-Smith, 1985). One of these constituents, cinnamamide, was ultimately identified as a potentially useful bird repellent (Crocker and Perry, 1990; Crocker *et al.*, 1993). Further collaboration between evolutionary ecologists and wildlife managers might reveal additional naturally occurring anti-herbivory compounds that could prove useful for crop protection.

Avian conservation and agriculture

There is increasing recognition that agricultural areas can be important to avian conservation (Johnson, 1997; Shahabuddin, 1997; Hobson, 1998), so a major challenge is to find ways for agriculture and birds to coexist amicably. Too often, attractive feeding opportunities in crop habitat are over-exploited by a few problem species, provoking responses by growers that are detrimental to all species using the resource. In certain situations, incentives from government and private sources might be provided for producers whose agricultural activity supports bird populations (Huner, 2000). Alternatively, perhaps coalitions of government and private conservation organizations can work with agricultural producers to establish and maintain alternative feeding sites for crop-depredating bird species. Whatever form it takes, increased communication between agricultural producers and avian conservationists is crucial so that the needs and expectations of all interests can be better understood and appreciated.

Acknowledgements

Paul Lyrene provided access to blueberry cultivars at the University of Florida Horticulture Unit, and Kelly Goocher was principally responsible for compiling field data on bird damage to blueberries. Arlene McGrane assisted with feeding trials and Kandy Roca supervised care of the captive birds. Doug Levey, Mark Tobin and Kirsten Silvius provided helpful comments on the manuscript.

References

Alcock, J. (1970) Punishment levels and the response of black-capped chickadees to three kinds of artificial seeds. *Animal Behaviour* 18, 592–599.

Avery, M.L., Daneke, D.E., Decker, D.G., Lefebvre, P.W., Matteson, R.E. and Nelms, C.O. (1988) Flight pen evaluation of eyespot balloons to protect citrus from bird depredations. In: Crabb, A.C. and Marsh, R.E. (eds) *Proceedings of Thirteenth Vertebrate Pest Conference.* University of California, Davis, California, pp. 277–280.

Avery, M.L., Goocher, K.J. and Cone, M.A. (1993) Handling efficiency and berry size preferences of cedar waxwings. *Wilson Bulletin* 105, 604–611.

Avery, M.L., Nelson, J.W. and Cone, M.A. (1994) Survey of bird damage to blueberries in North America. In: Curtis, P.D., Fargione, M.J. and Caslick, J.E. (eds) *Proceedings of Fifth Eastern Wildlife Damage Control Conference.* Cornell University, Ithaca, New York, pp. 105–110.

Avery, M.L., Decker, D.G., Humphrey, J.S., Aronov, E., Linscombe, S.D. and Way, M.O. (1995a) Methyl anthranilate as a rice seed treatment to deter birds. *Journal of Wildlife Management* 59, 50–56.

Avery, M.L., Decker, D.G., Humphrey, J.S., Hayes, A.A. and Laukert, C.C. (1995b) Color, size, and location of artificial fruits affect sucrose avoidance by cedar waxwings and European starlings. *Auk* 112, 436–444.

Avery, M.L., Primus, T.M., Defrancesco, J., Cummings, J.L., Decker, D.G., Humphrey, J.S., Davis, J.E. and Deacon, R. (1996) Field evaluation of methyl anthranilate for deterring birds eating blueberries. *Journal of Wildlife Management* 60, 929–934.

Avery, M.L., Humphrey, J.S. and Decker, D.G. (1997) Feeding deterrence of anthraquinone, anthracene, and anthrone to rice-eating birds. *Journal of Wildlife Management* 61, 1359–1365.

Bailey, P.T. and Smith, G. (1979) Methiocarb as a bird repellent on wine grapes. *Australian Journal of Experimental Agriculture and Animal Husbandry* 19, 247–250.

Blackwell, B.F., Seamans, T.W. and Dolbeer, R.A. (1999) Plant growth regulator (Stronghold™) enhances repellency of anthraquinone formulation (Flight Control™) to Canada geese. *Journal of Wildlife Management* 63, 1336–1343.

Brugger, K.E., Nol, P. and Phillips, C.I. (1993) Sucrose repellency to European starlings: will high-sucrose cultivars deter bird damage to fruit? *Ecological Applications* 3, 256–261.

Bullard, R.W. and York, J.O. (1996) Screening grain sorghums for bird tolerance and nutritional quality. *Crop Protection* 15, 159–165.

Clark, L. (1996) Trigeminal repellents do not promote conditioned odor avoidance in European starlings. *Wilson Bulletin* 108, 36–52.

Clark, L. (1998) Physiological, ecological, and evolutionary bases for the avoidance of chemical irritants by birds. *Current Ornithology* 14, 1–37.

Conover, M.R. (1982) Behavioral techniques to reduce bird damage to blueberries: methiocarb and a hawk-kite predator model. *Wildlife Society Bulletin* 10, 211–216.

Crabb, A.C. (1979) A report on efficacy of methiocarb as an avian repellent in figs and results of industry-wide bird damage assessments. In: Jackson, W.B. (ed.) *Proceedings Eighth Bird Control Seminar.* Bowling Green State University, Bowling Green, Ohio, pp. 25–30.

Crocker, D.R. and Perry, S.M. (1990) Plant chemistry and bird repellents. *Ibis* 132, 300–308.

Crocker, D.R., Perry, S.M., Wilson, M., Bishop, J.D. and Scanlon, C.B. (1993) Repellency of cinnamic acid derivatives to captive rock doves. *Journal of Wildlife Management* 57, 113–122.

Darnell, R.L., Cano-Medrano, R., Koch, K.E. and Avery, M.L. (1994) Differences in sucrose metabolism relative to accumulation of bird-deterrent sucrose levels in fruits of wild and domestic *Vaccinium* species. *Physiologia Plantarum* 92, 336–342.

Denslow, J.S., Levey, D.J., Moermond, T.C. and Wentworth, B.C. (1987) A synthetic diet for fruit-eating birds. *Wilson Bulletin* 99, 131–134.

Dolbeer, R.A., Avery, M.L. and Tobin, M.E. (1994) Assessment of field hazards to birds from methiocarb applications to fruit crops. *Pesticide Science* 40, 147–161.

Elliott, H.N. (1964) Starlings in the Pacific northwest. In: *Proceedings of Second Vertebrate Pest Conference.* University of California, Davis, California, pp. 29–39.

Fuller-Perrine, L.D. and Tobin, M.E. (1993) A method for applying and removing bird-exclusion netting in commercial vineyards. *Wildlife Society Bulletin* 21, 47–51.

Glahn, J.F., Palacios, J.D. and Garrison, M.V. (2000) Controlling great-tailed grackle damage to citrus in the lower Rio Grande valley, Texas. In: Parkhurst, J.A. (ed.) *Proceedings Eighth Eastern Wildlife Damage Control Conference,* Asheville, North Carolina.

Greig-Smith, P.W. (1985) The importance of flavour in determining the feeding preferences of bullfinches (*Pyrrhula pyrrhula*) for buds of two pear cultivars. *Journal of Applied Ecology* 22, 29–37.

Greig-Smith, P.W., Wilson, M.F., Blunden, C.A. and Wilson, G.M. (1983) Bud-eating by bullfinches, *Pyrrhula pyrrhula* in relation to the chemical constituents of two pear cultivars. *Annals of Applied Biology* 103, 335–343.

Hobson, K.A. (1998) Conservation in a cup: the neotropical migrant coffee connection. *Picoides* 11, 24–25.

Hothem, R.L., DeHaven, R.W. and Fairaizl, S.D. (1988) *Bird Damage to Sunflower in North Dakota, South Dakota, and Minnesota, 1979–1981.* Fish and Wildlife Technical Report 15, United States Department of the Interior, Fish and Wildlife Service, Washington, DC.

Huner, J.V. (2000) Crawfish and water birds. *American Scientist* 88, 301–303.

Johnson, A.R. (1997) Long-term studies and conservation of greater flamingos in the Camargue and Mediterranean. *Colonial Waterbirds* 20, 306–315.

Johnson, D.B., Guthery, F.S. and Koerth, N.E. (1989) Grackle damage to grapefruit in the lower Rio Grande valley. *Wildlife Society Bulletin* 17, 46–50.

Larsen, K.H. and Mott, D.F. (1970) House finch removal from a western Oregon blueberry planting. *Murrelet* 51, 15–16.

Martínez del Rio, C. (1990) Dietary, phylogenetic, and ecological correlates of intestinal sucrase and maltase activity in birds. *Physiological Zoology* 63, 987–1011.

Martínez del Rio, C., Karasov, W.H. and Levey, D.J. (1989) Physiological basis and ecological consequences of sugar preferences in cedar waxwings. *Auk* 106, 64–71.

Mason, J.R., Adams, M.A. and Clark, L. (1989) Anthranilate repellency to starlings: chemical correlates and sensory perception. *Journal of Wildlife Management* 53, 55–64.

Nelms, C.O., Avery, M.L. and Decker, D.G. (1990) Assessment of bird damage to early-ripening blueberries in Florida. In: Davis, L.R. and Marsh, R.E. (eds) *Proceedings of the Fourteenth Vertebrate Pest Conference.* University of California, Davis, California, pp. 302–306.

Palmer, T.K. (1970) House finch (linnet) control in California. In: Dana, R.H. (ed.) *Proceedings of the Fourth Vertebrate Pest Conference.* University of California, Davis, California, pp. 173–178.

Plesser, H., Omasi, S. and Yom-Tov, Y. (1983) Mist nets as a means of eliminating bird damage to vineyards. *Crop Protection* 2, 503–506.

Porter, R.E.R. (1982) Comparison of exclosure and methiocarb for protecting sweet cherries from birds, and the effect of washing on residues. *New Zealand Journal of Experimental Agriculture* 10, 413–418.

Pyke, G.H., Pulliam, H.R. and Charnov, E.L. (1977) Optimal foraging: a selective review of theory and tests. *Quarterly Review of Biology* 52, 137–154.

Rey, P.J. and Gutiérrez, J.E. (1996) Pecking of olives by frugivorous birds: a shift in feeding behaviour to overcome gape limitation. *Journal of Avian Biology* 27, 327–333.

Rogers, J.G., Jr (1974) Responses of caged red-winged blackbirds to two types of repellents. *Journal of Wildlife Management* 38, 418–423.

Rogers, J.G., Jr (1978) Some characteristics of conditioned aversion in red-winged blackbirds. *Auk* 95, 362–369.

Shahabuddin, G. (1997) Preliminary observations on the role of coffee plantations as avifaunal refuges in the Palni Hills of the Western Ghats. *Journal of the Bombay Natural History Society* 94, 10–21.

Sherburne, J.A. (1972) Effects of seasonal changes in the abundance and chemistry of fleshy fruits of northeastern woody shrubs on patterns of exploitation by frugivorous birds. PhD thesis, Cornell University, Ithaca, New York.

Skeate, S.T. (1987) Interactions between birds and fruits in a northern Florida hammock community. *Ecology* 68, 297–309.

Socci, A.M., Pritts, M.P. and Kelly, M.J. (1997) Potential use of sucrose as a feeding deterrent for frugivorous birds. *HortTechnology* 7, 250–253.

Tahon, J. (1980) Attempts to control starlings at roosts using explosives. In: Wright, E.N. (ed.) *Bird Problems in Agriculture.* British Crop Protection Council, Croydon, pp. 56–68.

Tillman, E.A., Van Doorn, A. and Avery, M.L. (2001) Bird damage to tropical fruit in south Florida. In: Brittingham, M.C. (eds) *Proceedings of the Ninth Eastern Wildlife Damage Control Conference,* University Park, Pennsylvania.

Tipton, A.R., Rappole, J.H., Kane, A.H., Flores, R.H., Johnson, D.B., Hobbs, J., Schulz, P., Beasom, S.L. and Palacios, J. (1989) Use of monofilament line, reflective tape, beach-balls, and pyrotechnics for controlling grackle damage to citrus. In: *Ninth Great Plains Wildlife Damage Control Workshop Proceedings.* USDA Forest Service General Technical Report RM-171, Fort Collins, Colorado, pp. 126–128.

Tobin, M.E., Woronecki, P.P., Dolbeer, R.A. and Bruggers, R.L. (1988) Reflecting tape fails to protect ripening blueberries from bird damage. *Wildlife Society Bulletin* 16, 300–303.

Tobin, M.E., Dolbeer, R.A., Webster, C.M. and Seamans, T.W. (1991) Cultivar differences in bird damage to cherries. *Wildlife Society Bulletin* 19, 190–194.

USDA (1999) *41.0 Million Dollars of Fruit Lost to Wildlife Damage.* Report SpCr3(5-99), US Department of Agriculture, Agricultural Statistics Board, National Agricultural Statistics Service, Washington, DC.

Weitzel, N.H. (1988) Nest-site competition between the European starling and native breeding birds in northwestern Nevada. *Condor* 90, 515–517.

32 Harvest and Management of Forest Fruits by Humans: Implications for Fruit–Frugivore Interactions

Susan M. Moegenburg*
Department of Zoology, University of Florida, Gainesville, FL 32611-8525, USA

Introduction

Ecological theory and empirical studies suggest that frugivores can influence the abundance and distribution of fruiting plants (Janzen, 1970; Howe and Smallwood, 1982; Fragoso, 1997; see also Terborgh *et al.*, this volume). Somewhat independently, anthropological and archaeological studies have documented the use, management and domestication of fruits and fruit-producing trees by humans, which has apparently altered the abundance and distribution of fruiting plants in forests worldwide (Posey, 1985; Denevan, 1992; Balée, 1993, 1994). To date, plant composition and structure in tropical forests has typically been attributed to either ecological or anthropological processes. Few studies have considered the interaction between humans and non-human frugivores (hereafter 'frugivores') in determining fruiting-plant distributions or the effects of fruit management and extraction by humans on forest frugivores.

For millennia people have been altering the abundance and distribution of fruits in forests by harvesting fruits from naturally occurring trees and by enriching forests with fruit-producing species. Archaeological evidence shows, for example, that people have utilized fruits from Amazonian and Malaysian palm-trees for thousands of years, and botanical evidence suggests that many tropical forest stands are relicts of human use and management for fruit and other resources (Posey, 1985; Mabberley, 1992; Balée, 1994; Roosevelt *et al.*, 1996; McCann, 1999). Most, if not all, of the fruit species harvested by people are also consumed by frugivores. Are these frugivores affected when their food resource is increased through management or decreased through harvest? It is difficult to know the ecological and evolutionary consequences of ancient fruit harvest and forest management for frugivory and seed dispersal; however, clues can be gleaned from current practices, many of which involve the same species used historically by forest-dwelling people. Examination of contemporary forest enrichment and fruit harvest may provide insight into the importance of both past and present human–fruit–frugivore interactions. Furthermore, it may aid in the

*Present address: Smithsonian Migratory Bird Center, National Zoological Park, Washington, DC 20008, USA.

development of sustainable management and harvest practices.

In this chapter I explore the historical and contemporary effects of fruit harvest and forest management for fruit-producing trees on frugivores and fruit–frugivore interactions. To do so, I combine a general overview of harvest and management activities with data from my work on *Euterpe oleracea*, a palm that produces one of the most intensively harvested and managed fruits in the eastern Amazon region. The *E. oleracea* system typifies the complexity of human–fruit–frugivore interactions: on the one hand, management increases fruit availability to frugivores; on the other, fruit harvest reduces availability. Furthermore, harvest and management are ancient practices that have been partially responsible for the creation of high-density, monodominant *E. oleracea* stands (Anderson *et al.*, 1995; Hiraoka, 1995). Although fruit harvest and forest management occur worldwide in many types of ecosystems, most of my examples and my case-study focus on fruit and tree species in neotropical forests. My goals are to demonstrate that: (i) humans substantially alter the abundance of fruit in forests through harvest and management; (ii) harvest and management affect frugivore behaviour and fruit–frugivore interactions; and (iii) fruit harvest and forest management have occurred for a very long time across very large areas, potentially influencing the evolution of fruit–frugivore interactions.

Fruit Harvest

Probably the most ancient human activity affecting fruit abundance in forests is simply the removal of fruit from trees. Archaeological data from Amazonian Brazil show that people utilized fruit from palm-trees as early as 10,000 years BP (Roosevelt *et al.*, 1996). During the past several centuries, European travellers to neotropical regions recorded long lists of native fruits extracted by people and used for food, beverages, medicines, dyes and other purposes (Gomez-Pompa and Kaus, 1990). Harvest of a plethora of forest fruits continues today in the region of Iquitos, Peru, where 120 regularly consumed species of fruits are harvested exclusively from forest trees (Vasquez and Gentry, 1989). In addition to involving many species, harvest also occurs over vast areas of forest. In the Colombian Amazon, for example, indigenous reserves encompass 18 million ha (Bunyard, 1989). In Brazil, 22,000 km^2 comprise a series of 'extractive reserves' – areas designated for long-term sustainable harvest of forest resources by residents (Fearnside, 1989). A primary source of income in extractive reserves is derived from sales of fruit, such as Brazil nuts (*Bertholettia excelsa, Lecythidaceae*) (Kainer *et al.*, 1998).

Although data on the amount of fruit extracted from tropical forests are rare, a few studies have demonstrated that harvests can be very high. For example, *c.* 50,000 tons of Brazil nuts are collected from wild trees in the Brazilian Amazon each year (Panayotou and Ashton, 1992). Additional harvests of Brazil nuts, which are eaten by agoutis (*Dasyprocta* spp.), other rodents (e.g. *Proechimys*) and macaws (*Ara* spp.) (Peres and Baider, 1997; Peres *et al.*, 1997), occur in other Amazonian and Guianan regions. Intensive harvest of forest fruits also occurs in palaeotropical regions. In Borneo, for example, extractions of edible illipe nuts (*Dipterocarpaceae*), which are eaten by bearded pigs (*Sus barbatus*) and rodents, range between 10 tons in non-masting years and > 28,000 tons in masting years (Panayotou and Ashton, 1992). For these high-demand species, the amount of fruit harvested often represents a significant portion of forest fruit production (Vasquez and Gentry, 1989). In one region of India, for instance, the annual extraction of 1251 tons of amla (*Phyllanthus emblica, Euphorbiaceae*) fruit, which are consumed by ungulates, represents 13% of the regional fruit production of this species (Shankar *et al.*, 1996).

While large volumes of some species of fruits and seeds are harvested and sold, small amounts of many others are collected and used primarily for subsistence consumption. Vasquez and Gentry (1989) found that 136 of the 193 (70%) species consumed near Iquitos, Peru, were not sold but were either utilized directly by the harvester or traded. In regions that are distant from markets, people tend to rely more heavily on forest resources, and the diversity of extracted species tends to be high. Ka'apor people of the Brazilian Amazon, for example, recognize 179 species of edible non-domesticated plants (Balée, 1994).

Palms are among the plant families most heavily used by people throughout tropical regions (Boom, 1988). Rural people in Brazil, Peru and South-East Asia all use fruits from > 12 species of palms (Balick, 1988; Davis, 1988; Mejia, 1988). In the region near Iquitos, Peru, at least 23 species of palm fruits are consumed and 14 are sold in markets (Table 32.1; Vasquez and Gentry, 1989). Not only is the diversity of marketed palm species high, but so is the volume moved through markets. An estimated 15 tons of *Mauritia flexuosa* palm fruits are sold daily in the Iquitos, Peru, market (Padoch, 1988) and up to 40 tons of *E. oleracea* fruits are sold daily in the Belém, Brazil, market (Clay, 1997). In Florida, USA, > 680 tons of fruits are harvested annually from saw palmetto (*Serenoa repens*) (Bennett and Kicklin, 1998). All these wild-harvested palm fruits are also consumed by a diverse array of frugivorous vertebrates (Table 32.1).

Impacts of Harvest

Plant populations

Because fruits and seeds play multiple roles in forests, their extraction may have multiple effects on populations, communities and ecosystems (Nepstad *et al.*, 1992; Murali *et al.*, 1996; Witkowski and Lamont, 1996). At the population and community levels, harvest of fruits and seeds may affect the survival and fitness of adult trees and reduce the food supply to fruit- and seed-eating animals. At the ecosystem level, harvest may affect the cycling of nutrients and energy.

The most direct effect of fruit and seed extraction is on individuals of species killed during the harvesting process. While fruit and seed harvest is typically not destructive to parent plants, trees of some species are felled to harvest their fruits. Clearly, such harvest can have an impact on tree populations. For example, palm populations from which trees are harvested for construction materials (*Iriartia deltoidea*), palm heart (*E. oleracea* and *Euterpe edulis*) and fruit (*Astrocaryum murumuru*) tend to have lower densities of adult trees than do non-harvested populations (Fig. 32.1; Kahn, 1988; Anderson *et al.*, 1991; Clark *et al.*, 1995; Pollack *et al.*, 1995; Clay, 1997; Galetti and Fernandez, 1998). Similar destructive harvesting has been noted for *Jessenia bataua* (*Palmae*) and *Parahancornia peruviana* (*Apocynaceae*) (Vasquez and Gentry, 1989). One of the most well-known examples of destructive fruit harvest is the dioecious palm *M. flexuosa*, which people avoid climbing because the stem contains skin-irritating silica (Padoch, 1988). Intensive cutting of female trees for > 50 years has led to regional declines of female trees in *M. flexuosa* populations and local extinction near Iquitos, Peru (Vasquez and Gentry, 1989). In one Peruvian reserve, destructive harvest has reduced the area covered by high-density stands of *M. flexuosa* from 37% to 15% (Achung *et al.*, 1995).

In most species, harvest is not destructive because fruits and seeds are either gathered from the floor after falling from the parent crown (e.g. Brazil nuts) or from the crown by harvesters who climb the trees (e.g. *E. oleracea*, *P. emblica*). This type of harvest may, however, have population-level effects. In particular, fruit and seed harvest may reduce regeneration in the harvested populations, because sexual regeneration depends upon the germination of viable seeds (Peters, 1990). Numerous studies have shown that, because the majority of seeds produced by tropical tree species are often destroyed by predators or pathogens either before or after germinating, a very large number of seeds is needed for a parent tree to replace itself (Janzen, 1970, 1971). When people remove a portion of the seed crop, they may therefore be reducing the ability of trees to produce offspring. This may be the case in some Brazil nut (*B. excelsa*) and amla (*P. emblica*) populations in Brazil and India, respectively, in which seedlings and advanced regeneration are rare or missing (Nepstad *et al.*, 1992; Murali *et al.*, 1996; Peres and Baider, 1997).

Several studies have used demographic data and matrix modelling to assess reduced regeneration in areas where fruits and seeds are harvested (Peters, 1990; Bernal, 1998). The goal of these studies is to quantify the percentage of the seed crop necessary to maintain stable populations. The remaining percentage may, according to these studies, be harvested.

Table 32.1. Selected tree species that produce fruits and/or seeds harvested by people, including geographical ranges, the frugivores that consume the fruits or seeds and the amounts harvested.

Fruit species	Range	Frugivores	Human use	Sources
Mauritia flexuosa (*Palmae*)	Pan-Amazonian	*Tapirus terrestris*, *Tayassu pecari*, *Mazama americana*, *Agouti paca*, *Dasyprocta fulginosa*, *Eira barbara*, *Mazama gouazoubira*, *Tayassu tajaca*, *Proechimys* spp., *Ara* spp.	6.1 tons ha^{-1} $year^{-1}$	Anderson *et al.*, 1991; Peters, 1992; Achung *et al.*, 1995; Fragoso, 1997
Euterpe edulis (*Palmae*)	Brazilian Atlantic Forest	30 bird and 13 mammal species	> 200 stems cut ha^{-1}	Galetti and Aleixo, 1998
Euterpe oleracea (*Palmae*)	Amazon River estuary	> 40 bird species and six mammal species	13,000 kg ha^{-1} $year^{-1}$; high-density stands cover > 10,000 km^2	Anderson *et al.*, 1995; Muñiz-Miret *et al.*, 1996; Moegenburg, 2000
Serenoa repens (*Palmae*)	Florida, USA	*Odocoileus virginianus*, *Ursus americanus*	680 tons annually	Bennett and Hicklin, 1998
Attalea speciosa (*Palmae*)	Eastern Amazonia	*Agouti paca*, *Dasyprocta punctata*, *Coendu prehensilis*, *Proechimys* spp., *Mesomys hispidus*, *T. tajaca* and *T. peccari*	1.5 ± 0.1 tons ha^{-1} $year^{-1}$; high-density stands cover > 200,000 km^2	Anderson *et al.*, 1991
Jessenia bataua (*Palmae*)	Western Amazonia	See *M. flexuosa* above	3.5 tons ha^{-1} $year^{-1}$	Peters, 1990
Brosimum alicastrum (*Moraceae*)	Mexico	Various rodents	6 tons ha^{-1} $year^{-1}$	Ricker *et al.*, 1999
Phyllanthus emblica (*Euphorbiaceae*)	India	Spotted deer, mouse deer, sambur	1895 kg ha^{-1} $year^{-1}$	Shankar *et al.*, 1996

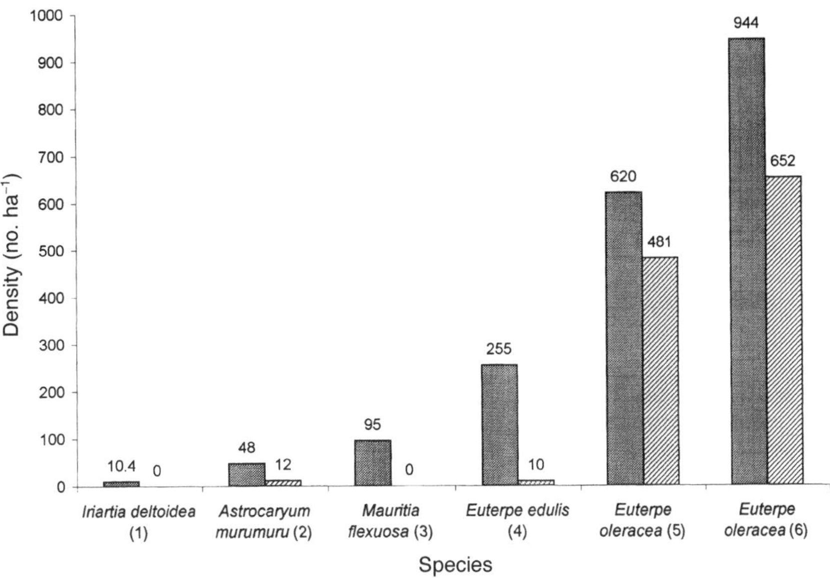

Fig. 32.1. Densities of palm-trees in non-harvested (solid bars) and harvested (striped bars) populations. Actual densities are indicated above bars and sources are noted in parentheses. Lower densities probably result from harvest. Sources: (1) Clark *et al.* (1995); (2) Anderson *et al.* (1995); (3) Kahn (1988); (4) Galetti and Aleixo (1998); (5) Pollack *et al.* (1995); (6) Moegenburg (2000).

Applying this method to several species in Peru and Mexico, Peters (1990) found that 80% of *Grias peruviana* (*Lecythidaceae*) seeds and 98% of *Brosimum alicastrum* (*Moraceae*) seeds could be harvested without leading to population declines. In a similar study of the vegetable-ivory palm, *Phytelephas seemannii*, in Colombia, Bernal (1998) projected that 86% seed removal could be sustained without reducing regeneration. These models are simplistic, however, because they fail to account for seed loss to predators and pathogens, which is often very high (Janzen, 1971). Furthermore, the models focus solely on the percentage of seeds required for plant regeneration; they do not consider that seeds and fruits may be necessary for the maintenance of frugivore populations.

Frugivores

The fleshy pulp and nutritious seeds harvested from many tropical tree species provide food for a substantial portion of the vertebrate community. In neotropical forests up to 40% of bird species and up to 64% of mammal species include fruits and seeds in their diets (Janson and Emmons, 1990; Karr *et al.*, 1990). In floodplain forests of the Amazon River, fruits of 38 genera and 28 families are an important dietary component for at least 200 species of fish, which swim under the canopy when the forest is flooded and subsist largely on fruits and seeds (Goulding, 1980). In fact, fruit abundance may regulate frugivore populations (Foster, 1982; Snyder *et al.*, 1987; Wright *et al.*, 1999) and determine frugivore distributions across space and time (Levey, 1988; Loiselle and Blake, 1991; Rey, 1995).

The apparent link between fruit and frugivore abundances implies that frugivore ecology and behaviour will be affected when humans harvest fruit. However, it is often assumed that fruit harvest has no measurable impacts on forest vertebrates (e.g. Peters, 1992). The disparity of opinions and lack of data about the effects of fruit harvest on fruit-eating animals have hindered the development of guidelines for fruit management and harvest at sustainable levels. To clarify this issue, I studied the harvest of fruit from the palm *E. oleracea* in the estuary region of the Amazon River. Across approximately 10,000 km^2 of flood-plain forests in the Amazon River estuary, *E. oleracea* forms monodominant stands, some of which are the result of historical or contemporary management (Kahn, 1988; Hiraoka, 1995). Individual trees produce up to 25 slender stems, which reach heights of 30 m. Reproductive stems produce infructescences bearing several thousand purple-black globose drupes, 1 cm in diameter (Moegenburg, 2000).

The ubiquity of *E. oleracea* across inhabited estuarine regions suggests its importance in the diet, culture and economy of the people in estuarine Amazonia (Anderson *et al.*, 1995; Hiraoka, 1995; Muñiz-Mirit *et al.*, 1996). *Euterpe oleracea* yields two of the region's most profitable non-timber forest products: heart of palm and fruit. People harvest *E. oleracea* fruit by climbing stems and removing infructescences. Fruits are processed into a drink consumed daily by many thousand people throughout the region. Where people have access to markets, they often manage forests to increase *E. oleracea* production so that fruit not consumed in the household can be sold. Commercial sales involve large volumes of fruit; harvest can exceed 13,000 kg ha^{-1} year^{-1} (Muñiz-Miret *et al.*, 1996).

To evaluate the effects of *E. oleracea* fruit harvest on frugivores, I conducted an experimental harvest of *E. oleracea* fruit in 1.8 ha plots in high-density palm forest (Moegenburg, 2000). In the study site (Caxiuanã National Forest, Pará, Brazil), *E. oleracea* forms high-density stands (> 300 trees ha^{-1}) and produces fruits that are consumed by at least 41 species of birds, five species of mammals and three species of fish (Moegenburg, 2000).

During 2 months of the *E. oleracea* fruiting season, I harvested 43–75% of ripe fruit (typical harvest levels near markets) from some areas, left other areas unharvested and compared frugivore visits to harvested and unharvested areas (Moegenburg, 2000). The frugivore community was clearly sensitive to the reduced availability of fruit resulting from harvest. Specifically, *E. oleracea* fruit removal reduced frugivorous bird biomass by 29% (Fig. 32.2a). This level of removal also reduced frugivorous bird-species richness, abundance and visit lengths by 25, 29 and 68%, respectively, and frugivorous mammal-species richness by 58% (Moegenburg, 2000). In addition, the abundance of a fruit-eating fish species (*Pyrrhulina* sp., Lebiasinidae) was reduced by 57% (Fig. 32.2b).

Frugivores that stopped visiting the sites of *E. oleracea* fruit removal moved to areas of higher fruit availability to forage; their populations were probably not affected by the relatively small scale of this experiment. However, *E. oleracea* fruits are harvested over a large portion of the Amazon River estuary, sometimes at very intensive levels (Muñiz-Miret *et al.*, 1996). Intensive harvest over such a large area may substantially reduce fruit availability to frugivores on the scale of their home ranges, having a potential impact on their populations. Indeed, fruit abundance on a large scale has been linked to adult survival and reproductive success in highly frugivorous animals (Foster, 1982; Adler, 1998; Wright *et al.*, 1999). In years of low community-wide fruit production, studies have documented high adult mortality and low reproductive success in several species of frugivorous mammals (Foster, 1982; Wright *et al.*, 1999). Moreover, Adler (1998) demonstrated that supplementing rodent populations with fruits and seeds increased population-level reproductive success. Similar responses in avian frugivores are suspected to occur in several psittacines, including the Puerto Rican parrot (*Amazona vittata*) (Snyder *et al.*, 1987) and Lear's macaw (*Anodorhynchus leari*) Abramson *et al.*, 1995).

In some palm species (e.g. *E. edulis*, *Euterpe precatoria*, *I. deltoidea*), reduction of fruit availability to frugivores occurs not through harvest of fruit but rather through harvest of stems (Galetti and Fernandez, 1998). How does reduction of fruit via stem harvest affect frugivores? Galetti and Aleixo (1998) compared the frugivorous bird communities in two forests in southern Brazil, one containing ten *E. edulis* stems ha^{-1} and from which *E. edulis* had been harvested 5–10 years previously, and a second containing 225 *E. edulis* stems ha^{-1}, from which none had been harvested. The authors found no difference in the abundance of frugivorous birds in harvested and unharvested forest and argue that, because *E. edulis* produces fruit concurrently with many other plant species, the 96% reduction in palm density in harvested forest represented a small overall

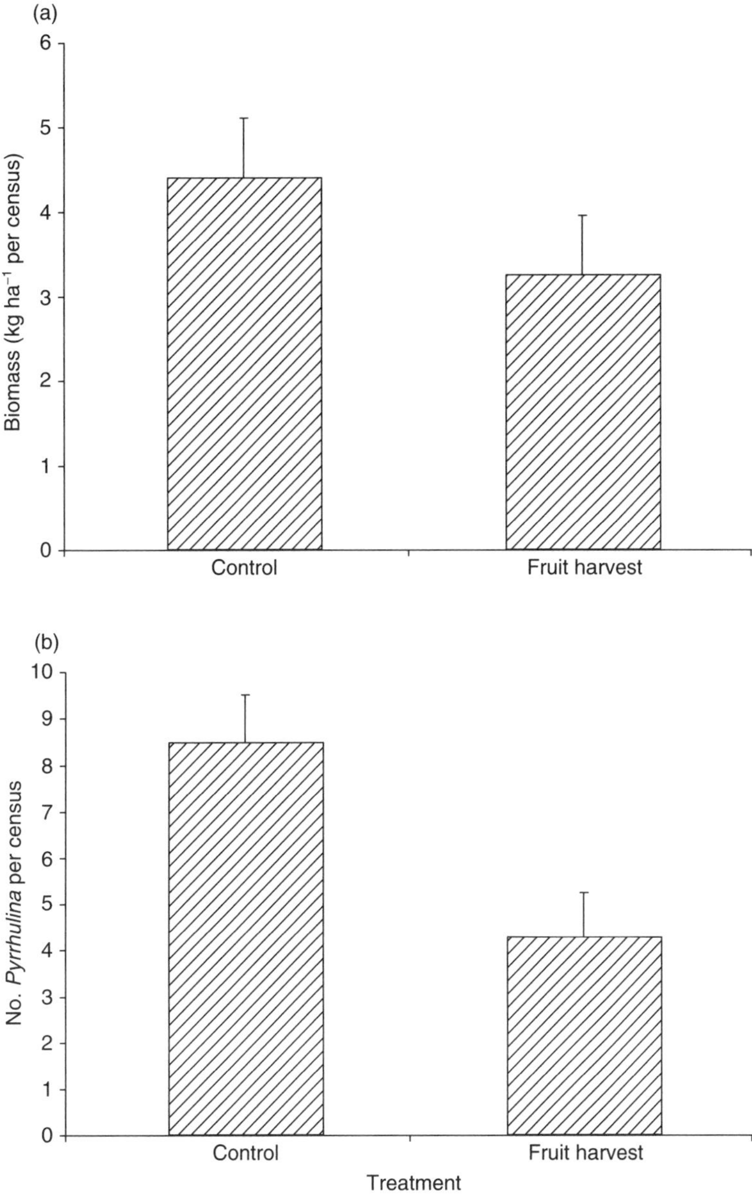

Fig. 32.2. Frugivorous bird biomass (a) and number of individuals of the fruit-eating fish *Pyrrhulina* sp. (b) in control plots and plots from which 43–75% of ripe *Euterpe oleracea* fruit had been harvested. Both biomass and number of *Pyrrhulina* were significantly lower ($P < 0.05$) in harvest plots than in control plots.

loss of fruit available to frugivores. Their conclusion suggests that effects of fruit harvest on frugivores depend on the identity of the harvested species. In addition, the effects may depend on the spatial and temporal scale at which they are measured. Galetti and Aleixo (1998) censused birds across large areas, which probably encompassed the home ranges of many bird species. Their results suggest that small-scale reduction of frugivore abundance due to fruit harvest, as shown by Moegenburg (2000), may not be indicative of frugivore responses at larger spatial scales. This idea is not supported, however, by the findings of Rey (1995), who showed that frugivore abundance was correlated with both local and regional fruit abundance in olive orchards. While these studies provide a good start, further research is required to understand the effects of fruit harvest by people on animal behaviour and ecology.

If fruit harvest reduces the abundance of frugivores, other forest processes, such as seed dispersal, may be disrupted (Strahl and Grahal, 1991). In many tropical forests the majority (50–80%) of tree species produce fruits whose seeds are dispersed by vertebrates (Gentry, 1982). If frugivore visits to a forest decline due to reduced fruit availability, then sympatric vertebrate-dispersed plants may suffer reduced seed dispersal. Reduced dispersal, in turn, could lead to reduced tree-species diversity, because fewer seeds will arrive in predator- and pathogen-free sites (see Terborgh *et al.*, this volume).

Nutrient cycles

In addition to affecting plant and frugivore individuals, populations and communities, fruit harvest by people may also affect ecosystem processes, such as nutrient cycling. Some fruits that fall to the forest floor are eaten by terrestrial herbivores; others decompose, releasing nutrients. In central Amazonia, fruits comprise only 7% of the total litter fall (Franken *et al.*, 1979), but are responsible for 14% of the total nitrogen (N) and 25% of the total phosphorus (P) recycled in floodplain forest (Franken, 1979). This amounts to 10.31 kg ha^{-1} $year^{-1}$ N and 0.35 kg ha^{-1} $year^{-1}$ P (Franken, 1979).

When people extract fruits and other materials, they may break the cycle of nutrient exchange between growing and decomposing material. Furthermore, extraction may permanently remove limiting nutrients, such as N and P, from the system, which can affect overall productivity (Vitousek, 1984). In a study of commercial bloom picking from *Banksia* trees in Australia, for example, Witkowski and Lamont (1996) found that trees from which blooms were harvested had 30% lower N and P content than did non-harvested trees. Moreover, the soil surrounding harvested populations contained 3103 and 152 g ha^{-1} less N and P, respectively, than did soil from unharvested populations.

Amazonian sites, where soils and rivers are nutrient-poor and the majority of nutrients are derived from decomposing organic material (Vitousek, 1984), may also lose nutrients when fruits and seeds are extracted. In my flooded-forest study site, annual production of *E. oleracea* fruit ranges from 2456 to 5236 kg ha^{-1} $year^{-1}$ (Moegenburg, 2000). When frugivores, such as parrots, feed on *E. oleracea* trees, they dislodge many whole fruits and drop many partially eaten fruits. These fruits fall in the water and decompose, losing nutrients rapidly (Fig. 32.3). In a fruit-decomposition study, I found that, if all *E. oleracea* fruits fell in the water and decomposed, then 11.22–23.93 kg N ha^{-1} and 0.76–1.62 kg P ha^{-1} would leach from them (Moegenburg, 2000). Of course, not all fruits fall into the water; an unknown proportion is consumed by vertebrates. If vertebrates removed 50% of the total ripe fruit and the remaining 50% were left to decompose, N and P leaching would total 1.61–3.44 and 0.11–0.23 kg ha^{-1} $year^{-1}$, respectively. Because P can limit productivity in Amazonian sites (Vitousek, 1984), P reduction caused by fruit harvest may indeed affect forest productivity.

In summary, the large volumes of fruit harvested from forests by people are likely to affect plant populations, frugivore populations and communities and nutrient cycling. These effects should be considered both in studies of fruit–frugivore interactions and during development of policies for sustainable forest use.

Forest Enrichment with Fruit-producing Species

Contemporary forest enrichment

As reviewed above, the fruits of forest trees are important in the cultures, diets and economies of forest-dwelling people throughout tropical regions (Balick, 1988; Boom, 1988; Anderson *et al.*, 1995; Hiraoka, 1995). This reliance on fruits has led people to increase

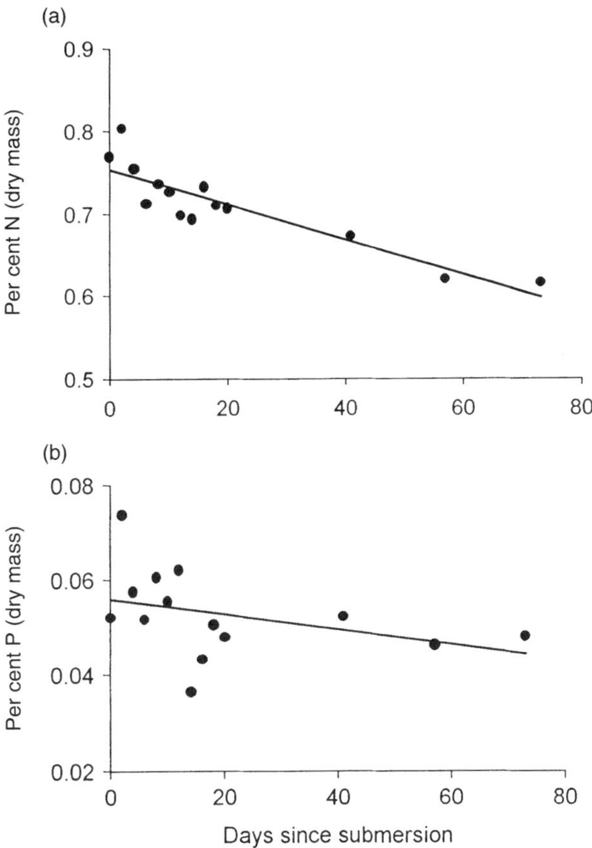

Fig. 32.3. Loss of nitrogen (a) and phosphorus (b) from submerged *Euterpe oleracea* fruit over 3.5 months. Plotted are means ± 1 SE of three replicates, each containing 47 fruits.

the densities of species from which fruit is harvested. Such forest management has the opposite effect of fruit harvest: it increases the amount of fruit that is produced and, perhaps, available to frugivores. The most typical method of increasing production is through enrichment planting, defined as: 'increasing the density of crop-producing species by planting in their natural habitats' (Schultze *et al.*, 1994, p. 582).

Many types of forest stands have enriched densities of fruit-producing trees. In some, called 'enriched forests', enrichment planting is the sole, or predominant, management activity. More intensive management is carried out in 'forest gardens' – stands that are managed at an intermediate intensity but that retain elements of non-managed or primary forest. In the Amazon basin, forest gardens tend to be located ≥ 100 m from homes and have a high diversity of tree species, which produce fruit at different times of the year (Anderson *et al.*, 1995; Pinedo-Vasquez and Padoch, 1996). The most intensively managed stands, usually located near human settlements, are variously termed 'orchards', 'home gardens', 'kitchen gardens' and agroforests (Gomez-Pompa and Kaus, 1990; Frumhoff, 1995; Moguel and Toledo, 1999). Most trees in these forests are planted and yield products harvested by people. Because most people who live in tropical forests maintain one or more of these types of enriched stands, forests over a very extensive area support elevated densities of fleshy fruit-producing trees. Moreover, many forests not currently under enrichment management were enriched in the past (Mabberley, 1992; Balée, 1993, 1994; Clark, 1996; McCann, 1999).

Few studies have compared tree densities in enriched versus non-enriched stands, but those that have show that enriched stands support significantly greater densities of useful plant species (Fig. 32.4). For example, stands enriched with *E. oleracea* in the Amazon River estuary contain approximately seven times the number of reproductive *E. oleracea* trees as do non-enriched stands (Muñiz-Miret *et al.*, 1996; Moegenburg, 2000). In other regions, densities of *Jacaratia spinosa, Attalea speciosa, Spondias mombin, Inga edulis, Hevea brasiliensis* and *B. excelsa,* among other harvested species, are higher in enriched stands (Balée, 1993; Anderson *et al.*, 1995; De Jong, 1996). The fruits produced by many of these species are consumed by both people and forest frugivores.

Because enrichment can greatly increase the economic value of tropical forests, it is encouraged in several types of forestry systems. For example, in extractive reserves in Brazil, residents plant Brazil-nut seedlings in forest gaps, pastures and regenerating fallows (Kainer *et al.*, 1998). A similar effort is under way in Los Tuxtlas, Mexico, where Ricker *et al.* (1999) recommend planting 40–200 seedlings of fruit-producing trees per hectare to increase the economic value of forests. In Brazil, Anderson *et al.* (1991) recommend a density of 100 mature *A. speciosa* palms per hectare in enriched secondary forests.

Impacts of enrichment on frugivores

Enriched forest stands, with their high density of fruit-producing trees, apparently attract large numbers of fruit-eating animals. In Brazilian forests enriched with *E. oleracea,* for example, all but four of 19 fruit-eating bird genera were detected more frequently in enriched stands than in non-enriched stands (Fig. 32.5). Most species in these genera were observed consuming *E. oleracea* fruit (Moegenburg, 2000). This implies a preference of fruit-eaters for forests with higher *E. oleracea* fruit availability, and supports the notion that frugivores track fruit abundance across space and time (Levey, 1988; Loiselle and Blake, 1991; Rey, 1995). Moreover, it shows that frugivore abundance is probably altered in the many types of forests enriched with

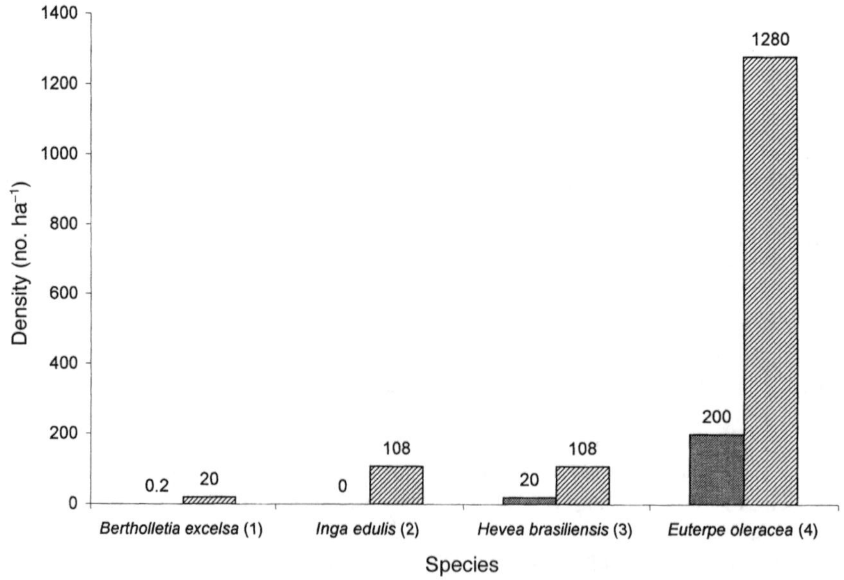

Fig. 32.4. Densities of tree species in non-enriched (solid bars) and enriched (striped bars) populations. Actual densities are indicated above bars and sources are noted in parentheses. Higher densities probably result from enrichment. Sources: (1) Kainer *et al.* (1998); (2, 3) Anderson *et al.* (1995); (4) Moegenburg (2000).

fruit-producing trees, including agroforests (Frumhoff, 1995; Moguel and Toledo, 1999).

The history of enrichment management

The practices of forest enrichment and fruit harvest that persist today are continuations of processes that have long occurred in tropical forests (Posey, 1985; Mabberley, 1992; Balée, 1993, 1994). Just as contemporary people rely heavily on forest fruits, ancient cultures subsisted on forest products, including many fruits (Gomez-Pompa and Kaus, 1990; Roosevelt *et al.*, 1996). In addition to simple extraction of fruit from forests, both archaeological and botanical data imply that humans employed management activities to increase fruit production in forests (Gomez-Pompa, 1990; Mabberley, 1992; Balée, 1993, 1994). Even where ancient civilizations have long ago disappeared, the 'footprint' of their forest management remains. For example, 16th-century explorers wrote of extensive forests managed by the Mayans of Mexico for fruits such as those from the tree *B. alicastrum* (*Moraceae*) (Gomez-Pompa and Kaus, 1990). In Mexico today, *B. alicastrum* is a forest dominant in numerous vegetation types of nearly all humid and subhumid forests (Gomez-Pompa and Kaus, 1990). In addition to producing edible fruits, which are still harvested today, *B. alicastrum* yields edible latex and bark, which are also used by people. Gomez-Pompa and Kaus (1990) cited at least 27 other edible, fleshy fruit-producing species that are abundant or dominant in mature forest in Mexico, and suggested that past management increased their densities. A similar pattern of higher densities of fruit-producing trees in formerly managed forests is seen in Amazonian Brazil, where up to 12% of the total forest area has apparently been modified by people (Balée, 1993).

Current distributions and densities of certain species of palms may be particularly related to past management. In the Brazilian Amazon, > 196,370 km^2 are estimated to consist of

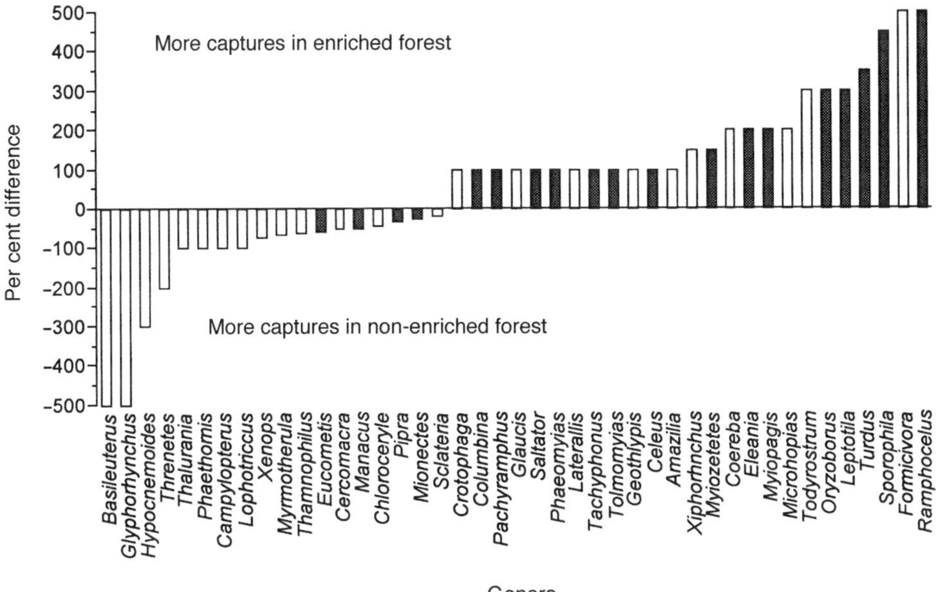

Fig. 32.5. Per cent difference in capture rates of bird genera in *E. oleracea*-enriched and non-enriched forest stands. Genera with positive values were captured more frequently in enriched stands, while those with negative values were captured more frequently in non-enriched stands. Black bars indicate genera that include fruit in their diet, while white bars represent genera with other diet compositions. All but four fruit-eating genera were captured more frequently in enriched forest stands.

so-called 'babassu forests' – forests dominated by the palm *A. speciosa*, which supplies nutritional pulp, oil-rich seeds and other food items (Anderson *et al.*, 1991). Anderson *et al.* (1991) posit that *A. speciosa* at one time had a much more restricted range, which was subsequently expanded through management. In another example, *Prestoa* palms were found to occur at higher densities in previously disturbed sites in Puerto Rico (García-Montiel and Scatena, 1994). Former management may have also reduced palm densities in Costa Rican forest, where Clark *et al.* (1995) suggest that lower densities of *I. deltoidea* than would be predicted by edaphic conditions are the result of past harvest of the high-quality wood.

Conclusions

Implications for studies of fruit–frugivore interactions

In summary, humans substantially alter the abundance and distribution of fruit and fruiting plants via harvest and management. In fact, harvest and management have occurred across vast areas for millennia. Even in forests where no humans now live, the footprint of prior human activities may persist and be the template on to which other biotic and abiotic processes are sketched. Indeed, the changes in fruit abundance resulting from harvest and management apparently affect the abundance of frugivores. What are the implications of this manipulation of fruit abundance by humans for the ecology and evolution of fruit–frugivore interactions?

Ecologists should recognize that fruiting-tree abundance and distribution in many forests result from a complex suite of factors, including not only frugivory, seed dispersal and predation, but also management and harvest by humans (Janzen, 1970; Howe and Smallwood, 1982; Posey, 1985; Balée, 1993, 1994; see also Terborgh *et al.*, this volume). Due to these complex human–fruit–frugivore interactions, caution must be used when interpreting patterns of present plant distributions, such as the clumped pattern exhibited by some species (*Maximilliana maripa, Palmae*) (Fragoso, 1997), as due only to non-human factors (McCann, 1999). That humans have altered plant-species abundance and diversity in some forests does not reduce the value of those forests for ecological studies or biological conservation (Clark, 1996). Rather, understanding how humans alter plant distribution and abundance will allow for better understanding of the actual role of frugivores and seed-dispersers in shaping tropical forests.

There are numerous ways in which fruit harvest and management may in the past have affected frugivores and forest composition. The concentration of fleshy fruits in managed forests, for example, may have affected the behaviour of frugivores, encouraging them to spend more time in managed forests and less time in forests with relatively low densities of fleshy fruits. If frugivore habitat choice has been affected in this way, other processes are likely to have been affected, too. In particular, if frugivores concentrated their activity in fruit-enriched stands, they may have dispersed more seeds into those stands than into other stand types, thereby further affecting forest composition (Mabberley, 1992).

Human–fruit–frugivore interactions may affect more, however, than just forest composition and tree-species diversity. The long time frame during which forest enrichment and fruit harvest have taken place – up to 10,000 years in Amazonia, for example (Roosevelt *et al.*, 1996) – implies that these processes may influence both ecological and evolutionary relationships between plants and animals. Consider the consequences of humans increasing the relative abundance of fruits with certain traits, such as low concentrations of secondary compounds or thin exocarps, and decreasing the relative abundance of fruits with other kinds of traits. Over time, frugivores may develop preferences for the types of fruits that are most abundant or most palatable. Alternatively, frugivores may develop mechanisms to detoxify secondary compounds found in some unripe fruits, if humans consistently harvest the most abundant fruits when they are ripe. The recognition of humans as a factor affecting fruit–frugivore interactions leads to many such avenues for interesting research.

Implications for conservation

As we have seen, fruit harvest and forest enrichment with fruit-producing species apparently affect forest composition, frugivores and nutrient cycling. Harvest and enrichment are therefore not entirely benign activities, as has been claimed (Peters, 1990, 1992; Anderson *et al.*, 1995). Rather, they should be added, along with habitat loss, fragmentation, hunting and capture for trade, to the list of human activities that affect frugivore populations and fruit–frugivore interactions. Extraction of fruits from wild-occurring trees and enrichment of forests with fruit-producing species have great potential for increasing incomes that can be generated from forests (Fearnside, 1989). Nevertheless, such forests should be seen as complements to, not substitutes for, more strictly protected areas.

Directions for future research

Much research is needed to understand the relative roles of humans and frugivores in shaping tropical forests. I offer some suggestions that would contribute to that goal.

1. Whenever possible, studies of frugivore effects on plant community composition and structure should include the history, especially human use history, of the site.
2. Studies that link forest use by humans with forest composition are needed to provide a greater understanding of the spatial extent of human-influenced forests.
3. Hundreds of species of fruits and seeds are harvested from tropical forests and yet almost nothing is known about the impacts of such harvest. Many more studies are needed on non-timber forest products to determine the effects of harvest on plant populations, fruit- and seed-eaters, fruit–frugivore interactions and other species in the community.
4. Most valuable would be studies that determine the amount of fruits and seeds that can be removed without perturbing the ecology of the system. Systems in which forest-dwelling people harvest fruits provide ideal opportunities for *in situ* experiments. Finally, effects of harvest should be measured over long time frames (multiyear) and at spatial scales that encompass the home ranges of fruit-eating animals.

Acknowledgements

I gratefully acknowledge field assistance from B. Ferreira, R. Newman and A. Castelo Branco Pina and logistical support from M. Jardim and P. Lisboa. My ideas and the manuscript have been greatly influenced by interactions with D. Levey, C. Uhl, J. Putz, J. Terborgh, S. Weinstein, M. Hiraoka, M. Galetti, J. Motta and the residents of the Cajari River Extractive Reserve, Amapá, Brazil. While writing this chapter, I was supported by a STAR fellowship from the US Environmental Protection Agency.

References

Abramson, J., Speer, B.L. and Thomson, J.B. (1995) *The Large Macaws: Their Care, Breeding, and Conservation.* Raintree Publications, Fort Bragg, California, USA, 534 pp.

Achung, F.R., Achung, M.R. and Ruestra, P.V. (1995) *Realidady perspectivas: La Reserva Nacional Pacaya-Samiria.* Grafica Bellido, Lima, Peru, 132 pp.

Adler, G.H. (1998) Impacts of resource abundance on populations of a tropical forest rodent. *Ecology* 79, 242–254.

Anderson, A.B., May, P.H. and Balick, M.J. (1991) *The Subsidy from Nature.* Columbia University Press, New York, USA, 233 pp.

Anderson, A.B., Magee, P., Gély, A. and Jardim, M.A.G. (1995) Forest management patterns in the flood-plain of the Amazon estuary. *Conservation Biology* 9, 47–61.

Balée, W. (1993) Indigenous transformation of Amazonian forests: an example from Maranhão, Brazil. *L'Homme* 126–128, 231–254.

Balée, W. (1994) *Footprints of the Forest: Ka'apor Ethnobotany.* Columbia University Press, New York, 396 pp.

Balick, M.J. (1988) The use of palms by the Apinayé and Guajajara Indians of northeastern Brazil. *Advances in Economic Botany* 6, 65–90.

Bennett, B.C. and Hicklin, J.R. (1998) Uses of saw palmetto (*Serenoa repens, Arecaceae*) in Florida. *Economic Botany* 52, 381–393.

Bernal, R.G. (1998) Demography of the vegetable ivory palm *Phytelephas seemannii* in Colombia, and the impact of seed harvesting. *Journal of Applied Ecology* 35, 64–74.

Boom, B.M. (1988) The Chácobo Indians and their palms. *Advances in Economic Botany* 6, 91–97.

Bunyard, P. (1989) Guardians of the Amazon. *New Scientist* 16 December, 38–41.

Clark, D.A., Clark, D.B., Sandoval, R. and Castro, C.M.V. (1995) Edaphic and human effects on landscape-scale distributions of tropical rainforest palms. *Ecology* 76, 2581–2594.

Clark, D.B. (1996) Abolishing virginity. *Journal of Tropical Ecology* 12, 735–739.

Clay, J.W. (1997) The impact of palm heart harvesting in the Amazon estuary. In: Freese, C.H. (ed.) *Harvesting Wild Species: Implications for Biodiversity Conservation.* Johns Hopkins University Press, Baltimore, Maryland, pp. 283–314.

Davis, T.A. (1988) Uses of semi-wild palms in Indonesia and elsewhere in south and southeast Asia. *Advances in Economic Botany* 6, 98–118.

De Jong, W. (1996) Swidden–fallow agroforestry in Amazonia: diversity at close distance. *Agroforestry Systems* 34, 277–290.

Denevan, W.M. (1992) The pristine myth: the landscape of the Americas in 1492. *Annals of the American Geographers* 83, 369–385.

Fearnside, P.M. (1989) Extractive reserves in Brazilian Amazonia. *Bioscience* 39, 387–393.

Foster, R.B. (1982) Famine on Barro Colorado Island. In: Leigh, E.G., Jr, Rand, A.S. and Windsor, D.M. (eds) *The Ecology of a Tropical Forest.* Smithsonian Institution Press, Washington, DC, pp. 201–212.

Fragoso, J.M.V. (1997) Tapir-generated seed shadows: scale-dependent patchiness in the Amazon forest. *Journal of Ecology* 85, 519–529.

Franken, M. (1979) Major nutrient and energy content of the litterfall of a riverine forest of central Amazonia. *Tropical Ecology* 2, 211–224.

Franken, M., Irmler, U. and Klinge, H. (1979) Litterfall in inundation, riverine, and terra firme forests of central Amazonia. *Tropical Ecology* 2, 225–235.

Frumhoff, P. (1995) Conserving wildlife in forests managed for timber. *BioScience* 45, 456–464.

Galetti, M. and Aleixo, A. (1998) Effects of palm heart harvesting on avian frugivores in the Atlantic rain forest of Brazil. *Journal of Applied Ecology* 35, 286–293.

Galetti, M. and Fernandez, J.C. (1998) Palm heart harvesting in the Brazilian Atlantic forest: changes in industry structure and the illegal trade. *Journal of Applied Ecology* 35, 294–301.

Galetti, M., Zipparro, V. and Morellato, L.P. (1999) Fruit phenology and frugivory on the palm *Euterpe edulis* in a lowland atlantic forest of Brazil. *Ecotropica* 5, 115–122.

García-Montiel, D.C. and Scatena, F.N. (1994) The effect of human activity on the structure and composition of a tropical forest in Puerto Rico. *Forest Ecology and Management* 63, 57–78.

Gentry, A.H. (1982) Patterns of neotropical plant species diversity. *Evolutionary Biology* 15, 1–84.

Gomez-Pompa, A. and Kaus, A. (1990) Traditional management of tropical forests in Mexico. In: Anderson, A.B. (ed.) *Alternatives to Deforestation.* Columbia University Press, New York, pp. 45–64.

Goulding, M. (1980) *The Fishes and the Forest.* University of California Press, Berkeley, California, 280 pp.

Hiraoka, M. (1995) Land use changes in the Amazon estuary. *Global Environmental Change* 5, 323–336.

Howe, H.F. and Smallwood, J. (1982) Ecology of seed dispersal. *Annual Review of Ecology and Systematics* 13, 201–228.

Janson, C.H. and Emmons, L.H. (1990) Ecological structure of the nonflying mammal community at Cocha Cashu biological station, Manu National Park, Peru. In: Gentry, A.H. (ed.) *Four Neotropical Rainforests.* Yale University Press, London, pp. 314–338.

Janzen, D.H. (1970) Herbivores and the number of tree species in tropical forests. *American Naturalist* 104, 501–528.

Janzen, D.H. (1971) Seed predation by animals. *Annual Review of Ecology and Systematics* 2, 465–492.

Kahn, F. (1988) Ecology of economically important palms in Peruvian Amazonia. *Advances in Economic Botany* 6, 42–49.

Kainer, K.A., Duryea, M.L., Costa de Macedo, N. and Williams, K. (1998) Brazil nut seedling establishment and autecology in extractive reserves of Acre, Brazil. *Ecological Applications* 8, 397–410.

Karr, J.R., Robinson, S., Blake, J.G. and Bierregaard, R.O., Jr (1990) Birds of four neotropical forests. In: Gentry, A.H. (ed.) *Four Neotropical Rainforests.* Yale University Press, London, pp. 237–272.

Levey, D.J. (1988) Spatial and temporal variation in Costa Rican fruit and fruit-eating bird abundance. *Ecological Monographs* 58, 251–269.

Loiselle, B.A. and Blake, J.G. (1991) Temporal variation in birds and fruits along an elevational gradient in Costa Rica. *Ecology* 72, 180–193.

Mabberley, D.J. (1992) *Tropical Rain Forest Ecology.* Chapman & Hall, New York, 285 pp.

McCann, J.M. (1999) Before 1492: the making of the pre-Columbia landscape. Part II: The vegetation, and implications for restoration for 2000 and beyond. *Ecological Restoration* 17, 107–119.

Mejia, C.K. (1988) Utilization of palms in eleven mestizo villages of the Peruvian Amazon (Ucayali River, Dept. of Loreto). *Advances in Economic Botany* 6, 130–136.

Moegenburg, S.M. (2000) Fruit–frugivore interactions in *Euterpe* palm forests of the Amazon River floodplain. PhD thesis, University of Florida, Gainesville, Florida.

Moguel, P. and Toledo, V.M. (1999) Biodiversity conservation in traditional coffee production systems of Mexico. *Conservation Biology* 13, 11–21.

Muñiz-Mirit, N., Vamos, R., Hiraoka, M., Montagnini, F. and Mendelson, R. (1996) The economic value of managing the açaí palm (*Euterpe oleracea* Mart.) in the floodplains of the Amazon estuary, Pará, Brazil. *Forest Ecology and Management* 87, 163–173.

Murali, K.S., Shankar, U., Shaanker, R.U., Ganeshiah, K.N. and Bawa, K.S. (1996) Extraction of non-timber forest products in the forests of Biligiri Rangan Hills, India. 2. Impact of NTFP extraction on regeneration, population structure, and species composition. *Economic Botany* 50, 252–269.

Nepstad, D.C., Brown, F., Luz, L., Alechandra, A. and Viana, V. (1992) Biotic impoverishment of Amazonian forests by rubber tappers, loggers, and cattle ranchers. In: Nepstad, D.C. and Schwartzman, S. (eds) *Non-Timber Products from Tropical Forests. Advances in Economic Botany* 9, 1–14.

Padoch, C. (1988) Aguaje (*Mauritia flexuosa*) in the economy of Iquitos, Peru. *Advances in Economic Botany* 6, 214–224.

Panayotou, T. and Ashton, P.S. (1992) *Not by Timber Alone.* Island Press, Washington, DC, 282 pp.

Peres, C.A. and Baider, C. (1997) Seed dispersal, spatial distribution and population structure of Brazil nut trees (*Bertholletia excelsa*) in southeastern Amazonia. *Journal of Tropical Ecology* 13, 595–616.

Peres, C.A., Schiesari, L.C. and Dias-Leme, C.L. (1997) Vertebrate predation of Brazil-nuts (*Bertholletia excelsa, Lecythidaceae*), an agouti-dispersed Amazonian seed crop: a test of escape hypothesis. *Journal of Tropical Ecology* 13, 69–79.

Peters, C.M. (1990) Population ecology and management of forest fruit trees in Peruvian Amazon. In: Anderson, A.B. (ed.) *Alternatives to Deforestation.* Columbia University Press, New York, pp. 86–98.

Peters, C.M. (1992) The ecology and economics of oligarchic forests. In: Nepstad, D.C. and Schwartzman, S. (eds) *Non-Timber Products from Tropical Forests. Advances in Economic Botany* 9, 15–22.

Pinedo-Vasquez, M. and Padoch, C. (1996) Managing forest remnants and forest gardens in Peru and Indonesia. In: Schelhas, J. and Greenberg, R. (eds) *Forest Patches in Tropical Landscapes.* Island Press, Washington, DC, pp. 327–342.

Pollack, H., Mattos, M. and Uhl, C. (1995) A profile of palm heart extraction in the Amazon Estuary. *Human Ecology* 23, 357–385.

Posey, D.A. (1985) Indigenous management of tropical forest ecosystems: the case of the Kayapó Indians of the Brazilian Amazon. *Agroforestry Systems* 3, 139–158.

Rey, P.J. (1995) Spatio-temporal variation in fruit and frugivorous bird abundance in olive orchards. *Ecology* 76, 1625–1635.

Ricker, M., Mendelsohn, R.O., Daly, D.C. and Ángeles, G. (1999) Enriching the rainforest with native fruit trees: an ecological and economic analysis in Los Tuxtlas (Veracruz, Mexico). *Ecological Economics* 31, 439–448.

Roosevelt, A.C., Llima da Costa, M., Lopes Machado, C., Michab, M., Mercier, N., Valladas, H., Feathers, J., Barnett, W., Imazio da Silveira, M., Henderson, A., Silva, J., Chernoff, B., Reese, D.S., Holman, J.A., Toth, N. and Schick, K. (1996) Paleoindian cave dwellers in the Amazon: the peopling of the Americas. *Science* 272, 373–384.

Schultze, P.C., Leighton, M. and Peart, D.R. (1994) Enrichment planting in selectively-logged rain forest: a combined ecological and economic analysis. *Ecological Applications* 4, 581–592.

Shankar, U., Murali, K.S., Shaanker, R.U., Ganeshaiah, K.N. and Bawa, K.S. (1996) Extraction of non-timber forest products in the forests of Biligiri Rangan Hills, India. 3. Productivity, extraction and prospects of sustainable harvest of amla *Phyllanthus emblica* (*Euphorbiaceae*). *Economic Botany* 50, 270–279.

Snyder, N.F., Wiley, J.W. and Kepler, C.B. (1987) *The Parrots of Luquillo: Natural History and Conservation of the Puerto Rican Parrot.* Western Foundation of Vertebrate Zoology, Los Angeles, California, 384 pp.

Strahl, S.D. and Grahal, A. (1991) Conservation of large avian frugivores and the management of Neotropical protected areas. *Oryx* 25, 50–55.

Vasquez, R. and Gentry, A.H. (1989) Use and misuse of forest-harvested fruits in the Iquitos area. *Conservation Biology* 3, 350–361.

Vitousek, P.M. (1984) Litterfall, nutrient cycling, and nutrient limitation in tropical forests. *Ecology* 65, 285–298.

Witkowski, E.T.F. and Lamont, B.B. (1996) Nutrient losses from commercial picking and cockatoo removal of *Banksia hookeriana* blooms at the organ, plant, and site levels. *Journal of Applied Ecology* 33, 131–140.

Wright, S.J., Carrasco, C., Calderón, O. and Paton, S. (1999) The El Niño southern oscillation, variable fruit production, and famine in a tropical forest. *Ecology* 80, 1632–1647.

Index

Page numbers in bold refer to illustrations and tables